EVOLUTIONARY COMPUTATION

Books of Related Interest from IEEE Press . . .

FUZZY SYSTEMS DESIGN PRINCIPLES: Building Fuzzy IF-THEN Rule Bases
Riza C. Berkan and Sheldon L. Trubatch
1997   Cloth   520 pp   IEEE Product No. PC5622   ISBN 0-7803-1151-5

WHAT'S SIZE GOT TO DO WITH IT? Understanding Computer Rightsizing
John E. Blyler and Gary A. Ray
1998   Paper   416 pp   IEEE Product No. PP4499   ISBN 0-7803-1096-9

EVOLUTIONARY COMPUTATION: Toward a New Philosophy of Machine Intelligence
David B. Fogel
1995   Cloth   288 pp   IEEE Product No. PC3871   ISBN 0-7803-1038-1

UNDERSTANDING NEURAL NETWORKS AND FUZZY LOGIC: Basic Concepts and Applications
Stamatios V. Kartalopoulos
1996   Paper   232 pp   IEEE Product No. PP5591   ISBN 0-7803-1128-0

EVOLUTIONARY COMPUTATION

The Fossil Record

Edited by

David B. Fogel

IEEE PRESS

A Selected Reprint Volume

IEEE Neural Networks Council, *Sponsor*

The Institute of Electrical and Electronics Engineers, Inc., New York

This book may be purchased at a discount from the publisher
when ordered in bulk quantities. Contact:

IEEE Press Marketing
Attn: Special Sales
445 Hoes Lane, P.O. Box 1331
Piscataway, NJ 08855-1331
Fax: (732) 981-9334

For more information on the IEEE Press,
visit the IEEE home page: http://www.ieee.org/

Printed in the United States of America

10   9   8   7   6   5   4   3   2   1

ISBN 0-7803-3481-7
IEEE Order Number: PC5737

Library of Congress Cataloging-in-Publication Data

Evolutionary computation : the fossil record / edited by David Fogel.
    p.   cm.
    Includes bibliographical references and index.
    "A selected reprint volume."
    "IEEE order number: PC5737"—T.p. verso.
    ISBN 0-7803-3481-7 (cloth)
    1. Evolutionary programming (Computer science)   I. Fogel, David
B.
QA76.618.E883   1998
005.1—dc21   98-2728
                                                          CIP

To the pioneers . . .

Contents

Preface

Evolutionary computation has a long history. Unfortunately, this history is virtually unknown to many practitioners in the field. This is a shame, not simply for the handicap that it imposes on the scientific advancement of the field, but also because the history of evolutionary computation is a truly fascinating tale of multiple independent beginnings, ingenious inventions, disappointments, and marvelous successes, often achieved in the face of limited, if not antiquated, computing equipment. It is a story that deserves to be told, heard, and repeated to others.

The word *story* appears in "history," and rightly so. History is not a mere cataloguing of events or a listing of independent facts and places and writings; and that is particularly so with science. The history of any science cannot be adequately represented by a simple index, perhaps organized chronologically or alphabetically—the most rudimentary of possible orderings. Such listings amount to nothing more than the inventory of a stamp collection and make no real contribution to a greater understanding of the science that underlies each effort.

Although the *Science Citation Index* does not suffice as a historical document, this and other listings of prior art do provide a record—a fossil record—of the endeavor of science. Like the fossil record of natural evolution, they also contain many missing links. An incomplete accounting, unfortunately, is unavoidable: Publications of limited distribution, particularly conference proceedings, often have short lifespans and then vanish with little trace. Eventually, their authors too vanish, and unfortunately the memories of their contribution to the advancement of knowledge die with them. This situation is just now beginning to occur in the field of evolutionary computation; it makes the timeliness of an accurate accounting of the history of this field even more important. Many pioneers in evolutionary computation can still provide firsthand information regarding their facts, data, processes, and the culture in which they worked. Two decades from now, circumstances will be very different and we will need to rely on a record such as that provided by the current volume.

Natural evolution is a historical process: "Anything that changes in time, by definition, has a history" (Mayr, 1982, p. 1). And the essence of evolution is change. Evolution is a dynamic, two-step process of random variation and selection (Mayr, 1988, pp. 97–98) that engenders constant change to individuals within a population in response to environmental dynamics and demands. Individuals who are poorly adapted to their current environment are culled from the population. The adaptations that allow specific individuals to survive are heritable, coded in a complex genetic milieu that is passed on to successive progeny. Thus, adaptations are a historical sequence of consecutive events, each one leading to the next. The converse, however, is also true: History itself is an evolutionary process, not just in the sense that it changes over time but also in the more strict sense that it undergoes mutation and selection.

The situation is not unlike the child's game of *telephone*. Several children are needed to play. An initial message, perhaps provided by a teacher, is sent from one child to another, whispered in the ear, then passed along again in sequence. Each child can only hear what the last child whispers. The message is subject to mutation at each step, occasionally by accident, sometimes with devious intent. At last, the final student divulges the message that he or she heard, as it was understood. This "mutant" message is then compared to the actual message, revealed by the teacher. In effect, the teacher applies a form of selection to restore the original message. And so it is with history: "Written histories, like science itself, are constantly in need of revision. Erroneous interpretations of an earlier author eventually become myths, accepted without question and carried forward from generation to generation" (Mayr, 1982, p. 1).

The aim of this volume, then, is to perform the function of selection on what has mostly been a process of mutation. Such error correction is surely needed for the conventional accounting of the history of evolutionary computation is more of a fable or science fiction than a factual record. Readers only casually acquainted with the field may be surprised to learn, for example, that the idea to use recombination in population-based evolutionary simulations did not arise in a single major innovation (cf. Mitchell, 1996, p. 3; Levy, 1992, p. 169) but in fact was commonly, if not routinely, applied in multiple independent lines of investigation in the 1950s and 1960s. It may also be of interest that the prospects of using computers to study "life-as-it-could-be" (i.e., artificial life) rather than "life-as-we-know-it" were plainly clear to population geneticists and evolutionary biologists in the same time frame (cf. Waldrop, 1992, p. 200). Many supposedly recent inventions (e.g., the use of more than two parents to generate offspring, the evolution of neural networks, the coevolution of individuals in a population) actually occurred more than 30 years ago. By unearthing these scientific fossils here, I hope both to revive interest in many of these long-forgotten works simply for the fact that they exist and to reawaken investigation into avenues of scientific inquiry that have been hibernating far too long.

All historical accounting is necessarily subjective. While compiling this record, I have made several judgments as to which papers merit reprinting. Ultimately, the papers selected met one or more of the following criteria: (1) They offered a first or very early attempt at a specific approach, (2) they had a significant impact on the future development of the field, (3) they could have had a significant impact had they received due attention, (4) they represented a key turning point in the field, or (5) I found them to be personally interesting.[1] With only two exceptions, no papers more recent than five years from the publication of this volume were chosen; I felt that later papers had not been allowed the time to "fossilize" yet. Assuredly, several such candidate papers will enter in future editions of this work. The two exceptions (Fogel, 1994; Back et al., 1997) are introductory surveys of evolutionary computation. These articles are included mostly for the sake of novice readers, but also for their historical significance: The journal issues in which the articles appeared marked pivotal events in the acceptance of evolutionary computation as a practical tool for solving real-world problems. All but two of the reprinted papers have been scanned in their original form. I gratefully acknowledge all the permissions-to-reprint granted by the publishers and authors.

With the exception of the information in the first chapter, the fossil record presented here begins in the 1950s and progresses forward through time, concluding with more contemporary research. This chronology reveals the multiple independent efforts in evolutionary computation in light of the state-of-the-art at the time those contributions were offered. In some cases of very closely related work, multiple papers from different periods have been reprinted in a single section; yet there are strong connections among several papers separated by many chapters and, thus, possibly decades of time. The index provided at the back of the volume should aid the reader in finding common threads among these various instances, while the original material that introduces each section places the reprinted works in the context of prior art and subsequent evolution.

I have had the good fortune to know many of the pioneers in evolutionary computation. Several related personal anecdotes and helped review the editorial introductions to their own contributions, as well as offered comments on the overall content. In addition, many colleagues reviewed portions of this work, while a few generous friends scrutinized it in its entirety. In particular, thanks are owed to Russell Anderson, Jim Antonisse, Wirt Atmar, Robert Axelrod, Thomas Bäck, Peter Bienert, William Bossert, George Box, George Burgin, Michael Conrad, Frank Cornett, Nichael Cramer, Jason Daida, Lawrence Davis, Don Dearholt, Bill Dress, George Dyson, Tom English, Larry Ersay, Lawrence Fogel, Anne Fraser, Alex Fraser, Richard Friedberg, George Friedman, Roman Galar, Joseph Goguen, David Goldberg, John Grefenstette, John Holland, Howard Kaufman, John Koza, Michael Lyle, Bob Marks, Zbigniew Michalewicz, Javier Montez, Heinz Mühlenbein, James North, Martin Nowak, Al Owens, Howard Pattee, Fred Petry, Martin Pincus, Bill Porto, Tom Ray, Ingo Rechenberg, Michel Rogson, Ron Root, Hans-Paul Schwefel, Michele Sebag, Rob Smith, Steve Smith, Robert Toombs, and Lotfi Zadeh. I'm especially grateful to Joe Felsenstein for providing an exhaustive listing of efforts to use computer simulation in genetic systems from the 1950s to the 1970s; to Antanas Zilinskas for providing several references to Russian literature along with English translations; to my brother, Gary, not only for his careful review of the text, but for spending many hours assisting me in digging up old papers in the library; to Jacquelyn Moore for helping me locate several pioneers in evolutionary computation; and to Pete Angeline, who not only helped improve the exposition but framed its concept with me over a period of two years. I also appreciate the assistance of John Griffin, Marilyn Giannakouros, Denise Phillip, Mark Morrell, and Karen Hawkins at IEEE Press.

Hans Bremermann, a recipient of the Lifetime Achievement Award from the Evolutionary Programming Society, died in early 1996, but I was fortunate that he was able to review the relevant material in his section before his death. I regret not being able to show him the finished product. I also regret not being able to learn firsthand about the efforts of several other pioneers, including Nils Barricelli, Woody Bledsoe, Bradford Dunham, Gordon Pask, and Jack Walsh. I hope they would be pleased.

David B. Fogel
La Jolla, California

---

[1] There were two other criteria: I had to be able to both find and read the paper. No doubt, despite the assistance of colleagues, significant research in simulated evolution from the former Soviet Union and other countries remains to be discovered.

References

T. Bäck, U. Hammel, and H.-P. Schwefel (1997) "Evolutionary computation: Comments on the history and current state," *IEEE Transactions on Evolutionary Computation,* Vol. 1:1, pp. 3–17.

D. B. Fogel (1994) "An introduction to simulated evolution," *IEEE Transactions on Neural Networks: Special Issue on Evolutionary Computation,* Vol. 5:1, pp. 3–14.

S. Levy (1992) *Artificial Life: A Report from the Frontier where Computers meet Biology,* Vintage Books, NY.

E. Mayr (1982) *The Growth of Biological Thought: Diversity, Evolution, and Inheritance,* Belknap, Harvard.

E. Mayr (1988) *Toward a New Philosophy of Biology: Observations of an Evolutionist,* Belknap, Harvard.

M. Mitchell (1996) *An Introduction to Genetic Algorithms,* MIT Press, Cambridge, MA.

M. M. Waldrop (1992) *Complexity: The Emerging Science at the Edge of Order and Chaos,* Simon & Schuster, NY.

Chapter 1
An Introduction to Evolutionary Computation

D. B. Fogel (1994) "An Introduction to Simulated Evolutionary Optimization," *IEEE Trans. Neural Networks,* Vol. 5:1, pp. 3–14.

T. Bäck, U. Hammel, and H.-P. Schwefel (1997) "Evolutionary Computation: Comments on the History and Current State," *IEEE Trans. Evolutionary Computation,* Vol. 1:1, pp. 3–17.

EVOLUTION is in essence a two-step process of random variation and selection (Mayr, 1988, pp. 97–98). A population of individuals is exposed to an environment and responds with a collection of behaviors. Some of these behaviors are better suited to meet the demands of the environment than are others. Selection tends to eliminate those individuals that demonstrate inappropriate behaviors. The survivors reproduce, and the genetics underlying their behavioral traits are passed on to their offspring. But this replication is never without error, nor can individual genotypes remain free of random mutations. The introduction of random genetic variation in turn leads to novel behavioral characteristics, and the process of evolution iterates. Over successive generations, increasingly appropriate behaviors accumulate within evolving phyletic lines (Atmar, 1994).

Evolution optimizes behaviors (i.e., the phenotype), not the underlying genetics per se, because selection can act only in the face of phenotypic variation. The manner in which functional adaptations are encoded in genetics is transparent to selection; only the realized behaviors resulting from the interaction of the genotype with the environment can be assessed by competitive selection. Useful variations have the best chance of being preserved in the struggle for life, leading to a process of continual improvement (Darwin, 1859, p. 130). Evolution may in fact create "organs of extreme perfection and complication" (Darwin, 1859, p. 171), but must always act within the constraints of physical development and the historical accidents of life that precede the current population. Evolution is entirely opportunistic (Jacob, 1977), and can only work within the variation present in extant individuals.

The process of evolution can be modeled algorithmically and simulated on a computer. In the most elementary of models, it may be summarized as a difference equation:

$$\mathbf{x}[t + 1] = s(v(\mathbf{x}[t]))$$

where the population at time, t, denoted as $\mathbf{x}[t]$, is operated on by random variation, v, and selection, s, to give rise to a new population $\mathbf{x}[t + 1]$. Natural evolution does not occur in discontinuous time intervals, but the use of a digital computer requires discrete events. Over successive iterations of variation and selection, an evolutionary algorithm can drive a population toward particular optima on a response surface that represents the measurable worth of each possible individual that might reside in a population. *Evolutionary computation* is the field that studies the properties of these algorithms and similar procedures for simulating evolution on a computer.

Although the term *evolutionary computation* was invented as recently as 1991, the field has a history that spans four decades. Many independent efforts to simulate evolution on a computer were offered in the 1950s and 1960s. Three broadly similar avenues of investigation in simulated evolution have survived as main disciplines within the field: evolution strategies, evolutionary programming, and genetic algorithms. Each begins with a population of contending trial solutions brought to a task at hand. New solutions are created by randomly varying the existing solutions. An objective measure of performance is used to assess the "fitness" of each trial solution, and a selection mechanism determines which solutions to retain as "parents" for the subsequent generation. The differences between the procedures are characterized by the typical data representations, the types of variations that are imposed on solutions to create offspring, and the methods employed for selecting new parents. Over time, however, these differences have become increasingly blurred, and will likely become of only historical interest.

The two papers reprinted here, Fogel (1994) and Bäck et al. (1997), provide surveys of evolutionary computation. Fogel (1994) offered an introduction to a special issue of the *IEEE Transactions on Neural Networks* devoted to evolutionary computation, while Bäck et al. (1997) offered the first paper of the *IEEE Transactions on Evolutionary Computation*. These two publications represent important milestones in the acceptance of evolutionary algorithms as practical tools for addressing complex problems in engineering. The papers include numerous references that will assist novice readers who are just entering the field.

References

[1] W. Atmar (1994) "Notes on the simulation of evolution," *IEEE Trans. Neural Networks,* Vol. 5:1, pp. 130–147.

[2] T. Bäck, U. Hammel, and H.-P. Schwefel (1997) "Evolutionary computation: comments on the history and current state," *IEEE Trans. Evolutionary Computation,* Vol. 1:1, pp. 3–17.

[3] C. Darwin (1859) *The Origin of Species by Means of Natural Selection or the Preservation of Favoured Races in the Struggle for Life,* Mentor Reprint, 1958, NY.

[4] D. B. Fogel (1994) "An introduction to simulated evolutionary optimization," *IEEE Trans. Neural Networks,* Vol. 5:1, pp. 3–14.

[5] F. Jacob (1977) "Evolution and tinkering," *Science,* Vol. 196, pp. 1161–1166.

[6] E. Mayr (1988) *Toward a New Philosophy of Biology: Observations of an Evolutionist,* Belknap, Harvard.

IEEE TRANSACTIONS ON NEURAL NETWORKS, VOL. 5, NO. 1, JANUARY 1994

An Introduction to Simulated Evolutionary Optimization

David B. Fogel, *Member, IEEE*

Abstract—Natural evolution is a population-based optimization process. Simulating this process on a computer results in stochastic optimization techniques that can often outperform classical methods of optimization when applied to difficult real-world problems. There are currently three main avenues of research in simulated evolution: genetic algorithms, evolution strategies, and evolutionary programming. Each method emphasizes a different facet of natural evolution. Genetic algorithms stress chromosomal operators. Evolution strategies emphasize behavioral changes at the level of the individual. Evolutionary programming stresses behavioral change at the level of the species. The development of each of these procedures over the past 35 years is described. Some recent efforts in these areas are reviewed.

I. INTRODUCTION

THE fundamental approach to optimization is to formulate a single standard of measurement—a cost function—that summarizes the performance or value of a decision and iteratively improve this performance by selecting from among the available alternatives. Most classical methods of optimization generate a deterministic sequence of trial solutions based on the gradient or higher-order statistics of the cost function [1, chaps. 8–10]. Under regularity conditions on this function, these techniques can be shown to generate sequences that asymptotically converge to locally optimal solutions, and in certain cases they converge exponentially fast [2, pp. 12–15]. Variations on these procedures are often applied to training neural networks (backpropagation) [3], [4], or estimating parameters in system identification and adaptive control applications (recursive prediction error methods, Newton-Gauss) [2, pp. 22–23], [5]. But the methods often fail to perform adequately when random perturbations are imposed on the cost function. Further, locally optimal solutions often prove insufficient for real-world engineering problems.

Darwinian evolution is intrinsically a robust search and optimization mechanism. Evolved biota demonstrate optimized complex behavior at every level: the cell, the organ, the individual, and the population. The problems that biological species have solved are typified by chaos, chance, temporality, and nonlinear interactivity. These are also characteristics of problems that have proved to be especially intractable to classic methods of optimization. The evolutionary process can be applied to problems where heuristic solutions are not available or generally lead to unsatisfactory results.

The most widely accepted collection of evolutionary theories is the neo-Darwinian paradigm. These arguments assert that the history of life can be fully accounted for by physical processes operating on and within populations and species [6, p. 39]. These processes are reproduction, mutation, competition, and selection. Reproduction is an obvious property of extant species. Further, species have such great reproductive potential that their population size would increase at an exponential rate if all individuals of the species were to reproduce successfully [7], [8, p. 479]. Reproduction is accomplished through the transfer of an individual's genetic program (either asexually or sexually) to progeny. Mutation, in a positively entropic system, is guaranteed, in that replication errors during information transfer will necessarily occur. Competition is a consequence of expanding populations in a finite resource space. Selection is the inevitable result of competitive replication as species fill the available space. Evolution becomes the inescapable result of interacting basic physical statistical processes ([9], [10, p. 25], [11] and others).

Individuals and species can be viewed as a duality of their genetic program, the genotype, and their expressed behavioral traits, the phenotype. The genotype provides a mechanism for the storage of experiential evidence, of historically acquired information. Unfortunately, the results of genetic variations are generally unpredictable due to the universal effects of pleiotropy and polygeny (Fig. 1) [8], [12], [13], [14, p. 224], [15]–[19], [20, p. 296]. *Pleiotropy* is the effect that a single gene may simultaneously affect several phenotypic traits. *Polygeny* is the effect that a single phenotypic characteristic may be determined by the simultaneous interaction of many genes. There are no one-gene, one-trait relationships in natural evolved systems. The phenotype varies as a complex, nonlinear function of the interaction between underlying genetic structures and current environmental conditions. Very different genetic structures may code for equivalent behaviors, just as diverse computer programs can generate similar functions.

Selection directly acts only on the expressed behaviors of individuals and species [19, pp. 477–478]. Wright [21] offered the concept of adaptive topography to describe the fitness of individuals and species (minimally, isolated reproductive populations termed *demes*). A population of genotypes maps to respective phenotypes (*sensu* Lewontin [22]), which are in turn mapped onto the adaptive topography (Fig. 2). Each peak corresponds to an optimized collection of phenotypes, and thus one or more sets of optimized genotypes. Evolution probabilistically proceeds up the slopes of the topography toward peaks as selection culls inappropriate phenotypic variants.

Others [11], [23, pp. 400–401] have suggested that it is more appropriate to view the adaptive landscape from an inverted

Manuscript received April 15, 1993; revised August 2, 1993.
The author is with Natural Selection, Inc., La Jolla, CA 92037.
IEEE Log Number 9213549.

Reprinted from *IEEE Transactions on Neural Networks*, Vol. 5:1, pp. 3-14, January, 1994.

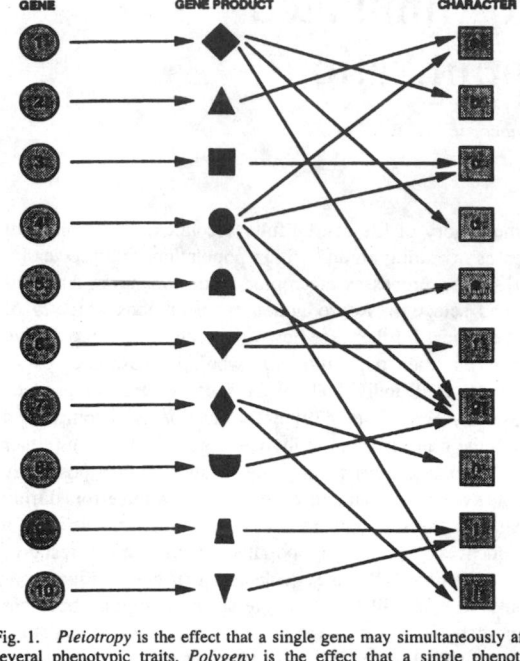

Fig. 1. *Pleiotropy* is the effect that a single gene may simultaneously affect several phenotypic traits. *Polygeny* is the effect that a single phenotypic characteristic may be determined by the simultaneous interaction of many genes. These one-to-many and many-to-one mappings are pervasive in natural systems. As a result, even small changes to a single gene may induce a raft of behavioral changes in the individual (after [18]).

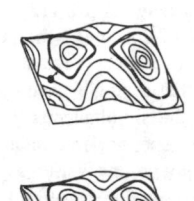

Fig. 2. Wright's adaptive topology, inverted. An adaptive topography, or adaptive landscape, is defined to represent the fitness of all possible phenotypes. Wright [21] proposed that as selection culls the least appropriate existing behaviors relative to others in the population, the population advances to areas of higher fitness on the landscape. Atmar [11] and others have suggested viewing the topography from an inverted perspective. Populations then advance to areas of lower behavioral error.

position. The peaks become troughs, "minimized prediction error entropy wells" [11]. Such a viewpoint is intuitively appealing. Searching for peaks depicts evolution as a slowly advancing, tedious, uncertain process. Moreover, there appears to be a certain fragility to an evolving phyletic line; an optimized population might be expected to quickly fall off the peak under slight perturbations. The inverted topography leaves an altogether different impression. Populations advance

rapidly, falling down the walls of the error troughs until their cohesive set of interrelated behaviors are optimized, at which point stagnation occurs. If the topography is generally static, rapid descents will be followed by long periods of stasis. If, however, the topography is in continual flux, stagnation may never set in.

Viewed in this manner, evolution is an obvious optimizing problem-solving process. Selection drives phenotypes as close to the optimum as possible, given initial conditions and environmental constraints. But the environment is continually changing. Species lag behind, constantly evolving toward a new optimum. No organism should be viewed as being perfectly adapted to its environment. The suboptimality of behavior is to be expected in any dynamic environment that mandates trade-offs between behavioral requirements. But selection never ceases to operate, regardless of the population's position on the topography.

Mayr [19, p. 532] has summarized some of the more salient characteristics of the neo-Darwinian paradigm. These include:

1) The individual is the primary target of selection.
2) Genetic variation is largely a chance phenomenon. Stochastic processes play a significant role in evolution.
3) Genotypic variation is largely a product of recombination and "only ultimately of mutation."
4) "Gradual" evolution may incorporate phenotypic discontinuities.
5) Not all phenotypic changes are necessarily consequences of *ad hoc* natural selection.
6) Evolution is a change in adaptation and diversity, not merely a change in gene frequencies.
7) Selection is probabilistic, not deterministic.

Simulations of evolution should rely on these foundations.

II. GENETIC ALGORITHMS

Fraser [24]–[28], Bremermann *et al.* [29]–[36], Reed *et al.* [37], and Holland [38], [39] proposed similar algorithms that simulate genetic systems. These procedures are now described by the term *genetic algorithms* and are typically implemented as follows:

1) The problem to be addressed is defined and captured in an objective function that indicates the fitness of any potential solution.
2) A population of candidate solutions is initialized subject to certain constraints. Typically, each trial solution is coded as a vector x, termed a *chromosome,* with elements being described as *genes* and varying values at specific positions called *alleles.* Holland [39, pp. 70–72] suggested that all solutions should be represented by binary strings. For example, if it were desired to find the scalar value x that maximizes:

$$F(x) = -x^2,$$

then a finite range of values for x would be selected and the minimum possible value in the range would be represented by the string $\{0 \ldots 0\}$, with the maximum value being represented by the string $\{1 \ldots 1\}$. The de-

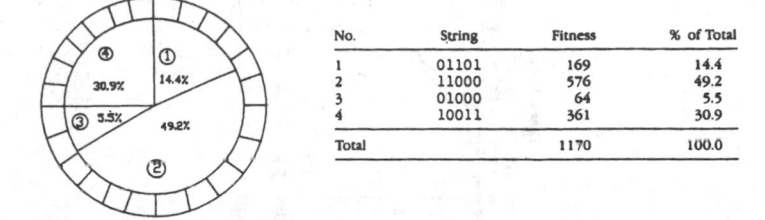

No.	String	Fitness	% of Total
1	01101	169	14.4
2	11000	576	49.2
3	01000	64	5.5
4	10011	361	30.9
Total		1170	100.0

Fig. 3. Roulette wheel selection in genetic algorithms. Selection in genetic algorithms is often accomplished via differential reproduction according to fitness. In the typical approach, each chromosome is given a probability of being copied into the next generation that is proportional to its fitness relative to all other chromosomes in the population. Successive trials are conducted in which a chromosome is selected, until all available positions are filled. Those chromosomes with above-average fitness will tend to generate more copies than those with below-average fitness. The figure is adapted from [143].

sired degree of precision would indicate the appropriate length of the binary coding.

3) Each chromosome, x_i, $i = 1, \ldots, P$, in the population is decoded into a form appropriate for evaluation and is then assigned a fitness score, $\mu(x_i)$ according to the objective.

4) Each chromosome is assigned a probability of reproduction, p_i, $i = 1, \ldots, P$, so that its likelihood of being selected is proportional to its fitness relative to the other chromosomes in the population. If the fitness of each chromosome is a strictly positive number to be maximized, this is often accomplished using *roulette wheel selection* (see Fig. 3).

5) According to the assigned probabilities of reproduction, p_i, $i = 1, \ldots, P$, a new population of chromosomes is generated by probabilistically selecting strings from the current population. The selected chromosomes generate "offspring" via the use of specific genetic operators, such as crossover and bit mutation. Crossover is applied to two chromosomes *(parents)* and creates two new chromosomes *(offspring)* by selecting a random position along the coding and splicing the section that appears before the selected position in the first string with the section that appears after the selected position in the second string, and vice versa (see Fig. 4). Other, more sophisticated, crossover operators have been introduced and will be discussed later. Bit mutation simply offers the chance to flip each bit in the coding of a new solution. Typical values for the probabilities of crossover and bit mutation range from 0.6 to 0.95 and 0.001 to 0.01, respectively [40], [41].

6) The process is halted if a suitable solution has been found, or if the available computing time has expired; otherwise the process proceeds to step (3) where the new chromosomes are scored and the cycle is repeated.

For example, suppose the task is to find a vector of 100 bits {0,1} such that the sum of all of the bits in the vector is maximized. The objective function could be written as:

$$\mu(x) = \sum_{i=1}^{100} x_i,$$

where x is a vector of 100 symbols from {0,1}. Any such vector x could be scored with respect to $\mu(x)$ and would

Crossover Point

Parent #1: 1 1 0 1 | 0 1 1 1 1 0 1   Offspring #1: 1 0 1 0 0 1 1 1 1 0 1

→

Parent #2: 1 0 1 0 | 0 0 0 0 1 0 0   Offspring #2: 1 1 0 1 0 0 0 0 1 0 0

Fig. 4. The one-point crossover operator. A typical method of recombination in genetic algorithms is to select two parents and randomly choose a splicing point along the chromosomes. The segments from the two parents are exchanged and two new offspring are created.

receive a fitness rating ranging from zero to 100. Let an initial population of 100 parents be selected completely at random and subjected to roulette wheel selection in light of $\mu(x)$, with the probabilities of crossover and bit mutation being 0.8 and 0.01, respectively. Fig. 5 shows the rate of improvement of the best vector in the population, and the average of all parents, at each generation (one complete iteration of steps 3–6) under such conditions. The process rapidly converges on vectors of all 1's.

There are a number of issues that must be addressed when using a genetic algorithm. For example, the necessity for binary codings has received considerable criticism [42]–[44]. To understand the motivation for using bit strings, the notion of a schema must be introduced. Consider a string of symbols from an alphabet A. Suppose that some of the components of the string are held fixed while others are free to vary. Define a wild card symbol, #, that matches any symbol from A. A string with fixed and variable symbols defines a schema. Consider the string {01##}, defined over the union of {#} and the alphabet $A = \{0,1\}$. This set includes {0100}, {0101}, {0110} and {0111}. Holland [39, pp. 66–74] recognized that every string that is evaluated actually offers partial information about the expected fitness of all possible schemata in which that string resides. That is, if the string {0000} is evaluated to have some fitness, then partial information is also received about the worth of sampling from variations in {0###}, {#0##}, {#00#}, {#0#0}, and so forth. This characteristic is termed *implicit parallelism*, as it is through a single sample that information is gained with respect to many schemata. Holland [39, p. 71] speculated that it would be beneficial to maximize the number of schemata being sampled, thus providing maximum implicit parallelism, and proved that this is achieved for

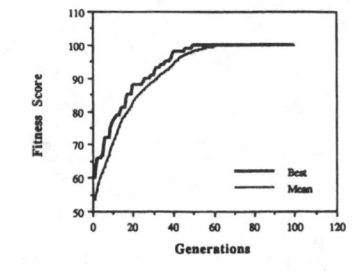

Fig. 5. The rate of optimization in a simple binary coding problem using a standard genetic algorithm. The curves indicate the fitness of the best chromosomes in the population and the mean fitness of all parents at each generation. The optimum fitness is 100 units.

$|A| = 2$. Binary strings were therefore suggested as a universal representation.

The use of binary strings is not universally accepted in genetic algorithm literature, however. Michalewicz [44, p. 82] indicates that for real-valued numerical optimization problems, floating-point representations outperform binary representations because they are more consistent, more precise, and lead to faster execution. But Michalewicz [44, p. 75] also claims that genetic algorithms perform poorly when the state space of possible solutions is extremely large, as would be required for high-precision numerical optimization of many variables that could take on real-values in a large range. This claim is perhaps too broad. The size of the state space alone does not determine the efficiency of the genetic algorithm, regardless of the choice of representation. Very large state spaces can sometimes be searched quite efficiently, and relatively small state spaces sometimes provide significant difficulties. But it is fair to say that maximizing implicit parallelism will not always provide for optimum performance. Many researchers in genetic algorithms have foregone the bit strings suggested by Holland [39, pp. 70–72] and have achieved reasonable results to difficult problems [44]–[47].

Selection in proportion to fitness can be problematic. There are two practical considerations: 1) roulette wheel selection depends upon positive values, and 2) simply adding a large constant value to the objective function can eliminate selection, with the algorithm then proceeding as a purely random walk. There are several heuristics that have been devised to compensate for these issues. For example, the fitness of all parents can be scaled relative to the lowest fitness in the population, or proportional selection can be based on ranking by fitness. Selection based on ranking also eliminates problems with functions that have large offsets.

One mathematical problem with selecting parents to reproduce in proportion to their relative fitness is that this procedure cannot ensure asymptotic convergence to a global optimum [48]. The best chromosome in the population may be lost at any generation, and there is no assurance that any gains made up to a given generation will be retained in future generations. This can be overcome by employing a heuristic termed *elitist selection* [49], which simply always retains the best chromosome in the population. This procedure guarantees asymptotic convergence [48], [50], [51], but the specific rates of convergence vary by problem and are generally unknown.

The crossover operator has been termed the distinguishing feature of genetic algorithms [52, pp. 17–18]. Holland [39, pp. 110–111] indicates that crossover provides the main search operator while bit mutation simply serves as a background operator to ensure that all possible alleles can enter the population. The probabilities commonly assigned to crossover and bit mutation reflect this philosophical view. But the choice of crossover operator is not straightforward.

Holland [39, p. 160], and others [53], [54], propose that genetic algorithms work by identifying good "building blocks" and eventually combining these to get larger building blocks. This idea has become known as the *building block hypothesis*. The hypothesis suggests that a one-point crossover operator would perform better than an operator that, say, took one bit from either parent with equal probability *(uniform crossover)*, because it could maintain sequences (blocks) of "good code" that are associated with above-average performance and not disrupt their linkage. But this has not been clearly demonstrated in the literature. Syswerda [55] conducted function optimization experiments with uniform crossover, two-point crossover and one-point crossover. Uniform crossover provided generally better solutions with less computational effort. Moreover, it has been noted that sections of code that reside at opposite ends of a chromosome are more likely to be disrupted under one-point crossover than are sections that are near the middle of the chromosome. Holland [39, pp. 106–109] proposed an inversion operator that would reverse the index position for a section of the chromosome, so that linkages could be constructed between arbitrary genes. But inversion has not been found to be useful in practice [52, p. 21]. The relevance of the building block hypothesis is presently unclear, but its value is likely to vary significantly by problem.

Premature convergence is another important concern in genetic algorithms. This occurs when the population of chromosomes reaches a configuration such that crossover no longer produces offspring that can outperform their parents, as must be the case in a homogeneous population. Under such circumstances, all standard forms of crossover simply regenerate the current parents. Any further optimization relies solely on bit mutation and can be quite slow. Premature convergence is often observed in genetic algorithm research ([40], [52, pp. 25, 26], [56], [57], and others) because of the exponential reproduction of the best observed chromosomes coupled with the strong emphasis on crossover. Davis [52, pp. 26, 27]

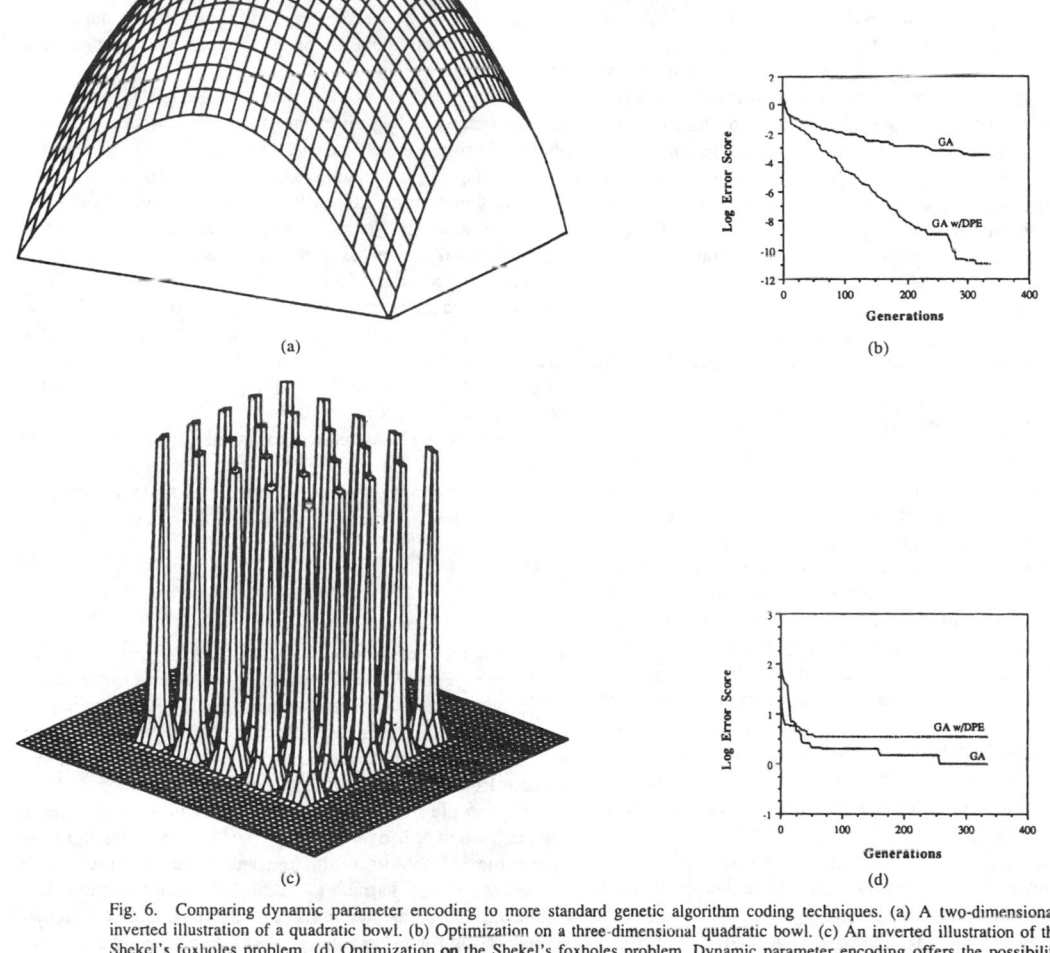

(a)

(b)

(c)

(d)

Fig. 6. Comparing dynamic parameter encoding to more standard genetic algorithm coding techniques. (a) A two-dimensional, inverted illustration of a quadratic bowl. (b) Optimization on a three-dimensional quadratic bowl. (c) An inverted illustration of the Shekel's foxholes problem. (d) Optimization on the Shekel's foxholes problem. Dynamic parameter encoding offers the possibility of increasing the precision of a solution on-line, but may also encounter problems with premature convergence.

recommends that when the population converges on a chromosome that would require the simultaneous mutation of many bits in order to improve it, the run is practically completed and it should either be restarted using a different random seed, or hill-climbing heuristics should be employed to search for improvements.

One recent proposal for alleviating the problems associated with premature convergence was offered in [41]. The method, termed *dynamic parameter encoding* (DPE), dynamically resizes the available range of each parameter. Broadly, when a heuristic suggests that the population has converged, the minimum and maximum values for the range are resized to a smaller window and the process is iterated. In this manner, DPE can zoom in on solutions that are closer to

the global optimum than provided by the initial precision. Schraudolph [58] has kindly provided results from experiments with DPE presented in [41]. As indicated in Fig. 6, DPE clearly outperforms the standard genetic algorithm when searching a quadratic bowl, but actually performs worse on a multimodal function (Shekel's foxholes). The effectiveness of DPE is an open, promising area of research. DPE only zooms in, so the initial range of parameters must be set to include the global optimum or it will not be found. But it would be relatively straightforward to include a mechanism in DPE to expand the search window, as well as reduce it.

Although many open questions remain, genetic algorithms have been used to successfully address diverse practical optimization problems [59]. While some researchers do not view

7

genetic algorithms as function optimization procedures *per se* (e.g., [60]), they are commonly used for precisely that purpose. Current research efforts include: 1) developing a stronger mathematical foundation for the genetic algorithm as an optimization technique [41], [48], [61], [62], including analysis of classes of problems that are difficult for genetic algorithms [63]–[66] as well as the sensitivity to performance of the general technique to various operator and parameter settings [42], [44], [67]–[70]; 2) comparing genetic algorithms to other optimization methods and examining the manner in which they can be enhanced by incorporating other procedures such as simulated annealing [71]–[73]; 3) using genetic algorithms for computer programming and engineering problems [74]–[79]; 4) applying genetic algorithms to machine learning rule-based classifier systems [80]–[84]; 5) using genetic algorithms as a basis for artificial life simulations [85], [86, pp. 186–195]; and 6) implementing genetic algorithms on parallel machines [87]–[89]. The most recent investigations can be found in [90].

III. EVOLUTION STRATEGIES AND EVOLUTIONARY PROGRAMMING

An alternative approach to simulating evolution was independently adopted by Schwefel [91] and Rechenberg [92] collaborating in Germany, and L. Fogel [93], [94] in the United States, and later pursued by [95]–[99], among others. These models, commonly described by the terms *evolution strategies* or *evolutionary programming,* or more broadly as *evolutionary algorithms* [87], [100] (although many authors use this term to describe the entire field of simulated evolution), emphasize the behavioral link between parents and offspring, or between reproductive populations, rather than the genetic link. When applied to real-valued function optimization, the most simple method is implemented as follows:

1) The problem is defined as finding the real-valued n-dimensional vector x that is associated with the extremum of a functional $F(x) : R^n \rightarrow R$. Without loss of generality, let the procedure be implemented as a minimization process.
2) An initial population of parent vectors, x_i, $i = 1, \ldots, P$, is selected at random from a feasible range in each dimension. The distribution of initial trials is typically uniform.
3) An offspring vector, x_i', $i = 1, \ldots, P$, is created from each parent x_i by adding a Gaussian random variable with zero mean and preselected standard deviation to each component of x_i.
4) Selection then determines which of these vectors to maintain by comparing the errors $F(x_i)$ and $F(x_i')$, $i = 1, \ldots, P$. The P vectors that possess the least error become the new parents for the next generation.
5) The process of generating new trials and selecting those with least error continues until a sufficient solution is reached or the available computation is exhausted.

In this model, each component of a trial solution is viewed as a behavioral trait, not as a gene. A genetic source for these phenotypic traits is presumed, but the nature of the linkage is not detailed. It is assumed that whatever genetic transfor-

mations occur, the resulting change in each behavioral trait will follow a Gaussian distribution with zero mean difference and some standard deviation. Specific genetic alterations can affect many phenotypic characteristics due to pleiotropy and polygeny (Fig. 1). It is therefore appropriate to simultaneously vary all of the components of a parent in the creation of a new offspring.

The original efforts in evolution strategies [91], [92] examined the preceding algorithm but focused on a single parent-single offspring search. This was termed a $(1 + 1) - ES$ in that a single offspring is created from a single parent and both are placed in competition for survival, with selection eliminating the poorer solution. There were two main drawbacks to this approach when viewed as a practical optimization algorithm: 1) the constant standard deviation (step size) in each dimension made the procedure slow to converge on optimal solutions, and 2) the brittle nature of a point-to-point search made the procedure susceptible to stagnation at local minima (although the procedure can be shown to asymptotically converge to the global optimum vector x) [101].

Rechenberg [92] defined the expected convergence rate of the algorithm as the ratio of the average distance covered toward the optimum and the number of trials required to achieve this improvement. For a quadratic function

$$ F(x) = \sum_{i=1}^{n} x_i^2, \tag{1} $$

where x is an n-dimensional vector of reals, and x_i denotes the ith component of x, Rechenberg [92] demonstrated that the optimum expected convergence rate is given when $\sigma \approx 1.224r/n$, where σ is the standard deviation of the zero mean Gaussian perturbation, r denotes the current Euclidean distance from the optimumn and there are n dimensions. Thus, for this simple function the optimum convergence rate is obtained when the average step size is proportional to the square root of the error function and inversely proportional to the number of variables. Additional analyses have been conducted on other functions and the results have yielded similar forms for setting the standard deviation [102].

The use of multiple parents and offspring in evolution strategies was developed by Schwefel [103], [104]. Two approaches are currently explored, denoted by $(\mu + \lambda) - ES$ and $(\mu, \lambda) - ES$. In the former, μ parents are used to create λ offspring and all solutions compete for survival, with the best being selected as parents of the next generation. In the latter, only the λ offspring compete for survival, and the parents are completely replaced each generation. That is, the lifespan of every solution is limited to a single generation. Increasing the population size increases the rate of optimization over a fixed number of generations.

To provide a very simple example, suppose it is desired to find the minimum of the function in (1) for $n = 3$. Let the original population consist of 30 parents, with each component initialized in accordance with a uniform distribution over $[-5.12, 5.12]$ (after [40]). Let one offspring be created from each parent by adding a Gaussian random variable with mean zero and variance equal to the error score of the parent divided

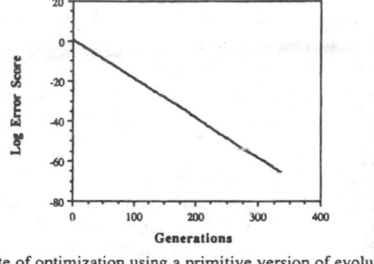

Fig. 7. The rate of optimization using a primitive version of evolution strategies on the three-dimensional quadratic bowl. Thirty parents are maintained at each generation. Offspring are created by adding a Gaussian random variable to each component.

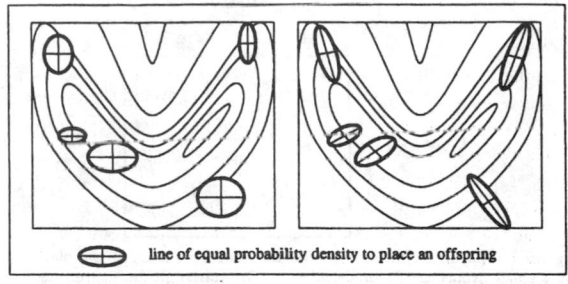

line of equal probability density to place an offspring

Fig. 8. Under independent Gaussian perturbations to each component of every parent, new trials are are distributed such that the contours of equal probability are aligned with the coordinate axes (left picture). This will not be optimal in general because the contours of the response are rarely similarly aligned. Schwefel [104] suggests a mechanism for incorporating self-adaptive covariance terms. Under this procedure, new trials can be distributed in any orientation (right picture). The evolutionary process adapts to the contours of the response surface, distributing trials so as to maximaize the probability of discovering improved solutions.

by the square of the number of dimensions ($3^2 = 9$) to each component. Let selection simply retain the best 30 vectors in the population of parents and offspring. Fig. 7 indicates the rate of optimization of the best vector in the population as a function of the number of generations. The process rapidly converges close to the unique global optimum.

Rather than using a heuristic schedule for reducing the step size over time, Schwefel [104] developed the idea of making the distribution of new trials from each parent an additional adaptive parameter (Rechenberg, personal communication, indicates that he introduced the idea in 1967). In this procedure, each solution vector comprises not only the trial vector x of n dimensions, but a perturbation vector σ which provides instructions on how to mutate x and is itself subject to mutation. For example, if x is the current position vector and σ is a vector of variances corresponding to each dimension of x, then a new solution vector (x',σ') could be created as:

$$\sigma'_i = \sigma_i \exp(\tau' \cdot N(0,1) + \tau \cdot N_i(0,1))$$
$$x'_i = x_i + N(0,\sigma'_i)$$

where $i = 1, \ldots, n$, and $N(0,1)$ represents a single standard Gaussian random variable, $N_i(0,1)$ represents the ith independent identically distributed standard Gaussian, and τ and τ' are operator set parameters which define global and individual step-sizes [102]. In this manner, the evolution strategy can self-adapt to the width of the error surface and more appropriately distribute trials. This method was extended again [104] to incorporate correlated mutations so that the distribution of new trials could adapt to contours on the error surface (Fig. 8).

Finally, additional extensions were made to evolution strategies to include methods for recombining individual solutions in the creation of new offspring. There are many proposed procedures. These include selecting individual components from either of two parents at random, averaging individual components from two parents with a given weighting, and so forth [102].

The original evolutionary programming approach was similar to that of Schwefel and Rechenberg but involved a more complex problem, that of creating artificial intelligence. Fogel [94] proposed that intelligent behavior requires the composite ability to predict one's environment coupled with a translation of the predictions into a suitable response in light of the given goal. To provide maximum generality, in a series

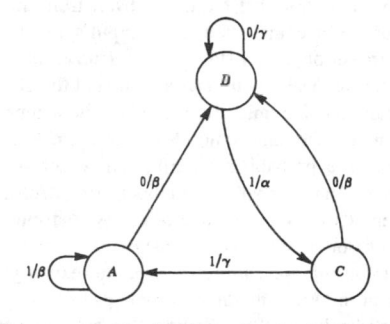

Fig. 9. A finite state machine (FSM) consists of a finite number of states. For each state, for every possible input symbol, there is an associated output symbol and next-state transition. In the figure, input symbols are shown to the left of the virgule, output symbols are shown to the right. The input alphabet is {0, 1} and the output alphabet is {α, β, γ}. The machine is presumed to start in state A. The figure is taken from [144].

of experiments, a simulated environment was described as sequence of symbols taken from a finite alphabet. The problem was then defined to evolve an algorithm that would operate on the sequence of symbols thus far observed in such a manner as to produce an output symbol that is likely to maximize the benefit to the algorithm in light of the next symbol to appear in the environment and a well-defined payoff function. Finite state machines (FSM's) [105] provided a useful representation for the required behavior (Fig. 9).

Evolutionary programming operated on FSM's as follows:

1) Initially, a population of parent FSM's is randomly constructed.

2) The parents are exposed to the environment; that is, the sequence of symbols that have been observed up to the current time. For each parent machine, as each input symbol is offered to the machine, each output symbol is compared to the next input symbol. The worth of this prediction is then measured with respect to the given payoff function (e.g., all–none, absolute error, squared error, or any other expression of the meaning of the symbols). After the last prediction is made, a function

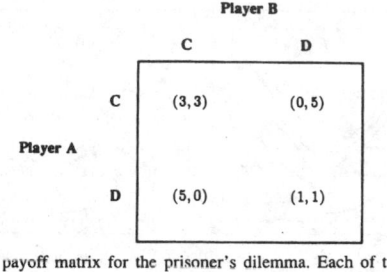

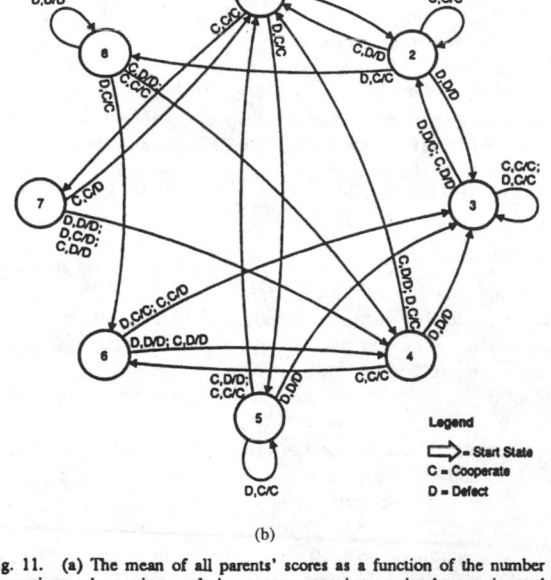

Fig. 10. A payoff matrix for the prisoner's dilemma. Each of two players must either cooperate (C) or defect (D). The entries in the matrix, (a,b), indicate the gain to players A and B, respectively. This payoff matrix was used in simulations in [106]–[108].

of the payoff for each symbol (e.g., average payoff per symbol) indicates the fitness of the machine.

3) Offspring machines are created by randomly mutating each parent machine. There are five possible modes of random mutation that naturally result from the description of the machine: change an output symbol, change a state transition, add a state, delete a state, or change the initial state. The deletion of a state and the changing of the start state are only allowed when the parent machine has more than one state. Mutations are chosen with respect to a probability distribution, which is typically uniform. The number of mutations per offspring is also chosen with respect to a probability distribution (e.g., Poisson) or may be fixed a priori.

4) The offspring are evaluated over the existing environment in the same manner as their parents.

5) Those machines that provide the greatest payoff are retained to become parents of the next generation. Typically, the parent population remains the same size, simply for convenience.

6) Steps 3)–5) are iterated until it is required to make an actual prediction of the next symbol (not yet experienced) from the environment. The best machine is selected to generate this prediction, the new symbol is added to the experienced environment, and the process reverts to step 2).

The prediction problem is a sequence of static optimization problems in which the adaptive topography (fitness function) is time-varying. The process can be easily extended to abstract situations in which the payoffs for individual behaviors depend not only on an extrinsic payoff function, but also on the behavior of other individuals in the population. For example, Fogel [106], [107], following previous foundational research by Axelrod using genetic algorithms [108], evolved a population of FSM's in light of the iterated prisoner's dilemma (Fig. 10). Starting with completely random FSM's of one to five states, but ultimately possessing a maximum of eight states, the simulated evolution quickly converged on mutually cooperative behavior (Fig. 11). The evolving FSM's essentially learned to predict the behavior (a sequence of symbols) of other FSM's in the evolving population.

Evolutionary programming has recently been applied to real-valued continuous optimization problems and is virtually

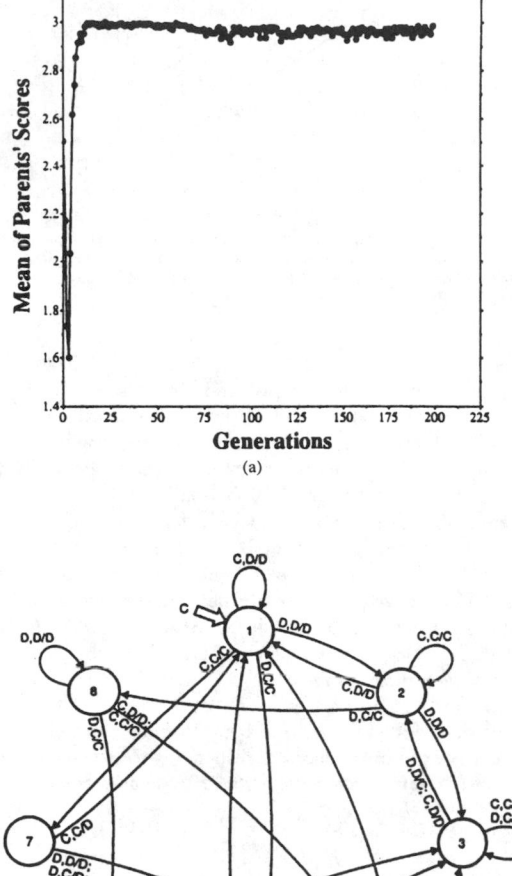

Fig. 11. (a) The mean of all parents' scores as a function of the number generations when using evolutionary programming to simulate an iterated prisoner's dilemma incorporating 50 parents coded as finite state machines (FSM's). The input alphabet consists of the previous moves for the current player and the opponent {(C,C), (C,D), (D,C), (D,D)}; the output alphabet consists of the next move {C,D}. Each FSM plays against every other FSM in the population over a long series of moves. The results indicate a propensity to evolve cooperative behavior even though it would appear more beneficial for an individual to defect on any given play. (b) A typical FSM evolved after 200 generations when using 100 parents. The cooperative nature of the machine can be observed by noting that (C,C) typically elicits further cooperation, and in states 2 and 3, such cooperation will be absorbing. Further, (D,D) typically elicits further defection, indicating that the machine will not be taken advantage of during an encounter with a purely selfish machine. These results appear in [107].

equivalent in many cases to the procedures used in evolution strategies. The extension to using self-adapting independent variances was offered in [109] with procedures for optimizing the covariance matrix used in generating new trials offered in [110]. These methods differ from those offered in [104] in that Gaussian perturbations are appl ied to the self-adaptive parameters instead of lognormal perturbations. Initial comparisons [111], [112] indicate that the procedures in [104] appear to be more robust than those in [110]. One possible explanation for this would be that it is easier for variances of individual terms to transition between small and large values under the method of [104]. Theoretical and empirical comparison between these mechanisms is an open area of research.

As currently implemented, there are two essential differences between evolution strategies and evolutionary programming.

1) Evolution strategies rely on strict deterministic selection. Evolutionary programming typically emphasizes the probabilistic nature of selection by conducting a stochastic tournament for survival at each generation. The probability that a particular trial solution will be maintained is made a function of its rank in the population.

2) Evolution strategies typically abstracts coding structures as analogues of individuals. Evolutionary programming typically abstracts coding structures as analogues of distinct species (reproductive populations). Therefore, evolution strategies may use recombination operations to generate new trials [111], but evolutionary programming does not, as there is no sexual communication between species [100].

The current efforts in evolution strategies and evolutionary programming follow lines of investigation similar to those in genetic algorithms: 1) developing mathematical foundations for the procedures [51], [111], [113], investigating their computational complexity theoretically and empirically [114], [115] and combining evolutionary optimization with more traditional search techniques [116]; 2) using evolutionary algorithms to train and design neural networks [117]–[121]; 3) examining evolutionary algorithms for system identification, control, and robotics applications [122]–[127], as well as pattern recognition problems [128]–[130], along with the possibility for synergism between evolutionary and fuzzy systems [131], [132]; 4) applying evolutionary optimization to machine learning [133]; 5) relating evolutionary models to biological observations or applications [107], [134]–[137]; and also 6) designing evolutionary algorithms for implementation on parallel processing machines [138], [139], [140]. The most recent investigations can be found in [141], [142].

IV. SUMMARY

Simulated evolution has a long history. Similar ideas and implementations have been independently invented numerous times. There are currently three main lines of investigation: genetic algorithms, evolution strategies, and evolutionary programming. These methods share many similarities. Each maintains a population of trial solutions, imposes random changes to those solutions, and incorporates the use of selection to determine which solutions to maintain into future generations and which to remove from the pool of trials. But these methods also have important differences. Genetic algorithms emphasize models of genetic operators as observed in nature, such as crossing over, inversion, and point mutation and apply these to abstracted chromosomes. Evolution strategies and evolutionary programming emphasize mutational transformations that maintain behavioral linkage between each parent and its offspring, respectively, at the level of the individual or the species. Recombination may be appropriately applied to individuals, but is not applicable for species.

No model can be a complete description of the true system. Each of the three possible evolutionary approaches described above is incomplete. But each has also been demonstrated to be of practical use when applied to difficult optimization problems. The greatest potential for the application of evolutionary optimization to real-world problems will come from their implementation on parallel machines, for evolution is an inherently parallel process. Recent advances in distributed processing architectures will result in dramatically reduced execution times for simulations that would simply be impractical on current serial computers.

Natural evolution is a robust yet efficient problem-solving technique. Simulated evolution can be made as robust. The same procedures can be applied to diverse problems with relatively little reprogramming. While such efforts will undoubtedly continue to address difficult real-world problems, the ultimate advancement of the field will, as always, rely on the careful observation and abstraction of the natural process of evolution.

ACKNOWLEDGMENT

The author is grateful to W. Atmar, T. Bäck, L. Davis, G. B. Fogel, L. J. Fogel, E. Mayr, Z. Michalewicz, G. Rudolph, H.-P. Schwefel, and the anonymous referees for their helpful comments and criticisms of this review.

REFERENCES

[1] M. S. Bazaraa and C. M. Shetty, *Nonlinear Programming*, New York: John Wiley, 1979.
[2] B. D. O. Anderson, R. R. Bitmead, C. R. Johnson, P. V. Kokotovic, R. L. Kosut, I. M. Y. Mareels, L. Praly, and B. D. Riedle, *Stability of Adaptive Systems: Passivity and Averaging Analysis*. Cambridge, MA: MIT Press, 1986.
[3] P. Werbos, "Beyond regression: new tools for prediction and analysis in the behavioral sciences," Doctoral dissertation, Harvard University, 1974.
[4] D. E. Rumelhart and J. L. McClelland, *Parallel Distributed Processing: Explorations in the Microstructures of Cognition*. vol. 1, Cambridge, MA: MIT Press, 1986.
[5] L. Ljung, *System Identification: Theory for the User*, Englewood Cliffs, NJ: Prentice-Hall, 1987.
[6] A. Hoffman, *Arguments on Evolution: A Paleontologist's Perspective*, New York: Oxford University Press, 1988.
[7] T. R. Malthus, *An Essay on the Principle of Population, as it Affects the Future Improvement of Society*, 6th ed., London: Murray, 1826.
[8] E. Mayr, *The Growth of Biological Thought: Diversity, Evolution and Inheritance*, Cambridge, MA: Belknap Press, 1988.
[9] J. Huxley, "The evolutionary process," in *Evolution as a Process*, J. Huxley, A. C. Hardy, and E. B. Ford, Eds. New York: Collier Books, pp. 9–33, 1963.
[10] D. E. Wooldridge, *The Mechanical Man: The Physical Basis of Intelligent Life*. New York: McGraw-Hill, 1968.

[11] W. Atmar, "The inevitability of evolutionary invention," unpublished manuscript, 1979.
[12] S. Wright, "Evolution in Mendelian populations," *Genetics*, vol. 16, pp. 97–159, 1931.
[13] S. Wright, "The evolution of life," Panel discussion in *Evolution After Darwin: Issues in Evolution*, vol. III, S. Tax and C. Callender, Eds. Chicago: Univ. of Chicago Press, 1960.
[14] G. G. Simpson, *The Meaning of Evolution: A Study of the History of Life and Its Significance for Man*. New Haven, CT: Yale Univ. Press, 1949.
[15] T. Dobzhansky, *Genetics of the Evolutionary Processes*. New York: Columbia Univ. Press, 1970.
[16] S. M. Stanley, "A theory of evolution above the species level," *Proc. Nat. Acad. Sci.*, vol. 72, no. 2, pp. 646–650, 1975.
[17] E. Mayr, "Where are we?" *Cold Spring Harbor Symp. Quant. Biol.*, vol. 24, pp. 409–440, 1959.
[18] E. Mayr, *Animal Species and Evolution*. Cambridge, MA: Belknap Press, 1963.
[19] E. Mayr, *Toward a New Philosophy of Biology: Observations of an Evolutionist*. Cambridge, MA: Belknap Press, 1988.
[20] R. Dawkins, *The Blind Watchmaker*. Oxford: Clarendon Press, 1986.
[21] S. Wright, "The roles of mutation, inbreeding, crossbreeding, and selection in evolution," *Proc. 6th Int. Cong. Genetics, Ithaca*, vol. 1, pp. 356–366, 1932.
[22] R. C. Lewontin, *The Genetic Basis of Evolutionary Change*. New York: Columbia University Press, NY, 1974.
[23] P. H. Raven and G. B. Johnson, *Biology*, St. Louis, MO: Times Mirror, 1986.
[24] A. S. Fraser, "Simulation of genetic systems by automatic digital computers. I. Introduction," *Australian J. of Biol. Sci.*, vol. 10, pp. 484–491, 1957.
[25] A. S. Fraser, "Simulation of genetic systems by automatic digital computers. II. Effects of linkage on rates of advance under selection," *Australian J. of Biol. Sci.*, vol. 10, pp. 492–499, 1957.
[26] A. S. Fraser, "Simulation of genetic systems by automatic digital computers. IV. Epistasis," *Australian J. of Biol. Sci.*, vol. 13, pp. 329–346, 1960.
[27] A. S. Fraser, "Simulation of genetic systems," *J. of Theor. Biol.*, vol. 2, pp. 329–346, 1962.
[28] A. S. Fraser, "The evolution of purposive behavior," in *Purposive Systems*, H. von Foerster, J. D. White, L. J. Peterson, and J. K. Russell, Eds. Washington, DC: Spartan Books, pp. 15–23, 1968.
[29] H. J. Bremermann, "The evolution of intelligence. The nervous system as a model of its environment," Technical Report No. 1, Contract No. 477(17), Dept. of Mathematics, Univ. of Washington, Seattle, 1958.
[30] H. J. Bremermann, "Optimization through evolution and recombination," in *Self-Organizing Systems*. M. C. Yovits, G. T. Jacobi, and G. D. Goldstine, Eds. Washington, DC: Spartan Books, pp. 93–106, 1962.
[31] H. J. Bremermann, "Quantitative aspects of goal-seeking self-organizing systems," in *Progress in Theoretical Biology*, vol. 1, New York: Academic Press, pp. 59–77, 1967.
[32] H. J. Bremermann, "Numerical Optimization Procedures Derived from Biological Evolution Processes," in *Cybernetic Problems in Bionics*, H. L. Oestreicher and D. R. Moore, Eds. New York: Gordon & Breach, pp. 543–562, 1968.
[33] H. J. Bremermann, "On the Dynamics and Trajectories of Evolution Processes," in *Biogenesis, Evolution, Homeostasis*. A. Locker, Ed. New York: Springer-Verlag, pp. 29–37, 1973.
[34] H. J. Bremermann and M. Rogson, "An Evolution-Type Search Method for Convex Sets," ONR Technical Report, Contracts 222(85) and 3656(58), UC Berkeley, 1964.
[35] H. J. Bremermann, M. Rogson, and S. Salaff, "Search by Evolution," in *Biophysics and Cybernetic Systems*. M. Maxfield, A. Callahan, and L. J. Fogel, Eds. Washington, DC: Spartan Books, pp. 157–167, 1965.
[36] H. J. Bremermann, M. Rogson, and S. Salaff, "Global Properties of Evolution Processes," in *Natural Automata and Useful Simulations*. H. H. Pattee, E. A. Edlsack, L. Fein, and A. B. Callahan, Eds. Washington, DC: Spartan Books, pp. 3–41, 1966.
[37] J. Reed, R. Toombs, and N. A. Barricelli, "Simulation of biological evolution and machine learning," *Journal of Theoretical Biology*, vol. 17, pp. 319–342, 1967.
[38] J. H. Holland, "Adaptive plans optimal for payoff-only environments," *Proc. of the 2nd Hawaii Int. Conf. on System Sciences*, pp. 917–920, 1969.
[39] J. H. Holland, *Adaptation in Natural and Artificial Systems*. Ann Arbor: Univ. Of Michigan Press, 1975.
[40] K. A. De Jong, "The analysis of the behavior of a class of genetic adaptive systems," Doctoral dissertation, Univ. of Michigan, Ann Arbor, 1975.

[41] N. N. Schraudolph and R. K. Belew, "Dynamic parameter encoding for genetic algorithms," *Machine Learning*, vol. 9, no. 1, pp. 9–21, 1992.
[42] G. A. Vignaux and Z. Michalewicz, "A genetic algorithm for the linear transportation problem," *IEEE Trans. on Systems, Man and Cybernetics*, vol. 21, no. 2, pp. 445–452, 1991.
[43] J. Antonisse, "A new interpretation of schema notation that overturns the binary encoding constraint," *Proc. of the Third International Conf. on Genetic Algorithms*, J. D. Schaffer, Ed. San Mateo, CA: Morgan Kaufmann Publishers, pp. 86–91, 1989.
[44] Z. Michalewicz, *Genetic Algorithms + Data Structures = Evolution Programs*. New York: Springer-Verlag, 1992.
[45] D. J. Montana, "Automated parameter tuning for interpretation of synthetic images," in *Handbook of Genetic Algorithms*. L. Davis, Ed. New York: Van Nostrand Reinhold, pp. 282–311, 1991.
[46] G. Syswerda, "Schedule optimization using genetic algorithms," in *Handbook of Genetic Algorithms*, L. Davis, Ed. New York: Van Nostrand Reinhold, pp. 332–349, 1991.
[47] A. H. Wright, "Genetic algorithms for real parameter optimization," *Foundations of Genetic Algorithms*, G. J. E. Rawlins, Ed. San Mateo, CA: Morgan Kaufmann Publishers, pp. 205–218, 1991.
[48] G. Rudolph, "Convergence properties of canonical genetic algorithms," *IEEE Trans. on Neural Networks*, vol. 5. no. 1, 1994.
[49] J. J. Grefenstette, "Optimization of control parameters for genetic algorithms," *IEEE Trans. Sys., Man and Cybern.*, vol. 16, no. 1, pp. 122–128, 1986.
[50] A. E. Eiben, E. H. Aarts, and K. M. Van Hee, "Global convergence of genetic algorithms: An infinite Markov chain analysis," *Parallel Problem Solving from Nature*, H.-P. Schwefel and R. Männer, Eds. Heidelberg, Berlin: Springer-Verlag, pp. 4–12, 1991.
[51] D. B. Fogel, "Asymptotic convergence properties of genetic algorithms and evolutionary programming: Analysis and experiments," *Cybernetics and Systems*, in press, 1994.
[52] L. Davis, Ed. *Handbook of Genetic Algorithms*, New York: Van Nostrand Reinhold, 1991.
[53] D. E. Goldberg, "Computer-aided gas pipeline operation using genetic algorithms and rule learning," Doctoral dissertation, Univ. of Michigan, Ann Arbor, 1983.
[54] J. J. Grefenstette, R. Gopal, B. Rosmaita, and D. Van Gucht, "Genetic algorithms for the traveling salesman problem," in *Proc. of an Intern. Conf. on Genetic Algorithms and Their Applications*, J. J. Grefenstette, Ed. Lawrence Earlbaum, pp. 160–168, 1985.
[55] G. Syswerda, "Uniform crossover in genetic algorithms," in *Proc. of the Third Intern. Conf. on Genetic Algorithms*, J. D. Schaffer, Ed. San Mateo, CA: Morgan Kaufmann, pp. 2–9, 1989.
[56] A. S. Bickel and R. W. Bickel, "Determination of near-optimum use of hospital diagnostic resources using the 'GENES' genetic algorithm shell," *Comput. Biol. Med.*, vol. 20, no. 1, pp. 1–13, 1990.
[57] G. Pitney, T. R. Smith, and D. Greenwood, "Genetic design of processing elements for path planning networks," *Proc. of the Int. Joint Conf. on Neural Networks 1990*, vol. III, IEEE, pp. 925–932, 1990.
[58] N. N. Schraudolph, personal communication, UCSD, 1992.
[59] J. H. Holland, "Genetic algorithms," *Scientific American*, pp. 66–72, July, 1992.
[60] K. A. De Jong, "Are genetic algorithms function optimizers?" *Proc. of the Sec. Parallel Problem Solving from Nature Conf.*, R. Männer and B. Manderick, Eds. The Netherlands: Elsevier Science Press, pp. 3–14, 1992.
[61] T. E. Davis and J. C. Principe, "A simulated annealing like convergence theory for the simple genetic algorithm," *Proc. of the Fourth Intern. Conf. on Genetic Algorithms*, R. K. Belew and L. B. Booker, Eds. San Mateo, CA: Morgan Kaufmann, pp. 174–181, 1991.
[62] X. Qi and F. Palmieri, "Adaptive mutation in the genetic algorithm," *Proc. of the Sec. Ann. Conf. on Evolutionary Programming*, D.B. Fogel and W. Atmar, Eds. La Jolla, CA: Evolutionary Programming Society, pp. 192–196, 1993.
[63] G. E. Liepins and M. D. Vose, "Deceptiveness and genetic algorithm dynamics," in *Foundations of Genetic Algorithms*, G. J. E. Rawlins, Ed. San Mateo, CA: Morgan Kaufmann, pp. 36–52, 1991.
[64] L. D. Whitley, "Fundamental principles of deception in genetic search," in *Foundations of Genetic Algorithms*, G. J. E. Rawlins, Ed. San Mateo, CA: Morgan Kaufmann, pp. 221–241, 1991.
[65] W. E. Hart and R. K. Belew, "Optimizing an arbitrary function is hard for the genetic algorithm," *Proc. of the Fourth Intern. Conf. on Genetic Algorithms*, R. K. Belew and L. B. Booker, Eds. San Mateo, CA: Morgan Kaufmann, pp. 190–195, 1991.
[66] S. Forrest and M. Mitchell, "What makes a problem hard for a genetic algorithm?" *Machine Learning*. vol. 13, no. 2–3, pp. 285–319, 1993.
[67] J. D. Schaffer and L. J. Eshelman, "On crossover as an evolutionarily

viable strategy," in *Proc. of the Fourth Intern. Conf. on Genetic Algorithms*, R. K. Belew and L. B. Booker, Eds. San Mateo, CA: Morgan Kaufmann, pp. 61–68, 1991.

[68] W. M. Spears and K. A. De Jong, "On the virtues of parameterized uniform crossover," in *Proc. of the Fourth Intern. Conf. on Genetic Algorithms*, R. K. Belew and L. B. Booker, Eds. San Mateo, CA: Morgan Kaufmann, pp. 230–236, 1991.

[69] D. E. Goldberg, K. Deb, and J. H. Clark, "Genetic algorithms: noise, and the sizing of populations," *Complex Systems*, vol. 6, pp. 333–362, 1992.

[70] V. Kreinovich, C. Quintana, and O. Fuentes, "Genetic algorithms—what fitness scaling is optimal," *Cybernetics and Systems*, vol. 24, no. 1, pp. 9–26, 1993.

[71] S. W. Mahfoud and D. E. Goldberg, "Parallel recombinative simulated annealing: A genetic algorithm," IlliGAL Report No. 92002, Univ. of Illinois, Urbana-Champaign, 1992.

[72] L. Ingber and B. Rosen, "Genetic algorithms and very fast simulated annealing—a comparison," *Math. and Comp. Model.*, vol. 16, no. 11, pp. 87–100, 1992.

[73] D. Adler, "Genetic algorithms and simulated annealing: A marriage proposal," in *IEEE Conference on Neural Networks 1993*, pp. 1104–1109, 1993.

[74] J. R. Koza, "A hierarchical approach to learning the boolean multiplexer function," in *Foundations of Genetic Algorithms*. G. J. E. Rawlins, Ed. San Mateo, CA: Morgan Kaufmann, pp. 171–192, 1991.

[75] J. R. Koza, *Genetic Programming*. Cambridge, MA: MIT Press, 1992.

[76] S. Forrest and G. Mayer-Kress, "Genetic algorithms, nonlinear dynamical systems, and models of international security," *Handbook of Genetic Algorithms*, L. Davis, Ed. New York: Van Nostrand Reinhold, pp. 166–185, 1991.

[77] J. R. Koza, "Hierarchical automatic function definition in genetic programming," *Foundations of Genetic Algorithms 2*, L. D. Whitley, Ed. San Mateo, CA: Morgan Kaufmann, pp. 297–318, 1992.

[78] K. Kristinsson and G. A. Dumont, "System identification and control using genetic algorithms," *IEEE Trans. Sys., Man and Cybern.*, vol. 22, no. 5, pp. 1033–1046, 1992.

[79] K. Krishnakumar and D. E. Goldberg, "Control system optimization using genetic algorithms," *Journ. of Guidance, Control and Dynamics*, vol. 15, no. 3, pp. 735–740, 1992.

[80] J. H. Holland, "Concerning the emergence of tag-mediated lookahead in classifier systems," *Physica D*, vol. 42, pp. 188–201, 1990.

[81] S. Forrest and J. H. Miller, "Emergent behavior in classifier systems," *Physica D*, vol. 42, pp. 213–227, 1990.

[82] R. L. Riolo, "Modeling simple human category learning with a classifier system," in *Proc. of the Fourth Intern. Conf. on Genetic Algorithms*, R. K. Belew and L. B. Booker, Eds. San Mateo, CA: Morgan Kaufmann, pp. 324–333, 1991.

[83] G. E. Liepins, M. R. Hilliard, M. Palmer and G. Rangarajan, "Credit assignment and discovery in classifier systems," *Intern. Journ. of Intelligent Sys.*, vol. 6, no. 1, pp. 55–69, 1991.

[84] S. Tokinaga and A.B. Whinston, "Applying adaptive credit assignment algorithms for the learning classifier system based upon the genetic algorithm," *IEICE Trans. on Fund. Elec. Comm. and Comp. Sci.*, vol. E75A, no. 5, pp. 568–577, 1992.

[85] D. Jefferson, R. Collins, C. Cooper, M. Dyer, M. Flowers, R. Korf, C. Taylor, and A. Wang, "Evolution as a theme in artificial life: The Genesys/Tracker system," in *Artificial Life II*, C. G. Langton, C. Taylor, J. D. Farmer, and S. Rasmussen, Eds. Reading, MA: Addison-Wesley, pp. 549–578, 1991.

[86] J. H. Holland, *Adaptation in Natural and Artificial Systems*. 2nd ed., Cambridge, MA: MIT Press, 1992.

[87] H. Mühlenbein, "Evolution in time and space—the parallel genetic algorithm," in *Foundations of Genetic Algorithms*, G. J. E. Rawlins, Ed. San Mateo, CA: Morgan Kaufmann, pp. 316–337, 1991.

[88] P. Spiessens and B. Manderick, "A massively parallel genetic algorithm: implementation and first analysis," in *Proc. of the Fourth Intern. Conf. on Genetic Algorithms*, R. K. Belew and L. B. Booker, Eds. San Mateo, CA: Morgan Kaufmann, pp. 279–286, 1991.

[89] H. Mühlenbein, M. Schomisch and J. Born, "The parallel genetic algorithm as function optimizer," *Parallel Computing*, vol. 17, pp. 619–632, 1991.

[90] S. Forrest, Ed., *Proc. of the Fifth Intern. Conf. on Genetic Algorithms*, San Mateo, CA: Morgan Kaufmann, 1993.

[91] H.-P. Schwefel, "Kybernetische evolution als strategie der experimentellen forschung in der strmungstechnik," Diploma thesis, Technical Univ. of Berlin, 1965.

[92] I. Rechenberg, *Evolutionsstrategie: Optimierung technischer systeme nach prinzipien der biolgischen evolution*. Stuttgart: Frommann-Holzboog Verlag, 1973.

[93] L. J. Fogel, "Autonomous automata," *Industrial Research*, vol. 4, pp. 14–19, 1962.

[94] L. J. Fogel, "On the organization of intellect," Doctoral dissertation, UCLA, 1964.

[95] M. Conrad, "Evolutionary learning circuits," *Journ. Theor. Biol.*, vol. 46, pp. 167–188, 1974.

[96] M. Conrad and H. H. Pattee, "Evolution experiments with an artificial ecosystem," *Journ. Theor. Biol.*, vol. 28, pp. 393–409, 1970.

[97] G. H. Burgin, "On playing two-person zero-sum games against nonminimax players," *IEEE Trans. on Systems Science and Cybernetics*, vol. SSC-5, no. 4, pp. 369–370, 1969.

[98] G. H. Burgin, "System identification by quasilinearization and evolutionary programming," *Journal of Cybernetics*, vol. 2, no. 3, pp. 4–23, 1974.

[99] J. W. Atmar, "Speculation on the evolution of intelligence and its possible realization in machine form," Doctoral dissertation, New Mexico State University, Las Cruces, 1976.

[100] D. B. Fogel, "On the philosophical differences between genetic algorithms and evolutionary algorithms," in *Proc. of the Sec. Ann. Conf. on Evolutionary Programming*, D. B. Fogel and W. Atmar, Eds. La Jolla, CA: Evolutionary Programming Society, pp. 23–29, 1993.

[101] F. J. Solis and R. J.-B. Wets, "Minimization by random search techniques," *Math. Operations Research*, vol. 6, pp. 19–30, 1981.

[102] T. Bäck and H.-P. Schwefel, "An Overview of Evolutionary Algorithms for Parameter Optimization," *Evolutionary Computation*, vol. 1, no. 1, pp. 1–24, 1993.

[103] H.-P. Schwefel, "Numerische optimierung von computer-modellen mittels der evoluionsstrategie," *Interdisciplinary systems research*, vol. 26, Basel: Birkhuser, 1977.

[104] H.-P. Schwefel, *Numerical Optimization of Computer Models*. Chichester, UK: John Wiley, 1981.

[105] G. H. Mealy, "A method of synthesizing sequential circuits," *Bell Sys. Tech. Journ.*, vol. 34, pp. 1054–1079, 1955.

[106] D. B. Fogel, "The evolution of intelligent decision making in gaming," *Cybernetics and Systems*, vol. 22, pp. 223–236, 1991.

[107] D. B. Fogel, "Evolving behaviors in the iterated prisoner's dilemma," *Evolutionary Computation*, vol. 1, no. 1, pp. 77–97, 1993.

[108] R. Axelrod, "The evolution of strategies in the iterated prisoner's dilemma," in *Genetic Algorithms and Simulated Annealing*, L. Davis, Ed. London: Pitman Publishing, pp. 32–41, 1987.

[109] D. B. Fogel, L. J. Fogel, and W. Atmar, "Meta-evolutionary programming," in *Proc. of the 25th Asilomar Conf. on Signals, Systems and Computers*, R. R. Chen, Ed. IEEE Computer Society, pp. 540–545, 1991.

[110] D. B. Fogel, L. J. Fogel, W. Atmar, and G. B. Fogel, "Hierarchic methods of evolutionary programming," in *Proc. of the First Ann. Conf. on Evolutionary Programming*, D. B. Fogel and W. Atmar, Eds. La Jolla, CA: Evolutionary Programming Society, pp. 175–182, 1992.

[111] T. Bäck, G. Rudolph, and H.-P. Schwefel, "Evolutionary programming and evolution strategies: similarities and differences," in *Proc. of the Second Ann. Conf. on Evolutionary Programming*, D. B. Fogel and W. Atmar, Eds. La Jolla, CA: Evolutionary Programming Society, pp. 11–22, 1993.

[112] N. Saravanan, "Learning of Strategy Parameters in Evolutionary Programming," *Proc. of Third Annual Conference on Evolutionary Programming*, A. V. Sebald and L. J. Fogel, Eds. RiverEdge, NJ: World Scientific, to appear, 1994.

[113] G. Rudolph, "On correlated mutations in evolution strategies," in *Parallel Problem Solving from Nature 2*, R. Männer and B. Manderick, Eds. The Netherlands: Elsevier Science Press, pp. 105–114, 1992.

[114] B. K. Ambati, J. Ambati, and M. M. Mokhtar, "Heuristic combinatorial optimization by simulated darwinian evolution: A polynomial time algorithm for the traveling salesman problem," *Biological Cybernetics*, vol. 65, pp. 31–35, 1991.

[115] D. B. Fogel, "Empirical estimation of the computation required to discover approximate solutions to the traveling salesman problem using evolutionary programming," in *Proc. of the Second Ann. Conf. on Evolutionary Programming*, D. B. Fogel and W. Atmar, Eds. La Jolla, CA: Evolutionary Programming Society, in press, 1993.

[116] D. Waagen, P. Diercks, and J. R. McDonnell, "The stochastic direction set algorithm: A hybrid technique for finding function extrema," in *Proc. of the First Ann. Conf. on Evolutionary Programming*, D. B. Fogel and W. Atmar, Eds. La Jolla, CA: Evolutionary Programming Society, pp. 35–42, 1992.

[117] R. Lohmann, "Structure evolution and incomplete induction," in *Proc. of the Sec. Parallel Problem Solving from Nature Conf.*, R. Männer and B. Manderick, Eds. The Netherlands: Elsevier Science Press, pp. 175–186, 1992.

IEEE TRANSACTIONS ON NEURAL NETWORKS, VOL. 5, NO. 1, JANUARY 1994

[118] J. R. McDonnell and D. Waagen, "Evolving neural network connectivity," *Intern. Conf. on Neural Networks 1993*, IEEE, pp. 863–868, 1993.

[119] P. J. Angeline, G. Saunders and J. Pollack, "An evolutionary algorithm that constructs neural networks," *IEEE Trans. Neural Networks*, vol. 5, no 1, 1994.

[120] D. B. Fogel, "Using evolutionary programming to create neural networks that are capable of playing tic-tac-toe," *Inern. Conf. on Neural Networks 1993*, IEEE, pp. 875–880, 1993.

[121] R. Smalz and M. Conrad, "Evolutionary credit apportionment and time-dependent neural processing," in *Proc. of the Second Ann. Conf. on Evolutionary Programming*, D. B. Fogel and W. Atmar, Eds. La Jolla, CA: Evolutionary Programming Society, pp. 119–126, 1993.

[122] W. Kuhn and A. Visser, "Identification of the system parameter of a 6 axis robot with the help of an evolution strategy," *Robotersysteme*, vol. 8, no. 3, pp. 123–133, 1992.

[123] J. R. McDonnell, B. D. Andersen, W. C. Page and F. Pin, "Mobile manipulator configuration optimization using evolutionary programming," in *Proc. of the First Ann. Conf. on Evolutionary Programming*, D. B. Fogel and W. Atmar, Eds. La Jolla, CA: Evolutionary Programming Society, pp. 52–62, 1992.

[124] W. C. Page, B. D. Andersen, and J. R. McDonnell, "An evolutionary programming approach to multi-dimensional path planning," in *Proc. of the First Ann. Conf. on Evolutionary Programming*, D. B. Fogel and W. Atmar, Eds. La Jolla, CA: Evolutionary Programming Society, pp. 63–70, 1992.

[125] A. V. Sebald, J. Schlenzig, and D. B. Fogel, "Minimax design of CMAC encoded neural controllers for systems with variable time delay," in *Proc. of the First Ann. Conf. on Evolutionary Programming*, D. B. Fogel and W. Atmar, Eds. La Jolla, CA: Evolutionary Programming Society, pp. 120–126, 1992.

[126] D. B. Fogel, *System Identification Through Simulated Evolution: A Machine Learning Approach to Modeling*. Needham, MA: Ginn Press, 1991.

[127] D. B. Fogel, "Using evolutionary programming for modeling: An ocean acoustic example," *IEEE Journ. on Oceanic Engineering*, vol. 17, no. 4, pp. 333–340, 1992.

[128] V. W. Porto, "Alternative methods for training neural networks," in *Proc. of the First Ann. Conf. on Evolutionary Programming*, D. B. Fogel and W. Atmar, Eds. La Jolla, CA: Evolutionary Programming Society, pp. 100–110, 1992.

[129] L. A. Tamburino, M. A. Zmuda and M. M. Rizki, "Applying evolutionary search to pattern recognition problems," in *Proc. of the Sec. Ann. Conf. on Evolutionary Programming*, D. B. Fogel and W. Atmar, Eds. Evolutionary Programming Society, La Jolla, CA, pp. 183–191, 1993.

[130] M. M. Rizki, L. A. Tamburino and M. A. Zmuda, "Evolving multi-resolution feature detectors," in *Proc. of the Sec. Ann. Conf. on Evolu-tionary Programming*, D. B. Fogel and W. Atmar, Eds. La Jolla, CA: Evolutionary Programming Society, pp. 108–118, 1993.

[131] D. B. Fogel and P. K. Simpson, "Evolving fuzzy clusters," *Intern. Conf. on Neural Networks 1993*, IEEE, pp. 1829–1834, 1993.

[132] S. Haffner and A. V. Sebald, "Computer-aided design of fuzzy HVAC controllers using evolutionary programming," in *Proc. of the Sec. Ann. Conf. on Evolutionary Programming*, D. B. Fogel and W. Atmar, Eds. La Jolla, CA: Evolutionary Programming Society, pp. 98–107, 1993.

[133] S. H. Rubin, "Case-based learning: A new paradigm for automated knowledge acquisition," *ISA Transactions*, Special Issue on Artif. Intell. for Eng., Design and Manuf., vol. 31, pp. 181–209, 1992.

[134] W. Atmar, "Notes on the simulation of evolution," *IEEE Trans. on Neural Networks*, vol. 5. no. 1, 1994.

[135] G. B. Fogel, "An introduction to the protein folding problem and the potential application of evolutionary programming," in *Proc. of the Sec. Ann. Conf. on Evolutionary Programming*, D. B. Fogel and W. Atmar, Eds. La Jolla, CA: Evolutionary Programming Society, pp. 170–177, 1993.

[136] M. Conrad, "Molecular computing: The lock-key paradigm," *Computer*, Special Issue on Molecular Computing, M. Conrad, Ed. Nov., pp. 11–20, 1992.

[137] J. O'Callaghan and M. Conrad, "Symbiotic interactions in the EVOLVE III ecosystem model," *BioSystems*, vol. 26, pp. 199–209, 1992.

[138] G. Rudolph, "Parallel approaches to stochastic global optimization," in *Parallel Computing: From Theory to Sound Practice*. W. Joosen and E. Milgrom, Eds. Amsterdam: IOS Press pp. 256–267, 1992.

[139] B. S. Duncan, "Parallel evolutionary programming," in *Proc. of the Sec. Ann. Conf. on Evolutionary Programming*, D.B. Fogel and W. Atmar, Eds. La Jolla, CA: Evolutionary Programming Society, pp. 202–208, 1993.

[140] F. Hoffmeister, "Scalable Parallelism by Evolutionary Algorithms," in *Parallel Comp. & Math. Opt.*, D. B. Grauer, Ed. Heidelberg, Berlin: Springer-Verlag, pp. 177–198, 1991.

[141] D. B. Fogel and W. Atmar Eds., *Proc. of the Sec. Ann. Conf. on Evolutionary Programming*, La Jolla, CA: Evolutionary Programming Society, 1993.

[142] R. Männer and B. Manderick, Eds., *Proc. of the Sec. Parallel Problem Solving from Nature Conf.* The Netherlands: Elsevier Science Press, 1992.

[143] D. E. Goldberg, *Genetic Algorithms in Search, Optimization and Machine Learning*. Reading, MA: Addison Wesley, 1989.

[144] L. J. Fogel, A. J. Owens, and M. J. Walsh, *Artificial Intelligence through Simulated Evolution*. New York: John Wiley, 1966.

Evolutionary Computation: Comments on the History and Current State

Thomas Bäck, Ulrich Hammel, and Hans-Paul Schwefel

Abstract— Evolutionary computation has started to receive significant attention during the last decade, although the origins can be traced back to the late 1950's. This article surveys the history as well as the current state of this rapidly growing field. We describe the purpose, the general structure, and the working principles of different approaches, including genetic algorithms (GA) [with links to *genetic programming* (GP) and *classifier systems* (CS)], *evolution strategies* (ES), and *evolutionary programming* (EP) by analysis and comparison of their most important constituents (i.e., representations, variation operators, reproduction, and selection mechanism). Finally, we give a brief overview on the manifold of application domains, although this necessarily must remain incomplete.

Index Terms— Classifier systems, evolution strategies, evolutionary computation, evolutionary programming, genetic algorithms, genetic programming.

I. EVOLUTIONARY COMPUTATION: ROOTS AND PURPOSE

THIS first issue of the IEEE TRANSACTIONS ON EVOLUTIONARY COMPUTATION marks an important point in the history of the rapidly growing field of evolutionary computation, and we are glad to participate in this event. In preparation for this summary, we strove to provide a comprehensive review of both the history and the state of the art in the field for both the novice and the expert in evolutionary computation. Our selections of material are necessarily subjective, and we regret any significant omissions.

Although the origins of evolutionary computation can be traced back to the late 1950's (see e.g., the influencing works of Bremermann [1], Friedberg [2], [3], Box [4], and others), the field remained relatively unknown to the broader scientific community for almost three decades. This was largely due to the lack of available powerful computer platforms at that time, but also due to some methodological shortcomings of those early approaches (see, e.g., Fogel [5, p. 103]).

The fundamental work of Holland [6], Rechenberg [7], Schwefel [8], and Fogel [9] served to slowly change this picture during the 1970's, and we currently observe a remarkable

Manuscript received November 13, 1996; revised January 23, 1997. The work of T. Bäck was supported by a grant from the German BMBF, Project EVOALG.
T. Bäck is with the Informatik Centrum Dortmund, Center for Applied Systems Analysis (CASA), D-44227 Dortmund, Germany, and Leiden University, NL-2333 CA Leiden, The Netherlands (e-mail: baeck@icd.de).

U. Hammel and H.-P. Schwefel are with the Computer Science Department, Dortmund University, D-44221 Dortmund, Germany (e-mail: hammel@LS11.informatik.uni-dortmund.de; schwefel@LS11.informatik.uni-dortmund.de).
Publisher Item Identifier S 1089-778X(97)03305-5.

and steady (still exponential) increase in the number of publications (see, e.g., the bibliography of [10]) and conferences in this field, a clear demonstration of the scientific as well as economic relevance of this subject matter.

But what are the benefits of evolutionary computation (compared to other approaches) which may justify the effort invested in this area? We argue that the most significant advantage of using evolutionary search lies in the gain of flexibility and adaptability to the task at hand, in combination with robust performance (although this depends on the problem class) and global search characteristics. In fact, evolutionary computation should be understood as a general adaptable concept for problem solving, especially well suited for solving difficult optimization problems, rather than a collection of related and ready-to-use algorithms.

The majority of current implementations of evolutionary algorithms descend from three strongly related but independently developed approaches: *genetic algorithms, evolutionary programming*, and *evolution strategies*.

Genetic algorithms, introduced by Holland [6], [11], [12], and subsequently studied by De Jong [13]–[16], Goldberg [17]–[21], and others such as Davis [22], Eshelman [23], [24], Forrest [25], Grefenstette [26]–[29], Koza [30], [31], Mitchell [32], Riolo [33], [34], and Schaffer [35]–[37], to name only a few, have been originally proposed as a general model of adaptive processes, but by far the largest application of the techniques is in the domain of optimization [15], [16]. Since this is true for all three of the mainstream algorithms presented in this paper, we will discuss their capabilities and performance mainly as optimization strategies.

Evolutionary programming, introduced by Fogel [9], [38] and extended in Burgin [39], [40], Atmar [41], Fogel [42]–[44], and others, was originally offered as an attempt to create artificial intelligence. The approach was to evolve finite state machines (FSM) to predict events on the basis of former observations. An FSM is an abstract machine which transforms a sequence of input symbols into a sequence of output symbols. The transformation depends on a finite set of states and a finite set of state transition rules. The performance of an FSM with respect to its environment might then be measured on the basis of the machine's prediction capability, i.e., by comparing each output symbol with the next input symbol and measuring the worth of a prediction by some payoff function.

Evolution strategies, as developed by Rechenberg [45], [46] and Schwefel [47], [48], and extended by Herdy [49], Kursawe [50], Ostermeier [51], [52], Rudolph [53], Schwefel [54], and

Reprinted from *IEEE Transactions on Evolutionary Computation*, Vol. 1:1, pp. 3-17, April, 1997.

others, were initially designed with the goal of solving difficult discrete and continuous, mainly experimental [55], parameter optimization problems.

During the 1980's, advances in computer performance enabled the application of evolutionary algorithms to solve difficult real-world optimization problems, and the solutions received a broader audience. In addition, beginning in 1985, international conferences on the techniques were offered (mainly focusing on genetic algorithms [56]–[61], with an early emphasis on evolutionary programming [62]–[66], as small workshops on theoretical aspects of genetic algorithms [67]–[69], as a genetic programming conference [70], with the general theme of problem solving methods gleaned from nature [71]–[74], and with the general topic of evolutionary computation [75]–[78]). But somewhat surprisingly, the researchers in the various disciplines of evolutionary computation remained isolated from each other until the meetings in the early 1990's [59], [63], [71].

The remainder of this paper is intended as an overview of the current state of the field. We cannot claim that this overview is close to complete. As good starting points for further studies we recommend [5], [18], [22], [31], [32], [48], and [79]–[82]. In addition moderated mailing lists[1] and newsgroups[2] allow one to keep track of current events and discussions in the field.

In the next section we describe the application domain of evolutionary algorithms and contrast them with the traditional approach of mathematical programming.

II. OPTIMIZATION, EVOLUTIONARY COMPUTATION, AND MATHEMATICAL PROGRAMMING

In general, an optimization problem requires finding a setting $\vec{x} \in M$ of free parameters of the system under consideration, such that a certain quality criterion $f: M \to \mathbb{R}$ (typically called the *objective function*) is maximized (or, equivalently, minimized)

$$f(\vec{x}) \to \max. \tag{1}$$

The objective function might be given by real-world systems of arbitrary complexity. The solution to the *global* optimization problem (1) requires finding a vector $\vec{x}^*$ such that $\forall \vec{x} \in M: f(\vec{x}) \leq f(\vec{x}^*) = f^*$. Characteristics such as *multimodality*, i.e., the existence of several *local maxima* $\vec{x}'$ with

$$\exists \varepsilon > 0: \forall \vec{x} \in M: \rho(\vec{x}, \vec{x}') < \varepsilon \Rightarrow f(\vec{x}) \leq f(\vec{x}') \tag{2}$$

(where ρ denotes a distance measure on M), *constraints*, i.e., restrictions on the set M by functions $g_j: M \to \mathbb{R}$ such that the set of *feasible* solutions $F \subseteq M$ is only a subset of the domain of the variables

$$F = \{\vec{x} \in M \mid g_j(\vec{x}) \geq 0 \,\forall j\} \tag{3}$$

and other factors, such as large dimensionality, strong nonlinearities, nondifferentiability, and noisy and time-varying

[1] For example, GA-List-Request@AIC.NRL.NAVY.MIL and EP-List-Request@magenta.me.fau.edu.

[2] For example, comp.ai.genetic.

objective functions, frequently lead to difficult if not unsolvable optimization tasks (see [83, p. 6]). But even in the latter case, the identification of an improvement of the currently known best solution through optimization is often already a big success for practical problems, and in many cases evolutionary algorithms provide an efficient and effective method to achieve this.

Optimization problems occur in many technical, economic, and scientific projects, like cost-, time-, and risk-minimization or quality-, profit-, and efficiency-maximization [10], [22] (see also [80, part G]). Thus, the development of general strategies is of great value.

In real-world situations the objective function f and the constraints g_j are often not analytically treatable or are even not given in closed form, e.g., if the function definition is based on a simulation model [84], [85].

The traditional approach in such cases is to develop a formal model that resembles the original functions close enough but is solvable by means of traditional mathematical methods such as linear and nonlinear programming. This approach most often requires simplifications of the original problem formulation. Thus, an important aspect of mathematical programming lies in the design of the formal model.

No doubt, this approach has proven to be very successful in many applications, but has several drawbacks which motivated the search for novel approaches, where evolutionary computation is one of the most promising directions. The most severe problem is that, due to oversimplifications, the computed solutions do not solve the original problem. Such problems, e.g., in the case of simulation models, are then often considered unsolvable.

The fundamental difference in the evolutionary computation approach is to adapt the method to the problem at hand. In our opinion, evolutionary algorithms should not be considered as off-the-peg, ready-to-use algorithms but rather as a general concept which can be tailored to most of the real-world applications that often are beyond solution by means of traditional methods. Once a successful EC-framework has been developed it can be incrementally adapted to the problem under consideration [86], to changes of the requirements of the project, to modifications of the model, and to the change of hardware resources.

III. THE STRUCTURE OF AN EVOLUTIONARY ALGORITHM

Evolutionary algorithms mimic the process of natural evolution, the driving process for the emergence of complex and well-adapted organic structures. To put it succinctly and with strong simplifications, evolution is the result of the interplay between the creation of new genetic information and its evaluation and selection. A single individual of a population is affected by other individuals of the population (e.g., by food competition, predators, and mating), as well as by the environment (e.g., by food supply and climate). The better an individual performs under these conditions the greater is the chance for the individual to live for a longer while and generate offspring, which in turn inherit the (disturbed) parental genetic information. Over the course of evolution, this leads to a penetration of the population with the genetic information of individuals of above-average fitness. The nondeterministic nature of reproduction leads to a permanent production of novel genetic information and therefore to the creation of differing offspring (see [5], [79], and [87] for more details).

This neo-Darwinian model of organic evolution is reflected by the structure of the following general evolutionary algorithm.

Algorithm 1:

```
t := 0;
initialize P(t);
evaluate P(t);
while not terminate do
    P'(t) := variation [P(t)];
    evaluate [P'(t)];
    P(t + 1) := select [P'(t) ∪ Q];
    t := t + 1;
od
```

In this algorithm, $P(t)$ denotes a population of μ individuals at generation t. Q is a special set of individuals that might be considered for selection, e.g., $Q = P(t)$ (but $Q = \emptyset$ is possible as well). An offspring population $P'(t)$ of size λ is generated by means of variation operators such as recombination and/or mutation (but others such as inversion [11, pp. 106–109] are also possible) from the population $P(t)$. The offspring individuals are then evaluated by calculating the objective function values $f(\vec{x}_k)$ for each of the solutions $\vec{x}_k$ represented by individuals in $P'(t)$, and selection based on the fitness values is performed to drive the process toward better solutions. It should be noted that $\lambda = 1$ is possible, thus including so-called *steady-state* selection schemes [88], [89] if used in combination with $Q = P(t)$. Furthermore, by choosing $1 \leq \lambda \leq \mu$ an arbitrary value of the *generation gap* [90] is adjustable, such that the transition between strictly generational and steady-state variants of the algorithm is also taken into account by the formulation offered here. It should also be noted that $\lambda > \mu$, i.e., a reproduction surplus, is the normal case in nature.

IV. DESIGNING AN EVOLUTIONARY ALGORITHM

As mentioned, at least three variants of evolutionary algorithms have to be distinguished: genetic algorithms, evolutionary programming, and evolution strategies. From these ("canonical") approaches innumerable variants have been derived. Their main differences lie in:

- the representation of individuals;
- the design of the variation operators (mutation and/or recombination);
- the selection/reproduction mechanism.

In most real-world applications the search space is defined by a set of objects, e.g., processing units, pumps, heaters, and coolers of a chemical plant, each of which have different parameters such as energy consumption, capacity, etc. Those parameters which are subject to optimization constitute the so-called *phenotype space*. On the other hand the genetic

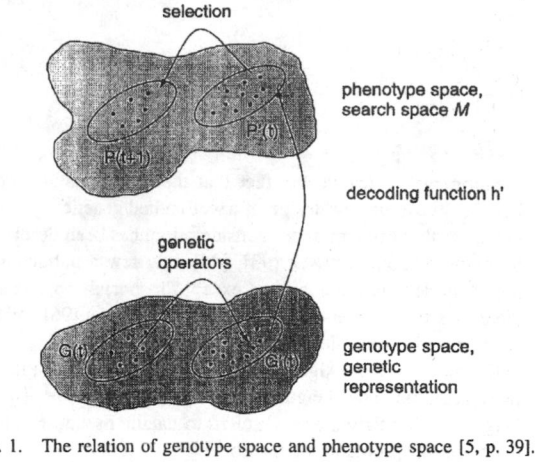

Fig. 1. The relation of genotype space and phenotype space [5, p. 39].

operators often work on abstract mathematical objects like binary strings, the *genotype space*. Obviously, a mapping or coding function between the phenotype and genotype space is required. Fig. 1 sketches the situation (see also [5, pp. 38–43]).

In general, two different approaches can be followed. The first is to choose one of the standard algorithms and to design a decoding function according to the requirements of the algorithm. The second suggests designing the representation as close as possible to the characteristics of the phenotype space, almost avoiding the need for a decoding function.

Many empirical and theoretical results are available for the standard instances of evolutionary algorithms, which is clearly an important advantage of the first approach, especially with regard to the reuse and parameter setting of operators. On the other hand, a complex coding function may introduce additional nonlinearities and other mathematical difficulties which can hinder the search process substantially [79, pp. 221–227], [82, p. 97].

There is no general answer to the question of which one of the two approaches mentioned above to follow for a specific project, but many practical applications have shown that the best solutions could be found after imposing substantial modifications to the standard algorithms [86]. We think that most practitioners prefer natural, problem-related representations. Michalewicz [82, p. 4] offers:

> It seems that a "natural" representation of a potential solution for a given problem plus a family of applicable "genetic" operators might be quite useful in the approximation of solutions of many problems, and this nature-modeled approach ... is a promising direction for problem solving in general.

Furthermore, many researchers also use hybrid algorithms, i.e., combinations of evolutionary search heuristics and traditional as well as knowledge-based search techniques [22, p. 56], [91], [92].

It should be emphasized that all this becomes possible because the requirements for the application of evolutionary heuristics are so modest compared to most other search techniques. In our opinion, this is one of the most important strengths of the evolutionary approach and one of the reasons for the popularity evolutionary computation has gained throughout the last decade.

A. The Representation

Surprisingly, despite the fact that the representation problem, i.e., the choice or design of a well-suited genetic representation for the problem under consideration, has been described by many researchers [82], [93], [94] only few a publications explicitly deal with this subject except for specialized research directions such as *genetic programming* [31], [95], [96] and the evolution of neural networks [97], [98].

Canonical genetic algorithms use a binary representation of individuals as fixed-length strings over the alphabet $\{0, 1\}$ [11], such that they are well suited to handle pseudo-Boolean optimization problems of the form

$$f: \{0, 1\}^{\ell} \to \mathbb{R}. \tag{4}$$

Sticking to the binary representation, genetic algorithms often enforce the utilization of encoding and decoding functions $h: M \to \{0, 1\}^{\ell}$ and $h': \{0, 1\}^{\ell} \to M$ that facilitate mapping solutions $\vec{x} \in M$ to binary strings $h(\vec{x}) \in \{0, 1\}^{\ell}$ and vice versa, which sometimes requires rather complex mappings h and h'. In case of continuous parameter optimization problems, for instance, genetic algorithms typically represent a real-valued vector $\vec{x} \in \mathbb{R}^n$ by a binary string $\vec{y} \in \{0, 1\}^{\ell}$ as follows: the binary string is logically divided into n segments of equal length ℓ' (i.e., $\ell = n \cdot \ell'$), each segment is decoded to yield the corresponding integer value, and the integer value is in turn linearly mapped to the interval $[u_i, v_i] \subset \mathbb{R}$ (corresponding with the ith segment of the binary string) of real values [18].

The strong preference for using binary representations of solutions in genetic algorithms is derived from *schema theory* [11], which analyzes genetic algorithms in terms of their expected schema sampling behavior under the assumption that mutation and recombination are detrimental. The term *schema* denotes a similarity template that represents a subset of $\{0, 1\}^{\ell}$, and the *schema theorem* of genetic algorithms offers that the canonical genetic algorithm provides a near-optimal sampling strategy (in terms of minimizing expected losses) for schemata by increasing the number of well-performing, short (i.e., with small distance between the left-most and right-most defined position), and low-order (i.e., with few specified bits) schemata (so-called building blocks) over subsequent generations (see [18] for a more detailed introduction to the schema theorem). The fundamental argument to justify the strong emphasis on binary alphabets is derived from the fact that the number of schemata is maximized for a given finite number of search points under a binary alphabet [18, pp. 40–41]. Consequently, the schema theory presently seems to favor binary representations of solutions (but see [99] for an alternative view and [100] for a transfer of schema theory to S-expression representations used in genetic programming).

Practical experience, as well as some theoretical hints regarding the binary encoding of continuous object variables [101]–[105], however, indicate that the binary representation has some disadvantages. The coding function might introduce an additional multimodality, thus making the combined objective function $f = f' \circ h'$ (where $f': M \to \mathbb{R}$) more complex than the original problem f' was. In fact, the schema theory relies on approximations [11, pp. 78–83] and the optimization criterion to minimize the *overall* expected loss (corresponding to the sum of all fitness values of all individuals ever sampled during the evolution) rather than the criterion to maximize the best fitness value ever found [15]. In concluding this brief excursion into the theory of canonical genetic algorithms, we would like to emphasize the recent work by Vose [106]–[109] and others [110], [111] on modeling genetic algorithms by Markov chain theory. This approach has already provided a remarkable insight into their convergence properties and dynamical behavior and led to the development of so-called *executable models* that facilitate the direct simulation of genetic algorithms by Markov chains for problems of sufficiently small dimension [112], [113].

In contrast to genetic algorithms, the representation in *evolution strategies* and *evolutionary programming* is directly based on real-valued vectors when dealing with continuous parameter optimization problems of the general form

$$f: M \subseteq \mathbb{R}^n \to \mathbb{R}. \tag{5}$$

Both methods have originally been developed and are also used, however, for combinatorial optimization problems [42], [43], [55]. Moreover, since many real-world problems have complex search spaces which cannot be mapped "canonically" to one of the representations mentioned so far, many strategy variants, e.g., for integer [114], mixed-integer [115], structure optimization [116], [117], and others [82, ch. 10], have been introduced in the literature, but exhaustive comparative studies especially for nonstandard representations are still missing. The actual development of the field is characterized by a progressing integration of the different approaches, such that the utilization of the common labels "genetic algorithm," "evolution strategy," and "evolutionary programming" might be sometimes even misleading.

B. Mutation

Of course, the design of variation operators has to obey the mathematical properties of the chosen representation, but there are still many degrees of freedom.

Mutation in genetic algorithms was introduced as a dedicated "background operator" of small importance (see [11, pp. 109–111]). Mutation works by inverting bits with very small probability such as $p_m = 0.001$ [13], $p_m \in [0.005, 0.01]$ [118], or $p_m = 1/\ell$ [119], [120]. Recent studies have impressively clarified, however, that much larger mutation rates, decreasing over the course of evolution, are often helpful with respect to the convergence reliability and velocity of a genetic algorithm [101], [121], and that even self-adaptive mutation rates are effective for pseudo-Boolean problems [122]–[124].

Originally, mutation in evolutionary programming was implemented as a random change (or multiple changes) of the description of the finite state machines according to five different modifications: change of an output symbol, change of a state transition, addition of a state, deletion of a state, or change

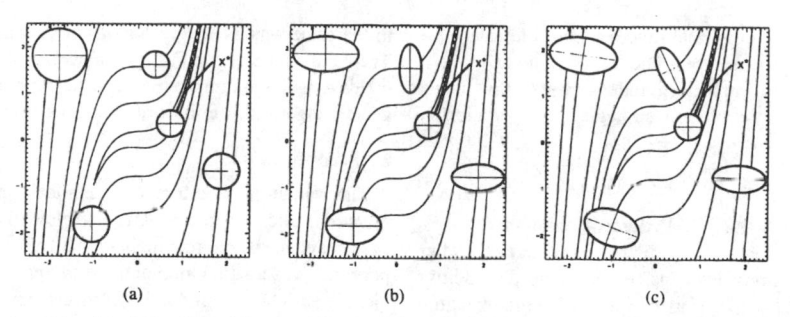

(a)   (b)   (c)

Fig. 2. Two-dimensional contour plot of the effect of the mutation operator in case of self-adaptation of (a) a single step size, (b) n step sizes, and (c) covariances. x^* denotes the optimizer. The ellipses represent one line of equal probability to place an offspring that is generated by mutation from the parent individual located at the center of the ellipses. Five sample individuals are shown in each of the plots.

of the initial state. The mutations were typically performed with uniform probability, and the number of mutations for a single offspring was either fixed or also chosen according to a probability distribution. Currently, the most frequently used mutation scheme as applied to real-valued representations is very similar to that of evolution strategies.

In evolution strategies, the individuals consist of object variables $x_i \in \mathbb{R}$ ($1 \leq i \leq n$) and so-called *strategy parameters*, which are discussed in the next section. Mutation is then performed independently on each vector element by adding a normally distributed random value with expectation zero and standard deviation σ (the notation $N_i(\cdot, \cdot)$ indicates that the random variable is sampled anew for each value of the index i)

$$x_i' = x_i + \sigma \cdot N_i(0, 1). \qquad (6)$$

This raises the question of how to control the so-called step size σ of (6), which is discussed in the next section.

C. Self-Adaptation

In [125] Schwefel introduced an endogenous mechanism for step-size control by incorporating these parameters into the representation in order to facilitate the evolutionary *self-adaptation* of these parameters by applying evolutionary operators to the object variables and the strategy parameters for mutation at the same time, i.e., searching the space of solutions and strategy parameters simultaneously. This way, a suitable adjustment and diversity of mutation parameters should be provided under arbitrary circumstances.

More formally, an individual $\vec{a} = (\vec{x}, \vec{\sigma})$ consists of object variables $\vec{x} \in \mathbb{R}^n$ and strategy parameters $\vec{\sigma} \in \mathbb{R}_+^n$. The mutation operator works by adding a normally distributed random vector $\vec{z} \in \mathbb{R}^n$ with $z_i \sim N(0, \sigma_i^2)$ (i.e., the components of $\vec{z}$ are normally distributed with expectation zero and variance σ_i^2).

The effect of mutation is now defined as

$$\sigma_i' = \sigma_i \cdot \exp\left[\tau' \cdot N(0, 1) + \tau \cdot N_i(0, 1)\right] \qquad (7)$$
$$x_i' = x_i + \sigma_i' \cdot N_i(0, 1) \qquad (8)$$

where $\tau' \propto (\sqrt{2n})^{-1}$ and $\tau \propto (\sqrt{2\sqrt{n}})^{-1}$.

This mutation scheme, which is most frequently used in evolution strategies, is schematically depicted (for $n = 2$)

in the middle of Fig. 2. The locations of equal probability density for descendants are concentric hyperellipses (just one is depicted in Fig. 2) around the parental midpoint. In the case considered here, i.e., up to n variances, but no covariances, the axes of the hyperellipses are congruent with the coordinate axes.

Two modifications of this scheme have to be mentioned: a simplified version uses just one step-size parameter for all of the object variables. In this case the hyperellipses are reduced to hyperspheres, as depicted in the left part of Fig. 2. A more elaborate *correlated mutation* scheme allows for the rotation of hyperellipses, as shown in the right part of Fig. 2. This mechanism aims at a better adaptation to the topology of the objective function (for details, see [79]).

The settings for the *learning rates* τ and τ' are recommended as upper bounds for the choice of these parameters (see [126, pp. 167–168]), but one should have in mind that, depending on the particular topological characteristics of the objective function, the optimal setting of these parameters might differ from the values proposed. For the case of one self-adaptable step size, however, Beyer has recently theoretically shown that, for the sphere model (a quadratic bowl), the setting $\tau_0 \propto 1/\sqrt{n}$ is the optimal choice, maximizing the convergence velocity [127].

The amount of information included into the individuals by means of the self-adaptation principle increases from the simple case of one standard deviation up to the order of n^2 additional parameters, which reflects an enormous degree of freedom for the *internal models* of the individuals. This growing degree of freedom often enhances the global search capabilities of the algorithm at the cost of the expense in computation time, and it also reflects a shift from the precise *adaptation* of a few strategy parameters (as in case of one step size) to the exploitation of a large *diversity* of strategy parameters. In case of correlated mutations, Rudolph [128] has shown that an approximation of the Hessian could be computed with an upper bound of $\mu + \lambda = (n^2 + 3n + 4)/2$ on the population size, but the typical population sizes $\mu = 15$ and $\lambda = 100$, independently of n, are certainly not sufficient to achieve this.

The choice of a logarithmic normal distribution for the modification of the standard deviations σ_i is presently also acknowledged in evolutionary programming literature

[129]–[131]. Extensive empirical investigations indicate some advantage of this scheme over the original additive self-adaptation mechanism introduced independently (but about 20 years later than in evolution strategies) in evolutionary programming [132] where

$$\sigma'_i = \sigma_i \cdot [1 + \alpha \cdot N(0, 1)] \qquad (9)$$

(with a setting of $\alpha \approx 0.2$ [131]). Recent preliminary investigations indicate, however, that this becomes reversed when noisy objective functions are considered, where the additive mechanism seems to outperform multiplicative modifications [133].

A study by Gehlhaar and Fogel [134] also indicates that the order of the modifications of x_i and σ_i has a strong impact on the effectiveness of self-adaptation: It appears important to mutate the standard deviations first and to use the mutated standard deviations for the modification of object variables. As the authors point out in that study, the reversed mechanism might suffer from generating offspring that have useful object variable vectors but poor strategy parameter vectors because these have not been used to determine the position of the offspring itself.

More work needs to be performed, however, to achieve any clear understanding of the general advantages or disadvantages of one self-adaptation scheme compared to the other mechanisms. A recent theoretical study by Beyer presents a first step toward this goal [127]. In this work, the author shows that the self-adaptation principle works for a variety of different probability density functions for the modification of the step size, i.e., it is an extremely robust mechanism. Moreover, [127] clarifies that (9) is obtained from the corresponding equation for evolution strategies with one self-adaptable step size by Taylor expansion breaking off after the linear term, such that both methods behave equivalently for small settings of the learning rates τ and α, when $\tau = \alpha$. This prediction was confirmed perfectly by an experiment reported in [135].

Apart from the early work by Schaffer and Morishima [37], self-adaptation has only recently been introduced in genetic algorithms as a mechanism for evolving the parameters of variation operators. In [37], *punctuated crossover* was offered as a method for adapting both the number and position of crossover points for a multipoint crossover operator in canonical genetic algorithms. Although this approach seemed promising, the operator has not been used widely. A simpler approach toward self-adapting the crossover operator was presented by Spears [136], who allowed individuals to choose between two-point crossover and uniform crossover by means of a self-adaptable operator choice bit attached to the representation of individuals. The results indicated that, in case of crossover operators, rather than adapting to the single best operator for a given problem, the mechanism seems to benefit from the existing diversity of operators available for crossover.

Concerning the mutation operator in genetic algorithms, some effort to facilitate self-adaptation of the mutation rate has been presented by Smith and Fogarty [123], based on earlier work by Bäck [137]. These approaches incorporate the mutation rate $p_m \in [0, 1]$ into the representation of individuals and allow for mutation and recombination of the mutation rate

in the same way as the vector of binary variables is evolved. The results reported in [123] demonstrate that the mechanism yields a significant improvement in performance of a canonical genetic algorithm on the test functions used.

D. Recombination

The variation operators of canonical genetic algorithms, mutation, and recombination are typically applied with a strong emphasis on recombination. The standard algorithm performs a so-called one-point crossover, where two individuals are chosen randomly from the population, a position in the bitstrings is randomly determined as the crossover point, and an offspring is generated by concatenating the left substring of one parent and the right substring of the other parent. Numerous extensions of this operator, such as increasing the number of crossover points [138], uniform crossover (each bit is chosen randomly from the corresponding parental bits) [139], and others, have been proposed, but similar to evolution strategies no generally useful recipe for the choice of a recombination operator can be given. The theoretical analysis of recombination is still to a large extent an open problem. Recent work on *multi-parent recombination*, where more than two individuals participate in generating a single offspring individual, clarifies that this generalization of recombination might yield a performance improvement in many application examples [140]–[142]. Unlike evolution strategies, where it is either utilized for the creation of all members of the intermediate population (the default case) or not at all, the recombination operator in genetic algorithms is typically applied with a certain probability p_c, and commonly proposed settings of the crossover probability are $p_c = 0.6$ [13] and $p_c \in [0.75, 0.95]$ [118].

In evolution strategies recombination is incorporated into the main loop of the algorithm as the first operator (see Algorithm 1) and generates a new intermediate population of λ individuals by λ-fold application to the parent population, creating one individual per application from ϱ $(1 \leq \varrho \leq \mu)$ individuals. Normally, $\varrho = 2$ or $\varrho = \mu$ (so-called global recombination) are chosen. The recombination types for object variables and strategy parameters in evolution strategies often differ from each other, and typical examples are *discrete recombination* (random choices of single variables from parents, comparable to uniform crossover in genetic algorithms) and *intermediary recombination* (often arithmetic averaging, but other variants such as geometrical crossover [143] are also possible). For further details on these operators, see [79].

The advantages or disadvantages of recombination for a particular objective function can hardly be assessed in advance, and certainly no generally useful setting of recombination operators (such as the discrete recombination of object variables and global intermediary of strategy parameters as we have claimed in [79, pp. 82–83]) exists. Recently, Kursawe has impressively demonstrated that, using an inappropriate setting of the recombination operator, the (15 100)-evolution strategy with n self-adaptable variances might even diverge on a sphere model for $n = 100$ [144]. Kursawe shows that the appropriate choice of the recombination operator not only depends on the objective function topology, but also on the dimension of the objective function and the number of strategy parameters incorporated into the individuals. Only recently, Rechenberg [46] and Beyer [142] presented first results concerning the convergence velocity analysis of global recombination in case of the sphere model. These results clarify that, for using one (rather than n as in Kursawe's experiment) optimally chosen standard deviation σ, a μ-fold speedup is achieved by both recombination variants. Beyer's interpretation of the results, however, is somewhat surprising because it does not put down the success of this operator on the existence of building blocks which are usefully rearranged in an offspring individual, but rather explains it as a *genetic repair* of the harmful parts of mutation.

Concerning evolutionary programming, a rash statement based on the common understanding of the contending structures as individuals would be to claim that evolutionary programming simply does not use recombination. Rather than focusing on the mechanism of sexual recombination, however, Fogel [145] argues that one may examine and simulate its functional effect and correspondingly interpret a string of symbols as a reproducing population or species, thus making recombination a nonissue (refer to [145] for philosophical reasons underlining this choice).

E. Selection

Unlike the variation operators which work on the genetic representation, the selection operator is based solely on the fitness values of the individuals.

In genetic algorithms, selection is typically implemented as a probabilistic operator, using the relative fitness $p(\vec{a}_i) = f(\vec{a}_i)/\sum_{j=1}^{\mu} f(\vec{a}_j)$ to determine the selection probability of an individual $\vec{a}_i$ (*proportional selection*). This method requires positive fitness values and a maximization task, so that *scaling functions* are often utilized to transform the fitness values accordingly (see, e.g., [18, p. 124]). Rather than using absolute fitness values, *rank-based selection* methods utilize the indexes of individuals when ordered according to fitness values to calculate the corresponding selection probabilities. Linear [146] as well as nonlinear [82, p. 60] mappings have been proposed for this type of selection operator. *Tournament selection* [147] works by taking a random uniform sample of a certain size $q > 1$ from the population, selecting the best of these q individuals to survive for the next generation, and repeating the process until the new population is filled. This method gains increasing popularity because it is easy to implement, computationally efficient, and allows for fine-tuning the selective pressure by increasing or decreasing the tournament size q. For an overview of selection methods and a characterization of their selective pressure in terms of numerical measures, the reader should consult [148] and [149]. While most of these selection operators have been introduced in the framework of a generational genetic algorithm, they can also be used in combination with the steady-state and generation gap methods outlined in Section III.

The (μ, λ)-evolution strategy uses a deterministic selection scheme. The notation (μ, λ) indicates that μ parents create $\lambda > \mu$ offspring by means of recombination and mutation, and the best μ offspring individuals are deterministically

selected to replace the parents (in this case, $Q = \emptyset$ in Algorithm 1). Notice that this mechanism allows that the best member of the population at generation $t + 1$ might perform *worse* than the best individual at generation t, i.e., the method is not *elitist*, thus allowing the strategy to accept temporary deteriorations that might help to leave the region of attraction of a local optimum and reach a better optimum. In contrast, the $(\mu + \lambda)$ strategy selects the μ survivors from the union of parents and offspring, such that a monotonic course of evolution is guaranteed [$Q = P(t)$ in Algorithm 1]. Due to recommendations by Schwefel, however, the (μ, λ) strategy is preferred over the $(\mu + \lambda)$ strategy, although recent experimental findings seem to indicate that the latter performs as well as or better than the (μ, λ) strategy in many practical cases [134]. It should also be noted that both schemes can be interpreted as instances of the general (μ, κ, λ) strategy, where $1 \leq \kappa \leq \infty$ denotes the maximum life span (in generations) of an individual. For $\kappa = 1$, the selection method yields the (μ, λ) strategy, while it turns into the $(\mu + \lambda)$ strategy for $\kappa = \infty$ [54].

A minor difference between evolutionary programming and evolution strategies consists in the choice of a probabilistic variant of $(\mu + \lambda)$ selection in evolutionary programming, where each solution out of offspring and parent individuals is evaluated against $q > 1$ (typically, $q \leq 10$) other randomly chosen solutions from the union of parent and offspring individuals [$Q = P(t)$ in Algorithm 1]. For each comparison, a "win" is assigned if an individual's score is better or equal to that of its opponent, and the μ individuals with the greatest number of wins are retained to be parents of the next generation. As shown in [79, pp. 96–99], this selection method is a probabilistic version of $(\mu + \lambda)$ selection which becomes more and more deterministic as the number q of competitors is increased. Whether or not a probabilistic selection scheme should be preferable over a deterministic scheme remains an open question.

Evolutionary algorithms can easily be ported to parallel computer architectures [150], [151]. Since the individuals can be modified and, most importantly, evaluated independently of each other, we should expect a speed-up scaling linear with the number of processing units p as long as p does not exceed the population size μ. But selection operates on the whole population so this operator eventually slows down the overall performance, especially for massively parallel architectures where $p \gg \mu$. This observation motivated the development of parallel algorithms using local selection within subpopulations like in *migration models* [53], [152] or within small neighborhoods of spatially arranged individuals like in *diffusion models* [153]–[156] (also called *cellular evolutionary algorithms* [157]–[159]). It can be observed that local selection techniques not only yield a considerable speed-up on parallel architectures, but also improve the robustness of the algorithms [46], [116], [160].

F. Other Evolutionary Algorithm Variants

Although it is impossible to present a thorough overview of all variants of evolutionary computation here, it seems appropriate to explicitly mention *order-based genetic algorithms* [18], [82], *classifier systems* [161], [162], and *genetic programming* [31], [70], [81], [163] as branches of genetic algorithms that have developed into their own directions of research and application. The following overview is restricted to a brief statement of their domain of application and some literature references:

- *Order-based genetic algorithms* were proposed for searching the space of *permutations* $\pi: \{1, \cdots, n\} \rightarrow \{1, \cdots, n\}$ directly rather than using complex decoding functions for mapping binary strings to permutations and preserving feasible permutations under mutation and crossover (as proposed in [164]). They apply specialized recombination (such as *order crossover* or *partially matched crossover*) and mutation operators (such as random exchanges of two elements of the permutation) which preserve permutations (see [82, ch. 10] for an overview).

- *Classifier systems* use an evolutionary algorithm to search the space of *production rules* (often encoded by strings over a ternary alphabet, but also sometimes using symbolic rules [165]) of a learning system capable of induction and generalization [18, ch. 6], [161], [166], [167]. Typically, the *Michigan* approach and the *Pittsburgh* approach are distinguished according to whether an individual corresponds with a single rule of the rule-based system (Michigan) or with a complete rule base (Pittsburgh).

- *Genetic programming* applies evolutionary search to the space of tree structures which may be interpreted as computer programs in a language suitable to modification by mutation and recombination. The dominant approach to genetic programming uses (a subset of) LISP programs (S expressions) as genotype space [31], [163], but other programming languages including machine code are also used (see, e.g., [70], [81], and [168]).

Throughout this section we made the attempt to compare the constituents of evolutionary algorithms in terms of their canonical forms. But in practice the borders between these approaches are much more fluid. We can observe a steady evolution in this field by modifying (mutating), (re)combining, and validating (evaluating) the current approaches, permanently improving the population of evolutionary algorithms.

V. APPLICATIONS

Practical application problems in fields as diverse as engineering, natural sciences, economics, and business (to mention only some of the most prominent representatives) often exhibit a number of characteristics that prevent the straightforward application of standard instances of evolutionary algorithms. Typical problems encountered when developing an evolutionary algorithm for a practical application include the following.

1) A suitable representation and corresponding operators need to be developed when the canonical representation is different from binary strings or real-valued vectors.
2) Various constraints need to be taken into account by means of a suitable method (ranging from penalty functions to repair algorithms, constraint-preserving operators, and decoders; see [169] for an overview).
3) Expert knowledge about the problem needs to be incorporated into the representation and the operators in order to guide the search process and increase its convergence velocity—without running into the trap, however, of being confused and misled by expert beliefs and habits which might not correspond with the best solutions.
4) An objective function needs to be developed, often in cooperation with experts from the particular application field.
5) The parameters of the evolutionary algorithm need to be set (or tuned) and the feasibility of the approach needs to be assessed by comparing the results to expert solutions (used so far) or, if applicable, solutions obtained by other algorithms.

Most of these topics require experience with evolutionary algorithms as well as cooperation between the application's expert and the evolutionary algorithm expert, and only few general results are available to guide the design of the algorithm (e.g., representation-independent recombination and mutation operators [170], [171], the requirement that small changes by mutation occur more frequently than large ones [48], [172], and a quantification of the selective pressure imposed by the most commonly used selection operators [149]). Nevertheless, evolutionary algorithms often yield excellent results when applied to complex optimization problems where other methods are either not applicable or turn out to be unsatisfactory (a variety of examples can be found in [80]).

Important practical problem classes where evolutionary algorithms yield solutions of high quality include engineering design applications involving continuous parameters (e.g., for the design of aircraft [173], [174] structural mechanics problems based on two-dimensional shape representations [175], electromagnetic systems [176], and mobile manipulators [177], [178]), discrete parameters (e.g., for multiplierless digital filter optimization [179], the design of a linear collider [180], or nuclear reactor fuel arrangement optimization [181]), and mixed-integer representations (e.g., for the design of survivable networks [182] and optical multilayer systems [115]). Combinatorial optimization problems with a straightforward binary representation of solutions have also been treated successfully with canonical genetic algorithms and their derivatives (e.g., set partitioning and its application to airline crew scheduling [183], knapsack problems [184], [185], and others [186]). Relevant applications to combinatorial problems utilizing a permutation representation of solutions are also found in the domains of scheduling (e.g., production scheduling [187] and related problems [188]), routing (e.g., of vehicles [189] or telephone calls [190]), and packing (e.g., of pallets on a truck [191]).

The existing range of successful applications is extremely broad, thus by far preventing an exhaustive overview—the list of fields and example applications should be taken as a hint for further reading rather than a representative overview. Some of the most challenging applications with a large profit potential are found in the field of biochemical drug design, where evolutionary algorithms have gained remarkable interest and success in the past few years as an optimization procedure to support protein engineering [134], [192]–[194]. Also, finance and business provide a promising field of profitable applications [195], but of course few details are published about this work (see, e.g., [196]). In fact, the relation between evolutionary algorithms and economics has found increasing interest in the past few years and is now widely seen as a promising modeling approach for agents acting in a complex, uncertain situation [197].

In concluding this section, we refer to the research field of *computational intelligence* (see Section VI for details) and the applications of evolutionary computation to the other main fields of computational intelligence, namely fuzzy logic and neural networks. An overview of the utilization of genetic algorithms to train and construct neural networks is given in [198], and of course other variants of evolutionary algorithms can also be used for this task (see e.g., [199] for an evolutionary programming, [200] for an evolution strategy example, and [97] and [201] for genetic algorithm examples). Similarly, both the rule base and membership functions of fuzzy systems can be optimized by evolutionary algorithms, typically yielding improvements of the performance of the fuzzy system (e.g., [202]–[206]). The interaction of computational intelligence techniques and hybridization with other methods such as expert systems and local optimization techniques certainly opens a new direction of research toward hybrid systems that exhibit problem solving capabilities approaching those of naturally intelligent systems in the future. Evolutionary algorithms, seen as a technique to evolve machine intelligence (see [5]), are one of the mandatory prerequisites for achieving this goal by means of algorithmic principles that are already working quite successfully in natural evolution [207].

VI. SUMMARY AND OUTLOOK

To summarize, the current state of evolutionary computation research can be characterized as in the following.

- The basic concepts have been developed more than 35 years ago, but it took almost two decades for their potential to be recognized by a larger audience.
- Application-oriented research in evolutionary computation is quite successful and almost dominates the field (if we consider the majority of papers). Only few potential application domains could be identified, if any, where evolutionary algorithms have not been tested so far. In many cases they have been used to produce good, if not superior, results.
- In contrast, the theoretical foundations are to some extent still weak. To say it more pithy: "We know that they work, but we do not know why." As a consequence, inexperienced users fall into the same traps repeatedly, since there are only few rules of thumb for the design and parameterization of evolutionary algorithms.
  A constructive approach for the synthesis of evolutionary algorithms, i.e., the choice or design of the representations, variation operators, and selection mechanisms is needed. But first investigations pointing in the direction of design principles for representation-independent operators

are encouraging [171], as well, as is the work on complex nonstandard representations such as in the field of genetic programming.

- Likewise, the field still lacks a sound formal characterization of the application domain and the limits of evolutionary computation. This requires future efforts in the field of complexity theory.

There exists a strong relationship between evolutionary computation and some other techniques, e.g., fuzzy logic and neural networks, usually regarded as elements of artificial intelligence. Following Bezdek [208], their main common characteristic lies in their numerical knowledge representation, which differentiates them from traditional symbolic artificial intelligence. Bezdek suggested the term *computational intelligence* for this special branch of artificial intelligence with the following characteristics[3]:

1) numerical knowledge representation;
2) adaptability;
3) fault tolerance;
4) processing speed comparable to human cognition processes;
5) error rate optimality (e.g., with respect to a Bayesian estimate of the probability of a certain error on future data).

We regard computational intelligence as one of the most innovative research directions in connection with evolutionary computation, since we may expect that efficient, robust, and easy-to-use solutions to complex real-world problems will be developed on the basis of these complementary techniques. In this field, we expect an impetus from the interdisciplinary cooperation, e.g., techniques for tightly coupling evolutionary and problem domain heuristics, more elaborate techniques for self-adaptation, as well as an important step toward machine intelligence.

Finally, it should be pointed out that we are far from using all potentially helpful features of evolution within evolutionary algorithms. Comparing natural evolution and the algorithms discussed here, we can immediately identify a list of important differences, which all might be exploited to obtain more robust search algorithms *and* a better understanding of natural evolution.

- Natural evolution works under dynamically changing environmental conditions, with nonstationary optima and even changing optimization criteria, and the individuals themselves are also changing the structure of the adaptive landscape during adaptation [210]. In evolutionary algorithms, environmental conditions are often static, but nonelitist variants are able to deal with changing environments. It is certainly worthwhile, however, to consider a more flexible life span concept for individuals in evolutionary algorithms than just the extremes of a maximum life span of one generation [as in a (μ, λ) strategy] and of an unlimited life span (as in an elitist strategy), by introducing an aging parameter as in the (μ, κ, λ) strategy [54].

---

[3]The term "computational intelligence" was originally coined by Cercone and McCalla [209].

- The long-term goal of evolution consists of the maintenance of *evolvability* of a population [95], guaranteed by mutation, and a preservation of diversity within the population (the term *meliorization* describes this more appropriately than optimization or adaptation does). In contrast, evolutionary algorithms often aim at finding a precise solution and converging to this solution.

- In natural evolution, many criteria need to be met at the same time, while most evolutionary algorithms are designed for single fitness criteria (see [211] for an overview of the existing attempts to apply evolutionary algorithms to multiobjective optimization). The concepts of *diploidy* or *polyploidy* combined with *dominance and recessivity* [50] as well as the idea of introducing two sexes with different selection criteria might be helpful for such problems [212], [213].

- Natural evolution neither assumes global knowledge (about all fitness values of all individuals) nor a generational synchronization, while many evolutionary algorithms still identify an iteration of the algorithm with one complete generation update. Fine-grained asynchronously parallel variants of evolutionary algorithms, introducing local neighborhoods for recombination and selection and a time-space organization like in cellular automata [157]–[159] represent an attempt to overcome these restrictions.

- The *co-evolution* of species such as in predator-prey interactions implies that the adaptive landscape of individuals of one species changes as members of the other species make their adaptive moves [214]. Both the work on competitive fitness evaluation presented in [215] and the co-evolution of separate populations [216], [217] present successful approaches to incorporate the aspect of mutual interaction of different adaptive landscapes into evolutionary algorithms. As clarified by the work of Kauffman [214], however, we are just beginning to explore the dynamics of co-evolving systems and to exploit the principle for practical problem solving and evolutionary simulation.

- The genotype-phenotype mapping in nature, realized by the *genetic code* as well as the *epigenetic apparatus* (i.e., the biochemical processes facilitating the development and differentiation of an individual's cells into organs and systems), has evolved over time, while the mapping is usually fixed in evolutionary algorithms (dynamic parameter encoding as presented in [218] being a notable exception). An evolutionary self-adaptation of the genotype-phenotype mapping might be an interesting way to make the search more flexible, starting with a coarse-grained, volume-oriented search and focusing on promising regions of the search space as the evolution proceeds.

- Other topics, such as multicellularity and *ontogeny* of individuals, up to the development of their own brains (individual learning, such as accounted for by the Baldwin effect in evolution [219]), are usually not modeled in evolutionary algorithms. The self-adaptation of strategy parameters is just a first step into this direction, realizing the idea that each individual might have its own internal strategy to deal with its environment. This strategy might be more complex than the simple mutation parameters presently taken into account by evolution strategies and evolutionary programming.

With all this in mind, we are convinced that we are just beginning to understand and to exploit the full potential of evolutionary computation. Concerning basic research as well as practical applications to challenging industrial problems, evolutionary algorithms offer a wide range of promising further investigations, and it will be exciting to observe the future development of the field.

ACKNOWLEDGMENT

The authors would like to thank D. B. Fogel and three anonymous reviewers for their very valuable and detailed comments that helped them improve the paper. They also appreciate the informal comments of another anonymous reviewer, and the efforts of the anonymous associate editor responsible for handling the paper submission and review procedure. The first author would also like to thank C. Müller for her patience.

REFERENCES

[1] H. J. Bremermann, "Optimization through evolution and recombination," in *Self-Organizing Systems,* M. C. Yovits *et al.,* Eds. Washington, DC: Spartan, 1962.

[2] R. M. Friedberg, "A learning machine: Part I," *IBM J.,* vol. 2, no. 1, pp. 2–13, Jan. 1958.

[3] R. M. Friedberg, B. Dunham, and J. H. North, "A learning machine: Part II," *IBM J.,* vol. 3, no. 7, pp. 282–287, July 1959.

[4] G. E. P. Box, "Evolutionary operation: A method for increasing industrial productivity," *Appl. Statistics,* vol. VI, no. 2, pp. 81–101, 1957.

[5] D. B. Fogel, *Evolutionary Computation: Toward a New Philosophy of Machine Intelligence.* Piscataway, NJ: IEEE Press, 1995.

[6] J. H. Holland, "Outline for a logical theory of adaptive systems," *J. Assoc. Comput. Mach.,* vol. 3, pp. 297–314, 1962.

[7] I. Rechenberg, "Cybernetic solution path of an experimental problem," Royal Aircraft Establishment, Library translation No. 1122, Farnborough, Hants., U.K., Aug. 1965.

[8] H.-P. Schwefel, "Projekt MHD-Staustrahlrohr: Experimentelle Optimierung einer Zweiphasendüse, Teil I," Technischer Bericht 11.034/68, 35, AEG Forschungsinstitut, Berlin, Germany, Oct. 1968.

[9] L. J. Fogel, "Autonomous automata," *Ind. Res.,* vol. 4, pp. 14–19, 1962.

[10] J. T. Alander, "Indexed bibliography of genetic algorithms papers of 1996," University of Vaasa, Department of Information Technology and Production Economics, Rep. 94-1-96, 1995, (ftp.uwasa.fi, cs/report94-1, ga96bib.ps.Z).

[11] J. H. Holland, *Adaptation in Natural and Artificial Systems.* Ann Arbor, MI: Univ. of Michigan Press, 1975.

[12] J. H. Holland and J. S. Reitman, "Cognitive systems based on adaptive algorithms," in *Pattern-Directed Inference Systems,* D. A. Waterman and F. Hayes-Roth, Eds. New York: Academic, 1978.

[13] K. A. De Jong, "An analysis of the behavior of a class of genetic adaptive systems," Ph.D. dissertation, Univ. of Michigan, Ann Arbor, 1975, Diss. Abstr. Int. 36(10), 5140B, University Microfilms no. 76-9381.

[14] _____, "On using genetic algorithms to search program spaces," in *Proc. 2nd Int. Conf. on Genetic Algorithms and Their Applications.* Hillsdale, NJ: Lawrence Erlbaum, 1987, pp. 210–216.

[15] _____, "Are genetic algorithms function optimizers?" in *Parallel Problem Solving from Nature 2.* Amsterdam, The Netherlands: Elsevier, 1992, pp. 3–13.

[16] _____, "Genetic algorithms are NOT function optimizers," in *Foundations of Genetic Algorithms 2.* San Mateo, CA: Morgan Kaufmann, 1993, pp. 5–17.

[17] D. E. Goldberg, "Genetic algorithms and rule learning in dynamic system control," in *Proc. 1st Int. Conf. on Genetic Algorithms and Their Applications.* Hillsdale, NJ: Lawrence Erlbaum, 1985, pp. 8–15.

[18] _____, *Genetic Algorithms in Search, Optimization and Machine Learning.* Reading, MA: Addison-Wesley, 1989.

[19] _____, "The theory of virtual alphabets," in *Parallel Problem Solving from Nature—Proc. 1st Workshop PPSN I.* (Lecture Notes in Computer Science, vol. 496). Berlin, Germany: Springer, 1991, pp. 13–22.

[20] D. E. Goldberg, K. Deb, and J. H. Clark, "Genetic algorithms, noise, and the sizing of populations," *Complex Syst.,* vol. 6, pp. 333–362, 1992.

[21] D. E. Goldberg, K. Deb, H. Kargupta, and G. Harik, "Rapid, accurate optimization of difficult problems using fast messy genetic algorithms," in *Proc. 5th Int. Conf. on Genetic Algorithms.* San Mateo, CA: Morgan Kaufmann, 1993, pp. 56–64.

[22] L. Davis, Ed., *Handbook of Genetic Algorithms.* New York: Van Nostrand Reinhold, 1991.

[23] L. J. Eshelman and J. D. Schaffer, "Crossover's niche," in *Proc. 5th Int. Conf. on Genetic Algorithms.* San Mateo, CA: Morgan Kaufmann, 1993, pp. 9–14.

[24] _____, "Productive recombination and propagating and preserving schemata," in *Foundations of Genetic Algorithms 3.* San Francisco, CA: Morgan Kaufmann, 1995, pp. 299–313.

[25] S. Forrest and M. Mitchell, "What makes a problem hard for a genetic algorithm? Some anomalous results and their explanation," *Mach. Learn.,* vol. 13, pp. 285–319, 1993.

[26] J. J. Grefenstette, "Optimization of control parameters for genetic algorithms," *IEEE Trans. Syst., Man Cybern.,* vol. SMC-16, no. 1, pp. 122–128, 1986.

[27] _____, "Incorporating problem specific knowledge into genetic algorithms," in *Genetic Algorithms and Simulated Annealing,* L. Davis, Ed. San Mateo, CA: Morgan Kaufmann, 1987, pp. 42–60.

[28] _____, "Conditions for implicit parallelism," in *Foundations of Genetic Algorithms.* San Mateo, CA: Morgan Kaufmann, 1991, pp. 252–261.

[29] _____, "Deception considered harmful," in *Foundations of Genetic Algorithms 2.* San Mateo, CA: Morgan Kaufmann, 1993, pp. 75–91.

[30] J. R. Koza, "Hierarchical genetic algorithms operating on populations of computer programs," in *Proc. 11th Int. Joint Conf. on Artificial Intelligence,* N. S. Sridharan, Ed. San Mateo, CA: Morgan Kaufmann, 1989, pp. 768–774.

[31] _____, *Genetic Programming: On the Programming of Computers by Means of Natural Selection.* Cambridge, MA: MIT Press, 1992.

[32] M. Mitchell, *An Introduction to Genetic Algorithms.* Cambridge, MA: MIT Press, 1996.

[33] R. L. Riolo, "The emergence of coupled sequences of classifiers," in *Proc. 3rd Int. Conf. on Genetic Algorithms.* San Mateo, CA: Morgan Kaufmann, 1989, pp. 256–264.

[34] _____, "The emergence of default hierarchies in learning classifier systems," in *Proc. 3rd Int. Conf. on Genetic Algorithms.* San Mateo, CA: Morgan Kaufmann, 1989, pp. 322–327.

[35] J. D. Schaffer, "Multiple objective optimization with vector evaluated genetic algorithms," in *Proc. 1st Int. Conf. on Genetic Algorithms and Their Applications.* Hillsdale, NJ: Lawrence Erlbaum, 1985, pp. 93–100.

[36] J. D. Schaffer and L. J. Eshelman, "On crossover as an evolutionary viable strategy," in *Proc. 4th Int. Conf. on Genetic Algorithms.* San Mateo, CA: Morgan Kaufmann, 1991, pp. 61–68.

[37] J. D. Schaffer and A. Morishima, "An adaptive crossover distribution mechanism for genetic algorithms," in *Proc. 2nd Int. Conf. on Genetic Algorithms and Their Applications.* Hillsdale, NJ: Lawrence Erlbaum, 1987, pp. 36–40.

[38] L. J. Fogel, "On the organization of intellect," Ph.D. dissertation, University of California, Los Angeles, 1964.

[39] G. H. Burgin, "On playing two-person zero-sum games against nonminimax players," *IEEE Trans. Syst. Sci. Cybern.,* vol. SSC-5, no. 4, pp. 369–370, Oct. 1969.

[40] _____, "Systems identification by quasilinearization and evolutionary programming," *J. Cybern.,* vol. 3, no. 2, pp. 56–75, 1973.

[41] J. W. Atmar, "Speculation on the evolution of intelligence and its possible realization in machine form," Ph.D. dissertation, New Mexico State Univ., Las Cruces, 1976.

[42] L. J. Fogel, A. J. Owens, and M. J. Walsh, *Artificial Intelligence Through Simulated Evolution.* New York: Wiley, 1966.

[43] D. B. Fogel, "An evolutionary approach to the traveling salesman problem," *Biological Cybern.,* vol. 60, pp. 139–144, 1988.

[44] _____, "Evolving artificial intelligence," Ph.D. dissertation, Univ. of California, San Diego, 1992.

[45] I. Rechenberg, *Evolutionsstrategie: Optimierung technischer Systeme nach Prinzipien der biologischen Evolution.* Stuttgart, Germany: Frommann-Holzboog, 1973.

[46] _____, *Evolutionsstrategie '94,* in *Werkstatt Bionik und Evolutionstechnik.* Stuttgart, Germany: Frommann-Holzboog, 1994, vol. 1.

[47] H.-P. Schwefel, *Evolutionsstrategie und numerische Optimierung* Dissertation, Technische Universität Berlin, Germany, May 1975.

[48] _____, *Evolution and Optimum Seeking.* New York: Wiley, 1995 (Sixth-Generation Computer Technology Series).

[49] M. Herdy, "Reproductive isolation as strategy parameter in hierarchically organized evolution strategies," in *Parallel Problem Solving from Nature 2.* Amsterdam, The Netherlands: Elsevier, 1992, pp. 207–217.

[50] F. Kursawe, "A variant of Evolution Strategies for vector optimization," in *Parallel Problem Solving from Nature—Proc. 1st Workshop PPSN I* (Lecture Notes in Computer Science, vol. 496). Berlin, Germany: Springer, 1991, pp. 193–197.

[51] A. Ostermeier, "An evolution strategy with momentum adaptation of the random number distribution," in *Parallel Problem Solving from Nature 2.* Amsterdam, The Netherlands: Elsevier, 1992, pp. 197–206.

[52] A. Ostermeier, A. Gawelczyk, and N. Hansen, "Step-size adaptation based on nonlocal use of selection information," in *Parallel Problem Solving from Nature—PPSN III, Int. Conf. on Evolutionary Computation.* (Lecture Notes in Computer Science, vol. 866). Berlin, Germany: Springer, 1994, pp. 189–198.

[53] G. Rudolph, "Global optimization by means of distributed evolution strategies," in *Parallel Problem Solving from Nature—Proc. 1st Workshop PPSN I* (Lecture Notes in Computer Science, vol. 496). Berlin, Germany: Springer, 1991, pp. 209–213.

[54] H.-P. Schwefel and G. Rudolph, "Contemporary evolution strategies," in *Advances in Artificial Life. 3rd Int. Conf. on Artificial Life* (Lecture Notes in Artificial Intelligence, vol. 929), F. Morán, A. Moreno, J. J. Merelo, and P. Chacón, Eds. Berlin, Germany: Springer, 1995, pp. 893–907.

[55] J. Klockgether and H.-P. Schwefel, "Two-phase nozzle and hollow core jet experiments," in *Proc. 11th Symp. Engineering Aspects of Magnetohydrodynamics,* D. G. Elliott, Ed. Pasadena, CA: California Institute of Technology, Mar. 24–26, 1970, pp. 141–148.

[56] J. J. Grefenstette, Ed., *Proc. 1st Int. Conf. on Genetic Algorithms and Their Applications.* Hillsdale, NJ: Lawrence Erlbaum, 1985.

[57] _____, *Proc. 2nd Int. Conf. on Genetic Algorithms and Their Applications.* Hillsdale, NJ: Lawrence Erlbaum, 1987.

[58] J. D. Schaffer, Ed., *Proc. 3rd Int. Conf. on Genetic Algorithms.* San Mateo, CA: Morgan Kaufmann, 1989.

[59] R. K. Belew and L. B. Booker, Eds., *Proc. 4th Int. Conf. on Genetic Algorithms.* San Mateo, CA, Morgan Kaufmann, 1991.

[60] S. Forrest, Ed., *Proc. 5th Int. Conf. on Genetic Algorithms.* San Mateo, CA: Morgan Kaufmann, 1993.

[61] L. Eshelman, Ed., *Genetic Algorithms: Proc. 6th Int. Conf.* San Francisco, CA: Morgan Kaufmann, 1995.

[62] D. B. Fogel and W. Atmar, Eds., *Proc 1st Annu. Conf. on Evolutionary Programming.* San Diego, CA: Evolutionary Programming Society, 1992.

[63] _____, *Proc. 2nd Annu. Conf. on Evolutionary Programming.* San Diego, CA: Evolutionary Programming Society, 1993.

[64] A. V. Sebald and L. J. Fogel, Eds., *Proc. 3rd Annual Conf. on Evolutionary Programming.* Singapore: World Scientific, 1994.

[65] J. R. McDonnell, R. G. Reynolds, and D. B. Fogel, Eds., *Proc. 4th Annu. Conf. on Evolutionary Programming.* Cambridge, MA: MIT Press, 1995.

[66] L. J. Fogel, P. J. Angeline, and T. Bäck, Eds., *Proc. 5th Annu. Conf. on Evolutionary Programming.* Cambridge, MA: The MIT Press, 1996.

[67] G. J. E. Rawlins, Ed., *Foundations of Genetic Algorithms.* San Mateo, CA: Morgan Kaufmann, 1991.

[68] L. D. Whitley, Ed., *Foundations of Genetic Algorithms 2.* San Mateo, CA: Morgan Kaufmann, 1993.

[69] M. D. Vose and L. D. Whitley, Ed., *Foundations of Genetic Algorithms 3.* San Francisco, CA: Morgan Kaufmann, 1995.

[70] J. R. Koza, D. E. Goldberg, D. B. Fogel, and R. L. Riolo, Eds., *Genetic Programming 1996. Proc. 1st Annu. Conf.* Cambridge, MA: MIT Press, 1996.

[71] H.-P. Schwefel and R. Männer, Eds., *Parallel Problem Solving from Nature—Proc. 1st Workshop PPSN I.* Berlin, Germany: Springer, 1991, vol. 496 of *Lecture Notes in Computer Science.*

[72] R. Männer and B. Manderick, Eds., *Parallel Problem Solving from Nature 2.* Amsterdam, The Netherlands: Elsevier, 1992.

[73] Y. Davidor, H.-P. Schwefel, and R. Männer, Eds., *Parallel Problem Solving from Nature—PPSN III, Int. Conf. on Evolutionary Computation.* (Lecture Notes in Computer Science, vol. 866) Berlin: Springer, 1994.

[74] H.-M. Voigt, W. Ebeling, I. Rechenberg, and H.-P. Schwefel, Eds., *Parallel Problem Solving from Nature IV. Proc. Int. Conf. on Evolutionary Computation.* Berlin, Germany: Springer, 1996, vol. 1141 of *Lecture Notes in Computer Science.*

[75] *Proc. 1st IEEE Conf. on Evolutionary Computation, Orlando, FL.* Piscataway, NJ: IEEE Press, 1994.

[76] *Proc. 2nd IEEE Conf. on Evolutionary Computation, Perth, Australia.* Piscataway, NJ: IEEE Press, 1995.

[77] *Proc. 3rd IEEE Conf. on Evolutionary Computation, Nagoya, Japan.* Piscataway, NJ: IEEE Press, 1996.

[78] *Proc. 4th IEEE Conf. on Evolutionary Computation, Indianapolis, IN.* Piscataway, NJ: IEEE Press, 1997.

[79] T. Bäck, *Evolutionary Algorithms in Theory and Practice.* New York: Oxford Univ. Press, 1996.

[80] T. Bäck, D. B. Fogel, and Z. Michalewicz, Eds., *Handbook of Evolutionary Computation.* New York: Oxford Univ. Press and Institute of Physics, 1997.

[81] K. E. Kinnear, Ed., *Advances in Genetic Programming.* Cambridge, MA: MIT Press, 1994.

[82] Z. Michalewicz, *Genetic Algorithms + Data Structures = Evolution Programs.* Berlin, Germany: Springer, 1996.

[83] A. Törn and A. Žilinskas, *Global Optimization* (Lecture Notes in Computer Science, vol. 350). Berlin: Springer, 1989.

[84] T. Bäck, U. Hammel, M. Schütz, H.-P. Schwefel, and J. Sprave, "Applications of evolutionary algorithms at the center for applied systems analysis," in *Computational Methods in Applied Sciences'96*, J.-A. Désidéri, C. Hirsch, P. Le Tallec, E. Oñate, M. Pandolfi, J. Périaux, and E. Stein, Eds. Chichester, UK: Wiley, 1996, pp. 243–250.

[85] H.-P. Schwefel, "Direct search for optimal parameters within simulation models," in *Proc. 12th Annu. Simulation Symp., Tampa, FL, Mar. 1979,* pp. 91–102.

[86] Z. Michalewicz, "A hierarchy of evolution programs: An experimental study," *Evolutionary Computation,* vol. 1, no. 1, pp. 51–76, 1993.

[87] W. Atmar, "Notes on the simulation of evolution," *IEEE Trans. Neural Networks,* vol. 5, no. 1, pp. 130–148, 1994.

[88] L. D. Whitley, "The GENITOR algorithm and selection pressure: Why rank-based allocation of reproductive trials is best," in *Proc. 3rd Int. Conf. on Genetic Algorithms.* San Mateo, CA: Morgan Kaufmann, 1989, pp. 116–121.

[89] L. D. Whitley and J. Kauth, "GENITOR: A different genetic algorithm," in *Proc. Rocky Mountain Conf. Artificial Intel.,* Denver, CO, 1988, pp. 118–130.

[90] K. A. De Jong and J. Sarma, "Generation gaps revisited," in *Foundations of Genetic Algorithms 2.* San Mateo, CA: Morgan Kaufmann, 1993, pp. 19–28.

[91] D. J. Powell, M. M. Skolnick, and S. S. Tong, "Interdigitation: A hybrid technique for engineering design optimization employing genetic algorithms, expert systems, and numerical optimization," in *Handbook of Genetic Algorithms.* New York: Van Nostrand Reinhold, 1991, ch. 20, pp. 312–321.

[92] J.-M. Renders and S. P. Flasse, "Hybrid methods using genetic algorithms for global optimization," *IEEE Trans. Syst., Man, Cybern. B,* vol. 26, no. 2, pp. 243–258, 1996.

[93] K. A. De Jong, "Evolutionary computation: Recent developments and open issues," in *1st Int. Conf. on Evolutionary Computation and Its Applications,* E. D. Goodman, B. Punch, and V. Uskov, Eds. Moskau: Presidium of the Russian Academy of Science, 1996, pp. 7–17.

[94] M. Mitchell and S. Forrest, "Genetic algorithms and artificial life," *Artificial Life,* vol. 1, no. 3, pp. 267–289, 1995.

[95] L. Altenberg, "The evolution of evolvability in genetic programming," in *Advances in Genetic Programming.* Cambridge, MA: MIT Press, 1994, pp. 47–74.

[96] R. Keller and W. Banzhaf, "Genetic programming using genotype-phenotype mapping from linear genomes into linear phenotypes," in *Genetic Programming 1996: Proc. 1st Annu. Conf.,* J. R. Koza, D. E. Goldberg, D. B. Fogel, and R. L. Riolo, Eds., 1996.

[97] F. Gruau, "Genetic synthesis of modular neural networks," in *Proc. 5th Int. Conf. on Genetic Algorithms.* San Mateo, CA: Morgan Kaufmann, 1993, pp. 318–325.

[98] M. Mandischer, "Representation and evolution of neural networks," in *Artificial Neural Nets and Genetic Algorithms,* R. F. Albrecht, C. R. Reeves, and N. C. Steele, Eds. Wien, Germany: Springer, 1993, pp. 643–649.

[99] H. J. Antonisse, "A new interpretation of schema notation that overturns the binary encoding constraint," in *Proc. 3rd Int. Conf. on Genetic Algorithms.* San Mateo, CA: Morgan Kaufmann, 1989, pp. 86–91.

[100] U.-M. O'Reilly and F. Oppacher, "The troubling aspects of a building block hypothesis for genetic programming," in *Foundations of Genetic Algorithms 3.* San Francisco, CA: Morgan Kaufmann, 1995, pp. 73–88.

[101] T. Bäck, "Optimal mutation rates in genetic search," in *Proc. 5th Int. Conf. on Genetic Algorithms,* S. Forrest, Ed. San Mateo, CA: Morgan Kaufmann, 1993, pp. 2–8.

[102] L. J. Eshelman and J. D. Schaffer, "Real-coded genetic algorithms and interval-schemata," in *Foundations of Genetic Algorithms 2.* San Mateo, CA: Morgan Kaufmann, 1993, pp. 187–202.

[103] C. Z. Janikow and Z. Michalewicz, "An experimental comparison of binary and floating point representations in genetic algorithms," in *Proc. 4th Int. Conf. on Genetic Algorithms.* San Mateo, CA, Morgan Kaufmann, 1991, pp. 31–36.

[104] N. J. Radcliffe, "Equivalence class analysis of genetic algorithms," *Complex Systems,* vol. 5, no. 2, pp. 183–206, 1991.

[105] A. H. Wright, "Genetic algorithms for real parameter optimization," in *Foundations of Genetic Algorithms.* San Mateo, CA: Morgan Kaufmann, 1991, pp. 205–218.

[106] A. Nix and M. D. Vose, "Modeling genetic algorithms with markov chains," *Ann. Math. Artif. Intell.,* vol. 5, pp. 79–88, 1992.

[107] M. D. Vose, "Modeling simple genetic algorithms," in *Foundations of Genetic Algorithms 2.* San Mateo, CA: Morgan Kaufmann, 1993, pp. 63–73.

[108] M. D. Vose and A. H. Wright, "Simple genetic algorithms with linear fitness," *Evolutionary Computation,* vol. 2, no. 4, pp. 347–368, 1994.

[109] M. D. Vose, "Modeling simple genetic algorithms," *Evolutionary Computation,* vol. 3, no. 4, pp. 453–472, 1995.

[110] G. Rudolph, "Convergence analysis of canonical genetic algorithms," *IEEE Trans. Neural Networks,* Special Issue on Evolutionary Computation, vol. 5, no. 1, pp. 96–101, 1994.

[111] J. Suzuki, "A Markov chain analysis on simple genetic algorithms," *IEEE Trans. Syst., Man, Cybern.,* vol. 25, no. 4, pp. 655–659, Apr. 1995.

[112] K. A. De Jong, W. M. Spears, and D. F. Gordon, "Using Markov chains to analyze GAFO's," in *Foundations of Genetic Algorithms 3.* San Francisco, CA: Morgan Kaufmann, 1995, pp. 115–137.

[113] L. D. Whitley, "An executable model of a simple genetic algorithm," in *Foundations of Genetic Algorithms 3.* San Francisco, CA: Morgan Kaufmann, 1995, pp. 45–62.

[114] G. Rudolph, "An evolutionary algorithm for integer programming," in *Parallel Problem Solving from Nature—PPSN III, Int. Conf. on Evolutionary Computation* (Lecture Notes in Computer Science, vol. 866). Berlin, Germany: Springer, 1994, pp. 139–148.

[115] M. Schütz and J. Sprave, "Application of parallel mixed-integer evolution strategies with mutation rate pooling," in *Proc. 5th Annu. Conf. on Evolutionary Programming.* Cambridge, MA: MIT Press, 1996, pp. 345–354.

[116] B. Groß, U. Hammel, A. Meyer, P. Maldaner, P. Roosen, and M. Schütz, "Optimization of heat exchanger networks by means of evolution strategies," in *Parallel Problem Solving from Nature IV. Proc. Int. Conf. on Evolutionary Computation.* (Lecture Notes in Computer Science, vol. 1141). Berlin: Springer, 1996, pp. 1002–1011.

[117] R. Lohmann, "Structure evolution in neural systems," in *Dynamic, Genetic, and Chaotic Programming,* B. Soucek and the IRIS Group, Eds. New York: Wiley, 1992, pp. 395–411.

[118] J. D. Schaffer, R. A. Caruana, L. J. Eshelman, and R. Das, "A study of control parameters affecting online performance of genetic algorithms for function optimization," in *Proc. 3rd Int. Conf. on Genetic Algorithms.* San Mateo, CA: Morgan Kaufmann, 1989, pp. 51–60.

[119] H. J. Bremermann, M. Rogson, and S. Salaff, "Global properties of evolution processes," in *Natural Automata and Useful Simulations,* H. H. Pattec, E. A. Edelsack, L. Fein, and A. B. Callahan, Eds. Washington, DC: Spartan, 1966, ch. 1, pp. 3–41.

[120] H. Mühlenbein, "How genetic algorithms really work: I. Mutation and hillclimbing," in *Parallel Problem Solving from Nature 2.* Amsterdam: Elsevier, 1992, pp. 15–25.

[121] T. C. Fogarty, "Varying the probability of mutation in the genetic algorithm," in *Proc. 3rd Int. Conf. on Genetic Algorithms.* San Mateo, CA: Morgan Kaufmann, 1989, pp. 104–109.

[122] T. Bäck and M. Schütz, "Intelligent mutation rate control in canonical genetic algorithms," in *Foundations of Intelligent Systems, 9th Int. Symp., ISMIS'96* (Lecture Notes in Artificial Intelligence, vol. 1079), Z. W. Ras and M. Michalewicz, Eds. Berlin, Germany: Springer, 1996, pp. 158–167.

[123] J. Smith and T. C. Fogarty, "Self adaptation of mutation rates in a steady state genetic algorithm," in *Proc. 3rd IEEE Conf. on Evolutionary Computation.* Piscataway, NJ: IEEE Press, 1996, pp. 318–323.

[124] M. Yanagiya, "A simple mutation-dependent genetic algorithm," in *Proc. 5th Int. Conf. on Genetic Algorithms.* San Mateo, CA: Morgan Kaufmann, 1993, p. 659.

[125] H.-P. Schwefel, *Numerical Optimization of Computer Models.* Chichester: Wiley, 1981.

[126] ———, *Numerische Optimierung von Computer-Modellen mittels der Evolutionsstrategie,* vol. 26 of *Interdisciplinary Systems Research.* Basel, Germany: Birkhäuser, 1977.

[127] H.-G. Beyer, "Toward a theory of evolution strategies: Self-adaptation," *Evolutionary Computation*, vol. 3, no. 3, pp. 311–348, 1995.

[128] G. Rudolph, "On correlated mutations in evolution strategies," in *Parallel Problem Solving from Nature 2.* Amsterdam, The Netherlands: Elsevier, 1992, pp. 105–114.

[129] N. Saravanan and D. B. Fogel, "Evolving neurocontrollers using evolutionary programming," in *Proc. 1st IEEE Conf. on Evolutionary Computation.* Piscataway, NJ: IEEE Press, 1994, vol. 1, pp. 217–222.

[130] _____, "Learning of strategy parameters in evolutionary programming: An empirical study," in *Proc. 3rd Annu. Conf. on Evolutionary Programming.* Singapore: World Scientific, 1994, pp. 269–280.

[131] N. Saravanan, D. B. Fogel, and K. M. Nelson, "A comparison of methods for self-adaptation in evolutionary algorithms," *BioSystems*, vol. 36, pp. 157–166, 1995.

[132] D. B. Fogel, L. J. Fogel, and W. Atmar, "Meta-evolutionary programming," in *Proc. 25th Asilomar Conf. Sig., Sys. Comp.*, R. R. Chen, Ed. Pacific Grove, CA, 1991, pp. 540–545.

[133] P. J. Angeline, "The effects of noise on self-adaptive evolutionary optimization," in *Proc. 5th Annu. Conf. on Evolutionary Programming.* Cambridge, MA: MIT Press, 1996, pp. 433–440.

[134] D. K. Gehlhaar and D. B. Fogel, "Tuning evolutionary programming for conformationally flexible molecular docking," in *Proc. 5th Annu. Conf. on Evolutionary Programming.* Cambridge, MA: MIT Press, 1996, pp. 419–429.

[135] T. Bäck and H.-P. Schwefel, "Evolutionary computation: An overview," in *Proc. 3rd IEEE Conf. on Evolutionary Computation.* Piscataway, NJ: IEEE Press, 1996, pp. 20–29.

[136] W. M. Spears, "Adapting crossover in evolutionary algorithms," in *Proc. 4th Annu. Conf. on Evolutionary Programming.* Cambridge, MA: MIT Press, 1995, pp. 367–384.

[137] T. Bäck, "Self Adaptation in Genetic Algorithms," in *Proceedings of the 1st European Conference on Artificial Life*, F. J. Varela and P. Bourgine, Eds. Cambridge, MA: MIT Press, 1992, pp. 263–271.

[138] L. J. Eshelman, R. A. Caruna, and J. D. Schaffer, "Biases in the crossover landscape," in *Proc. 3rd Int. Conf. on Genetic Algorithms.* San Mateo, CA: Morgan Kaufmann, 1989, pp. 10–19.

[139] G. Syswerda, "Uniform crossover in genetic algorithms," in *Proc. 3rd Int. Conf. on Genetic Algorithms.* San Mateo, CA: Morgan Kaufmann, 1989, pp. 2–9.

[140] A. E. Eiben, P.-E. Raué, and Zs. Ruttkay, "Genetic algorithms with multi-parent recombination," in *Parallel Problem Solving from Nature—PPSN III, Int. Conf. on Evolutionary Computation.* Berlin: Springer, 1994, vol. 866 of *Lecture Notes in Computer Science*, pp. 78–87.

[141] A. E. Eiben, C. H. M. van Kemenade, and J. N. Kok, "Orgy in the computer: Multi-parent reproduction in genetic algorithms," in *Advances in Artificial Life. 3rd Int. Conf. on Artificial Life*, F. Morán, A. Moreno, J. J. Merelo, and P. Chacón, Eds. Berlin: Springer, 1995, vol. 929 of *Lecture Notes in Artificial Intelligence*, pp. 934–945.

[142] H.-G. Beyer, "Toward a theory of evolution strategies: On the benefits of sex—the $(\mu/\mu, \lambda)$-theory," *Evolutionary Computation*, vol. 3, no. 1, pp. 81–111, 1995.

[143] Z. Michalewicz, G. Nazhiyath, and M. Michalewicz, "A note on usefulness of geometrical crossover for numerical optimization problems," in *Proc. 5th Annu. Conf. on Evolutionary Programming.* Cambridge, MA: The MIT Press, 1996, pp. 305–312.

[144] F. Kursawe, "Toward self-adapting evolution strategies," in *Proc. 2nd IEEE Conf. Evolutionary Computation, Perth, Australia.* Piscataway, NJ: IEEE Press, 1995, pp. 283–288.

[145] D. B. Fogel, "On the philosophical differences between evolutionary algorithms and genetic algorithms," in *Proc. 2nd Annu. Conf. on Evolutionary Programming.* San Diego, CA: Evolutionary Programming Society, 1993, pp. 23–29.

[146] J. E. Baker, "Adaptive selection methods for genetic algorithms," in *Proc. 1st Int. Conf. on Genetic Algorithms and Their Applications.* Hillsdale, NJ: Lawrence Erlbaum, 1985, pp. 101–111.

[147] D. E. Goldberg, B. Korb, and K. Deb, "Messy genetic algorithms: Motivation, analysis, and first results," *Complex Syst.*, vol. 3, no. 5, pp. 493–530, Oct. 1989.

[148] T. Bäck, "Selective pressure in evolutionary algorithms: A characterization of selection mechanisms," in *Proc. 1st IEEE Conf. on Evolutionary Computation.* Piscataway, NJ: IEEE Press, 1994, pp. 57–62.

[149] D. E. Goldberg and K. Deb, "A comparative analysis of selection schemes used in genetic algorithms," in *Foundations of Genetic Algorithms.* San Mateo, CA: Morgan Kaufmann, 1991, pp. 69–93.

[150] M. Dorigo and V. Maniezzo, "Parallel genetic algorithms: Introduction and overview of current research," in *Parallel Genetic Algorithms: Theory & Applications, Frontiers in Artificial Intelligence and Applications,*

J. Stender, Ed. Amsterdam, The Netherlands: IOS, 1993, pp. 5–42.

[151] F. Hoffmeister, "Scalable parallelism by evolutionary algorithms," in *Parallel Computing and Mathematical Optimization*, (Lecture Notes in Economics and Mathematical Systems, vol. 367), M. Grauer and D. B. Pressmar, Eds. Berlin, Germany: Springer, 1991, pp. 177–198.

[152] M. Munetomo, Y. Takai, and Y. Sato, "An efficient migration scheme for subpopulation-based asynchronously parallel genetic algorithms," in *Proc. 5th Int. Conf. on Genetic Algorithms.* San Mateo, CA: Morgan Kaufmann, 1993, p. 649.

[153] S. Baluja, "Structure and performance of fine-grain parallelism in genetic search," in *Proc. 5th Int. Conf. on Genetic Algorithms.* San Mateo, CA: Morgan Kaufmann, 1993, pp. 155–162.

[154] R. J. Collins and D. R. Jefferson, "Selection in massively parallel genetic algorithms," in *Proc. 4th Int. Conf. on Genetic Algorithms.* San Mateo, CA, Morgan Kaufmann, 1991, pp. 249–256.

[155] M. Gorges-Schleuter, "ASPARAGOS: An asynchronous parallel genetic optimization strategy," in *Proc. 3rd Int. Conf. on Genetic Algorithms.* San Mateo, CA: Morgan Kaufmann, 1989, pp. 422–427.

[156] P. Spiessens and B. Manderick, "Fine-grained parallel genetic algorithms," in *Proc. 3rd Int. Conf. on Genetic Algorithms.* San Mateo, CA: Morgan Kaufmann, 1989, pp. 428–433.

[157] V. S. Gordon, K. Mathias, and L. D. Whitley, "Cellular genetic algorithms as function optimizers: Locality effects," in *Proc. 1994 ACM Symp. on Applied Computing*, E. Deaton, D. Oppenheim, J. Urban, and H. Berghel, Eds. New York: ACM, 1994, pp. 237–241.

[158] G. Rudolph and J. Sprave, "A cellular genetic algorithm with self-adjusting acceptance threshold," in *Proc. 1st IEE/IEEE Int. Conf. Genetic Algorithms in Eng. Sys.: Innovations and Appl.* London: IEE, 1995, pp. 365–372.

[159] L. D. Whitley, "Cellular genetic algorithms," in *Proc. 5th Int. Conf. on Genetic Algorithms.* San Mateo, CA: Morgan Kaufmann, 1993, p. 658.

[160] M. Gorges-Schleuter, "Comparison of local mating strategies in massively parallel genetic algorithms," in *Parallel Problem Solving from Nature 2.* Amsterdam: Elsevier, 1992, pp. 553–562.

[161] J. H. Holland, K. J. Holyoak, R. E. Nisbett, and P. R. Thagard, *Induction: Processes of Inference, Learning, and Discovery.* Cambridge, MA: MIT Press, 1986.

[162] R. Serra and G. Zanarini, *Complex Systems and Cognitive Processes.* Berlin: Springer, 1990.

[163] M. L. Cramer, "A representation for the adaptive generation of simple sequential programs," in *Proc. 1st Int. Conf. on Genetic Algorithms and Their Applications.* Hillsdale, NJ: Lawrence Erlbaum, 1985, pp. 183–187.

[164] J. C. Bean, "Genetics and random keys for sequences and optimization," Department of Industrial and Operations Engineering, The Univ. of Michigan, Ann Arbor, Tech. Rep. 92-43, 1993.

[165] D. A. Gordon and J. J. Grefenstette, "Explanations of empirically derived reactive plans," in *Proc. Seventh Int. Conf. on Machine Learning.* San Mateo, CA: Morgan Kaufmann, June 1990, pp. 198–203.

[166] L. B. Booker, D. E. Goldberg, and J. H. Holland, "Classifier systems and genetic algorithms," in *Machine Learning: Paradigms and Methods*, J. G. Carbonell, Ed. Cambridge, MA: MIT Press/Elsevier, 1989, pp. 235–282.

[167] S. W. Wilson, "ZCS: A zeroth level classifier system," *Evolutionary Computation*, vol. 2, no. 1, pp. 1–18, 1994.

[168] F. D. Francone, P. Nordin, and W. Banzhaf, "Benchmarking the generalization capabilities of a compiling genetic programming system using sparse data sets," in *Genetic Programming 1996. Proc. 1st Annu. Conf.* Cambridge, MA: MIT Press, 1996, pp. 72–80.

[169] Z. Michalewicz and M. Schoenauer, "Evolutionary algorithms for constrained parameter optimization problems," *Evolutionary Computation*, vol. 4, no. 1, pp. 1–32, 1996.

[170] N. J. Radcliffe, "The algebra of genetic algorithms," *Ann. Math. Artif. Intell.*, vol. 10, pp. 339–384, 1994.

[171] P. D. Surry and N. J. Radcliffe, "Formal algorithms + formal representations = search strategies," in *Parallel Problem Solving from Nature IV. Proc. Int. Conf. on Evolutionary Computation.* (Lecture Notes in Computer Science, vol. 1141) Berlin, Germany: Springer, 1996, pp. 366–375.

[172] N. J. Radcliffe and P. D. Surry, "Fitness variance of formae and performance prediction," in *Foundations of Genetic Algorithms 3.* San Francisco, CA: Morgan Kaufmann, 1995, pp. 51–72.

[173] M. F. Bramlette and E. E. Bouchard, "Genetic algorithms in parametric design of aircraft," in *Handbook of Genetic Algorithms.* New York: Van Nostrand Reinhold, 1991, ch. 10, pp. 109–123.

[174] J. Périaux, M. Sefrioui, B. Stoufflet, B. Mantel, and E. Laporte, "Robust genetic algorithms for optimization problems in aerodynamic design," in *Genetic Algorithms in Engineering and Computer Science*, G. Winter,

J. Périaux, M. Galán, and P. Cuesta, Eds. Chichester: Wiley, 1995, ch. 19, pp. 371–396.

[175] M. Schoenauer, "Shape representations for evolutionary optimization and identification in structural mechanics," in *Genetic Algorithms in Engineering and Computer Science*, G. Winter, J. Périaux, M. Galán, and P. Cuesta, Eds. Chichester: Wiley, 1995, ch. 22, pp. 443–463.

[176] E. Michielssen and D. S. Weile, "Electromagnetic system design using genetic algorithms," in *Genetic Algorithms in Engineering and Computer Science*, G. Winter, J. Périaux, M. Galán, and P. Cuesta, Eds. Chichester: Wiley, 1995, ch. 18, pp. 345–369.

[177] B. Anderson, J. McDonnell, and W. Page, "Configuration optimization of mobile manipulators with equality constraints using evolutionary programming," in *Proc. 1st Annu. Conf. on Evolutionary Programming*. San Diego, CA: Evolutionary Programming Society, 1992, pp. 71–79.

[178] J. R. McDonnell, B. L. Anderson, W. C. Page, and F. G. Pin, "Mobile manipulator configuration optimization using evolutionary programming," in *Proc 1st Annu. Conf. on Evolutionary Programming*. San Diego, CA: Evolutionary Programming Society, 1992, pp. 52–62.

[179] J. D. Schaffer and L. J. Eshelman, "Designing multiplierless digital filters using genetic algorithms," in *Proc. 5th Int. Conf. on Genetic Algorithms*. San Mateo, CA: Morgan Kaufmann, 1993, pp. 439–444.

[180] H.-G. Beyer, "Some aspects of the 'evolution strategy' for solving TSP-like optimization problems appearing at the design studies of a 0.5 TeV e^+e^--linear collider," in *Parallel Problem Solving from Nature 2*. Amsterdam: Elsevier, 1992, pp. 361–370.

[181] T. Bäck, J. Heistermann, C. Kappler, and M. Zamparelli, "Evolutionary algorithms support refueling of pressurized water reactors," in *Proc. 3rd IEEE Conference on Evolutionary Computation*. Piscataway, NJ: IEEE Press, 1996, pp. 104–108.

[182] L. Davis, D. Orvosh, A. Cox, and Y. Qiu, "A genetic algorithm for survivable network design," in *Proc. 5th Int. Conf. on Genetic Algorithms*. San Mateo, CA: Morgan Kaufmann, 1993, pp. 408–415.

[183] D. M. Levine, "A genetic algorithm for the set partitioning problem," in *Proc. 5th Int. Conf. on Genetic Algorithms*. San Mateo, CA: Morgan Kaufmann, 1993, pp. 481–487.

[184] V. S. Gordon, A. P. W. Böhm, and L. D. Whitley, "A note on the performance of genetic algorithms on zero-one knapsack problems," in *Proc. 1994 ACM Symp. Applied Computing*, E. Deaton, D. Oppenheim, J. Urban, and H. Berghel, Eds. New York: ACM, 1994, pp. 194–195.

[185] A. Olsen, "Penalty functions and the knapsack problem," in *Proc. 1st IEEE Conf. on Evolutionary Computation*. Piscataway, NJ: IEEE Press, 1994, pp. 554–558.

[186] S. Khuri, T. Bäck, and J. Heitkötter, "An evolutionary approach to combinatorial optimization problems," in *Proc. 22nd Annu. ACM Computer Science Conf.*, D. Cizmar, Ed. New York: ACM, 1994, pp. 66–73.

[187] R. Bruns, "Direct chromosome representation and advanced genetic operators for production scheduling," in *Proc. 1st Annu. Conf. on Evolutionary Programming*. San Diego, CA: Evolutionary Programming Society, 1992, pp. 352–359.

[188] H.-L. Fang, P. Ross, and D. Corne, "A promising genetic algorithm approach to job-shop scheduling, rescheduling, and open-shop scheduling problems," in *Proc. 1st Annu. Conf. on Evolutionary Programming*. San Diego, CA: Evolutionary Programming Society, 1992, pp. 375–382.

[189] J. I. Blanton and R. L. Wainwright, "Multiple vehicle routing with time and capacity constraints using genetic algorithms," in *Proc. 5th Int. Conf. on Genetic Algorithms*. San Mateo, CA: Morgan Kaufmann, 1993, pp. 452–459.

[190] L. A. Cox, L. Davis, and Y. Qiu, "Dynamic anticipatory routing in circuit-switched telecommunications networks," in *Handbook of Genetic Algorithms*. New York: Van Nostrand Reinhold, 1991, ch. 11, pp. 109–143.

[191] K. Juliff, "A multi-chromosome genetic algorithm for pallet loading," in *Proc. 5th Int. Conf. on Genetic Algorithms*. San Mateo, CA: Morgan Kaufmann, 1993, pp. 467–473.

[192] S. Schulze-Kremer, "Genetic algorithms for protein ternary structure prediction," in *Parallel Genetic Algorithms: Theory & Applications*, J. Stender, Ed. Amsterdam: IOS, 1993, *Frontiers in Artificial Intelligence and Applications*, pp. 129–150.

[193] R. Unger and J. Moult, "A genetic algorithm for 3D protein folding simulation," in *Proc. 5th Int. Conf. on Genetic Algorithms*. San Mateo, CA: Morgan Kaufmann, 1993, pp. 581–588.

[194] D. C. Youvan, A. P. Arkin, and M. M. Yang, "Recursive ensemble mutagenesis: A combinatorial optimization technique for protein engineering," in *Parallel Problem Solving from Nature 2*. Amsterdam: Elsevier, 1992, pp. 401–410.

[195] R. F. Walker, E. W. Haasdijk, and M. C. Gerrets, "Credit evaluation using a genetic algorithm," in *Intelligent Systems for Finance and Business*. Chichester: Wiley, 1995, ch. 3, pp. 39–59.

[196] S. Goonatilake and P. Treleaven, Eds., *Intelligent Systems for Finance and Business*. Chichester: Wiley, 1995.

[197] P. G. Harrald, "Evolutionary algorithms and economic models: A view," in *Proc. 5th Annu. Conf. on Evolutionary Programming*. Cambridge, MA: MIT Press, 1996, pp. 3–7.

[198] L. D. Whitley, "Genetic algorithms and neural networks," in *Genetic Algorithms in Engineering and Computer Science*, G. Winter, J. Périaux, M. Galán, and P. Cuesta, Eds. Chichester, UK: Wiley, 1995, ch. 11, pp. 203–216.

[199] P. J. Angeline, G. M. Saunders, and J. B. Pollack, "An evolutionary algorithm that constructs recurrent neural networks," *IEEE Trans. Neural Networks*, vol. 5, no. 1, pp. 54–65, 1994.

[200] W. Wienholt, "Minimizing the system error in feedforward neural networks with evolution strategy," in *Proc. Int. Conf. on Artificial Neural Networks*, S. Gielen and B. Kappen, Eds. London: Springer, 1993, pp. 490–493.

[201] M. Mandischer, "Genetic optimization and representation of neural networks," in *Proc. 4th Australian Conf. on Neural Networks*, P. Leong and M. Jabri, Eds. Sidney Univ., Dept. Elect. Eng., 1993, pp. 122–125.

[202] A. Homaifar and E. McCormick, "Full design of fuzzy controllers using genetic algorithms," in *Neural and Stochastic Methods in Image and Signal Processing*, S.-S. Chen, Ed. The International Society for Optical Engineering, 1992, vol. SPIE-1766, pp. 393–404.

[203] C. L. Karr, "Genetic algorithms for fuzzy controllers," *AI Expert*, vol. 6, no. 2, pp. 27–33, 1991.

[204] S. B. Haffner and A. V. Sebald, "Computer-aided design of fuzzy HVAC controllers using evolutionary programming," in *Proc. 2nd Annu. Conf. on Evolutionary Programming*. San Diego, CA: Evolutionary Programming Society, 1993, , pp. 98–107.

[205] P. Thrift, "Fuzzy logic synthesis with genetic algorithms," in *Proc. 4th Int. Conf. on Genetic Algorithms*. San Mateo, CA, Morgan Kaufmann, 1991, pp. 514–518.

[206] P. Wang and D. P. Kwok, "Optimal fuzzy PID control based on genetic algorithm," in *Proc. 1992 Int. Conf. on Industrial Electronics, Control, and Instrumentation*. Piscataway, NJ: IEEE Press, 1992, vol. 2, pp. 977–981.

[207] D. C. Dennett, *Darwin's Dangerous Idea*, New York: Touchstone, 1995.

[208] J. C. Bezdek, "What is computational intelligence?" in *Computational Intelligence: Imitating Life*, J. M. Zurada, R. J. Marks II, and Ch. J. Robinson, Eds. New York: IEEE Press, 1994, pp. 1–12.

[209] N. Cercone and G. McCalla, "Ten years of computational intelligence," *Computational Intelligence*, vol. 10, no. 4, pp. i–vi, 1994.

[210] J. Schull, "The view from the adaptive landscape," in *Parallel Problem Solving from Nature—Proc. 1st Workshop PPSN I*, Berlin: Springer, 1991, vol. 496 of *Lecture Notes in Computer Science*, pp. 415–427.

[211] C. M. Fonseca and P. J. Fleming, "An overview of evolutionary algorithms in multiobjective optimization," *Evolutionary Computation*, vol. 3, no. 1, pp. 1–16, 1995.

[212] J. Lis and A. E. Eiben, "Multi-sexual genetic algorithm for multiobjective optimization," in *Proc. 4th IEEE Conf. Evolutionary Computation, Indianapolis, IN* Piscataway, NJ: IEEE Press, 1997.

[213] E. Ronald, "When selection meets seduction," in *Genetic Algorithms: Proc. 6th Int. Conf.* San Francisco, CA: Morgan Kaufmann, 1995, pp. 167–173.

[214] S. A. Kauffman, *The Origins of Order. Self-Organization and Selection in Evolution*. New York: Oxford Univ. Press, 1993.

[215] P. J. Angeline and J. B. Pollack, "Competitive environments evolve better solutions for complex tasks," in *Proc. 5th Int. Conf. on Genetic Algorithms*. San Mateo, CA: Morgan Kaufmann, 1993, pp. 264–270.

[216] W. D. Hillis, "Co-evolving parasites improve simulated evolution as an optimization procedure," in *Emergent Computation. Self-Organizing, Collective, and Cooperative Phenomena in Natural and Artificial Computing Networks*. Cambridge, MA: MIT Press, 1990, pp. 228–234.

[217] J. Paredis, "Coevolutionary life-time learning," in *Parallel Problem Solving from Nature IV. Proc. Int. Conf. on Evolutionary Computation*. (Lecture Notes in Computer Science, vol. 1141). Berlin, Germany: Springer, 1996, pp. 72–80.

[218] N. N. Schraudolph and R. K. Belew, "Dynamic parameter encoding for genetic algorithms," *Machine Learning*, vol. 9, pp. 9–21, 1992.

[219] R. W. Anderson, "Genetic mechanisms underlying the Baldwin effect are evident in natural antibodies," in *Proc. 4th Annu. Conf. on Evolutionary Programming*. Cambridge, MA: MIT Press, 1995, pp. 547–564.

[220] S. Forrest, Ed., *Emergent Computation. Self-Organizing, Collective, and Cooperative Phenomena in Natural and Artificial Computing Networks*. Cambridge, MA: MIT Press, 1990.

Chapter 2
Evolving Control Circuits
for Autonomous Robots

G. J. Friedman (1956) "Selective Feedback Computers for Engineering Synthesis and Nervous System Analogy," master's thesis, UCLA.

FRIEDMAN (1956) offered perhaps the earliest systematic proposal for generating autonomous behavior in a robot using the principles of variation and selection. The thesis proposed that a series of control circuits, much like contemporary neural networks, could be evolved through "selective feedback" in a manner akin to natural selection. The hypothesized environment was a field of varying temperature and light sources, and a robot was required to navigate in the field so as to maintain a desired temperature based on sensed information.

The control circuits were composed of "basic neural elements" that could perform excitatory and inhibitory functions. Groups of these elements could be gathered to implement decayed integrators, attenuators, amplifiers, and so forth. Friedman (1956) argued that versatile behaviors could be constructed by connecting such groups, and offered a mechanism for automatically constructing, testing, and evaluating these circuits using random mutation and selection.

Under some simplifying assumptions, Friedman (1956) showed that the time required to evolve suitable circuits for some complex behaviors was expected to be on the order of the logarithm of the time required to design the circuits by a completely random search. For moderately complex behaviors, the evolutionary search required minutes of computation, compared to billions of years for random search. Some of the assumptions were very optimistic, such as decomposing a complex behavioral task into a sequence of intermediate behaviors which could be accomplished in a geometric series of increasing difficulty, but even if the effectiveness of evolutionary search were diminished by several orders of magnitude it would still have an enormous advantage over random search.

The experiments in autonomous robotics suggested in Friedman (1956) were never actually implemented in practice. But the work is strongly suggestive of Braitenberg's "Vehicles" which appeared 28 years later (Braitenberg, 1984) and it also anticipates the subsequent, independently formulated work in evolving neural networks (e.g., Montana and Davis, 1989; and others), electronic circuits (e.g., Koza et al., 1996; and others), and robotics (e.g., Brooks, 1992; and others), as well as the now common concept of fitness landscapes (which was also proposed earlier in Wright, 1932), as well as the idea that complex systems can arise from the interaction of simple rules. A peer reviewed section of the thesis was later published as Friedman (1959), and no further extensions were made.

This paper was typeset especially for this edited volume and does not appear in its original form. The layout of the master's thesis incorporated wide margins with double spacing; the original document is 78 pages. All of the figures have been reproduced by electronic scanning and appear as they do in the original manuscript. Other handwritten marks, for equations and other notation, have been typeset in the version offered here. Friedman (1959) was used to verify the interpretation of the handwritten equations in Friedman (1956), but there are some discrepancies between the equations in the two documents; the attempt was to be consistent to Friedman (1956) while also being correct. The font was chosen to be as close to the original as possible, and misspellings were not corrected. It is hoped that no other misspellings or errors were introduced during this typesetting.

References

[1] V. Braitenberg (1984) *Vehicles: Experiments in Synthetic Psychology*, MIT Press, Cambridge, MA.

[2] R. A. Brooks (1992) "Artificial life and real robots," *Toward a Practice of Autonomous Systems: Proc. of the 1st European Conf. on Artificial Life*, F. J. Varela and P. Bourgine (eds.), MIT Press, Cambridge, MA, pp. 3–10.

[3] G. J. Friedman (1956) "Selective feedback computers for engineering synthesis and nervous system analogy," master's thesis, UCLA.

[4] G. J. Friedman (1959) "Digital simulation of an evolutionary process," *General Systems: Yearbook of the Society for General Systems Research*, Vol. 4, pp. 171–184.

[5] J. R. Koza, F. H. Bennett, D. Andre, and M. A. Keane (1996) "Four problems for which a computer program evolved by genetic programming is competitive with human performance," *Proc. of the 1996 IEEE Conf. on Evolutionary Computation*, IEEE Press, NY, pp. 1–10.

[6] D. J. Montana and L. Davis (1989) "Training feedforward neural networks using genetic algorithms," *Proc. of the 11th Intern. Joint Conf. on Artificial Intelligence*, N. S. Sridharan (ed.), Morgan Kaufmann, San Mateo, CA, pp. 762–767.

[7] S. Wright (1932) "The roles of mutation, inbreeding, crossbreeding, and selection in evolution," *Proc. of the 6th Intern. Cong. on Genetics*, Vol. 1, Ithaca, NY, pp. 356–366.

UNIVERSITY OF CALIFORNIA, LOS ANGELES

Selective Feedback Computers for Engineering
Synthesis and Nervous System Analogy

A thesis submitted in partial satisfaction of the
requirements for the degree of Master of Science
in Engineering

by

George J. Friedman

February, 1956

ABSTRACT

"Selective Feedback Computers for Engineering
    Synthesis and Nervous System Analogy"

MS Thesis by George J. Friedman

Many similarities of behavior exist between modern feedback
and computer systems, and the nervous system in animals. In an
effort to investigate new areas of similarity, this paper develops a model for Darwin's Law of Natural Selection, called a Selective Feedback Computer, which designs engineering circuits
and unifies previous models of learning behavior.

The Selective Feedback Computer imposes a series of increasingly severe "environments" upon a variable circuit. Each circuit is automatically tested and evaluated, and new circuit selections are made on the basis of past successes and failures.
The time required for the Selective Feedback Computer to attain
a solution in the form of a final circuit design is calculated
as a function of the probability of the circuit, and is small
compared to the time required by a random process.

Five control circuits in a series which could have resulted
from the application of the Selective Feedback Computer are described schematically. The behavior of these circuits is analogous to tropism, differentiated stimulus-response, adaptive, and
associative behavior. Two of the circuits operate in a similar
manner to two already constructed models for learning behavior.

The Selective Feedback Computer offers a possible tool for the
future development of control and computer systems, and for psychological research.

ACKNOWLEDGMENTS

Appreciation is expressed to Mr. George Mount for his patience
and guidance in the formulation of the concepts in this thesis;
to Mr. Joseph Gengerelli of the Psychology Department, Mr. Thomas Rogers of the Engineering Department, and Mr. Waldo Furgason
of the Zoology Department for their valuable comments, to my
wife, Ruthanne, for typing and proofreading the manuscript, and
to Messrs. Ephraim Cohen and Richard Hepburn for their services
as professional laymen.

                                        GJF

[Table of Contents and List of Illustrations omitted]

I INTRODUCTION

Many similarities of behavior have been shown to exist between living and non-living systems. Modern feedback control systems exhibit the same goal-seeking tendencies as the nervous system of an animal. Giant electronic "brains" imitate the computational and logical activity of the human mind. Special models have been built which simulate still other types of behavior, such as adaptation and the conditioned response.[1,3]

The theme of this paper is to investigate the possibility of finding new and perhaps more basic areas of similarity between the behavior of the living and non-living worlds.

1.1 The Law of Natural Selection

One of the most fundamental phenomena of the living world is the complex process called "evolution", which in theory relates all life forms. The core of the evolutionary process was expressed by Darwin in his "Law of Natural Selection" and can be formulated in the following steps which consist of four observed facts and a general conclusion:

Observed Facts:

a) A very large variety of organisms exist, and a very large number of each type are born.

b) Under a given environment, only a limited number of organisms survive and reproduce.

c) Random mutations often occur and one or more of an offspring's characteristics are different from those of its parent(s).

d) The survivors reproduce themselves and pass most of their characteristics, including the mutations, to their offspring.

General Conclusion:

e) Assuming the above facts and a constant environment, a series of individuals is established which converges upon a type better fitted for survival.

1.2 Objective

The general objective of this thesis is to develop a model for the evolutionary process which will establish a series of models for learning behavior in a manner similar to that by which the process of natural selection establishes a series of individuals on an evolutionary scale. It will be attempted to show that this model for the evolutionary process, which will be called a "Selective Feedback Computer", not only furnishes a logical relationship between all members of a series of learning models, but also presents a possible methods for engineering circuit design.

The specific objectives are outlined in Part II, STATEMENT OF THE PROBLEM.

II STATEMENT OF THE PROBLEM

2.1 Problem for Part III

To describe the organization of a machine composed of simple components whose overall operation has the following similarities to Darwin's Law of Natural Selection: (see section 1.1)

a) The ability to produce a large variety of distinct circuits.

b) The ability to subject any circuit to a series of tests of varying difficulty, and to detect a "success" or "failure" at each test according to predetermined criteria.

c) The ability to record each successful circuit and to reproduce it in slightly altered form so that it, in turn, can be subjected to the series of tests.

d) The ability to favor the circuits which are successful in most of the test by reproducing and retesting them most frequently.

(If these steps can be realized in a machine, then a conclusion parallel to the general conclusion of 1.1e can be made:

e) A series of circuits is established which converges upon a type which passes the greatest number of tests.)

2.2 Problem for Part IV

To estimate the advantage that the method of the Selective Feedback Computer has over a random method of finding or "designing" a circuit which meets specific needs (i.e., pass specific tests).

2.3 Problem for Part V

To demonstrate the operation of the Selective Feedback Computer by applying it to this particular problem:

Given:

A mobile, temperature-sensitive device in a temperature-varying field. It is desired to maintain the temperature of the device's interior within five degrees of 0° F. The device has seven transducers (See Fig. 1)

Four of the transducers are sensors and detect the following physical variables:

a) the temperature at the interior; symbol Ⓔ

b) " " " " left side; " Ⓛ

c) " " " " right side; " Ⓡ

d) a flash of light in immediate vicinity; Ⓕ

The remaining three of the transducers are effectors, and when energized by an informational input, perform the following physical duties:

a) push on the surroundings on the left; ⊡L

b) " " " " " " right; ⊡R

c) " " " " in any direction; ⊡A

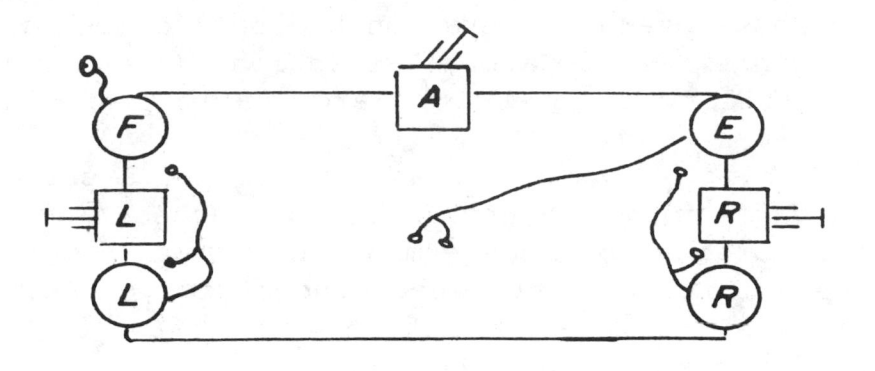

a) PHYSICAL DIAGRAM

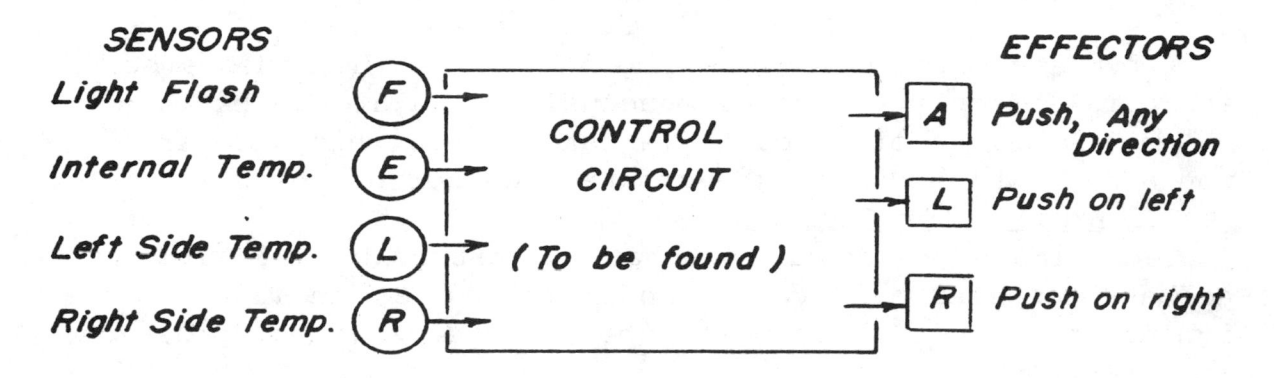

SENSORS			EFFECTORS
Light Flash	F	A	Push, Any Direction
Internal Temp.	E	CONTROL CIRCUIT	
Left Side Temp.	L	(To be found)	L Push on left
Right Side Temp.	R	R	Push on right

b) CONTROL CIRCUIT SCHEMATIC

1. The internal temperature must not diverge more than five degrees from zero degrees.

2. If:
$$\left\{ \begin{array}{cccc} External\ Temp: & 10° & 20° & 30° & 30° \\ Time: & 4 & 2 & 1 & 0 \end{array} \right\},\ \begin{array}{l} Internal\ Temp. \\ will\ exceed\ 5° \end{array}$$

3. One "push" moves the DEVICE its own length in one sec.

4. The sensors activate when $\Delta t = 3°$

c) TEMPERATURE,   HEAT TRANSFER, AND
MOBILITY DATA

Figure 1.   MOBILE, TEMPERATURE-SENSITIVE DEVICE

The heat transfer characteristics and mobility of the device are listed in Fig. 1.

The temperature-varying field (Fig. 2) is divided into five areas possessing the following modes of change:

a) Small areas of sudden high temperature;

b) Large areas of slowly increasing temperature;

c) Very narrow area of sudden high temperature;

d) A sudden, very high temperature area on the right, always preceded by a flash of light;

e) A sudden, very high temperature area on the left 10% of the time, and on the right 90% of the time, in each case preceded by a flash of light.

To Find:

A circuit with the sensors as inputs and the effectors as outputs which will control the movements of the device so that it will have a high probability of maintaining the controlled variable (its internal temperature) within the prescribed limits while in any area of the field.

III ORGANIZATION OF THE SELECTIVE FEEDBACK COMPUTER

3.1 Overall Operation

Assume that it is desired to design a circuit which will perform a certain function or operation. In order to determine whether a trial circuit meets the requirements of this function, it must be subjected to a test and its performance must be evaluated. If the function to be performed is a complex one, it may be expressed as the sum of other simpler functions, each one having its own test and performance evaluation. A series of tests can thus be established and the assumption is made that a circuit which satisfies any of the tests will be similar in some way to the circuit which satisfies all the tests. The possibility of arranging the tests in order of increasing difficulty is also assumed.

Stated very generally, the overall operation of the Selective Feedback Computer involves the application of a series of tests to a trial circuit and the evaluation of the circuit's performance at each test. New circuit selections are then made on the basis of the successes and failures of previously tested trial circuits.

The components of the Selective Feedback Computer, described in detail in section 3.2, are, in general, as follows: (Fig. 3)

a) A Variable Circuit whose network and parameters can be altered through a Switchbox;

b) An Exciter which applies a series of tests to the Variable Circuit, and a Criterion Detector which evaluates the per-

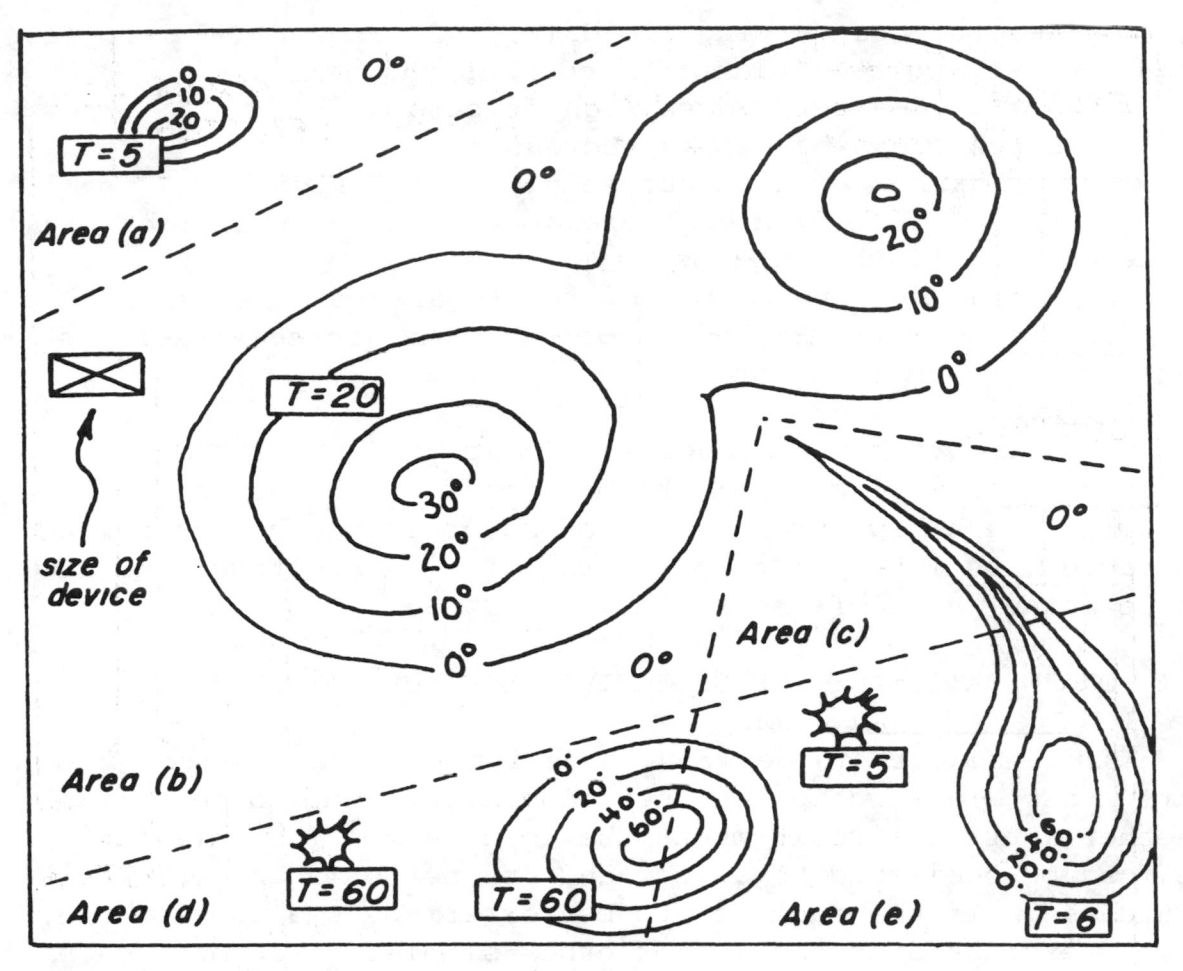

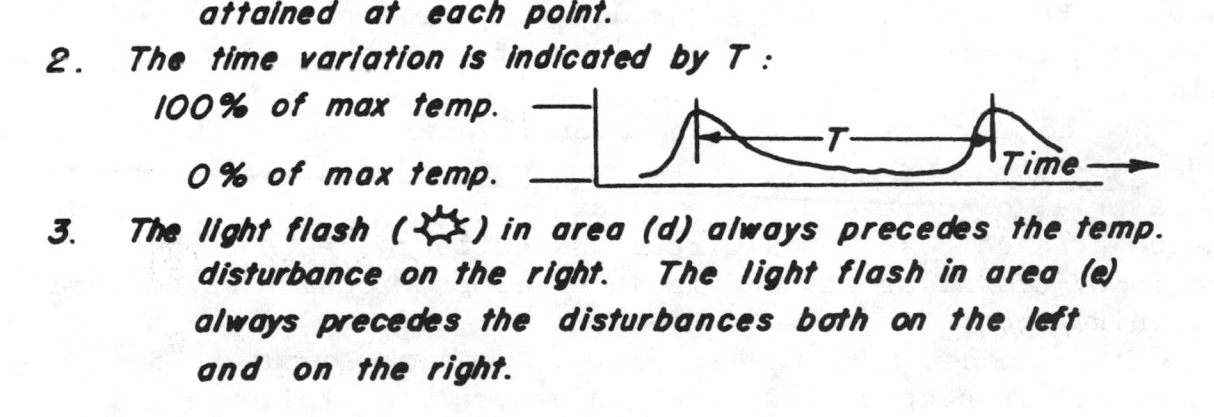

NOTES:

1. The contours indicate the MAXIMUM temperatures attained at each point.

2. The time variation is indicated by T :

100% of max temp.

0% of max temp.

3. The light flash (☼) in area (d) always precedes the temp. disturbance on the right. The light flash in area (e) always precedes the disturbances both on the left and on the right.

Figure 2. THE ENVIRONMENT OF THE DEVICE

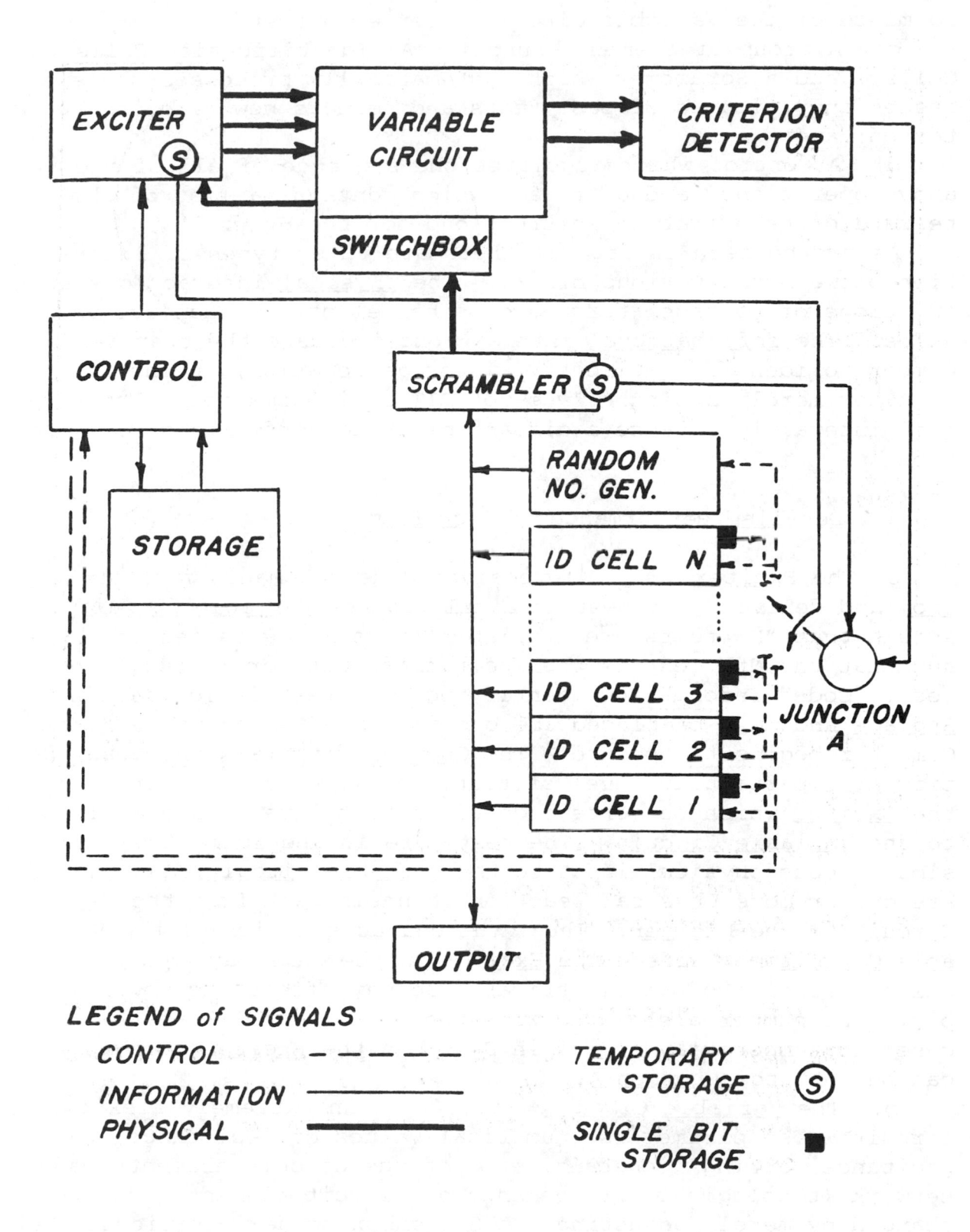

Figure 3.   SELECTIVE FEEDBACK COMPUTER

formance of the Variable Circuit after each test;

    c)   A group of elements—Junction A, Identification Cells (ID Cells), and a Scrambler—which systematically processes the successes and failures of past tests and selects new circuits to be tested; and

    d)   A Control which regulates the sequence of all of the above operations, and a Storage which contains a list of the tests and the operating instructions for the Control.

    As can be seen in Fig. 3, there are three types of interaction between the various blocks. The _physical_ interactions are functions of voltage, torque, or other electro-mechanical variables. The _informational_ interactions indicate the transfer of a number to identify a circuit, a storage location, an instruction, or merely a single "yes-no" BIT of information. The _control_ interactions cause a distant block to perform a given function.

3.2 _Detailed Performance of Each Item_

    a. The _Exciter_ (Fig. 4) receives coded commands from the _Control_ and delivers one test at a time to the _Variable Circuit_. Assume that "N" tests are available to form the series of tests, and that the Nth test is the least difficult, or lowest level of test. Coded information identifying each test is located in _Storage_ and must be placed there before the Selective Feedback Computer begins to operate. The _Control_ withdraws the code identifying a particular level of test, say level 3, and sends it to the _Exciter_ which converts code #3 to test level 3 and applies it to the _Variable Circuit_. The tests are in the form of several simultaneous physical input to the _Variable Circuit_, and there are one or more physical reaction channels back from the _Variable Circuit_ to the _Exciter_. Thus, the characteristics of the _Variable Circuit_ may affect the _Exciter_ in the same way that the characteristics of an electrical load may affect its power supply. The number associated with the test (number 3 in this case) is temporarily stored in S within the _Exciter_ so that it can be sent to _Junction A_.

    b. The _Variable Circuit_ (Fig. 4) is an extremely flexible circuit whose parameters (numerical values of resistance, capacitance, spring, constant, etc. of the circuit elements) and network (topological arrangements of circuit elements) can be changed by merely adjusting the _Switchbox_ to new positions. New circuit selections enter the _Switchbox_ as physical inputs from the _Scrambler_. At any given setting of the switches, a unique circuit is set up, and the settings of all the switches, taken

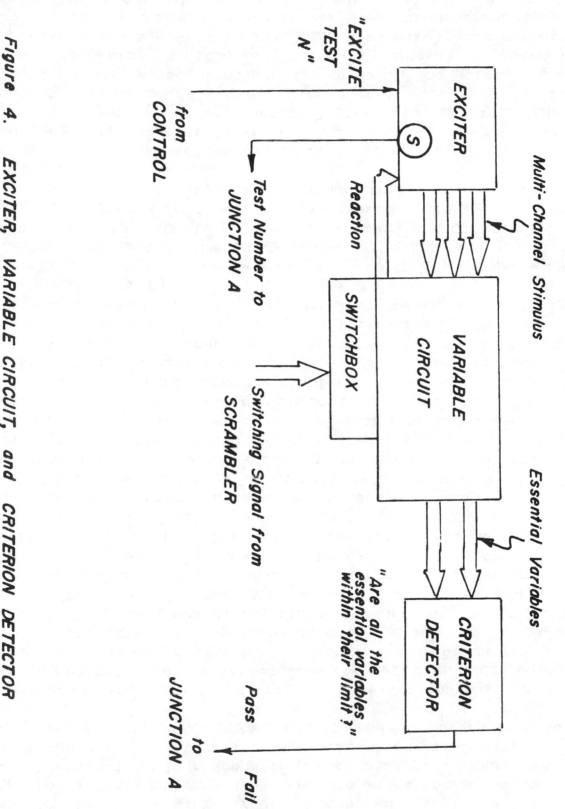

Figure 4. EXCITER, VARIABLE CIRCUIT, and CRITERION DETECTOR

in order, form an "Identification Number" (ID Number) (say 555555) for the particular circuit. Thus, when test level 3 is applied to circuit number 555555, certain important variables within the circuit will respond in a unique manner. Measurements of these important variables, called "essential variables", are sent to the <u>Criterion</u> <u>Detector</u> where they are evaluated.

c. The <u>Criterion</u> <u>Detector</u> (Fig. 4) receives the physical output of the essential variables from the <u>Variable</u> <u>Circuit</u> and checks each variable to determine if it is inside or outside previously prescribed limits. The criterion which the <u>Criterion</u> <u>Detector</u> applies when evaluating the performance of the <u>Variable</u> <u>Circuit</u> under a certain test is as follows: If all essential variables are within their limits, the particular circuit will have passed the particular test; if any essential variable exceeds its limits, it will have failed the test. Assume, for example, that it is desired to maintain the voltage at location P between 0 and 10 volts and the temperature at location Q between 50° and 55°. This particular voltage and temperature would be the "essential variables" and if, when circuit 555555 responds to test 3, they exceed the specified limits, circuit 555555 will have failed test 3. If the circuit was successful in resisting the unstabilizing effects of the physical inputs of test 3 so as to maintain these variables within their limits, then circuit 555555 will have passed test 3. This variable-limited criterion has wider application than what may be thought at first glance, and its choice is justified in section 3.5.

The single BIT of information regarding "pass" or "fail" is sent to <u>Junction</u> <u>A</u>.

d. <u>Junction</u> <u>A</u> (Fig. 5) is a logical circuit with inputs from the <u>Criterion</u> <u>Detector</u>, the <u>Scrambler</u>, and the <u>Exciter</u>, and with one output to the <u>ID</u> <u>Cells</u>. Assume the input from the <u>Criterion</u> <u>Detector</u> indicates a "pass" on the last test. Then the ID Number of the successful circuit, which is temporarily stored in Ⓢ of the <u>Scrambler</u>, will be delivered to the proper <u>ID</u> <u>Cell</u>. The proper ID Cell is determined by the number of the test just passed, which is temporarily stored in Ⓢ of the <u>Exciter</u>. Thus, if circuit 555555 passes test number 3, then the <u>Criterion</u> <u>Detector</u> output will indicate a "pass", the circuit's number, 555555, will be supplied by the temporary storage Ⓢ in the Scrambler, and the information that the last test was number 3, supplied by the temporary storage Ⓢ in the <u>Exciter</u>, will direct the number 555555 to <u>ID</u> <u>Cell</u> number 3 where it will be stored permanently.

If the input from the <u>Criterion</u> <u>Detector</u> indicates a "fail" on the last test, then the identification number in the Scram-

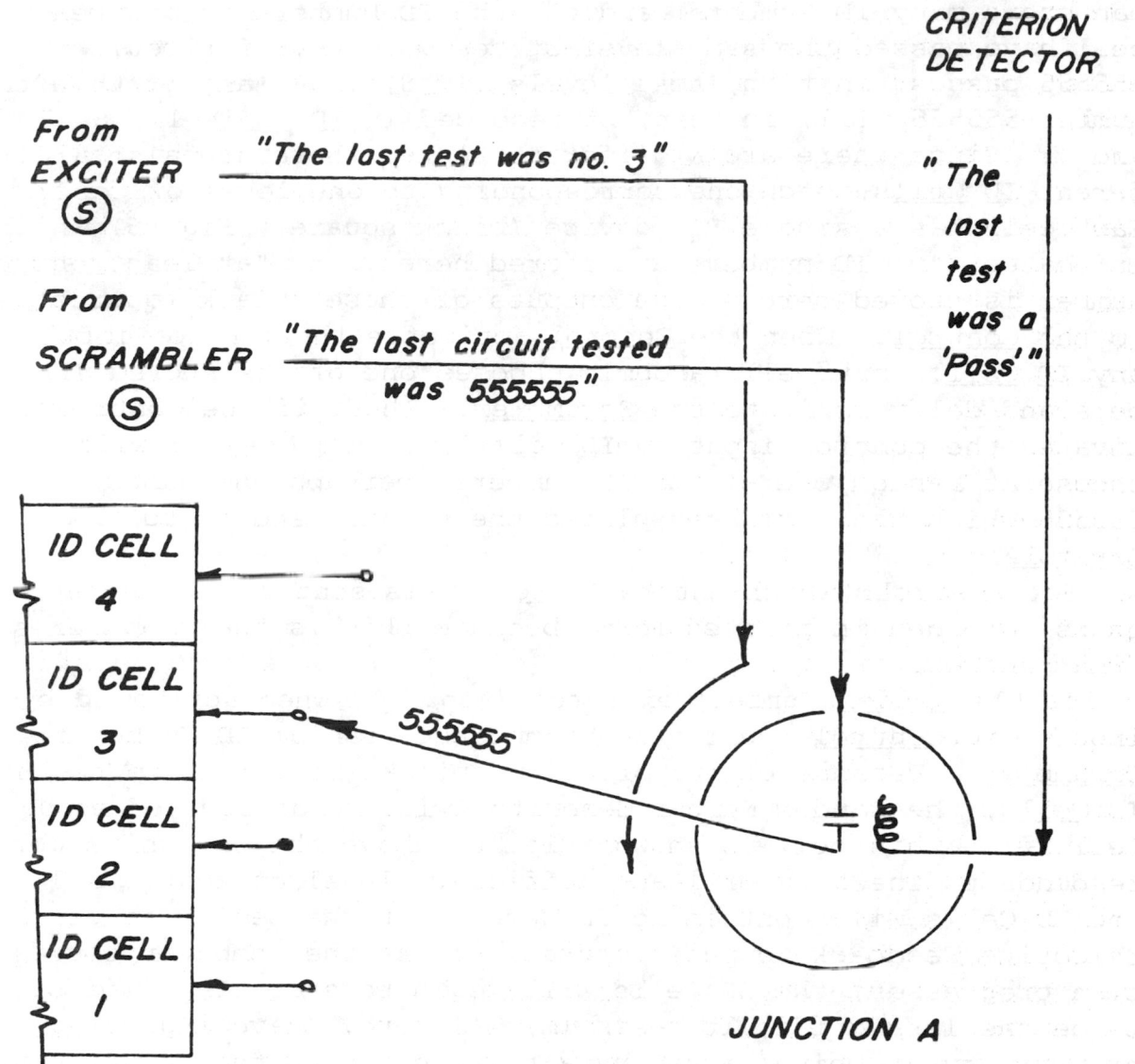

Figure 5.   JUNCTION A

bler will not be allowed to reach any of the ID Cells.

    e. The ID Cells (Fig. 6) are groups of registers which store
the identification numbers of successful circuits. Each cell
can store many ID Numbers, and all the ID Numbers in a given
cell have passed the same level of test. Thus, if circuit
555555 passes, in turn, test levels 11, 8, 7, 4, and 3, then the
number 555555 will, in turn, be recorded in ID Cells 11, 8, 7, 4,
and 3. Since there are N different tests, there are also N different ID Cells, each one corresponding to one level of test.
Each cell has a single BIT device (black squares, Fig. 6) which
indicates, "No ID numbers are stored here", or, "at least one ID
number is stored here". The outputs of these "black squares" go
to the Control. When the Control activates the control input to
any ID Cell, that cell randomly chooses one of its stored ID Numbers and delivers it to the Scrambler. Thus, if the Control activates the control input to ID Cell 3, then ID Cell 3 will
choose at random one of its ID Numbers (perhaps the number
555555 which has just been placed there) and send it to the
Scrambler.

    Every number which enters ID Cell 1 is sent to Output where
it is recorded in printed form, because this is the computer's
final answer.

    f. The Random Number Generator (Fig. 6), when activated by an
input from Control, delivers a completely random ID Number to the
Scrambler. Because of its close functional relationship to the
ID Cells, the Random Number Generator will be designated as ID
Cell N+1. This cell is shown (Fig. 6) above the one which corresponds to the Nth, or least difficult, level of test. All of
the ID Cells will contain no ID Numbers at the beginning of the
Selective Feedback Computer operation. As the computer opera-
tion progresses, the ID Cells will begin to store ID Numbers.
Since the less difficult tests (N, N-1, etc.) have a greater
probability of being passed by a trial circuit, then ID Cells N,
N-1, etc., corresponding to these easy tests, will start receiving ID Numbers earlier and at a greater frequency than the ID
Cells, ...3, 2, 1.

    However, the Random Number Generator is, by definition, never
"empty", and will always be able to supply an ID Number to the
Scrambler when activated by the Control.

    g. The Scrambler (Fig. 6), when supplied with an ID Number
from any of the ID Cells, adds or subtracts one small random
digit (1, 2, or 3) in a random place in the ID Number. (Thus,
555555 may become 557555 or 535555, but not 535655 or 550555).
The altered number is then sent to the Switchbox of the Variable
Circuit and is also stored temporarily in (S) within the Scrambler

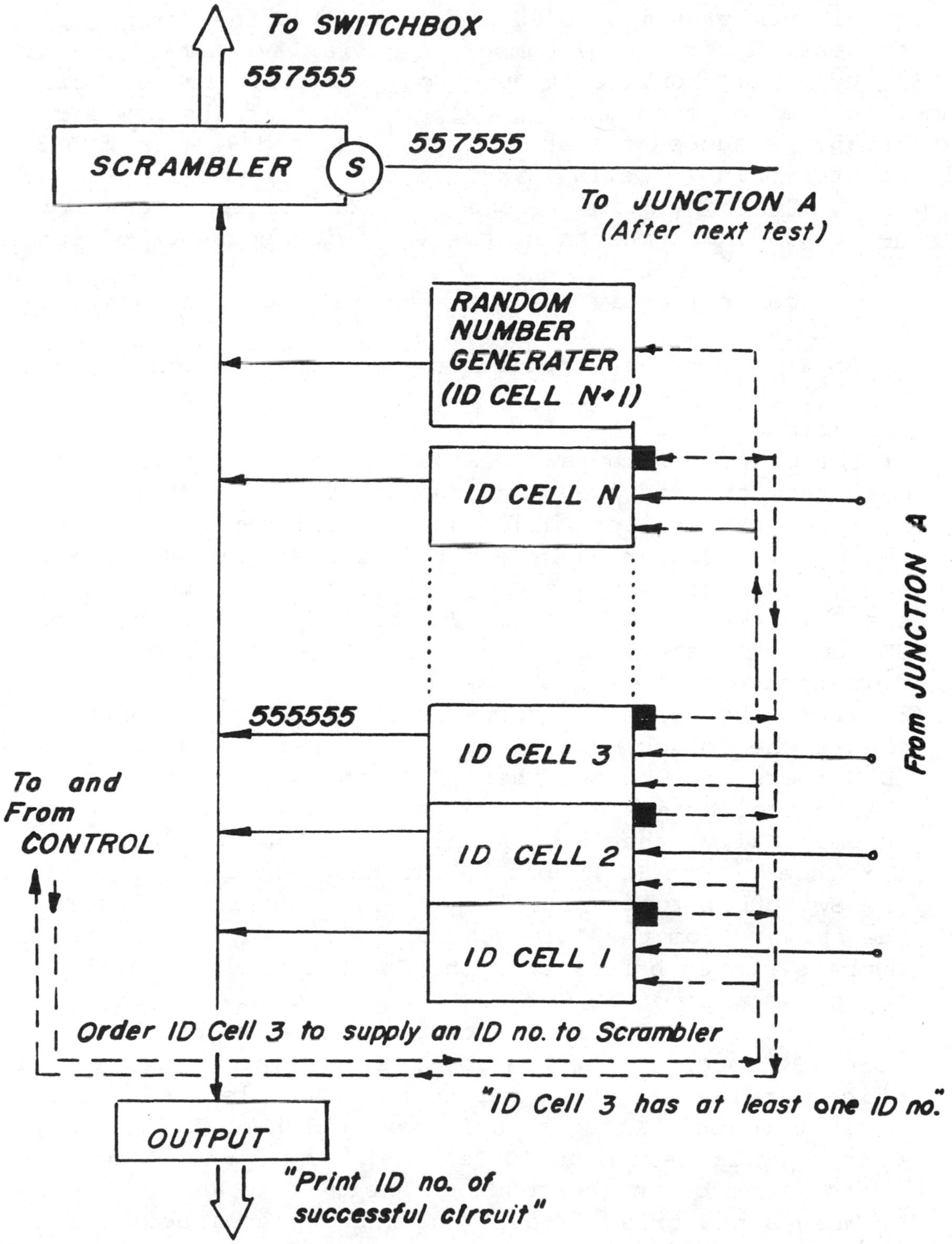

To SWITCHBOX
557555

SCRAMBLER (S) 557555
To JUNCTION A
(After next test)

RANDOM
NUMBER
GENERATER
(ID CELL N+1)

ID CELL N

ID CELL 3

555555

To and
From
CONTROL

ID CELL 2

ID CELL 1

From JUNCTION A

Order ID Cell 3 to supply an ID no. to Scrambler

"ID Cell 3 has at least one ID no."

OUTPUT

"Print ID no. of
successful circuit"

Figure 6. ID CELLS, SCRAMBLER, and OUTPUT

so that it can be sent to <u>Junction A</u> after the next test. If the new circuit is successful in any of its forthcoming tests, then its circuit number temporarily stored in Ⓢ in the <u>Scrambler</u>, and each successful test level number, temporarily stored in Ⓢ in the <u>Exciter</u>, will combine at <u>Junction A</u> to store the new circuit number in the corresponding <u>ID</u> <u>Cells</u>. Thus, if the new circuit, say 575555, is successful at tests 13, 10, and 9, then 575555 will be stored in ID Cells 13, 10, and 9.

    h. The <u>Control</u> (Fig. 7) receives coded instructions from <u>Storage</u> (Fig. 7) and maintains the following sequence of operation:

    (1) The control activates the Random Number Generator, (ID Cell N+1).

       An ID Number is supplied to the Scrambler which, through the Switchbox, establishes the first trial circuit in the Variable Circuit.

    (2) The Control withdraws the series of coded tests, one by one, from the Storage, and sends them to the Exciter.

       Then, the Exciter subjects the trial circuit to the tests. The Criterion Detector evaluates the performance of the circuit at each test. Finally, the Ⓢ in the Exciter, the Ⓢ in the Scrambler, and Junction A operate to place the ID Number of the trial circuit in those ID Cells which correspond to the tests passed.

    (3) After the series of tests is completed, the Control checks the "black square" of ID Cell N to determine if any ID Numbers are located there. If the answer is "yes", then the Control activates the control input to ID Cell N and, automatically, that cell supplies an ID Number to the Scrambler. The ID Number is slightly altered and sent to the Switchbox to be established as a new trial circuit. If the answer from the "black square" is, "no, there are no ID Numbers stored here", then the Control will check the "black square" of ID Cell (N-1), and so on through (N-2),...,3, 2, 1, until a cell is found which has one or more ID Numbers. The Control then activates this ID Cell, which supplies an ID Number to the Scrambler, establishing a new circuit. After reaching and checking ID Cell 1, the Control will return to ID Cell N+1 (the Random Number Generator) and begin the sequence anew.

    (4) When a new trial circuit has been established in the Variable Circuit, the Control will return to step (2) and subject the new circuit to the series of tests.

    (5) At the conclusion of each series of tests, the Control repeats step (3) (goes to the next ID Cell, checks its

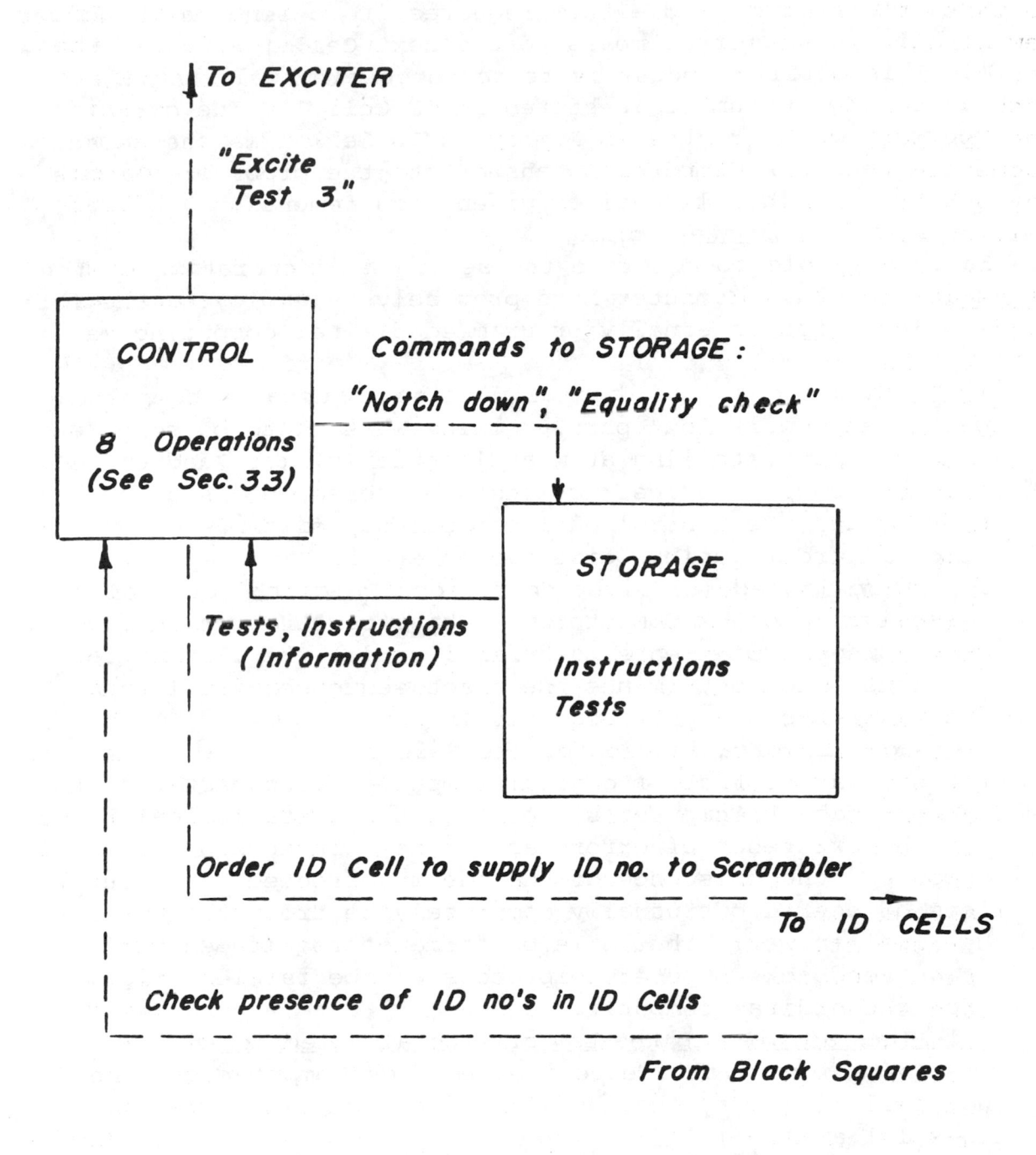

Figure 7. CONTROL and STORAGE

"black square", and either requires it to send an ID Number to the Scrambler or moves to the next Cell.)

(6) This entire process is repeated continuously until at least one ID Number is stored in ID Cell 1. The circuit corresponding to the ID Number in ID Cell 1 is the Selective Feedback Computer's "answer" to the problem: "Find a circuit which will satisfy given requirements".

3.3 Digital Computer Format

It is possible to describe the sequence of operation of the Selective Feedback Computer more precisely by employing the format developed for internally programmed digital computing machines.

An "instruction" in a digital machine is usually composed of two parts: an "operation" part, telling what is to be done, and an "address" part, telling at what location(s) (or "addresses") it is to be done. The instructions are stored in a sequence in the Storage and the Control will select and perform each instruction in order, unless directed otherwise.

Below are listed (a) eight operations which the Control of the Selective Feedback Computer will have to "understand", (b) certain Storage Assignments, and finally, (c) the list of 16 instructions which determines the exact sequence of all the operations: (Charts a, b, c respectively)

Although the organization of the Selective Feedback Computer is very similar to that of digital computers, the fact that the Selective Feedback Computer's operation has a statistical basis permits the existence of errors and noise. In an ordinary digital computer, the existence of but one uncorrected error out of millions of operations usually completely destroys the validity of the final answer. Thus, the performance requirements of the Selective Feedback Computer components can be far less rigid than for the digital computer.

3.4 Similarities to the Law of Natural Selection

The operation of the Selective Feedback Computer has the following similarities to the Law of Natural Selection: (Refer to sections 1.1 and 2.1)

a) "The ability to produce a large variety of distinct circuits". This is the defined performance of the Variable Circuit (section 3.2b). If there are d switches in the Switchbox and each switch has A settings, then the number of possible circuits is A^d.

b) "The ability to subject any circuit to a series of tests of varying difficulty, and to detect a 'success' or 'failure' at each test according to predetermined criteria". This is the function of the Exciter and Criteria Detector (sections 3.2a and

Operation	Explanation
Supply ID __M__	Cause the ID Cell, which is listed in Storage location M, to send an ID Number to the Scrambler.
Excite test __N__	Cause the Exciter to subject the Variable Circuit to the test whose address is listed at storage location N.
Notch down __O__	Subtract 0001 from the number in storage location O and place the result back in O.
Equality check __P Q__	If the numbers in storage locations P and Q are equal, record a "yes" in the single BIT register located in the Control. If they are unequal, record a "no".
Check Presence __R__	Go to the black square of the ID Cell listed in storage location R, and if there are one or more ID Numbers stored there, record "yes" in the single BIT register; if there are none, record a "no".
Conditional transfer __S__	If the single BIT register indicates "yes", go directly to storage location S for the next instruction. If "no", continue taking instructions sequentially.
Unconditional transfer __T__	Go directly to storage location T for the next instruction.
Place __U V__	Place the contents of storage location U in storage location V. The former contents of V are lost, but the former contents of U remain in U.

CHART (a)   LIST AND EXPLANATIONS OF OPERATIONS

Location:	Contents:
1-99	The list of instructions.
100-(N+99)	The series of tests.
200	(N+1), the ID Cell which supplies the first ID Number.
201	(N+99), the address of the first test.
202	Working ID Cell number, (initially N+1).
203	Working test number, (initially N+99).
204	The number: zero.
205	The number: 99.
ID Cell N	The ID Numbers of circuits which passed test N.
....	....
....	....
ID Cell 2	The ID Numbers of circuits which passed test 2.
ID Cell 1	The ID Numbers of circuits which passed test 1.

CHART (b)   STORAGE ASSIGNMENTS

3.2c).

 c) "The ability to record each successful circuit and to re-produce it in slightly altered form so that it, in turn, can be subjected to the series of tests". The ID Cells, Scrambler, Switchbox and Control perform these functions.

 d) "The ability to favor the circuits which are successful in most of the tests by reproducing and retesting them most frequently". The circuit which passes the greatest number of tests will have its number recorded in the greatest number of ID Cells. Since numbers for new circuits are taken from the ID Cells in sequence, the ID Numbers recorded at the most ID Cells will be "reproduced" and retested most frequently (section 3.2h).

 e) THEREFORE, a series of circuits is established which converges upon a type which passes the greatest number of tests.

3.5 Justification of the "Variable-Limited" Criterion

The choice of the "pass" or "fail" criterion (section 3.2c) which selects certain important or "essential" variables and studies their deviation from a desired position was in accordance with W. Ross Ashby's work on the Homeostat[1]. In order to emphasize the wide application of this type of criterion, a few specific physical systems will be described:

 a) The purpose of a simple pressure controller in a water pipe is to receive water at a high, variable pressure and discharge it at a lower, constant pressure. If the upstream pressure (input function) varies slowly, the pressure controller (stability-seeking circuit) can successfully maintain the discharge pressure (output variable) within the specified limits, and the system is stable for the given input. However, if the input function is a large magnitude pressure impulse, the stability-producing tendency of the circuit will not be strong or quick enough to keep the output variable from exceeding its limits and the system has failed its purpose.

 b) The purpose of a heating and air-conditioning control system is to maintain each room of a building within its own distinct temperature and humidity limits, and in addition, maintain variables necessary to its own function (steam or oil pressure, water temperature) within their particular limits. The input function is, therefore, the outside temperature, humidity, and wind velocity which are all functions of time and are of various magnitudes. If these are severe enough, either the room conditions may stray too far from the "thermostat" settings, or a mechanical shutdown may occur. Either of these cases of instability of the output variables would mean that the control

Location	Instruction	Explanation
001	Supply ID 202	Establishes a circuit in Variable Circuit.
002	Excite test 203	Runs through the series
003	Notch down 203	of tests, and when all
004	Equality check 203,205	have been used, adjusts
005	Cond'l transfer 007	203 back to initial
006	Uncond'l transfer 002	value again.
007	Place 201,203	
008	Notch down 202	Adjusts to next ID Cell Number.
009	Equality check 202,204	If all the ID Cells have
010	Cond'l transfer 012	been used, adjusts 202
011	Uncond'l transfer 014	back to initial value
012	Place 200,202	and returns to instruc-
013	Uncond'l transfer 001	tion 001.
014	Check presence 202	If an ID Number is in
015	Cond'l transfer 001	the ID Cell, goes to
016	Uncond'l transfer 008	001 and tests it. If not, goes to 008 and adjusts to next ID Cell.

CHART (c)   LIST OF INSTRUCTIONS (PROGRAM)

system or circuit had not satisfied the test imposed upon it be the input functions.

c) A mammalian organism living in a variable air temperature and oxygen supply must maintain (at least!) his blood temperature and pressure within certain narrow limits in order to survive. His "circuit" reacts in such a way to moderate changes in input function that his output variables remain well within healthy limits. If he is unable to maintain stability, he has not satisfied the criterion imposed upon him by the environment, and death is the usual result.

As an individual, his "circuit" is capable of great change which may be manifested as learning, whereby identical input functions may produce vastly different reactions as the individual grows older. However, the capacity for change is fixed with an individual, and in order to increase this capacity or flexibility, "learning" on a more basic, or evolutionary plane must be accomplished.

3.6 Evolutionary Engineering Synthesis

Consider the fundamental difference in the two methods of producing a new organism:

a) Selective Breeding: Here, the breeder has in mind certain characteristics he wishes the final organism to possess. He attempts to mate individuals, each with a portion of these characteristics, and hopes to obtain an individual with all of them.

b) Natural Selection: Here, the environment imposes a criterion of survival on a race of species. By constant application of the rule, "The dead shall not breed", a new race will evolve which is better adapted to survival under that environment.

Now consider the methods which an engineer could employ in the design of a new device:

(1) Ordinary Synthesis: He decides which characteristics he wishes for his device in order for it to perform its intended purpose. He then takes the usually well known, desired characteristics observed from many other machines and puts them together.

(2) Evolutionary Synthesis: He starts with a very flexible evolutionary machine and impresses the intended purpose upon it as a criterion. This involves the forcing to the surface of unknown, subtle characteristics which become evident only after they have survived a series of tests, all other ways having failed.

The first type of design of is the planned, "forward loop" type. The second might be termed the "correction feedback" type of design, involving a continuous pragmatic testing of the validity of the solution or partial solution.

This second method is inherent in the Selective Feedback Computer.

3.7 "Thinking Machine" Evolution

The method of natural selection might also be applicable to the design of a more flexible calculating machine. Present analogue and digital calculators handle mathematical problems fabulously well and can handle problems in logic if they are presented symbolically and can be solved by algebraic-type operations. In all these cases, the method of solution, the operational schedules, the programming, etc. were all previously built into the machine. The designer, so to speak, knew much more about the fundamental nature of the problem than did the machine.

On the other hand, mathematics and symbolic logic are far too cumbersome and even inapplicable in the solution of many classes of problems. An example of an extremely well defined, yet highly complicated problem is a chess game.

The design of chess playing machines has been a very fertile field, and, within the knowledge of the author, has always been attempted on the same basis as the design of the machine which solves the relatively simple problem of five simultaneous differential equations. That is to say, the theoretical aspects of the problem are first thoroughly analyzed, isomorphic analogues are designed and built into the elements and operations of the machine, and the problem is "cranked" out.

The "design" of a chess playing machine could possibly be effected by means of the natural selection principles of the Selective Feedback Computer. A basic, flexible learning circuit which possessed no built-in chess analogues could be first "taught" the moves of the game, and then by being allowed to play a series of increasingly difficult games, could abstract "strategies" and other "principles", which perhaps are even unknown to the designer. Thus, as Ashby[2] claims to be possible with all machines which have random inputs and methods of selection, there will be more design", or BITS of information, existing in the final machine than was actually available during its construction.

IV EVALUATION OF SELECTIVE FEEDBACK COMPUTER PERFORMANCE

4.1 Time Required for Random Searching

In discussing the possible methods of searching for a satisfactory circuit, it will be convenient to refer to a spatial diagram of the variables under consideration (Fig. 8).

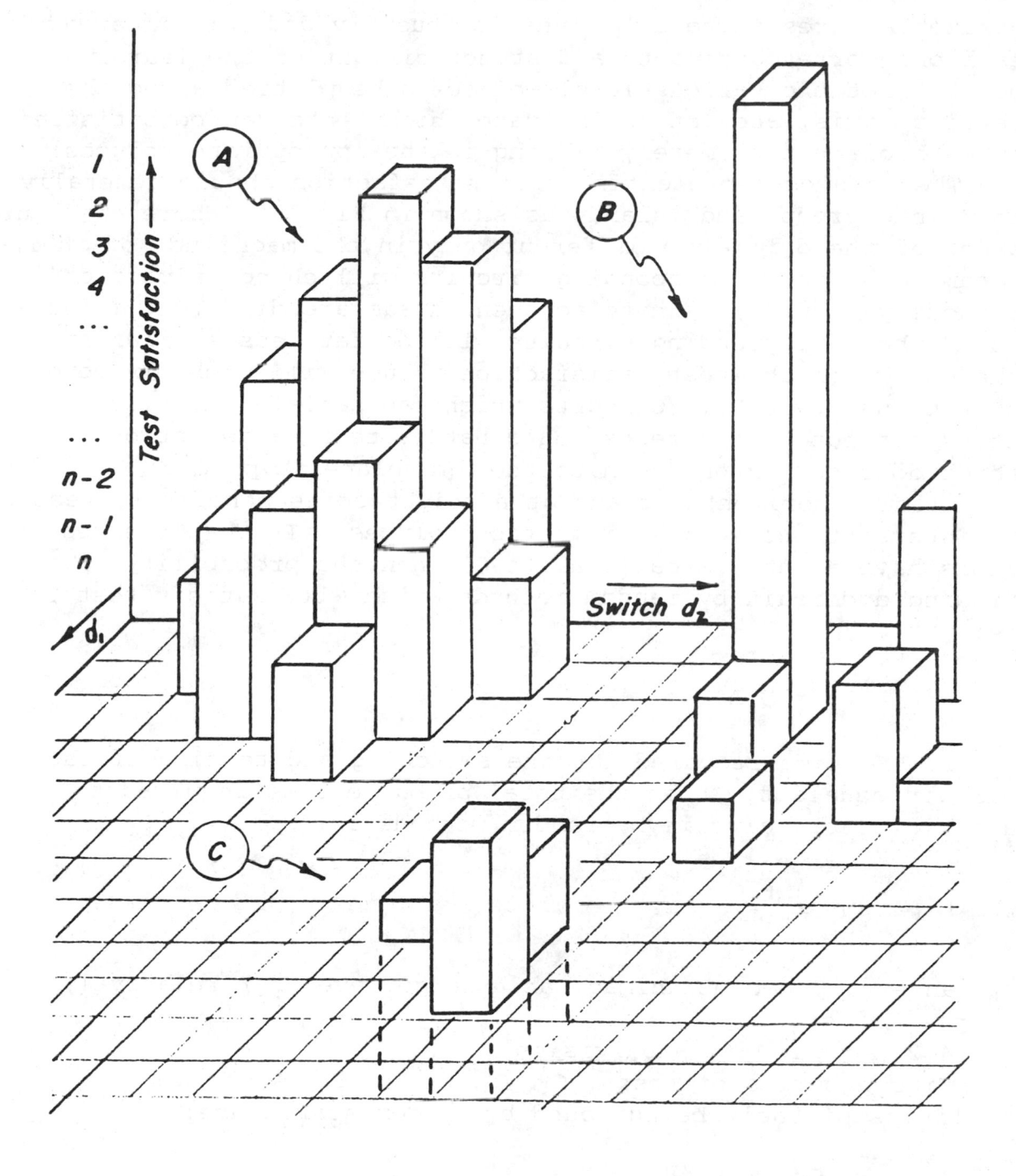

Figure 8. PLOT of TEST SATISFACTION vs SWITCH SETTING

Assume for a moment that the number of switches, d, equals two, and that the switch positions, A, are measured along the horizontal axes. The $d_1 d_2$ plane is thus divided into A^2 areas, each one corresponding to a distinct circuit of the learning model. Let the various levels of tests be plotted along the vertical axis, and let us indicate which tests can be satisfied by each circuit by merely filling in the appropriate volumes.

The volumes representing test satisfaction should generally form into groups and islands as shown in Fig. 8. Where adjacent areas of the $d_1 d_2$ plane different only in the magnitude of their parameters, the corresponding circuits will change almost continuously (type A). Where adjacent areas are different in network, the corresponding circuits will be far less similar and the change in the test satisfaction volume will probably be discontinuous (type B). Circuits which can satisfy "difficult" tests but not "easy" tests would have a test satisfaction volume which does not extend down to the $d_1 d_2$ plane (type C).

Allow a horizontal plane at a height corresponding to test C_1 to intersect the test satisfaction volumes. If the Intersections have a total area equal to Q, then the probability of finding a circuit by random methods which will satisfy test C_1 is:

$$P_1 = Q/A^2 \qquad (1)$$

If the time required for the switching and testing of each circuit equals t_1, then the total probable time required to find a satisfactory circuit by random methods is:

$$T_R = \frac{t_1}{P_1} = t_1 \frac{A^2}{Q} \qquad (2)$$

T_R can easily become unimaginable large, even for relatively simple criteria.

4.2 Probability Subdivision

Find a series of tests, C_1, C_2, C_3 ... C_n, such that the probabilities of their being found by random search are:

$$
\begin{aligned}
P_2 &= K P_1 \\
P_3 &= K^2 P_1 \\
&\cdots\cdots\cdots \\
P_n &= K^{n-1} P_1
\end{aligned}
\qquad (3)
$$

As before, have horizontal planes corresponding to the above tests intersect the test satisfaction volumes. If the curves of intersection are projected down to the d_1d_2 plane, a series of closed test contours are formed (Fig. 9). In searching for a circuit by the method outlined in section 2.3, first the outermost contour is found, and then each succeeding inner contour is found in turn.

The time to find a point within the outermost contour, P_n, is:

$$T = \frac{t_n}{P_1 K^{n-1}} \qquad (4)$$

Assume, for the present, that the probability of finding a P_{i-1} contour once a P_i contour has been found is equal to $1/k$. This assumption is prohibitively optimistic for a realistic calculation for the Selective Feedback Computer; however, it should be of value to briefly investigate this case as the ideal, limiting once. Then, by ideal probability subdivision, the total time to find a circuit which will satisfy test C_1 is:

$$T_{IPS} = \frac{t_n}{P_1 K^{n-1}} + \sum_{i=1}^{n-1} t_i K \qquad (5)$$

Assuming that all the t_i's = t, and that the probability subdivision is of a fineness which will allow:

$$K^n P_1 = 1 \qquad (6)$$

then the total time becomes:

$$T_{IPS} = tnP_1^{\frac{-1}{n}} \qquad (7)$$

where: t is the unit switching and testing time,
         n is the number of probability subdivisions,
         P_1 is the probability of the final circuit.

There exists an optimum n for each P_i which will allow T_{ips} to go to a minimum. By Differentiating Eq. (7) with respect to n and setting the result equal to zero, the optimum number of subdivisions is found to be:

$$n = -\log_e P_1 \qquad (8)$$

Substituting back into Eq. (7), a value is obtained for the total solution time of an ideal probability subdivided system

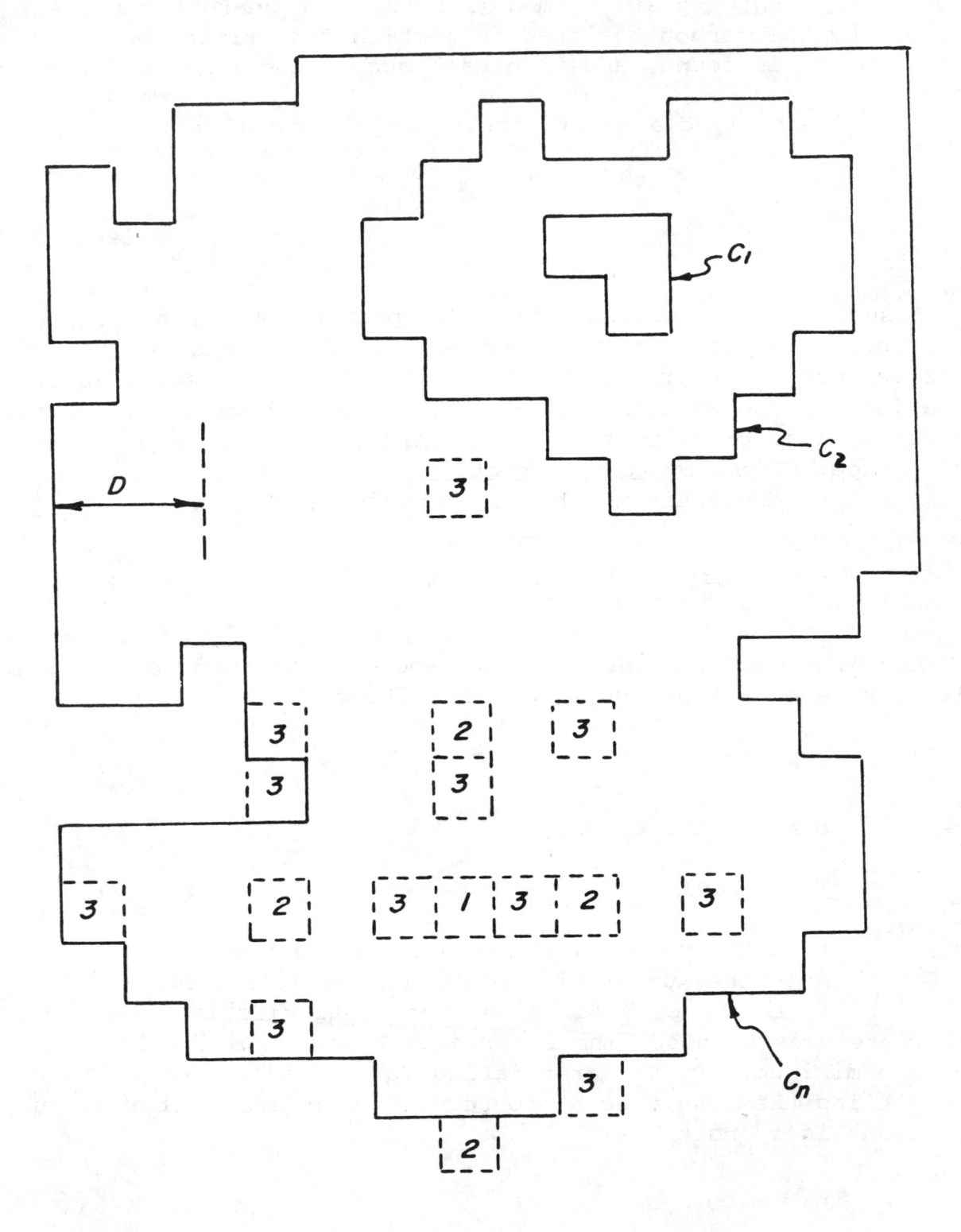

Figure 9. PROBABILITY CONTOURS and SEARCHING PATTERNS

with the optimum number of subdivisions:

$$\left(T_{IPS}\right)_{min} = -\log_e P_1^e \qquad\qquad (9)$$

As can be seen from Fig. 10, finding a circuit of probability of 10^{-18} by this method takes but a few seconds, compared to the billions of years necessary with a fully random method.

4.3 Selective Feedback Computer Dispersive Effect

Reconsider the test contours of Fig. 9. In the preceeding section regarding an ideal search, it was assumed that, once any point within the outermost contour C_n was found, the search for the next innermost contour would consist entirely of choices from within contour C_n. This essentially requires that the searching device know the complete location of contour C_n by merely hitting one point within it. Actually, the Selective Feedback Computer has no information about the location of contour after hitting within it just once.

Let us examine more closely how the Selective Feedback Computer searches a contour once it has found a point within it. Assume that the circuit corresponding to point 1 (Fig. 9) were established in the Variable Circuit. According to the operational schedule outlined in section 3.3, Junction A would recognize point 1 as belonging with contour C_n and would feed it back to the ID Cells and Scrambler, where small random changes in one variable at a time are made. The next points to be tried in the Learning Model are, therefore, those labeled "2". These will be tested and only those still falling within C_n will be fed back, forming the third "generation". Thus, the automatic tendency is to favor the points inside of the contour since any trial which wanders outside is not fed back.

However, a certain amount of dispersion is unavoidable, and it can be quantitatively described as follows:

Allow the searching process to continue until every point inside the contour has been tried, and, for the sake of simplicity, assume that no point has been tried twice. There will exist a dispersion of trials outside the contour, and the ratio of the number of dispersed trials to the total number of circuits within C_n is what interests us now.

Let the average switching distance that comprises a "mutation" equal D. If a dotted line were drawn "D" switchings inside contour C_n (Fig. 9), then it could be assumed that the "parents" of all circuits which fall outside of contour C_n lie in the area between the dotted line and contour C_n. Therefore, the number of dispersed, useless trials is proportional to the number

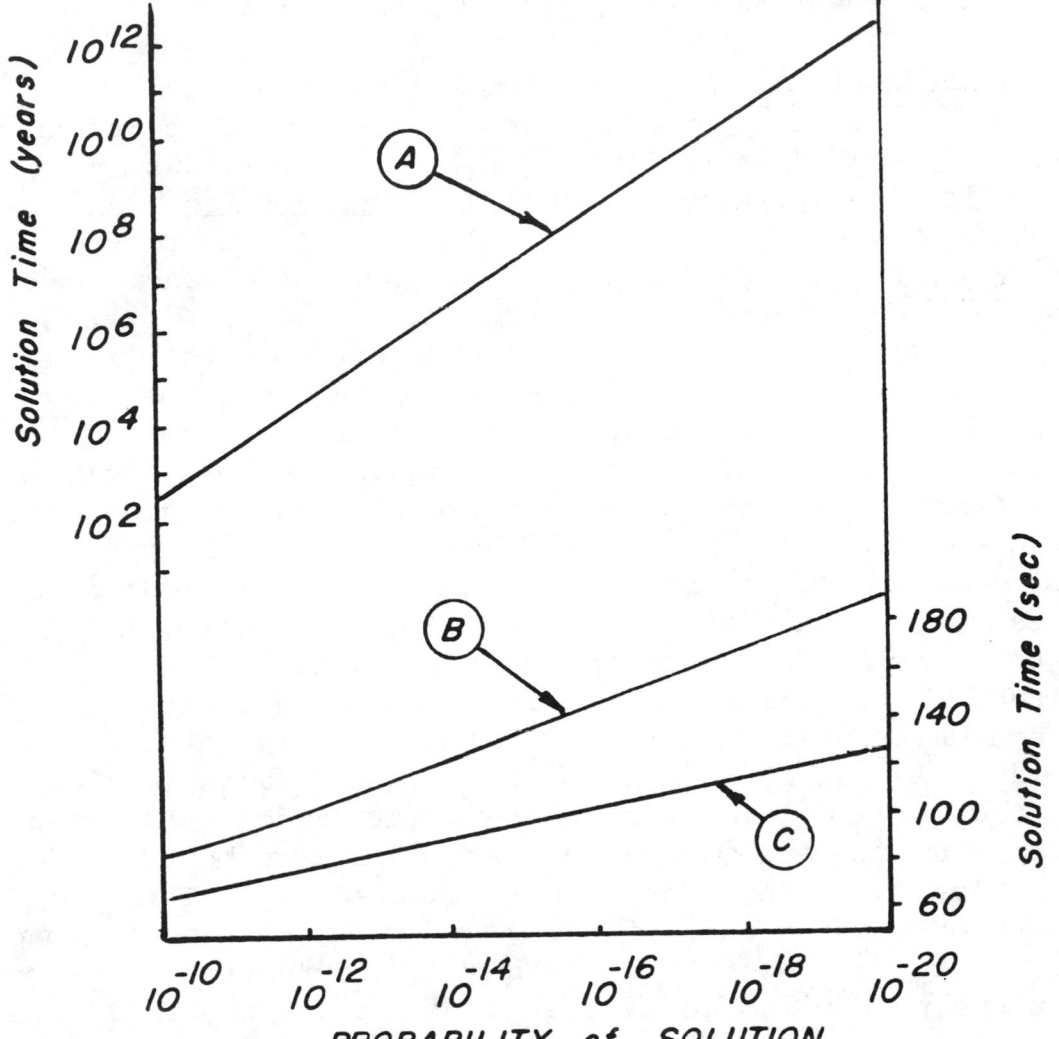

METHOD A: Random Search; $T = P^{-1}$

METHOD B: Selective Feedback Computer; $T = \left[\dfrac{\ln P}{2e\sqrt{a}}\right]^2 \left[1 + \dfrac{4DS}{A\ln P}\dfrac{d_d^2}{e^{2/d}}\right]$

METHOD C: Ideal Probability Subdivision; $T = -\ln P^e$

Figure 10. OPTIMUM SOLUTION TIMES (3 Methods)

PURELY RANDOM SYSTEM: $T_R = tP^{-1}$

IDEAL PROBABILITY SUBDIVISION:

$$T_{IPS} = tnp^{-\frac{1}{n}}$$

$$(T_{IPS})_{min} = -\log_e P^e \quad (\text{when } n = -\log_e P)$$

SELECTIVE FEEDBACK COMPUTER:

$$T_{SFC} = \frac{t}{k^{n-1}P} + \frac{nkt}{a}\sum_{i=1}^{n-1}\left(1 + S_d D \frac{d^2}{r_{i+1}}\right)$$

$$(T_{SFC})_{min} \cong \left[\frac{\log_e P_1}{2ea^{1/2}}\right]^2\left[1 + \frac{4S_d D d^2}{A\log_e Pe^{2/d}}\right]$$

Where:

T_R = Total solution time for random searching

T_{IPS} = Total solution time using Ideal Probability Subdivision

T_{DM} = Total solution time for the Selective Feedback Computer

t = Switching time per trial

P = Probability of finding solution

M = Number of probability subdivisions (or contours)

K = Ratio of probabilities between successive contours

$\ln$ = Logarithm to the base e

e = 2.718...

a = Factor due to random pairings

S_d = Shape factor for "perimeter-area" ratio in d dimensions

d = Number of switches (or dimensions)

A = Number of possible positions on each switch

r = Number of switching positions from the center to the periphery of a hypercontour

D = Average switching distance per "mutation".

CHART (d)   PARTIAL SUMMARY OF EQUATIONS

number of possible "points of escape" multiplied by the number of directions of escape from each point. A good approximation of the area between the dotted line and C_n is: (perimeter of C_n) (D).

So, instead of having to search through only the contour's area for a small inner contour, we must search the area plus the dispersed points or:

$$f = \left(1 + 2D\frac{\text{Perimeter}}{\text{Area}}\right) \tag{10}$$

times as much as before, where f equals the dispersion factor and 2 is the number of directions of escape.

At this time it should be remembered that, in the general case, the number of switches (d) that control the circuits of the Learning Model is far greater than 2, and therefore, the $d_1 d_2$ plane of Fig. 8 becomes a hyperspace of d dimensions and the contours of Fig. 9 become hypersurfaces of d-1 dimensions. The ratio of "surface" to "volume" of a hypersphere of d dimensions is simply:

$$\frac{\text{Surface}}{\text{Volume}} = \frac{d}{r} \tag{11}$$

This value is for "spheres" and is minimum. A suitable "shape factor" should be applied for more general shapes.

Since the number of directions of escape from a point on the perimeter of a hypercontour is merely equal to d, the dispersion factor for a contour of d dimensions is:

$$f = \left(1 + DS_a\frac{d^2}{r}\right) \tag{12}$$

where S_d = Shape factor for d dimensional space,
      r = Number of switching points from the center
          to the periphery of the d-1 dimensional contour.

Eq. 12, in giving the dispersion factor, indicates the unavoidable inefficiency inherent in this method of search.

4.4 <u>Selective</u> <u>Feedback</u> <u>Computer</u> <u>Mixing</u> <u>Effect</u>

The dispersive effect described in the previous section could be decreased somewhat by requiring that, instead of an ID Cell sending an ID Number to the Scrambler, it choose two of its number and send their average to the Scrambler. For simply shaped contours, this method would decrease the dispersion to almost

towards the "center of gravity" of the contour where the inner contours will most likely be found. (See Appendix "C"). In many instances, however, this mixing method would have detrimental effects and because of the initial complication it introduces, it was left out of the Selective Feedback Computer of section 3.3.

For the sake of completeness, the effects of the mixing can be lumped in a single variable, a, defined as the ratio of the probabilities of finding the next inner contour with mixing and without mixing.

4.5 Selective Feedback Computer Solution Time

By reasoning similar to that leading to Eq. 5, an expression can be written for the solution time, or total time necessary for the Selective Feedback Computer to search all the preliminary contours and find the final contour:

$$T_{SFC} = \frac{t}{K^{n-1}P_1} + \frac{nKt}{a} \sum_{i=1}^{n-1} \left(1 + DS_d \frac{d^2}{r_{i+1}} \right) \tag{13}$$

In order to get Eq. 13 into a closed form, it was necessary to assume that d and n were much greater than 1, and that the shape of each of the contours is roughly hyperspherical. The following approximate result was obtained:

$$T_{SFC} \approx \frac{tn^2}{a} P^{\frac{-1}{n}} \left[1 + 2\frac{S_d Dd^2}{An} P^{\frac{-1}{n}} \right] \tag{14}$$

In comparing this equation to Eq. 7 for the ideal probability subdivision, three differences are seen:

a) An extra "n" in the coefficient, due to the simultaneous operation of all n ID Cells.

b) The "a" in the coefficient's denominator, due to the mixing effect of section 4.4.

c) The entire expression inside the brackets, due to the dispersion effect discussed in section 4.3.

If Eq. 14 is now differentiated with respect to n, the optimum number of probability subdivisions is approximately:

$$n \approx \frac{-1}{2} \log_e P_1 \tag{15}$$

which, when substituted back in Eq. 14:

$$\left(T_{SFC}\right)_{min} \approx \left[\frac{\log_e P_1}{2ea^{1/2}}\right]^2 \left[1 + \frac{4S_d Dd^2}{A\log_e P_1 e^{2/d}}\right] \qquad (16)$$

4.6 Comparison and Summary of Equations

A comparison between the total times required for Random Search, for Ideal Probability Subdivision and for the Selective Feedback Computer is shown in Fig. 10.

The errors introduced by the approximations in sections 4.3 to 4.5 may run over several hundred percent, especially if the contour perimeters are much larger than the contour radii. However, the emphasis in this discussion was on the exponent of the time, since reference to Fig. 10 will demonstrate that any error would have to be extreme in order to affect the comparison of solution times shown.

V APPLICATION OF THE SELECTIVE FEEDBACK COMPUTER

5.1 Review of the Problem

According to the operational sequence of the Selective Feedback Computer (section 3.3), a circuit meeting complex requirements is attained by means of a searching process involving a series of simpler circuits which approach progressively closer to the final requirements. The sections below describe such a series of circuits which could have been operated upon by the Selective Feedback Computer.

These circuits correspond to special cases of the Variable Circuit and they form the control systems for the mobile, temperature-sensitive device described in section 2.3, as it is subjected to the various tests shown in Fig. 2.

5.2 Basic Informational Elements

Only two basic irritable (information-carrying) elements will be employed to build the control circuit of the device. Each type, inhibitory and excitatory, shall be assumed to have the following simple characteristics (see Fig. 11):

a) A minimum threshold for an input to cause activation.

b) An impulse output of constant magnitude.

c) A recovery, or relaxation time, immediately following the impulse output when the unit is insensitive.

The basic units can be combined in many ways to perform more complex functions (Fig. 12). Thus, a large or small group of excitatory elements are amplitude and form sensitive as well as time sensitive; large and small excitatory groups in "endless circle" arrangement can prolong and integrate impulse inputs;

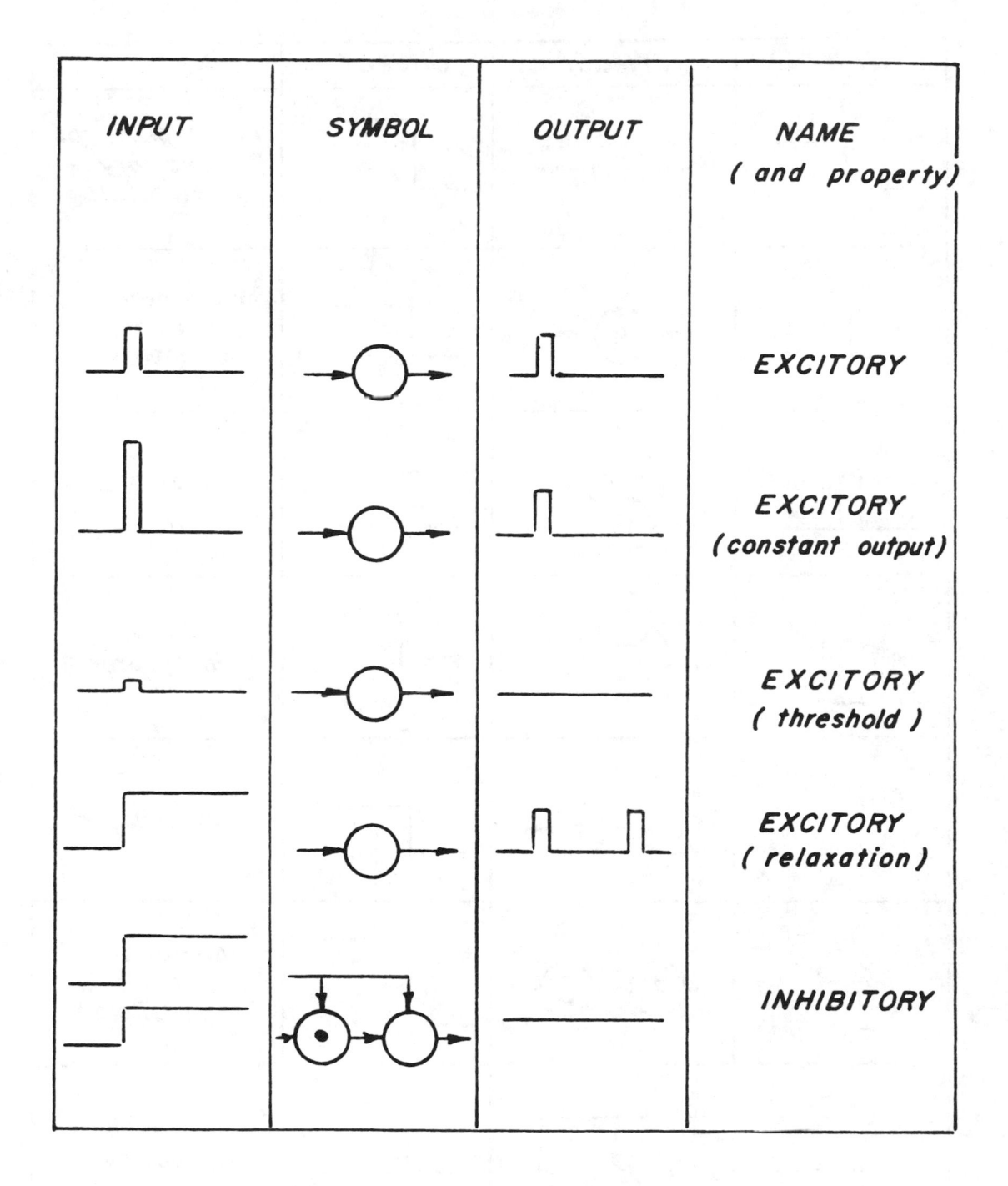

Figure 11. BASIC "Neural" ELEMENTS

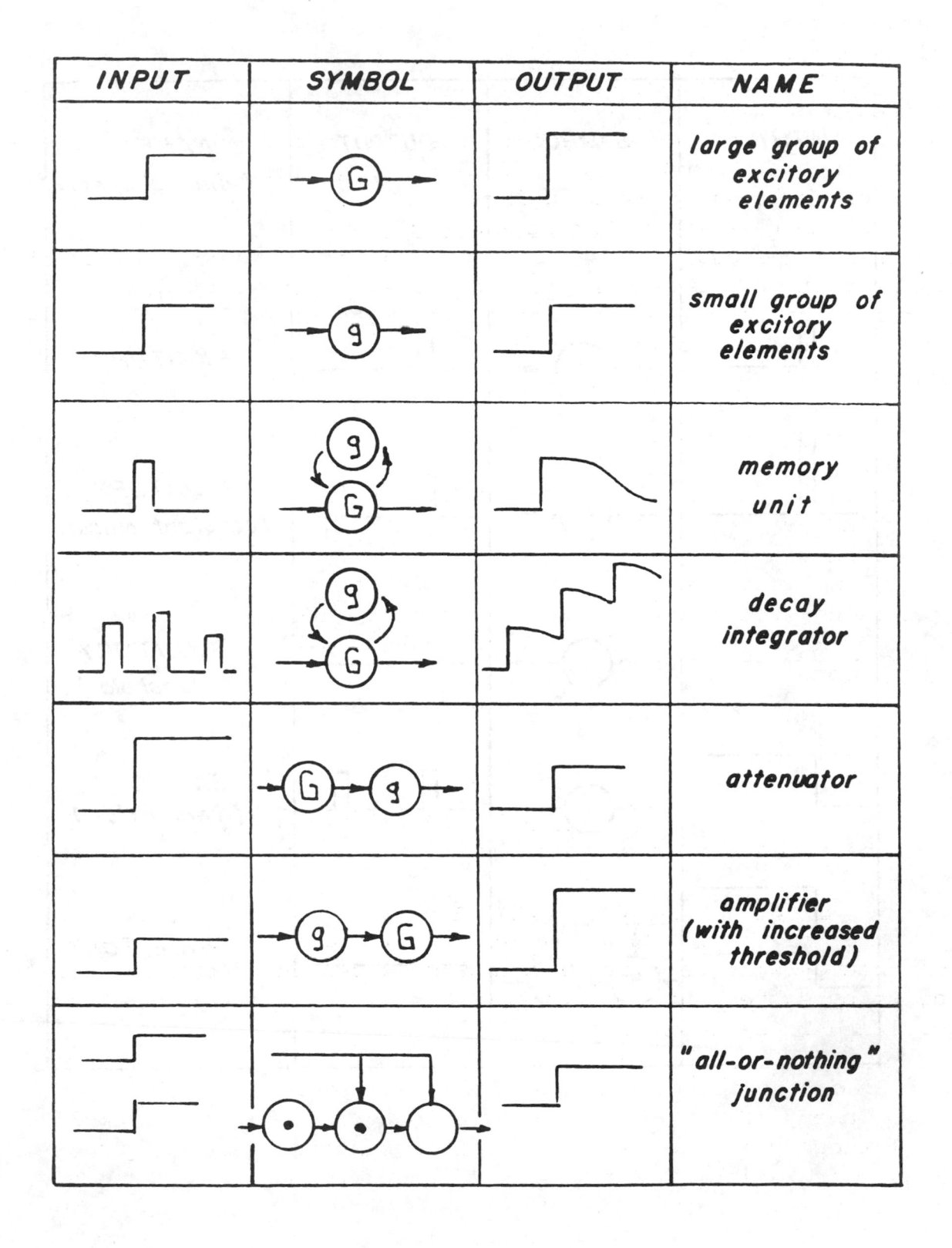

Figure 12. BASIC "Neural" GROUPS

large and small excitatory groups in series can attenuate, amplify and increase the threshold to an input; and, in conjunction with inhibitory units, an "all-or-nothing" junction is attained.

In order to reduce the complication in the schematic drawings, special symbols are introduced (Fig. 13) to substitute for some of the basic unit groupings.

5.3 Arrangement of Variable Circuit

Using the basic elements described above, "arbitrary" decisions must now be made regarding the switching arrangement, proportions of each element, and flexibility of the Variable Circuit for this problem. The basic elements will be used in the proportions indicated in Fig. 14: eight single excitatory elements, two inhibitory elements, five "all-or-nothing" junctions, five "decay integrator" groups, and two "pronounced threshold groups" (Fig. 13). Although the total number of excitatory and inhibitory elements in the above list is in the neighborhood of one hundred, the switching arrangement will be such that the above 22 groups will act as units.

On the average, it will be assumed that the circuit has the following switching flexibility: each of the 22 units may have 0, 1 or 2 outputs to any of three other units. The number of variations for each unit is:

Outputs	Variations
0	1
1	3
2	3×2
Total:	10 variations/unit

The total number of variations in this circuit is, therefore, 10^{22}.

5.4 Series of Control Circuits

With the basic structure of the Variable Circuit determined, an attempt will be made to establish a series of control circuits by subjecting the Variable Circuit to the series of tests outlined in section 2.3.

Control Circuit (a)

Place the temperature-sensitive device in Area (a) of its environment (Figs. 2 and 15). All possible control circuits will succeed if the device is in the constant zero temperature region (location X, Fig. 15). If the device is between the 0° and 3° contour of the pulsating "warm spot" (location Y), the internal temperature would quickly rise above 5°, and a purely random circuit would "fail". (Defined in sections 3.2 and 3.5).

However, consider Control Circuit (a) shown in Fig. 15. Ei-

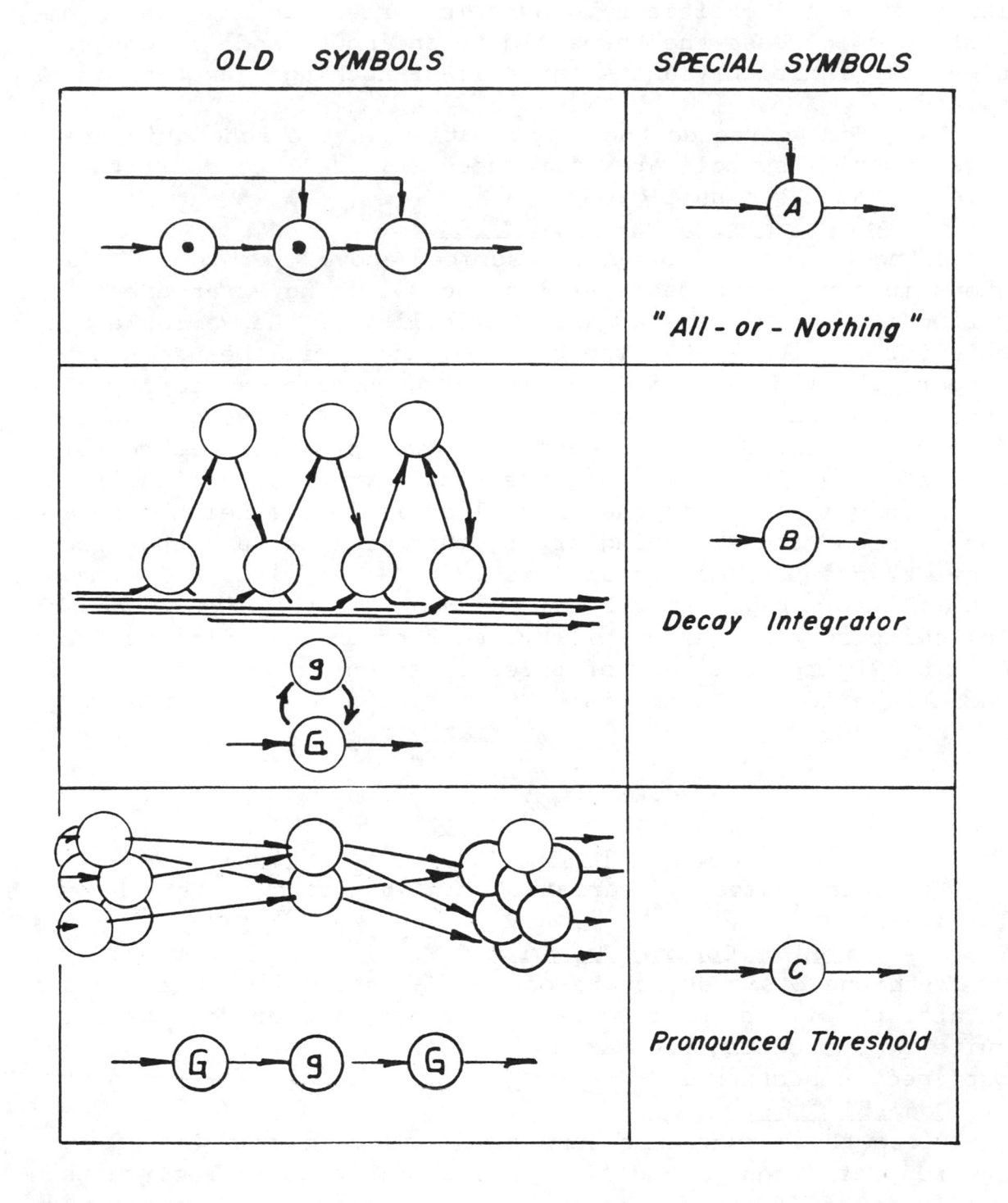

Figure 13. SPECIAL SYMBOL for "Neural" GROUPS

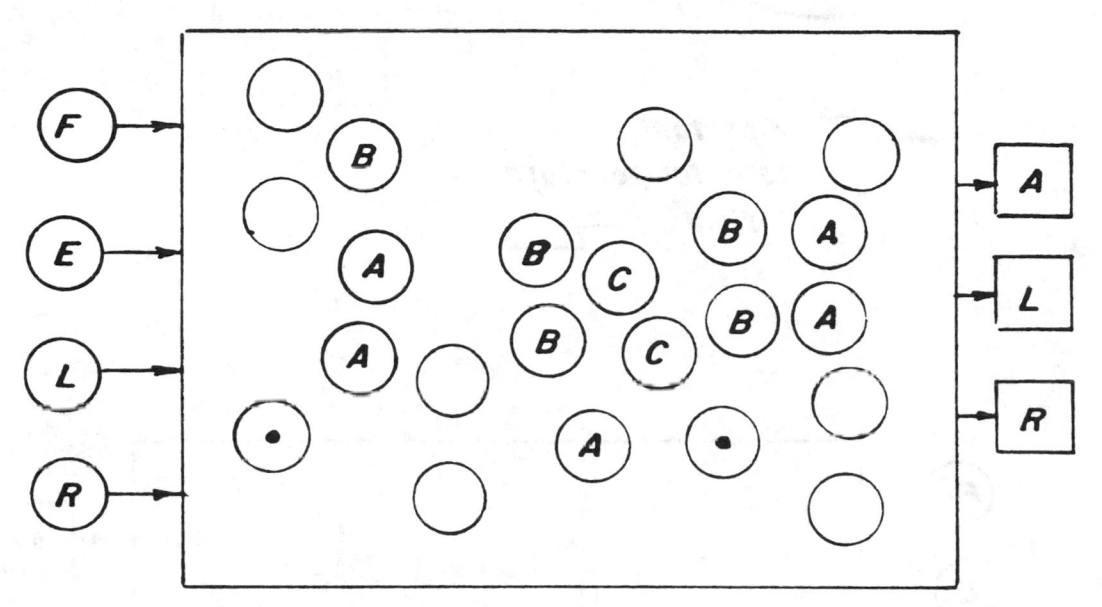

COMPONENTS:

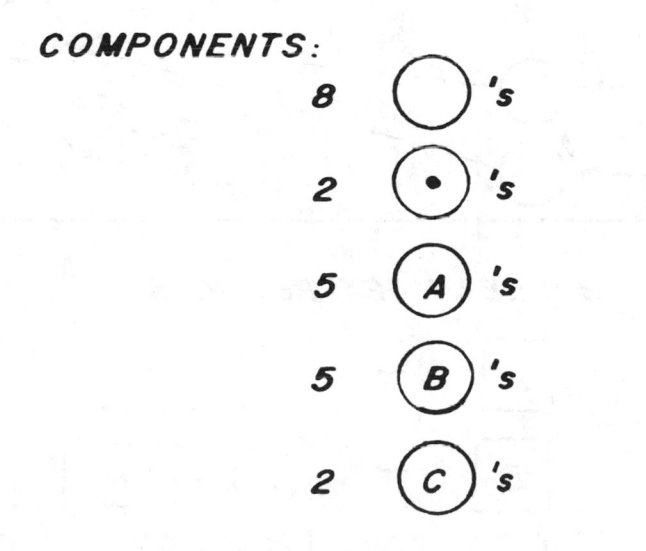

8 ◯ 's

2 ◉ 's

5 Ⓐ 's

5 Ⓑ 's

2 Ⓒ 's

Figure 14. ARRANGEMENT OF VARIABLE CIRCUIT

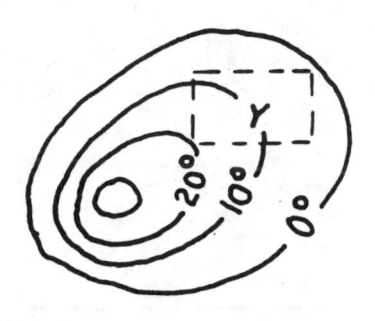

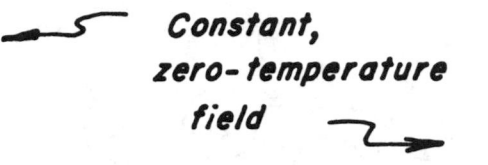

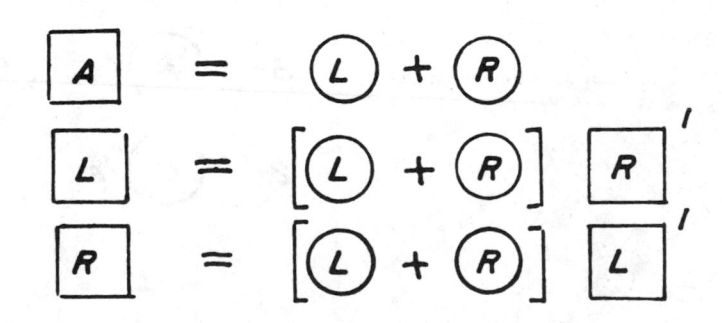

Constant,
zero-temperature
field

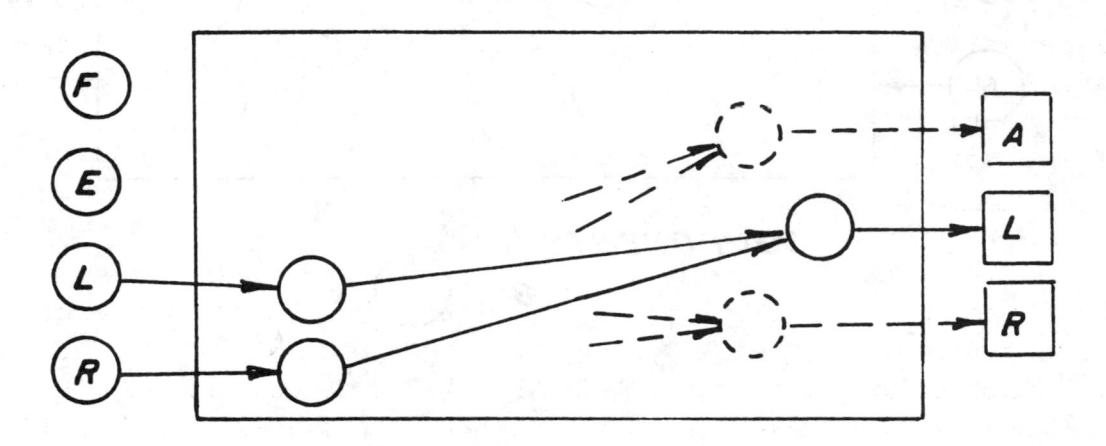

BOOLEAN EXPRESSIONS

$$A = L + R$$

$$L = [L + R] \, R'$$

$$R = [L + R] \, L'$$

Figure 15. AREA (a), and CONTROL CIRCUIT (a).

ther of the sensors, Ⓛ or Ⓡ, will activate any of the effectors
⬜A⬜ , ⬜L⬜ , or ⬜R⬜ , with exception that ⬜L⬜ and ⬜R⬜ may not be activated
simultaneously. Using the definitions:

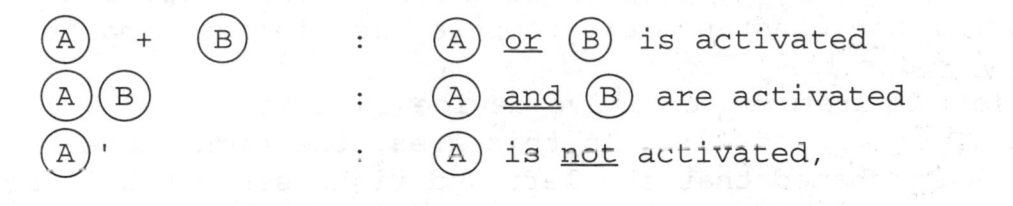

Ⓐ + Ⓑ   :   Ⓐ or Ⓑ is activated

ⒶⒷ   :   Ⓐ and Ⓑ are activated

Ⓐ'   :   Ⓐ is not activated,

these relationships may be stated with the following Boolean
expressions:

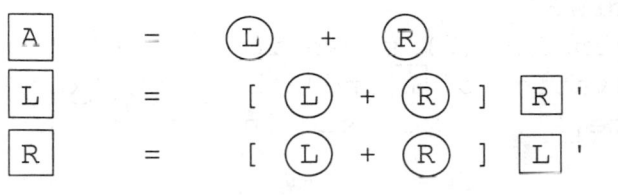

⬜A⬜   =   Ⓛ + Ⓡ

⬜L⬜   =   [Ⓛ + Ⓡ] ⬜R⬜'

⬜R⬜   =   [Ⓛ + Ⓡ] ⬜L⬜'

Thus, as any of the sensors detect a change in temperature
from 0°, any of the pushing effectors will be activated, and will
continue to be activated until the sensors no longer detect a
temperature different from 0°. Because of the small size of this
warm spot, the probability is very high that the device will
escape before its internal temperature exceeds 5°. (See Fig. 1).

Control Circuit (b)

Place the temperature-sensitive device in Area (b) of its
environment (Figs. 2 and 16). In this case, the temperature
field has a much greater area, and while Circuit (a) might have
about a 40% chance of escaping from location X, it would have a
far smaller probability of escaping from location Y.

However, consider Control Circuit (b) shown in Fig. 16. The
left or right effector will be activated if the proper sensor is
activated, except that if both those sensors are activated, then
the ⬜A⬜ effector is activated and both ⬜L⬜ and ⬜R⬜ are inhibited. Expressed in Boolean Algebra:

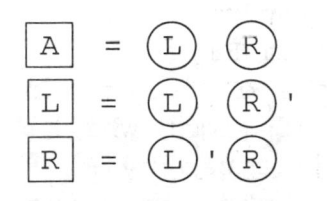

⬜A⬜ = Ⓛ Ⓡ

⬜L⬜ = Ⓛ Ⓡ'

⬜R⬜ = Ⓛ' Ⓡ

Thus, whenever Control Circuit (b) is in a gradual sloping
temperature gradient, the sensor on one side will detect a
higher temperature than the other and the appropriate effector

will push the device in a direction away from the high temperature. When both sides detect high, but equal temperatures (location Z, Fig. 16), then the action of the random effector will be such as to move the device perpendicular to the temperature gradients until the constant zero temperature field is reached.

Control Circuit (c)

Place the temperature-sensitive device in Area (c) of its environment (Figs. 2 and 17). In this area, the high temperature field is so shaped that the left and right sensors will not detect a temperature difference even though the device is "on" a high temperature field (location X). Both circuits (a) and (b) would fail in this situation, because Ⓛ and Ⓡ are the only sensors which activate the effectors.

However, consider Control Circuit (c) shown in Fig. 17. Activation of Ⓔ inhibits the activation of ⟦L⟧ and ⟦R⟧ by Ⓛ and Ⓡ respectively, but it does activate ⟦A⟧. Expressed in Boolean Algebra:

$$\boxed{A} = \bigcirc\!\!\!\!E \ + \ \bigcirc\!\!\!\!L \ \bigcirc\!\!\!\!R$$

$$\boxed{L} = \bigcirc\!\!\!\!L \ \bigcirc\!\!\!\!R \ ' \ \bigcirc\!\!\!\!E \ '$$

$$\boxed{R} = \bigcirc\!\!\!\!L \ ' \ \bigcirc\!\!\!\!R \ \bigcirc\!\!\!\!E \ '$$

Thus, if the operation of the previous circuitry is insufficient to keep the internal temperature down, then the internal sensor Ⓔ will detect the high temperature, will inhibit the previous course of action, and the system will revert to another, random course which pushes it out of danger.

Control Circuit (d)

Place the temperature-sensitive device in Area (d) of its environment (Figs. 2 and 18). The pulsating temperature field in this area is much higher than in the previous areas. A circuit caught in the field above the 30° contour would fail before any control action could be initiated (Fig. 1). Control Circuit (c), if caught in location Y, could probably escape, however, if it were caught in location X in the future, it would fail.

However, consider Control Circuit (d) shown in Fig. 18. (Only the elements added this time are shown; the circuitry for (a), (b) and (d) is still present). The flash of light which always preceeds the high temperature field is detected by Ⓕ. This activates the all-or-nothing junction Ⓐ and the decay integrator group Ⓑ in such a way that, in the future, activation of Ⓕ will have the same effect on ⟦R⟧ as will activation of Ⓡ.

Thus, Control Circuit (d), once having experienced a coinci-

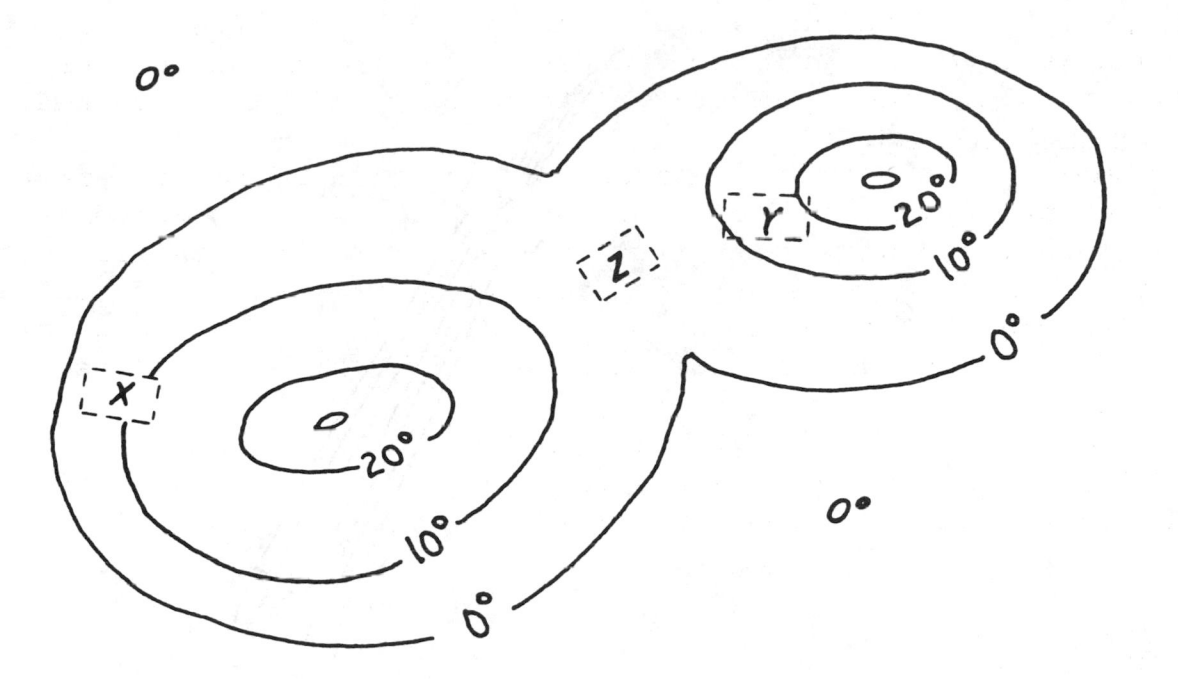

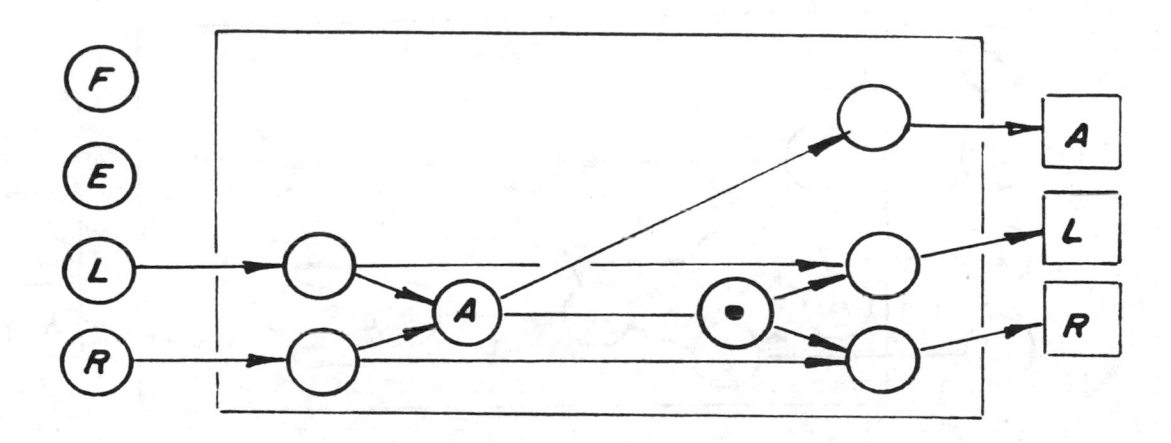

BOOLEAN EXPRESSIONS

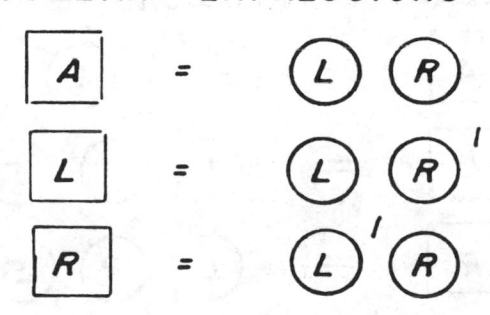

Figure 16.   AREA (b),   CONTROL CIRCUIT (b).

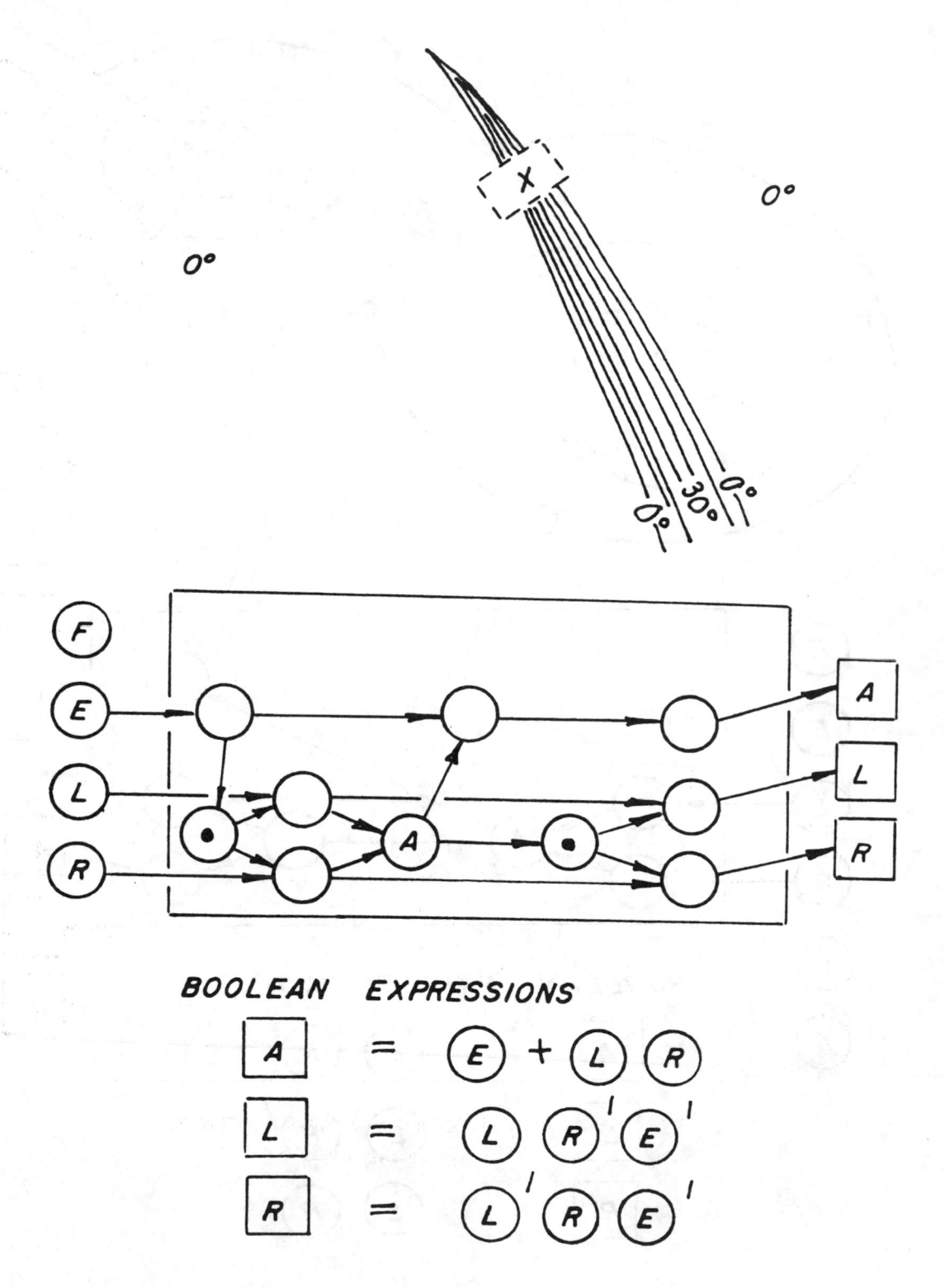

BOOLEAN EXPRESSIONS

A = E + L R

L = L R' E'

R = L' R E'

Figure 17. AREA (c), CONTROL CIRCUIT (c).

0°

Light Flash

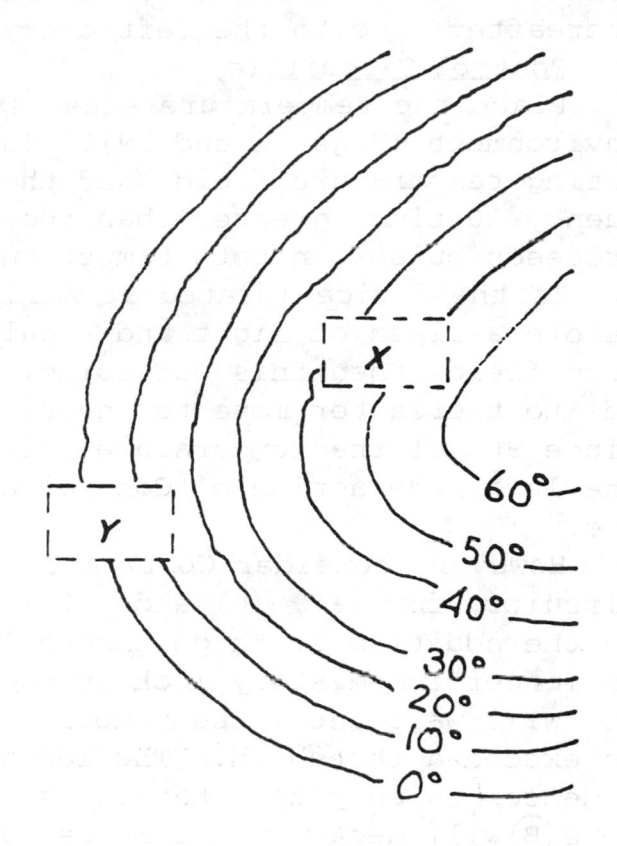

X

Y

60°
50°
40°
30°
20°
10°
0°

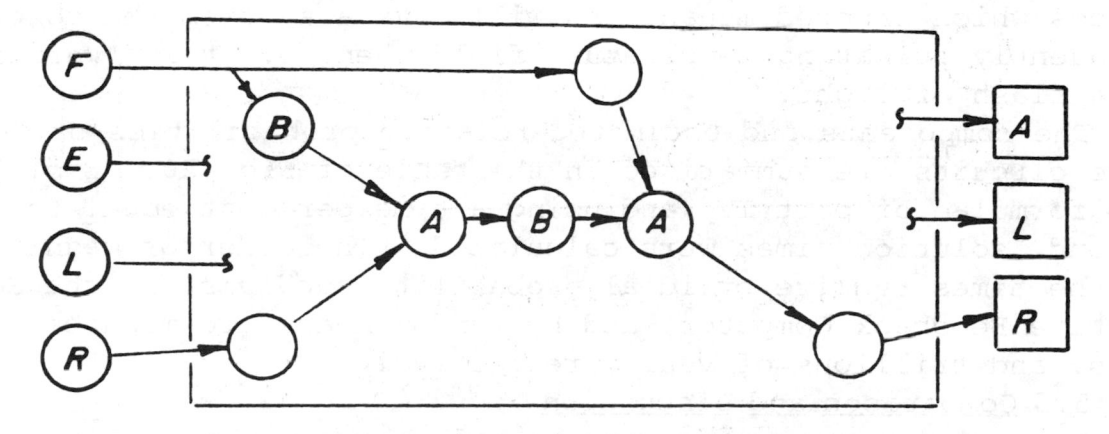

NOTE: This circuit is superimposed
on circuit (c).

Figure 18.   AREA (d), CONTROL CIRCUIT (d).

dence between a flash of light and a danger on the right, will
thereafter move to the left every time the light flashes.

Control Circuit (e)

Place the temperature-sensitive device in Area (e) of its
environment (Figs. 2 and 19). This area has two very high, pulsating temperature field, and the one on the right has a frequency 10 times greater than the one on the left. A light flash
preceeds pulses in both temperature fields.

If the device located at X is controlled by Circuit (d) just
before a flash of light and a pulse occur at the left temperature field, then this one coincidence will cause Control Circuit
(d) to thereafter move to the right after every flash of light.
Since 90% of the temperature pulses are on the right instead of
the left, the action of Control Circuit (d) will lead to failure.

However, consider Control Circuit (e) shown in Fig. 19. (The
circuitry for (a), (b) and (c) was omitted again for clarity).
By the addition of "high threshold groups" $\textcircled{C}$, the circuit will
no longer immediately lock in the cross-over from $\textcircled{F}$ to $\boxed{R}$ or $\boxed{L}$,
but will wait until the number of coincidences is great enough
to exceed a threshold. The length of time between $\textcircled{F}$ and $\textcircled{L}$ coincidences is so great that the signals through the decay integra-
tors $\textcircled{B}$ will decay with time as rapidly as they increase with
each new coincidence.

Thus, Control Circuit (e) will select the series of coincidences which is predominant and will move away from the most
frequently pulsating temperature field whenever it is "warned"
by a flash of light.

The components and estimated relative probabilities of the
five circuits are summarized in the table of Fig. 20. Employing
the formulae of part IV, and using a time per test equal to 1
second, solution times were calculated. The order of magnitude
of the times require by ideal probability subdivision, the Selective Feedback Computer, and by random means are minutes,
days, and trillions of years, respectively.

5.5 Comparison and Discussion

A crude comparison may now be made between the series of control circuits just developed and behavior in the living world.

Circuit (a) is similar to a simple tropism wherein the set
stimulus-response patterns have an urge to do "anything!" at the
first sign of danger.

Circuit (b) has, in addition, a degree of differentiation
wherein stimulii from separate sources are distinguishable and
lead to different responses.

Circuit (c) has the characteristics of "Adaptation" as de-

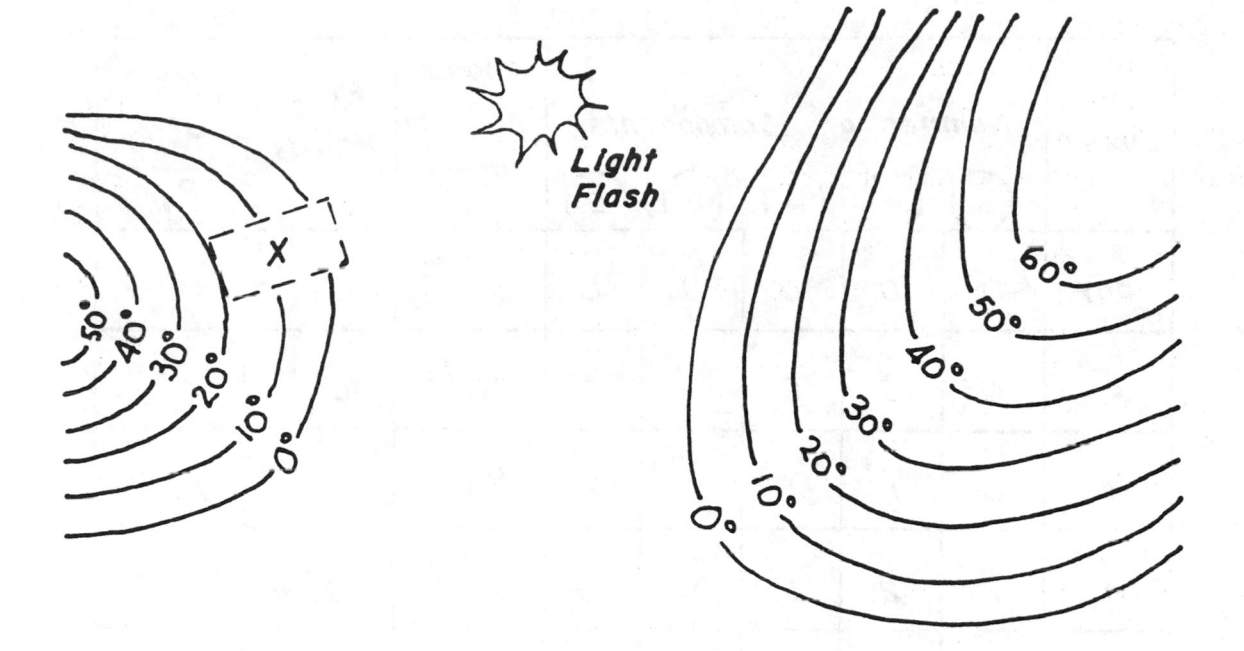

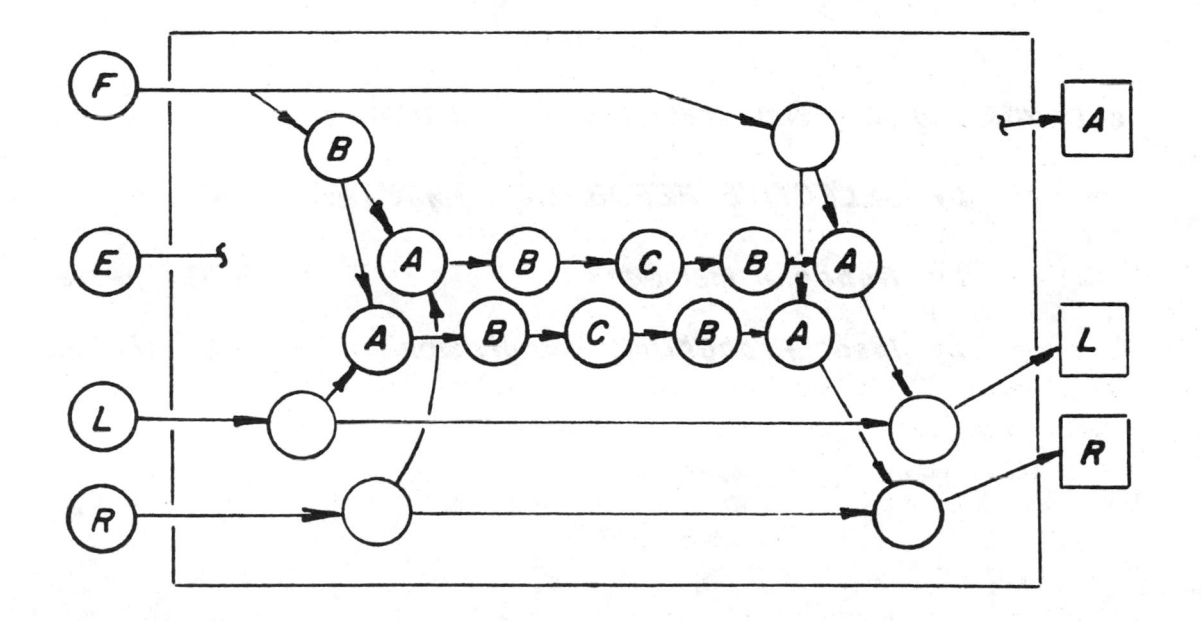

Figure 19.   AREA (e),   CIRCUIT (e).

Circuit	Number of Components					Approx. no. of Ways	Prob- ability	$K = \dfrac{P_{n+1}}{P_n}$
	◯	⊙	Ⓐ	Ⓑ	Ⓒ			
any	0	0	0	0	0	10^{22}	1	
a	4	0	0	0	0	10^{18}	10^{-4}	10^{-4}
b	6	1	1	0	0	10^{14}	10^{-8}	10^{-4}
c	7	2	1	0	0	10^{12}	10^{-10}	10^{-2}
d	8	2	3	2	0	10^{7}	10^{-15}	10^{-5}
e	8	2	5	5	2	10^{2}	10^{-20}	10^{-5}

Estimates of the time required for solution

by SELECTIVE FEEDBACK COMPUTER   4 days

by Random Methods 10^{13} years

by Ideal Probability Subdivision   2 minutes

Figure 20. PROBABILITY and OCCURANCE CHART for CONTROL CIRCUIT SERIES

fined by Ashby[1]: The ability to change from one course of action to another whenever it is sensed that one of the essential variables is diverging too far. Circuit (c) behaves like a simplified Homeostat (see Appendix A).

Circuit (d) is similar to a thresholdless conditioned response behavior, which "locks in" after being presented with but a single coincidence.

Circuit (e) has more of the characteristics of the conditioned response, such as threshold activation, memory, and forgetfulness. Its behavior is almost identical with CORA, the Conditioned Response Analogue, built by Walter[3] (see Appendix B).

Given a sufficient Variable Circuit, and an appropriate series of environments, the circuit, acting according to the law of natural selection, will evolve itself into one capable of very complex functions. However, the choice of environments seems to be infinite and there is the disquieting possibility that it may take more BITS of information to set up a proper series of environments that it would require to design the final circuit.

VI CONCLUSIONS

The Selective Feedback Computer, if supplied with the proper series of tests or "environments" can design a circuit to perform a given operation. The method employed is analogous to Darwin's Law of Natural Selection. Where as the solution times for random methods of circuit design are on the order of the reciprocal of the probability of the circuit, the solution time for the Selective Feedback Computer is on the order of the _logarithm_ of the reciprocal of the probability.

The concepts and schematic illustrations in this paper, while not conclusively demonstrating the usefulness of Selective Feedback Computers for engineering synthesis or biological analogy, did at least indicate a possible area for further investigation.

BIBLIOGRAPHY

1. Ashby, W. Ross: <u>Design</u> <u>For</u> <u>A</u> <u>Brain</u>, Wiley, New York, 1952.

2. Ashby, W. Ross: <u>Can</u> <u>a</u> <u>Chess</u> <u>Machine</u> <u>Outplay</u> <u>its</u> <u>Maker</u>? <u>British</u> <u>Journal</u> <u>for</u> <u>the</u> <u>Philosophy</u> <u>of</u> <u>Science</u>, May 1952.

3. Walter, W. Gray: <u>The</u> <u>Living</u> <u>Brain</u>, W. W. Norton, New York, 1953.

4. Rashevsky, N.: <u>Mathematical</u> <u>Biophysics</u>, Ch. 46, Boolean Algebra of Neural Nets., University of Chicago Press, 1948.

APPENDIX "A"

THE HOMEOSTAT AND ULTRASTABILITY

In his book, <u>Design</u> <u>For</u> <u>a</u> <u>Brain</u>[1], Ashby rigorously develops his concept of "ultrastability" in animals and machines. Briefly, an ultrastable system is one which can (a) detect instability within itself, and (b) then change itself by a "random" process until it does become stable.

It is assumed that there are certain essential variables associated with every organism. These variables have given limits, which if exceeded, cause death. (Two examples are blood temperature and pressure in a mammal). Survival occurs when the organism's behavior takes no essential variable outside these limits. An organism's desire or drive for <u>survival</u> could be recognized as a tendency for its behavior to <u>stabilize</u> its essential variables within their physiological limits.

Ashby's studies of simple life forms revealed that, whenever an environment became dangerous, the organism would attempt a series of methods to extricate itself. If any method is seen to be failing, the switch to the next one is done very rapidly, as if certain internal parameters had effected a step-function change. The behavior of Ashby's machine, the Homeostat, is analogous to this.

Functionally, the Homeostat consists of four needles swinging across four dials which represent its essential variables (Fig. 21). All four needles are magnetically and electronically interconnected through an adjustable circuit so that each one affects all the others. The essential variable limits are arbi-

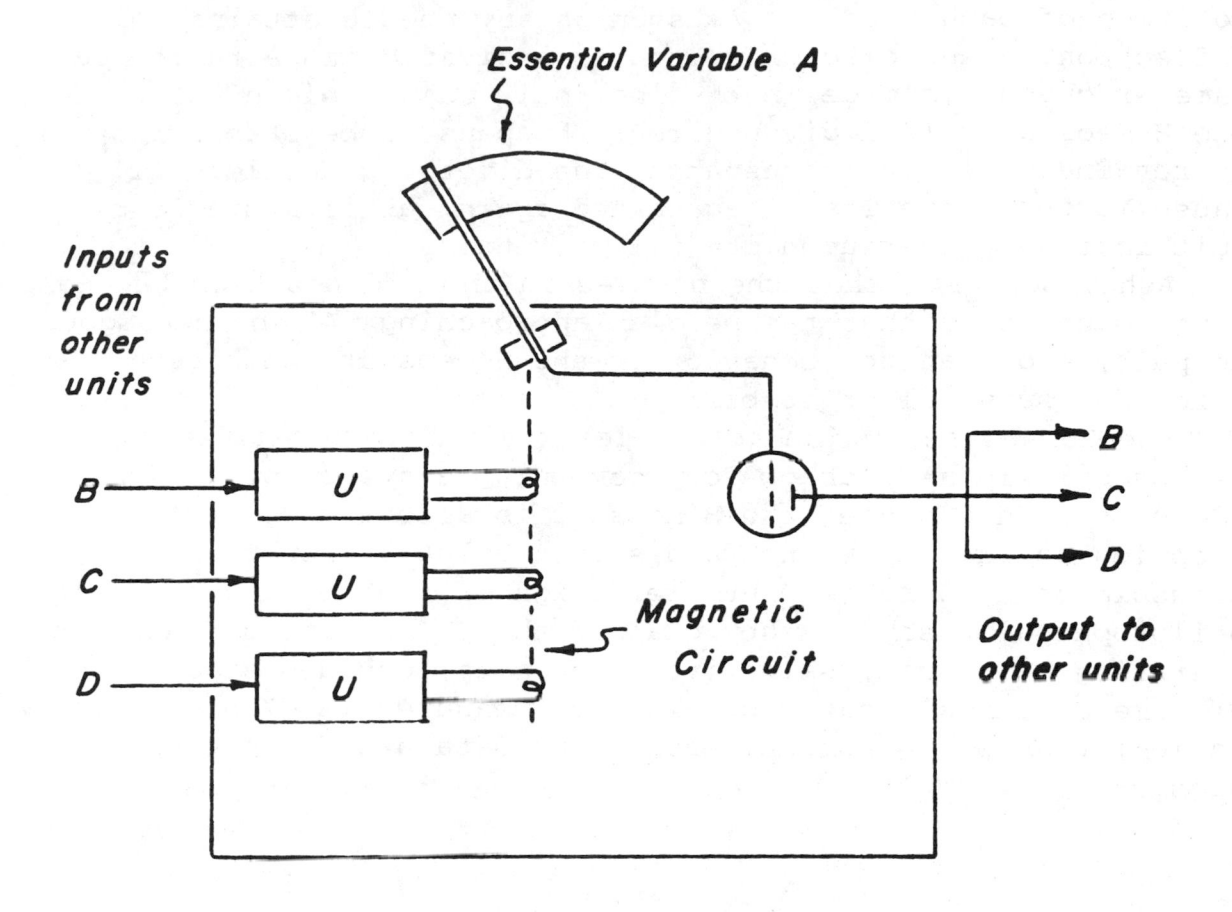

HOMEOSTAT; UNIT A

(One of four identical units)

Figure 21.   HOMEOSTAT

trarily defined to be at a deflection of ±45° from the center
position of each needle. As soon as any needle attains this
deflection, a uniselector switch is activated and a random parameter change is made in the internal, adjustable circuit of
the Homeostat. If a given circuit happens to be stable, it will
be retained. If it is unstable, the diverging needles will
cause it to change itself again and again, until a stable circuit is found, whereupon it will be retained.

Ashby suggests that the needles in the mid position be analogous to comfort, that the needles approaching ±45° be analogous
to pain, and that the behavior of the Homeostat exhibits a "desire" for survival or stability.

The Homeostat, which acts selectively toward stable circuits,
is essentially a feedback loop governing a system of feedback
elements. In a sense, the Homeostat is also a computer which
supplies as an output, not a discrete value or functional relationship of a variable, but the design of stable circuits. In a
philosophical essay[2], Ashby claimed that a machine of this type,
operating on random, selective principles, could over a period
of time attain a greater complexity (measured in BITS of information) than what was inherent in the data available in its design.

CORA; CONDITIONED RESPONSE ANALOGUE

W. Grey Walter's CORA$_3$, is an analogue of the classical conditioned response. The problem of circuit design was attacked
by the "black box method" (Fig. 2), and seven rather distinct
steps "from chance to meaning", as Walter puts it, are outlined.

The seven steps shown in Fig. 22 are easily synthesized by
electronic means, and the system thus formed has the following
characteristics in common with real animals: (Classical example
of Pavlov's salivating dog and the ringing of a bell).

1) Differentiation: dog salivates at the beginning of presentation of food, stops at the end.

2) Extension: Dog's memory of the bell decays approximately
exponentially with time.

3) Area of coincidence (or association) is inversely proportional to elapsed time between bell and food. If bell is rung
after or during food presentation, there is no association.

4) If the time between coincidence is too great, the memory
decays to zero. Repeated coincidences have no <u>conditioned</u> <u>re-</u>
<u>sponse</u> effect until a threshold is reached.

5) Once the threshold is reached, the <u>conditioned</u> <u>response</u>
abruptly comes in full force.

6) Preservation of conditioned response also fades away with
time unless recharged. If the time constant of (6)'s decay is
large with respect to the time constant of (4)'s increase, the
conditioned response will be kept at near maximum amplitudes
unless coincidences no longer appear, whereupon it will die out.

CORA works somewhat like a statistical analyzer, predicting
the probability of future coincidence on the basis of its past
experience. For a primitive animal, this process helps to lend
order to an apparently chaotic environment. On a higher level
it is very similar to mental association and semantic identification.

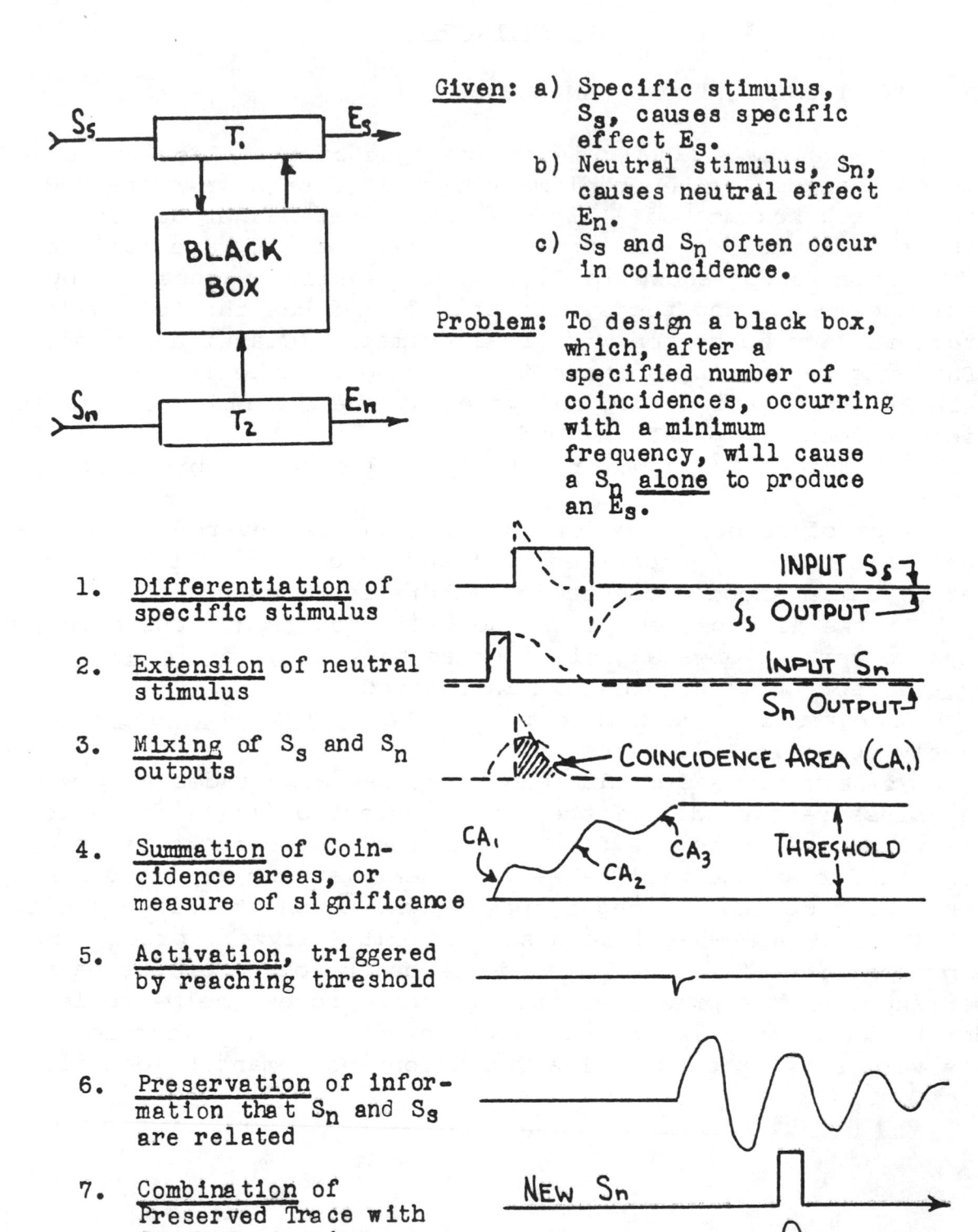

Given: a) Specific stimulus, S_s, causes specific effect E_s.
b) Neutral stimulus, S_n, causes neutral effect E_n.
c) S_s and S_n often occur in coincidence.

Problem: To design a black box, which, after a specified number of coincidences, occurring with a minimum frequency, will cause a S_n <u>alone</u> to produce an E_s.

1. <u>Differentiation</u> of specific stimulus

2. <u>Extension</u> of neutral stimulus

3. <u>Mixing</u> of S_s and S_n outputs

4. <u>Summation</u> of Coincidence areas, or measure of significance

5. <u>Activation</u>, triggered by reaching threshold

6. <u>Preservation</u> of information that S_n and S_s are related

7. <u>Combination</u> of Preserved Trace with fresh S_n to give conditioned response

Figure 22. CORA Analysis

APPENDIX "C"

FREQUENCY DISTRIBUTION OF A SERIES OF POINTS (a)
MADE UP FROM THE MIXING OF PAIRS OF POINTS (b)
WHICH ARE ALL EQUALLY PROBABLE OVER THE AREA OF A
CIRCLE

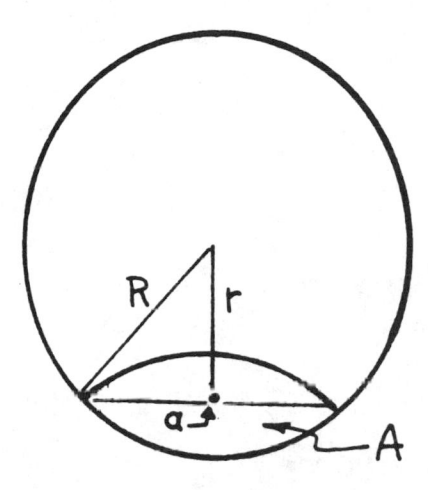

Only the "b" points which
are within the area, A, can
contribute to the point "a"
at a distance r from the
center.

The relative frequency
distribution of any "a"
point would be directly
proportional to its "A" area.

Relative Frequency Distribution $= A_r$

$$A_r = R^2 \cos^{-1}\left(\frac{r}{R}\right) - r\sqrt{R^2 - r^2}$$

or: $A_r = \cos^{-1}k - k\sqrt{1-k^2} \quad \left[\text{Letting } R^2 = 1, \frac{r}{R} = k\right]$

The relative frequency distribution is now normalized
by dividing by the total frequency distribution volume,
T.

$$T = \int_0^1 (\cos^{-1}k - k\sqrt{1-k^2})2\pi k\, dk = \frac{\pi^2}{8}$$

Normalized Freq. Dist. $= \frac{8}{\pi^2}\left(\cos^{-1}k - k\sqrt{1-k^2}\right)$

For the unmixed case (b); Norm Freq Dist $= \frac{1}{\pi}$

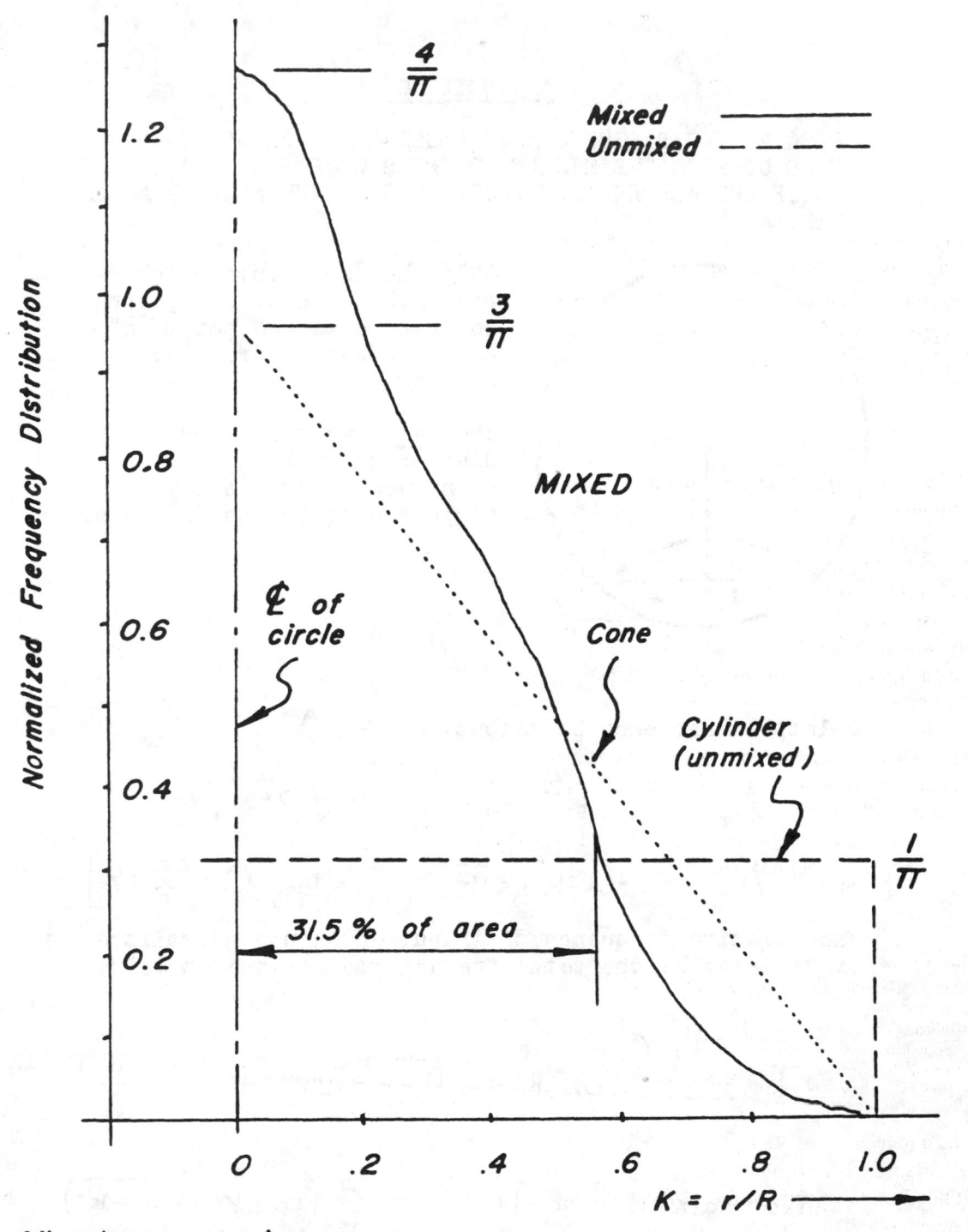

All volumes equal one

Figure 23. NORMALIZED FREQUENCY DISTRIBUTIONS

84

Chapter 3
Simulating the Evolution
of Genetic Systems

A. S. Fraser (1957) "Simulation of Genetic Systems by Automatic Digital Computers. I. Introduction," *Australian J. Biological Sciences,* Vol. 10, pp. 484–491.

J. L. Crosby (1967) "Computers in the Study of Evolution," *Sci. Prog. Oxf.,* Vol. 55, pp. 279–292.

A. S. Fraser (1968) "The Evolution of Purposive Behavior," *Purposive Systems,* H. von Foerster, J. D. White, L. J. Peterson, and J. K. Russell (eds.), Spartan Books, Washington D.C., pp. 15–23.

FRASER (1957a) (reprinted here) was perhaps the first to employ computers to study aspects of genetic systems. Simulations were conducted using diploid organisms represented by binary strings of a given length, say n. Each bit in a string represented an allele (either dominant or recessive) and the phenotype of each organism was determined by its genetic composition. Reproduction was accomplished using an n-point crossover operator where each position along an organism's genetic string was assigned a probability of breaking for recombination. Interactions between genes could be addressed nominally by forming *linkage groups*. This was accomplished by varying the probability of a crossover occurring at each locus along the strings. For example, assigning probabilities of zero to all positions except one would mandate linkages between all neighboring genes on either side of the crossover point (Figure 3.1).

Location	0	1	2	3	4	5
$P(r)$		0.1	0.1	0.5	0.1	0.1

Fig. 3.1. Fraser (1957a) adopted an approach to assign arbitrary probabilities of recombination, $P(r)$, in between locations along a simulated chromosome. Linkage groups could be formed by adjusting these probabilities. This form of recombination anticipated the one- and multi-point crossover operators which received attention in subsequent research in genetic algorithms (e.g., Holland, 1975; Goldberg, 1989; and others).

The general procedure was for a population of P parents to give rise to P' offspring via recombination. Selection would then eliminate all but P of the offspring (as well as all of the parents) for the next generation based on a function of the assigned phenotypic value.[1] In general, the rule for determining phenotypes could be arbitrary, but Fraser (1957a) offered some specific possibilities to model the effects of dominance and recessiveness. Selection could be applied toward the extreme values of the phenotype (essentially performing function maximization or minimization) or the mean values (stabilizing selection against

extremes). The possibility for varying the number of progeny per parent based on the parents' phenotypes was also introduced.

Fraser and colleagues studied this general system in a series of papers published over a decade (Fraser, 1957b, 1958; Barker, 1958a, 1958b; Fraser, 1960a, 1960b, 1960c, 1962; Fraser and Hansche, 1965; Fraser et al., 1966; Fraser and Burnell, 1967a, 1967b) culminating in the book *Computer Models in Genetics* (Fraser and Burnell, 1970). Attention was given to the varying effects of linkage, epistasis, rates of reproduction, and additional factors on the rates of advance under selection, as well as the genetic variability of a population and other statistics.

The use of computer models in population genetics became widespread in the late 1950s and 1960s (Martin and Cockerham, 1958, 1960; Crosby, 1960; Gill, 1963, 1965a, 1965b, 1965c; Ewens and Ewens, 1966; Bliss and Gates, 1968; Gates, 1968; Justice and Gervinski, 1968; and others) because computers could be used to overcome the simplifying assumptions required for mathematical treatment of questions of interest (Gill, 1965a; and others). All of the above-cited simulations relied on recombination and mating, and it is clear that this was routine. Crosby (1963), Crosby (1967) (reprinted here), and Levin (1969) provided reviews of these and other efforts (also see Crosby, 1973).

In retrospect, the computational procedure of Fraser (1957a, and others) was essentially equivalent to the canonical genetic algorithm later popularized in Holland (1975), Goldberg (1989), and others. A population of binary strings was subjected to discrete recombination and selection in light of fitness criteria, and the number of progeny per parent could be made a function of the parental fitness. By the time of Fraser's later contributions (Fraser and Burnell, 1967a), the models also incorporated inversion to build arbitrary linkages between alternative genes (cf. Holland, 1975, pp. 106–109) and were described in clear terms as "purposive" and "learning" (Fraser, 1968) (reprinted here).

---

[1]This anticipated the (μ, λ) selection method of evolution strategies (see Chapter 8).

References

[1] J. S. F. Barker (1958a) "Simulation of genetic systems by automatic digital computers. III. Selection between alleles at an autosomal locus," *Australian J. Biological Sciences,* Vol. 11, pp. 603–612.

[2] J. S. F. Barker (1958b) "Simulation of genetic systems by automatic digital computers. IV. Selection between alleles at a sex-linked locus," *Australian J. Biological Sciences,* Vol. 11, pp. 613–625.

[3] F. A. Bliss and C. E. Gates (1968) "Directional selection in simulated populations of self-pollinated plants," *Australian J. Biological Sciences,* Vol. 21, pp. 705–719.

[4] J. L. Crosby (1960) "The use of electronic computation in the study of random fluctuations in rapidly evolving populations," *Phil. Trans. Royal Soc. B,* Vol. 242, pp. 551–573.

[5] J. L. Crosby (1963) "Evolution by computer," *New Scientist,* Vol. 17, pp. 415–417.

[6] J. L. Crosby (1967) "Computers in the study of evolution," *Sci. Prog. Oxf.,* Vol. 55, pp. 279–292.

[7] J. L. Crosby (1973) *Computer Simulation in Genetics,* John Wiley, NY.

[8] W. J. Ewens and P. M. Ewens (1966) "The maintenance of alleles by mutation—Monte Carlo results for normal and self-sterility populations," *Heredity,* Vol. 21, pp. 371–378.

[9] A. S. Fraser (1957a) "Simulation of genetic systems by automatic digital computers. I. Introduction," *Australian J. Biological Sciences,* Vol. 10, pp. 484–491.

[10] A. S. Fraser (1957b) "Simulation of genetic systems by automatic digital computers. II. Effects of linkage or rates of advance under selection," *Australian J. Biological Sciences,* Vol. 10, pp. 492–499.

[11] A. S. Fraser (1958) "Monte Carlo analyses of genetic models," *Nature,* Vol. 181, pp. 208–209.

[12] A. S. Fraser (1960a) "Simulation of genetic systems by automatic digital computers. 5–linkage, dominance and epistasis," *Biometrical Genetics,* O. Kempthorne (ed.), Pergamon Press, NY, pp. 70–83.

[13] A. S. Fraser (1960b) "Simulation of genetic systems by automatic digital computers. VI. Epistasis," *Australian J. Biological Sciences,* Vol. 13:2, pp. 150–162.

[14] A. S. Fraser (1960c) "Simulation of genetic systems by automatic digital computers. VII. Effects of reproductive rate, and intensity of selection, on genetic structure," *Australian J. Biological Sciences,* Vol. 13, pp. 344–350.

[15] A. S. Fraser (1962) "Simulation of genetic systems," *J. Theoret. Biol.,* Vol. 2, pp. 329–346.

[16] A. S. Fraser (1968) "The evolution of purposive behavior," *Purposive Systems,* H. von Foerster, J. D. White, L. J. Peterson, and J. K. Russell (eds.), Spartan Books, Washington, D.C., pp. 15–23.

[17] A. S. Fraser and D. Burnell (1967a) "Simulation of genetic systems. XI. Inversion polymorphism," *Am. J. of Human Genetics,* Vol. 19:3, pp. 270–287.

[18] A. S. Fraser and D. Burnell (1967b) "Simulation of genetic systems. XII. Models of inversion polymorphism," *Genetics,* Vol. 57, pp. 267–282.

[19] A. S. Fraser and D. Burnell (1970) *Computer Models in Genetics,* McGraw-Hill, NY.

[20] A. S. Fraser, D. Burnell, and D. Miller (1966) "Simulation of genetic systems. X. Inversion polymorphism," *J. Theoret. Biol.,* Vol. 13, pp. 1–14.

[21] A. S. Fraser and P. E. Hansche (1965) "Simulation of genetic systems. Major and minor loci," *Proc. of the 11th Intern. Cong. on Genetics,* Vol. 3, S. J. Geerts (ed.), Pergamon Press, Oxford, pp. 507–516.

[22] C. E. Gates (1968) "Increasing efficiency of genetic simulation," (abstract) *Biometrics,* Vol. 24, pp. 1031–1032.

[23] J. L. Gill (1963) "Simulation of genetic systems," (abstract) *Biometrics,* Vol. 19, p. 654.

[24] J. L. Gill (1965a) "Effects of finite size on selection advance in simulated genetic populations," *Australian J. Biological Sciences,* Vol. 18, pp. 599–617.

[25] J. L. Gill (1965b) "A Monte Carlo evaluation of predicted selection response," *Australian J. Biological Sciences,* Vol. 18, pp. 999–1007.

[26] J. L. Gill (1965c) "Selection and linkage in simulated genetic populations," *Australian J. Biological Sciences,* Vol. 18, pp. 1171–1187.

[27] D. E. Goldberg (1989) *Genetic Algorithms in Search, Optimization, and Machine Learning,* Addison-Wesley, Reading, MA.

[28] J. H. Holland (1975) *Adaptation in Natural and Artificial Systems,* Univ. of Michigan Press, Ann Arbor, MI.

[29] K. E. Justice and J. M. Gervinski (1968) "Electronic simulation of the dynamics of evolving biological systems," *Cybernetics Problems in Bionics,* H. L. Oestreicher and D. R. Moore (eds.), Gordon and Breach, NY, pp. 205–228.

[30] B. R. Levin (1969) "Simulation of genetic systems," *Computer Applications in Genetics,* N. E. Morton (ed.), University of Hawaii Press, Honolulu, pp. 38–48.

[31] F. G. Martin and C. C. Cockerham (1958) "Adaptation of high speed computing machines for empirical selection studies," (abstract) *Biometrics,* Vol. 14, p. 571.

[32] F. G. Martin and C. C. Cockerham (1960) "High speed selection studies," *Biometrical Genetics,* O. Kempthorne (ed.), Pergamon Press, London, pp. 35–45.

SIMULATION OF GENETIC SYSTEMS BY AUTOMATIC DIGITAL COMPUTERS

I. INTRODUCTION

By A. S. Fraser*

[*Manuscript received June 26, 1957*]

Summary

Methods of setting automatic digital computers to simulate the algebraic aspects of reproduction, segregation, and selection are discussed. The application of these methods to the problem of the importance of linkage in multifactorial inheritance is illustrated by results from the SILLIAC.

I. Introduction

In recent years a field of mathematics has become prominent which is based on the simulation of stochastic processes. This field has been termed the Monte Carlo method and its prominence can be directly attributed to the introduction of automatic electronic digital computers. The Monte Carlo method involves in most of its applications several hundreds of thousands of arithmetic steps and therefore would be impractical without the speed of automatic computers.

The majority of genetic problems depend for their resolution on the algebra of repetitive sequences, and it is relatively easy to apply the Monte Carlo method to these sequences. The general problem of applying the Monte Carlo method to genetic problems using an automatic digital computer (the SILLIAC) is discussed in this paper.

II. Binary Representation of Genetic Formula

If we consider two alleles, $+^a$ and a, at a locus, the genotypes are represented as $+^a+^a$, $+^a a$, and aa respectively, and the gametes as $+^a$ and a. In this system $+$ and *not-plus* specify the two alleles, a specifies the locus. The symbols 1 and 0 can

TABLE 1
NORMAL AND BINARY REPRESENTATION OF TWO HAPLOID GENOTYPES

Normal	$+^a$	b	c	d	$+^e$	f	g	$+^h$	$+^i$	$+^j$	$+^k$	l	m
Binary	1	0	0	0	1	0	0	1	1	1	1	0	0

Normal	$+^a$	$+^b$	$+^c$	$+^d$	$+^e$	f	g	h	i	j	k	$+^l$	$+^m$
Binary	1	1	1	1	1	0	0	0	0	0	0	1	1

be substituted for $+$ and *not-plus*, and the position of the symbols in a register can specify the locus. This is illustrated in Table 1 for two haploid genotypes.

Since it is possible to manipulate the individual digits of a register, this binary arithmetical representation of genetic formulae is the first step in simulating genetic systems.

*Animal Genetics Section, C.S.I.R.O., University of Sydney.

III. "LOGICAL" ALGEBRA

A special aspect of the circuits of the ILLIAC family of computers is the ability to perform the operations of "logical" algebra. These are illustrated below:

$$0 \ 1 \ 1 \ \& \ 0 \ 0 \ 1 = 0 \ 0 \ 1 \qquad \text{Logical product}$$
$$0 \ 1 \ 1 \equiv 0 \ 0 \ 1 = 0 \ 1 \ 0 \qquad \text{Logical equivalent}$$
$$0 \ 1 \ 1 \wedge 0 \ 0 \ 1 = 1 \ 0 \ 0 \qquad \text{Logical not-sum}$$

These operations can be used to allow identification of the genetic nature of an individual at each locus, i.e. whether the individual is $+^i+^i$, $+^i i$, or ii at a locus i. Given two haploid genotypes A (paternal) and B (maternal), then

$$A \ \& \ B \text{ identifies the } ++ \text{ loci,}$$
$$A \equiv B \text{ identifies the } +i \text{ loci,}$$
$$A \wedge B \text{ identifies the } ii \text{ loci.}$$

This is illustrated for a diploid genotype of 10 loci:

A (paternal genotype)	1 0 1 1 0 0 0 1 1 0
B (maternal genotype)	1 1 0 1 1 0 0 1 0 1
$A \ \& \ B$	= 1 0 0 1 0 0 0 1 0 0
$A \equiv B$	= 0 1 1 0 1 0 0 0 1 1
$A \wedge B$	= 0 0 0 0 0 1 1 0 0 0

The importance of these transformations is most evident in the determination of the phenotypic value corresponding to a specific genotype.

IV. DETERMINATION OF PHENOTYPIC VALUE

If the contribution of the homozygous recessive loci to the phenotype is taken as zero it is only necessary to specify the phenotypic component of the homozygous dominant loci and the dominance term. If a is the phenotypic component and h is the dominance term, then the phenotypic contribution of a locus is 0, ah, or a, depending on whether it is 0/0, 0/1, or 1/1.

In the general case each locus may have a different phenotypic component and dominance term, and the vectors $\{a_i\}$ and $\{h_i\}$ need to be specified. If these be considered as diagonal matrices, and the logical product and logical sum of A and B be also considered as diagonal matrices (where A is the maternal and B the paternal genotype), then the phenotype of the individual is given by

$$\text{diag } A \ \& \ B \ . \ \text{diag } \{a_i\} + \text{diag } A \equiv B \ . \ \text{diag } \{a_i\} \ . \ \text{diag } \{h_i\} = \text{diag } [P_i],$$

and

$$P_{AB} = \Sigma p.$$

This is illustrated for a genotype of four loci, in which $\{a_i\}$ is $\{1, 2, 3, 4\}$ and $\{h_i\}$ is $\{1 \cdot 0, 0 \cdot 5, 0 \cdot 25, 0 \cdot 0\}$. Using binary notation, $A = 1 \ 0 \ 1 \ 0$, and $B = 1 \ 1 \ 0 \ 0$. Then diag $A \ \& \ B \ . \ \text{diag } \{a_i\} =$

$$\begin{bmatrix} 1 & . & . & . \\ . & 0 & . & . \\ . & . & 0 & . \\ . & . & . & 0 \end{bmatrix} \begin{bmatrix} 1 & . & . & . \\ . & 2 & . & . \\ . & . & 3 & . \\ . & . & . & 4 \end{bmatrix} = \begin{bmatrix} 1 & . & . & . \\ . & 0 & . & . \\ . & . & 0 & . \\ . & . & . & 0 \end{bmatrix} = C,$$

and diag $A \equiv B$. diag $\{a_i\}$. diag $\{h_i\} =$

$$\begin{bmatrix} 0 & . & . & . \\ . & 1 & . & . \\ . & . & 1 & . \\ . & . & . & 0 \end{bmatrix} \begin{bmatrix} 1 & . & . & . \\ . & 2 & . & . \\ . & . & 3 & . \\ . & . & . & 4 \end{bmatrix} \begin{bmatrix} 1{\cdot}0 & . & . & . \\ . & 0{\cdot}5 & . & . \\ . & . & 0{\cdot}25 & . \\ . & . & . & 0{\cdot}0 \end{bmatrix} = \begin{bmatrix} 0 & . & . & . \\ . & 1 & . & . \\ . & . & 0{\cdot}75 & . \\ . & . & . & 0 \end{bmatrix} = D,$$

and

$$C + D = \begin{bmatrix} 1 & . & . & . \\ . & 1 & . & . \\ . & . & 0{\cdot}75 & : \\ . & . & . & 0 \end{bmatrix} = [P_i],$$

from which

$$P_{AB} = \Sigma p = 2{\cdot}75.$$

The steps of this sequence can easily be programmed and the required space in the memory is small: two sets of n registers where n is the number of loci.

V. INTER-LOCUS INTERACTIONS

It is necessary for the completely general case to include inter-locus inter-actions. This can be done by specifying three matrices of order n, each matrix specify-ing the occurrence and order of interactions for each genetic state, i.e. separate matrices are required for the 0/0, 0/1, and 1/1 states. The space required to store these matrices would severely restrict the usefulness of a programme except for very few loci. This restriction can be avoided by restricting the number of degrees of interaction to a reasonable number, say 3. Here the required storage space would be 27 registers for the three matrices, and $2n^2$ digits for storage of the matrix specifying whether interaction occurs, and if so, of what type.

In the following matrix the rows specify the locus whose phenotype is modified by interaction, and the columns specify the locus performing the interaction. The elements of the matrix are modulo$_4$, i.e. 0, 1, 2, or 3, and specify, if 0, that no inter-action occurs, if 1, 2, 3, that interaction does occur and is of type 1, 2, or 3. If n is 20, then only 20 registers are required to store this matrix which is a minor demand on the memory space:

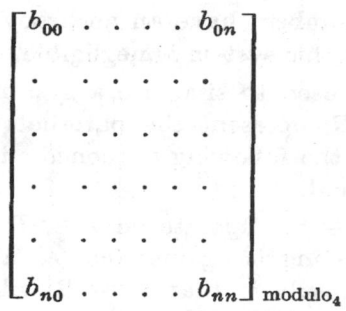

These sequences produce a single-valued phenotype. It is possible, by paralleling to produce a multi-valued phenotype, thus allowing simulation of genetic systems involving several characters.

VI. ENVIRONMENTAL EFFECTS

The above method of determining the phenotype corresponding to a specific genotype does not include any effect of non-genetic factors. Since the majority of the unsolved problems of mathematical genetics occur in systems with environmental modification of the phenotype it is necessary to simulate the occurrence of an environmental component of the phenotype which is independent of the genotype.

This can be accomplished by specifying a function $r = f(x)$ such that if r is a random number in the range 0 to 1, then x is a random normal deviate also in the range -1 to $+1$. Hastings (1955) has devised several functions which, using linear combinations of r, produce values corresponding closely to random normal deviates.

Given that P_i is the potential phenotype of the ith genotype, then by generating a random number, r_i, and finding x_i, the transformation of the potential phenotype into the actual phenotype is given by

$$[P_i]_{actual} = P_i \pm x_i . P_i.$$

It is clearly possible, by specifying different forms of $r = f(x)$, to simulate any degree of environmental modification of the potential phenotype. It is also possible to specify relations between the genotype and $r = f(x)$, i.e. to simulate genetic control of environmental stability. This is likely to be an important feature of programmes designed to examine systems showing "homeostasis" (Lerner 1954).

VII. SEGREGATION

Gametes produced by an individual heterozygous for a single locus are of two types, both occurring with equal probability: e.g. the gametes produced by the heterozygote 0/1 are of the types 0 and 1 with equal probability. This process can be simulated by generating a random number, r_i, which lies in the range

$$0 \leqslant r_i \leqslant 1.$$

If this number is tested against 0·5, then, in a set of q random numbers, the occurrence of tests which exceed 0·5 will have the same probability as those which are less than 0·5, allowing for slight bias introduced by the accuracy of the random number. If the random numbers have an accuracy of 1×2^{-38}, as is usual in the SILLIAC, then any bias of this system is negligible.

This method can be used to simulate a system of n genetically independent loci. Given that A and B represent the paternal and maternal genotypes of an individual as before, then the following sequence will simulate the production of a gamete by such an individual.

Form $A \, \& \, B$ and $A \equiv B$. Operate on $A \equiv B$ in digital sequence, generating a random number, and testing it against 0·5 (as above) for each digital position which contains 1. The result is that $A \equiv B$ will be transformed into a term

$<A \equiv B>$ in which the digital positions containing 0 are unaffected, whereas those containing 1 are left alone or changed to 0 with equal probabilities which are independent between positions. If we represent the operation of the random transform by $\bar{R}$, then this is expressed by

$$\bar{R} : A \equiv B \to <A \equiv B>.$$

Then $A \& B + <A \equiv B>$ gives a number whose digital conformation is such that (i) it has 0 wherever the AB configuration was 0/0, (ii) it has 1 wherever the AB configuration was 1/1, and (iii) it has 1 or 0 at equal probability wherever the AB configuration was 0/1 or 1/0.

VIII. RECOMBINATION

The simulation of recombination can be accomplished given the vector of frequencies of recombinant and non-recombinant classes. The vector is illustrated below for the gametes produced by an individual heterozygous at three loci, in coupling, where r_1 is the recombination between the first and second loci, and r_2 is the recombination between the second and third loci.

	Vector of Types of Gametes	Frequencies	
	0 0 0	$\frac{1}{2}(1-r_1)(1-r_2)$	$= f_{000}$
	0 0 1	$\frac{1}{2}(1-r_1)r_2$	$= f_{001}$
	0 1 0	$\frac{1}{2}r_1 r_2$	$= f_{010}$
$\frac{1\ 1\ 1}{0\ 0\ 0}$ produces	0 1 1	$\frac{1}{2}r_1(1-r_2)$	$= f_{011}$
(individual's	1 0 0	$\frac{1}{2}r_1(1-r_2)$	$= f_{100}$
genotype)	1 0 1	$\frac{1}{2}r_1 r_2$	$= f_{101}$
	1 1 0	$\frac{1}{2}(1-r_1)r_2$	$= f_{110}$
	1 1 1	$\frac{1}{2}(1-r_1)(1-r_2)$	$= f_{111}$

This illustration has been given for a triple heterozygote, but $\{f_i\}$, the vector of frequencies of recombinants and non-recombinants, is not restricted to a particular genotype. The same distribution of frequencies of recombinants and non-recombinants occurs for all genotypes. The effect of considering a different genotype is to change the vector of types of gametes. This is illustrated for three genotypes:

	Vector of Types of Gametes				
	1 0 1		1 1 0		0 0 0
	1 0 0		1 1 1		0 0 0
	1 1 1		1 0 0		0 0 0
$\frac{1\ 0\ 1}{0\ 1\ 0}$ produces	1 1 0	$\frac{1\ 1\ 0}{1\ 0\ 0}$ produces	1 0 1	$\frac{0\ 0\ 0}{0\ 0\ 0}$ produces	0 0 0
	0 0 1		1 1 0		0 0 0
	0 0 0		1 1 1		0 0 0
	0 1 1		1 0 0		0 0 0
	0 1 0		1 0 1		0 0 0

The sequence of operations necessary to simulate the production of gametes by an individual of any genotype is shown below, given that the genotype of the individual is *abc/def* and that $\{f_i\}$ is the vector of frequencies of recombinants and non-recombinants. The first step is to form the vector of types of gametes as shown:

Vector of Types of Gametes

$a\ b\ c$

$a\ b\ f$

$a\ e\ c$

$\dfrac{abc}{def}$ produces $\quad a\ e\ f$

$d\ b\ c$

$d\ b\ f$

$d\ e\ c$

$d\ e\ f$

The second step is to transform $\{f_i\}$ by sequential summation to give

$$\{f_{000};\ f_{000}+f_{001};\ f_{000}+f_{001}+f_{010};\ \ldots\ldots\ldots\ldots\ldots \Sigma f\}.$$

It is convenient to set $\{f_i\}$ such that $\Sigma f = 1$. The vector produced by sequential summation will be termed the $\{F_i\}$ vector.

A random number, r_i, is then generated such that

$$0 \leqslant r_i \leqslant 1.$$

This number is then tested across $\{F_i\}$ until

$$F_i < r_i < F_{i+1}.$$

Then the ith term in the vector of types of gametes is taken as the gamete produced. Repetition of this sequence will produce a number of gametes in which the various types occur with probabilities corresponding to the frequencies of recombinants and non-recombinants.

The simulation of the formation of a gamete, g, by an individual of genetic constitution A/B is represented as

$$T : A/B \rightarrow g,$$

where T represents the operation of forming the vector of types of gametes, testing a random number against $\{F_i\}$, and then selecting the ith type of gamete from the vector of types of gametes.

This method of simulating recombination can be set to simulate the independent assortment of linkage groups by specifying one or more of the values of r to be 0·5. If the genetic system to be simulated is of six loci then the following values of r will simulate two linkage groups:

Location	0	1	2	3	4	5
r		0·1	0·1	0·5	0·1	0·1

A major limitation of this method is that 2^n registers, where n is the number of loci, are required to store $\{F_i\}$. This limitation can be reduced by simulating the formation of each linkage group separately. If linkage groups are numbered from 0 to N, then the operation can be represented as

$$T : A_i/B_i \to g_i,$$

where $i = 0$ to N. Then, since each of the transforms of this transformation are independent, the vector of "chromosomes", i.e. $\{g_i\}$, simulates the independent assortment of linkage groups and the occurrence of linkage within each linkage group. A major advantage of this system is that the number of loci is Nn', where n' is the number of loci per linkage group. The required space in the memory is then $N2^{n'}$, which is considerably less than $2^{Nn'}$; further, if the restriction that the linkage relations are the same for each linkage group be accepted, then the required storage is $2^{n'}$, and any restrictions of number of loci are imposed by the time necessary for calculation rather than by the size of the memory.

The sequences discussed above allow the simulation of (i) the formation of a set of genotypes, $\{A/B\}_{\text{progeny}}$, from a set of parental genotypes, $\{A/B\}_{\text{parents}}$, and (ii) the formation of a set of phenotypes, $\{p\}_{\text{progeny}}$, from the genotypes of the progeny, $\{A/B\}_{\text{progeny}}$.

IX. SELECTION

There are several methods of simulating selection, the majority being variations of the following sequence. The first step is to re-order the phenotypes of the progeny, i.e. $\{p\}_{\text{progeny}}$, in ascending or descending sequence, re-ordering the genotypes of the progeny correspondingly. Then, if there are N' progeny and the programme specifies that N'' of these be retained as parents, it is possible to take (i) the N'' genotypes of $\{A/B\}_{\text{progeny}}$ with the greatest phenotypes, (ii) the N'' genotypes of $\{A/B\}_{\text{progeny}}$ with the least phenotypes, or (iii) the N'' genotypes of $\{A/B\}_{\text{progeny}}$ with phenotypes closest to the mean. These three alternatives simulate selection for one or other extreme or against extremes.

An important aspect of re-ordering the set of genotypes of progeny is that, where phenotypes are identical, position in that section of the store be kept at random, since, where the programme specifies the formation of many progeny per mating, it is possible, without such a precaution, for selection to favour a specific mating.

A deficiency of this method of simulating the operation of selection is that no account is taken of the variation of the number of progeny produced per individual. This can be included by a variation of the method of random normal deviates discussed above. If r_i is a random number and $r_i = (x)$, where x is a random normal deviate in the range 0 to 1, then, given q, the potential number of progeny, it is possible to modify this by

$$q \cdot x_i = Q_i,$$

where Q_i is the actual number of progeny for the ith individual.

Clearly many variations are possible. An example is to relate q to the potential or actual phenotype. This is accomplished by setting $q = \mathrm{f}\,(P_i)_{\text{potential}}$, or $q = \mathrm{f}(P_i)_{\text{actual}}$. A limitation of all these more sophisticated models is that the time necessary for calculation is increased, sometimes markedly, over the simpler models.

X. Conclusions

The sequences discussed above, when formulated in a programme suitable for an automatic computer, allow simulation of genetic systems. There are various problems in which this approach may be useful. These are (i) the effects of linkage on the efficiency of selection, (ii) the construction of tables relating the competitive efficiencies of alleles to the parameters of population size, intensity of selection, etc., and (iii) the comparison of the efficiencies of different breeding plans for varying degrees of inter-locus interactions. In subsequent papers of this series results gained from running various programmes covering these problems will be discussed.

XI. Acknowledgments

I am deeply indebted to Dr. P. J. Claringbold for introducing me to the Monte Carlo method, and for his patient tuition in various aspects of programming. The staff of the Adolph Basser Computing Laboratory were extremely helpful, particularly Dr. J. C. Bennett, Dr. B. Chartres, and Mr. J. Butcher. I am grateful to Miss J. Ogilvie for her preparation and meticulous checking of the various programmes.

XII. References

Hastings, C. (Jr.) (1955).—"Approximations for Digital Computers." (Princeton University Press.)

Lerner, M. (1954).—"Genetical Homeostasis." (Oliver & Boyd: Edinburgh.)

Sci. Prog., Oxf. (1967) **55**, 279–292.

Computers in the study of evolution

J. L. Crosby

This paper gives an account of the principles of computer simulation of genetic systems and its application to the theoretical study of processes of evolution. The scope of this technique is illustrated by reference to a number of examples.

Introduction

The study of population genetics and evolution processes suffers from two substantial disadvantages. The more obvious of these is the impracticability, in the great majority of systems, of conducting experiments. Except in special cases, rates of evolution are too slow and organisms too difficult to manage as experimental material. *Drosophila* provides a shining exception, but an evolution theory based largely on *Drosophila* would be even more unsatisfactory than a similarly based genetics. In general, it is usually impossible or impracticable to test hypotheses about evolution in a particular species by the deliberate setting up of controlled experiments with living organisms of that species.

We can attempt partially to get round this difficulty by constructing models representing the evolutionary system we wish to study, and using these to test at least the theoretical validity of our ideas. Such models may be inanimate, in which case they will involve a great deal of time and tedious manipulation. Or we can use tractable living organisms as the material for our models, deliberately treating a specific organism as though it were a generalized one, and designing the experiments on the assumption that the results will have a wider validity than simple application to the experimental organism. Many experiments with *Drosophila* seem to have been designed from this broader view rather than as experiments specifically intended as investigations of the evolutionary biology of *Drosophila*.

Both kinds of approach are valid, provided their limitations are clearly recognized. For example, where one organism is used as a model for a generality of organisms, we can never have full knowledge of the biology of the experimental system. There may be many things happening, and many variables

Dr Crosby is Reader in Genetics in the Botany Department, University of Durham.

(some specific to the experimental organism and perhaps invalidating the concept of generality) of which we have no knowledge and over which we can therefore have no control. An inanimate model does not (or at least it should not) suffer from this defect.

Just as the experimental approach to population genetics faces difficulties, so does the theoretical, but for quite different reasons. It is easy to see that, in principle, if we know the genetical constitution of a population in one generation, and have sufficient information about such things as its breeding system, selection pressures, mutation rates, etc., we should be able to produce mathematical formulae by which we could calculate its genetical constitution for the next generation. By iteration, we could calculate an evolutionary course for the population. This is the basic idea of the mathematical approach to evolution theory, and this sort of operation has been carried out many times for relatively simple systems. But there are many evolutionary systems which seem to be far too complex to be capable of mathematical representation without simplification to a degree which would render validity doubtful, and many others which would far overtax the limited mathematical ability of most biologists interested in evolution.

This kind of mathematical approach has one major defect. It is deterministic. That is, once theoretical generating equations have been established, the course of evolution is predetermined and invariable. This is quite untrue of real systems, where the actual genetical constitution of a population may (and usually will) differ to a greater or lesser extent from that which could have been predicted from the preceding one, because of the random variability which is inherent in systems of living organisms. A simple genetical experiment will rarely give precisely the ratios predicted by mendelian theory.

Evolution is a stochastic process. Prediction is of limited reliability—limited both in terms of accuracy and in the range of time over which it retains any value. Depending very much on the system which is being considered, prediction may have some value over many generations, or may be quite worthless extended even over very few.

Some evolutionary systems have fairly consistent negative feed-back characteristics, if we may introduce jargon from cybernetics. That is to say, selective forces tend to correct any divergence from the deterministic path and, on the whole, evolutionary progress will approximate fairly well to prediction; if the system is in approximate equilibrium, this condition will be maintained.

Many systems may vary in their feed-back properties at different times in their evolutionary history. When there is no feed-back reaction, the population will drift unpredictably and there will be no equilibrium. A system which develops positive feed-back (disruptive selection and speciation processes provide examples) will tend for a while to diverge with increasing rapidity from any predicted path; equilibria in systems liable to such development will be metastable. Some of the most interesting evolutionary processes fall into this class and would seem to be quite outside the reach of mathematical treatment, since the development of positive feed-back characteristics is often very largely a matter of chance, triggered off perhaps by some unusual and unpredictable combination of genetic circumstances.

Speaking generally, we may say that a deterministic system has one endpoint which it reaches by one route, whereas a stochastic system may travel by many possible routes to many possible endpoints of various degrees of probability, or with no endpoint at all. Even a highly improbable endpoint may still be a possible one.

Stochastic processes and computers

To treat an evolutionary system stochastically, unless it is an extremely simple one, would seem to be beyond mathematics. We have only to compare the genetical simplicity of the hypothetical systems dealt with by Wright[1] in his classic theoretical investigations on random fluctuations in mendelian populations with the complexity of almost any system we may be interested in, to see this point.

It is here that the electronic computer comes into the picture, and provides us with an extremely powerful tool for the theoretical investigation of complex evolutionary processes.

Computers allow great increase in the complexity of mathematical models of evolutionary systems whose evaluation without their aid would be impracticable. As examples of this we may cite the investigation by Allan & Robertson[5] of the effect of an initial period of reverse selection on the subsequent response to forward selection, and Lewontin's[6] use of a mathematical model for the investigation of the interaction of selection and linkage. But whatever the importance of this work, it is of little interest to us here, for the treatment was purely deterministic and only the scale of operations had been enhanced, without the introduction of anything really new. The computer was used only as a fast and powerful calculating machine. Computers are much more versatile than that. Besides, the use of computers merely to permit increase in mathematical complexity tends to lead to the production of papers even more incomprehensible to the average geneticist than those of the pre-computer age.

The possibility of a much wider exploitation of computers in genetics becomes apparent when it is realized that there is no difficulty in treating models of evolutionary systems stochastically, and that the basic objection to models that they are slow to manipulate disappears when we further realize that we can set up abstract models in a computer, because these can be manipulated with great rapidity. Such models, which are much easier to understand than the algebraic operations of biomathematicians and which may simulate real or theoretical systems of great complexity well beyond the range of mathematical representation, can easily be treated stochastically through processes in the computer which produce reliable approximations to random variability. These processes involve the iterative calculation of huge cycles of pseudo-random numbers which may, if care is taken in the choice of generating functions, safely be treated as though they were truly random numbers.[2]

The potentialities of computer simulation seem to have been recognized independently by several people. The quite different approaches to the problem by Fraser[3] and by Crosby[4] will serve to illustrate the principles involved.

Algebraic simulation

Crosby's technique retained a substantial though simple mathematical basis, and has not been so widely used as Fraser's more thoroughgoing technique of digital simulation, which Crosby himself soon adopted. It has, however, a number of points of interest which are worth considering, and may well find wider application in the future. It will be discussed first, so that a more coherent account can be given of Fraser's method and of the work which developed from it.

Crosby developed his technique specifically for application to a long-standing research project on evolution of the breeding system of the primrose (*Primula vulgaris*) from heterostyly to homostyly, which was at the same time an evolution from outbreeding to inbreeding. This was a relatively simple system to treat mathematically, and the course of evolution could be predicted from fairly simple equations. In general, studies of the populations in the field seemed to support the theoretical treatment, but there was much more irregularity in the distribution of homostyly than would have been expected.

Although the system was one of negative feed-back, it was of such a kind that random fluctuations would be expected to cause substantial acceleration or deceleration of the evolutionary process, but without significantly affecting the route or the endpoint. The irregularities found in the distribution of homostyly might therefore be the result of different rates of evolution in different populations, resulting simply from random fluctuations.

The problem was to find a way by which this supposition could be theoretically justified, and it was decided to explore the potentialities of electronic computation.

In the primrose populations under consideration, four different genotypes are possible. If for one population in any one generation the frequencies of these genotypes were represented respectively as p_0 q_0 r_0 s_0, then their expected frequencies p_1 q_1 r_1 s_1 in the next generation could easily be calculated. By repetition of this process, a predicted course of evolutionary change could be calculated for as many generations as desired. But this was a deterministic calculation, and it had to be transformed into a stochastic one. That is to say, the required result of the first calculation was not $p_1\ q_1\ r_1\ s_1$ for the frequencies of the four genotypes, but some other values for these frequencies which might have been arrived at by chance when $p_1\ q_1\ r_1\ s_1$ were the expected frequencies. And it would be these other values which would be the basis for the calculation of the following generation, and so on.

This kind of problem is more difficult than it might appear to be at first glance. There does not seem to be any practicable way by which this transformation could be achieved in a single operation, except in small populations with only two variables.

After the calculation of $p_1\ q_1\ r_1\ s_1$, the method used to make the transformation was to choose plants one at a time by generating a pseudo-random number in the range 0 to $n-1$, and subtracting from it successively $np_1\ nq_1\ nr_1$ and ns_1 until it became negative; the plant was then considered to be of the genotype corresponding to the last number subtracted. This was repeated until the number of plants required for the population had been obtained. In effect, this is the same as making a random selection of the required number of plants from a very large population in which the four genotypes occur in the proportions $p_1\ q_1\ r_1\ s_1$. This is obviously a slow process, very much slower than the initial calculation of the expected frequencies, and the time required to produce each generation is dependent on the population size. This places a practical limitation on the size of population which can be simulated, a disability which also appears in Fraser's technique; however, store capacity, which is a further limiting factor there, does not have any limiting effect on population size in the technique under discussion.

This treatment of the primrose problem made it clear that random fluctuations could lead to very substantial differences in rates of evolution, as had been expected, but the technique was capable of much greater elaboration. Evolution of homostyly is not merely intra-populational—it involves centrifugal spread of homostyly resulting from gene flow between populations.

It was possible to construct a computer model which considered simultaneous evolution in a number of hypothetical populations of various sizes scattered in an imaginary countryside and between which there was gene flow. The magnitude of this gene flow depended on the distance apart of the populations and on the nature of the imaginary terrain; it was also subject to random fluctuations.

The results from a simple form of this model[4] were amply confirmed in a much more elaborate model (unpublished) which simulated a section of countryside of very heterogeneous topography containing 150 populations. Two hundred years of evolution with this complex model produced a pattern of distribution strikingly similar to that found in the naturally occurring populations in the West of England.

In a study of a curious situation in American mouse populations in which certain sub-lethal alleles persisted in surprisingly high frequencies, apparently due to a combination of heterozygote advantage and a large deviation from an expected mendelian segregation ratio, Lewontin & Dunn[7] used a similar method of stochastic transformation but applied it to the choice of sperm and ova from pre-calculated gamete pools, each individual offspring being formed from a sperm and an ovum so chosen. But instead of including the process of natural selection in the initial calculation, they performed a selection test on each individual as produced, determining its fitness by referring its genotype to a pre-set table and retaining or rejecting the individual according to the relative magnitudes of the fitness value and a pseudo-random number. This method of simulating natural selection is more typical of the digital simulation technique described later.

This approach to the problem gave a more adequate account of it than purely mathematical analysis had achieved, and indicated that chance played a large part in determining the elimination or establishment of the sub-lethal alleles in the various mouse populations.

This technique of genetic simulation suffers from the disadvantage that it requires faithful mathematical representation of the evolutionary system we are studying. If the system is a complicated one, the degree of mathematical expertise required may be too much for the average biologist and may also involve simplifying assumptions of doubtful validity. The full biological implications of any mathematical treatment must be understood, and this is not always easy, either for the mathematician or the biologist. Although this technique certainly has important potentialities which may lead to its wider adoption in the future, it has not so far been greatly exploited.

Simulation by direct digital representation

The simulation method developed by Fraser[3] is fundamentally non-mathematical. That is to say, while mathematical operations occur in the programmes, their general function is manipulative. The basis of simulation is not mathematical. Instead, symbolic models of organisms, in terms of their genotypes, are set up in the computer and made to undergo such processes as reproduction, genetic segregation and selection in a way which is intended to be strictly analogous to the natural processes.

Genetic segregation, at least in a diploid nucleus, is a binary process in the sense that in a heterozygote *Aa* there is a choice of two, and only two, possibilities as to which allele will enter any one product of meiosis. The numbers in a computer are binary numbers, and for each digit position there are two choices only, 0 and 1, as to which digit will actually be present. Fraser[3] pointed out that this correspondence between the two systems could be exploited to give an excellent method of simulating within a computer the genotype of an organism.

For example, we may represent the heterozygote *Aa* by the digits 01 in a word of the computer ('word' is the term used for the basic units in the computer store, whose digits may represent instructions, numbers or some other kind of information). The gametes produced from this heterozygote will each be represented by a single digit, and by reference to a pseudo-random number we can for any one gamete decide whether that digit is to be the 0 or the 1, with equal chances as in the ordinary mendelian segregation of which the operation is a model. Self-fertilization is simulated by choosing a second gamete in the same way, and combining the two digits to produce a zygote. For cross-fertilization we have two simulated organisms and produce one gamete from each. In either case repetition of segregation and fertilization will produce a family.

We are not limited to representation of a single gene. For instance, an organism of genotype *AaBBCcdd*, which in practice is more conveniently written as *ABCdaBcd*, may be represented by 00011011. If the genes are taken to be unlinked, gamete formation will be simulated by random choices between the first and fifth digits, the second and sixth, the third and seventh, and the fourth and eighth. If we wish to represent a gene with more than two alleles, we simply use more than one digit—two will give four different alleles, and three will give eight.

The digits in a computer word can be manipulated in a number of different ways, arithmetic or non-arithmetic, and this provides us with a wide repertoire of genetical processes which can be simulated; these processes do not necessarily have to follow the natural 'laws' of genetics, but may be governed by any rules of genetical behaviour which we may care to invent.

A simple illustration of digit manipulation is given by the determination of the phenotype of the organism that has just been used for illustration, *ABCdaBcd*, where the capital letters represent completely dominant alleles; these should always be represented by zeros. To obtain the phenotype of *Aa*, we simply multiply together its digits 01; this gives us 0 which represents *A*, the required phenotype. To obtain the phenotype of *ABCdaBcd* we multiply the first and fifth digits of 00011011, the second and sixth, and so on; this gives us the answer 0001 corresponding to the phenotype *ABCd*. This operation is easily carried out by the computer, for after a simple preliminary relocation of the digits, a single instruction for logical multiplication determines the operation for all genes simultaneously. Of course, where there is no dominance, the genotype digits also describe the phenotype.

The simulated genes in an example such as this may be supposed to have quite unrelated phenotypic effects, or may interact in various ways. If we store in the computer a table of pre-set parameters giving for each possible phenotype such information as fertility, viability, its breeding characteristics or more 'visible' properties, then quick reference to this table by use of the phenotypic digits will enable the computer to find out how the particular organism is to be dealt with. Sometimes, by tricks of programming, the phenotypic (or genotypic) digits may be used directly to determine the course taken by the programme, without need for reference to a table of parameters.

Unless we wish to be really excessive, there is no limit in principle to the number of gene loci we can simulate for an organism; we can string computer words together if one word has insufficient digits for our purpose. In practice, store capacity and the time required for computation provide limits. However, provided we do not need more than one word for each gamete, and provided our genes are all unlinked, we can simulate segregation and fertilization for all gene loci simultaneously;[8, 9] that is, increase in the number of independent genes under consideration does not, within the limits of the capacity of the manipulative unit of the computer (usually the word) cause any increase in computation time.

Simulation of evolution

The logical step which allows the simulation of processes of evolution is from reproduction of a family to reproduction of a population, and it is quite an easy step.

In the simplest case, if an organism is represented by one word, then a number of words will simulate a population. If this population is supposed to be panmictic, then we can produce one offspring for the next generation by choosing two parents at random and making them reproduce as previously described. Reproduction of the whole population is achieved by repetition of this process until the required number of offspring are produced. From this generation, the next one is derived, and so on. If the population is not panmictic, the process is essentially the same except that random selection of parents has to conform to limitations set by the breeding system.

Natural selection can be superimposed on the process quite easily. If we imagine that selection pressure against an organism acts through a low fertility, then each time one is chosen to be a parent a test is made against a pseudo-random number in such a way that the chance of that organism being used for reproduction on that occasion is equal to its fertility relative to a fully fertile organism; if it fails the test, it is rejected this time though it may still be chosen again later. If selection pressure acts through low viability, then a similar test is employed immediately after reproduction, failure again leading to rejection, permanently this time. No difficulty arises when both fertility and viability are involved. A method which has been used by some workers[3, 15] is to select the fittest individuals to form a parent pool, choosing parents from the pool at random for each reproduction. This truncation method may save time, but is open to the serious objection that it does not relate an individual's reproductive potential sufficiently closely to its fitness, while at the same time it turns the merely improbable into the impossible, so reducing the stochastic element. Other criteria of selection may be taken into consideration, and there are situations such as those involving the breeding system where selection forces are inherent in the system and require no separate action by the programme.

Two serious limitations on this technique of simulation will be apparent. The computation time required for the production of one offspring is considerably greater than that required by Crosby's method, and this provides one limit on the population size we can work with. Secondly, store capacity also provides a limitation on population size.

So far, therefore, this technique has been restricted in its application to rather small populations, of a few thousand individuals at most. But its advantages provide more than adequate compensation. It allows precise and often elegant analogues to be made of quite complex genetic systems, avoiding the imprecision and dubious simplification of mathematical treatment where this would be possible, and allowing analysis in the many cases beyond the reach of mathematics. It also provides a good discipline, for the use of this technique enforces more precise thinking about the system. For instance, a statement such as '30 per cent inbreeding' may be adequate for a mathematical analysis, but the computer requires precise programme instruction on how this is to be achieved, in terms making biological sense. Because we have to think more deeply and precisely about the system we are simulating, a better understanding of it may be obtained even before a programme is successfully run.

It could be said that this technique avoids mathematics by making the electronic organisms do the mathematics themselves. At least, the method avoids arguments about the validity of mathematical and statistical methods, which one sometimes suspects to be of more interest to the biomathematician than are the biological systems they are supposed to elucidate.

An example of Crosby's which illustrates this concerns the breeding system of *Oenothera organensis*,[10] which had been reduced to about 500 plants in New Mexico with between them forty-five different alleles of the incompatability gene *s*. These alleles have no selective value apart from the breeding system, which would work perfectly well with only three. Random fluctuations would be expected to reduce their number rapidly from forty-five in a population of that size by chance elimination of many of the alleles, and it was difficult to see how an equilibrium number of forty-five could be maintained. A subtlety which caused all the mathematical trouble is that because of the nature of the breeding system any *s* allele has a selective advantage when it is scarce, so elimination is not as fast as might be expected. Wright dealt with this mathematically, but not to Fisher's satisfaction, and a mathematical argument developed involving Wright, Fisher and Moran in which the breeding system of *Oe. organensis* almost disappeared from view.

Computer simulation of this system was very simple, and had the added advantage that the simulation conformed precisely to Wright's view of the natural system when he treated it mathematically; the selective advantage of rare alleles was inherent in the model as in nature and required no special consideration. It was also very much quicker to write the programme than to try to understand the arguments in the various contentious papers.

The mathematical analysis concluded that the equilibrium number of alleles should be far less than forty-five, but gave no effective explanation of the observable discrepancy. Computer analysis agreed broadly with the conclusion, but went further in suggesting and strongly supporting an explanation. There was good circumstantial evidence that the species had once been abundant and widespread, and had been almost eliminated during the nineteenth century. As with the primrose populations it was possible to simulate spatial distribution, and this showed that even with very restricted gene flow most of the large number of incompatibility alleles which so big a population could carry would be well distributed throughout it. After the catastrophe, each of the few surviving pockets of the species should therefore have contained a high number of alleles. *Oe. organensis* is a long-lived perennial, and there have been insufficient generations since the catastrophe for this number to have fallen very much. The mathematical arguments may have been interesting, but they had little relevance to the biology of the problem. Wright had also attempted to consider spatial distribution; this was necessarily rudimentary because of the limitations of mathematics, and had led him to a conclusion quite different from that derived from the more realistic computer model.

Some applications of digital simulation

Much of the work using digital simulation has tended to concentrate on problems which were more elaborate versions of those dealt with by more orthodox mathematical methods. One obvious field for attack was that of polygenic systems, where many genes may contribute to the expression of one trait, and quite different assortments of genes may produce very similar phenotypes. Such systems may thus have simultaneously a high genotypic variability and a low phenotypic variability. Fraser[11] investigated the problem of how the former could be maintained and stabilized while the latter is retained, and whether such a situation is one which could be brought about by natural selection, since it has many advantages from an evolutionary point of view.

By using a digital simulation model involving twenty genes, he was able to show that where a character was controlled basically by a few genes whose expression could be modified by other genes, it was possible to select for low phenotypic variability with little change in genotypic variability, selection acting by changing the frequencies and patterns of the modifying genes.

Another problem which has attracted attention from several workers is that of the role of linkage in evolutionary processes,[9, 12, 13, 15–17] but this is much less easy to tackle. Unfortunately, while the computer finds many processes easy to simulate, linkage is one which is much more difficult for the computer than for the living organism. Simulation of more than two or three linked genes presents serious difficulties of time or storage space. In one of his programmes designed for demonstration to university classes in genetics, Crosby[14] dealt with linkage by simulating the behaviour of chromosomes in crossing-over in meiosis, and was able to take into account localization of chiasmata and interference.

The first method used by Fraser[9] employed the same principle as the stochastic transformation already described for non-digital simulation. The frequencies of the possible different gene sequences (recombinant and non-recombinant) were calculated and for each gamete a sequence was chosen by serial subtraction of the frequencies from a pseudo-random number until this gave a negative result. A much faster method is to store representations ('masks') of the different patterns of recombination or non-recombination in quantities proportional to their expected frequencies, and to select one from the store by single reference to a pseudo-random number; this demands a very great deal of store space. Both methods suffer from the disadvantage that if we are dealing with n genes, the number of possible gene sequences is 2^n.

A quite different method of dealing with linkage is by a 'random walk' technique.[18] The two homologous chromosomes are imagined lying side by side, and the gene sequence for a gamete is determined by starting at the end of one chromosome chosen at random, and 'walking' along it, changing chromosomes or not at each gene by referring a pseudo-random number to a pre-set parameter for the frequency of crossing-over at that point. It would be difficult, but not impossible, to take chiasma interference into account in this method; there does not seem to have been any attempt to do this.

Several workers have used digital models in order experimentally to test preconceived ideas derived from mathematical treatments. A recent example of this is an investigation by Young[15] of directional selection with different levels of selection intensity, heritability and recombination probability.

Gill & Clemmer[13] introduced the ingenious idea of 'tagging' genes. By having only five genes to a word, each could be labelled with a number in such a way that the descendants of any particular gene in any individual could be followed for several generations. This enabled an estimate to be made of a coefficient of inbreeding for any generation by determining the proportion in the population of allelic pairs whose two alleles were descended from a common ancestral allele of the first generation. Five genes were used on each of five pairs of chromosomes, with combinations of three degrees of linkage and three intensities of selection. Selection and linkage each significantly increased the degree of inbreeding, linkage the more effectively. But the system was somewhat unrealistic to the extent that no generation began with more than eight parents, which seems far too few. One might think that it was a pity to spoil a good idea in this way.

The broadening repertoire of simulation techniques

The models which have so far been referred to deal with problems which have a general thematic resemblance to one another, in that they are extensions of familiar problems of population genetics. The next two models to be described, which have not yet been published, are of interest because they break new ground and indicate directions in which computer genetics may develop.

Crosby has extended the simulation of spatial arrangement of populations to that of individual plants within a population. In the models which have so far been discussed, the breeding range of an individual extended uniformly through-out the population, but now it was localized in a gaussian manner to the immediate vicinity of that individual. This model also introduced genetical discontinuity, since it considered the population genetics of the region of contact between two subspecies (necessarily genetically different) which could initially interbreed but whose hybrids had low fertility.

This situation seems not to have been treated mathematically, though it had been discussed by Dobzhansky[19] and others who had suggested that selection could be expected of genes which would produce barriers to hybridization, since this involves gamete wastage and has therefore a selective disadvantage.

The model genotype gave the two subspecies the opportunity of evolving differences in flowering period, which would be an effective barrier to hybridization. The combined population size was 2500. So long as flowering was approximately simultaneous, there was extensive hybridization in the region of contact and the two subspecies overlapped each other's territory very little. As soon as marked differences in flowering period appeared, the amount of hybridization decreased and the two subspecies began to interpenetrate. Eventually, after some 500 generations, they were growing thoroughly intermingled with little coincidence of flowering and very little hybridization. The model had successfully demonstrated the evolution of genetic isolation barriers, and a change in status from subspecies to species.

This model was elaborated by the introduction of ecological differences, and the technique clearly has considerable potential for development as a tool in genecological research.

One very interesting way in which the range of possibilities in computer simulation is being extended is shown by a project which is being developed by Rosenberg in the University of Michigan, who has kindly allowed me to refer to it.

This is based on the simulation of an idealized biological population consisting of a set of unicellular individuals each having one pair of chromosomes with up to twenty genes. Each gene may have as many as sixteen alleles. Each cell has a set of chemicals whose reactions are controlled by idealized enzymes, and the presence of any particular enzyme depends on the presence of the appropriate allele. The cells are able to communicate with their environment through the diffusion of chemicals.

For each individual at any particular time, a fitness function may be estimated which depends upon the extent to which a subset of its chemicals matches predetermined concentration requirements. Selection operates by making the number of offspring resulting from a mating proportional to the sum of the fitness functions of the parents.

Selection is thus strictly phenotypic, with a highly complex non-linear time-dependent relationship between genotype and phenotype. The development of a model of this order of complexity will allow computer techniques to be exploited in a wide range of interesting experiments.

Future development of computer genetics

Computer simulation techniques for the study of genetic systems face one severe inherent restriction. There is a practical limit on the size of population which can be simulated, and this arises from two distinct causes.

Firstly, size of the simulated population has been restricted by the available capacity of direct access store. In a problem of even moderate genetic complexity this limits population size to a few thousands. This will be a quite inadequate figure when complex population problems involving spatial distribution and genetical and ecological heterogeneity are involved. For emphasis, it may be noted that Young[15] refers to a population of a thousand as 'large'.

Use of magnetic tape to increase population size by extending store capacity is of little value, and quite impossible where random access to the whole population is required, because access time is far too slow. But the new technique of magnetic disc storage, with its much faster access time, has radically altered this position and there is no reason to suppose that for time-sharing (multiprogramming) machines with disc storage facilities there would be any great difficulty in producing efficient programmes which would allow the simulation of very much larger populations maintained essentially in disc storage.

But here the second limiting factor comes into consideration. Basically, a computer can only do one thing at a time, and this includes reproduction of an individual. It is a matter of simple arithmetic to see that an experiment on a population of 2000 which takes an hour will, on blowing up to 50,000 by utilizing extra storage capacity, take about a day. It is most unlikely that there will be, in the foreseeable future, enough computer time available for many such experiments.

More efficient programming will not solve this problem, for it can only make a fractional difference to the running speed. Faster computers would help, but there is good reason for believing that we may not be very far from the limit in computer speed. More computers, providing more computer time, would also help, but the dream of a fast machine all to oneself seems just now to be not even a remote possibility.

The answer may have to be found in a reconsideration of simulation methods. A return to the kind of technique originally used by Crosby might be useful for some kinds of problems, but it will need a much more effective and faster method of stochastic transformation.

But in spite of its limitations, computer simulation is a valuable technique which deserves to be much more widely exploited than it has been. Geneticists have been rather slow to take advantage of its possibilities, perhaps because most of them are still rather afraid of computers. But it is not difficult to learn to programme a computer. The difficulty lies in the genetics.

References

1. WRIGHT S. (1931) *Genetics* **16**, 97.
2. HAMMERSLEY J. M. & HANDSCOMB D.C. (1964) *Monte Carlo Methods*. Methuen, London.
3. FRASER A.S. (1958) *Aust. J. biol. Sci.* **10**, 484.
4. CROSBY J.L. (1960) *Phil. Trans. R. Soc.* B, **242**, 551.
5. ALLAN J.S. & ROBERTSON A. (1964) *Genet. Res.* **5**, 68.
6. LEWONTIN R.C. (1964) *Genetics* **49**, 49.
7. LEWONTIN R.C. & DUNN L.C. (1960) *Genetics*, **45**, 705
8. FRASER A.S. & HANSCHE P.E. (1965) *Genetics Today*, vol. 3. p. 507. Pergamon Press, Oxford.
9. FRASER A.S. (1958) *Aust. J. biol. Sci.* **10**, 492.
10. CROSBY J.L. (1966) *Evolution* **20**, 567.
11. FRASER A.S. (1960) *Aust. J. biol. Sci.* **13**, 150.
12. GILL J.L. (1965) *Aust. J. biol. Sci.* **18**, 1171.
13. GILL J.L. & CLEMMER B.A. (1966) *Aust. J. biol. Sci.* **19**, 307.
14. CROSBY J.L. (1961) *Heredity* **16**, 255.
15. YOUNG S.S.Y. (1966) *Genetics* **53**, 189.
16. MARTIN F.G. & COCKERHAM C.C. (1960) *Biometrical Genetics* (Ed. by O. Kempthorne), p. 34. Pergamon Press, Oxford.
17. FRASER A.S. (1962) *J. theoret. Biol.* **2**, 329.
18. FRASER A.S. (1960) *Biometrical Genetics* (Ed. by O. Kempthorne), p. 70. Pergamon Press, Oxford.
19. DOBZHANSKY TH. (1940) *Am. Nat.* **74**, 312.

The Evolution
of Purposive Behavior

ALEXANDER S. FRASER

University of Cincinnati

Cincinnati, Ohio

Purposive behavior has two aspects, the ability to specify an intrinsic purpose and the ability to initiate and maintain a pattern of behavior that is directed towards achievement of an intrinsic purpose. The evolution of purposive behavior, therefore, has two aspects and their examination requires models of evolution based on known parameters of the genetic basis of living things. These models will then allow us to determine the constraints on the evolution of purposive behavior.

The essentials of a model of organic evolution are

1. a system of stored information. This will consist of molecules of DNA, i.e., the information will consist of a vector of quarternary bits.

$$\text{Information} = [\dots wxyz \dots]_4$$

2. a system of transfer of information from one generation to the next. This can be by simple replication, but hereditary transfer in the majority of living things involves combination and recombination. The rules for combination and recombination involve the alignment in parallel of two genetic vectors and the formation of a new vector by recombination along this parallel

alignment. The formation of the new vector is analogous to a random walk along the length of the two vectors, with a specified probability of changing from one vector to another between each element and the next.

$$[\ldots wxyz \ldots]_i \qquad [\ldots \dot{w}\dot{x}\dot{y}\dot{z} \ldots]_j$$

$$\downarrow \text{ Combination}$$

$$[\ldots wxyz \ldots]_i$$
$$[\ldots \dot{w}\dot{x}\dot{y}\dot{z} \ldots]_j$$

$$\downarrow \text{ Recombination}$$

$$[\ldots \overline{w}|x \ y|z \ldots]_i$$
$$[\ldots \dot{w} \ \dot{x} \ \dot{y} \ \dot{z} \ldots]_j$$

$$\downarrow$$

$$[\ldots w\dot{x}\dot{y}z \ldots]_k$$

If we assume that the probability of recombination is constant over all pairs of adjacent elements, then the system of recombination involves an operator, 0_r, specifying that the probability of changing vector in the random walk is r (where r lies in the range of $0 < r \leqslant 0.5$).

3. a system of mutation of stored information. This will introduce a random element such that the genetic vectors will be random sets of elements, if no other factors are operative. Known rates of mutation are very low, and most computer models of evolution have been formed without including a mutation factor. The effects of mutation have been introduced by assuming that the genetic vectors of the initial population are completely random, having resulted from a long-continued mutation process.

4. a system of translation of the genetic information into a form against which value judgments are made. Such value judgments will have the effect of determining the probability of transmission of a vector of genetic information from one generation to the next.

The interaction of these four systems has the consequence of modifying the pool of genetic variability in such a way that individual genetic vectors have a greater probability of surviving value judgments. This can be described as a process of learning. A particular genetic vector with a high probability of surviving value judgments could occur at a very low frequency in the initial population, but over a period of time the frequency of this vector will increase. This could result in the population eventually consisting only of the particular genetic vector. The population will have *rejected genetic information of low survival value and retained only genetic information of high survival value.*

The simplest genetic system is one in which the genetic vectors have only a single element which may have two values, e.g., 0 and 1. If one of these elements has a higher probability of survival than the other, then the probability of the population acquiring increased survival follows fairly simple algebraic statements. However, the genetic system of any organism does not involve just a single pair of alleles; instead, a population will be genetically variable at many, if not all, of its hundreds of genetic loci. The problem is to devise methods of describing and analyzing the changes that will occur in such systems of extensive variability, in which many different genetic vectors have effectively the same survival value. Evolution becomes a question of whether a meaningful genetic change is possible from a basis of almost infinite genetic diversity.

An analogy can be drawn to the derivation of language, given the existence of an alphabet. An endless variety of letter combinations needs to be refined into a meaningful system of information. The analogy is not one that can be pushed too far—there is implicit meaning in some genetic variation—a gene causes the production of a protein, the form of which relates to a specific substrate, whereas letter combinations have no such possibility of intrinsic meanings.

Computer simulation has been used to examine a particular genetic system to determine whether a particular mode of value judgments of genetic vectors can impose genetic meaning on an initially meaningless mélange of genetic variation. Computer runs were made with systems of up to 30 loci, i.e., the genetic vectors had up to 30 elements, but the essentials of the system can be understood best by first considering a system of two loci, i.e., genetic vectors having two elements. The elements were specified to have two states: 0 and 1. Normal genetic symbology uses letters for genetic elements, specifying states (alleles) by upper and lower case. There are ten possible genetic combinations in this system for the diploid state. (The diploid state is one in which an individual is specified by two genetic vectors, one inherited from each parent:)

$$
\begin{bmatrix} (ab) \\ (ab) \end{bmatrix} \quad \begin{bmatrix} (Ab) \\ (ab) \end{bmatrix} \quad \begin{bmatrix} (Ab) \\ (Ab) \end{bmatrix} \quad \begin{bmatrix} (AB) \\ (Ab) \end{bmatrix} \quad \begin{bmatrix} (AB) \\ (AB) \end{bmatrix}
$$

$$
\begin{bmatrix} (aB) \\ (ab) \end{bmatrix} \quad \begin{bmatrix} (AB) \\ (ab) \end{bmatrix} \quad \begin{bmatrix} (AB) \\ (aB) \end{bmatrix}
$$

$$
\begin{bmatrix} (Ab) \\ (aB) \end{bmatrix}
$$

$$
\begin{bmatrix} (aB) \\ (aB) \end{bmatrix}
$$

| 0 | 1 | 2 | 3 | 4 |

Genetic value

If we specify that lower case elements have a value of 0 and upper case have a value of 1, then the genetic values can be specified by addition over the genetic specification. Some genetic values are unique to just one genotype (genetic specification), e.g., (ab)(ab) is the only genotype with a genetic value of 0, and (AB)(AB) is the only genotype with a genetic value of 4. Other genetic values are given by more than one genotype, e.g., the genetic value of 2 is specified by four genotypes.

We can specify survival value as some function of genetic value. We have concentrated on what is termed the optimum model, in which genetic survival is maximal for individuals with an intermediate genetic value, and minimal for individuals with extreme genetic values. We have been examining the importance of being mediocre. This model can be referred to patterns of behavior. Suppose that an individual's reaction to light is specified by its genetic value, having a negative phototaxis for low genetic values, a positive phototaxis for high genetic values, and a zero phototaxis for intermediate genetic values. Then, if survival is maximal for individuals who do not react to light, selection will favor intermediate genetic values. Here, the intrinsic purpose is specified—do not react to light, and we are examining the evolution of a pattern of behavior to satisfy the intrinsic purpose. An extreme form of selection for intermediate values is that in which only individuals with the most intermediate value survive; individuals with other values do not survive, e.g., only individuals showing no phototaxis survive. This is illustrated below with the substitution of a binary notation for the letter case notation.

$$
\begin{bmatrix} (00) \\ (00) \end{bmatrix} \quad \begin{bmatrix} (01) \\ (00) \end{bmatrix} \quad \begin{bmatrix} (10) \\ (10) \end{bmatrix} \quad \begin{bmatrix} (01) \\ (01) \end{bmatrix} \quad \begin{bmatrix} (11) \\ (10) \end{bmatrix} \quad \begin{bmatrix} (11) \\ (11) \end{bmatrix}
$$

$$
\begin{bmatrix} (10) \\ (00) \end{bmatrix} \quad \begin{bmatrix} (11) \\ (00) \end{bmatrix} \quad \begin{bmatrix} (10) \\ (01) \end{bmatrix} \quad \begin{bmatrix} (11) \\ (01) \end{bmatrix}
$$

Do not survive	Survive	Do not survive

Imposition of this type of value judgment can have three possible effects on the genetic structure of the population.

1. All individuals become (10)(10). This is genetic fixation for this homozygous genotype. Such individuals can only produce (10) gametes regardless of recombination.

2. All individuals become (01)(01). This is genetic fixation for this homozygous genotype. Such individuals can only produce (01) gametes regardless of recombination.

3. Parents are of four types in exactly equal proportions.

$$
\begin{bmatrix} (01) \\ (01) \end{bmatrix} \quad , \quad \begin{bmatrix} (01) \\ (10) \end{bmatrix} \quad , \quad \begin{bmatrix} (11) \\ (00) \end{bmatrix} \quad , \quad \begin{bmatrix} (10) \\ (10) \end{bmatrix}
$$

In this case recombination in the second and third of these genotypes will change the proportions of gametes produced, but the changes will be complementary if the frequencies of these two genotypes are equal. This situation is metastable—any increase in frequency of the first or fourth genotypes, i.e.,

(01)(01) or (10)(10) will result in the population moving to the state of fixation for the genotype that was in excess.

There is a considerable complexity to the nature of the metastable state, dependent on the degree of recombination that occurs. If recombination is at a maximum of 0.5, then the above statement holds, whereas if recombination does not occur then the metastable state involves only three genotypes in the proportions as shown.

$$\begin{bmatrix} (01) \\ (01) \end{bmatrix} \qquad\qquad \begin{bmatrix} (01) \\ (10) \end{bmatrix} \qquad\qquad \begin{bmatrix} (10) \\ (10) \end{bmatrix}$$
$$1 \quad : \quad 2 \quad : \quad 1$$

It is obvious that the words purpose and evolution are rather difficult to apply. If the purpose is to achieve a population with an intermediate phenotype, then there are three separate ways of achieving this purpose. If, on the other hand, the aim is to achieve a single genetic solution, then the actual aim that is achieved will be decided by stochastic factors if the initial genetic variability is at a maximum. Random factors will determine whether the population moves towards the (10)(10) homozygous state or to the (01)(01) homozygous state. Achieving either state is an increase of the survival status of the population—it will be genetically erudite in the matter of survival, but its manner of achieving this aim was decided by an initial random perturbation. The random factors impose genetic direction; these factors give genetic purpose to an otherwise indeterminate situation.

This model can be extended to involve more loci. For four loci, there are six possible fixation states.

(0101)(0101)
(0110)(0110)
(0011)(0011)
(1001)(1001)
(1010)(1010)
(1100)(1100)

For six loci the number of fixed intermediate states is greater yet: 20. The number of possible intermediate states increases geometrically. A genetic system approaching reality involves tens of loci, and the number of intermediate states in such genetic systems is approaching an effectively infinite dimension. Computer simulation has been used to examine the effectiveness of selection on such systems, with a genetic model of 30 loci. Any specific genetic combination has a probability of occurrence of $1/3^{30}$; there are 3^{30} possible genotypes. The number of genotypes that satisfy the requirement of having the intermediate phenotype is less than 3^{30}, but it is so large that the indeterminacy of the system is extremely marked. Selection in the absence of random perturbations will be ineffective if the genes are not linked. Selection cannot impose genetic purpose, and stochastic perturbations need to be fairly large to impose direction. The genetic system has no inherent purpose, and it is intractable to the imposition of a purpose by stochastic factors.

In a situation such as this, it is pertinent to consider the effects of introduction of unlikely modifications since, over the time periods involved in the evolutionary systems, even extremely rare possibilities become feasible. Such an unlikely system is that of an inversion. Chromosomal mutations are known in which a segment becomes inverted, i.e., a segment is effectively removed and reinserted in the alternate sequence.

$$A \ B \ C \ D \ E \ F \ G \ H \ I \ J$$
$$\downarrow \quad \text{Inversion}$$
$$A \ B \cdot H \ G \ F \ E \ D \ C \cdot I \ J$$
Inverted segment

Inversions have the effect of reducing recombination amongst the genes of the inverted segment to near zero. An inversion, therefore, isolates a particular genetic combination from the normal process of recombination. A large inversion can result in a loosely linked set of genes becoming extremely tightly linked. The occurrence of such a large inversion could have marked effects in a population that is indeterminate because the genes are freely recombining but are spaced along a chromosome. The inversion by causing the genes to be tightly linked can affect the situation. This has been examined by computer modeling.

A chromosome that has become inverted contains a particular set of 0 and 1 alleles. It will most probably be unbalanced, i.e., the sum of its genetic constitution will not be intermediate; there will be more 0's than 1's or vice versa. We have used a model involving 30 loci and a balanced chromosome would have 15 zero alleles and 15 unit alleles—this can be expressed by the formula, $0^{15}1^{15}$. An unbalanced chromosome would then have a value such that 0^i1^j, $\neq$ j. Values such as $0^{14}1^{16}$, $0^{13}1^{17}$, $0^{12}1^{18}$, etc. are probable in the system we have examined, and we have examined the effect of introducing an inverted chromosome with a particular unbalanced conformation into a population which is indeterminate, as described above. The results of computer simulation of this model have shown that (under certain constraints) the frequency of the inversion rapidly increases to a maximum of 50 percent. This is at first sight surprising—the inversion is unbalanced yet the selective system favors a balanced constitution. Why does the inversion increase in frequency? The answer lies in the stability of the inversion in heredity. The suppression of recombination results in the inverted chromosome being transmitted in heredity unchanged, which is in marked contrast to an uninverted chromosome in which the free recombination results in its unchanged transmission in heredity being extremely improbable. This gives the inversion such an advantage that the disadvantage of genetic unbalance is overridden, with the result that the inversion increases in frequency. In a system where purpose cannot be defined, then an unlikely event which introduces even an inefficient purpose sets the population on that path. However, one of the constraints that was discovered from the computer simulation was that the frequency of the inversion needs to exceed a certain value before the process of moving in a defined purpose begins—if the inversion occurs at a lower frequency, it does not increase and the inversion is lost from the population. Even with an inversion there is some need for stochastic variation to attain purpose.

The point that emerges from these simulations is that genetic learning can proceed in a deterministic sequence of increase once a stochastic process has set a basically indeterminate system into one of the many possible pathways. In our original separation of the two aspects of purpose, that of the ability to specify an intrinsic purpose and that of elaborating a pattern of behavior that is directed towards achievement of a specified purpose, it appears that the evolution of the latter can easily be accounted for by the action of selection, but the evolution of the former can and probably has required stochastic fluctuations to break the impasse of indeterminacy.

Chapter 4
Evolving Online Productivity

G. E. P. Box (1957) "Evolutionary Operation: A Method for Increasing Industrial Productivity,"
Applied Statistics, Vol. 6:2, pp. 81–101.

THE well-known statistician G. E. P. Box proposed an evolutionary approach to optimizing industrial production, particularly in the chemical industry. The method, termed "evolutionary operation," or "EVOP," was first published in archival literature in Box (1957) (reprinted here), following earlier presentation to the International Conference on Statistical Quality Control in Paris, July 1955, after its introduction at Imperial Chemical Industries in England in 1954.[1] In essence, the technique was to make small modifications to current production settings in two or three parameters at a time, note the observed change in plant performance, then select the best settings for which there was sufficient statistical evidence of an improvement and repeat the process. This was in contrast to the still typical reluctance of production managers to change any settings in a working factory: The conventional view holds that when the plant is "working" it is best not to effect any change.

Box (1957) was explicit about the analogy between his proposed method and the process of evolution. Modifications to processes constituted mutations, while adjustment of the production parameters to their observed best levels constituted selection. The imposed modifications were not chosen at random, but rather in a factorial design[2] where each parameter was varied in a low and high setting. These settings, however, represented only minor changes to the current parameter profile; changes were kept relatively small so as not to induce a significant chance of loss due to the production of unsatisfactory material.[3] The changes were implemented online, while the plant was in operation, not in any computer simulation. This requirement constrained the possible "mutations" that could be imposed. The choice of modifications (size and type) was to be made by a committee of plant personnel, with the suggestion that they meet periodically (perhaps once a month) (Box, 1957). After several cycles through parameter level combinations in the factorial design, if there were sufficient evidence to identify positive change in the plant's performance, the nominal setting of the appropriate parameter would be modified and a new factorial experiment would be started. Thus the technique essentially used a single parent parameter profile to generate multiple offspring and selected, based on statistical evidence, the best of the offspring and parent to serve as a parent for further exploration.

Although the EVOP procedure, as originally described, required a human (or humans) in the loop, Box (1957) wrote of the possibility for making the process completely automatic. This was beyond the instrumentation capability of most industrial plants at the time, however. Efforts were made instead to

[1]The first written account of the EVOP technique was Box (1954), an internal report from Imperial Chemical Industries, Ltd. (Box, 1997).

[2]A factorial design is a statistical method in which multiple parameter effects are assessed by experimenting at all levels of a given factor in combination with all levels of every other factor in the experiment (Hicks, 1982, pp. 88–89).

[3]The small modifications that were imposed may have hindered EVOP from escaping local minima, but this was likely a sacrifice that was to be tolerated in favor of gaining some improvement over initial suboptimal parameter settings without inducing any significant probability of generating catastrophic results. Goldberg (1989, p. 104) offered a criticism of Box's analogy of the variation of parameters to genetic mutation: " . . . under this definition, just about anything that changes a structure qualifies as a mutation. This is much too loose a definition, and we must be more careful that essential parts of the analogy are preserved." But any genetic change does indeed constitute a mutation (Hartl and Clark, 1989, p. 96; Dobzhansky, 1970, p. 65, defined mutation as not including changes by recombination, but this would not alter the argument for the present circumstance because EVOP did not use recombination). Box (1997) also stressed that his use of the term *mutation* applied more properly to entirely new routes to manufacture particular organic chemicals, rather than simple variations on a theme. Goldberg (1989, p. 104) also offered a criticism of the regular pattern of possible variations; however, Satterthwaite (1959a) suggested a randomized EVOP (REVOP) (detailed in Lowe, 1964) following previous work on random balance experiments (Satterthwaite, 1959b). It should be noted that Box and Draper (1969, pp. 176–179) did not endorse the REVOP procedure, nor another modification termed the "simplex EVOP" (Spendley et al., 1962).

simplify the calculations required for the EVOP committees (Box and Hunter, 1959). Box (1960) and Box and Chanmugam (1962) suggested the use of EVOP for optimizing time-varying plants where a particular parameter (e.g., a catalyst) might vary continuously in time in an unknown manner, requiring online adjustment to these fluctuations. Box (1966) formed a problem in decision theory regarding when to make a parameter adjustment based on the magnitude of the observed change in plant response. Box and Draper (1969) provided a comprehensive review of the process and developed procedures.

EVOP attracted considerable attention and was subsequently put in practice in several chemical companies in the United States (Koehler, 1959; Hunter, 1960; and others,[4] also see Barnett, 1960, for an early review). It remains in limited use (e.g., Banerjee and Bhattacharyya, 1993).

References

[1] R. Banerjee and B. C. Bhattacharyya (1993) "Evolutionary operation (EVOP) to optimize 3-dimensional biological experiments," *Biotechnology and Bioengineering,* Vol. 41:1, pp. 67–71.

[2] E. H. Barnett (1960) "Introduction to evolutionary operation," *Industrial and Engineering Chemistry,* Vol. 52:6, pp. 500–503.

[3] G. E. P. Box (1954) "Evolutionary operation: A proposed technique for process development," Internal report, Imperial Chemical Industries, Ltd., Dyestuffs Division, Manchester, U.K.

[4] G. E. P. Box (1957) "Evolutionary operation: A method for increasing industrial productivity," *Applied Statistics,* Vol. 6:2, pp. 81–101.

[5] G. E. P. Box (1960) "Some general considerations in process optimization," *J. Basic Engineering,* March, pp. 113–119.

[6] G. E. P. Box (1966) "A simple system of evolutionary operation subject to empirical feedback," *Technometrics,* Vol. 8:1, pp. 19–26.

[7] G. E. P. Box (1997) personal communication, Center for Quality and Productivity Improvement, Madison, WI.

[8] G. E. P. Box and J. Chanmugam (1962) "Adaptive optimization of continuous processes," *Industrial and Engineering Chemistry Fundamentals,* Vol. 1:1, pp. 2–16.

[9] G. E. P. Box and N. R. Draper (1969) *Evolutionary Operation: A Statistical Method for Process Control,* John Wiley, NY.

[10] G. E. P. Box and J. S. Hunter (1959) "Condensed calculations for evolutionary operation programs," *Technometrics,* Vol. 1:1, pp. 77–95.

[11] T. Dobzhansky (1970) *Genetics of the Evolutionary Process,* Columbia Univ. Press, NY.

[12] D. E. Goldberg (1989) *Genetic Algorithms in Search, Optimization and Machine Learning,* Addison-Wesley, Reading, MA.

[13] D. L. Hartl and A. G. Clark (1989) *Principles of Population Genetics,* 2nd ed., Sinauer, Sunderland, MA.

[14] C. R. Hicks (1982) *Fundamental Concepts in the Design of Experiments,* 3rd ed., Holt, Rinehart and Winston, NY.

[15] J. S. Hunter (1960) "Optimize your chemical process with evolutionary operations," *Chemical Engineering,* September, pp. 193–202.

[16] T. Koehler (1958) "Evolutionary operation, some actual examples," *Proc. 2nd Stevens Symp.,* co-sponsored by the Chemical Division and Metropolitan Section, American Society for Quality Control.

[17] T. L. Koehler (1959) "Evolution operation: A program for optimizing plant operation," *Chem. Eng. Prog.,* Vol. 55:10, pp. 76–79.

[18] C. W. Lowe (1964) "Some techniques of evolutionary operation," *Trans. Instn. Chem. Engrs.,* Vol. 42, pp. T334–344.

[19] F. S. Riordan (1958) "Problems in the administration of evolutionary operation," paper read to the 1958 Conference, Chemical Division, American Society for Quality Control, Buffalo, NY.

[20] F. S. Riordan (1959) "EVOP-revolution by evolution," paper presented at the all-day conference on statistical quality control and methods in industry, co-sponsored by South Texas Section ASQC and the Texas Manufacturer's Assn.

[21] F. E. Satterthwaite (1959a) "Author's response to discussions," *Technometrics,* Vol. 1:2, pp. 184–192.

[22] F. E. Satterthwaite (1959b) "Random balance experimentation," *Technometrics,* Vol. 1:2, pp. 111–137.

[23] W. Spendley, G. R. Hext, and F. R. Himsworth (1962) "Sequential application of simplex designs in optimization and evolutionary operation," *Technometrics,* Vol. 4:4, pp. 441–461.

---

[4]Two publications (Box and Hunter, 1959; Hunter, 1960) cite Koehler (1958), Riordan (1958, 1959).

EVOLUTIONARY OPERATION:*
A METHOD FOR INCREASING INDUSTRIAL PRODUCTIVITY

GEORGE E. P. BOX†

Imperial Chemical Industries Limited
Dyestuffs Division Headquarters, Manchester

The rate at which industrial processes are improved is limited by the present shortage of technical personnel. Dr Box describes a method of process improvement which supplements the more orthodox studies and is run in the normal course of production by plant personnel themselves. The basic philosophy is introduced that industrial processes should be run so as to generate not only product, but also information on how the product can be improved.

Introduction

Much scientific effort in industry is directed on the one hand to the discovery of new products and processes and on the other to their development and improvement. This paper is concerned with a particular aspect of the problem of *improving* industrial processes.

Industrial organisations usually have specialist groups of scientific workers in research, development, and experimental departments, permanently occupied with improving manufacturing processes, who employ a wide variety of techniques, ranging from the fundamental study of reaction mechanisms to the purely empirical assessment of the effects of changes in variables. Associated experimentation may be conducted in the laboratory, on the pilot plant, and on the full scale process, and in particular may involve the use of statistical techniques having a fairly high degree of sophistication.[1-7] As a result of the application of this variety of specialised effort a steady rate of increase in productivity is usually attained.

Ultimately the rate of improvement is limited by the shortage of technical personnel. This shortage can be expected to become more rather than less severe, and in searching for further ways of attaining greater process efficiency one must look for methods which are sparing in their use of scientific manpower. The object of this paper is to outline one such device, which has been applied with considerable success over the past few years.

This is called 'Evolutionary Operation'. It is a method of process operation which has a 'built-in' procedure to increase productivity. It uses some simple statistical ideas and is run during normal routine

* Based on a paper given at the International Conference on Statistical Quality Control organised by the European Productivity Agency of OEEC in Paris in July 1955.
† Now Director of the Statistical Techniques Group at Princeton University.

production largely by plant personnel themselves. Its basic philosophy is that it is nearly always inefficient to run an industrial process to produce product alone. A process should be run so as to generate product *plus information on how to improve the product.*

The technique is in no sense a substitute for the more fundamental investigations referred to above. On the contrary, the effects discovered by the application of evolutionary operation, particularly those which are of an unexpected kind, help to indicate new areas where fundamental research might be rewarding. Although the method has been specifically developed as a production technique for the chemical industry, it is believed that it has more general applicability.

Plant-scale and Small-scale Experiments

In the chemical industry the plant process will usually have been arrived at after considerable experimentation on the small scale. Now the optimum conditions of operation on the small scale usually provide no more than a good first approximation to the full-scale optimum. Because of this it is commonly found that considerable modification of the conditions arrived at from small-scale work is necessary before a comparable result can be obtained on the plant itself.

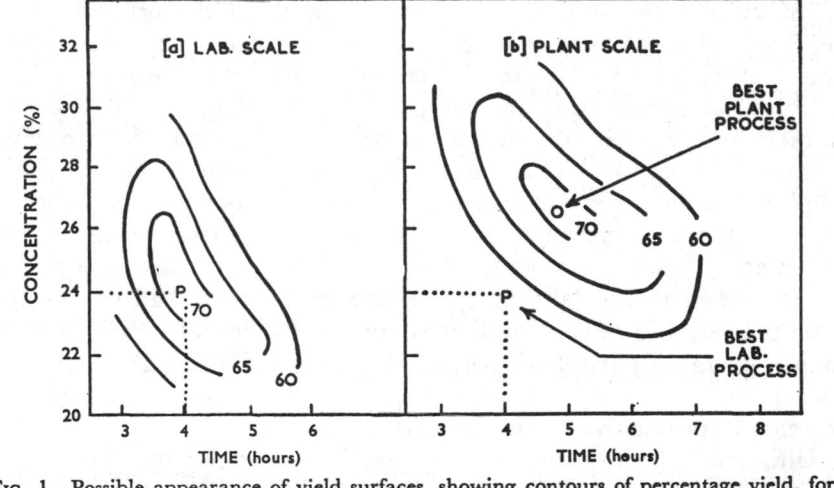

Fig. 1. Possible appearance of yield surfaces, showing contours of percentage yield, for a process conducted (*a*) on the laboratory scale and (*b*) on the plant scale.

Fig. 1 shows how, in translating the process from the laboratory to the plant, the yield surface can become distorted and displaced due to scale-up effects. It will be seen that when this happens one might expect the best laboratory conditions to give disappointingly low yields on the full scale. The effort to move from the laboratory maximum-yield conditions at P to the plant maximum-yield conditions at O must evidently be exerted *on the plant itself* and not in the laboratory—investigation in the laboratory can only lead back to P.

For those unfamiliar with the representation adopted in Fig. 1 it should be explained that the relationship between the response ('yield') and the process variables ('time of reaction' and 'concentration of one of the reactants') is imagined to be represented by a solid graph or 'response surface'. In the neighbourhood of maximum yield such a surface may have the appearance of a mound. The height of the mound at any particular point represents the yield at some set of reaction conditions. To allow representation in two dimensions the yield is shown by contours in the same way that the height of land is represented by contours on a map. Although for simplicity the above discussion is conducted in terms of yield, it should be understood that conditions giving highest yield would often not represent the optimum process. There would, for example, be no advantage in obtaining a higher yield if to do so involved the use of a disproportionate amount of some expensive starting material. The principal response usually considered, therefore, is 'the cost of producing unit quantity of product under the specified manufacturing conditions', or some other measure of productivity which takes account of the cost of running the process.

In addition to improvements made possible by adjustment of process conditions already studied on the small scale, further progress is usually possible by the introduction of new modifications not considered—and often not capable of being studied—at the small-scale stage of development.

Adjustments are made when the plant is first installed, but these seldom result in the location of the ultimate plant optimum, and as a result of special experimental campaigns, chance discoveries, and new ideas, improvement usually continues over many years. The object of evolutionary operation is to speed up this process.

Analogy with Evolutionary Process

The method used to speed the improvement is illustrated by the following analogy. Living things advance by means of two mechanisms:

(i) Genetic variability due to various agencies such as mutation.

(ii) Natural selection.

Chemical processes advance in a similar manner. Discovery of a new route for manufacture corresponds to a mutation. Adjustment of the process variables to their best levels, once the route is agreed, involves a process of natural selection in which unpromising combinations of the levels of process variables are neglected in favour of promising ones.

Fig. 2 illustrates diagrammatically the possible evolution of a species of lobster. It is supposed that a particular mutation produces a type of lobster with 'length of claws' and 'pressure attainable between claws' corresponding to the point P on the diagram and that in a given environment the contours of 'percentage surviving long enough to reproduce' are like those shown in the figure.

The dots around P indicate offspring produced by the initial type of lobster. Since those in the direction of the arrow have the greatest chance of survival, over a period of time the scatter of points representing succeeding generations of lobsters will automatically move up the survival surface. This automatic process of natural selection ensures, without any special effort on the part of the lobsters, that optimum-type lobsters exist. It also ensures that if the environment changes so that the survival surface is altered, the lobsters will change correspondingly to the new point of maximum survival.

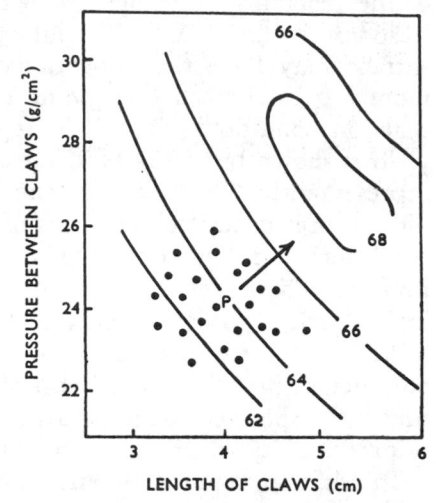

FIG. 2. Evolution of a species of lobster. Contours show the percentage surviving long enough to reproduce in a given environment.

What we have to do is to imitate this process. That is to say, we have to institute a set of rules for *normal plant operation* so that (without serious danger of loss through manufacture of unsatisfactory material) an evolutionary force is at work which steadily and automatically moves the process towards its optimum conditions if it is not operating there already. Such a technique will gradually nudge the operating procedure into the form that is ideally suited to the particular piece of equipment which happens to be available. The two essential features of the evolutionary process are:

(i) *Variation*.

(ii) *Selection* of 'favourable' variants.

Static and Evolutionary Operation

Routine production is normally conducted by running the plant at rigidly defined operating conditions called the 'works process'. The works process embodies the best conditions of operation known at the time. The manufacturing procedure, in which the plant operator aims always to reproduce exactly this same set of conditions, will be called the method of *Static Operation*. Although this method of operation, if strictly adhered to, clearly precludes the possibility of evolutionary development, yet the *objectives* which it sets out to achieve are nevertheless essential to successful manufacture, for in practice we are interested not only in the productivity of the process, but also in the physical properties of the product which is manufactured. These physical properties might fall outside specification limits if arbitrary deviations from the works process were allowed. Our modified method of operation must therefore include safeguards which will ensure that the risk of producing appreciable amounts of material of unsatisfactory quality is acceptably small.

In the method of Evolutionary Operation a carefully planned cycle of minor variants on the works process is agreed. The routine of plant operation then consists of running each of the variants in turn and continually repeating the cycle. The cycle of variants follows a simple pattern, the persistent repetition of which allows evidence concerning the yield and physical properties of the product in the immediate vicinity of the works process to accumulate during routine manufacture. In this way we use routine manufacture to generate not only the product we require but also the information we need to improve it.

Controlled variation having thus been introduced into the manufacture, the effect of selection is introduced by arranging that the results are continuously presented to the plant manager in a way which is easily comprehended. This allows him to see what changes ought to be made to improve manufacture. The stream of information concerning the products from the various manufacturing conditions is summarised on an *Information Board* prominently displayed in the plant manager's office. This is continuously brought up to date by a clerk to whom the duty is specifically assigned. The information is set out in such a way that the plant manager can at any time see what weight of evidence exists for moving the centre of the scheme of variants to some new point, what types of change are undesirable from the standpoint of producing material of inferior quality, how much the scheme is costing to run, and so on.

In making a permanent change in the routine of plant operation the situation is very different from that which we meet in running specialised experiments on the plant. The latter will last a limited time, during which special facilities can be made available. Furthermore some manufacture of substandard material is to be expected and will be budgeted for. Evolutionary operation, however, is virtually a *permanent* method of running the plant and cannot therefore demand special facilities and concessions. For this reason only small changes in the levels of the variables can be permitted, and only techniques simple enough to be run continuously by works personnel themselves under actual conditions of manufacture can be employed.

The effects of the deliberate changes in the variables will usually be masked by large errors customarily found on the full scale. *However, since production will continue anyway, a cycle of variants which does not significantly effect production can be run almost indefinitely, and because of constant repetition the effect of small changes can be detected.*

An Example

To illustrate the procedure we consider one phase of evolutionary operation for a particular batch process. At this stage of development two process factors—the percentage concentration of one of the reactants, and the temperature at which the reaction was conducted—

were being studied following the scheme of variants shown in Fig. 3. The works process is labelled (1) and the four variants are labelled (2), (3), (4), and (5). One batch of product was made at each set of conditions, which were run successively in the order 1, 2, 3, 4, 5; 1, 2, 3, 4, 5; and so on. Three responses were recorded:

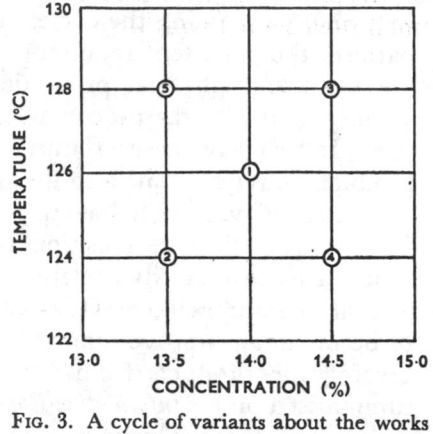

FIG. 3. A cycle of variants about the works process.

(i) The cost of manufacturing unit weight of product. This was obtained by dividing 'the cost of running at the specified conditions' by 'the observed weight yield at those conditions'. It was desired to bring this cost to the smallest value possible, subject to certain restrictions listed in (ii) and (iii) below.

(ii) The percentage of a certain impurity. It was desired that this should not exceed 0·5.

(iii) A measure of fluidity. It was preferred that this should lie between the limits 55 and 80.

The information coming from the experiment was recorded by writing in chalk on an ordinary blackboard. Alternatively, wax pencil on a white plastic board or magnetic letters and numbers on a steel board could have been used. The essential thing is that it should be a simple matter to erase or remove one number and replace it by another. The scheme set out in Fig. 4 is not the only one which could have been adopted, but is intended to show the sort of calculations and layout of the results which have been found useful.

The phase number at the top left-hand corner of the board indicates that two previous phases of evolutionary operation have already been completed on this process. In general these might have involved other variables or the same variables at other levels. In order that the new results may be considered in proper relation to those obtained previously, the final average values recorded in previous phases should also be available (for example on sheets of paper pinned to the board). The cycle number in the top right-hand corner indicates that 16 cycles of this third phase of operation have been completed. There follows a plan of the cycle of variants being run.

The table below this summarises the current situation. First are shown the requirements which it is desired to satisfy. These are followed by the running (i.e. up-to-date) averages at the various manufacturing conditions set out so as to follow the plan of the cycle of variants. This arrangement makes it easy to appreciate the general implications of the results. A measure of the reliability of the individual running averages is supplied by the '95% error limits'. These are simply the quantities

126

$\pm ts/\sqrt{n}$ appropriate to the calculation of fixed sample-size confidence limits for individual means. In this expression s is the standard deviation, n the number of cycles, and t the appropriate significance point of Student's t distribution. The scheme can be run until these limits for the means of the principal response have been reduced to an acceptably small preassigned width.[8]

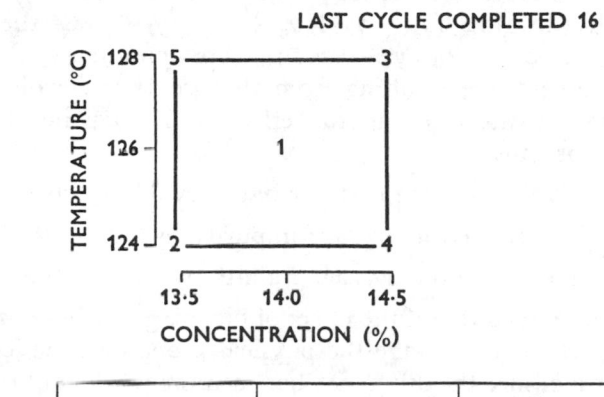

PHASE 3   LAST CYCLE COMPLETED 16

		Cost	Impurity (%)	Fluidity
Requirement		Minimum	Less than 0·50	Between 55 and 80
Running Averages		32·6 33·9 32·8 32·3 33·4	0·29 0·35 0·27 0·17 0·19	73·2 76·2 71·3 60·2 67·6
95% Error Limits		±0·7	±0·03	±1·1
Effects with 95% Error Limits	Conc.	1·2 ± 0·7	0·04 ± 0·03	5·2 ± 1·1
	Temp.	0·4 ± 0·7	0·14 ± 0·03	10·8 ± 1·1
	C × T	0·1 ± 0·7	0·02 ± 0·03	−2·2 ± 1·1
	Change in Mean	0·2 ± 0·6	−0·02 ± 0·03	−1·6 ± 1·0
Standard Deviation 95% Error Limits		1·44 1·22 1·76	0·059 0·050 0·072	2·12 1·80 2·59
Prior Estimate		2·71	0·054	3·22

FIG. 4. Appearance of information board at the end of cycle 16.

Below this are shown the 'effects' of the variables and their 95% error limits. The concentration and temperature effects are each calculated in the usual way as the difference in the average values of the response at the higher and the lower level of the variable. The value at the centre conditions does not enter the calculations except in computing the 'change in mean' effect. If y_1, y_2, y_3, y_4, and y_5 are the running averages of one of the responses after n cycles of operation, the effects and their limits of error for this particular example are as follows:

Effect	Value	Limits of Error
Concentration ..	$(y_3 + y_4 - y_2 - y_5)/2$	$\pm ts/\sqrt{n}$
Temperature ..	$(y_3 + y_5 - y_2 - y_4)/2$	$\pm ts/\sqrt{n}$
Interaction ..	$(y_2 + y_3 - y_4 - y_5)/2$	$\pm ts/\sqrt{n}$
Change in mean ..	$(y_2 + y_3 + y_4 + y_5 - 4y_1)/5$	$\pm 2ts/\sqrt{(5n)}$

The effect referred to as the 'change in mean' is simply the grand mean for all the runs $(y_1 + y_2 + y_3 + y_4 + y_5)/5$ less the mean for the 'standard' conditions y_1. It is therefore an estimate of the difference in average response resulting from the use of the evolutionary scheme. In the present example the effect of introducing the evolutionary scheme is thus:

(i) To increase the cost per batch by $0 \cdot 2 \pm 0 \cdot 6$ units.

(ii) To reduce the average impurity by $0 \cdot 02 \pm 0 \cdot 03\%$.

(iii) To reduce the average fluidity by $1 \cdot 6 \pm 1 \cdot 0$ units.

(It will be seen that if the effect of blending batches of slightly different qualities was to average the physical properties, the 'change in mean' would measure the difference between the product of the works process and a blend of the products from evolutionary operation. In many actual examples partial or complete blending of the products does naturally occur at the later stages of manufacture, so that where physical properties behave approximately additively there may in fact be remarkably little overall change in the manufactured product due to evolutionary operation. In some cases, especially if the effect of introducing small variations in the levels of the variables was unexpectedly large, blending could be deliberately introduced to produce an acceptable product.)

If the variants cover a region of the response surface which is a sloping plane, then it is not difficult to see that the true 'change in mean' will be zero. For a convex surface such as that near a maximum it will be negative, whereas for a concave surface such as that near a minimum it will be positive. It can be shown that, on certain plausible assumptions, the 'change in mean' is proportional to the sum of the quadratic constants which measure curvature in the directions of the variables. When the interaction effect and the change in mean effect are not small compared with the single effect of the variables, this indicates that a maximum or minimum is being approached and for exact location and exploration a technique fully set out elsewhere[4-6] is adopted.

For the cost response the change in mean supplies a continuous measure of the cost of obtaining information by the process of evolutionary operation. In the present example this is estimated at $0 \cdot 2 \pm 0 \cdot 6$ units of cost per batch. If the cost surface were locally planar, evolutionary operation would cost nothing. In practice, concavity of the cost surface is to be expected, since a minimum is usually being approached, so that there will usually be a small cost associated with running the evolutionary process. Except when the process has been brought very close to its ultimate optimum this cost will be redeemed many times over by the value of the permanent process improvements that occur from time to time as a result of the information generated by the evolutionary scheme.

After the calculated effects, the experimental error standard deviations calculated from the observations themselves are shown. Except in the initial stages of the scheme these are used in computing the limits of error for the running means and for the 'effects' of the variables. The normal theory fixed sample-size 95% confidence limits for these estimates of the standard deviations are also shown. The final items are the estimates obtained from prior data used in initiating the scheme.

In general, by inspecting the results set out on the information board in the light of the requirements it is desired to satisfy and from expert knowledge of other factors which affect plant operation, the plant manager decides at any particular stage whether

(a) to wait for further information;

(b) to modify operation.

Under (b) some of the alternatives open to him are:

 (i) To adopt one of the variants as the new 'works process' and to recommence the cycle about this new centre point.

 (ii) To explore an indicated favourable direction of advance and recommence the cycle around the best conditions found. (This exploration may be done, for example, by making a series of tentative advances in the indicated direction, at each stage running the new conditions and the previous best conditions alternately, until sufficient evidence has been gathered.)

(iii) To change the pattern of variants to one in which the levels are more widely spaced.

(iv) To substitute new variables for one or more of the old variables.

In the example actually discussed it is seen that a decrease in concentration would be expected to result in reduced cost, reduced impurity, and reduced fluidity. The effect on cost of a decrease in temperature is uncertain but is more likely to be favourable than not, and would almost certainly result in marked reductions in impurity and fluidity. These facts have to be considered, bearing in mind that a fluidity of less than 55 is undesirable and that although a further large reduction in impurity is welcome it is not necessary in order to meet the specification. It was decided in the event to explore the effects of reducing concentration alone. Phase 3 was terminated, and in the next phase the three processes (13%, 126°), (13·5%, 126°), (14%, 126°) were compared. The first of these gave a mean cost of 32·1 with an impurity of 0·25 and a fluidity of 60·7, and was adopted as a base for further development.

A geometrical display of the results on the information board may also be used when three variables are jointly considered. In this case perspective drawings of the sort illustrated in Fig. 5 are used. The

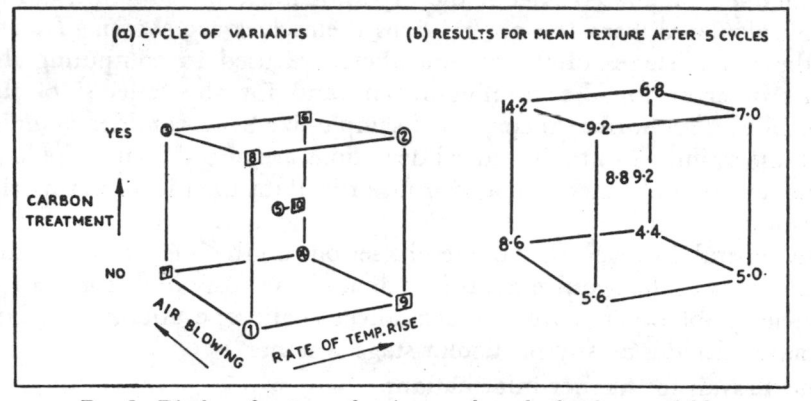

FIG. 5. Display of pattern of variants and results for three variables.

cycle of variants for a three-variable scheme is shown in Fig. 5(a), while 5(b) shows the results from one of the responses after five cycles of operation. This method may, of course, be applied to responses which cannot be measured directly. In the case illustrated in Fig. 5(b) an important property of the product was a somewhat esoteric quality referred to as 'texture'. A set of artificial standard samples were prepared which were judged by experts to have a range of 'textures' in approximately uniform steps, and these were arbitrarily scored. The texture of a sample of each manufactured material was then matched against the standards and an appropriate score given to it. In a similar way a scheme to evolve conditions which would give a product less inclined to 'cake' has been run, in which caking was judged by visual inspection and scored by comparison with a verbally defined scale.

Selection of the Variants

The technique outlined differs from the natural evolutionary process in one vital respect. In nature the variants occur spontaneously, but in our artificial evolutionary process we have to introduce them. Variants involving the levels of temperature, concentration, pressure, etc., are natural choices, but there are usually an almost unlimited number of less obvious ways in which manufacturing procedure can be tentatively modified. Frequent instances of marked improvement due to some innovation never previously considered in a process which has been running for many years testify to the existence of valuable modifications waiting to be thought of.

To make our artificial evolutionary process really effective, therefore, one more circumstance is needed—we must set up a situation in which useful ideas are continually forthcoming. An atmosphere for the generation of such ideas is perhaps best induced by bringing together at suitable intervals a group of people with special, but different, technical backgrounds. In addition to plant personnel themselves, obvious candidates for such a group are, for example, a research man with an intimate knowledge of the chemistry of the similar reactions and a chemical engineer with special knowledge of the type of plant in question. The intention should be to have complementary rather than common disciplines represented.

These people should form the nucleus of a small evolutionary operation committee, meeting perhaps once a month, whose duty it is to help and advise the plant manager in the performance of evolutionary operation. The major task of such a group is to discuss the implications of current results and make suggestions for future phases of operation. Their deliberations will frequently lead to the formulation of theories which in turn suggest new modifications that can be tried with profit.

Since questions of modification of certain physical properties of the manufactured product may arise, a representative of the department responsible for the quality of manufacture should also be on the evolutionary operation committee. Rather more may be got from the results and more ambitious techniques adopted if a statistician is also present at the meetings.

With the establishment of this committee all the requirements for an efficient evolutionary method of production are satisfied and the 'closed loop' illustrated in Fig. 6 is obtained. We are thus provided with a practical method of 'automatic optimisation' which requires no special equipment and which can be applied to almost any manufacturing process, whether the plant concerned is simple or elaborate.

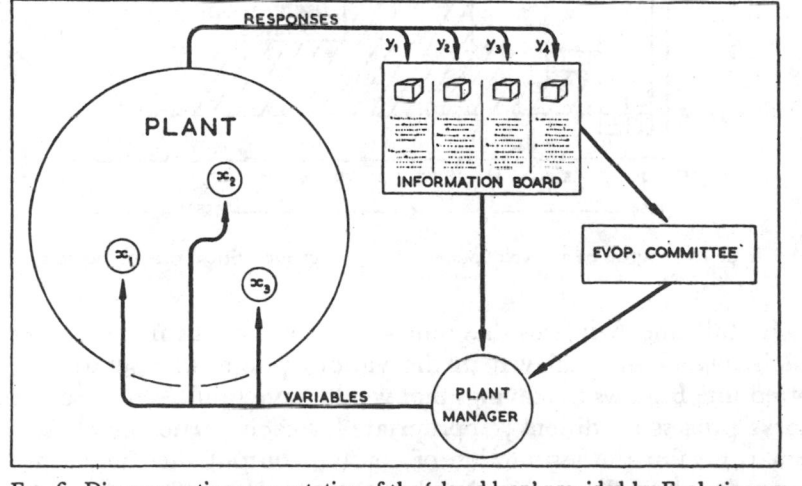

FIG. 6. Diagrammatic representation of the 'closed loop' provided by Evolutionary Operation.

At the beginning of this article I spoke of evolutionary operation as being run largely by plant personnel themselves rather than by specialists. The use of some specialists as advisers on the evolutionary operation committee does not seriously vitiate this principle. In practice, the time spent by the specialists is perhaps one afternoon a month, and the ultimate responsibility for running the scheme still rests with the plant manager and not with the specialists.

When not to Stop

With an alert team of workers new ideas should be continually forthcoming and the evolutionary method becomes virtually a permanent mode of operation and should be so regarded. Only if it seemed that more would be lost than gained from the evolutionary procedure would the reintroduction of static operation be justified. In practice it is found that even very small gains will justify the continual operation of the evolutionary method. The situation at any given time can be appraised by the use of a pictorial log like that shown in Fig. 7.

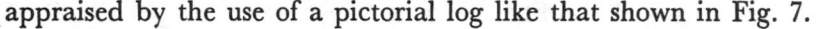

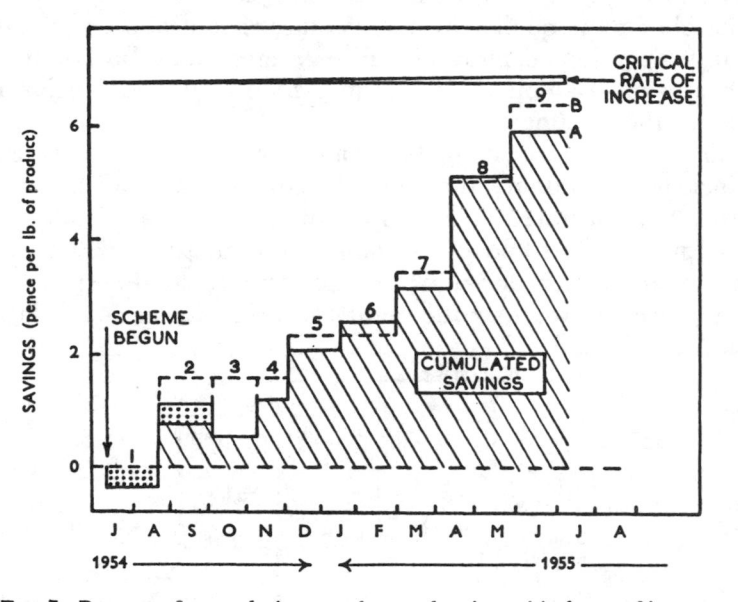

FIG. 7. Progress of an evolutionary scheme, showing critical rate of increase.

The full line A in this diagram shows the savings in pence per lb. which have been achieved in the various phases of operation. The dotted line B shows the savings that would have resulted if the centre or 'works process conditions', appropriate for each particular phase, had been run. On the assumption of constant output the shaded area is proportional to the accumulated savings resulting from the scheme, while each of the rectangular areas between the dotted and full lines shows the accumulated 'expenses' of running the scheme during that phase. The speckled area in phase 1 represents the cumulated expenses for the scheme during this phase and is debited from the cumulated savings in phase 2.

Whereas each phase of evolutionary operation, and consequently the expenses associated with it, lasts for only a limited time, any improvements which result go on for as long as the process is used. Suppose it is assumed that process improvements will go on earning money for p years after they are discovered and that the running of the evolutionary scheme adds c pence per lb. to the cost of the product. Then the question of whether or not, at any instant of time, evolutionary operation should be continued may be resolved by comparing the rate of improvement r which it is expected may be produced by the evolutionary process (measured in pence per lb. per year) with the critical rate of improvement r_0 given by $r_0 = c/p$. For if it is expected that the evolutionary scheme will need to be run for time t years to produce an improvement at the end of that period of rt pence per lb., and if k lb. of product is made per year, then the total saving during the p years for which the discovery is used will be $rtkp$ pence. During this time, kt lb. of product will be made and the loss due to running the evolutionary scheme will be ckt pence. Thus the scheme will pay off if $rtkp$ is greater than ckt, that is if r is greater than $r_0 = c/p$.

As an example consider the situation in Fig. 7. Should the scheme there shown be continued or not? Let us suppose (very conservatively) that improvements on this process are expected to go on earning money for 3 years after their discovery, so that p is put equal to 3. Suppose also that c is taken to be the average of the values experienced in the 9 previous phases. This gives the value $c = 0\cdot3$ pence per lb. We then find for the critical rate $r_0 = 0\cdot3/3 = 0\cdot1$ pence per lb. per year. Thus so long as the rate of improvement due to the evolutionary process is expected to be at least as great as $0\cdot1$ pence per lb. per year the evolutionary scheme should be continued. This critical rate of increase is shown diagrammatically at the top of Fig. 7. It will be seen that the actual rate of improvement which had been experienced over the previous year was about 6 pence per lb. per year (about 60 times the critical rate), and there is no evidence as yet of any flagging in this rate of improvement. There is therefore no doubt whatever that this scheme should be continued.

The example given is by no means atypical, which explains my insistence that the evolutionary method should be regarded as virtually a permanent mode of operation. It is psychologically wrong to talk of production under such a scheme as 'experimental manufacture', since an experiment is something which is done for a limited period and is not part of the normal run of things.

Some Questions and Tentative Answers

Like statistical quality control, evolutionary operation is designed for application in the factory itself. Its aims are different and more ambitious than those of quality control, however, since it is directed to ensure not a more uniform product but a cheaper and better product. I believe that this technique, if applied sufficiently widely, can have a marked affect in achieving greater productivity in industry by ensuring that the plant that is available, whether old or new, is operated in the best possible manner. The outline above is intentionally general because it is the general attitude and philosophy that is important here, and not the particular details of application. A full account is nevertheless being prepared in which a number of technical questions are discussed. In the present paper I shall do no more than indicate some of these questions and some tentative replies, which it is hoped to amplify and to justify in the later discussion.

Q. How many variables should be included at one time in an evolutionary scheme?

A. Usually two or three variables can be handled satisfactorily under the normal production conditions with which I am familiar. It should perhaps be emphasised once again that what is being discussed is the normal production situation in which evolutionary operation is applied. For specialist short-period investigations (which, as has been explained in the introduction, are *not* the subject of this account but which nevertheless play an extremely important part in the general scheme of process development to which evolutionary operation is a supplement) the situation is entirely different. In these specialist investigations where, *for a limited period*, it is permissible to interfere with production, and where special supervision and other facilities are made available, the object should be to saturate (or following Satterthwaite[9] possibly 'super-saturate') the experiment with as many factors as possible.

By studying the variables in groups of three or so at a time we forgo the possibility of detecting dependence (interaction) between variables not in the same group. The effect of this limitation should be minimised as much as possible by examining in the same group sets of variables which are expected to be interrelated. Periodically variables whose effects have been found to be important in different phases of operation should be tested together.

Q. What patterns of variants are of most value?

A. A variety of patterns of variants are useful for particular purposes. Among the most valuable for initial use are those based on two-level factorial designs with one or more added points at the centre conditions. These are the arrangements shown for two and three variables in Figs. 3 and 5. They have the advantages that:

(i) They are simple to comprehend, perform and analyse.

(ii) The added centre points allow continual reference to the 'standard' process and permit the 'cost' of the evolutionary scheme to be assessed.

(iii) Complexity of the surface is easily detected by considering the relative magnitudes of the simple 'main effects' on the one hand and the 'change in mean' and interaction effects on the other.

(iv) They can be made the nucleus of more elaborate designs (in particular of composite second-order rotatable designs[10]) by which complexity of the surface may be elucidated.

(v) They lend themselves conveniently to 'blocking arrangements' whereby extraneous disturbances due to such uncontrolled factors as time trends may be reduced. In Fig. 5(a), for example, the circles and squares indicate two sub-cycles into which the complete cycle may be divided. A general change in mean occurring between sub-cycles will not bias the estimation of effects.

Q. How should past plant-records be used in planning the evolutionary scheme?

A. When, as is often the case, past plant-records covering long periods of normal operation are available, the planning of an evolutionary scheme should always begin with a careful study of these records. In particular they may be used to determine the approximate magnitude and nature of the uncontrolled variation in the various responses, and consequently the number of repetitions of the cycle likely to be needed to detect effects of a given size.

From these records we can find out whether the errors in the principal response can be regarded as effectively independent and, if not, we can determine the nature of the dependence that exists. This is of some importance in choosing the period for which each variant should be run before changing to the next variant. In practice, of course, this period depends partly on convenience of operation and, for continuous processes, on the time it takes for the plant to settle down after a change in reaction conditions. The time for running each variant which gives the maximum amount of information for a given total period of production can be shown to depend on the nature of the dependence between the observations, and may be determined by a fairly simple use of the correlogram along the lines considered by Jowett[11] or, equivalently, by considering the spectral properties of the record.

Q. Should the variants be run in random order?

A. Faced with the possibility of serial correlation between successive observations, the statistician would normally wish to perform the variants in random order within each cycle, thus guaranteeing the validity of the simple type of analysis used in the example of Fig. 4. However, in some cases, particularly where the time for running each cycle is short, it is much simpler to run a systematic routine of variants on the plant than a random one. In these circumstances randomisation is usually abandoned (as was in fact done in the example considered). The observations after n cycles of k variants can be written in a table having n rows and k columns, and we are only concerned with comparison of *column* means. Now the major part of the dependence occurs *within* rows, and in this situation, as was pointed out by R. A. Fisher,[14,15] the simple analysis of the type we have considered above will usually not be seriously invalidated. Where the correlogram is available it is possible to determine how far dependence between observations will affect the simple analysis and what remedial measures, if any, need be taken, although this refinement would seldom be worth while.

Q. How should multiple responses be considered?

A. Although it is theoretically possible to equate all responses to a single criterion such as profitability, this usually presents great practical difficulties. As a general rule it is best to represent the problem as one of improving a *principal* response (for example the cost per lb. of product) subject to satisfying certain conditions on a number of *auxiliary* responses. These auxiliary responses usually measure the quality and important physical properties of the product.

Very careful thought in the selection of the principal response is essential. The vital question to ask is: 'If this response is improved will it mean *necessarily* that the process is improved?'

In the example of Fig. 4 the reconciliation of the requirements for the various responses in the light of the experimental results was done intuitively and led to the decision to reduce concentration alone. This intuitive approach has the virtue of simplicity and allows background information not emanating from the experimental results themselves to be taken into account. It is fairly satisfactory in the situation specifically dealt with here when there are only two or three variables to consider. However, as has been pointed out elsewhere,[12] the problem is really one of programming, in the sense of linear programming, with the added complications that the problems are not always approximately linear and that the restraints are not known exactly but must be estimated. In the fuller account we show how certain calculations can help with the more difficult cases.

Q. How best can the stream of information coming from the plant during the evolutionary process be presented to those responsible for deciding what to do?

A. Two things are necessary: first, to show how much weight ought to be attached to the results, and second, to present them in such a way that their interpretation is facilitated as much as possible.

To convey a sense of the degree of reliability which the plant manager should associate with the results, a number of ideas have been tried. In the original schemes various types of sequential charts and significance tests were used. It is now felt, however, that the problem is not one of significance testing and that what is needed is a presentation of the information contained in the data unweighted by external features subsequently injected into the situation. For example, the particular choice of the risks α and β and of the hypotheses 'which it is desired to test' (subtleties not readily comprehended by plant managers) can completely alter the apparent implications of a set of data when these are plotted on a sequential chart. If the observations are roughly normally distributed, are independent, and have constant variance, then all the information they contain is included in the mean, the standard deviation, and the number of observations. These statistics seem best comprehended in the form of a mean with its 95% confidence limits. In appraising the results prior information about the importance of different sorts of effect must be used. It seems best, however, to separate this from the presentation of the results, which then refer to information supplied by the observations and to nothing else. This problem is regarded as being far from solved. It involves many intricacies which cannot be discussed in the present account and is probably best considered in terms of stochastic approximation[16] and servo-mechanism theory. All that would be claimed for the present method is that it does allow a satisfactory evolutionary process to go on.

To show the implications of the mean results once it becomes apparent that these are determined sufficiently accurately, there is no doubt that for two or three variables geometrical representation, such as that shown in Figs. 4 and 5, is ideal. It allows the general trend in the responses and their relationship to each other to be appreciated in a manner not possible in any other way.

The plant manager should run evolutionary operation in much the same way as he would play a card game. The information board shows him his 'hand' at any given time, and depending on that hand there are a number of actions he can take (including drawing a further card and deciding what to do then).

Where one or more of the variants is clearly better than the works process or where clear-cut trends in the results exist, the plant manager will have no difficulty in following the indications of the information board. Where the results indicate that complexity exists he will be able to obtain the help of the statistician on the evolutionary operation committee in elucidating the results and, where necessary, in augmenting and modifying[4-6] the cycle of variants in the next phase of operation so as to resolve the complexity.

A duplicate information board may be kept on the plant itself and its significance explained to process workers. This provides added interest and is an incentive to accurate operation, which itself can result in general improvement in productivity.

Q. In what way can the results from small-scale experimental studies be used in planning the evolutionary scheme, and how should this affect the way in which these small studies are conducted?

A. The complexity mentioned above arises principally because the variables studied fail to behave independently in their effects on the response; that is, they interact. The plant process will usually have been arrived at as the result of a small-scale investigation of at least some of the variables. This investigation should have culminated in a study of the local 'geography' of the response surfaces in the neighbourhood of the proposed operating conditions. The principal features of the laboratory response surfaces will normally be preserved on the plant scale even though some distortion occurs. If the characteristics of the laboratory surfaces have been determined in the manner mentioned above, it is frequently possible to discover transforms of the variables originally considered which act approximately independently, at least for the principal response. By working in terms of these new variables, difficulties due to complexity of the surface can be greatly reduced in the plant-scale investigation.

Q. In practice it is impossible to attain truly static operation. Small variations in the process conditions are bound to occur from one run to the next. In cases where these changes are recorded, why should one bother to carry out a special pattern of variants? Why not use the 'pattern of variants' supplied by the natural variation of the process to supply information on which evolutionary improvement can be based?

A. It is true, of course, that for data generated by natural variation the simple type of analysis of the results which has been used above would no longer be applicable. However, this itself is no reason why the natural pattern of variants should not be used. Suppose that the level of response is denoted by y and that there are k variables whose levels are denoted by $x_1, x_2, \ldots, x_k$; suppose that the works process is defined by the particular set of conditions $x_{10}, x_{20}, \ldots, x_{k0}$ and that, owing to imperfect control, fluctuations about those levels occur and are recorded. Then we can, for example, assume a local relationship of the form

$$y = b_0 + b_1 x_1 + b_2 x_2 + \ldots + b_k x_k$$

and estimate the coefficients continuously by the method of least squares (multiple regression). If our assumptions were correct, these coefficients would measure the individual effects of the variables. The calculations required to fit the equations are laborious but, as has been suggested by Professor Goodman,[13] could in principle be done mechanically (e.g. by an electronic computer).

At first sight such a method appears attractive, for here we seem to have an evolutionary scheme in which we do not need to bother about introducing variants deliberately. On closer examination, however, its value seems much more doubtful. Many investigations have been made by statisticians over the years in which plant records have been analysed by multiple regression in an attempt to determine the 'effects' of the variables and so to improve the process. In my experience the results of such investigations are nearly always disappointing. The reasons are not far to seek:

1. Many of the factors that may vitally affect the efficiency of the process are not in the normal course of events altered at all.

2. Those factors which vary naturally do so, not over the ranges we should like, but over ranges dictated by the degree of control which happens to exist. The more control is improved, the less information we get.

3. The fluctuations that naturally occur in the variables are often heavily correlated. This results in poor precision of the estimates when we try to disentangle the effects of the variables one from the other.

4. Accidental modifications often tend to happen in 'phases' and so become spuriously correlated with causally unrelated time-trends in response. Such effects can lead to completely wrong conclusions. Attempts to eliminate time-trends computationally usually eliminate the effects of the factors at the same time.

What this all amounts to is that a naturally occurring scheme of variants is not very likely to provide a good, or even passable, 'design' and consequently that the amount of information generated by natural variation may be scarcely worth salvaging.

Q. Can evolutionary operation be made automatic?

A. The procedure of evolutionary operation so far described is a 'manual' one. It requires no special facilities and can be immediately applied in one form or another to a very large proportion of industrial processes. This is so whether the available plant is of the crudest kind or whether it includes such refinements as automatic controllers and recorders. The plant manager is himself a part of the 'closed loop', thus ensuring that sensible action will be taken even in unforeseen circumstances.

With a sufficiently instrumented plant the evolutionary procedure is, of course, capable of being made completely automatic. Thus variables whose levels are regulated by a controller can be automatically changed at regular intervals so as to follow a cycle of variants, and a response such as cost per lb. can be automatically computed from the readings of instruments which measure the properties of the product. The cumulated differences in response at the various process conditions can be used to trigger off adjustments in the location of the pattern of variants, so completing the evolutionary process.

In continuous processes (where there is a continuous input of starting materials and a continuous output of product) the 'pattern of variants', instead of being a discrete set of points as in Fig. 3, can consist of a continuous locus. The problem of detecting the effects of the variables is then precisely that which arises in communication theory, of detecting a signal of known form in a noisy channel.

The introduction of automatic evolutionary operation would usually be worth while only if the response surface itself was changing in some way and it was desirable to attempt to follow that change. For many chemical processes the response surfaces are reasonably stable. In some, however, unpredictable but steady changes can occur owing to slow changes in raw material (such as crude oil) or in catalyst activity. Here unpredictable differences in the position of the optimum conditions may occur between batches of catalyst and also within the life of a single catalyst batch. In these cases automatic evolutionary operation may be effective in keeping the plant operating near its best performance, but only if the rate at which information is generated is sufficiently large compared with the rate at which the optimum conditions are changing. This is essentially a problem in the theory of servo-mechanisms.

Discussion

The device I have described is of course only a more powerful and concentrated form of the naturally occurring evolutionary process which goes on during all manufacture. In the ordinary course of events once the favourable effect of a deliberate or accidental modification is *recognised*, that modification is included in the works process. Unfortunately, because of a high level of variation, which usually obscures all but very large effects, favourable modifications frequently go unrecognised unless they are forced to reveal themselves by the device used here of constant repetition and consequent averaging-out of errors.

That many of the problems touched on in the later part of this paper are still the subject of active investigation should not obscure the fact that evolutionary operation, as set out in earlier sections, is a practical and immediately available method which ought to be more widely applied.

Both practical experience and theoretical consideration show that very little can be lost and a great deal can be gained by application of the technique, and for this reason evolutionary operation should be adopted as a *normal production method*. Static operation should be tolerated only if good reasons for not using the evolutionary procedure can be advanced.

REFERENCES

[1] YOUDEN, W. J. (1954 onwards). Bimonthly articles on 'Statistical Design', *Industrial and Engineering Chemistry*.

[2] DANIEL, C. and RIBLETT, E. W. (1954). 'A multifactor experiment', *Industrial and Engineering Chemistry*, **46**, 1465.

[3] VAURIO, V. W. and DANIEL, C. (1954). 'Evaluation of several sets of constants and several sources of variability', *Chemical Engineering Progress*, **50**, 81.

[4] BOX, G. E. P., CONNOR, L. R., COUSINS, W. R., DAVIES, O. L. (Editor), HIMSWORTH, F. R., and SILLITO, G. P. (1954). *The Design and Analysis of Industrial Experiments*. Oliver and Boyd, Edinburgh and London.

[5] BOX, G. E. P. and WILSON, K. B. (1951). 'On the experimental attainment of optimum conditions', *J. R. Statist. Soc.*, B, **13**, 1.

[6] BOX, G. E. P. (1954). 'The exploration and exploitation of response surfaces: some general considerations and examples', *Biometrics*, **10**, 16.

[7] BOX, G. E. P. and YOULE, P. V. (1955). 'The exploration and exploitation of response surfaces: an example of the link between the fitted surface and the basic mechanism of the system', *Biometrics*, **11**, 287.

[8] ANSCOMBE, F. J. (1954). 'Fixed-sample-size analysis of sequential observations', *Biometrics*, **10**, 89.

[9] SATTERTHWAITE, F. E. (1956). (Unpublished communication.)

[10] Box, G. E. P. and HUNTER, J. S. (1956). 'Multifactor experimental designs for exploring response surfaces', *Ann. Math. Statist.* (In the press.)

[11] JOWETT, G. H. (1955). 'The comparison of means of industrial time series', *Applied Statistics*, **4**, 32.

[12] Box, G. E. P. (1955). Discussion at Symposium on Linear Programming, *J. R. Statist. Soc.*, B, **17**, 198.

[13] WILKES, M. V. (1956). Discussion of paper 'Application of digital computers in the exploration of functional relationships' by G. E. P. Box and G. A. COUTIE at the Convention on Digital Computer Techniques. *Proceedings of the Institution of Electrical Engineers*, **103**, Part B, Supplement No. 1, 108.

[14] FISHER, R. A. (1941). *Statistical Methods for Research Workers*, 8th edition, p. 226. Oliver and Boyd, Edinburgh and London.

[15] Box, G. E. P. (1954). 'Some theorems on quadratic forms applied in the study of analysis of variance problems. II: Effects of inequality of variance and of correlation between errors in the two-way classification'. *Ann. Math. Statist.*, **25**, 484.

[16] ROBBINS, H. and MONRO, S. (1951). 'A stochastic approximation method', *Ann. Math. Statist.*, **22**, 400.

Chapter 5
Evolving Computer Programs

R. M. Friedberg (1958) "A Learning Machine: Part I," *IBM J. Research and Development,* Vol. 2:1, pp. 2–13.

R. M. Friedberg, B. Dunham, and J. H. North (1959) "A Learning Machine: Part II," *IBM J. Research and Development,* Vol. 3, pp. 282–287.

FRIEDBERG (1958) and Friedberg et al. (1959) were among the first efforts to evolve computer programs. Although the word "evolution" does not appear in either paper, the intent to simulate evolution was plainly in the minds of the researchers (North, 1997), and this was the common perspective of the work (Jackson, 1974, pp. 376–377). Moreover, a subsequent paper by Dunham and North (Dunham et al., 1963), the coauthors of Friedberg et al. (1959), modeled evolution explicitly as an optimization process (also see Dunham et al., 1959). The task adopted by Friedberg was to generate a set of machine language (i.e., binary encoded) instructions that could perform relatively simple calculations (e.g., add the numbers in two data locations).[1]

The speed of the computing machinery available in the late 1950s demanded several procedural shortcuts in order to test the devised methods. In particular, there was a hope that structurally similar programs could be grouped together in classes. Students of genetic algorithms may recognize the comments: "Methods that resemble one another must be associated in classes, and a record must be kept on each class. In this way, the success or failure of a method will be interpreted as a reflection not only on that method but all its classmates. Thus a large universe of methods may be sifted in a relatively small number of trials . . ." as being a forerunner of the notion of *intrinsic parallelism* proposed in Holland (1975, p. 71). Specifically, a *class* was defined to consist of all programs having a certain instruction at a certain location. A learning procedure was intended to compare the performance of two nonoverlapping classes of programs, essentially evaluating alternative instructions that might occur at a given location. Again, this is similar to the notion of *schemata* also offered later in Holland (1975, p. 68).

A credit assignment algorithm was invented to partition the influence of individual instructions. Friedberg (1958) recognized that the behavior resulting from executing each instruction was not independent of the other instructions, but the amount of time required to attempt credit assignment to pairs or higher-order combinations of instructions was prohibitive. A record was kept of the number of successes and failures for each instruction (by way of determining the success or failure of the program in which it resided). A separate program (the learner) operated on the success numbers associated with various instructions to attempt to include more "good" instructions rather than "bad" instructions. In retrospect, it is not clear that the assignment procedure utilized was sufficient for the chosen tasks. Moreover, specific details of the implementation resulted in a cycling effect in which similar unsuccessful programs were constructed at periodic intervals. Thus, although the credit assignment procedure was intended to increase the efficiency of the search, it may have unintentionally had the opposite effect.

Friedberg et al. (1959) assayed the efficiency of their procedures in comparison with versions of a completely random search. It was observed that in several comparisons, the random search outperformed previously suggested methods of learning to solve the given problems. In the end, the intrinsically parallel credit assignment method was abandoned in favor of partitioning the given problems into subproblems to be handled in a chosen order of difficulty. This resulted in searches that compared favorably to completely random generate-and-test, but the method exhibited a tendency to stagnate at programs that were incapable of solving all of the subproblems associated with an overall problem.

Friedberg (1958) reported the results as demonstrating "limited success," but Friedberg et al. (1959) suggested that "where we should go from here is not entirely clear." Thus the authors' opinion of the research was somewhat uncertain. In a widely read review, Minsky (1961) offered an unfavorable opinion of the potential for this method of program generation, calling it "a comparable failure. . . . The machine did learn to solve some extremely simple problems. But it took on the order of 1,000 times longer than pure chance would expect." It is somewhat difficult to determine from where this figure was derived. An initial set of experiments with an *unprimed* version of the procedure

---

[1]Friedberg (1997) remarked that earlier unpublished work by Marvin Minsky on variable logic networks served as his inspiration for the ideas presented in Friedberg (1958) and Friedberg et al. (1959).

required an average of 100,000 trials to solve a particular bit manipulation problem. When knowledge was applied in priming the initial program and the procedures were then compared to versions of random search, the best methods of the learning technique compared equally to the performance of the best random search methods. The average number of runs required per success for the random methods was on the order of 10,000, and for Minsky's description to be correct, it would need to be on the order of 100. Moreover, the procedures for learning to solve problems in subtasks showed demonstrable improvement over the random search methods.

Minsky (1961) also criticized this latter approach of partitioning the problem, although he wrote "in the last section of [Friedbcrg et al., 1959] we see some real success obtained by breaking the problem into parts and solving them sequentially." Criticism was directed toward the stagnation of this approach at what essentially were local optima in the space of possible programs. Minsky attributed this to the "mesa phenomena" (Minsky and Selfridge, 1961): "In changing just one instruction at a time the machine had not taken large enough steps in its search through the program space" (Minsky, 1961). Careful reexamination of the methods of Friedberg et al. (1959) indicate, however, that the procedure being criticized did not alter just one instruction at a time. Instead, potentially many (i.e., even all) instructions could be altered depending on which instructions had been involved in the successful completion of previous subtasks. Regardless, the stagnation effect was real and was not able to be overcome.

The research of Friedberg (1958) and Friedberg et al. (1959) may be viewed as an early precursor to the proposals of Holland (1975) involving genetic algorithms and classifier systems: the learning procedure operated on binary strings and attempted to apportion credit to individual instructions despite having their functionality depend on multiple-instruction interactions.

Two additional tangential insights can be gleaned from Friedberg's research: (1) the use of domain knowledge can accelerate an optimization procedure (in this case priming and resetting procedures were used to set up the programs for each run), and (2) randomly generated programs may have a tendency to generate irrelevant instructions, an effect that has been observed in genetic programming research (e.g., Kinnear, 1993). Dunham et al. (1963) followed with two additional

recognitions: (1) suspending selection and accepting small changes that result in worse solutions can be useful for ameliorating local stagnation (see Chapter 14 on soft selection), and (2) it is sometimes efficient to vary the granularity of the evaluation function, such that rough estimates of a solution's quality can be obtained more rapidly. Dunham et al. (1963) reported success in evolving the wiring of a 1400-terminal black box built in four layers (the box was actually manufactured from the evolved designs and provided the inertial guidance system for a missile [North, 1997]) and offered,"We feel the 'natural selection' approach is more powerful than commonly recognized, and we hope to see it more widely exploited." Dunham and North pursued this line of research within IBM through the 1970s (Dunham et al., 1974ab) and 1980s up until their retirements in 1992 and 1991, respectively.

References

[1] B. Dunham, D. Fridshal, R. Fridshal, and J. H. North (1963) "Design by natural selection," *Synthese*, Vol. 15, pp. 254–259.

[2] B. Dunham, H. Lewitan, and J. H. North (1974a) "Simultaneous solution of multiple problems by natural selection," *IBM Techncial Disclosure Bulletin*, Vol. 17:7, pp. 2191–2192.

[3] B. Dunham, H. Lewitan, and J. H. North (1974b) "Introduction of artificial data to enable natural selection scoring mechanism to handle more complex cases," *IBM Techncial Disclosure Bulletin*, Vol. 17:7, p. 2193.

[4] B. Dunham, D. Middleton, J. H. North, J. A. Sliter, and J. W. Weltzien (1959) "The multipurpose bias device. Part II: The efficiency of logical elements," *IBM J. Research and Development*, Vol. 3:1, pp. 46–53.

[5] R. M. Fricdberg (1958) "A learning machine: Part I," *IBM J. Research and Development*, Vol. 2:1, pp. 2–13.

[6] R. M. Friedberg (1997) personal communication, Barnard Univ., NY.

[7] R. M. Friedberg, B. Dunham, and J. H. North (1959) "A learning machine: Part II," *IBM J. Research and Development*, Vol. 3, pp. 282–287.

[8] J. H. Holland (1975) *Adaptation in Natural and Artificial Systems*, Univ. of Michigan Press, Ann Arbor, MI.

[9] P. C. Jackson (1974) *Introduction to Artificial Intelligence*, Dover Publications (reprint), NY.

[10] K. E. Kinnear (1993) "Evolving a sort: Lessons in genetic programming," *Proc. of the 1993 Intern. Conf. on Neural Networks*, Vol. 2, IEEE Press, NY, pp. 881.

[11] M. L. Minsky (1961) "Steps toward artificial intelligence," *Proc. of the IRE*, Vol. 49:1, pp. 8–30.

[12] M. L. Minsky and O. G. Selfridge (1961) "Learning in random nets," *Proc. of the 4th Lond. Symp. on Information Theory*, C. Cherry (ed.), Butterworths, London.

[13] J. H. North (1997) personal communication, Yorktown Heights, NY.

R. M. Friedberg

A Learning Machine: Part I

Abstract: Machines would be more useful if they could learn to perform tasks for which they were not given precise methods. Difficulties that attend giving a machine this ability are discussed. It is proposed that the program of a stored-program computer be gradually improved by a learning procedure which tries many programs and chooses, from the instructions that may occupy a given location, the one most often associated with a successful result. An experimental test of this principle is described in detail. Preliminary results, which show limited success, are reported and interpreted. Further results and conclusions will appear in the second part of the paper.

Introduction

We are seldom satisfied to have assigned our more laborious tasks to machinery. We turn with impatience to whatever still occupies our time and ask whether ingenuity cannot bring it, too, into the domain of automation. Although modern electronic computers have relieved us of many tedious calculations, we are still faced with difficult tasks in which the slowness of our thoughts and the shortness of our memory limit us severely, but for which present machines are less adequate than we because they lack judgment. If we are ever to make a machine that will speak, understand or translate human languages, solve mathematical problems with imagination, practice a profession or direct an organization, either we must reduce these activities to a science so exact that we can tell a machine precisely how to go about doing them or we must develop a machine that can do things without being told precisely how. This paper explores the second possibility.

If a machine is not told *how* to do something, at least some indication must be given of *what* it is to do; otherwise we could not direct its efforts toward a particular problem. It is difficult to see a way of telling it *what* without telling it *how*, except by allowing it to try out procedures at random or according to some unintelligent system and informing it constantly whether or not it is doing what we wish. The machine might be designed to gravitate toward those procedures which most often elicit from us a favorable response. We could teach this machine to perform a task even though we could not describe a precise method for performing it, provided only that we understood the task well enough to be able to ascertain whether or not it had been done successfully.

Such a machine, it may be objected, would only follow precise orders just as present computers do. Even if it acted sometimes at random, we should have to give it a method for generating random numbers. We should have to give it a method for correlating its behavior with our responses and for adjusting its behavior accordingly. In short, although it might learn to perform a task without being told precisely how to perform it, it would still have to be told precisely how to learn.

This is true, but it does not lessen the desirability of such a learning machine. On the one hand, even the simplest feedback devices do things without being told exactly how. A thermostat is not told at what level to keep the furnace running, although it is told how to readjust the furnace if the room temperature is too high or too low. On the other hand, even the most powerful learning device known may well follow a precise program at a very elementary level. It appears that each neuron of the human brain follows laws of cause and effect, but the organization of the brain is so complex that a determinism is not manifest in its activities as a whole. Between the thermostat and the brain there may be no gulf in principle. Yet in practice there is a gulf so wide that bridging it would be an enormous achievement. When we look at the mechanism of a thermostat, we can see in detail how the thermostat does its job. When we examine the parts of the brain, we are at a loss to understand, from their properties, how the brain does what it does, except in a vague way. What we want, then, is to equip a machine with a learning procedure by which it can develop methods that cannot, at least, be deduced *trivially* from an examination of the learning mechanism.

Certain difficulties confront us immediately. If, as

suggested earlier. the machine is to try out methods and select the better ones, we must. present it *a priori* with a well-defined universe of methods from which it must choose those to be tried. If this universe is small, then the "inventiveness" of the machine is severely limited and the value of the methods that it develops depends more on our astuteness in choosing a universe containing good methods than on the ability of the learning procedure to pick the best methods from among those in the universe. For example, we might design a method that used several parameters and cause the learning procedure to vary the parameters until it found the most successful set of values. The universe of permissible methods would then consist of all methods combining the form we devised with arbitrary values of the parameters. While this might yield excellent results for some problems, for others we probably could not devise any general form which did not exclude some methods much superior to any it included. In order really to give the learner a "free hand," we should present it with a universe which, although well-defined, is so large and varied that we are not even acquainted with the forms of all the methods it contains. Of course we must expect that in any universe so uncensored the majority of methods are useless.

This raises another difficulty in turn. If the universe is very large, the learning procedure cannot practically try out each permissible method repeatedly in order to evaluate it. Methods that resemble one another must be associated in classes, and a record must be kept on each class. In this way the success or failure of a method will be interpreted as a reflection not only on that method but on all its classmates. Thus a large universe of methods may be sifted in a relatively small number of trials, provided that the criterion by which two methods are classed together is a good one. So again the effectiveness of the learner may be limited by the inadequacy of whatever principle we devise.

Our experience of stored-program computers suggests a scheme which may fit the requirements. Let the universe of methods consist of all programs that can possibly be written for a given computer. This universe is well-defined, yet presumably it excludes no conceivable method except by reason of the computer's size, and even for a small computer it includes a great many of the methods that ingenuity might discover, although senseless programs are naturally in the majority. Let a class consist of all programs having a certain instruction in a certain location. Thus each program is a member of as many different classes as there are locations, and the learning procedure, in comparing the performance of two non-overlapping classes of programs, really evaluates one instruction against another that might occupy the same location.

At first thought it seems that not much can be expected from this plan of classification. Surely, having individual instructions in common is only a superficial resemblance between programs. Programmers know all too well that two programs may have almost identical

form, differing only in one or two instructions, and yet have entirely different *intent*, one carrying out the programmer's wishes and the other producing "garbage." On the other hand, a very slight change in intent may require a drastic change in form, as when an instruction is inserted and a whole block of instructions must be displaced, so that no location in that block contains the same instruction as before.

Nevertheless, the scheme can be defended. Form and intent, to be sure, are related quite discontinuously in the compact, economical programs that programmers write, but a learning machine would probably develop much more inefficient programs in which many irrelevant instructions were scattered among the instructions that were essential to the intent. Among such programs, slight changes in form might well correspond to slight changes in intent, so that programs falling into the same classes tended to perform similar acts.

The versatility of the scheme is in its favor. In order to make the learner turn its attention from one problem to another, one need only change the criterion by which one informs it of success or failure. Moreover, we wish our machine not merely to learn to solve one isolated problem after another, but to develop an ability to handle whole classes of related problems. Programmers have found that certain sequences of instructions, or subroutines, occur again and again in many of their programs. It is as though any sensible program, no matter what its purpose, must rest on the same basic fabric of program organization. From this point of view, it is quite plausible that the learner, by including certain subroutines in a program, could improve greatly its chances of adjusting the rest of the program so as to perform a task successfully, regardless of just what task was assigned to it. The instructions composing such a subroutine ought to acquire good records, since the class of programs having these instructions in common would contain a particularly high concentration of successful programs. If the learner were to improve its general performance by attaching good records to the instructions composing a number of valuable subroutines, we might justifiably say that it had acquired not a mere *habit* of answering a certain problem correctly, but a general *ability* to do well on a large class of problems.

It is true that a subroutine usually consists of several instructions, and we propose here to evaluate only single instructions, since keeping a record even on all pairs of instructions would require enormously more time and storage. Perhaps the scheme could be improved by giving the learner a flexible way of reassigning its bookkeeping space. A special record might be kept for a pair of instructions if programs containing the pair did considerably better than programs containing either instruction without the other. On the other hand, the record on a single instruction might be dropped if programs containing it did neither better nor worse, on the average, than other programs. However, the scheme as it stands may well suffice for learning subroutines. Suppose that those programs which contain a special pair of instructions tend very often to be successful. Each member of this pair should enjoy from the outset a slight statistical advantage over its competitors, because among the (admittedly rare) programs tested that contain the other member of the pair, those that also contain the first member are more often successful than those that do not. This statistical advantage should cause the learning procedure more and more often to select programs for trial that contain one or both of the pair. The more often each member of the pair is used, the greater advantage does the other enjoy over its competitors. Eventually both members of the pair should have good records and be used often. The same process is conceivable for subroutines of arbitrary length.

Plausible though the foregoing arguments may sound to sympathetic ears, the critical mind notes that they depend more on far-fetched assumptions and less on demonstrable premises, the further they proceed. One may doubt seriously that a machine can really accomplish anything by trying out many programs and keeping a record in which each instruction is associated with the successes and failures of programs containing it. Supposing that this procedure did lead to some progress, one may ask whether even the simplest problem would not require the trial of an astronomical number of programs, especially if progress were to depend on the gradual influence of very small statistical differences. Therefore, an experiment was begun to test a learning procedure of this type. A hypothetical computer was designed for this purpose and called Herman, the letters of which stand for nothing in particular. Herman has a very simple logic such that every number of 14 bits is a meaningful instruction and every sequence of 64 instructions is a performable program. An outside agent called the Teacher causes Herman's program to be performed many times and examines Herman's memory each time to see whether a desired task has been performed successfully in that trial. The Teacher's announcements of success and failure enable a third element, the Learner, to evaluate the different instructions which, on different occasions, appear in Herman's program. Basing its acts on this evaluation, the Learner tries to include "good" instructions in the program rather than "bad" ones. The experiment is run by simulation of these three elements on the IBM 704 Electronic Data Processing Machine.

The remainder of Part I contains a description of the experiment and some early results. The experiment is unfinished at the time of the present writing. Part II will appear later with additional results and conclusions drawn from them.

Experimental methods

• Computer

Herman is a sequential stored-program computer with a program of 64 instructions in locations numbered I_0 to I_{63}. During the running of this program, the instruc-tions are not modified, but they may be modified between runs by the learning procedure. The program itself acts upon the data in 64 locations numbered D_0 to D_{63}. Each data location D_n contains one bit. Each instruction location I_n contains a 14-bit instruction. When an instruction is executed, its first two bits are interpreted as an operation code; its next six bits, which form a number a, are interpreted either as a data address D_a or as an instruction address I_a, depending on the operation code; and its last six bits are also interpreted either as a data address D_b or as an instruction address I_b. The way in which the instruction is executed depends not only on the operation code and on the two numbers a and b, but also on the number n of the location I_n in which the instruction is stored. If

$$\overbrace{n}^{\text{6 bits}}$$ is the number of a location I_n containing

$$\underbrace{\text{"bits 1-2}}_{op}, \underbrace{\text{bits 3-8}}_{a}, \underbrace{\text{bits 9-14"}}_{b},$$

then the instruction in I_n is executed as follows:

If $op = 0$, take the next instruction
   from I_a if D_n contains 0,
   from I_b if D_n contains 1.

If $op = 1$, put into D_n, 0 if either D_a or D_b contains 0,
                          1 if both D_a and D_b contain 1.
Take the next instruction from I_{n+1}.

If $op = 2$, put into D_b the bit that appears in D_a. Take the next instruction from I_{n+1}.

If $op = 3$, put into D_n and into $D_{a(I_b)}$ the complement of the bit that appears in D_a. (The number $a(I_b)$ is found in positions 3 to 8 of the instruction location I_b.) Take the next instruction from I_{n+1}.

The choice of these particular operations was partly arbitrary and partly based on thought. The operations are powerful enough so that any procedure can in principle be programmed. (Actually the finite size of this computer makes sufficiently complex problems unprogrammable, but it was anticipated that the 64 instructions allowed would be more than were needed to program any of the problems submitted to the computer in the course of the present limited project.) The operations are simple enough not to impose any plan of organization on the data handled by the computer in the way that the structure of the IBM 704, for example, naturally groups the bits in storage into 36-bit binary numbers.

The peculiar use of addresses deserves explanation. As suggested above, the validity of keeping a separate record of success and failure for each instruction may be questioned on the ground that an instruction might be particularly well suited to play a part in an organized program and might thus tend to acquire a good record *as long as a certain other instruction was in the program;* when this other instruction was removed from the program, the former instruction might cease to have any virtue, and its good record would be misleading. For example, if Instruction 31 were to place the result of a calculation in a data location x, and if Instruction 32, which followed it, were to use the datum in location x to perform a calculation, these two instructions would be related in a sensible way and might be expected to contribute to the chances of success of a program containing them. In a conventional machine this relationship could not be viewed as a property of either instruction alone, for it depends on their both using the same data location x. If Instruction 31 were changed, Instruction 32 would lose its virtue unless (unlikely occurrence) the new instruction happened also to place data in location x.

In Herman, Instruction 32 might be "3, 31, 33." If Instruction 31 has an *op* code of 1 or 3, it places the result of the operation in D_{31}, the very location from which a datum is taken by the instruction "3, 31, 33." This is true no matter what the address bits in Instruction 31 are. Similarly, if Instruction 33 has an *op* code of 1, 2, or 3, it takes a datum from location D_a, where a is the number appearing in bits 3 to 8 of Instruction 33. In this same location is placed the result of executing the instruction "3, 31, 33." This is also true no matter what the number a is, or what the number in bits 9 to 14 of Instruction 33 is. Therefore, if the instruction "3, 31, 33" appears in location I_{32} and Instructions 31 and 33 are varied at random, the probability is ½ that Instructions 31 and 32 are related in the "sensible" way described above and ¾ that Instructions 32 and 33 are so related. If the instruction "3, 31, 33" in I_{32} acquires a good record, it may be expected therefore to continue to justify this record even if Instructions 31 and 33 are altered frequently.

It was for this reason that D_n was used in *ops* 0, 1 and 3 and that $a(I_b)$ was used in *op* 3. These features were not believed to eliminate the dependence of the virtue of an instruction on the presence of another instruction in the program, but they were expected to reduce it. Objections can be made. There may be "sensible" relationships other than that discussed above, relationships which depend on more than one instruction in spite of the special features of Herman. Or, if a learning procedure such as the one envisioned achieves success, it may do so by means of a program which has no characteristics that we would consider "sensible." Nevertheless, certain of the results to be described indicate that the special address features of Herman may have contributed to such success as was achieved.

● *Operation*

Before each trial of Herman, the Teacher places bits chosen at random in certain of the data locations (the input locations). The contents of the remaining data locations are left as they are from the preceding trial. Herman is started at Instruction 0 (that is, the first instruction to be executed is taken from I_0). If Instruction 63 is executed and is not a transfer instruction, then (there being no Instruction 64) Herman's program is considered to have *finished*. If this happens, the Teacher examines the contents of certain of the data locations (the output locations) and decides whether the bits in these locations satisfy a certain relation with the bits placed in the input locations at the beginning of the trial. If the relation is satisfied, the Teacher notifies the Learner of a success; otherwise, of a failure. This notification is the only information that the Learner receives about what the Teacher is doing. Neither the Learner nor Herman's program is "told" which of the locations D_0 to D_{63} are input locations, which are output locations, or what the Teacher's criterion of success is. This is primarily because no way was seen of making any of this information useful to the program or to the Learner without imposing one's own preconceptions on the way in which Herman might attack a problem.

Because of the transfer instruction (*op* 0), it is quite possible for Herman's program either to finish by reaching Instruction 63 after executing only a few instructions or to run for a long time or forever without finishing. Hence an arbitrary upper bound is set on the length of time a program may run on one trial. If, after the length of time required to execute 64 instructions, the program has not finished, Herman is stopped and the Learner is notified of a failure. It was believed that the problems to be presented to Herman could easily be solved by programs that would finish in considerably fewer than 64 instructions.

The choice of a subset of the 64 data locations to serve as input locations, the choice of a subset to serve as output locations, and the choice of a criterion by which the Teacher judges between success and failure together determine a single problem. It was intended that a single problem be presented to Herman for many (e.g., 50,000) successive trials, so that the input locations, the output locations, and the criterion of success would be fixed, while the bits placed in the input locations would be chosen anew at random before each trial. For example, the first problem that was given to Herman, called Problem 1, has D_0 as the only input location, D_{63} as the only output location, and identity between the output bit and the input bit as the criterion of success. Before each trial in which this problem is presented, the Teacher generates a random bit (0 or 1), records it as the input bit, and places it in D_0. If the program finishes in the time permitted, the Teacher examines the bit in D_{63} and notifies the Learner of success or failure according as this bit is the same as the input bit or not. After this has been done for many trials, the Learner should have evolved a program for Herman which will reproduce in D_{63} the bit presented to it in D_0, if not infallibly, at least in a large fraction of the trials.

● *Learning Procedure*

There are 2^{14} different instructions that could possibly occupy a single location I_n. It would be impractical to keep a record on each of these. Instead, two instructions (chosen initially at random) are "on record" at any time for each location I_n, so that there are altogether 128 instructions on record. For each I_n, one of the two instruc-

148

tions on record is "active" and the other is "inactive." In any trial the program executed by Herman consists of the 64 active instructions. The Learner has two ways of altering the program. It frequently interchanges the two instructions on record for a single location, so that first one and then the other becomes the active instruction. This process may be called "routine change." Occasionally the Learner makes a "random change"; that is, it erases one of the 128 instructions from the record and replaces it with a new 14-bit number chosen at random. The routine changes enable the Learner to accumulate data on the relative success of the two instructions on record for each location and gradually to favor the more successful instruction. The random changes are made in order that the Learner not be restricted to the 2^{64} programs that can be made from the instructions on record at any one time.

Both the routine and the random changes are governed largely by a number associated with each instruction on record, called its "success number." The success number is supposed to indicate how well an instruction has served over many thousands of previous trials. Each time a success is reported, the success number of every active instruction is increased by 1. (If the program finished the successful trial before executing more than 32 instructions, the success numbers are increased by 2 instead of 1. This is done in order to encourage the development of programs that do not take a long time to finish, because it was anticipated that the success of the project might depend on the number of trials that could be simulated in the limited computer time available.) When a new instruction is placed on record by a random change, its success number is set initially to a constant S_i. When any success number becomes equal to or greater than a constant S_m, all 128 success numbers are scaled down, i.e., multiplied by a constant r less than 1. The original design used $S_m = 2^{15}$, $S_i = \frac{7}{8}S_m$, $r = \frac{63}{64}$. There are two reasons for scaling. One reason is to keep the success numbers at a roughly constant average size, so that they are comparable with S_i. The other reason is to diminish slowly the importance attached to the relative success that various instructions enjoyed a long time ago, compared with the importance of their more recent relative performance. For example, if one instruction achieves 64 more successes than another and thereby acquires a success number that exceeds the other's by 64, scaling will preserve the ratio between the two success numbers but will lessen the difference so that it can be made up by only 63 successes of the second instruction.

Each instruction location I_n has at any time a "state number" which plays a part in determining routine changes. Each time a failure is reported, a certain location I_n is subjected to "criticism." The absolute difference between its state number and the success number of its inactive instruction is taken as the new state number. If the old state number was less than the success number of the inactive instruction, the inactive instruction becomes active and the active instruction becomes inactive. Otherwise the two instructions are left as they are found.

Each instruction location in turn, $I_0, I_1, \ldots, I_{63}, I_0, I_1, \ldots$, is subjected to criticism after successive failures. The reason that all the locations are not subjected to criticism after each failure is partly to save computer time and partly to ensure, by making routine changes one at a time, that a large number of different programs will be tried out.

The effect of this method of criticism is that the ratio between the success numbers of the two instructions governs the relative frequency with which each instruction emerges as active. Thus, if the success number of one instruction is roughly twice that of the other, a routine change will ordinarily be made whenever a criticism finds the latter instruction in the active position, but then it will usually require two criticisms to dislodge the former instruction from the active position. But, since exactly 64 failures must occur between successive criticisms of a single instruction location, an instruction may remain active for many more trials than its rival even though its success number is lower, simply by being less often associated with a failure. Thus the frequency with which each instruction on record is active depends partly on how well it is performing currently and partly on its long-term record, represented by its success number. When a certain set of instructions has been found to be successful in one problem and the Teacher now commences to pose another problem, it is intended that the frequent failures of the established program to perform the new problem will induce the Learner to alter the program and to use most frequently the instructions that are most often successful at the new problem. At the same time the instructions that were successful at the old problem ought not to be "forgotten," but should (at least for some time) retain their high success numbers, so that if the old problem is presented again the "memory" of these instructions will aid the Learner to arrive at a successful program. It should be emphasized that a change of problem is not signaled explicitly to the Learner but makes itself felt solely through the report of success or failure after each trial. The ability of the Learner to associate an instruction with a highly favorable long-term record, even while that instruction is currently inactive because it does not serve well in the problem at hand, is felt to be essential to the retention of things once learned.

A random change is made after every 64th failure. The instruction to be replaced is chosen from among one of four groups: the active and the inactive instructions in odd-numbered locations, and the active and the inactive instructions in even-numbered locations. These four groups are considered in turn, one at each random change. Of the group to be considered, an instruction with the lowest success number is replaced by a random instruction, which is given the success number S_i. The reason for dividing the 128 instructions on record into four groups is that considerable computer time is required to find the lowest of 128 numbers. The purpose of timing random changes every so many failures is to make them infrequent when the program is doing well.

The random instruction is obtained from a multiplicative random-number generator. A 35-bit binary random number is multiplied by $23 \times 10^{10}+1$ and divided by 2^{35}. The remainder of this division, a 35-bit number, is taken as the new random number. The quotient yields a random instruction. If we call the lowest-order bit of the quotient bit 1, the *op* code is taken from bits 29 and 28, the first address is taken from bits 25 to 20, and the second address is taken from bits 7 to 2. The way of extracting an instruction was determined by the requirements of simulating Herman on the IBM 704. This random-number generator was chosen because it takes little time and storage and was known not to give zero for many more generations than the project would require. It was not considered necessary that the random instructions used should pass any particular sophisticated test of randomness. The starting random number was 10987654321 (decimal).

The parameters had to be adjusted by guesswork. Random changes should be made often enough so that a variety of programs is available to the Learner, but not so often that "good" instructions are erased from record before they can establish their superiority. The rate at which the importance of ancient successes is diminished by the scaling of success numbers may be estimated as $-\log r/(1-r)S_m$, and is therefore roughly independent of r [since $(1-r) \ll 1$] and inversely proportional to S_m. This rate should, perhaps, be made comparable to the rate at which the program is renewed by random changes. S_i should be lower than the success numbers of some instructions, so that an instruction may, by acquiring a high success number, preserve itself indefinitely from random changes, but not lower than the success numbers of all instructions, for then a new random instruction would almost surely be removed from record by the next random change before it had a chance to establish its worth. If r is too low, the mass of success numbers will undergo large fluctuations that disturb their relationship to S_i. If r is too high, the rounding error in scaling will distort the ratios of success numbers.

● *Simulation*

Herman, the Learner, and the Teacher are simulated together in the IBM 704. The program runs from 5,000 to 10,000 trials of Herman each minute, including the intervening acts of the Teacher and Learner. The actual execution of Herman's program is the most time-consuming part of each trial. The part of the program that simulates the Teacher is rewritten or altered from day to day so as to present different problems or introduce modifications into the Learner. At the end of each day's run the IBM 704 punches out binary cards representing the state of Herman and the Learner. At the start of a later run, these cards can be read in so that the run will continue as though the 704 had not stopped, with the same active and inactive instructions, success numbers, and state numbers as at the end of the previous run. If desired, the day's run may begin with randomly chosen instructions, success numbers, and state numbers. In the course of each day's run a printed record is produced which indicates whether a previous run is being continued, identifies that run, identifies the problem being presented and any modifications in Herman or the Learner, and lists the number of successes achieved by Herman in each block of 10,000 trials.

Results

At the time of the present writing, only a few preliminary results have been obtained. These do not present a complete or conclusive picture, but they do indicate roughly the capabilities and limitations of Herman, and they suggest avenues of further exploration. It is intended that a more exhaustive set of experiments will be performed and published as Part II of this paper.

Some of the experiments were begun with "random initialization"—that is, random values were assigned to the success numbers, the state numbers, and the instructions in Herman's program. Other experiments were begun with a "history"—that is, these numbers were all given the values they had had at the end of some previous experiment.

● *Experiment 1*

After random initialization, Herman was presented with "Problem 1." In this problem, D_0 is the input location, D_{63} is the output location, and the criterion of success is that the output bit should be identical to the input bit.

The number of successes obtained by Herman in each block of 10,000 trials is shown in Table 1. Herman's progress on the problem may be divided into three stages. In Stage 1 (the first 60,000 trials), the frequency of success climbed steadily from almost 0 to slightly less than $\frac{1}{2}$. Since even a random program may be expected to succeed in Problem 1 in 50% of the trials in which it finishes within the time limit, it is fairly certain that during Stage 1 the time limit was being exceeded in a large fraction of the trials. The fact that the frequency of success stopped rising rather abruptly just before it would have reached $\frac{1}{2}$ indicates strongly that the rise in Stage 1 represents a gradual elimination of time failures, and that at the end of Stage 1 Herman's program was finishing within the time limit in about 90% of the trials and obtaining successes in roughly half of the 90%. This conclusion is supported by the fact that the simulation of each 10,000 trials at the end of Stage 1 required only about half as much running time on the IBM 704 as at the beginning of Stage 1.

In Stage 2 (the next 90,000 trials), the frequency of success fluctuated around roughly 45%. Apparently Herman was making no progress toward achieving the desired relationship between the output bit in D_{63} and the input bit in D_0. In Stage 3 (the last 50,000 trials) Herman suddenly "hit the jackpot." Since the Learner changes Herman's program only after a failure, it is obvious that if Herman hits on a program that is certain of success that program will remain unchanged as long as the same problem is presented. The program used by Herman during Stage 3 is reproduced in Chart 1. A careful examination reveals that it is certain to succeed indefinitely at

Problem 1, no matter what sequence of input bits it is given.

Table 1 *Problem 1*

Block of 10,000 Trials	Number of Successes
1st	26
2nd	511
3rd	1,822
4th	3,057
5th	3,853
6th	4,648
7th	4,741
8th	4,601
9th	4,387
10th	4,623
11th	4,123
12th	2,488
13th	4,246
14th	4,554
15th	4,382
16th	10,000
17th	10,000
18th	10,000
19th	10,000
20th	10,000

It appears that the Learner accomplished nothing in Stage 2 except to cast about at random until it hit upon a perfect program. However, one may contend that during Stage 2 the Learner was improving Herman's program in a way which did not increase immediately the frequency of success but which gradually increased the probability that further modifications would result in a perfect program. This contention receives some support from Experiments 2 and 3.

• *Experiment 2*

Problem 1 was presented after random initialization. The same Learner was used as in Experiment 1, but Herman was replaced by a slightly different computer which we may call Sherman. The latter is exactly like Herman except for some modifications in the way an instruction is executed. When $op = 0$, the conditional transfer depends on the bit in D_a instead of on that in D_n. When $op = 1$, the result of the operation is placed in D_{b+1} instead of in D_n. (If $b = 63$, $b + 1$ is taken as 0.) When $op = 3$, the result of the operation is placed in D_b and in $D_{b\pm32}$ instead of in D_n and in $D_{a(I_b)}$.

Sherman is about as powerful a computer as Herman, but it lacks the two features—the use of the instruction location as a third address and the indirect address $a(I_b)$—which were intended to increase the likelihood that a meaningful performance record for a single instruction could be kept independently of other instructions in the program. Experiment 2 was designed to show whether these two features actually contribute to Herman's performance.

Sherman passed through Stage 1 and Stage 2 much as

Herman did. Stage 1 took about 70,000 trials. During Stage 2, Sherman seemed to achieve greater average success than did Herman. The most striking result of Experiment 2 is that Stage 3 never arrived. Although the experiment was run for 800,000 trials, Sherman never succeeded in more than half of any 20,000 successive trials, whereas Herman, in Experiment 1, acquired a perfect program in 150,000 trials. One might suppose that Herman succeeded because the random initialization at the beginning of Experiment 1 was carried out with a "fortunate" set of random numbers. This hypothesis was tested by the next experiment.

• *Experiment 3*

Experiment 1 was rerun nine times, and random initialization was carried out with a different set of random numbers before each rerun. If the advent of Stage 3 in Experiment 1 were due merely to a lucky choice of random numbers for initialization, Stage 3 would probably not occur in the reruns.

The course of a rerun was not always marked by a clear division between Stage 1 and Stage 2. Sometimes the initial rise in frequency of success leveled off below 40%. Sometimes Stage 2 was so short that it could not be distinguished from Stage 1. Sometimes the initial rise was irregular instead of being smooth, as in Experiment 1.

Stage 3 arrived in every rerun except the last, during which the time allotted to the experiment ran out after 220,000 trials. In the other 8 reruns, the number of trials required for Herman to acquire a perfect program varied from 30,000 to 210,000, averaging about 100,000.

The IBM 704 was instructed in this experiment to discontinue each rerun as soon as at least 8,000 successes were obtained in a block of 10,000 trials. Four of the reruns ended with a block of 10,000 straight successes. Four ended with a block of 10,000 trials of which more than 8,000 but fewer than 10,000 were successes. It seems a safe inference that in each of the latter four a perfect program was obtained during the first 4,000 trials of the block and that the next block would have consisted of 10,000 successes. The results of every one of the 8 reruns were consistent with the supposition that Herman continued to succeed in fewer than 50% of the trials until a perfect program was found.

These results show that Herman's discovery of a perfect program in Experiment 1 was not a lucky accident. There are two ways to explain Herman's superiority over Sherman in finding programs perfect for Problem 1. Herman may surpass Sherman either in the number of perfect programs possible or in the efficiency with which the Learner can progress toward them. The features by which Herman differs from Sherman were actually designed with the latter possibility in mind, but the former cannot be ruled out.

Several more complicated problems were presented to Herman.

• *Experiment 4*

Starting from the end of Experiment 1 (that is, setting

Chart 1   Program obtained in Experiment 1. Input datum from Location 0 transferred to Location 63.

N	D	OP	A	B	N	D	OP	A	B
0	Input	0	21	5	32	0	3	3	19
1	1	0	22	60	33	1	2	33	44
2	0	0	53	12	34	1	1	39	61
3	0	0	46	4	35	1	0	27	18
4	0	1	53	27	36	1	3	23	56
5	0	2	63	22	37	0	0	11	63
6	0	2	26	3	38	1	0	19	58
7	1	3	19	37	39	1	0	24	42
8	0	2	0	44	40	0	0	19	62
9	0	0	28	22	41	0	2	58	38
10	0	2	10	18	42	0	3	7	28
11	1	2	19	12	43	1	2	13	3
12	1	1	20	5	44	1	2	24	62
13	1	3	27	55	45	0	3	54	13
14	1	3	17	13	46	1	3	45	59
15	1	2	63	32	47	0	1	32	19
16	0	3	36	29	48	1	3	23	59
17	0	0	56	63	49	0	1	38	22
18	1	0	27	44	50	1	0	41	15
19	1	2	63	3	51	1	0	61	38
20	1	3	2	8	52	0	2	1	21
21	1	2	29	26	53	0	1	6	50
22	0	1	16	28	54	0	3	30	32
23	0	0	24	60	55	0	3	57	4
24	1	3	41	63	56	0	1	53	3
25	0	3	43	45	57	1	3	4	8
26	0	1	1	61	58	0	3	18	42
27	0	0	3	17	59	0	1	31	7
28	0	0	13	24	60	0	1	62	2
29	1	2	5	42	61	0	3	44	46
30	0	2	14	20	62	1	3	25	42
31	1	0	8	47	63	Output	3	11	18

Herman's program, the success numbers, and the state numbers as they were then), Herman was presented with Problem 2, which is the same as Problem 1 except that the output bit in D_{63} must be the *complement* of the input bit in D_0. Obviously a program that is perfect for Problem 1 must be phenomenally unsuccessful for Problem 2. Problem 2 was continued until a perfect program was attained (as inferred from at least 7,000 successes in a block of 10,000 trials). Then Problem 1 was given until a perfect program was attained. This was continued, with the hope that the Learner would presently be able to adapt quickly to whichever of the two problems was presented. Then it could be considered to have "learned" not just a solution to a single problem but a generalized ability to handle problems in which the input location is D_0 and the output location is D_{63}.

As may be seen from Table 2, the result was more or less as desired. It is not understood why the first adaptation was made so quickly or why the next few took longer.

Table 2   **Approximate number of trials required before perfect program was found (starting from the end of Experiment 1).**

Problems	Number of Trials
Problem 2	400 (est.)
Problem 1	80,000
Problem 2	140,000
Problem 1	230,000
Problem 2	20,000
Problem 1	500 (est.)
Problem 2	500 (est.)
Problem 1	200 (est.)

● *Experiment 5*

Starting from the end of Experiment 4, Herman was presented with Problem 3. In this problem, D_0 and D_5 are the input locations, D_{62} and D_{63} are the output locations, and the criterion of success is that the two-bit binary number formed by the output bits (taking the low-order bit from D_{63}) be the sum of the two input bits.

It was intended that the choice of input and output locations in the various problems follow a consistent plan, so that the Learner, faced with a new problem, would have to adapt only to the features of the problem that were really new, and not also to an arbitrary rearrangement of input and output locations. In the expectation that some problems might involve numbers as many as five bits long, the policy was laid down of letting the first input number occupy locations D_0 onward, starting with the low-order bit; letting the second input number occupy locations D_5 onward, starting with the low-order bit; and letting the output number *end* with the low-order bit in D_{63}. This policy was to be followed even in problems which, like all those discussed in this paper, involved numbers of fewer than five bits.

In 2,420,000 trials, Herman obtained 612,063 successes, or slightly more than one success in four trials, which is the expected average for a random program

exclusive of time failures. The frequency of success fluctuated widely during the experiment, going below 14,000 out of 100,000 successive trials and above 33,000 out of 100,000 successive trials. It would be rash to infer from the data that there was a steady secular upward trend. The slight excess of the over-all average over ¼ should not be taken very seriously in view of the large short-term fluctuations. If the last 250,000 trials had been omitted, the over-all average would have fallen as far short of ¼ as it actually exceeded it. However, the closeness of the average to ¼ suggests that through all the fluctuations, which probably reflected changes in the program, Herman retained the habit of finishing usually within the time limit.

● *Experiment 6*

Starting from the end of Experiment 4, Herman was presented with Problem 4, in which a success is recorded if D_{63}, the only output location, finally contains the low-order bit of the sum of the input bits placed in D_0 and D_5, regardless of the final content of D_{62}.

The results followed the same pattern as Experiment 1. A perfect program, reproduced in Chart 2, was obtained after 940,000 trials.

● *Experiment 7*

Starting from the end of Experiment 6, Herman was presented with Problem 5, in which a success is recorded if D_{62}, the only output location, finally contains the high-order bit of the sum of the input bits placed in D_0 and D_5, regardless of the final content of D_{63}. It was hoped that if Herman acquired the ability to handle both Problems 4 and 5, it would not be too great a leap thence to progress to Problem 3, which combines 4 and 5.

No perfect program was obtained for this problem, although 2,740,000 trials were run. As in Experiment 5, there were large fluctuations in frequency of success. The total number of successes was 1,367,321, which falls short of half the number of trials by an amount insignificant in view of the short-term fluctuations.

It is supposed that the order in which problems are presented affects the learning process, although none of the experiments reported in this paper (Part I) show the effect clearly. Thus, the alternation of Problems 1 and 2 in Experiment 4 presumably encouraged the development of programs which were meaningful if D_0 was an input location and D_{63} an output location. This development presumably aided the subsequent learning of Problem 4; the only new location to be identified was D_5 as an input location. Once Problem 4 had been learned, the only new location to be identified in Problem 5 was D_{62} as an output location. The fact that Herman achieved more success in Problem 4 than in Problem 5 suggests three explanations:

1. that the logical function to be performed in Problem 5 (logical AND) is more difficult for Herman than that in Problem 4 (addition modulo 2);

2. that it is more difficult to identify a new output location than a new input location;

153

Chart 2 **Program obtained in Experiment 6. The addition modulo 2 of input data from Locations 0 and 5 is obtained in Location 63.**

N	D	OP	A	B	N	D	OP	A	B
0	Input	0	58	57	32	1	3	32	34
1	1	0	14	8	33	0	2	32	42
2	0	3	47	22	34	0	2	38	43
3	1	0	29	54	35	0	3	59	12
4	1	2	52	5	36	1	0	26	63
5	Input	0	45	24	37	0	2	37	18
6	0	0	23	11	38	0	2	11	45
7	0	3	4	37	39	1	3	5	56
8	0	0	18	24	40	1	0	18	29
9	0	3	18	57	41	1	3	49	23
10	0	0	2	21	42	0	0	2	33
11	1	0	10	34	43	0	2	63	9
12	0	0	32	58	44	0	0	54	14
13	1	0	40	56	45	1	1	33	47
14	0	2	53	5	46	0	3	1	16
15	0	0	22	35	47	1	3	57	19
16	0	0	14	25	48	1	3	16	34
17	1	2	41	59	49	0	1	14	29
18	1	2	23	30	50	0	3	38	47
19	0	3	45	51	51	1	0	43	39
20	0	3	41	26	52	1	0	41	9
21	0	1	32	44	53	1	0	2	21
22	1	1	38	13	54	1	2	1	25
23	0	3	5	56	55	1	1	63	23
24	1	0	46	53	56	0	1	5	8
25	1	0	26	63	57	0	0	46	24
26	1	2	34	43	58	1	2	35	26
27	1	1	62	5	59	0	1	18	20
28	0	2	10	19	60	1	3	15	29
29	0	1	58	61	61	0	2	27	24
30	0	1	17	33	62	0	3	11	58
31	1	0	63	32	63	Output	1	5	51

3. that Herman's experience with Problem 4 established D_5 less firmly as an input location than the previous alternation of Problems 1 and 2 had established D_0 and D_{63} as input and output locations, respectively.

The first explanation seems less promising than the other two. In any of these problems, learning to make the contents of certain locations depend on the contents of certain other locations seems a greater task than learning, once the input and output locations are identified, to make this dependence obey a certain logical function.

• Experiment 8

If a large group of problems were learned, it would become cumbersome to return repeatedly to each one of the group in order to retain the ability to perform it. This could be done more easily if Herman could learn to perform different problems on successive trials. For example, if Herman could learn to perform Problem 1 in every other trial and Problem 2 in the intervening trials, the ability to perform both problems might be renewed, when necessary, by presenting this alternation of them. There is no reason why a single program could not perform a different act in successive trials, for the effect of executing a program depends not only on the instructions of which it is composed but also on the content of the data locations other than the input locations. These contents, in turn, were determined by the action of the program in the preceding trial.

As preparation, Problem 6 was presented, starting from the end of Experiment 4. This problem has no input location and one output location, D_{63}, and a success is recorded if the output bit is a 1 in an odd-numbered trial or a 0 in an even-numbered trial.

Herman achieved a perfect program for Problem 6 in fewer than 20,000 trials. This is noteworthy since the learning of time-dependent behavior is a particularly interesting phenomenon in its own right.

After a perfect program had been obtained for Problem 6, Herman was presented with the alternation of Problems 1 and 2. That is, the output bit in D_{63} was required to be the same as the input bit placed in D_0 in even-numbered trials and to be its complement in odd-numbered trials.

The frequency of success did not exceed 50% in any block of 10,000 trials, although the experiment was run for 2,130,000 trials. No perfect program was obtained. In all, 863,447 trials were successful.

• Experiment 9

In Experiments 5 and 7, the frequency of success often stayed considerably above the expected fraction for a random program (¼ in Experiment 5, ½ in Experiment 7) for as many as 100,000 successive trials. It may safely be inferred that during those successful periods, programs were in use that tended to achieve success, although they did not achieve it infallibly. Since the Learner is supposed to retain the instructions comprising such programs, it is disturbing that these periods of success were often followed by periods in which the frequency of success was distinctly *below* the expected fraction. This indicates that "good" instructions were replaced by "bad" ones. In an effort to find out whether the Learner is capable at all of holding on to a "good" program which does not *always* achieve success, Problem 1 was presented to Herman after random initialization, and every tenth trial was ruled a failure no matter what Herman did. Thus even a perfect program would achieve success only in 9 trials out of 10.

Under these conditions Herman failed to retain a frequency of success higher than 45% for more than about 50,000 trials at any one time, although the experiment was run for 1,380,000 trials and often a single block of 10,000 trials yielded more than 7,000 successes. In all, 560,618 successes were obtained.

These results suggest that the Learner is seriously deficient in the ability to retain instructions that are statistically advantageous but not infallibly successful. This deficiency might be due to the random changes in the program. The Learner has features that were designed to protect "good" instructions from random changes, but perhaps the features did not work. To test this possibility, the Learner was modified so as to make no random changes. The preceding experiment was then repeated *starting from the end of Experiment 1*, so that there was at least one perfect program among the 2^{64} programs attainable by routine changes alone.

This experiment was run for 2,670,000 trials. During the first million trials, the frequency of success stayed fairly close to the expected random 45%. Although occasionally 7,000 or even 8,000 successes appeared in a single block of 10,000 trials, no three consecutive blocks each contained more than 6,000 successes. The last million trials included several successful periods, from 30,000 to 100,000 trials long, during which more than 8,000 and frequently just 9,000 successes were obtained in each block of 10,000. These successful periods were separated by normal periods, from 50,000 to 150,000 trials long, during which the frequency of success approximated 45%, as during the first million trials. The data do not suffice to show whether or not the successful periods would have grown longer and more frequent, had the experiment been prolonged. But the fact that the successful periods were absent at first and appeared after 1,700,000 trials indicates that the Learner, restricted to routine changes, is able gradually to favor particularly successful instructions even though they do not succeed more than 90% of the time.

The random changes cannot easily be dispensed with, however, for an unfavorable choice of a few key instructions can make it impossible to obtain a successful program from a given set of 128 instructions. The influential role of certain key instructions is perhaps an undesirable feature of the present scheme. For example, whenever Herman's program has been examined after a long string of successes, *op* 0 has been found in I_0 and *op* 1 or 3 in I_{63}, and reflection shows that these characteristics are almost necessary for a successful program if D_0 is used for input and D_{63} is used for output. An example of the

Learner's failure to protect "good" instructions from random changes is the fact that the key instructions in I_0 and I_{63} were changed in the course of Experiment 7.

The results obtained thus far, although fragmentary, show that in a practical number of trials a learning machine of the type described can achieve enough success to be suitable for informative experimentation. By continuing this kind of investigation, we may grow to understand the factors that influence the behavior of such a machine.

Received November 14, 1957

R. M. Friedberg
B. Dunham
J. H. North

A Learning Machine: Part II*

Abstract: An effort is made to improve the performance of the learning machine described in Part I, and the over-all effect of various changes is considered. Comparative runs by machines without the scoring mechanism indicate that the grading of individual instructions can aid in the learning process. A related study is made in which automatic debugging of programs is taken as a special case of machine search. The ability to partition problems and to deal with parts in order of difficulty proves helpful.

Introduction

The experiment described in Part I was continued after an interruption of several months. Two immediate objectives were set: (1) an explicit measure of Herman's learning efficiency, and (2) a better understanding of the factors which govern that efficiency. We are interested in Herman because elements which help or hinder his small-scale performance can well influence more substantial learning machines.

But how is efficiency to be measured? Suppose changes are made in Herman's program randomly, without the benefit of success numbers. For any reasonable problem, a correct program will be hit upon eventually. The question is *how much faster* Herman, with the aid of the scoring mechanism, will develop correct programs than he would merely by random, trial-and-error search.

Not all random searches are alike. Suppose we discover two men on a lake: Samson and Homer. Each is blind and cannot fix his precise location. Nevertheless, by dropping a lead line, each can determine exact depths. Samson is somewhat stolid. He is blown by chance winds about the lake, but he finds pleasure in dropping his lead line every five seconds and recording the depth. Since one drop follows so immediately upon another, his successive positions are close together and do not generally differ much in depth. Homer, on the other hand, takes great pleasure in surprise. He too is bandied about by chance winds, but he waits a full hour between drops. In this way, his successive positions tend to be rather remote; and he has no idea what will come next.

Suppose Samson and Homer compare the first 10,000 depths each has recorded. If we assume the lake remained unchanged, there is no reason why either man should have scored more *deep* depths than the other. Samson, of course, obtains his extreme readings somewhat in bunches. Suppose now we penalize Samson for his stolidity. Every time he records a deep depth of a certain magnitude, he must wait an hour before his next drop. His percentage of deep drops will become smaller, since he cannot take full advantage from having reached a deep section of the lake.

With this fantasy in mind, we set up two different random machines. One was like the penalized Samson, and the other was like Homer. When the Samson machine failed, only one or two of the 64 active instructions were changed. When a successful program emerged, the whole machine was started from scratch. The Homer machine, on the other hand, underwent total revision after each failure. As one would expect, Homer far surpassed Samson in the average speed with which he obtained correct programs. Nevertheless, Samson serves as a more valid basis for appraising Herman's success-number mechanism, because Samson is almost exactly like Herman except for lacking such a mechanism.

Teddy

Before we attempted to measure Herman's learning efficiency, we made two changes in his mode of operation by *priming* and *reset*. Both were calculated to improve his performance. In this way, production runs on the IBM 704 could be cut down and very much better statistics obtained.

As noted in Part I, a successful program is likely to contain *op* 0 in the initial location and *op* 1 or 3 in the final location, when the two locations are used for input-output purposes. We *prime* Herman by guaranteeing

*"A Learning Machine: Part I" by R. M. Friedberg, appeared in the *IBM Journal*, 2, No. 1, 2-13 (January 1958).

that this is the case. Priming is thus an *ad hoc* adjustment to Herman's particular characteristics; *reset,* on the other hand, is more fundamental. Although the mechanism of success numbers enables us to "criticize" and modify the instructions in the various locations, it in no way affects the data bits, which carry over unchanged from one run to another. This means there is a certain "dark area" in the experiment, in that a part of the machinery which most influences the outcome of a problem is almost totally independent of the success-number bookkeeping. Further, an important element which could easily be kept constant from run to run, thereby reducing the over-all complexity of the situation, is permitted to vary. Herman is *reset,* therefore, after every run by inserting ZERO in all data locations which do not represent the selected inputs. From here onward, we shall assume machines to be primed and reset, unless otherwise specified.

The efficiency with which variant machines dealt with Problem 1 was used as a rough touchstone of their learning potential. In this problem, D_0 is the input location, D_{63} is the output location, and the criterion of success is that the output bit should be identical to the input bit. Table 1 shows the general effect of priming and reset. At least 2,000,000 trial runs were made to obtain the statistics for each of the four possibilities.

Table 1 **Average number of trials required to achieve a perfect program for Problem 1.**

	Reset	No Reset
Herman primed	15,197	57,281
Herman unprimed	78,829	477,019

Herman's performance on Problem 1 was then compared to that of Samson and Homer, mentioned earlier. Samson was so set up that one active instruction was replaced by its inactive counterpart after every failure. After every 64 failures, one of the 128 instructions, chosen randomly, was replaced by a new random one. The interchange of active and inactive instructions was done both systematically and by random choice. The latter mode of operation is more efficient, since there is less likelihood that the machine will *cycle.* When changes are made in systematic order ($I_0, I_1, \ldots, I_{63}, I_0, I_1, \ldots$), almost duplicate programs will occur every 128 failures. Hence, programs that have already proved failures may be run again, which will inhibit the speed of learning. Samson was, therefore, set up in both ways for Problem 1. Both machines were given the usual 2,000,000 trials. On the average, the "systematic" Samson, which most resembled Herman, required 34,829 runs to achieve success. The "randomized" Samson required 9744. Homer was then tried. Set up in the manner earlier described, he achieved over 1000 perfect programs at an average of 356 trial runs.

The difference between the two Samsons suggested a possible change in Herman. "Criticism" of individual lo-

cations after failure might no longer be made in systematic order, but randomly. The "randomized" Herman was in fact set up, and (2,000,000 trials, Problem 1) averaged 9959 runs per success. Figure 1 shows graphically the comparative performance of the two Samsons and Hermans. The fact that the "randomized" Herman does not maintain the supremacy over the corresponding Samson which was shown by the "systematic" Herman is not surprising. Because random success numbers are assigned at the start, a problem must be run for some time before the success mechanism can take full effect. If average runs are sufficiently short, the success mechanism can in fact inhibit performance, since, in the initial stages, instructions with the lowest success numbers may well be ones which have had no effect whatever on the problem.

Because it was felt that Herman's performance could be substantially improved, it was decided to modify the general scoring mechanism; and a new machine, Teddy, was put together.

Teddy differs from Herman only in the way the associated Learner functions. Two general motivations governed his design: (1) elimination of *dark areas,* and (2) reduction of *traffic jams. Dark areas* arise when the critical mechanism does not attack those elements which have in fact had most to do with past performance. It

Figure 1 **Record of performance on Problem 1, random input.**

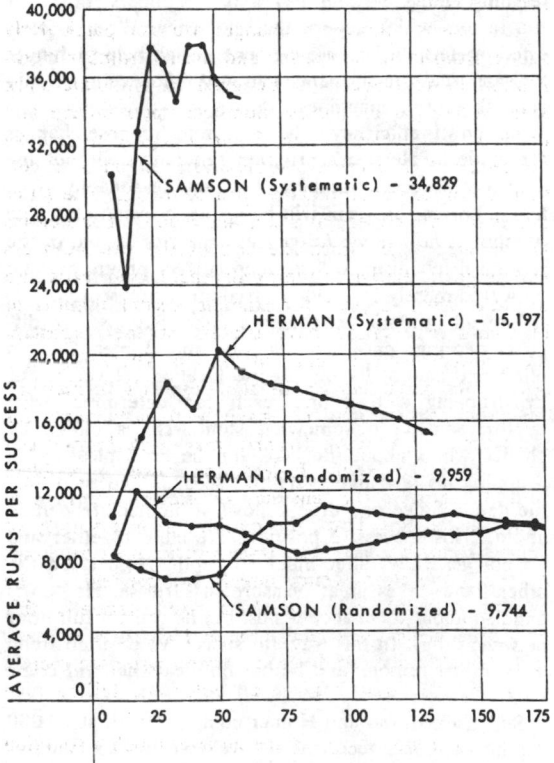

SUCCESSFUL PROGRAMS ACHIEVED

may either punish and reward innocent bystanders, or leave unnoticed true heroes and villains. *Traffic jams arise* when the machine gets itself into a difficult situation from which it can emerge only very slowly, if at all. For example, on a given production run where Herman learned Problem 1 45 times, in 42 cases it never took him more than 30,000 trials. In the remaining three cases, he did not take less than 86,000.

What then are the basic differences between Teddy's scoring mechanism and Herman's? *First, success numbers are modified only for participants.* A *participant* is an active instruction which was either executed in the run in question or referred to by the *b* part of an executed *op* 3. *Second, success numbers are both raised and lowered.* After every success the appropriate success numbers are increased by two. After every failure, the appropriate success numbers are reduced by two. *Third, the method of assigning success numbers is changed.* Upon initialization, all of the instructions are assigned a success number of 1000. The S_i of an instruction later introduced is the mean of the current success numbers of the other instructions. *Fourth, a new system for handling maximum and minimum numbers is introduced.* No number is taken as S_m, and scaling is eliminated. An instruction whose success number drops to 256 is replaced by a new random instruction. *Fifth, there is insured modification of the program after failure.* Only the participants are subjected to "criticism," and this process continues until one of the inactive instructions becomes active.

The reasons for these changes are not particularly subtle. Restricting "criticism" and success-number modification to participants is designed to eliminate dark areas. Lowering of success numbers upon failure and the new treatment of S_i and S_m are designed to remove the tendency of success numbers, after a run of a certain length, to bunch together just below S_m. The latter makes the interchange of active and inactive instructions almost automatic (since the size of the two numbers will be great relative to their difference), and leads also to traffic jams. Suppose, for example, a small number of Herman's instructions have success numbers substantially lower than the bunch at the top. These few may receive the great burden of the Learner's "criticism," even though they may not have been participants for some time. As a result, Herman may work himself out of a jam quite slowly. Indeed, the longer a problem runs, the greater the tendency of success numbers to bunch at the top. Thus, the machine (for a variety of reasons) may become more-or-less static in its behavior, its basic operations being largely independent of the changes made when it fails. As a manifestation of this phenomenon, an unprimed, unreset Herman was made to arrive at and retain for some time a program merely to finish, even though given an automatic failure nine times in ten.

Because Teddy's over-all design is so much a function of earlier experience with Herman, it may be of some interest to set forth a few details of the result just men-tioned. In a given lesson, we presented Herman (unprimed and unreset) with a new problem in which the only criterion of success is that the program finish in time. The Teacher, however, was made to report only one of every ten true successes to the Learner as a success, and the other nine as failures. The frequency of true success rose gradually from a small initial percentage to almost 100 per cent. It then remained well above 90 per cent for better than 400,000 trials, during which a good many of the 128 locations were affected by random changes. We examined several programs that arose during the latter 280,000 trials. They all lacked any *op* 0 instructions in the last ten pairs of active and inactive instructions, and possessed a number of *op* 0 instructions elsewhere with addresses designating these last locations. In fact, the empirical check indicated that the last 11 pairs of instructions were not replaced by any new random instruction in the final 100,000 runs. Both of these features are obviously prone to favor a success for the problem in question without necessarily ensuring it. When Samson (unprimed and unreset) was substituted for Herman in this experiment, the frequency of true successes did not rise above 15 per cent. Thus, it can be seen that the success-number mechanism is very effective for this problem, both in developing a high frequency of success and in maintaining it despite random changes in the program. This contrasts with the inability of Herman to maintain an almost successful program for Problem 1 when given a systematic failure only one true success in ten (Part I, Experiment 9).

Teddy's record of performance is somewhat better than Herman's. In a total run of 500,000, he was able to solve Problem 1 after an average of 1360 programs tried. It should be noted, however, that the way in which the problem is posed by the Teacher and the method of counting runs are both different from that described in Part I. Since Teddy is reset after every run, he will obtain an identical output given an identical input, provided no instructions have been changed. Hence, it is simpler to know when he has arrived at a perfect program—he need only have tried all of the possible input circumstances and obtained correct outputs. With this in mind, we set up the following procedure: a given program tries out all the input conditions before a change is made. Success numbers are appropriately modified, and one set of "criticisms" is made per failure. The resulting program is then tried, and so on. In tabulating the result, it is more convenient to count the programs tried than the individual runs. With this new mode of bookkeeping in effect, Herman (primed, reset, "randomized") required on the average 2890 program trials to solve Problem 1. The "randomized" Samson needed 4603 trials, and Homer, 321. Figure 2 shows these and other results.

Teddy and the "randomized" Herman were also tried on a few two-variable problems, in which D_0 and D_5 were taken as input locations, and D_{63} as the output. Table 2 provides a record of their achievement. The functions indicated are the familiar truth-functional ones.

Table 2 **Comparative performance of Teddy and Herman (randomized) on two-variable problems.**

Function	Average programs tried before success and number of successes obtained			
	Teddy		Herman	
AND	24,896	15	225,508	3
INCLUSIVE-OR	44,539	9	97,978	6
NOT-IF-THEN	18,633	8	141,306	3

One of the major objectives of Part I was to determine whether the grading of individual instructions would aid in the learning process. Samson, Herman, and Teddy, in their various forms, are all inhibited in that changes are made more or less one at a time. The generally superior performance of those machines with a success-number mechanism over those without does serve to indicate that such a mechanism can provide a sound basis for constructing a learning machine.

Another aim of Part I was to render more explicit just how a machine, in a progressive sequence of operations, could discover order in the midst of apparent

Figure 2 **Record of performance on Problem 1, systematic input.**

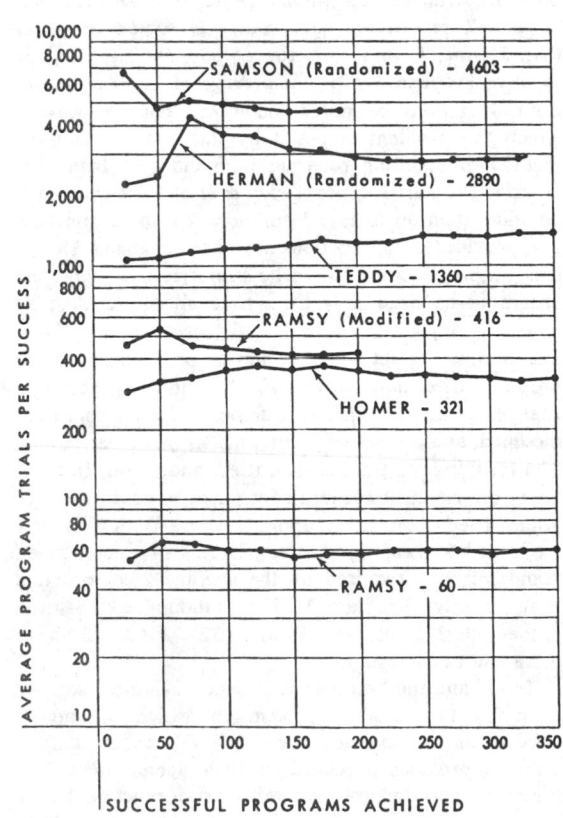

chaos. As we have seen, dark areas and traffic jams are likely to occur and do have a definite effect on the efficiency of the machine. Hence, in setting up future learning machines, we need to consider how these problems are to be managed in the given case. There is, of course, the fascinating prospect that a learning machine might so adjust itself as to eliminate traffic jams and dark areas progressively, as it gains experience.

Finally, there is the problem that making changes one at a time can very much inhibit a learning machine. Homer far outstrips Teddy in performance. The particular way in which success numbers function for the machines we have considered makes it difficult to avoid this inhibition. Hence, the question was next raised how an elementary learning machine, which followed reasonably simple but general directions, might be set up without the use of success numbers.

Ramsy

It was remarked in Part I that, although a learning machine might indeed "learn to perform a task without being told precisely how to perform it, it would still have to be told precisely how to learn." For the machines we have been considering, the ability to arrive at a program for solving problems simply from seeing whether trial runs succeed or fail depends in large part on having some effective way of selecting one imperfect program over another. Naturally, it is easy to recognize a perfect program when one comes along; and one might, as Homer does, simply try out one new program after another. Still, this method, though suggestive, does not seem very promising. The harder the problem, the less likely it will help us. If we could arrive at programs not yet perfect which have, nevertheless, a certain figure of merit in their favor; and, if we could use such programs as a decent basis for obtaining new programs with a higher figure of merit, perhaps we might develop a reasonable learning machine after all. But how is all this to be done?

Suppose we simplify the problem somewhat. Let us say we want a *self-debugging* machine capable of making an efficient, progressive search for correct programs, starting from scratch in each case. What exactly would such a search be like? Let us go back to the lake-bottom fantasy described in the Introduction. We now postulate Thales, a third man. Thales is neither blind nor blown about by chance winds. He can always return to a spot just left. To determine depths, he too must drop a lead line; but he does have the capacity to try out nearby positions before moving on to them. In this way, he can always move to a deeper part of the lake until he reaches a position of desired depth unless, of course, he is stopped at some point which is deep relative to its immediate surroundings but not to the lake as a whole. Under the latter circumstances, he must accept new positions which are not necessarily better, in order to get out of the immediate dead end.

Now, although the "Thales" technique of directed search may seem simple and straightforward, we do not in fact understand it or its applications fully. Nevertheless, we do know that it is a powerful method, which we have already used with considerable success on a number of occasions where other problem-solving techniques failed. It is typically applied to problems with answers easy to recognize but difficult to calculate. Hence, the analogy to the machine-learning experiment is helpful. On the one hand, we may receive some hints as to how the learning machine can be set up more efficiently. On the other hand, we are provided with an additional motivation for the experiment. A learning machine based on the principle of directed machine search would provide a rather illuminating example of this technique. After all, our basic objective in building and studying computers is to obtain greater problem-solving capacity. For the latter, we need not only better machines, but also a better understanding of how to use them.

Most problems we encounter can be broken down into parts, and often these parts are not difficult in themselves. All of us have dealt with outwardly hard problems which became easy when reduced to a set of simpler subproblems. Suppose two imperfect methods for solving a problem are at hand. One does no good at all. The other manages a certain segment of the problem. We would obviously attach a higher figure of merit to the second. If there were some way of leaving undisturbed those features of the better method which contributed to its partial success, while modifying other features which led to its partial failure, we should also have a decent basis for obtaining a new method with an even higher figure of merit. In this way, we could proceed to the solution by a sequence of definite steps. Our scheme of operation would then resemble Thales, in that a proposed step would be accepted only when in the right direction. Suppose the various parts, however, though much easier than the total problem itself, are not alike in difficulty. If the solution of one part does not help that of another, it would be advantageous to attack the more difficult parts first. Since we hope to leave undisturbed those features of our method which contribute to whatever partial success has been obtained, the farther on we get with a problem, the less of the method we have available to modify.

The machine Ramsy is based primarily upon two edicts: (1) partition the problem into parts, and (2) deal with the more difficult parts first. An added principle of operation is derived from Homer's superiority over the penalized Samson. When a purely random search is made, wholesale eradication should follow failure. Needless to say, dark areas and traffic jams are to be avoided.

It is convenient to retain the three basic blocks of the learning machines already discussed, but to modify their mode of interaction. In Part I, the machine was broken down into Teacher, Learner, and Slave (that is, Herman). We sometimes use the name of the Slave loosely in describing the whole machine, but no confusion should result from this. To this family of three, we provided certain information, namely the possible inputs and related outputs of the problem in question. No other information was given. Specifically, we did not adjust the family externally from problem to problem. In setting up a new family machine, we wish the same rules to apply for posing problems.

The Slave Ramsy is identical to Teddy and Herman in kinds of instruction and number of locations. The associated Teacher and Learner, however, have somewhat reversed roles. The Teacher is more active, the Learner more passive, and more information passes between them. The partitioning and ordering of the problem are done by the Teacher. The Learner keeps a limited set of records and provides a random-number generator. There are no success numbers or inactive instructions.

How is a problem partitioned? If we restrict our attention to "bit-manipulation" problems (which seem sufficient to our purpose), partitioning is straightforward. Suppose we wish Ramsy to solve the familiar problem AND for two variables, the input and output locations being D_0, D_5, and D_{63} as before. There are four possible input cases (two ONES, two ZEROS, et cetera), and these define the four parts of the problem.

How are the parts to be ordered? Assume that Ramsy is primed with an op 1 or 3 in the output location and will retain that operation. With op 1, it is easier to produce ZEROS than ONES; with op 3, the reverse. If a problem produces more ZERO outputs than ONES, it is better to have op 1 in the appropriate location; otherwise, op 3.

The Ramsy family's mode of operation can now be briefly summarized. A problem is posed as always. The Teacher primes the Slave appropriately and orders the parts in terms of difficulty. A most difficult part is given the Slave to solve first. If he fails on a given run, all of the participating instructions (with the exception of the bits determining the primed operations) are replaced by new random instructions. If he succeeds, all of the participants become *bound;* and the next part of the problem is undertaken. Bound instructions are exempted from later eradications unless Ramsy becomes *stuck,* which occurs when no participants can be replaced after a failing run because all are bound. Under such conditions, Ramsy is filled with new instructions and the problem begun again.

Ramsy proved far more successful than Homer. On Problem 1, for example, he required, on the average, only 60 trial programs in obtaining 500 solutions. Figure 2 shows graphically his performance. This superiority was also maintained on a number of two-variable problems, as indicated in Table 3.

To test the efficiency of the Homer-Samson principle, some limited runs were also made on a modified Ramsy in which never more than one participant was replaced by a random instruction. As expected, the more stolid machine proved less efficient. On Problem 1, for example, the modified Ramsy generated 200 perfect programs at an average of 416 trials each, as shown in Figure 2. A brief run was also made in which Ramsy was presented with the easier of the two problem conditions first. He averaged 144 program trials to solve Problem 1.

Table 3 Comparative performance of Ramsy and Homer on two-variable problems.

Function	Average programs tried before success and number of successes obtained			
	Ramsy		Homer	
EXCLUSIVE-OR	12,934	31	199,910	0
AND	1758	32	19,445	5
INCLUSIVE-OR	216	60	7208	42
NOT-IF-THEN	6820	35	14,247	16

Another variant of Ramsy included a built-in mechanism for shortening programs. This was done by a rather simple system of address modification which served to eliminate redundant participants. The technique worked smoothly but was of only moderate interest to the over-all experiment. Programs were very much shortened, but the speed of learning was not affected.

Still another variant included an elementary method for making D_{63} a function of D_5, when the latter served as an input. Learning efficiency was increased, but it was not easy to generalize the technique to deal with more involved cases.

The chief drawback to Ramsy is, of course, his growing tendency to become stuck as more and more complicated problems are encountered. Undoubtedly, an improved mechanism can be devised for dealing with this situation. The problem is analogous to that of Thales when he reaches a depth which is maximum for a given area, but not yet deep enough. Without taking a totally fresh start, he must "back-up" slightly in order to get out of the immediate dead end. Traffic-jam techniques appropriate for Thales may prove workable for Ramsy.

Added experience has revealed, however, an additional drawback to Ramsy, which would seem to make it profitable to start over with an entirely new Slave. Since the basic method of learning is quite different from that introduced in Part I, the particular operations selected for Herman need not be retained. These require, on the average, 260 microseconds to execute. If the tiny computer which we simulate on the large machine had an order code more closely akin to some of the actual instructions already available, the machine

time required to carry out the experiment could be substantially reduced. For a variety of reasons, this would very much facilitate the conduct of the experiment.

Conclusion

In setting up an experiment in which we study the properties of small, easily controlled machines as a guide to larger machines, we face special problems in understanding our results. Those characteristics of the small machine which obtain simply because it *is* small must be differentiated from more fundamental properties which may hold in the larger case. We have seen that the amount of change made when a program fails has a definite bearing upon the average speed with which successful programs are obtained. Homer makes large-scale changes upon failure, and surpasses Samson for this reason. Thales, on the other hand, undertakes only small changes; but those changes made are likely to be in the right direction. Hence there is a definite tie-in between the size of change made and our capacity to compare related positions. The results obtained do indicate that a simple reinforcement of individual instructions can aid in the learning process; but we have not wholly succeeded in setting up a machine based upon this principle. The difficulty is that, although small-scale changes are made upon failure, our scoring mechanism is admittedly loose. The importance of dark areas and traffic jams has also been emphasized. As we have seen, there are special problems in recognizing, avoiding, and getting out of the latter. The results with Ramsy indicate that the ability to partition problems and to deal with the parts in order of difficulty does prove helpful. In this connection, we found it worthwhile to note the analogy between machine learning and the problem-solving technique of directed machine search.

Where we should go from here is not entirely clear. Perhaps the experiment, with a radically different Slave, could be set up in a closer analogy to Thales. We have a somewhat vague but quite persistent sentiment that the methods we have used to bring about learning are too passive. Some scheme of ensured referencing or execution of inputs, such as that briefly suggested near the end of the last section, might well be introduced. In all events, we find the unanswered questions as fascinating as they are difficult.

Received March 24, 1959

Chapter 6
Artificial Life and Evolving Strategies

N. A. Barricelli (1962) "Numerical Testing of Evolution Theories. Part I: Theoretical Introduction and Basic Tests," *Acta Biotheoretica*, Vol. 16, No. 1–2, pp. 69–98.

N. A. Barricelli (1963) "Numerical Testing of Evolution Theories. Part II: Preliminary Tests of Performance, Symbiogenesis and Terrestrial Life," *Acta Biotheoretica*, Vol. 16, No. 3–4, pp. 99–126.

PERHAPS the earliest archival record of an evolutionary simulation is due to Barricelli (1954). Nils Aall Barricelli worked on the high-speed computer at the Institute for Advanced Study in Princeton, New Jersey, in 1953 and conducted experiments that were similar to some of the artificial life experiments later popularized in Langton (1987a).[1] Barricelli's original research was published in Italian, but due to some translation errors and additions, this was subsequently republished as Barricelli (1957) in English along with novel results. Barricelli had a long interest in the origin of life and attempted to answer some fundamental questions regarding natural evolution using simulations. In particular, his investigations concerned the conditions that genes must fulfill in order to give rise to the development of higher forms of life. He postulated that the most important conditions for genes to satisfy included: (1) the capacity to reproduce, (2) the capacity for changing to alternative forms (by mutation), and (3) a need of symbiosis (e.g., parasitism) with other genes or organisms (Barricelli, 1957). The emphasis on symbiosis was pervasive in his early work.

His initial experiments involved a relatively simple simulation of numeric elements in a grid. These elements propagated by means of local rules specified for each possible number. Assume that there is a vector of N cells. A new vector of N cells is determined by rules applied to the previous vector of N cells. Let the entry at the ith cell for generation g be denoted as x_{ig}. If $x_{ig} = 0$ then the ith cell is empty.

(1) A number $x_{ig} = n$ shifts in one generation n cells to the right if it is positive, or $|n|$ cells to the left if is it negative, unless this results in a "collision" with another number (i.e., the same rules applied to another number in the vector at generation g would have that number also propagate to the same cell).

(2) The same number $x_{ig} = n$ may reproduce m cells to the right (or $|m|$ to the left) if $x_{i+n,g} = m$, again with the exception of a collision.

(3) Reproduction may occur more than once if $x_{i+m,g} = r$ (where $r \neq 0$), then $x_{i+r,g+1} = n$, again with the exception of a collision.

(4) Should two numeric elements ("genes") collide, if they are equal, then one copy of the number is placed in the cell where the collision occurred, resulting in no mutation. If the numbers are not equal, a variety of rules that alter the numeric elements in the vector of N cells at generation $g + 1$ may be employed (see Barricelli, 1957, for details).

Figure 6.1 shows the propagation of an initial pattern involving the numbers 5, 1, and -3 under a mutation rule that zeroes out any cell when two or more numbers collide. The result of the propagation is an "emergent" pattern of $(5, -3, 1, -3, 0, -3, 1)$, which appears at least once in every generation after its creation in generation 4. The figure depicts a "flat" grid, but Barricelli (1957) indicated executing experiments with 512 cells in a tubular

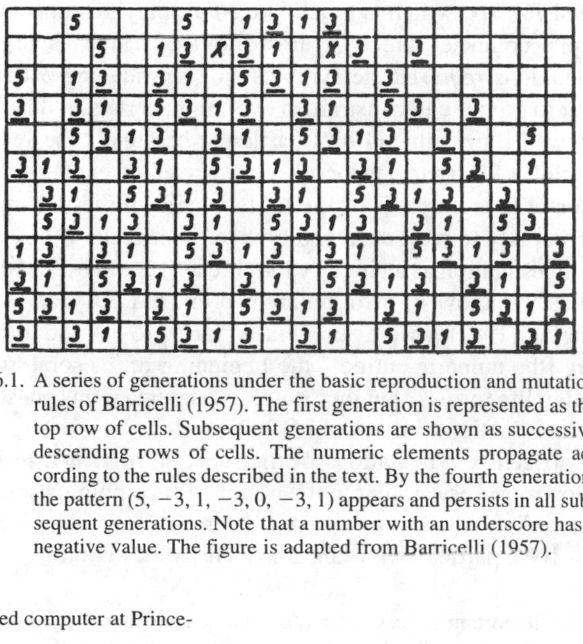

Fig. 6.1. A series of generations under the basic reproduction and mutation rules of Barricelli (1957). The first generation is represented as the top row of cells. Subsequent generations are shown as successive descending rows of cells. The numeric elements propagate according to the rules described in the text. By the fourth generation, the pattern $(5, -3, 1, -3, 0, -3, 1)$ appears and persists in all subsequent generations. Note that a number with an underscore has a negative value. The figure is adapted from Barricelli (1957).

---

[1]Barricelli (1957) noted that he executed some experiments prior to using the high-speed computer at Princeton. It is not clear if these experiments were conducted by hand or by electronic means.

fashion, in which the left and right edges were assumed to be adjacent. The experiments involved random initialization of numbers in the first generation (determined by using a set of playing cards). For each of the mutation rules, the use of only a single rule led to uniform conditions in which a single pattern expanded to fill the whole grid. The use of multiple rules was also investigated, with different rules being used in different parts of the grid. This was observed to prevent uniformity.

Barricelli (1957) examined some of the patterns that emerged as "organisms." He defined them to be *independent* if they could reproduce without requiring the presence of organisms with a different pattern ("another species"). Patterns were defined to be *dependent* (a "parasite") if they were unable to reproduce all of their elements alone, and needed a continuous supply of elements from other patterns for reproduction. Dependent parasitic patterns were observed, as well as other patterns characteristic of natural evolution (e.g., hitchhiking genes). Barricelli also observed patterns of recombination, including a multiple-point operation in which two patterns would collide and the result would be a new self-sustaining pattern with numeric elements chosen from each of the two "parents." The experimental design was also expanded to involve a two-dimensional grid at each generation, with suitably modified rules for numeric propagation. In retrospect, there is an obvious similarity in this work to Conway's Game of Life (for a review, see Dewdney, 1987).

Barricelli (1957) noted that the mutation and propagation rules were the prime determinant of the patterns that emerged and suggested a different set of experiments designed to develop patterns that were associated with specific functions. For example, it was suggested that the organisms play a game and have their performance determine their potential for survival.

The two papers reprinted here (Barricelli, 1962, 1963) offer results on emergent properties of numeric values propagated by specific rules beyond those of Barricelli (1957), as well as some of the earliest efforts at coevolutionary gaming (using a game called *Tac Tix,* which is much like *Nim*), in which one strategy for play competes directly against another. The mechanics of the simulations *required* the symbiosis of reproducing patterns in order to promote unconstrained evolution; certainly, this design followed Barricelli's (1962) particular bias that "the selection principle of Darwin's theory is not sufficient to explain the evolution of living organisms if one starts with entities having only the property to reproduce and mutate." Barricelli (1962) set up the investigation as an inquiry into life as it could be, rather than life as we know it, a formulation promoted 25 years later in Langton (1987b). He even went so far as to ask the question: "Are [the numeric entities] the beginning of, or some sort of, foreign life forms?" but then quickly discounted this question in the introduction of Barricelli (1963).

Barricelli's explanations of the function of sexual recombination were very similar to the notion of recombining schemata as building blocks, as offered later in Holland (1975, pp. 103–104), particularly when Barricelli (1962) wrote:

"An important success of the symbiogenesis theory [that complex organisms arise from combinations of genetic material from lesser organisms] is that it can explain in a simple way how sexual recombination originated, a phenomenon whose interpretation was a complete mystery before this theory occurred. In fact, if the genes are associated into larger organisms by symbiotic phenomena, they had from the beginning the capacity to change host and associate with new organisms as parasites and symbionts usually do. This capacity, already at the outset, conferred to the gene-association the adaptivity of organisms with sexual reproduction which is based on the fact that a large number of useful mutations—one in each gene—can expand in the population contemporaneously. This will sooner or later provide for the necessary reassociation of all the positively selected genes into the final organism. A great number of phenotypic problems can in this way be solved contemporaneously without delaying one another, a fact which permits an increase in the rate of evolutionary adaptation by a factor of several thousand (R. A. Fisher, p. 122–123, 1930)."

Barricelli (1962) "tested" the symbiogenetic theory by implementing reproduction rules that would promote the symbiosis of different numerical entities and determine if processes similar to those in natural evolution would occur. Indeed, symbiotic cooperation between such different entities was "rendered necessary for reproduction" in his simulations. As such, it is not clear that any real test of the effects of symbiosis had been conducted. Later, as coauthor of Reed et al. (1967) (see Chapter 11), Barricelli tempered his view on the importance of recombination by indicating its ability to accelerate evolution for phenotypic characters that were not polygenic (i.e., not based on the interaction of multiple genes) but to provide no apparent advantage when phenotypic characters were polygenic.

Barricelli (1962) clearly recognized the "programming" aspect of his evolutionary simulations: "The genetic structure of any symbioorganism can rightly be considered as a coded survival strategy program created and developed during its past evolutionary history." His insights also included the Markovian nature of the simulations, a facet which has been used to advantage to analyze the convergence properties of evolutionary algorithms (Rudolph, 1994; De Jong et al., 1995; and many others), as well as the potential for computing with DNA or other chemical methods (as has been promoted by Joyce, 1992; Conrad, 1992; Adelman, 1994; and others).

References

[1] L. M. Adelman (1994) "Molecular computation of solutions to combinatorial problems," *Science,* Vol. 266, pp. 1021–1024.

[2] N. A. Barricelli (1954) "Esempi numerici di processi di evoluzione," *Methodos,* pp. 45–68.

[3] N. A. Barricelli (1957) "Symbiogenetic evolution processes realized by artificial methods," *Methodos,* Vol. IX, No. 35–36, pp. 143–182.

[4] N. A. Barricelli (1962) "Numerical testing of evolution theories. Part I: Theoretical introduction and basic tests," *Acta Biotheoretica,* Vol. 16, No. 1–2, pp. 69–98.

[5] N. A. Barricelli (1963) "Numerical testing of evolution theories. Part II: Preliminary tests of performance, symbiogenesis and terrestrial life," *Acta Biotheoretica,* Vol. 16, No. 3–4, pp. 99–126.

[6] M. Conrad (1992) "Molecular computing: The lock-key paradigm," *Computer,* Vol. 25:11, pp. 11–20.

[7] K. A. De Jong, W. M. Spears, and D. F. Gordon (1995) "Using Markov chains to analyze GAFOs," *Foundations of Genetic Algorithms 3,* L. D Whitley and M. D. Vose (eds.), Morgan Kaufmann, San Mateo, CA, pp. 115–137.

[8] A. K. Dewdney (1987) "Computer recreations: The game of life acquires some successors in three dimensions," *Scientific American,* Vol. 256:2, pp. 16–24.

[9] R. A. Fisher (1930) *The Genetical Theory of Natural Selection,* Oxford University Press, London.

[10] J. H. Holland (1975) *Adaptation in Natural and Artificial Systems,* Univ. of Michigan Press, Ann Arbor, MI.

[11] G. F. Joyce (1992) "Directed molecular evolution," *Scientific American,* Vol. 267:6, pp. 90–97.

[12] C. G. Langton (ed.) (1987a) *Artificial Life: The Proc. of an Interdisciplinary Workshop on the Synthesis and Simulation of Living Systems,* Addison-Wesley, Reading, MA.

[13] C. G. Langton (1987b) "Artificial life," *Artificial Life: The Proc. of an Interdisciplinary Workshop on the Synthesis and Simulation of Living Systems,* C. G. Langton (ed.), Addison-Wesley, Reading, MA, pp. 1–47.

[14] J. Reed, R. Toombs, and N. A. Barricelli (1967) "Simulation of biological evolution and machine learning. I. Selection of self-reproducing numeric patterns by data processing machines, effects of hereditary control, mutation type and crossing," *J. Theoret. Biol.,* Vol. 17, pp. 319–342.

[15] G. Rudolph (1994) "Convergence analysis of canonical genetic algorithms," *IEEE Trans. on Neural Networks,* Vol. 5:1, pp. 96–101.

NUMERICAL TESTING OF EVOLUTION THEORIES

PART I

THEORETICAL INTRODUCTION AND BASIC TESTS [1]

by

NILS AALL BARRICELLI

(Vanderbilt University, Nashville 5, Tennessee)

(Rec. 27. XI. 1961)

NOTE BY THE AUTHOR

A few introductory remarks may be helpful to avoid possible misinterpretations when a relatively new subject is presented.

It is not the intention of the author to face the reader with a new type of life or new living forms. The numerical symbioorganisms presented in this paper are not living, not in the sense commonly attributed to this term. It is, however, the purpose of this paper to present a part of the rapidly growing evidence to the effect that, what we are used to consider living beings (terrestrial life forms) are a particular type of symbioorganisms and are subject to the same evolutionary improvements and to many other "biophenomena" which are common to a large class of symbioorganisms.

There is nothing more peculiar about this statement, except for its more recent discovery, than about statements like the following:

1) living organisms are physical bodies and are subject to the same mechanical and physical laws followed by all physical bodies.

2) living organisms are formed by chemical compounds and are subject to the same kind of chemical reactions, which are the object of biochemistry.

The investigation of physical, chemical and symbiogenetical phenomena are all important for the understanding of different aspects of the process which is designated by the common term "LIFE".

If the reader at any stage should fall for the temptation to attribute to the numerical symbioorganisms a little too many of the properties of living beings, please do not make the mistake of believing that this was the intention of the author. The properties of symbioorganisms are clearly and specifically described and they should be understood s t a t e m e n t   b y   s t a t e m e n t the way they are presented.

As to the terminology used the following principles have been followed: concepts presenting some form of similarity or analogy with a biological concept have been designated by the same or a similar term. For example: "symbioorganism" corresponding to the biologic term "organism"; or "symbiosis" (also termed "utility association") corresponding to the biologic term "symbiosis"; and "mutation" to disignate hereditary

1) The investigation presented in this paper was supported in part by research grants RG-6980 and C-2306 from the National Institutes of Health, United States Public Health Service. The first part of the investigation was performed while the author was a temporary member of the Institute for Advanced Study in Princeton, New Jersey, and the evolution experiments and competition tests were performed by the electronic computer of the Institute.

changes which are not normal crossing results, the same term used for such hereditary changes in biology. This terminology has the advantage of being easy to remember and of making the analogies with biological concepts immediately clear to the reader without requiring tedious explanations. However, the reader should be careful not to confuse a term used in connection with symbioorganisms with the same, or a similar term used in biology. The reader who is aware of this will have no difficulty in keeping the record straight. The terms and concepts used in connection with symbioorganisms are in no case identical to biological concepts; they are mathematical concepts (see appendix) usually applicable to a large variety of symbioorganisms. The biological terms are often selected by the common use of the language and they usually refer to empirical objects and observed biologic phenomena.

Empirical models of symbioorganisms, such as the numerical symbioorganisms developed in a particular experiment are to be considered as empirical representations of a mathematical entity in the same sense as a circular object can be used to represent a circle as a mean of investigation or description. The same terminology used in the mathematical entity investigated will of course also be used in its empirical models.

I. INTRODUCTION

This is the first of two papers the purpose of which is to give an orientation on the present stage and recent developments in the investigation of numeric evolution phenomena. The second paper (BARRICELLI, 1962) will appear in this journal later.

The Darwinian idea that evolution takes place by random hereditary changes and selection has from the beginning been handicapped by the fact that no proper test had been found to decide whether such evolution was possible and how it would develop under controlled conditions. A test using living organisms in rapid evolution (viruses or bacteria) would have the serious drawback that the causes of adaptation or evolution would be difficult to state unequivocally, and Lamarckian or other kinds of interpretation would be difficult to exclude. However if, instead of using living organisms, one could experiment with entities which, without any doubt could evolve exclusively by "mutations" and selection, then and only then would a successful evolution experiment give conclusive evidence; the better if the environmental factors also are under control as for example if they are limited to some sort of competition between the entities used.

The author has shown in several publications (BARRICELLI, 1954 and 1957) that evolution processes in many respects similar to biological evolution can now be obtained using self-reproducing entities of numerical nature whose properties are completely under control. Mutation and selection alone, however, proved insufficient to explain evolutionary phenomena. Another fundamental idea (to be explained in sections 2 and 3) had to be added to the theory before it was possible to obtain evolutionary processes and other phenomena characteristic of living organisms, which are listed in section 6.

It may appear that the properties one would have to assign to a population of self-reproducing elements in order to obtain Darwinian evolution are of a spectacular simplicity. The elements would only have to:

1. Be selfreproducing and
2. Undergo hereditary changes (mutations) in order to permit evolution by a process based on the survival of the fittest.

Such evolution would actually have the characteristics of a purely statistical phenomenon which may well occur in other than biological contexts.

In fact, examples of self-reproducing entities are well known and it is no problem to define and construct more such entities artificially. It is for instance well known that some chemical compounds, called "autocatalysts," have the ability to catalyze a chemical reaction between other compounds, a reaction whose result is, among other things, an increase of the catalyst itself. Such an autocatalyst will therefore become more and more abundant as the reaction proceeds, or in other words, it reproduces itself. At the same time, some compounds (called substrates) originally present will be consumed, while other compounds (called by-products) may, by the same reaction, be produced together with the autocatalyst.

Also in atomic physics, autocatalytic phenomena similar to the one above described are well known. For instance a neutron may start a chain reaction by hitting a uranium U_{235} atom producing among other things two or more neutrons, and thus reproduces itself by a process having the characteristics of what may be considered atomic autocatalysis. The Uranium U_{235} atoms act as the substrate of the nuclear reaction, which, in addition to neutrons also produces some by-products (some elements of atomic weight roughly half the weight of Uranium).

It is easy to construct artificial objects or define entities, for instance of arithmetic nature, which are able to reproduce themselves. In this case one may also choose the definition in such a way that the factors which are not indispensable, conceptually, such as substrates and by-products, will not be included among the elements to be operated upon. For instance, one may use as self-reproducing entities a group of numbers written in the first line of a crossection paper — see fig. 1, where negative numbers are underlined — and one may choose arbitrarily a reproduction rule for these numbers. The reproduction rule used in fig. 1 is the following: in one time unit (generation) a positive number m is reproduced m squares to the right and a negative number n is reproduced $-n$ squares to the left. The result obtained from the first line by following this reproduction rule is recorded in the second line. Applying the same reproduction rule to the second line, one obtains the third line, etc. Of course, in order to prosecute the operation, one would have to state some rule to decide what to do in the cases in which two different numbers happen to fall (collide) in the same square. This will be done below.

The other property essential for Darwinian evolution, *viz.*, undergoing hereditary changes, is probably not as common outside the living organisms as is the faculty to reproduce. There is little point in trying to enumerate phenomena which might be interpreted as examples of hereditary change. For the purposes of this paper, it is sufficient to recall that the genes of a cell and the viruses are examples of self-reproducing elements with the ability to undergo hereditary changes. In both cases the ability to reproduce and to undergo hereditary changes seems to be a property of the nucleic acid which is an important constituent of these structures.

There is no difficulty in defining mathematical entities which besides the faculty to reproduce have the property of undergoing hereditary changes. In the numerical entities defined above, one may for instance choose some mutation rules to apply when two numbers collide in the same square. The number to be put in the collision square may be different from the two colliding numbers and may therefore represent a mutation.

The following mutation rule has been applied in fig. 1 : two numbers which

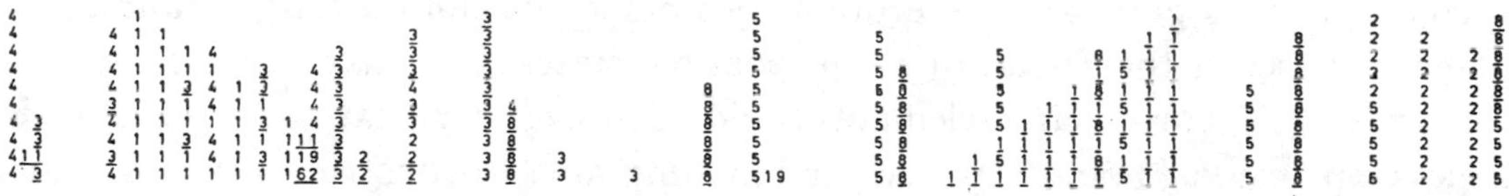

Fig. 1. Selfreproducing numbers (see text). Adaptive selection but no extensive evolution phenomena are possible.

collide in a square are added together, and from the result one subtracts the content of the square above the collision square (except when the square above the collision square is empty, in which case nothing will be subtracted). If three numbers collide in the same square, they are added together, and from the result one subtracts twice the content of the square above the collision place, and so on.

In this manner we have created a class of numbers which are able to reproduce and to undergo hereditary changes. The conditions for an evolution process according to the principle of Darwin's theory would appear to be present. The numbers which have the greatest survival in the environment created in fig. 1 by the rules stated above, will survive. The other numbers will be eliminated little by little. A process of adaptation to the environmental conditions, that is, a process of Darwinian evolution, will take place.

2. CRITICISM OF THE DARWINIAN EVOLUTION PRINCIPLE

The last example of Darwinian evolution clearly shows that something more is needed to understand the formation of organs and properties with a complexity comparable to those of living organisms. No matter how many mutations occur, the numbers of fig. 1 will never become anything more complex than plain numbers.

One may object that the situation is different for genes and viruses because of the greater complexity of the nucleic acid molecules which might be expected to present far greater evolutionary possibilities.

If the situation is different for the simplest self-reproducing elements of hereditary material, that must be for some other reason, but certainly not for the evolutionary possibilities of each individual element. If one considers the recombinational genes — which are the smallest fragments of hereditary material which can be separated from adjacent hereditary material by crossing over — as the self-reproducing elements from which the hereditary material is formed, one finds the following situation. In most recombinational genes only two allelic states are found; more rarely one finds 3 or more allelic states. A recombinational gene with n allelic states is a gene which can exist only in n different varities which can be reached by one or several consecutive mutations. The evolutionary possibilities of such a recombinational gene are therefore very limited. Evolution can do nothing better than select the fittest of the n possible states. It is hard to visualize how such a selfreproducing element could, simply by mutations and selection, develop into a "homo sapiens" or anything able to construct a homo sapiens even if allowance is made for the fact that the number of allelic states may be generally or very often underestimated.

As far as evolutionary possibilities are concerned, the situation does not seem to be any better for the recombinational genes, in spite of their apparent complexity, than it is for the numbers recorded in fig. 1.

If the Darwinian evolutionary principle is to explain the complexity and efficiency reached by living organisms, it must be applicable to the simplest elements which are able to reproduce and mutate. One cannot start with entities of considerable complexity and efficiency, as for example cells, even if relatively simple unicellular organisms, such as amoebae, are chosen. It is obvious that amoebae have a complexity, efficiency and evolutionary possibility far greater than the numbers in fig. 1. But these faculties of the amoeba are part of the properties which evolution theory is supposed to explain rather than to postulate as a prerequisite for evolution.

In conclusion, the selection principle of Darwin's theory is not sufficient to explain the evolution of living organisms if one starts with entities having only the property to reproduce and mutate. At least one more theoretical principle is needed, a principle which would explain how self-reproducing entities could give rise to organisms with the variability and evolutionary possibilities which characterize living organisms.

3. THE SYMBIOGENESIS THEORY

A solution of this difficulty is provided by a theory which can be introduced as a possible interpretation of the startling analogies existing between intracellular viruses and genes. Intracellular viruses and genes are probably both segments of nucleic acid with autocatalytic properties and with the ability to acquire hereditary changes. Examples of intermediate stages between viruses and genes have been found (*cf*. plasmagenes in *Paramecium*, SONNEBORN, 1949). Often the only way to distinguish a symbiotic virus from a gene is provided by the virus' ability to be transmitted by infection, because both are inherited in a similar way (*cf*. LEDERBERG & LEDERBERG, 1952). In recent years the study of bacterial genetics and the relation between bacteria and temperate phages has shown that an appreciable section of genetic material of the *E. Coli* bacterium K_{12} consists of symbiotic viruses. An interesting theory, "the symbiogenesis theory," provides a possible interpretation of the mentioned analogies between viruses and genes. Originally the term "symbiogenesis" was used to qualify any theory implying that the first cell were formed by a symbiotic association of several entities (MERESCHKOWSKY, 1910). A more recent version of symbiogenesis theory, which has proved most fruitful assumes that the genes and possibly other selfreproducing entities in the cell spring from originally independent viruses or virus-like organisms which by symbiosis have been united with the rest of the cell (KOZO-POLIANSKY, 1924; BARRICELLI, 1947, 1952, 1955). According to the symbiogenesis theory the evolution process which led to the formation of the cell was initiated by a symbiotic association between several virus-like organisms. Little by little an increasing number of symbionts of the same nature may have joined the group, and the evolution process thus initiated may have led to the formation of the various organs and properties of the cell.

This theory solves the dilemma in which we were placed by pushing Darwin's evolutionary principle to its ultimate consequences. Even if every recombinational gene were assumed to have only two allelic stages, the association of n genes (or primitive viruses) would yield an organism with 2^n possible varieties. Among these many varieties natural selection undoubtedly has a good choice. If the number n of genes is in the usual range of several thousand, it may not appear so incredibly fantastic to imagine that among the

2^n possible combinations there may be some which correspond to organisms as complex as *Drosophila* or possibly as complex as Man.

In short, the situation is completely reversed, and the limited number of allelic states in each self-reproducing element does not represent an important limitation if the symbiogenesis theory is correct. [2]

Until recently the possibilities of testing the theory were very limited. With the development of bacterial and virus genetics testing possibilities previously unsuspected became available. On a theoretical basis the use of high speed computers makes it possible today to decide what an evolution process would look like when developing in accordance to the symbiogenesis theory.

4. THE SYMBIOGENETIC INTERPRETATION OF CROSSING

An important success of the symbiogenesis theory is that it can explain in a simple way how sexual reproduction originated, a phenomenon whose interpretation was a complete mystery before this theory appeared. In fact, if the genes associated into larger organisms by symbiotic phenomena, they had from the beginning the capacity to change hosts and associate with new organisms as parasites and symbionts usually do. This capacity, already at the outset, conferred to the gene-association the adaptivity of organisms with sexual reproduction which is based on the fact that a large number of useful mutations — one in each gene — can expand in the population contempora-

---

2) This fact remains a fundamental difficulty in all non symbiogenetic attempts to explain the origin of life; namely, the enormous variability, and consequently, the enormous complexity of the first selfreproducing entity required by such explanations. This difficulty is partly recognized by the authors of non symbiogenetic explanations, since they all imply that life on this planet may have arisen only once in a period of several billion years. The possibility that life in a primitive form (microorganisms) might have been transmitted from other planets, by different types of space diffusion phenomena, has also been considered.

When the variability of the selfreproducing entities is limited to very few allelic states, the origin of such entities is no longer a problem. As a matter of fact, no variability is strictly required in each individual self-reproducing entity. The presence or the absence of a self-reproducing entity with the respective phenotypic effects these two alternatives may have on the organism, are in themselves two allelic hereditary characters.

Nobody knows at what rate new autocatalysts are being created by chance on this planet. But, to put it mildly, every chemist would probably suggest that the rate per second, or even better per microsecond, should be used as a unit rather than the rate per billion years.

The origin of the selfreproducing entities is, therefore, no problem. The problems are : (1) Whether a symbiotic association of a large group of autocatalysts is likely to be formed. (2) Whether the symbiotic association of a large group of selfreproducing entities is possible on statistical and mathematical grounds.

The second question is the one which is being directly answered in this paper. However, the nature of the answer and the results which will be presented are strongly suggestive that also the answer to the first question would be affirmative, and that terestrinal life is one result of such symbiotic associations.

neously. This will sooner or later provide for the necessary reassociation of all the positively selected genes into the final organism. A great number of phenotypic problems can in this way be solved contemporaneously without delaying one another, a fact which permits an increase in the rate of evolutionary adaptation by a factor of several thousand (R. A. Fisher, p. 122-123, 1930).

According to the symbiogenesis theory sexual reproduction is the result of an adaptive improvement of the original ability of the genes to change host organisms and recombine. Examples will be shown in this paper of crossing mechanisms and crossing phenomena observed during a series of artificial evolution phenomena developed as a test of the symbiogenesis theory (see also BARRICELLI, 1957). On this theory crossing mechanisms can be expected to exist in all organisms, even the most primitive ones.

Among the experimental facts in support of this interpretation of the origin of crossing and sexual reproduction one may cite:

1. Crossing in viruses. Crossing mechanisms are found in many viruses, particularly in bacterial viruses (phages). ADAMS, 1959, page 377).

2. Composite viruses. Examples are known of a virus which can carry the genetic material of one or even two different viruses in the same protein coat (JACOB, 1955; ADAMS, 1959). This is a startling example of symbiosis which might easily lead to the formation of a single virus with the genetic material of two or several different viruses.

3. Transduction in the E. Coli bacterium K_{12}. This form of transduction is due to the fact that a genetic segment containing several hereditary characters of the bacterium is still able, under certain conditions, to act as a virulent phage (parasite). Under other conditions it will act as a symbiont which is able to change the host organism and bring with it the hereditary characters which are localized in the genetic segment it represents. (ADAMS, 1959; BARRICELLI, 1956; various papers relevant to this point by LEDERBERG, JACOB, CAMPBELL, FUERST and others, are quoted by ADAMS, 1959).

4. Provirus region in K_{12}. An appreciable segment of genetic material in K_{12} bacteria, the provirus region, is formed by symbiotic viruses of the kind just described.

5. Defective phages. By mutation a virus may become unable to act as a parasite and hence in order to be transmitted to another host organism it may often be dependent on some mechanism different from the kind of transduction described above. Two such mechanisms (*Sal-*

monella transduction and regular cross) are described below. Usually a phage-genetic segment adapted to perform a function in the bacterium is no longer apt to perform its original function in the phage (too many mutations and substitutions). Transducing phages are therefore defective.

6. T r a n s d u c t i o n   i n   *S a l m o n e l l a .* Instead of carrying another virus or several other viruses in its protein coat, a phage may carry bacterial genes or genetic segments which no longer are able to act as parasites or individual organisms. This crossing mechanism is very common in the genus *Salmonella.*

7. R e g u l a r   c r o s s   i n K_{12}. In some cases two bacteria may attach together in order to permit the transmission of a genetic segment which would be unable to perform the operation without such assistance from the bacterial hosts. The genetic segment is transmitted from a "donor" (male) to a "recipient" (female) bacterium. In K_{12} the transmitted segment is usually different for different strains of donor bacteria used. The transmitted segment will, moreover, be shorter if the mating is interrupted. By interrupting the mating it is therefore possible to find out which part of the genetic segment is transmitted first and may be considered as the beginning or "origin" of the segment and which part represents the end of the segment. In all donor strains studied so far, the origin was represented by genetic material which at present is unable to act as a parasite and according to present knowledge has no autonomous means of transmitting itself to other bacteria. The ends of the transmitted segments, on the contrary, are all in or at the edge of the provirus region. This region consists of genetic segments which have their own mechanisms by which they may be transmitted to other bacteria (by the transduction mechanism of K_{12} bacteria). Therefore, they do not need the assistance of the bacterium in order to perform this operation.

The function of regular crossing in K_{12} is apparently to assist in the transmission of a genetic segment (probably a conglomerate of many defective viruses) which no longer has its own mechanism of transmission to other bacteria.

Several of the phenomena described above were predicted (BARRICELLI, 1947, 1952, 1955) by the symbiogenesis theory. The same phenomena give a fairly good picture of the way in which the relation between symbiotic viruses and cells may have developed from infection to a crossing mechanism. From regular crossing in K_{12}, with large genetic segments exchanged between copulating cells, a small step farther would apparently lead to a crossing mechanism like that of *Paramecium* with exchange of whole gametic nuclei between the copulating cells.

This is the symbiogenetic interpretation of the origin of sexual reproduction.

One may now ask: What about the origin of biologic evolution? Can the symbiogenesis theory explain a phenomenon like this?

5. THE FORMATION OF SYMBIOORGANISMS BY NUMERICAL ELEMENTS [3]

To test the symbiogenesis theory and find out whether it can explain how an evolution process like the biological one may arise, one may once again make use of numerical entities. All one has to do it to modify the reproduction rules used in fig. 1 in such a way that some kind of cooperation or "symbiosis" between different numerical entities will be promoted. This way one will find out whether it is true that by putting together elements with the fundamental properties of self-reproduction, mutability, and reciprocal (symbiotic) interdependency one can initiate an evolutionary process similar to biological evolution.

In order to promote symbiosis one may change in the following way the reproduction rules of fig. 1:

a) A number n will not be repeated in the square below it, as in fig. 1, but only n squares to the right (if positive) or $-n$ squares to the left (if negative) in the next row (see fig. 2 and 3). This operation will be called "translation to square (n) of next row without reproduction."

b) If the new position (n) happens to be below a square occupied by a different number m then a second n will be placed in position (m), which means m squares to the right (if m is positive) or $-m$ squares to the left (if m is negative) of the original n, but in the next row. This way a number n can reproduce if another number m different from n is present (see fig. 4).

If in this process the number n happens to come several times below different numbers it may reproduce several times (see fig. 5, where the number 4 from the second row appears 3 times in the third row).

The above reproduction rules are made with the purpose of permitting reproduction only when a numerical entity is together with other entities different from itself. A symbiotic cooperation between different numerical entities is thereby rendered necessary for reproduction.

---

3) All terms borrowed from biology when used in connection with symbiogenetic phenomena represent mathematical, not biological concepts (see appendix and note by the author at the beginning of this paper). They are in no case identical to the biologic concepts designated by the same or a similar term.

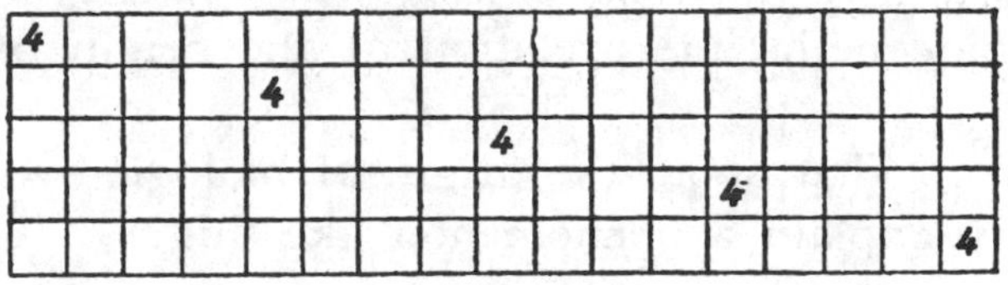

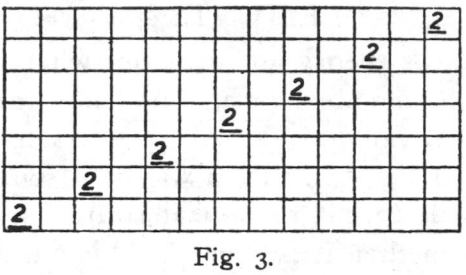

Fig. 2.

Fig. 3.

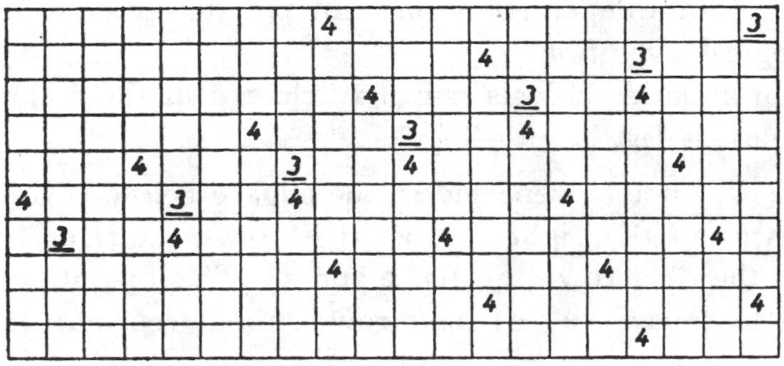

Fig. 4.

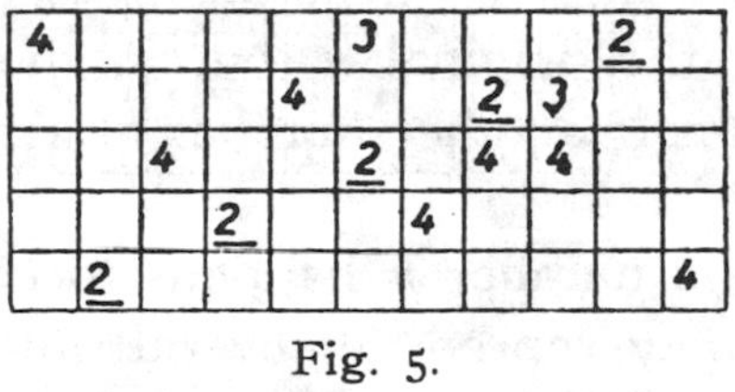

Fig. 5.

Fig. 2, 3, 4, 5. Reproduction rules requiring symbiosis (see text).

To the above reproduction rules one may add arbitrary mutation rules, for instance by taking advantage of the cases in which two different numbers collide in the same square. To begin with, a rule will be used in which no mutation can occur. The square in which two different numbers collide will be left empty, only marked by a cross (X) or "collision sign." If two identical numbers collide, one of them remains in the collision square.

In fig. 6 the absence of mutations is clearly manifested by the fact that the original numbers 5, 1 and -3 are present everywhere and no new number appears. However, in a few generations the numbers organize themselves into a stable configuration (5, -3, 1, -3, 0, -3, 1, 0 where an 0 marks an empty square) which is present everywhere in the figure. This kind of stable configurations will be called "numerical symbioorganisms." In the rest of this paper it will be shown that symbioorganisms have many properties similar to those observed in living structures. Some of the life-like properties of symbioorganisms which will be presented below are termed:

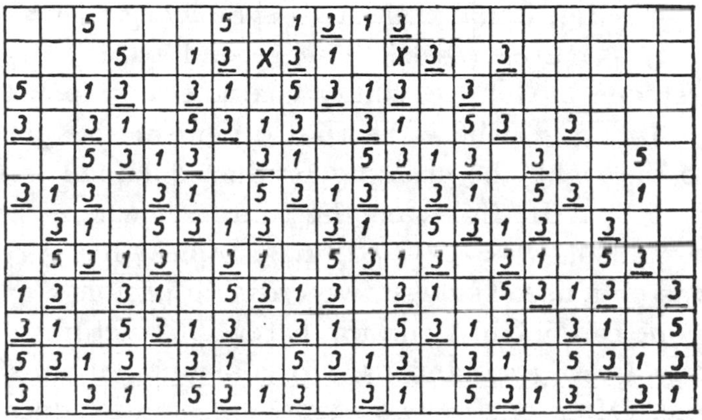

Fig. 6.

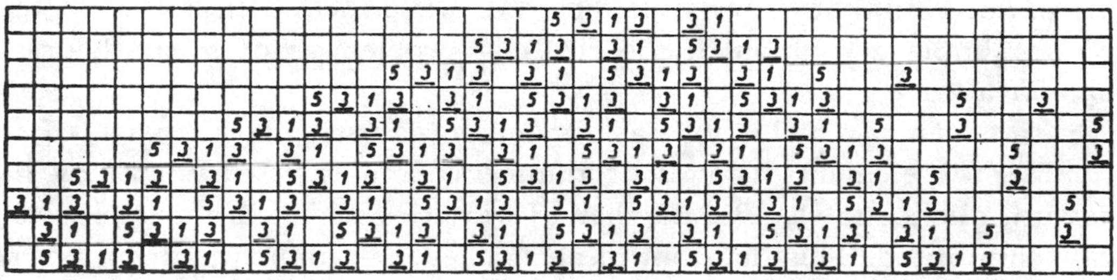

Fig. 7.

Fig. 6 and 7. Formation of a symbioorganism (6) and its reproduction characteristics (7).

(A) Selfreproduction; (B) Crossing; (C) Great variability; (D) Mutation (if the rules stated above are changed in order to permit mutation); (E) Spontaneous formation; (F) Parasitism; (G) Repairing mechanisms; (H) Evolution.

6. GENERAL PROPERTIES OF SYMBIOORGANISMS

Symbioorganisms do not have to be of numerical nature. Any elements with the property of self-reproduction and symbiotic cooperation may associate into symbioorganisms of some sort, no matter wheter they are of numerical, chemical, or any other nature. According to the symbiogenesis theory, living organisms are the particular kind of symbioorganisms which arise when organic molecules are used as self-reproducing elements.

Some general properties which one may expect to find very often in symbioorganisms and all of which are to be found in numerical symbioorganisms are the following:

A) S e l f r e p r o d u c t i o n. In fig. 7 it is shown that the symbioorganism (5, -3, 1, -3, 0, -3, 1, 0) — which arose in fig. 6 — is able to reproduce itself.

B) C r o s s i n g. In fig. 8 and 9 the two symbioorganisms (9, -11, 1, -7) and (5, -11, 1, -3) are crossed. The second parent organism differs from the first one by two hereditay characters, 5 instead of 9 and -3 instead of -7. In fig. 8, where the first organism was placed to the left, the second to the right, the crossing product obtained was the recombinant (5, -11, 1, -7). In fig. 9 on the contrary, where the order of the two parent organisms was reversed, the complementary recombinant (9, -11, 1, -3) was obtained. If these two crossing products are crossed with one another, one can obtain again the parental structures of the previous crosses as shown in fig. 10 and 11. Not all symbioorganisms do cross as neatly as these. Often one or two of the crossing products are not competetive in the presence of the parent organisms and can simply not arise, nor could they survive if they did. Nevertheless, the very existence of so simple a solution of the crossing problem is of great theoretical significance.

C) G r e a t v a r i a b i l i t y. Each symbioorganism may consist of any number of elements (genes or numbers) and each element may have several allelic states. The number of varieties which can arise is practically unlimited. This does not mean that every symbioorganism, no matter how primitive will show a great variability. But under proper conditions many symbioorganisms will (see fig. 24, generations 600, 1000, 1200, 1300, 1400, 1500, 1600, 1700 and 1800).

D) M u t a t i o n. To obtain mutations it is sufficient to change the rules for collision of different numerical elements. The change will have to be rather gentle if one wishes to keep the properties listed above.

All mutation rules which will be used in this paper consist in replacing some of the collision signs X (marking squares in which two different numbers have collided) by a new number or mutation.

One example of such mutation rules is the following (Rule A):

1) An X (collision sign) which happens to be under an occupied square remains.

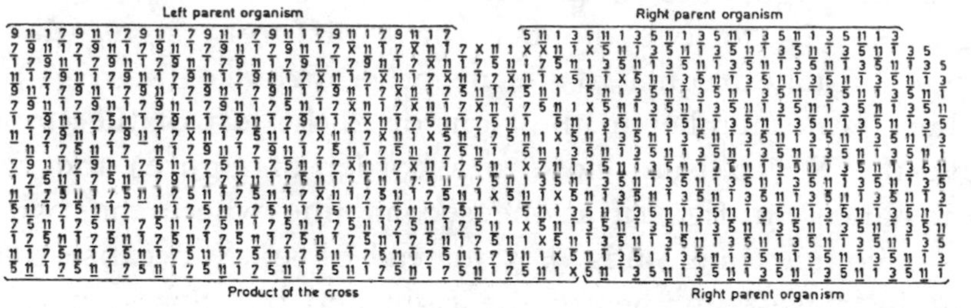

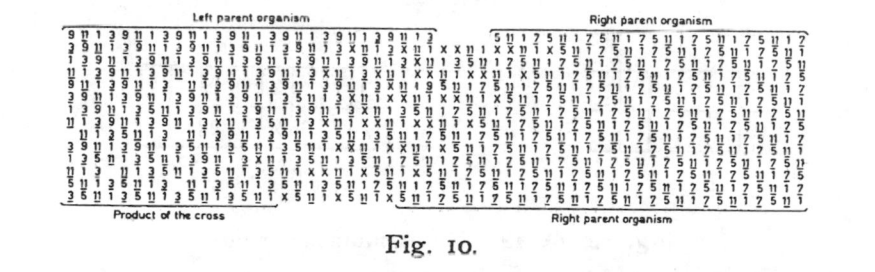

Fig. 8.

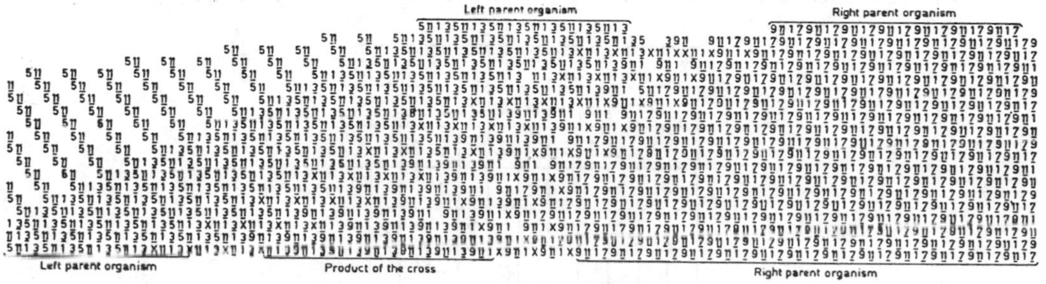

Fig. 9.

Fig. 10.

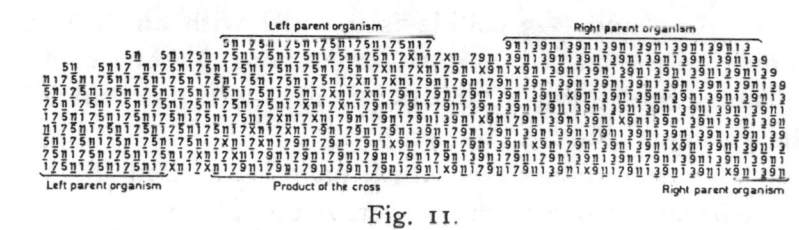

Fig. 11.

Fig. 8. 9. 10, 11. Crossbreeding in numeric symbioorganisms (see text). Crossing rules similar to those of haploid organisms.

2) An X which happens to be under an empty square or under another X is replaced by a number M (mutation) whose absolute value is equal to the distance between the closest number to the left and the closest number to the right of the empty square in question (the distance is measured in number of squares). If the two numbers have

the same sign, the mutation *M* will be the positive distance (sign +); if the two numbers have different signs, the mutation *M* will be taken equal to the negative distance (sign —).

3) If there is no number to the right or to the left of the empty square above the X (collision close to a border of figure) the X remains.

For instance, in fig. 12 at the second line where the two numbers 3 and —4 collide, one finds + 5 instead of an X because the distance between the closest number to the right and the closest number to the left in the line above is 5 squares and because they have the same sign (both are positive). On the other hand in fig. 13 the place where the

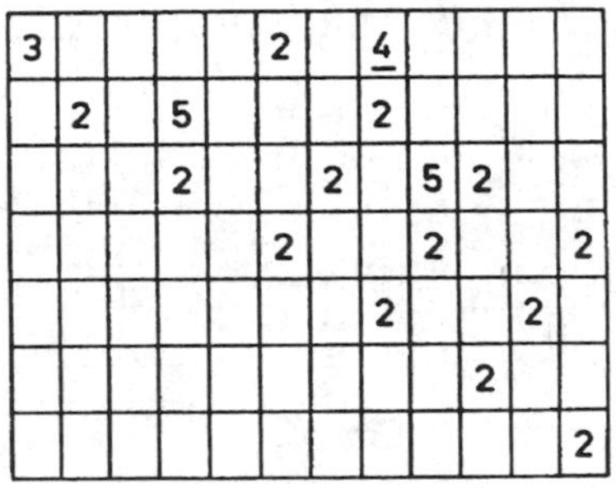

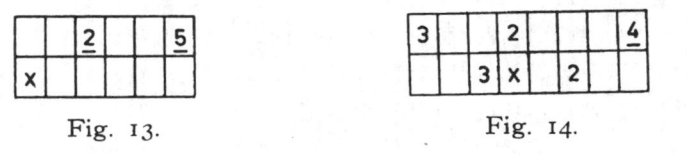

Fig. 12.

Fig. 13.   Fig. 14.

Fig. 12, 13, 14. The A-mutation rule.

two numbers —2 and —5 collide is marked with an X because there is no number to the left of X in the line above. Likewise in fig. 14 the X remains where the two numbers 3 and —4 collide since the square above the collision place is not empty.

Several other mutation rules which have been used in evolution experiments will be defined in this paper. Although the mutation rule may influence the kind of symbioorganisms which arise, the general character of the phenomena remains the same. [4]

---

4. In earlier experiments mutations based on random numbers have also been tried with comparable results. Some of the methods used in this paper will also yield fairly random mutations when a symbioorganism — as very often is the case — collides with invading numbers generated in an adjourned disorganized region. The frontier between two different symbioorganisms is usually disorganized.

E) Spontaneous formation. In fig. 15 the first row contains only the numbers 1, —1, or empty squares selected by a random procedure (heads and tails was played with two coins; two heads indicate 1, two tails —1, one head and one tail indicates an empty square). In the following generations other numbers than 1 and — 1 arose by mutation and various organisms recorded in the figures 16, 17, 18, 19, 20, 21, and 22 were formed.

This experiment shows that under favourable conditions symbioorganisms can be formed quite frequently and the formation of a symbioorganism is not a rare event. But of course if the environment is modified by the presence of other symbioorganisms such spontaneous formation can be prevented.

F) Parasitism. One may note that one of the symbioorganisms (1, —2, 1, 1, —2, 0) a so called tregener which arose near the upper right border of fig. 15 in the above experiment (the one which is recorded in fig. 17) does not reproduce completely. In fig. 17 the symbioorganism loses its elements one by one and at the end nothing remains. However, If the same symbioorganism is sowed together with the symbioorganism (1, —1) which acts as its host it will reproduce normally as shown in fig. 18. At the same time, the host is destroyed little by little. This relationship between two symbioorganisms is very similar to the phenomenon which in biology is called parasitism. The same terminology may be used here. The symbioorganism (1, —2, 1, 1, —2, 0) shall be called a parasite of (1, —1).

Parasitism is a very common phenomenon among symbioorganisms and it constitutes one of the major difficulties in the performance of numerical evolution experiments. Often an evolution process will end with the destruction of the species by a parasite which also will die out once the host is destroyed.

G) Repairing mechanism. Damages are usually repaired by cooperation among several neighboring symbioorganisms. The repairing process may often be partially or completely successful even if all the symbioorganisms present are damaged by removing (cancelling) some of their genes. In this case the repairing is more likely to succeed the greater the number of symbioorganisms cooperating.

This phenomenon presents a startling analogy with the repairing mechanism known in bacterial viruses as "reactivation by multiple infection with inactivated (for ex. irradiated) phages."

In all the 5 odd symbioorganisms of line 9 in fig. 23, a large fraction of the numerical elements, selected by a random procedure, are can-

Fig. 15. Spontaneous formation of symbioorganisms in an experiment started with random numbers

Fig. 16

Fig. 17

Fig. 18

Fig. 19

Fig. 20

Fig. 21

Fig. 22

Fig. 16, 17, 18, 19, 20, 21, 22. Reproduction of symbioorganisms which arose in the experiment of fig. 15. One organism (parasite Fig. 17) does not reproduce succesfully when alone. However in association with its host organism (Fig. 18) it reproduces normally.

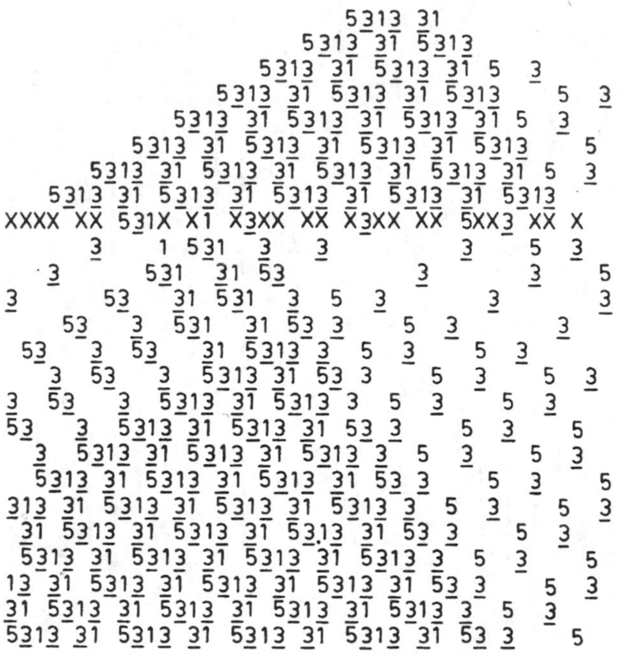

Fig. 23. Repairing mechanism by cooperation of several seriously damaged symbio-organisms. Removed numbers (damages) in the ninth line are marked by X's.

celled (replaced by an X in the fig.). The damaged symbioorganisms left reconstitute the original pattern in a few generations. None of the 5 odd damaged symbioorganisms of line 9 would in itself have been able either to reconstitute the original pattern or else to generate another pattern able to survive and reproduce.

Recovery by the above repairing mechanism is not always complete, and the reconstituted symbioorganism may often show deficiencies (missing genes). A deficiency is not always lethal, and the symbio-organism may therefore survive as a mutant.

H) Evolution. In fig. 15 the only example of hereditary change in a symbioorganism is the change which transformed (1, —1) into (1, —3) at the left side of the figure around generation 60. This change was induced by an external element (—3) entering the symbioorganism. Using high speed computers, evolution experiments have been per-formed which involved a large number of hereditary changes. Some of these evolution experiments will be described in this paper. The result of such experiments can be summarized as follows:

Symbioorganisms can be completely modified. Their complexity may drastically increase, and they can branch into different species which are unable to interbreed.

The evolution leads to a better adaptation to the artificial environmental conditions and a greater ability to compete with other symbioorganisms (see competition tests, Section 9).

Both mutation and crossing phenomena played a central part in evolution. To begin with, most of the crossing was performed by single elements (genes) which left one symbioorganism and entered another symbioorganism. Later the "regular" crossing mechanism described in fig. 8, 9, 10, and 11 became predominant. Such regular crossing is very similar to the crossing in haploid species of living organisms.

7. EVOLUTION EXPERIMENTS

In order to observe evolutionary phenomena, the experiment of fig. 15 would have to be repeated on a much larger scale and for a much larger number of generations. Such experiments were done in 1953, 1954, and 1956 in Princeton, New Jersey, using the electronic computer of the Institute for Advanced Study. Instead of 100 lines (generations) with 80 numbers each as in fig. 15, the Princeton experiments were continued for more than 5,000 generations using universes of 512 numbers. Moreover, the actual size of the universe was usually increased far beyond 512 numbers by running several parallel experiments with regular interchanging of several (50 to 100) consecutive numbers between two universes every 200 or every 500 generations (see BARRICELLI, 1957). A technical difficulty which showed up at once was the development of a phenomenon which will be called "homogeneity." Within a few hundred generations a single primitive variety of symbioorganism invaded the whole universe. After that stage was reached no collisions leading to new mutations occurred and no evolution was possible. The universe had reached a stage of "organized homogeneity" which would remain unchanged for any number of following generations, and the "final" symbioorganism occupying such universe would remain unchanged.

Attempts to modify the mutation rule did not remove this difficulty, but they often led to a different type of final symbioorganism better adapted to the new mutation rule. In many instances a new mutation rule would lead to a complete disorganization of the whole universe, apparently due to the death by starvation of a parasite, which in this case was the last surviving organism. After this stage of "disorganized homogeneity" was reached, some new organisms could arise again by chance once in a while. But they were always promptly destroyed and the stage of disorganized homogeneity was maintained. The reasons for development of disorganized homogeneity are not entirely understood. But a mutation rule which encourages parasitism

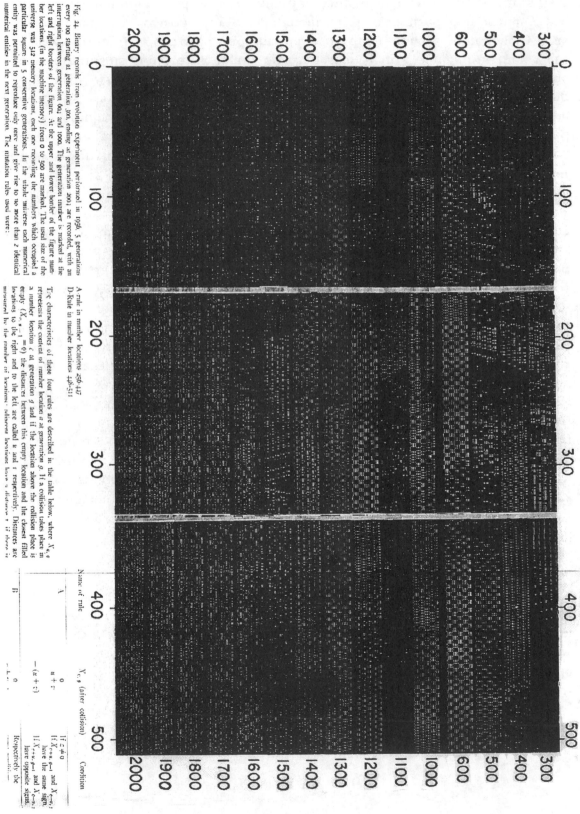

Fig. 24. Binary records from evolution experiment performed in 1956. 5 generations every 100 starting at generation 300, ending at generation 300; are recorded, with an interruption between generation 600 and 1000. The generation number is marked at the left and right borders of the figure. At the upper and lower border of the figure number locations (in the machine memory) from 0 to 500 are marked. The usual size of the universe was 512 memory locations, each one recording the numbers which occupied a particular square in 5 consecutive generations. In the whole universe each numerical entity was permitted to reproduce only once and give rise to no more than 2 identical numerical entities in the next generation. The initiation rules used were:

A rule in number locations 256-447
D-Rule in number locations 448-511

The characteristics of these four rules are described in the table below, where $X_{c,g}$ represents the content of number location c at generation g. If a collision takes place in a number location c at generation g and if the location above the collision place is empty ($X_{c,g-1} = 0$) the distances between this empty location and the closest filled locations to the right and to the left are called x and z respectively. Distances are measured by the number of locations; adjacent locations have a distance 1; if there is

Name of rule	$X_{c,g}$ (after collision)	Condition
A	0	if $z \geq x \neq 0$
	$x + z$	if $X_{c+x, g-1}$ and $X_{c-z, g-1}$ have the same sign
	$-(x + z)$	if $X_{c+x, g-1}$ and $X_{c-z, g-1}$ have opposite signs, Respectively the
D	0	

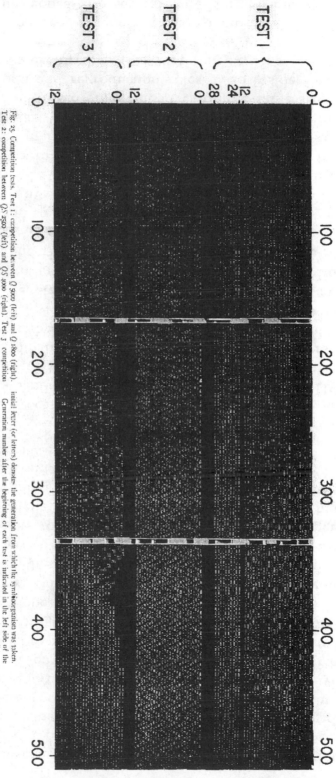

Fig. 25. Competition tests. Test 1: competition between Q 3000 (left) and Q 1800 (right).
Test 2: competition between QS 2500 (left) and QS 4000 (right). Test 3 competition
between QS 4000 (left), R 1900 (center) and Q 5000 (right). The number following the

initial letter (or letters) denotes the generation from which the symbioorganism was taken.
Generation number after the beginning of each test is indicated in the left side of the
figure. Number locations are marked at the lower and upper borders of the figure.

followed by starvation of the parasites once the host organisms are used up, could be expected to give a result like this.

Both homogeneity problems were eventually overcome by using different mutation rules in different sections of each universe. Also slight modifications of the reproduction rule were used in different universes to create different types of environment (BARRICELLI, 1957). Furthermore, by running several parallel experiments and by exchanging segments between two universes every 200 or 500 generations it was possible to break homogeneity whenever it developed in one of the universes.

Successful evolution experiments lasting for several thousand generations have been performed by these methods. However the reasons for the difficulties and for the success of the procedure used are not yet sufficiently understood. Later attempts to apply the same idea (see next paper of this series, BARRICELLI, 1962) have shown that this method is more likely than not to lead to a failure unless the same combination of mutation and reproduction rules is applied again. The rules used are indicated in the legend of fig. 24.

8. RESULTS

Part of an evolution process developed in Princeton in 1956 is described in fig. 24. The figure is obtained by a photographic method from IBM cards punched by the computer and represents a part of the memory of the computer at various stages during the evolution process. In a universe of 512 number locations (the number locations, 0, 100, 200, 300, 400, and 500 are marked in the upper and lower border of fig. 24), 5 generations every 100 are recorded in the figure, starting at generation 300 and ending at generation 2000 with an interruption between generation 600 and 1000. For example, the number 300 at the right border of the figure marks the beginning of the 5 generations 300-304, the location 400 marks the beginning of the 5 generations 400-404 and so forth. The number location and the generation can be used to identify any spot in the figure. The data of the figure are given in binary numbers. In a few cases the binary numbers will be translated into decimal numbers to identify some of the organisms for the reader.

The universe described in fig. 24 represents one of 5 parallel experiments performed by the computer. At regular intervals segments of the universe in fig. 24 were exchanged with two other universes of the same size but with slightly different mutation and reproduction rules. Also inside fig. 24, four different mutation rules were used in four different regions (see legend of fig. 24).

A set of numbers introduced at generation 200 from another universe of a parallel experiment, developed into several symbioorganisms in successive generations as recorded in the figure. At generation 600 a single species represented by a few different varieties able to interbreed, occupied the whole universe. At generation 1000 the same species and its parasites are still evolving by mutation, crossing, and selection. Its varieties which can be seen in the figure at generation 1000 are:

Symbioorganism	Type	Period	Number Locations
3 3 3 3 3 3	Independ.	2	0-80 and 490-511
3 3 3 3 6 3 3 6 3 3 3 3 3 6 3 3 6 3	Parasite	18	80-250
3 3 3 3 0 3 3 6 3 3 3 3 3 3 0 3 3 6 3	Parasite	18	250-400
3 3 3 3 9 3	Independ.	6	420-490

Symbioorganisms like these which are formed by multiples of 3 will be called *R*-organisms.

At generation 1100 the universe was almost entirely disorganized, but the *R*-organisms either survived around number location 200 or were reintroduced by an exchange with one of the other two universes. At generation 1200, three *R*-organisms occupied the region between the number locations 180 and 400. After generation 1200 the evolution of *R*-organisms, as can be seen in the figure, leads to a greater and greater complexity. Every number (gene) in the original symbioorganism is replaced several times by new mutations. After generation 1600 all genes are multiples of 6. This is a characteristic property of a type of the symbioorganisms which are called *Q*-organisms (BARRICELLI, 1957). After generation 1800 the *Q*-organisms acquire a periodicity of 72 genes, which will remain their characteristic period to the end of the experiment. After this characteristic periodicity was reached the evolution proceeded at a slower rate until generation 5000 when the experiment was discontinued. The last 3000 generations are not recorded in fig. 24, but the end result, or the *Q*-organism which occupied the same universe at generation 5000 will be shown in fig. 25.

A large number of other symbioorganisms arose after generation 2000. One of these, which will be called *QS*-organism, survived for a long time in competition with the *Q*-organisms. After generation 4000 the *Q*-organisms invaded all 5 universes and the *QS*-organisms were eradicated. The *QS*-organisms arose as a result of a symbiotic association of a *Q*-organism with another species (*S*-organism) whose genes are even numbers but mostly not multiples of 6. Every other gene in a *QS*-organism belongs to a *Q*-organism, and the remaining genes belong to an *S*-organism. Two examples of *QS*-organisms are recorded in fig. 25 in the second competition test. The *QS-organism* to the right has a period 16 and 4 alternate generations. Its numeric elements translated into decimals are in the 4 generations.

Generation

1	36,	14,	24,	10,	12,	30,	24,	34,	36,	14,	24,	10,	12,	30,	24, 34
2	12,	10,	24,	30,	36,	34,	24,	14,	12,	10,	24,	30,	36,	34,	24, 14
3	36,	30,	24,	34,	12,	14,	24,	10,	36,	30,	24,	34,	12,	14,	24, 10
4	12,	34,	24,	14,	36,	10,	24,	30,	12,	34,	24,	14,	36,	10,	24, 30

Several Q-organisms also show a comparable degree of complexity (for a more detailed description of some evolution processes see BARRICELLI, 1957). Both in the Q- and the QS-organisms every gene was replaced several times during the evolution processes. The number of different varieties which arose either by viable mutations, by crossing, or by invasion of foreign elements must have been at least several thousand in each of the two types of symbioorganisms. Several hundred-thousand mutations (judging from the amount of machine time used and other criteria) arose during the 5 parallel experiments, but less than 1 % of these mutations gave rise to competitive varieties.

9. COMPETITION TESTS

Three experiments made in order to compare the competitivity or fitness reached by several symbioorganisms are recorded in fig. 25. In the first experiment (Test 1) two Q-organisms are compared. To the right (number locations 240-511) is the Q-organism (Q 1800) which occupied the same region at generation 1800 in fig. 24. To the left (number locations 0-240) is a Q-organism (Q 5000) which occupied this region at generation 5000 in the same experiment which is partly recorded in fig. 24. The mutation and reproduction rules used and their distribution according to number location are the same as in fig. 24.

The experiment clearly shows that Q 5000 is much more able to compete under the conditions of fig. 24 and 25 than the Q 1800 organism. In the first 12 generations recorded in the figure Q 1800 is rapidly being destroyed and Q 5000 is rapidly invading its area. At the generations 24-27, which are also recorded in fig. 25, Q 5000 has invaded the whole universe and nothing is left of Q 1800. The reader may have noticed that Q 5000 penetrated into the region of Q 1800 not only from the left but also from the right border. This is because the reproduction rules have been applied considering the universe as circular, or in other words considering number location 0 as identical with 512 and number location 511 as identical with —1. The same has been done in fig. 24, and the reader may notice that the symbioorganism close to the left border is very often the same variety as the symbioorganism close to the right border in the same generation.

In the second experiment (Test 2) two QS-organisms are compared; a QS 2500 to the left (loc. 0-240) with a QS 4000 to the right (loc. 240-511).

Both symbioorganisms appeared in the continuation of the evolution experiment of fig. 24 in generations 2500 and 4000 respectively. In this case the two symbioorganisms are still able to interbreed and may be considered as belonging to the same species, in spite of the 1500 generation interval which separates them. Nevertheless a kind of frontier between the two organisms can easily be seen in the figure, marked by the empty spots due to collisions of different genes. It is evident that the genes of QS 4000 are rapidly invading the area of QS 2500. After 35 generations the entire universe of fig. 25 (Test 2) was occupied by the genes of QS 4000 and every gene of QS 2500 had disappeared. Also in this case the variety which had endured a longer period of evolution in the universe of fig. 24 was better fit to compete in this universe.

In the third experiment (Test 3) three symbioorganisms, Q 5000, QS 4000, and an R-organism which appeared in a parellel experiment at generation 1900, are compared. Again the symbioorganism (Q 5000), which had experienced the longest period of evolution, was the one which survived and destroyed the other two.

In any particular universe the survival ability of a symbioorganism is evidently determined by its genetic information, and it is clear that the ability to survive is improved by evolution. The genetic structure of any symbioorganism can rightly be considered as a coded survival strategy program created and developed during its past evolutionary history.

10. ERADICATION OF PRIMITIVE SYMBIOORGANISMS

The tests performed in fig. 25 show one reason why the Q-organisms are the only ones which survived at generation 5000: all other symbioorganisms had evidently been eradicated by the more competitive Q-varieties.

A glance at the conditions at generation 5000 or even at generation 4000 could easily give the mistaken impression that a symbioorganism had been formed by a lucky accident some time during the experiment, and that this had given rise to all the Q and QS varieties developed later. The reader who has followed the experiments of fig. 15 and fig. 24 from the beginning is certainly aware that a large number of unrelated symbioorganisms usually arise in such experiments. The only reason why they do not arise also at a later stage is that they have no chance to do so in the presence of far more competitive symbioorganisms developed later. The strong competition is also the reason why all symbioorganisms which survive at the end are related. In any given number of generations there is a finite probability that one or several competitors will be eradicated. At the end only one type of symbioorganisms (which in the present case happened to be the Q-organisms) will survive.

If this lesson can be applied to living organisms on Earth, the fact, that present life forms seem to be more or less related, is no argument for claiming that symbiogenesis occurred only once on this planet. Nor is the fact that new life forms do not seem to arise any longer an argument for claiming that symbiogenesis could not be frequent under favorable conditions and without competition from preadapted species (see also "oxygen as antibiotic agent," BARRICELLI, 1962, section 12).

11. THE EARLY CROSSING MECHANISMS

Another lesson which can be learned from these investigations is that crossing is a phenomenon which appears very early in a symbiogenetic evolution process and certainly before the symbioorganisms reach the complexity we find in living beings (see next paper of this series BARRICELLI, 1962). All symbioorganisms investigated showed some kind of crossing phenomena. In some cases a crossing process could have the character of an infection or infiltration by a symbiont or a numeric element delivered by another symbioorganism. However in all evolution processes the regular crossing mechanism described in the figures 8, 9, 10, and 11 became rapidly predominant with evolution (for a closer description and classification of the crossing mechanisms observed see BARRICELLI, 1957). The common idea that living beings may have existed for a long time before crossing mechanisms appeared is hardly consistent with a symbiogenetic interpretation. The discovery of crossing in viruses and bacteria support the idea that life may never have developed without crossing phenomena. The accidental loss of crossing ability observed in some cases is usually equivalent to the loss of more than 99.9 % of the potential evolution rapidity (FISHER. 1930). The extinction of the species seems the inevitable result, as soon as possible competitors take advantage of the situation.

12. CONCLUSION

It is clear that many phenomena which were considered peculiar for living beings are common to a large class of symbioorganisms. A question which apparently has troubled many inquirers in these problems is that of the nature of symbioorganisms and their relationship to living organisms. What are the numeric symbioorganisms presented in this paper, and what are all the other types of symbioorganisms, artificial or natural (see next paper of this series BARRICELLI, 1962)? Are they the beginning of, or some sort of, foreign life forms? Are they only models through which one is able to investigate some properties of living beings and some aspects of biological evolution?

The answers are: (1) They are not models, not any more than living organisms are models. They are a particular class of selfreproducing structures already defined. (2) It does not make sense to ask whether symbioorganisms are living as long as no clearcut definition of "living" has been given (see also: "Note by the author" at the beginning of the present paper). It makes better sense to ask whether the so-called living beings which populate this planet are symbioorganisms, and if so, what kind of symbioorganisms they are. The definition of symbioorganism has already been given and can be summarized as follows: A symbioorganism is a selfreproducing structure constructed by symbiotic association of several selfreproducing entities of any kind.

The numeric symbioorganisms presented in this paper have been developed in order to test a specific prediction of symbiogenesis theory. The prediction is that the phenomena listed in section 6 (general properties of symbioorganisms) and possibly other phenomena until recently considered characteristic of living organisms (biophenomena) are common to a large class of symbioorganisms. The test has given an unequivocal, positive answer. A large group of biophenomena has been detected in the numeric symbioorganisms developed. In the next paper of this series (BARRICELLI, 1962) two more biophenomena (the evolutionary improvement of performance in a specific task and an infection process by a parasite which produces limited injuries but does not destroy the host) will be described.

The answer obtained is also an answer to the question whether symbiogenesis theory can explain the presence of the previously listed biophenomena in living organisms. If living beings are symbioorganisms, they obviously may have the same chance as other symbioorganisms to develop these particular biophenomena.

The properties a class of self-reproducing entities would need in order to build up symbioorganisms and start a symbiogenetic evolution process are extremely simple. There is no reason to believe that billions of years would be needed to develop the biophenomena described. What is needed is not primarily time but a favorable chemical environment in which a large number of interconnected (symbiotic) autocatalytic reactions takes place. The results presented in this paper show that under proper conditions even in the very limited memory of a high speed computer a large number of symbioorganisms can arise by chance in a few seconds. It is only a matter of minutes before all the biophenomena described can be observed. Given enough time, there is no *a priori* reason why a proper chemical environment could not do the same job.

There are very few important limitations concerning temperature and other environmental conditions required for development of symbioorganisms. Neither the low temperature of the moons of Uranus nor the high temperature of the Sun-side face of Mercury are sufficient arguments to exclude the possibility of symbiogenetic phenomena. The experiment recorded in fig. 24 is a clear demonstration that symbiogenesis can not only take place on a planet or satellite but even in the memory of a high speed computer. To maintain that only conditions similar to those prevailing on Earth could permit symbiogenetic processes would obviously be too great a pretention.

The symbioorganisms evolving in a sufficiently large environment for a sufficiently long period of time may develop properties and organs with any degree of sophistication which could be useful in their particular environment. The number of allelic states and varieties which are possible in a symbioorganism (2^n or more, where n is the number of genes, see section on symbiogenesis theory) is not a serious limitation to the evolutionary possibilities. Unless some other severe limitation is imposed by the conditions of the experiment or the type of universe in which the organism exists (computer, planet, or test tube), there is no *a priori* reason for assuming that other classes of symbioorganisms could not reach the same complexity and efficiency characteristic for living organisms on this planet.

Whether life of the particular form (based on nucleic acid-protein association) which has developed on this planet is frequent or rare or even a unique phenomenon is difficult to say. On the other hand, symbiogenesis in one form or another is likely to have occurred on every planet or satellite where a large number of interconnected autocatalytic reactions are possible or have been possible in the past. The conditions required are too simple for considering symbiogenesis as a unique or even an infrequent phenomenon. However, as shown in the experiments of fig. 24 and 25, once a symbiogenetic process has established itself, it would probably prevent or promptly eradicate any other symbiogenetic activity which could possibly compete or interfere with it. Under these conditions such new symbiogenetic activities may indeed not only be infrequent but for all practical purposes impossible.

APPENDIX

Readers who are familiar with the notion of stochastic time-series will probably remember the definition of a Markovian time-series. An obvious extention of this concept is represented by the notion of a Markovian vector time series, in which each term is a vector V_t with several components (or a set of numbers) instead of a single number.

Definition:

If the probability distribution $P(V_{t+1})$ of V_{t+1} exclusively depends on the last preceding term V_t (*i.e.* on the values of the V_t components) the vector-time series is called Markovian. A degenerative case of a Markovian vector time-series is the so called deterministic case in which not just $P(V_{t+1})$ *but* V_{t+1} itself is entirely determined when V_t is given. This deterministic case, which would obviously be trival and uninteresting in a normal (numerical) time-series, is neither trival nor uninteresting in a vector time-series, as the reader will have the opportunity to experience. As a matter of fact the so called time t does not necessarily have to be time and could be substituted by any dimension (called pseudo-time) implying the same relationships between consecutive terms of the series. The numeric evolution processes presented in this paper are examples of a Markovian vector time-series, in which every vector (or term) has, as its components, the numbers (inclusive zeros or empty squares) constituting a particular generation of the process. Successive generations are successive terms (vectors) of a Markovian series. In fact every generation V_{t+1} or at least the probability distribution of its terms is completely determined when the preceding generation V_t is given. The Markovian series will be deterministic if the mutation rules applied specify the mutations in every single collision. It will not be deterministic if random numbers are used in the determination of at least part of the mutations.

As a matter of fact it is not necessary to consider only series consisting of vectors or sets of numbers. Instead of numbers one may consider any kind of entities (for example molecules or other gadgets) to be called "Markovian elements". If every term or set of elements V_{t+1} (or at least the probability distribution for the possible alternatives of V_{t+1}) exclusively depends on the preceding term V_t the sample of consecutive sets will be called a timed Markovian universe or shortly a universe. The notions of "deterministic" and "non-deterministic" universes are obvious from previous definitions.

Readers who are familiar with the complexity of time-series problems will be aware that Markovian universes can hardly have been the object of farreaching mathematical investigation. However, the use of high speed data processing computers makes possible the study of particular examples of Markovian phenomena such as the numeric evolution phenomena investigated in this paper.

The notion of Markovian universes and Markovian entities gives the basis for a mathematical definition of the concepts used in symbiogenesis theory. We shall, however, not afflict the reader by presenting a long list of definitions. It will be sufficient to point out that the notions of self-reproducing

Markovian entities, self-reproducing sets of Markovian entities, utility association (symbiosis), symbioorganism, hereditary changes by crossing, mutation *etc.* can be defined in terms of Markovian concepts and are to be considered as mathematical notions.

The numeric evolution processes and probably biologic evolution, if the symbiogenesis theory is correct, are empirical examples of two different symbiogenetic evolution processes. However, this statement does not tell what life is, and can only describe one aspect (the symbiogenetic and evolutionary aspect) of this complex phenomenon.

ZUSAMMENFASSUNG

Das Problem der Erforschung von Evolutionsphänomenen und Theorien unter Verwendung von künstlichen selbst-reproduzierenden Einheiten wird besprochen. Die Notwendigkeit eines theoretischen Prinzips zur Lösung des Variabilitätsproblems wird betont. Es wird gezeigt, dass das Variabilitätsproblem gelöst werden kann unter Annahme, dass selbst-reproduzierende Modelle (,,Symbio-organismen") von jeder Komplexität gebildet werden können durch eine symbiotische Assoziation von verschiedenen selbst-reproduzierenden Einheiten, jede mit sehr geringer oder gar keiner Variabilität (Symbiogenese-theorie). Weiter wird gezeigt, dass Symbio-organismen entworfen werden können, welche imstande sind, Kreuzungsphänomene verschiedener Art zu erfahren. Schlussfolgerungen, Voraussagungen und empirische Beobachtungen die eine symbiogenetische Interpretation der Entstehung von Kreuzungen in lebendigen Organismen stützen, werden besprochen.

Als einmal das Variabilitätsproblem überwunden war, ergab sich die Möglichkeit, künstliche selbst-reproduzierende Einheiten zu entwerfen, welche imstande sind, eine Anzahl verschiedener Evolutionsphänomene zu entwickeln. Die Symbio-organismen, entwickelt durch die elektronische Rechenmaschine von dem "Institute for Advanced Study, Princeton, N.J.", zeigten auch Kreuzungsphänomene, welche sich schnell zu einem Kreuzungsmechanismus entwickelten, der den Vererbungsgesetzen ähnlich ist, wie sie bei lbendigen Organismen beobachtet werden.

Eine Mannigfaltigkeit von anderen Phänomenen, welche in einer imposanten Weise mit biologischen Phänomenen übereinstimmen, sind in den Abschnitt über allgemeine Eigenschaften von Symbio-organismen aufgenommen. Ein konkurrenz Versuch zwischen mehreren Symbio-organismen, in verschiedenen Stadien eines Evolutionsexperiments gewählt, wurde angestellt. Der Versuch zeigte eine dramatische Verbesserung von Eignung und eine Zunahme von Konkurrenzfähigkeit während des Evolutionsprozesses. Verschiedene Folgerungen von evolutionären Fortgang und Aussterben von untereinander konkurrierenden Symbio-organismen werden besprochen. Die Einfachheit der Bedingungen für Symbiogenese und die grosse Wahrscheinlichkeit, dass symbiogenetische Phänomene verschiedener Art auf anderen Planeten und Satelliten in Entwicklung sein könnten, wird betont.

SUMMARY

The problem of testing evolution phenomena and theories by using artificial self-reproducing entities is discussed. The need of a theoretical principle which can permit the solution of the variability problem is emphasized. It is shown that the variability problem can be solved assuming that self-reproducing patterns (called symbio-organisms) of any complexity can be formed by a symbiotic association of several self-reproducing entities, each with very low variability or no varability at all (symbiogenesis theory).

Furthermore it is shown that symbio-organisms able to undergo crossing phenomena of various types can be designed. Consequences, predictions, and empirical observations supporting a symbiogenetic interpretation of the origin of crossing in living organisms are discussed.

Once the variability problem was overcome, it was possible to design artificial (e.g. numerical) selfreproducing entities able to develop a variety of evolutionary phenomena. The symbio-organisms developed by the electronic computer of the Institute for Advanced Study, Princeton, New Jersey, also showed crossing phenomena which rapidly developed into a crossing mechanism following laws of heredity similar to those observed in living organisms. A variety of other phenomena presenting impressive analogies with biological phenomena are listed in the section on general properties of symbio-organisms.

A competition test between several symbio-organisms selected in different stages of an evolution experiment was performed. The test showed a dramatic improvement of fitness and increase of competitivity during the evolution process. Several consequences of evolutionary improvement and eradication of competitors are discussed.

The simplicity of the conditions for symbiogenesis and the high probability that symbiogenetic phenomena of various kinds may have been developing on other planets and satellites is emphasized.

REFERENCES

ADAMS, M. H. (1959). Bacteriophages. — New York & London, Int. Publ., xviii + 592 p.

BARRICELLI, N. (1947). The hypothesis of the gene-symbiosis and its importance in the explanation of the origin of the gamic reproduction. — Oslo, Cammermeyers Boghandel, 9 p.

—— (1952). Mikroorganismenes genetikk. — Naturen 1952, Nr. 6, p. 162-191.

—— (1954). Esempi numerici di processi di evoluzione. — Methodos 1954, p. 45-68.

—— (1955). On the manner in which crossbreeding takes place in bacteriophages and bacteria. — Acta biotheor., Leiden XI, p. 75-84.

—— (1957). Symbiogenetic evolution processes realized by artificial methods. — Methodos IX, Nr 35-36.

—— (1962). Numerical testing of evolution theories: 2. Preliminary tests of performance. Symbiogenesis and terrestrial life. — Acta Biotheor. (in press).

FISHER, R. A. (1930). The genetical theory of natural selection. — London, Oxford Univ. Press, xiv + 272 p.

JACOB, F. (1955). Transduction of lysogeny in E. Coli. — Virology I, p. 207-220.

KOZO-POLIANSKY, B. (1924). Outline of a theory of symbiogenesis. — Selkhozgiz.

LEDERBERG, E. & J. (1952). Genetic studies of lysogenicity in Escherichia Coli. — Genetics XXXVIII, p. 51-64.

MERESCHOWSKY, C. (1910). Theorie der zwei Plasmaarten als Grundlage der Symbiogenesis, einer neuen Lehre von der Entstehung der Organismen. — Biol. Zbl. XXX, p. 278-303, 321-347, 353-367.

OPARIN, A. I. (1938). The origin of life. (Transl. with annotations by S. Morgulis) — New York, Macmillan, xii + 270 p.

SONNEBORN, T. M. (1949). Beyond the gene. — Amer. Scient. XXXVII, p. 33-59.

VON NEUMANN, J. (1951). The general and logical theory of automata. — In: L. A. Jeffress, ed., Cerebral mechanisms in behavior, The Hixon Symposium, p. 1-32. — New York, J. Wiley; London, Chapman & Hall.

NUMERICAL TESTING OF EVOLUTION THEORIES

Part II

PRELIMINARY TESTS OF PERFORMANCE. SYMBIOGENESIS AND TERRESTRIAL LIFE. [1]

by

NILS AALL BARRICELLI

(Department of Biology, Division of Molecular Biology)
(Vanderbilt University, Nashville, Tennessee)
(Rec. 27.XI.1961)

NOTE BY THE AUTHOR

In the latter part of this paper the nature of the relationship or similarities between living beings and other symbioorganisms is discussed. Some of the conclusions may be surprising to the reader. However, it must be pointed out that nothing which is presented in this paper can justify the conclusion that any other type of symbioorganism except the so called "Terrestrial life forms", which populate this planet, are alive. As a matter of fact this question has no meaning as long as there is no agreement on a definition of "living being". However, the reciprocal question "whether the objects we are used to call living beings are a particular class of symbioorganisms" has a meaning. This is the question we have been trying to answer in this paper and the preceeding paper in this series (BARRICELLI, 1962). If the nature of the answer and its consequences should make the reader feel somewhat disoriented an advise which may prove useful for science readers as it is for mountain climbers is "Hold on solid ground". Proven facts and rigorous deduction are the solid ground on which scientific knowledge can be based. Feelings and opinions and any form of instinctive resistancy to new ideas are not.

Everything which is said in this paper should be understood, statement by statement, the way it is presented. The author takes no responsibility for inferences and interpretations which are not rigorous consequences of the facts presented. As in the previous paper of this series (BARRICELLI, 1962) the terms used in connection with symbiogenetic phenomena do not have the same meaning they have in biology. They refer to mathematical concepts whose relation to the corresponding biological concepts is a matter of investigation.

---

1) This investigation was supported by research grant RG-6980 from the Division of General Medical Sciences of the National Institutes of Health, U.S.A. Public Health Service. The first part of this investigation was performed in the fall, 1959, while the author was visitor to the A.E.C. Computing Center, N.Y.U. The investigation was continued in the summer, 1960, while the author was Visiting Research Associate at Brookhaven National Laboratory, L.I.N.Y. and after his return to Vanderbilt University, Nashville, Tennessee

1. INTRODUCTION

In the first paper of this series (BARRICELLI, 1962) the results of numeric evolution experiments performed in Princeton, N. J. were presented. In one of the experiments, the evolutionary improvement was verified by competition tests between symbioorganisms at different stages of evolution. The tests clearly showed that symbioorganisms at a more advanced stage of evolution (with a longer evolution history behind them) easily eliminated more primitive organisms belonging to the same or to a different species (see fig. 25, BARRICELLI, 1962). Evidently the ability of the various symbioorganisms to perform operations necessary or useful for their survival was improved during the evolutionary process.

A question which arises in this connection is whether it would be possible to select symbioorganisms able to perform a specific task assigned to them. The task may be any operation permitting a measure of the performance reached by the symbioorganisms involved; for example, the task may consist in deciding the moves in a game being played against a human or against another symbioorganism. Evidently if a measurable improvement in a specific performance can be obtained by selection, this would open exciting possibilities. The evolutionary development of specialized structures with a specific function and a specific survival value could be open to investigation.

The problem of testing the improvement in a specific performance will be the primary subject of the first part of this paper.

A related problem should be mentioned here even though its investigation has not yet reached a stage where it can give fruitful results. A peculiar characteristic of the symbioorganisms developed so far is that they consist exclusively of selfreproducing entities, which perform the function of genetic material. These selfreproducing entities are permitted to interact exclusively with other selfreproducing entities. No other structures formed or modified or rearranged by the selfreproducing entities are involved. There is no parallel to what may be called somatic or non-genetic structures of living organisms.

This peculiarity is evidently due to the reproduction and mutation norms used. To save labor, computing time, and machine memory, the norms used did not involve entities which were not selfreproducing and could be dispensed of in the first evolution experiments. In the tests of performance to be reported below, the answers or decisions yielded by each symbioorganism will be expressed by a set of numbers. This will involve the formation of non-genetic numerical patterns characteristic for each symbioorganism. Such numerical patterns may present unlimited possibilities for developing structures and organs of any kind to perform the tasks for which they are designed. However, since computer time and memory still is a limiting factor, the non-genetic patterns of each numeric symbioorganism are constructed only when they are needed and are removed from the memory as soon as they have performed their task. This situation is in some respects comparable to the one which would arise among living beings if the genetic material got into the habit of creating a body or a somatic structure only when a situation arises which requires the performance of a specific task (for instance a fight with another organism), and assuming that the body would be disintegrated as soon as its objective had been fulfilled.

The experiments are not yet in a stage where the non-genetic patterns can be expected to yield important information. Only the results of the preliminary tests of performance and its evolutionary improvement will be discussed to some extent.

The last part of this paper will be dedicated to a discussion of the possibilities of obtaining symbiogenetic evolution processes by using a different set of reproduction and mutation rules (or "norm of action"). Particularly the use of rules applying the reproduction pattern of *DNA*-molecules (*DNA*-norm) and the implications this possibility may have with respect to the origin and history of terrestrial life are discussed.

2. PERFORMANCE TESTS

As already stated in the previous paper of this series (BARRICELLI, 1962), the genetic pattern of a symbioorganism performs the function of a survival strategy program developed during the past evolutionary history of the symbioorganism. The specific operations performed to bring the survival strategy into action are determined by the norm of action specifying the reproduction and mutation rules. This norm is the interpretation of the survival strategy programs. After this interpretation has been chosen arbitrarily to begin with, the various symbioorganisms develop their respective survival strategy programs based on the choice which has been made. Nothing prevents modifying the interpretation or adding a new interpretation to be used in special cases, for example, in the case of collision between two different numeric entities moving into the same location.

The last course is the one which will be followed in order to make it possible for a symbioorganism to perform a specific task. In case of collision between genes of two different symbioorganisms, the genetic patterns (or a part of the genetic patterns) of the two symbioorganisms will be interpreted as programs for the performance of a specific task according to a particular code designed for this purpose. The genetic patterns of the two symbioorganisms will consist of numbers, and numbers in the machine memory can be interpreted as instructions according to any arbitrary code which can be established by writing an interpretive program.

The operations to be performed by the two colliding symbioorganisms will consist in selecting the moves in a game to be played between them. The game which is used in the performance tests described below is a simple one denominated "Tac Tix." The rules of the game were published in Scientific American, February, 1958, p. 104-111, and will be explained below. The result of the game will decide which one of the two colliding genes will be permitted to occupy the collision place, namely the gene of the winner. Except for the cases of collision, which are often decided by games, the norms for reproduction and in many instances for mutation remain the same as in previous experiments (BARRICELLI, 1962).

By this procedure, the game strategy becomes part of the survival strategy of the competing symbioorganisms, and an evolutionary improvement in game performance can be expected.

3. RULES OF THE GAME

The rules of the game (Tac Tix) played between competing symbio-organisms are: In a square of $6 \times 6 = 36$ coins the two players remove alternatively one or several coins in a single row or a single column. Only

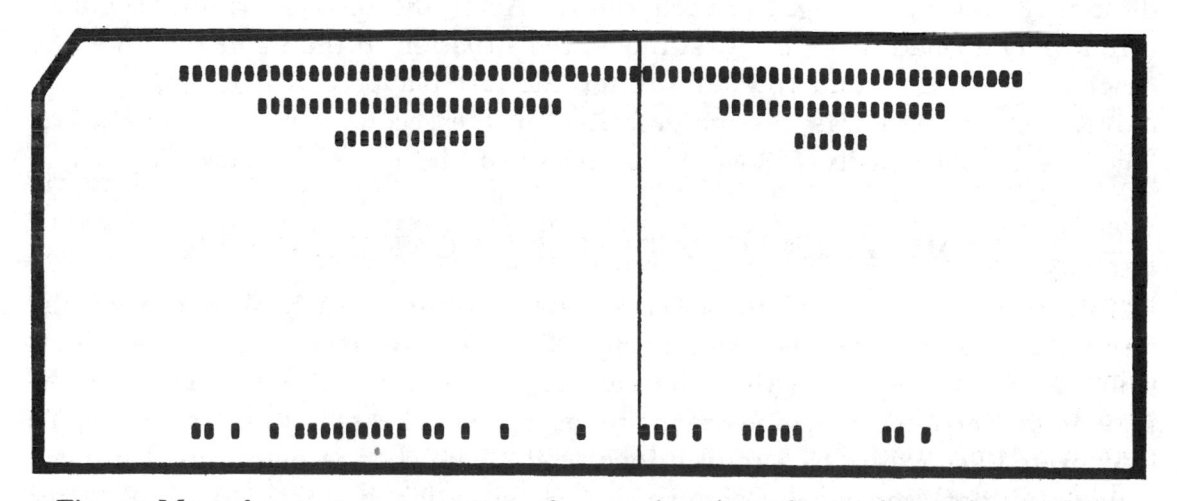

Fig. 1. Most frequent game pattern for unselected or damaged symbioorganisms.

an uninterrupted sequence of coins in a column or a row can be removed each time. For instance, when a row (or a column) is removed, the player who has his turn cannot take coins on both sides of the removed row (or column). The player who gets the last coin on the table loses the game.

To save space, in fig. 1 and 2 the 36 coins, which are represented by holes in an IBM card, are placed in a single row in the upper left side of each figure. But the game rules are applied assuming that the first 6 coins (holes) from the left represent the first row of a 6 × 6 square, while the next 6 coins (holes) represent the second row, etc. After the first symbioorganism

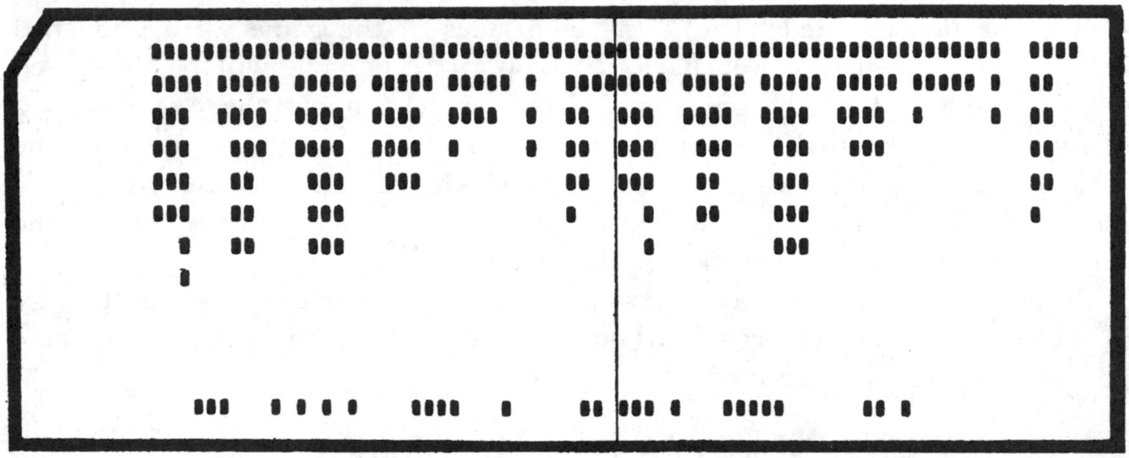

Fig. 2. Game played between two symbioorganisms after 1500 generations of selection for game performance.

(left player) had made his first move, the situation was the one described by the upper row to the right in each figure. After the second symbioorganism (right player) had made his first move, the situation in the game was the one described in the second row left. After the left player had made his second move, the situation was the one described in the second row to the right *etc*. The reader may easily follow the progress of the games in both figures.

4. FROM GENETIC PATTERN TO GAME STRATEGY

The procedure applied to select the next move in each situation of the game will now be described summarily. The idea consists in using the situation in the game before the move as d a t a   o f   t h e   p r o b l e m, and the genetic pattern of a symbioorganism as a s t r a t e g y   p r o g r a m to be interpreted according to a conventional code. The code is determined in such a manner that any genetic pattern, in any situation in the game, will, after a limited number of machine operations lead to a legal move of the game.

The computing machine used was [2]) the IBM 704 of the A. E. C. Computing Center, New York University, New York. Each memory location in

---

2) Readers who are not familiar with machine programming may skip the rest of this section and start with the next section.

this machine has 36 binary digits. This permits describing the situation in the game (36 coins present or removed) by a single binary number which can be stored in a single memory location. In other words, the problem is reduced to calculate at the nth move a 36 bit (no more than 36 digits binary) number s_{n+1} as a function $S_p(s_n)$ of another 36 bit number s_n: where s_n represents the present situation in the game (after n^{th} move) while s_{n+1} is the situation after next move. The function $S_p(X)$ shall be called decision-function. The decision-function $S_p(X)$ must be determined by the genetic pattern of the player p in such a way that only legal moves are possible. Apart from this restriction (only legal moves of the Tac Tix game) there are no other requirements to the decision-function $S_p(X)$. It would also be desirable not to insert other restrictions during the programming and rather leave to the symbioorganisms themselves the greatest possible liberty to choose the decision-function and to choose any particular decision-function in the largest possible number of ways by modifying their genetic patterns. Any restriction to the choice of decision-function would limit the game strategies available and may present a potential danger for preventing or delaying certain evolutionary improvements.

The solution of the problem is simplified by the fact that any 36 bit number can, by a simple operation, be transformed into a game situation s_{n+1} which can be reached by a l e g a l move from a preceding game situation s_n. This makes it possible to solve the problem in two steps: An unrestricted determination of a 36 bit number U_{n+1} by any function $U_p(s_n)$ of the present game situation s_n:

(1) $U_{n+1} = U_p(s_n)$

Transformation of the number U_{n+1} into a legal situation s_{n+1} following s_n by a legalizing operation $L(s_n, U_{n+1})$:

(2) $s_{n+1} = L(s_n, U_{n+1})$

s_n is supposed not to be zero; otherwise, the game would already be decided.

The first step calls for a method to select a section of the genetic pattern of each player to be used as game strategy program. The method should identify the gene to be used as first instruction in the game strategy program. It would be an advantage if always the same gene in a symbioorganism is used as first instruction in the game strategy program, since this would permit the specialization of a particular segment of the genetic pattern as an organ for game strategy operations. The method which has been used in the performance tests presented below consists in identifying the largest positive number in the genetic pattern and locating the first game strategy instruction in relation to this largest number. The method may lead to considerable ambiguity if the largest number is repeated several times in every period of the genetic pattern. For this reason, most symbioorganisms developed several game strategy programs with different characteristics and qualities. No steps have been taken against this sort of ambiguity in the performance tests described below. But the symbioorganisms themselves would have the possibility of preventing any ambiguity by a single mutation if an advantage in developing a single game strategy should be recognized.

The decision to play a game is taken every time a collision takes place in certain game-competition areas. Collisions very frequently occur between two symbioorganisms one to the right (right player) and one to the left (left player) of the collision place. But there may be disorganized areas on either side (disorganized right player and/or left player). Whatever the situation, the game strategy programs of the left player and right player will be identified in relation to the largest positive number in a certain region to the left and respectively to the right of the collision place. Each one of the two game strategy programs is arbitrarily identified with 16 consecutive numbers contained in 16 consecutive memory locations of the computer. Both sequences of 16 numbers are copied into memory locations reserved for the game strategy programs of the two players. The last 8 of the 16 numbers are used as parameters while the other 8 are copied in 8 new locations to be used as instructions. In these new location digits which could give meaningless or unwanted instructions are masked away by logic operations. Also digits which could give addresses outside the 16 locations of the game strategy program are removed in the same operation. A couple of digits needed are inserted and the instructions are executed. This process which is called a round of operations is repeated 50 times. Each time (or in each round of operations) a part of the 16 numbers may be replaced or changed and each one of the 8 instructions may be modified accordingly. At the end of this operation, the original genetic pattern has been transformed into a body of 16 numbers which will be called player body. The same two player bodies, one for the left and one for the right player, will be used to make all the moves of the game. Before each move, the last 4 of the 16 numbers of the player body which is supposed to make the move are replaced by 4 numbers (game data) derived from the present situation s_n of the game. The 4 numbers constituting the game data are the game situation s_n and 3 other numbers derived from s_n by one or two horizontal and/or vertical mirror substitutions in the game board. After inserting the game data 20 more rounds of operation are performed. Once again a part of the 16 numbers may be modified or substituted in each round of operation. The number to be used as U_{n+1} will be the last of the 16 numbers after the 20 rounds of operation are completed.

This solves the first part of the problem, the determination of U_{n+1} on the basis of the game situation s_n. The operations which determine U_{n+1} and therefore define the function U_p in relation (1) are to a large extent selected by the genes of the player, with the possibility (and, one may be tempted to say, the liberty) to select and program in many different ways almost any function U_p which can be calculated with the very limited machine time and memory space devoted to this purpose.

The legalizing operation $L(s_n, U_{n+1})$, which permits transformation of a 36 bit number U_{n+1} into a legal game situation s_{n+1} following the present situation s_n according to formula (2), is defined by logical operations which starting from the non-zero bits common to s_n and U_{n+1} (or, if there are none, starting from s_n) identify the lowest non-zero bit in this pattern. This bit is removed from s_n and if it is preceded by a consecutive set of non-zero bits common to s_n and U_{n+1} in the same row, or subsidiarely in the same column these bits are also removed.

5. SELECTION PROCEDURE

When two or more different numbers collide, the first and the last colliding numbers are recorded. The first one is treated as a gene of the left player, the last one as a gene of the right player (which is usually but not always correct). The game strategy programs of the left player and right player are identified. From each strategy program a set of numbers called player-body (respectively right player-body for the right player and left player-body for the left player) is determined. The first move is determined by using the player body of the first player as a set of instructions and the initial situation of the game as data of the problem. The next move is similarly determined by using the player body of the second player as a set of instructions and the situation in the game after the first move as data. The operation is repeated using alternately the right and left player bodies until all coins (holes) are removed and the issue is decided.

The gene of the winner—first or last colliding number both of which have been recorded for this purpose—enter the collision place. By this procedure, the best player will invade the contested areas and the other symbioorganisms will be obliged to retreat. One of the players can be, and often is, a damaged symbioorganism or a completely disorganized set of numbers. The performance of a symbioorganism can therefore also be tested against random or disorganized sets of numbers.

6. GAME QUALITY AND COMPETITIVITY

In games between symbioorganisms, a large fraction of the games are lost, as in fig. 1, by a stupid move of the loser, who takes all of several remaining coins instead of leaving one of them on the table. These games are called stupid games or games of quality 0. Some other games are decided by one or several correct final moves of the winner which leave no favorable choice to the loser. Favorable choice is here defined as a choice which would give the possibility of winning the game without a mistake from the opponent. For example, in the game of fig. 2, the winner's last and next to last moves are both correct decisions which leave no favorable choice for the loser.

The quality of a game which has been played can be measured by the number of correct final decisions of the winner. For human beginners—except those beginners who already know a similar game called Nim—the mean number of correct final decisions by the winner in the first 5 games played is between 1 and 2 when both players are beginners.

Unfortunately, the game quality does not only depend on the genetic strategy program but also on the condition of the players. Damaged symbio-organisms and disorganized sets of numbers are likely to play bad games. A quality record of the games played during an evolution process would, therefore, not only reflect the possible evolutionary improvement but also the extent of the damages caused by the competition between the symbio-organisms. For this and other reasons presented below, large fluctuations of game quality, unrelated to the evolutionary development, are observed.

A record of games won by the left player and by the right player in relation to the position of the various symbioorganisms and disorganized regions will therefore be given in fig. 4 and 6 to permit a better evaluation of the various situations.

7. RESULTS

Attempts to measure evolutionary progress in game performance were made in 1959 at the A. E. C. Computing Center in New York University with an IBM 704 computer and repeated in 1960 with the same computer while the author was visiting the Brookhaven National Laboratory in Long Island. In the first attempt, no successful evolution process was obtained and no measure of evolutionary progress in game performance could therefore be made. In 1960 the attempt was repeated using primarily the same combinations of mutation and reproduction norms which had been successfully used in Princeton before. However in some regions of the universe, which in this case had a size of 3072 numbers, mutations were replaced by game competitions. With this procedure two successful evolution processes (respectively called A and B) were obtained and the development of game performance could be observed and measured. In the later stages, the two experiments were linked together to observe game competitions between the two types of symbioorganisms developed. In each experiment (experiment A and experiment B) the evolution process started with inefficient game competition due to unsuccessful coding producing low variability or no variability at all in the game pattern. In the experiment A efficient game competition started at generation 1024. In experiment B efficient game competition started at generation 2560. In fig. 4, upper diagram, the games played at generation 1024, in the experiment A are marked by dots. Each dot is approximately at the number location (given by upper scale) where the collision occurred (+ or — 10 number locations). Above the 1024-line (at the level +0 and +1) are the games won by the right player, below the 1024-line at the level —0) are the games won by the left player. At the levels

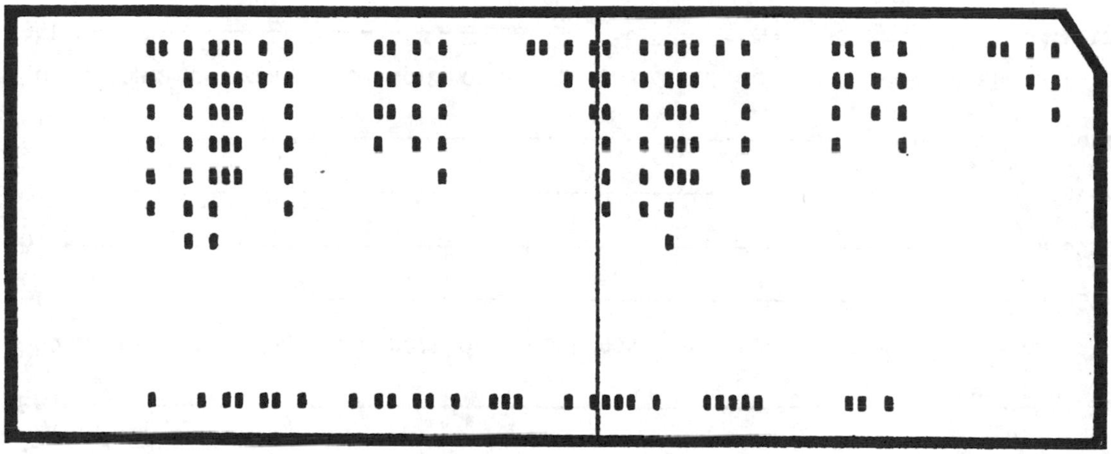

Fig. 3. Game of quality 4 played between two symbioorganisms after a preselected game-starting.

+0 and —0 are the dots corresponding to games of quality 0 (stupid games). At the level +1 is marked a game of quality 1 which was won by the right player. The same convention is used in the following diagrams of fig. 4 and in fig. 6, where dots at the levels +0 and —0 mark games of quality 0, dots at the levels +1 and —1 mark games of quality 1, dots at the levels +2 and —2 mark games of quality 2. Games of quality 3 or higher are all marked at the levels +3 or —3. The number of games (dots) in each diagram is marked in the right margin, the generation in which the games were played is marked in the left margin. In each diagram the most prominent symbioorganisms are indicated by horizontal segments (with arrow heads) which mark the regions in the universe occupied by the respective symbioorganisms. Tregeners (a type of symbioorganisms of periods 3 and 6) and tregener-derivatives (see fig.4) are related to the parasite recorded in the previous paper (BARRICELLI, 1962, fig. 17 and 18) "A" is the name of the type of

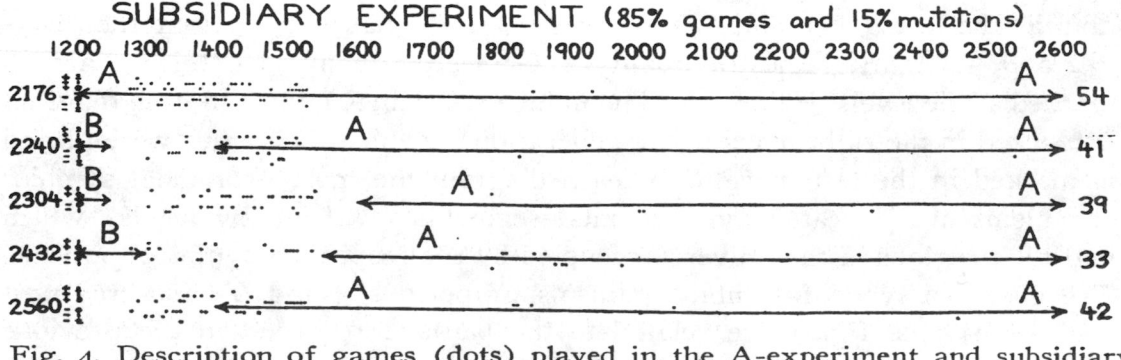

Fig. 4. Description of games (dots) played in the A-experiment and subsidiary experiment.

symbioorganism (of period 50 or 50-geners) which becomes predominant during the A-experiment. "B" is the name of the type of symbioorganism (60-geners and 72-geners) which becomes predominant during the B-experiment.

Games were played only in certain areas which were changed several times during the experiments, and can be approximately identified by the positions of the dots in fig. 4 and 6. Between number locations 1536 and 2048 all the games started regularly like the games in fig. 1 and 2. On both sides of this region, various preselected game startings, as for example in the game of fig. 3, were used to promote the development of more universal game strategies, the variety of the preselected game startings being an efficient method to prevent specialisation.

At generation 1024 in the A-experiment, when efficient game selection started, only one game out of 40 (dot in location 1450) was of quality 1. The other 39 were stupid games, mostly repetitions of the game in fig. 1. These games are won by the right player which is always the beginner in these experiments. The reader will, therefore, notice that all but 3 games were won by the right player (dots above the line) of generation 1024. Evidently, as long as the quality of the games is low, the right player has advantages in the region of regular game startings. The A-symbioorganism to the left of location 1800 has therefore no chance to penetrate the game area to the right, before its game performance is improved. Only at generation 1664, after A had invaded the game region to the left (between 1280 and 1536) thus improving its game strategy by adaptation to this game area, it started making some progress in the game region to the right.

At generation 2048 and every 256 generations after that, B-symbioorganisms from the parallel B-experiment were introduced in the region 0-1007. A large number of game competitions (dots) in the game region to the right of location 1024 is the result (see fig. 4 generations 2048, 2816, and 3072). In generations 2304 and and 2560 there was no game region in the proximity of location 1007 and only locations above 1200 are recorded in the figure for these generations.

A glance at fig. 4 shows that the quality of the games and the per cent of the games won by the right or the left player are very different in different regions and, at least in some places (like location 1900 at generation 1664 and location 1100 at generation 2176), are strongly related to the respective positions of competing symbioorganisms and/or disorganized regions. A closer inspection of fig. 4 can often tell in which positions the symbioorganisms are who win the games, or who are likely to play good games at least against a particular opponent whose tactics they have learned by mutation and selection. On the other hand, there is no clear evidence that the nature of game startings (regular or preselected startings) has any direct influence on the quality of the games or on the fraction of left (or right) victories after A-organisms had invaded the whole universe. No discontinuity in the quality or type of games can be seen in position 1536 which marks the frontier between preselected game startings (to the left) and regular game startings (to the right) at the generations 1792, 1920, 2048, 2176, 2816, 2944 etc. A difference can, however, be observed in the number of games played to the left of location 1536, compared with the number of games played to the right of this location. The difference, which is manifest at the generations 1792, 1920, 2048, 2944, etc. indicates a greater variability of the symbioorganisms in the preselected game starting's sector to the left of location 1536 than in the regular game starting's sector to the right of this location. The large variety of preselected game startings seems to promote variability. Usually variability will produce changes and permit evolution. But there is one case in which a stable situation with some variability in the region between location 1370 and 1500 was maintained for 128 generations without further changes. A glance at the games (dots) in this region on fig. 4 shows that the same games were played with the same results at generations 2944 and 3072.

The quality of the games, measured by the per cent of dots above the +o level and below the —o level, showed a tendency to increase during the A-experiment, but also large fluctuations. In fig. 5 the solid line shows the per cent of games with quality not lower than 1 for each generation represented in fig. 4. The significance of each value can be judged from the number of games on which it is calculated. This number is represented by vertical solid lines in the lower part of fig. 5. For example, the very low value (based on 5 games) at generation 2432 and the very high values (based respectively on 6 and 7 games) at generations 2954 and 3328 are not very significant. Some of the fluctuations are, however, significant and their interpretation is still a matter of investigation. In spite of the mentioned fluctuations, the solid line of fig. 5 suggests an increase in the percentage of games with quality 1 or higher by 10% to 15% every 1000 generations. From generation 1792 (or 768 generations after efficient game-selection had been started) disorganized areas appeared only in places where two symbioorganisms competed. A disorganized set of numbers had no longer any chance against symbioorganisms which had improved their game strategy for that many generations.

After generation 2048 a different continuation (subsidiary experiment) of the above experiment was attempted. The purpose of this subsidiary experiment was to keep the B-symbioorganism alive for a large number of generations rather than re-introducing it once every 256 generations in the region 0-1007 of the A-universe. In the game areas, the B-symbioorganisms are at a disadvantage in the competition with A-symbioorganisms to the right, as long as the game quality is low, for the reasons explained above. However, a reduction of the game frequency obtained by leaving about 15 % of the

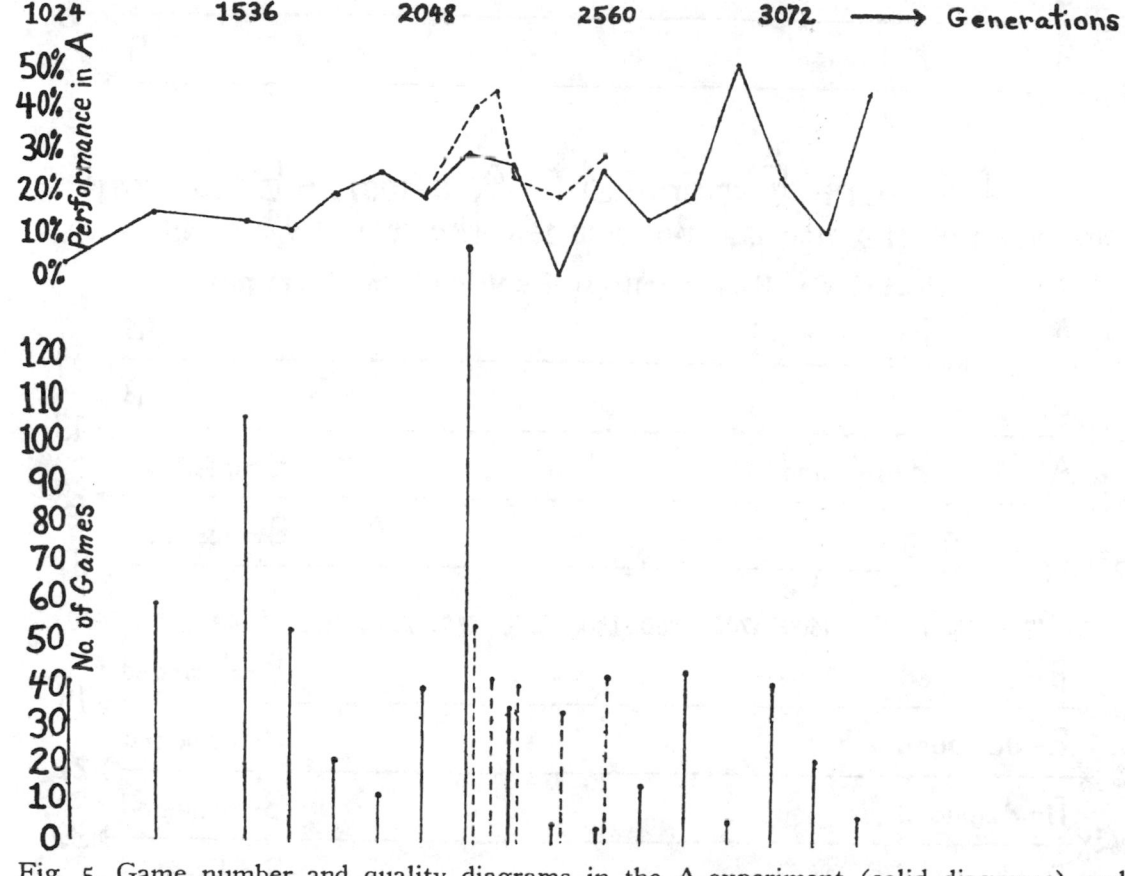

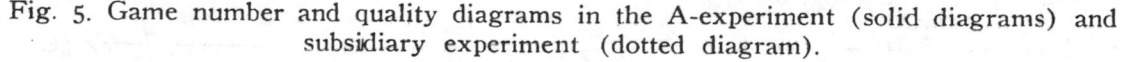

Fig. 5. Game number and quality diagrams in the A-experiment (solid diagrams) and subsidiary experiment (dotted diagram).

collision places empty (while the remaining 85 % of the collision places were occupied by the winner of a game) proved sufficient to give the B-symbio-organism a fighting chance. In the subsidiary experiment (see lower part of fig. 4) the B-symbioorganism was able to survive in a region below location 1300 which is mostly outside range of fig. 4. The quality of the games is represented by the dotted curve of fig. 5. The lower quality at generations 2304 and 2423 coincides with the situation in which most games are played in a disorganized area between the two symbioorganisms A and B (see fig. 4 lower part). The players are, therefore, mostly damaged or disorganized.

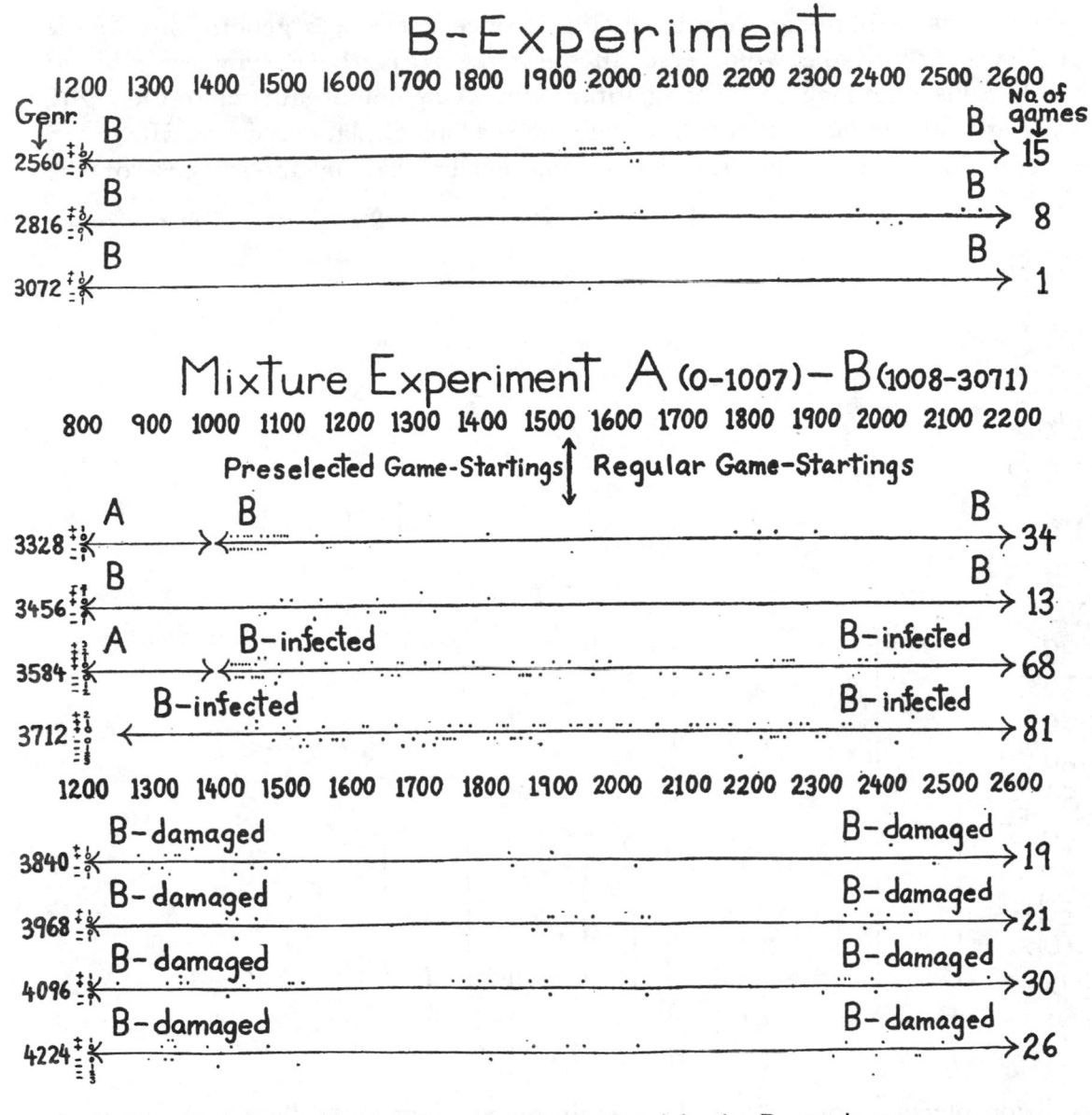

Fig. 6. Description of games (dots) played in the B-experiment.

The B-experiment is recorded in fig. 6 and the quality of the games in this experiment is recorded in fig. 7. Efficient game selection started at generation 2560 which is the first generation recorded in fig. 6 and 7. At generation 3328 and every 256 generations after that A-organisms from the parallel A-experiment was introduced in the region 0-1007. The collisions between A and B-genes are the cause of the large number of games (dots) to the right of location 1024 at generations 3328 and 3584 in fig. 6.

Around generation 3456, the B-symbioorganism developed an infection by a parasite which apparently damaged but did not kill the host. In fig. 8A and B (in which binary numbers—or genes—are recorded vertically, each column representing a binary number) two specimens of B-symbioorganisms are presented. The first specimen (fig. 8A) is taken from generation 3328 locations 1224-1296 before the infection developed. The second specimen

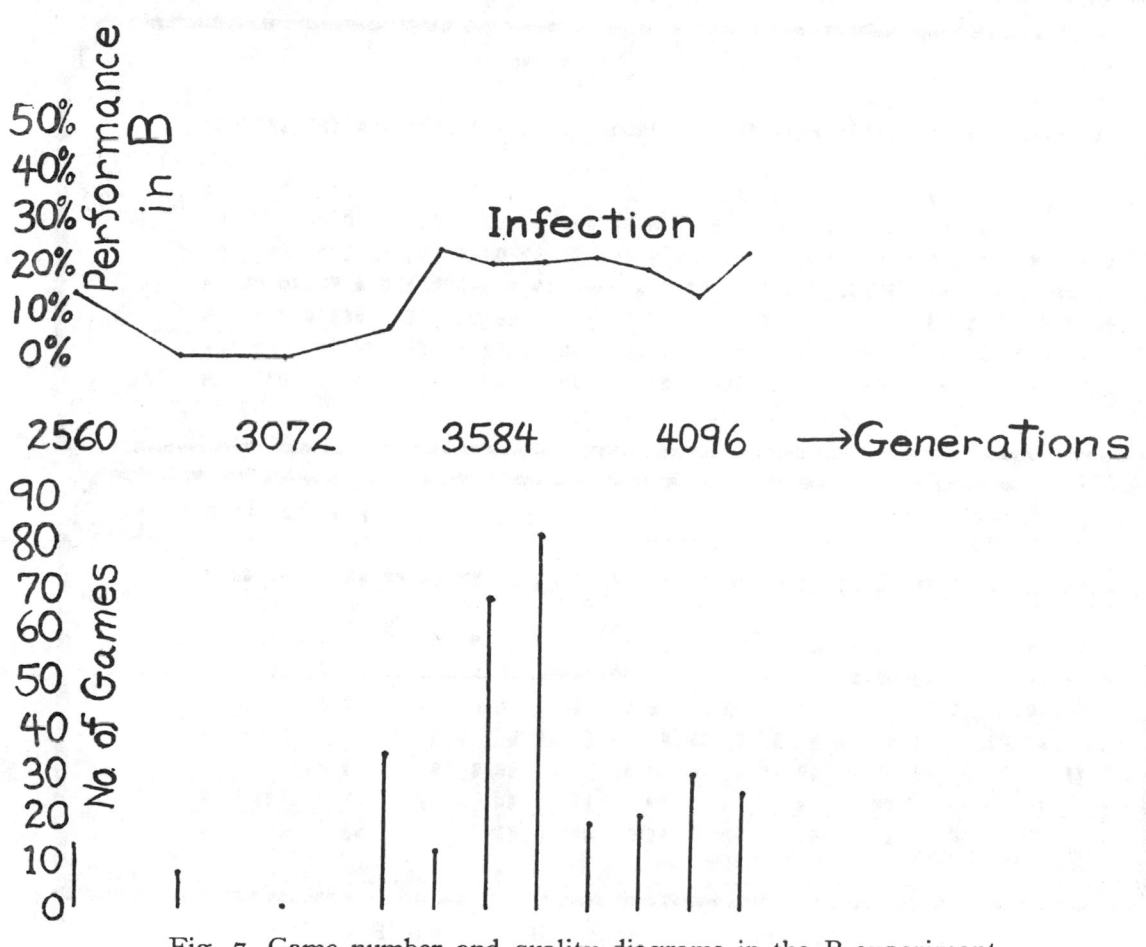

Fig. 7. Game number and quality diagrams in the B-experiment.

(fig. 8B) is taken from generation 3712 locations 2232-2304 at a stage of advanced infection. At this stage every third gene (see fig. 8B) of the host had been removed or replaced by the parasite. The genes of the parasite can easily be distinguished from those of the host because they do not have a 1 (or hole) in the upper row (second row from top which is filled in the uninfected host—fig. 8A—while every third place in this row is empty in the infected one—fig. 8B). Some cases of partial recovery have been observed, but it is not known whether the B-symbioorganisms can recover completely or whether this spectacular infection will eventually develop into a symbiotic relationship.

In the early acute phase of infection, a large number of games are played (see dots in fig. 6, generations 3584 and 3712, and vertical lines in fig. 7 same generations) as a result of the large number of collisions between parasite and host-genes. In the later stages of infection (generations 3840, 3968, 4096, 4224) the collision frequency is greatly reduced.

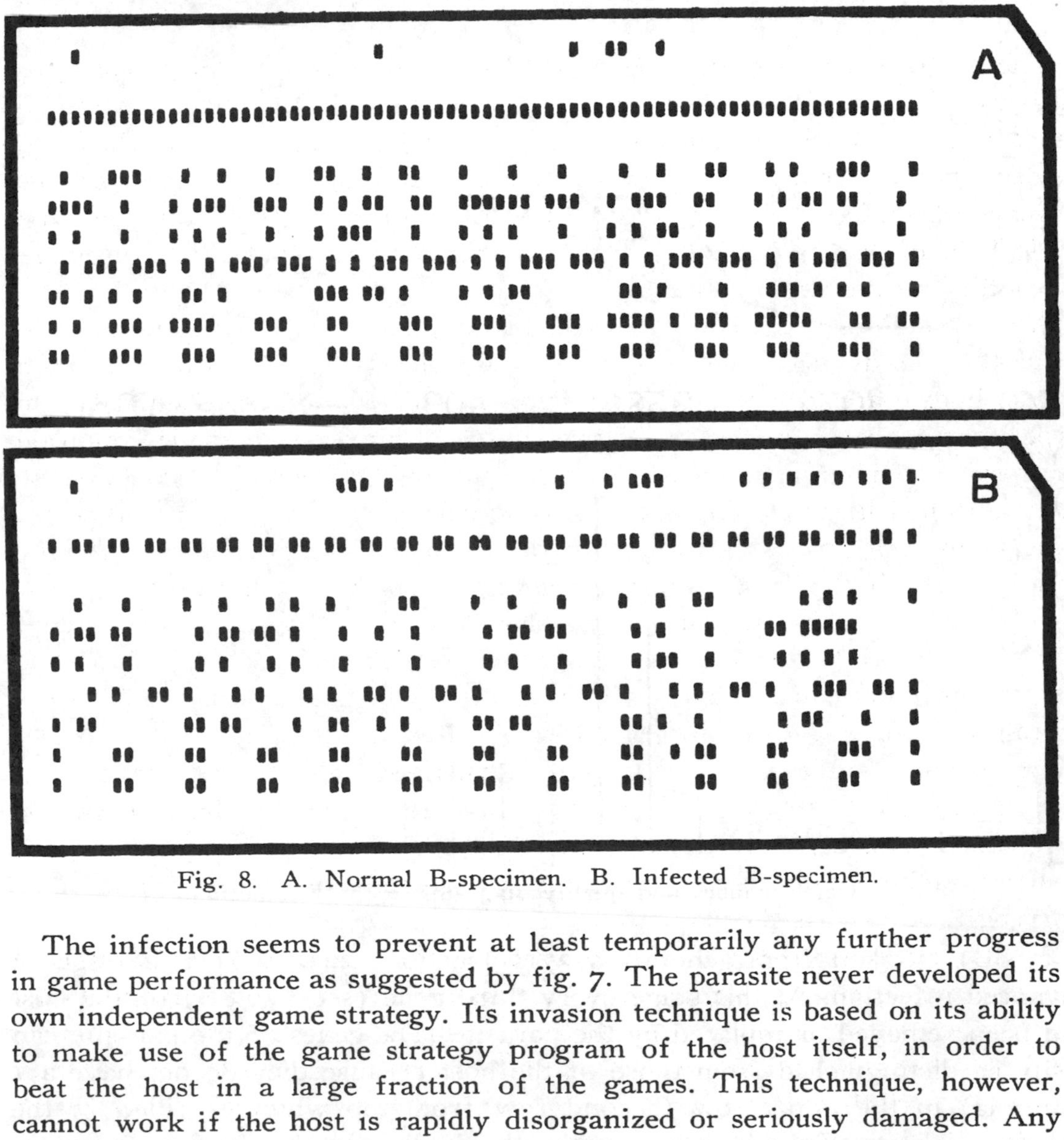

Fig. 8. A. Normal B-specimen. B. Infected B-specimen.

The infection seems to prevent at least temporarily any further progress in game performance as suggested by fig. 7. The parasite never developed its own independent game strategy. Its invasion technique is based on its ability to make use of the game strategy program of the host itself, in order to beat the host in a large fraction of the games. This technique, however, cannot work if the host is rapidly disorganized or seriously damaged. Any drastic reduction of game performance in an infected area would prevent further progress of the parasite from this area, and may locally produce a partial recovery of the host. This way a balance between host and parasite is obtained and the destruction of the host is prevented in game areas. Outside game areas a much more pronounced disorganization of the host is occasionally observed.

8. SIGNIFICANCE OF THE PERFORMANCE TESTS

The performance tests described in the previous section cannot give a true measure of the rapidity of improvement obtainable by the selection method used. These performance tests are the first combined experiment which succeeded at all. The choice of experimental conditions and parameters has primarily been based on a guess, and there is no information available which could help in deciding whether and how much the result could be improved by modifying the experiment.

A new combined experiment in which the game areas have been extended while the rounds of operations before each game (after the formation of the player body) are reduced from 20 to 10 has recently been started. This experiment is being carried on in two different universes and the conditions for periodical interchanging of symbioorganisms between the two universes are radically different from previous experiments. It is hoped that this experiment can give some of the information necessary to decide whether and how much the evolutionary progress of performance can be speeded up. With present speed, it may take 10,000 generations (about 80 machine hours on the IBM 704 or between 5 and 10 machine hours on the Atlas-Ferranti machine) to reach an average game quality higher than 1. The best averages obtained so far are around 0.4. Human beginners in the first 5 games show averages between 1 and 2. All these data must be taken with reservation due to the large fluctuations and the irregular character of the progress. However, there is no doubt that the progress is significant and that the symbioorganisms are "learning" the game by a sort of "evolutionary learning process" based on mutation, crossing and selection.

A fundamental question for practical application is: how would the evolutionary learning process work out in a more complicated game like, for instance, chess? It would seem that in a complicated game the evolutionary learning process would be much slower. However, the evolutionary learning process is very different from human learning. The learning of a seemingly simple operation like leaving a single pawn on the board in the Tac Tix game at the first given opportunity, may take a rather long time. On the other hand, the crossing mechanisms which play a fundamental role in this process make evolutionary learning extremely well suited to make progress in many different directions at the same time (cf. FISHER's law and the independent spreading of many different mutations). This may be rather difficult for humans who largely prefer to learn one thing at a time. There is no evidence that human standards can be used to evaluate machine time for evolutionary learning processes. The only way to decide the question is to try. The programming difficulties are greater with the chess game than with the game used above. On the other hand, chess-programming for high speed computers has already been successfully accomplished (BERNSTEIN, 1958). There is no doubt that the difficulties can be overcome.

At any rate, the value of the results presented does not primarily rest on the possibilities for practical applications, but on their biotheoretical significance. It has been shown not only that the symbioorganisms can improve by evolution, but how the improvement takes place in a particular set of operations necessary for their survival. It has been shown that given a chance to act on a set of pawns or toy bricks of some sort the symbioorganisms will "learn" how to operate them in a way which increases their chance for survival. This tendency to act on any thing which can have importance for survival is the key to the understanding of the formation of complex instruments and organs and the ultimate development of a whole body of somatic or non-genetic structures.

9. THE CHOICE OF REPRODUCTION NORMS

The development in game performance presented in this paper was made possible by a minor change in the reproduction and mutation rules (or "norm of action") used. This norm of action was designated as a "shift norm" (BARRICELLI, 1957) and this designation will still be maintained in spite of the change performed. A question which arises is: what would happen if more radical changes were applied or if a completely different norm were used.

As already pointed out in the previous paper (BARRICELLI, 1962) a norm which does not require symbiosis as a condition for reproduction or survival would not lead to the formation of symbioorganisms and hardly to any structure of complexity or evolutionary possibilities comparable to living organisms. Only norms of action requiring symbiosis (symbionorms) shall therefore be considered.

An extension of the shift norms to a two-dimensional universe—in which each generation consisted of vectors (pairs of numbers) scattered on a cross-section paper instead of a single row—has been tried in Princeton (BARRI-

CELLI, 1957). The experiment was not carried very far because of the much larger machine time and machine memory requirements of two-dimensional experiments. The phenomena observed were, however, of the same nature as those which have been described in the one-dimensional experiments.

Of fundamental interest for the understanding of the nature of terrestrial life is the fact that the well-known reproduction rules for *DNA* molecules are an example of a symbio-norm. In this norm, the nucleotides play the role of elementary selfreproducing entities (numbers); the polynucleotide chains or *DNA* molecules play the role of symbioorganisms. Proteins and other molecules which are constructed, rearranged, or modified by the catalytic action of the *DNA* and the associated enzymes have apparently the role of non-genetic objects, instruments, and products of the action of symbio-organisms (example: pawns in a game and numerical devices constructed or used to act upon them).

It is easy to show that the *DNA* reproduction rules constitute a symbio-norm. Each nucleotide would be unable to reproduce alone. In this respect a single nucleotide behaves like a single number in the numeric experiments already presented. It is the association of several nucleotides into polynucleotide chains (symbioorganisms) which confer to them the catalytic abilities necessary both for the construction (by the intermediate action of enzymes and other molecules) of new nucleotides and their association into new polynucleotide chains identical to the parental chains. The *DNA*-norm requires a symbiotic association into specific groups of nucleotides as a condition for reproduction.

The simplest known *DNA*-symbioorganisms (viruses) show the biophenomena listed in the previous paper (BARRICELLI, 1962, section 6), except spontaneous formation which cannot be observed in nature today (BARRICELLI, 1962, section 10).

It is important that the biophenomena observed are a consequence of the symbio-norm followed by the *DNA*-molecules. If the nucleotides, the amino acids, and other essential features involved were substituted by numbers in the memory of a computer instructed to apply the same symbio-norm (*DNA*-norm), one would observe exactly the same biophenomena, assuming successful programming. Obviously there may be technical difficulties, and our present knowledge is hardly sufficient to construct a true numerical model of the chemical phenomena involved in any particular *DNA* or polynucleotide duplication. An incomplete model describing only some fundamental aspects of *DNA*-duplication is the best one might be able to do for the time being. The question of present practical feasibility is however of no consequence for the argument, as long as no theoretic principle (like

Heisenberg's uncertainty principle, Einstein's critic of simultaneity) is involved, which would preclude the feasibility of the experiment. The question whether numbers or nucleotides or magnetic charges or any other gadgets in the memory of a high speed computer are used as selfreproducing elements has no influence on the result.

A lesson which can be learned from the performance tests described in this paper is that the nature of the biophenomena observed is not radically changed by a modification of the symbionorm which does not directly interfere with the reproduction mechanism. Neither A nor B nor any other symbioorganism developed in the performance tests show properties or biophenomena which are fundamentally different from those observed in the symbioorganisms developed during the Princeton experiments of 1954 and 1956. The infection developed by the B-symbioorganisms (fig. 8B) may seem more spectacular, but not fundamentally different from some of the other parasitic phenomena observed in previous experiments.

It seems likely that the nature of the biophenomena appearing in a symbiogenetic experiment is primarily, if not completely, determined by the reproduction norm used irrespective of possible interactions with other entities. If this applies also for the *DNA* reproduction norm, the interaction between *DNA* and proteins or other molecules may have little influence on the general features of the biophenomena, irrespective of the influence they may have on the selection of symbioorganisms. It might therefore be possible already with the scanty knowledge presently available to attempt numerical evolution experiments based on the *DNA*-norm in order to gather information on the manner in which a *DNA*-evolution process would develop.

10. *DNA*-SYMBIOGENESIS AND CROSSING

As a first step toward the use of a *DNA*-norm as a basis for a symbiogenetic evolution experiment, one may attempt to find in advance some of its characteristics and prospects. Some questions which were investigated before the shift norms were used and which may be worth investigating before a *DNA*-norm is used are the following: (1) what are the prospects of developing complex symbioorganisms; (2) what are the prospects of developing a crossing mechanism early in the evolution process. Without a crossing mechanism evolution would proceed at an extremely slow rate, and it is doubtful whether the experiment would be worth while.

The interpretation of crossing phenomena (particularly of the crossover mechanism) on a polynucleotide basis is a problem which has puzzled many investigators. The simple solution of the problem presented below may therefore have particular biotheoretic interest.

Both complexity and crossing would rapidly develop in *DNA*-symbio-organisms if the phenomenon, termed "complementary association," which is described below, can be expected to take place during the duplication of a polynucleotide chain. It is unknown whether complementary association would be possible when *DNA* duplicates under the action of an enzyme like polymerase. It may be necessary to assume that the duplication process is taking place under more primitive conditions perhaps similar to those existing before life orginated (possibly with a catalyst of non-biologic origin). The consecutive steps in the duplication process are assumed to be: (1) Separation of the two *DNA* strands. (2) Attachment of single nucleotides to each strand in the places where they fit with the complementary nucleotide. However, under this condition it is conceivable that not only single nucleotides, but occasionally a complementary single stranded polynucleotide-segment may be incorporated in the double stranded molecule which is formed (hypothesis of complementary association). If the two single stranded polynucleotide chains are complementary only in a border segment (overlap) shorter than either of the two chains (fig. 9B) the result can be a longer double stranded polynucleotide (fig. 9C). The process may, for example, start with a longitudinal association (fig. 9A). This condition is unstable since only a low percentage of the nucleotides facing one another are complementary but may end up in a complementary association (fig. 9B) which is stable since all nucleotides facing one another are complementary.

This complementary association mechanism provides a possible interpretation of evolutionary growth and increase of complexity in polynucleotides. A fact of considerable interest is, however, that it also provides a primitive

```
(A)        A G A A C A A
           A T A T C T T

(B)                                        G
           A  A G A A C A A
           A T A T C T T     T      T
           T
                        T

(C)        T A T A G A A C A A
           A T A T C T T G T T
```

Fig. 9. (A) Longitudinal association of two single stranded polynucleotides with a complementary segment (AGAA complementary to TCTT). (B) Complementary association and insertion of single nucleotides. (C) Formation of a double-stranded polynocleotide longer than both original chains (permitting evolutionary growth in size and complexity).

crossing mechanism for polynucleotides. In fact the two single-stranded polynucleotides in fig. 9A and B could be the result of incomplete duplication (partial replicas) of larger polynucleotides identical or homologous to the double-stranded polynucleotide in fig. 9C. Such incomplete replicas or partial replicas could for instance be the result of possible damages or too early separation of the two strands after duplication (separation before all nucleotides are filled in). In this case, the processes represented in fig. 9 will only restore the original size of the polynucleotides and can be considered a repairing mechanism (like multiplicity reactivation in viruses) rather than a growth mechanism. On the other hand if the two single-stranded poly-nucleotides in fig. 9A contained genetic markers (as a result of mutations or copying mistakes), the double-stranded polynucleotide of fig. 9C may be recombinant (containing copying mistakes inherited from both parents). Evidently the process described in fig. 9 can operate as a crossing mechanism. Partial replica models for virus crossing and reproduction (DOERMANN, 1953; DOERMANN & BOEHNER, 1961; BARRICELLI, 1952, 1955, 1960; BARRICELLI & DOERMANN, 1960, 1961) might be based on some mechanism of this or similar nature.

The above picture of a polynucleotide-crossing mechanism may or may not be the answer one would find by *DNA*-norm symbiogenesis experiments. But it shows at least that some simple crossing mechanisms for *DNA* molecules can be constructed on the basis of the *DNA*-reproduction model and might have a possibility to develop if a *DNA*-symbiogenesis experiment were attempted.

11. CHEMO-ANALOGICAL AND DIGITAL COMPUTERS

A question one may ask is: Why use a computer, rather than a chemical method to perform an evolution experiment based on the *DNA*-norm? The answer is: One method does not exclude the other and the distinction between the two methods may not be as fundamental as one would be inclined to believe. As a matter of fact, if *DNA*-norm experiments should become a frequent procedure, the question would arise whether it would be possible and convenient to construct an analogical computer especially designed for this type of experiments. Such a computer could essentially consist of an automatic, programmed chemical laboratory with read-in and read-out devices and other gadgets to perform the following operations: Interpret and transform information contained in IBM cards or magnetic tape into a specific arrangment of nucleotides and other molecules. Perform the chemical operations specified by the program (also contained in IBM cards or magnetic tape). Punch or read out the results into IBM cards or magnetic tape.

This "chemo-analogical computer" would probably do nothing which could not be done by a correctly programmed digital computer, if sufficient information were available to write a correct program. However, its value at least as a check for correct programming and a check for the theories used is evident.

A fact which emerges from this type of consideration is that the distinction between an evolution experiment performed by numbers in a computer or by nucleotides in a chemical laboratory is a rather subtle one. As a matter of fact, it is conceivable that the use of a digital computer and a chemo-analogical computer could be alternated in the same evolution experiment depending on which computer is available at any particular moment.

These considerations will make it clear for the reader that the fundamental difference between various types of symbioorganisms is the difference in the norms used. The question whether one type of symbioorganism is developed in the memory of a digital computer while another type is developed in a chemical laboratory or by a natural process on some planet or satellite *does not add anything fundamental to this difference.*

12. SYMBIOGENESIS AND TERRESTRIAL LIFE

It is doubtful whether a symbiogenetic evolution experiment based on *DNA*-norm could be carried far enough to see polynucleotides develop the ability to act on proteins. As a matter of fact, there is no assurance that the control of protein formation would be among the inventions of the symbio-organisms developed during an evolution experiment based on *DNA*-norm, no matter how far the experiment were carried on. Probably, in order to survive, the *DNA*-symbioorganisms would have to develop some means of controlling their chemical environment. But whether this would have to be done by enzymes or whether some other catalysts might be used and possibly developed into chemical instruments of comparable power, is still an open question [3]).

---

3) The problem of programming a norm permitting action of polynucleotides on protein formation (or any other action by a symbioorganism) is in several respects similar to the problem, already handled in this paper, to program a norm permitting symbioorganisms to act in the determination of the moves in a game. There is however the following fundamental difference; the rules for game-action were chosen arbitrarily and the symbioorganisms were purposefully given a large number of ways in which they could act on the game pattern or modify their game strategy. On the contrary, a programmed *DNA*-norm, if it shall have anything to do with *DNA*, must be a true copy of the reactions occurring in a particular chemical environment realizable in a hypothetical experiment. We cannot choose arbitrarily the ways in which polynucleotides, would act on protein synthesis. The possibilities of interfering with the phenomena would have to be restricted to the possibilities which would exist in a true chemical experiment.

What else an evolution process based on *DNA*-norm would show, besides the usual biophenomena already quoted, is hard to predict. However, it is obvious that a successful experiment might give results of fundamental theoretic interest.

The numeric evolution experiments which have been presented do not give information about the origin and history of terrestrial life. Nevertheless, a few fundamental notions have been established which may give some leads on the nature of the processes involved.

The very fact that the *DNA*-norm is a symbionorm (section 9) with characteristics suggesting a simple solution of the hybridization (crossing) problem, strongly supports the idea that the symbiogenesis of terrestrial life forms may have started by an association of nucleotides into polynucleotide chains. The only natural environment in which polynucleotides (viruses or cellular genetic material) reproduce nowadays is the interior, primarily the nucleus, of living cells. In view of the conservative nature of biologic systems (*cf.* chemical similarities between blood and sea water) it is tempting to assume that the environmment in which terrestial symbiogenesis occurred may have presented considerable chemical similarities to the nuclei of living cells (protoplasmic environment) (*cf.* RAPOPORT & RAPOPORT, 1958). The chemical similarities between nuclei of many different cells support this notion. The possibility for synthesis of several complex organic compounds in the absence of living organisms has already been established (UREY & MILLER, 1959; FOX, 1960).

One of the first steps in the evolutionary process leading toward the formation of cells, may have been a membrane (prototype of the present nuclear membranes). The function of this membrane may have been to protect a small fraction (protonucleus) of the medium in which a nucleic acid structure performed its activity, from chemical changes produced by external conditions or other nucleic acids (competitors and/or parasites). At the same time the membrane would prevent the dispersion of enzymes and other catalysts produced in the protonucleus. The membrane may originally have been formed and dissolved according to necessity at various stages of nucleic acid duplication and crossing.

After this stage, a consecutive chain of successful inventions may have transformed the protonuclei into the large variety of cells existing today. A large proportion of the early steps in evolution must have been related to the biosynthesis of compounds needed for nucleic acid duplication or for maintaining a favorable protoplasmic environment under deteriorating external conditions (HOROWITZ, 1945). In the meantime, the radical changes of environment caused by the chemical competition between living organisms has made all reproduction and all normal chemical activities of nucleic acids impossible outside cells. The most dramatic change caused by living organisms is probably the formation of atmospheric oxygen by chlorophyll carrying organisms. Oxygen is still a powerful poison for a large class of anaerobic organisms. Defense mechanisms against oxygen seem to be an important factor for survival in living organisms. In the reducing environment which seems to have characterized the earliest stage of biological evolution, oxygen may have been a powerful antibiotic weapon (BARRICELLI, 1962, section 10). The external cell walls were probably developed as an adaptive response to the changes of environment, in order to permit the maintenance of an environment (cytoplasmic environment) favorable to the chemical activity of the nucleus and its external (cytoplasmic) organelles. The cytoplasms of various types of cells (animal, vegetable, bacterial, *etc.*) may give information on the particular chemical environment in which each type of cell lived at the time when it developed its external cell wall.

After the cell walls were formed, crossing was still possible by diffusion and absorption of nucleic acids (transformation) and by virus infection (transduction). The cell-fusion or regular crossing mechanism must have been developed later (see BARRICELLI, 1962, section 4).

This tentative picture of the evolutionary development from polynucleotide to cell may or may not have something to do with the true history of terrestrial life. Other schemes, based for example on a polynucleotide infection of a selfreproducing protein nucleus or a nucleus which is made selfreproducing by the nucleic acid itself, are also conceivable. The symbionorm characterizing polynucleotide duplication seems however very well fit to start a symbiogenetic evolution process and must undoubtedly have played a fundamental role in the origin of terrestrial life whatever the nature of the other features involved in the process.

Whatever picture is preferred, there is no need for assuming that terrestrial life originated by anything comparable in complexity with a living cell: nearly a statistical impossibility. The first symbioorganisms may rather have been comparable in complexity to short polynucleotide chains. Large numbers of such symbioorganisms may have been created and destroyed over and over again. This is the stage in which biological evolution may have started, while the cell must have been a later and much more sophisticated product of the biological world.

SUMMARY

An interpretive system for the IBM 704 computer permitting interpretation of the genetic pattern of a numeric symbioorganism as a game strategy has been developed. Selection for best performance in a simple game (the game called Tac Tix published by GARDNER, 1958) has been applied in a preliminary experiment. An objective method to measure the quality of a game played is described. The results presented in the article show a small but significant improvement of game quality during a period of 2300 generations.

The general characteristics of the phenomena observed are similar to those of preceding evolution experiments (BARRICELLI, 1962 and 1957). However, a startling infection process caused by a parasite whose behaviour was influenced by the game competition is described. The parasite never developed an independent game strategy to any degree of efficiency and was entirely dependent on its host organism for game competitions with, and transmission of the infection to, uninfected hosts.

The consequences of substituting the reproduction and mutation rules (or "norm of action") used in the preceding evolution experiments are discussed. Particularly, the use of rules (*DNA*-norm) applying the reproduction pattern of *DNA*-molecules is considered, and the theoretic aspects of symbiogenetic evolution experiments based on *DNA*-norm are discussed. A few inferences concerning the origin and history of terrestrial life suggested by the results of the symbiogenesis experiments are presented.

ZUSAMMENFASSUNG

Ein interpretierendes System für die IBM 704 Rechenmaschine, dass Interpretation des genetischen Modells von einem numerischen Symbio-organismus wie eine Spiel Strategie, erlaubt, wurde entwickelt. Eine Forführung für Fähigkeit in einem einfachen Spiel (genannt Tac Tix veröffentlicht durch Gardner, 1958) wurde angewandt in einem preliminarisches Experiment. Eine objektive Methode zur Messung der Qualität des betreffenden Spiels wurde beschrieben. Die in dem Artikel angegebenen Resultate zeigen eine kleine, aber bedeutende Verbesserung der Spiel-Qualität während einr Periode von 2300 Generationen.

Die allgemeinen Kennziechen von den beobachteten Phänomenen sind denen von vorhergehenden Evolutions-Experimenten ähnlich (Barricelli 1962 und 1957). Jedoch, ein überraschender Infektions-Prozess, verursacht durch einen Parasit, wessen Verhalten beinflusst wurde durch die Spiel-Konkurrenz, wurde beschrieben. Der Parasit entwickelte niemals eine unabhängige Spiel-Strategie zu irgendeinem Masze von Fähigkeit und war ganz abhängig von seinem Gastherr-Organismus für Spiel-Konkurrenz mit, und Überbringung von der Infektion auf, uninfektierte Gäste.

Die Folgerungen von dem Ersetzen der Fortpflanzung und Mutations-Regeln (oder „Aktions-Norm"), angewandt in vorhergehenden Evolutions-Experimente wurden besprochen. Besonders der Gebrauch von Regeln (DNA-Norm) welche das Fortpflanzungs-Modell von DNA-Molekülen anwenden wurde betrachtet, und die theorethischen Aspekte von Symbiogenetischen Evolutions-Experimenten basiert auf DNA-Norm wurden besprochen. Einige Schlussfolgerungen in Bezug auf die Entstehung und Geschichte von dem Erdenleben suggeriert durch die Resultate von den Symbiogenese-Experimenten wurden gezeigt.

POST EDITORIAL NOTE

After this paper was submitted C. Bresch (Institut für Genetik der Universität zu Köln) has proposed (January 20, 1962) a phage-genetic theory making use of the principle of complementary association presented in section 10 of this paper. The same principle has been used by the author since the spring 1961 in several phage-genetic theories which are being tested by a high speed computer, IBM 7090.

REFERENCES

BARRICELLI, N. A. (1952). Mikroorganismenes genetikk. — Naturen 1952, Nr. 6, p. 162-191.

—— (1955). On the manner in which crossbreeding takes place in bacteriophages and bacteria. — Acta biotheor., Leiden XI, p. 75-84.

—— (1957). Symbiogenetic evolution processes realized by artificial methods. — Methodos IX, Nr. 35-36.

—— (1960). An analytical approach to the problems of phage recombination and reproduction: 1. Multiplicity reactivation and the nature of radiation damages. — Virology XI, p. 99-135.

—— (1962). Numerical testing of evolution theories: 1. Theoretical introduction and basic tests. — Acta biotheor., Leiden XVI, p. 69-98.

BARRICELLI, N. A. & A. H. DOERMANN (1960). An analytical approach to the problems of phage recombination and reproduction: 2. High negative interference. — Virology XI, p. 136-155.

—— (1961). An analytical approach to the problems of phage recombination and reproduction: 3. Crossreactivation. — Virology XIII, p. 460-476.

BERNSTEIN, A. & DE V. ROBERTS, M. (1958). Computer vs. chess-player. — Sci. Amer. cxcviii, p. 96-105.

DOERMANN, A. H. (1953). The vegetative state in the life cycle of bacteriophage. — Cold Spr. Harb. Symp. quant. Biol. XVIII, p. 3-11.

DOERMANN, A. H. & L. BOEHNER (1963). An experimental analysis of bacteriophage heterozygotes: 1. Mottled plaques from crosses involving six rII loci. — Virology (In press).

FOX, S. W. (1960). How did life begin? Recent experiments suggest an integrated origin of anabolism, protein, and cell boundaries. — Science CXXXII, p. 200-208.

GARDNER, M. (1958). Mathematical games. — Sci. Amer. cxcviii, p. 104-111.

HOROWITZ, N. H. (1945). On the evolution of biochemical syntheses. — Proc. nat. Acad. Sci., Wash. XXXI, p. 153-157.

KOZO-POLIANSKY, B. (1924). Outline of a theory of symbiogenesis. Selkhozgiz.

OPARIN, A. I. (1938). The origin of life. (Transl. with annotations by S. Morgulis) — New York, Macmillan, xii + 270 p.

UREY, H. C. & E. L. MILLER (1959). Organic compound synthesis on the primitive earth. — Science CXXX, p. 245-251.

Chapter 7
Artificial Intelligence through Simulated Evolution

L. J. Fogel, A. J. Owens, and M. J. Walsh (1965) "Artificial Intelligence through a Simulation of Evolution," *Biophysics and Cybernetic Systems: Proc. of the 2nd Cybernetic Sciences Symp.,* M. Maxfield, A. Callahan, and L. J. Fogel (eds.), Spartan Books, Washington D.C., pp. 131–155.

L. J. Fogel and W. S. McCulloch (1970) "Natural Automata and Prosthetic Devices," *Aids to Biological Communication: Prosthesis and Synthesis,* Vol. 2, D. M. Ramsey-Klee (ed.), Gordon and Breach, NY, pp. 221–262.

LAWRENCE Fogel served at the National Science Foundation (on leave from Convair) as special assistant to the associate director (research), Richard Bolt, in 1960. He was tasked to prepare a report on investing in basic research, including artificial intelligence. At that time, artificial intelligence primarily concerned heuristic programming and the simulation of primitive neural networks. Fogel considered both of these approaches to be limited because they modeled humans, rather than the underlying essential process that generally produces creatures of increasing intellect: evolution. He viewed intelligence in terms of adaptive behavior, and suggested that instead of modeling specific artifacts of evolution (e.g., humans), artificial intelligence should focus on modeling evolution itself (Fogel, 1962, 1963).

Intelligent behavior was viewed as a composite ability to predict one's environment coupled with a translation of each prediction into a suitable response in light of a given goal (to maximize a payoff function). Thus prediction is a prerequisite for intelligent behavior. For the sake of simplicity and generality, an environment was described as a sequence of symbols taken from a finite alphabet. The evolutionary problem was to devise an algorithm that would operate on the sequence of symbols thus far observed in such a manner as to produce an output symbol that is likely to maximize the algorithm's performance in light of both the next symbol to appear in the environment and a well-defined payoff function describing the worth of all of the possible correct and incorrect predictions. Finite state (Mealy)

machines (FSMs) provided a convenient representation for such an algorithm.

Specifically, the procedure for this *evolutionary programming* was as follows. A population of FSMs[1] is exposed to the environment—that is, the sequence of symbols that has been observed up to the current time. For each parent machine, as each input symbol is presented to the machine, the corresponding output symbol is compared with the next input symbol. The worth of this prediction is then measured with respect to the payoff function (e.g., all-none, absolute error, squared error, or any other expression that indicates the meaning of the symbols). After the last prediction is made, a function of the payoff for the sequence of symbols (e.g., average payoff per symbol) indicates the fitness of the machine. Offspring machines are created by randomly mutating the parents (i.e., through variations of adding and deleting states, changing output or next-state transitions, or changing the starting state)[2] and are scored in a similar manner. Those machines that provide the greatest payoff are retained to become parents of the next generation, and the procedure iterates. When new symbols are to be predicted, the best available machine serves as the basis for making such a prediction and the new observation is added to the available database.

This general procedure was applied to problems in prediction, system identification, and automatic control in a series of investigations (Fogel, 1964; Fogel et al., 1964, 1965a,[3] 1965b; and others). Fogel et al. (1965c) (reprinted here) is typical of

---

[1]The prose in several of the early papers on evolutionary programming emphasized a single parent giving rise to a single offspring, when in fact the reported experiments used a population of at least three parents (see Fogel et al., 1966, pp. 27–38). This choice of wording was unfortunate. Occasional reviews inappropriately described the work as using a population of size one (e.g., Holland, 1975, p. 163; Kieras, 1976; Rada, 1981; and others). The experiments in Fogel et al. (1965c) (reprinted here) used populations of at least three parents (Fogel, 1997).

[2]Fogel et al. (1966, pp. 21–23) also suggested the possibility of recombining finite state machines, particularly in a majority logic operation.

[3]This was also reprinted as Fogel et al. (1965b).

these studies, which were summarized in Fogel et al. (1966). Subsequent efforts extended evolutionary programming for predicting and classifying time series (Walsh, 1967; Lutter, 1968; Fogel and Moore, 1968; Lutter and Huntsinger, 1969; Burgin, 1974; Dearholt, 1976; and others), and also included the evolution of gaming strategies (Burgin, 1969) and the coevolution of FSMs where the fitness assigned to one strategy was determined in light of the opposing strategy (Fogel and Burgin, 1969). These studies were among the earliest to (1) use simulated evolution in forecasting, (2) include variable-length encodings, (3) use representations that take the form of a sequence of instructions, (4) incorporate a population of candidate solutions, and (5) coevolve evolutionary programs.

Research in evolutionary programming was also conducted independently at New Mexico State University[4] (Root, 1970; Cornett, 1972; Lyle, 1972; Holmes, 1973; Trellue, 1973; Montez, 1974; Atmar, 1976; Vincent, 1976; Williams, 1977; and others), primarily involving the recognition of handwritten characters and the induction of binary languages using Moore machines (also see Takeuchi, 1980; Tomita, 1982). Holmes (1973), following Lyle (1972), offered a procedure for dynamically adjusting the probabilities for alternative mutations as a function of their success or failure (or at random). Atmar (1976) used Mealy machines in a simulation of ecological tropisms where FSMs migrated in a grid representing a geographical area. Rates of learning were shown to vary as a function of population diversity, environmental heterogeneity, the size of the competitive arena, and the rate of mutation. Atmar (1976), following Wiener (1961), also argued that there are three forms of learning: (1) phylogenetic (arising within the phyletic line), (2) ontogenetic (arising within the individual), and (3) sociogenetic (arising within the social group), and that all of these were fundamentally evolutionary processes. Earlier, Campbell (1960) had argued that blind-variation-and-selective-survival is fundamental to all learning processes, and Fogel (1964; Fogel et al., 1966, pp. 108–111) also had argued a fundamental similarity between natural evolution and the scientific method, comparing organisms to hypotheses which are tested by their environment, with survival demonstrating suitability.

The evolutionary programming approach to artificial intelligence received significant attention from the artificial intelligence community,[5] and notable resistance from those following more traditional avenues of symbolic processing and heuristics (for a review, see Fogel, 1991, pp. 34–36). Fogel and McCulloch (1970) (reprinted here) offers a transcription of discussions between Fogel and scientific colleagues[6] from the 1966 conference on *Aids to Biological Communication: Prosthesis and Synthesis.* There was a clear reluctance to use randomness in advancing the search for superior solutions. Fogel concluded the discussion by pleading for logic and experiment in the face of *vox populi.*

Within the last decade, evolutionary programming has been extended to operate on arbitrary data structures, with no restrictions on the form of mutations that may be utilized (see Fogel and Atmar, 1992; and others). Whereas selection was often deterministic in early efforts (Fogel, 1964; and others), it is now standard to apply a probabilistic tournament that offers the possibility for lesser scoring solutions to survive (see Chapter 14 on soft selection). Further, the use of self-adaptation to autonomously update parameters controlling mutation in a second level of mutation and selection has been incorporated since Fogel et al. (1991).

References

[1] M. N. Anisimov and V. G. Indler (1988) "Use of evolutionary programming to control nonstationary agricultural objects," *Elektrotekhnika,* Vol. 59:4, pp. 31–35.

[2] J. W. Atmar (1976) "Speculation on the evolution of intelligence and its possible realization in machine form," Ph.D. diss., New Mexico State University, Las Cruces, NM.

[3] G. H. Burgin (1969) "On playing two-person zero-sum games against nonminimax players," *IEEE Trans. Systems Science and Cybern.,* Vol. SSC-5, pp. 369–370.

[4] G. H. Burgin (1974) "System identification by quasilinearization and evolutionary programming," *J. Cybernetics,* Vol. 2:3, pp. 4–23.

[5] D. T. Campbell (1960) "Blind variation and selective survival as a general strategy in knowledge-processes," *Self-Organizing Systems,* M. C. Yovits and S. Cameron (eds.), Pergamon, NY, pp. 205–231.

[6] F. N. Cornett (1972) "An application of evolutionary programming to pattern recognition," master's thesis, New Mexico State University, Las Cruces, NM.

[7] D. W. Dearholt (1976) "Some experiments on generalization using evolving automata," *Proc. 9th Intern. Conf. on Systems Sciences,* Honolulu, HI, pp. 131–133.

[8] B. S. Fleishman and I. L. Bukatova (1974) "Some analytical evaluations of evolutionary simulation parameters," Avtomatika i Vychislitel'naya Tekhnika, July-Aug., no.4, pp. 34–39. [Russian]

[9] D. B. Fogel (1991) *System Identification through Simulated Evolution: A Machine Learning Approach to Modeling,* Ginn Press, Needham, MA.

[10] D. B. Fogel and W. Atmar (eds.) (1992) *Proc. 1st Annual Conference on Evolutionary Programming,* Evolutionary Programming Society, La Jolla, CA.

[11] D. B. Fogel, L. J. Fogel, and J. W. Atmar (1991) "Meta-evolutionary programming," *Proc. of 25th Asilomar Conf. on Signals, Systems, and Computers,* R. R. Chen (ed.), Maple Press, San Jose, CA, pp. 540–545.

[12] L. J. Fogel (1962) "Autonomous automata," *Industrial Research,* Vol. 4, pp. 14–19.

[13] L. J. Fogel (1963) *Biotechnology: Concepts and Applications,* Prentice-Hall, Englewood Cliffs, NJ.

[4]This was primarily under the direction of Donald W. Dearholt.

[5]In particular, the approach was well known in the Soviet Union (Zilinskas, 1997) following the Russian translation of Fogel et al. (1966) as Fogel et al. (1969), and other papers. Russian studies following Fogel et al. (1966) include Fleishman and Bukatova (1974), Yakobson (1975), Anisimov and Indler (1988), and others.

[6]The discussion is reprinted in its entirety for completeness, although the subject matter relevant to evolutionary computation begins 17 pages into the paper. In the order of appearance, the listed participants are: Lawrence Fogel, Warren S. McCulloch, Marvin Minsky, Allen Newell, Edwin W. Paxon, John McCarthy, V. A. Kozhevnikov, Walter A. Rosenblith, E. E. David, Jr., Julian Bigelow, Alexander M. Letov, Frank Fremont-Smith, Walter L. Wasserman, Francis O. Schmitt, Michael Arbib, Donald M. MacKay, and Seymour Papert.

[14] L. J. Fogel (1964) "On the organization of intellect," Ph.D. diss., UCLA.

[15] L. J. Fogel (1997) personal communication, Natural Selection, Inc., La Jolla, CA.

[16] L. J. Fogel and G. H. Burgin (1969) "Competitive goal-seeking through evolutionary programming," final report, Contract AF 19(628)-5927, Air Force Cambridge Research Laboratories.

[17] L. J. Fogel and W. S. McCulloch (1970) "Natural Automata and Prosthetic Devices," *Aids to Biological Communication: Prosthesis and Synthesis,* Vol. 2, D. M. Ramsey-Klee (ed.), Gordon and Breach, NY, pp. 221–262.

[18] L. J. Fogel and R. A. Moore (1968) "Modeling the human operator with finite-state machines," final report, Contract no. NAS 1–6739, NASA Langley Research Center.

[19] L. J. Fogel, A. J. Owens, and M. J. Walsh (1964) "On the evolution of artificial intelligence," *Proc. 5th National Symp. on Human Factors in Engineering,* IEEE, San Diego, CA, pp. 63–76.

[20] L. J. Fogel, A. J. Owens, and M. J. Walsh (1965a) "Intelligent decision-making through a simulation of evolution," *IEEE Trans. Human Factors in Electronics,* Vol. HFE-6:1, pp. 13–23.

[21] L. J. Fogel, A. J. Owens, and M. J. Walsh (1965b) "Intelligent decision making through a simulation of evolution," *Behavioral Science,* Vol. 11:4, pp. 253–272.

[22] L. J. Fogel, A. J. Owens, and M. J. Walsh (1965c) "Artificial intelligence through a simulation of evolution," *Biophysics and Cybernetic Systems: Proc. of the 2nd Cybernetic Sciences Symp.,* M. Maxfield, A. Callahan, and L. J. Fogel (eds.), Spartan Books, Washington, D.C., pp. 131–155

[23] L. J. Fogel, A. J. Owens, and M. J. Walsh (1966) *Artificial Intelligence through Simulated Evolution,* John Wiley, NY.

[24] L. Fogel, I. Owens, and M. Uolsh (1969) *Artificial Intelligence and Evolutionary Modeling,* Mir, Moscow. [Russian]

[25] J. H. Holland (1975) *Adaptation in Natural and Artificial Systems,* Univ. of Michigan, Ann Arbor, MI.

[26] V. P. Holmes (1973) "Recognizing prime numbers with an evolutionary program," master's thesis, New Mexico State University, Las Cruces, NM.

[27] D. E. Kieras (1976) "Automata and S-R models," *J. Mathematical Psychology,* Vol. 13, pp. 127–147.

[28] B. E. Lutter (1968) "The application of artificial intelligence through the evolutionary programming technique to the control of chemical engineering processes," master's thesis, South Dakota School of Mines and Technology, Rapid City, SD.

[29] B. E. Lutter and R. C. Huntsinger (1969) "Engineering applications of finite automata," *Simulation,* Vol. 13, pp. 5–11.

[30] M. R. Lyle (1972) "An investigation into scoring techniques in evolutionary programming," master's thesis, New Mexico State University, Las Cruces, NM.

[31] J. Montez (1974) "Evolving automata for classifying electrocardiograms," master's thesis, New Mexico State University, Las Cruces, NM.

[32] R. Rada (1981) "Evolution and gradualness," *BioSystems,* Vol. 14, pp. 211–218.

[33] R. Root (1970) "An investigation of evolutionary programming," master's thesis, New Mexico State University, Las Cruces, NM.

[34] A. Takeuchi (1980) "Evolutionary automata—Comparison of automaton behavior and Restle's learning model," *Information Sciences,* Vol. 20, pp. 91–99.

[35] M. Tomita (1982) "Dynamic construction of finite automata from examples using hill-climbing," *Proc. 4th Ann. Cognitive Science Conf.,* Cognitive Science Society, Ann Arbor, MI, pp. 105–108.

[36] R. E. Trellue (1973) "The recognition of handprinted characters through evolutionary programming," master's thesis, New Mexico State University, Las Cruces, NM.

[37] R. W. Vincent (1976) "Evolving automata used for recognition of digitized strings," master's thesis, New Mexico State University, Las Cruces, NM.

[38] M. J. Walsh (1967) "Evolution of finite automata for prediction," final report, Contract no. RADC-TR-67 555, Rome Air Development Center, Griffiss AFB, NY.

[39] N. Wiener (1961) *Cybernetics,* Part 2, MIT Press, Cambridge, MA.

[40] G. L. Williams (1977) "Recognition of hand-printed numerals using evolving automata," master's thesis, New Mexico State University, Las Cruces, NM.

[41] B. M. Yakobson (1975) "Modelling evolutionary processes when designing engineering system," *Pribory i Sistemy Upravleniya,* no. 5, pp. 9–11. [Russian]

[42] A. Zilinskas (1997) personal communication, Institute of Mathematics and Informatics, Vilnius, Lithuania.

ARTIFICIAL INTELLIGENCE THROUGH A SIMULATION OF EVOLUTION*

LAWRENCE J. FOGEL, ALVIN J. OWENS and MICHAEL J. WALSH

General Dynamics/Astronautics, San Diego, California

INTRODUCTION

Both the bionic and heuristic programming approaches toward artificial intelligence attempt to model the information processing characteristics of that intelligent creature—Man. The immense complexity of the central nervous system, coupled with our incomplete knowledge of the neural and molecular mechanisms, limits our ability to replicate the biological entity which provides human intellect. Networks of threshold elements may simulate arrays of neurons, but this is a far cry from providing behavior at higher levels of abstraction. In short, the case for replicating nature in terms of physical correspondence stands on weak ground (as witness the fact that modern aircraft are not ornithoptors).†

Taking a broader view, heuristic programming focuses attention upon psychological aspects of human decision making in an attempt to devise logic which will overcome specific problems in a manner similar to that of an intelligent man. A variety of game playing and theorem proving programs have been successfully demonstrated, but this mimicry falls short of

---

* This research was in part supported by the Office of Naval Research (contract Nonr 4539 (00)) and in part by Goddard Space Flight Center, NASA (contract NAS 5-3907).

† These comments should not be taken as an attack on bionics. Such modeling may well provide valuable insight into biophysical functioning.

Reprinted with permission from *Biophysics and Cybernetic Systems: Proceedings of the 2nd Cybernetic Sciences Symposium,* L. J. Fogel, A. J. Owens, and M. J. Walsh, "Artificial intelligence through a simulation of evolution" M. Maxfield, A. Callahan, and L. J. Fogel (eds.), Spartan Books, Washington, D.C., pp. 131-155. © 1965 by L. J. Fogel and Spartan Books.

providing the flexibility which is essential to intelligent behavior; nor does it furnish an insight into the fundamental logic which makes intellect possible. Success in the field of artificial intelligence should require that inanimate machines solve problems which still remain to be solved by man, not because of their sheer speed, accuracy, or greater memory, but because they discover new techniques for solving the problem at hand.

Still another approach toward artificial intelligence is possible. Man may be recognized to be but a single artifact of the natural experiment called evolution; an experiment which has rather consistently produced creatures at higher and higher levels of intelligence. Might it not be far wiser to model the process of evolution—iterative mutation and selection—in order to discover successively better logic for seeking the given goal under the constraint imposed by the environment?

Intelligent behavior can result from an ability to predict the environment coupled with the selection of an algorithm which permits the translation of each prediction into a suitable response. For the sake of clarity, the following discussion will primarily be concerned with the problem of predicting the behavior of the observed environment. More specifically, the problem at each point in time is to devise an algorithm which will operate on the sequence of symbols thus far observed in order to produce an output symbol which will agree with the next symbol to emerge from the environment. Simulated evolution provides a means towards this end.

DISCUSSION

An arbitrary finite-state machine is exposed to the sequence of symbols which have thus far emerged from the environment. As this occurs, each output symbol from the machine is compared with the next input symbol. The percent correct score is a measure of the ability of this machine to predict the already experienced environment on the basis of the preceding symbols. An *offspring* of this machine is then produced through mutation, that is, through a single modification of the *parent* machine in accordance with some mutation noise distribution. Thus the offspring is made to differ from its parent either in an output symbol,* a state-transition, the number of states, or the initial state. †

---

* In the case of a binary environment, a deterministic procedure can be used to replace this type of mutation. As each symbol from the environment is predicted on the basis of the preceding symbols score is maintained of the relative frequency of success of each state-transition. A predictive fit score of greater than 0.5 can then be ensured by the reversal of output symbols on those state-transitions which were *more often wrong than right.*

† That state the machine is in when it receives the first symbol of its experience.

The offspring is now exposed to the same sequence of symbols which were experienced by the parent-machine and its prediction capability is similarly scored. If this score is found to equal or exceed that of the parent, the offspring survives to become the new parent. If not, it is discarded and a new offspring is generated. In this manner nonregressive evolution proceeds through successive finite-state machines which individually evidence increased ability to predict the already experienced sequence of symbols. At any point in time the remaining machine can be used for actual prediction; that is, it can be exercised by the last symbol to emerge from the environment thus producing an actual prediction of the next symbol to be experienced. The same machine is then used to parent succeeding offspring which are evaluated over the now longer recall. Thus the evolution continues in fast time in preparation for the next required actual prediction. Such predictions may take place periodically, aperiodically, or on request. They may be made whenever a specified predictive fit score has been attained, when some prechosen number of offspring have been evaluated, or when an appropriate number of generations* have occurred. Of course, in general, the longer the time interval between successive predictions, the greater the expectation of success. Similarly, the greater the speed of the computer facility (increase in the number of evaluated offspring) or the larger the available memory (increase of their permissible size) the greater the evolutionary prediction capability.

The evolutionary technique offers distinct versatility. For example, the desire to predict each second symbol in the future can be satisfied simply by scoring each offspring in terms of the correspondence between its output symbols and those symbols which emerge from the environment two symbols later. By the same token, appropriate scoring of the offspring permits the prediction of any particular future symbol, the average of some set of future symbols, or indeed, any well-defined function over the future. The desire for minimum error prediction may be satisfied by using a *magnitude of the difference* error matrix. Minimum *rms* error prediction results if each term of this error matrix is squared. If "a miss is as good as a mile" the error matrix should have equal nonzero off-diagonal terms and zero on the diagonal.

But, the purpose of the simulated evolution need not be restricted to prediction in any sense. The input symbols of the evolving machines may be individually associated with the set of possible stimuli, the output symbols with the set of alternative response, the goal: to achieve any well-defined function over the future sequence of responses. Here there is no longer a distinction between prediction and the algorithm which translates

---

* Defined later in this section.

prediction into response. The evolutionary program recommends each action in the light of its expected overall worth with respect to the given goal.

At the same time, it is reasonable in the interest of economy, to desire that the offspring be of minimum complexity. The maxim of parsimony may be directly incorporated into the evolutionary procedure by reducing the score of each machine in proportion to a measure of its complexity. The amount of this penalty may be influenced by the particular characteristics of the computer facility upon which the simulation is to be carried out. Thus, at each point in time, the evolutionary technique provides a non-regressive search through the domain of finite-state hypotheses for that logic which best satisfies the given goal under the constraint imposed by the available computation capability.

Efficiency of the simulated evolution can be improved in a number of ways. Any available information concerning the underlying logic of the environment can be incorporated in the form of the initial machine. If this *hint* is reasonably correct, the evolution should require fewer generations to attain the same score. If it is incorrect, this introduction of *false* information in no way precludes solution of the problem, although it may be expected to reduce the efficiency of the procedure.

Suitable choice of the mutation noise can increase the efficiency of evolution. For example, an increase in the probability of adding a state generates a wider selection of larger machines which should benefit evolution against a complex environment. In fact, the probability over the modes of mutation can be made to depend upon the evidence acquired within the evolutionary process itself. Thus, an experienced greater relative frequency of success for, say, changing the initial state might be made to increase the probability of this mode of mutation. Although this procedure may benefit the prediction of independent environments, it can offer a danger if the environment is interactive . . . an intelligent adversary might discover the specific dependency and use this knowledge to construct an obverse strategy.

The evolutionary search may be viewed as a selective random walk, a *hill climbing* procedure, in a hyperspace defined to include the finite-state machines and an additional coordinate on which is measured the predictive fit score.* The danger of becoming trapped on a secondary peak can be overcome by permitting multiple mutation, with the multiplicity being a function of the difference in predictive-fit score of successive generations. Thus, as the search nears a peak greater and greater *attention* is devoted

---

* The term "predictive-fit" is used in place of the more general "functional-fit" in view of the more immediate concern with the problem of prediction.

to generating more radical offspring in the hope of striking a point which may lie higher on the slope of another peak.

The evolutionary technique may be expected to predict nonstationary environments because of its continual search for a *best* logic. But selection of only the single best logic may be an overly severe policy. Certainly those offspring which have predictive-fit scores also characterize the logic of the environment in some meaningful manner. Why not mimic nature and save the best few machines at each point in time? In general, the highest scoring offspring has the greatest probability of giving rise to an even better offspring, thus it should receive most attention in terms of mutative reproduction. Lower ranked offspring may be regarded as insurance against gross nonstationarity of the environment. The distribution of mutative effort may well be in proportion to the normalized predictive-fit scores. Evaluated offspring are inserted into the rank order table of retained offspring and a generation is said to occur whenever an offspring is found which has a score equal to or greater than the score of the best machine.

All of the retained machines need not lie on the slopes of the peak which is identified by the best machine. Thus, saving the best few offspring may maintain a *cognizance* over several peaks, with the relative search effort being distributed in proportion to the expectation of significant new discoveries. The greater the number of surviving offspring, the larger the number of possible peaks. Of course, saving a greater number of offspring decrease the efficiency against a well-behaved environment. Here again, if properly chosen, the number of retained offspring and the distribution of mutative attention can improve the efficiency of evolution, but, at the same time any such restriction offers a danger if the environment takes the form of an intelligent adversary.

The efficiency of natural evolution is enhanced by the recombination of individuals of opposite sex. By analogy, why not retain worthwhile *traits* which have survived separate evalution ·by combining the best surviving machines through some genetic rule, mutating the product to yield offspring? Note that there is no need to restrict this mating to the two best surviving *individuals*. In fact the most obvious genetic rule, majority logic, only becomes meaningful with the combination of more than two machines†. Clearly, this opens the door to many new possibilities. For example, it may be fruitful to explore the combining of the best machines of several different generations in the hope of finding a model of the models

---

† It is always possible to express a finite set of machines which operate through a majority logic element as a single machine. Each state of the majority machine is a composite of a state from each of the given machines. Each transition is described by the input symbol which caused the respective transition in the given machines and by that output symbol which results when majority logic is applied to the output symbols from the given machines.

which had thus far been most successful. It is tempting to speculate on an obvious extension of this procedure which would operate simultaneously at several levels of abstraction, thus recombining the best machines over various levels.

EXPERIMENTS

In the interest of brevity only some of the series of experiments which were conducted to explore the predictive capability of the evolutionary technique will be reported, these being numbered consecutively for ease of reference. The original evolutionary program was written in Fortran II for the IBM-7094 to permit prediction of 2-symbol environments. This set of experiments demonstrated the feasibility of predicting cyclic environments, stationary environments, and the primeness of each next number in the sequence of positive increasing integers. [1,2]

A second program was then written to permit the prediction of 8-symbol environments. Unless otherwise indicated all of the following experiments started with the same arbitrary five-state machine. The recall was permitted to grow with experience starting with 40 symbols before the first prediction. The penalty for complexity was chosen to be 0.01 times the number of states in that machine. Single, double, or triple mutation of each parent machine occurred with equal probability and a maximum of 40 offspring or 10 generations were permitted before each successive actual prediction.

The first set of experiments concerned the prediction of an environment composed of a cyclic signal created by repetition of the simple pattern 13576420 which was disturbed by increasing levels of noise. With the environment consisting only of the undisturbed signal (Experiment 1) the evolutionary technique discovered a perfect one-state predictor machine within the first 18 evaluated offspring. The environment for Experiment 2 was generated by corrupting this signal by the equally-likely addition of $+1$ or -1 to certain symbols, these being identified by skipping a number of symbols from the last disturbed symbol in accordance with the next digit drawn from a uniformly distributed random number table. Quite arbitrarily, addition to the symbol 7 and subtraction from the symbol 0 were assumed to leave these symbols undisturbed. Thus, 82.5% of the symbols were left undisturbed. As shown in Fig. 1, 59.3% of the first 81 predictions were correct, there being only 6 errors in the last 30 predictions. During the evolution 3,241 different offspring were evaluated, the predictor machines growing in size to eight states.

The environment of Experiment 3 was obtained by disturbing the environment used in Experiment 2 once again in the same manner. Thus,

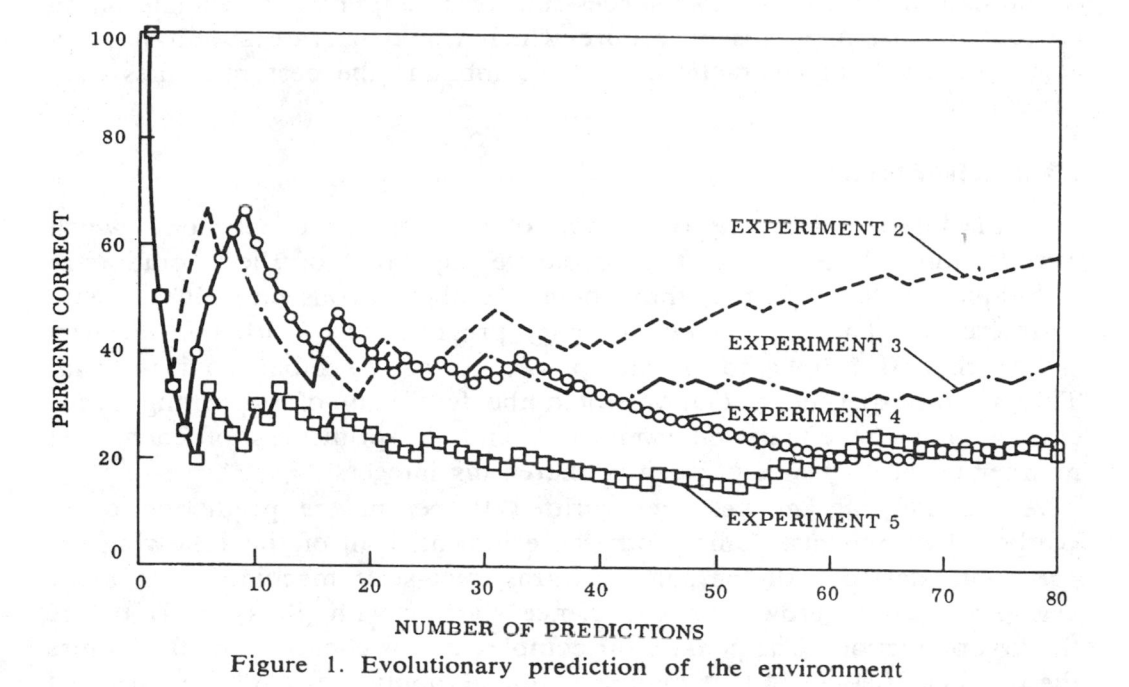

Figure 1. Evolutionary prediction of the environment

28.9% of the symbols were disturbed by $\pm$ 1, 1.5% were disturbed by $\pm$ 2; leaving 69.6% undisturbed. As shown in Fig. 1, 39.5% of the first 81 predictions were correct, there being a general increase in score in the last 20 predictions. During the evolution 3,236 different offspring were evaluated, the predictor machines growing in size to 15 states.

The environment for Experiment 4 was obtained by disturbing the environment used in Experiment 3 again in the same manner. Thus, 37.0% of the symbols were disturbed by $\pm$ 1, 2.5% were disturbed by $\pm$ 2, and 0.1% were disturbed by $\pm$ 3; leaving 60.4% of the symbols undisturbed. As shown on Fig. 1, 23.5% of the first 81 predictions were correct. During the evolution 3,214 different offspring were evaluated, the predictor machines growing in size rather steadily to 19 states.

The environment of Experiment 5 was obtained by disturbing a randomly chosen 50% of the symbols in the signal. Thus, 43.8% of the symbols were disturbed by $\pm$ 1 (the difference being due to the adopted rule concerning addition to 7 and substraction from 0); leaving 56.2% undisturbed. As shown in Fig. 1, 22.2% of the first 81 predictions were correct. During the evolution 3,195 different offspring were evaluted, the predictor machines growing in size somewhat irregularly to 17 states. It would appear that this last increase in the noise level (from 39.67% in Experiment 4 to

43.8% in Experiment 5) resulted in significantly degraded prediction of the environment for only the short recalls.

Figure 2 indicates the degree of correspondence between the sequence of predictions and the signal in these experiments. Note that after the first 76 predictions the signal was predicted in Experiment 5 as well as

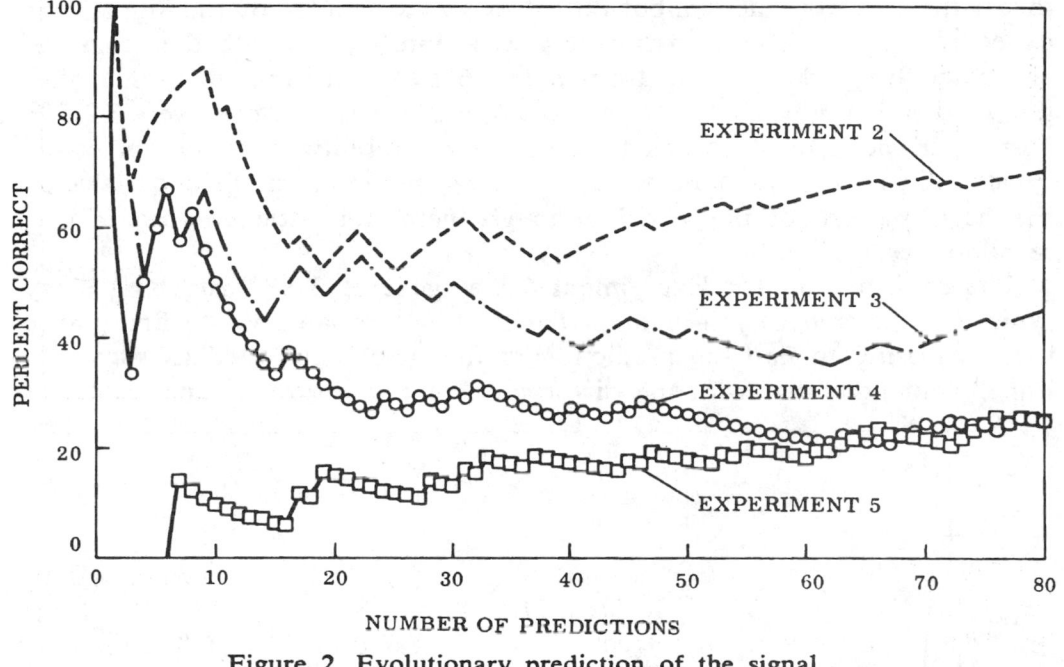

Figure 2. Evolutionary prediction of the signal.

it was in Experiment 4 in spite of the fact that a larger percentage of the signal symbols had been disturbed. This may be due to the fact that in the last experiment the symbols remained closer to the original signal. Such "consideration" for the magnitude of deviations is a result of using a distance-weighted error matrix, in this case the weighting being the magnitude of the symbol difference. In essence this choice converts the nominal scale of symbols to an ordinal scale.

It is of interest to examine each of the predictor-machines as representations of the periodic properties of the environment. The characteristic cycle for any finite-state machine is found by starting it in its initial state together with the first symbol of the recall then driving it by each of its successive output symbols until the output sequence is periodic. All of the characteristic cycles in Experiment 2 were eight symbols in length. The first 73 corresponded perfectly with the pattern of the signal but this *insight* was lost in the later predictions which were in error by one or two symbols. After the 14th prediction, the characteristic cycle remained 13576430.

The higher noise level of the third experiment resulted in characteristic cycles of varying length until the 26th prediction. From then on until the 70th prediction the characteristic cycle remained 1357643113576430, this being in error one symbol out of every two cycles of the signal. As expected, the result of Experiment 4 was more erratic with the length of characteristic cycle jumping from 8 to 16 and remaining the same after the 73rd prediction. Each of the last 24 characteristic cycles were 62.5% correct. Experiment 5 revealed even greater varability in the characteristic cycles. A majority of these were 8 or 16 symbols in length and reflected the basic pattern of the signal although there was little *one-to-one* correspondence.

The environment for Experiment 6 was generated by disturbing every symbol of the signal by $+ 1$ or $- 1$ with equal probability. At first glance it is surprising to find the prediction of the signal improved as shown in Fig. 3, but note that with the disturbance of every symbol one aspect of

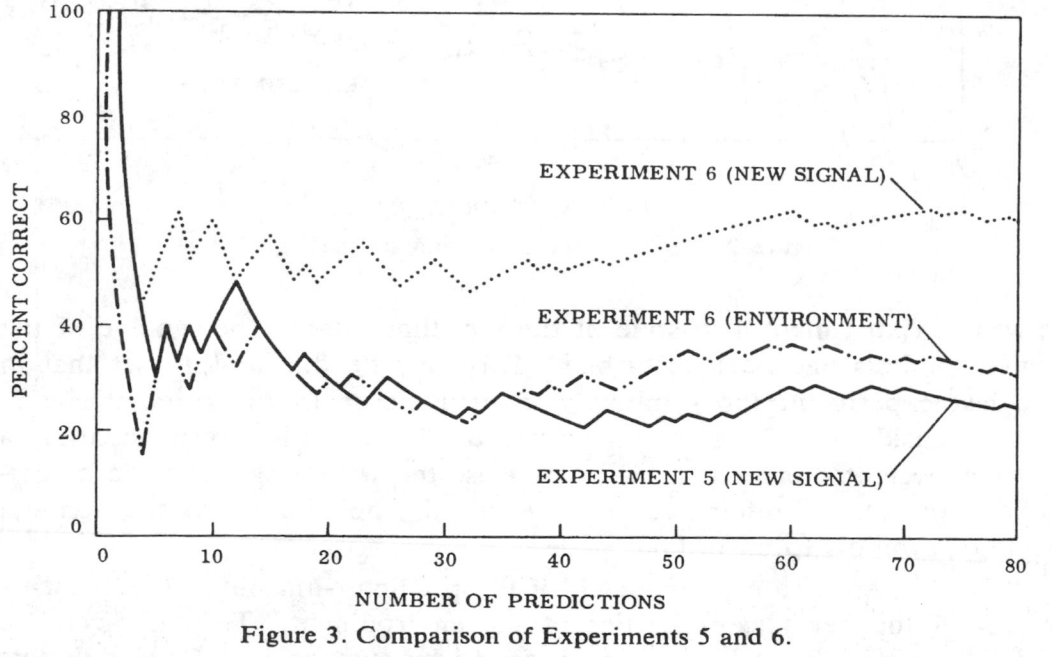

Figure 3. Comparison of Experiments 5 and 6.

the randomness of the environment was removed. In essence the signal had taken on a new form, the boundaries of the original signal 24677531 or 02465310, each having equal probability at each point in time. In the first 81 predictions this new signal was properly identified 70.4% of the time. In fact, the characteristic cycle of the last predictor-machine was 02267731102665331. This can be seen to lie on the boundaries of the original signal except for one symbol of every 17. In order to provide a basis for comparison the percent correct prediction of the environment is also shown as well as the result of Experiment 5 with respect to the new signal.

It is natural to inquire as to the extent the prediction capability will be degraded by *wild noise* (each disturbed symbol being replaced by a randomly chosen symbol from the input alphabet). The environments of Experiments 7 and 8 were generated by imposing this kind of disturbance on the original signal once and twice, respectively. As expected, the ability of the evolutionary program to predict the environment, as shown in Fig. 4, was somewhat poorer than in the comparable Experiments 2 and 3.

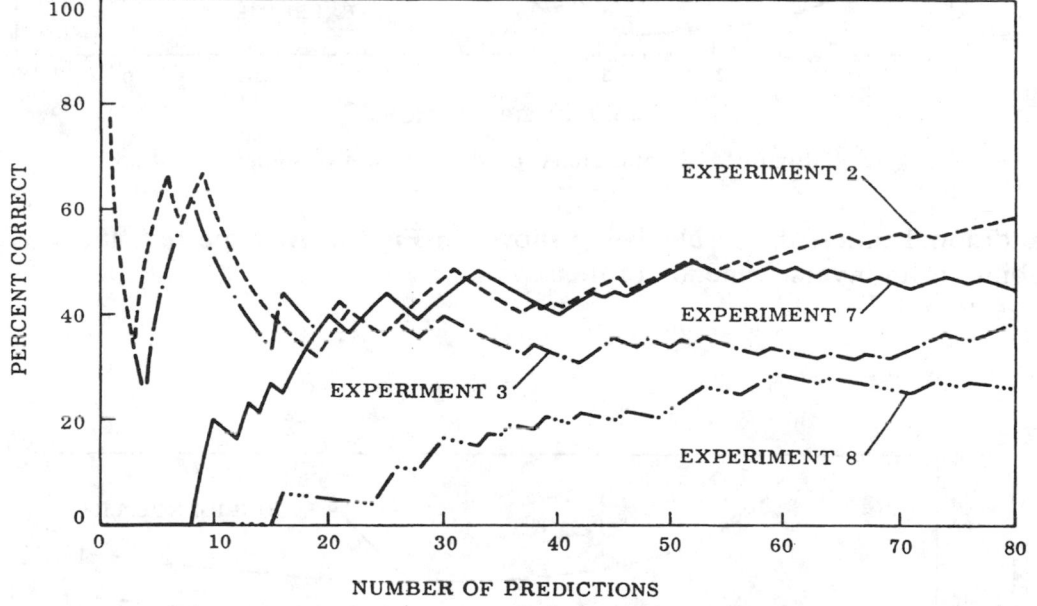

Figure 4. Evolutionary prediction of the environment.

Figure 5 indicates the degree of correspondence between the sequence of predictions and the signal in these same experiments. Here again the additional degree of randomness within the noise degrades the performance. Carrying this noise to the extreme results in a perfectly random environment. Experiment 9 revealed no significant ability of the evolutionary program to predict this environment.

The introduction of randomness always introduces questions of repeatability. In order to examine this point Experiment 2 was repeated nine

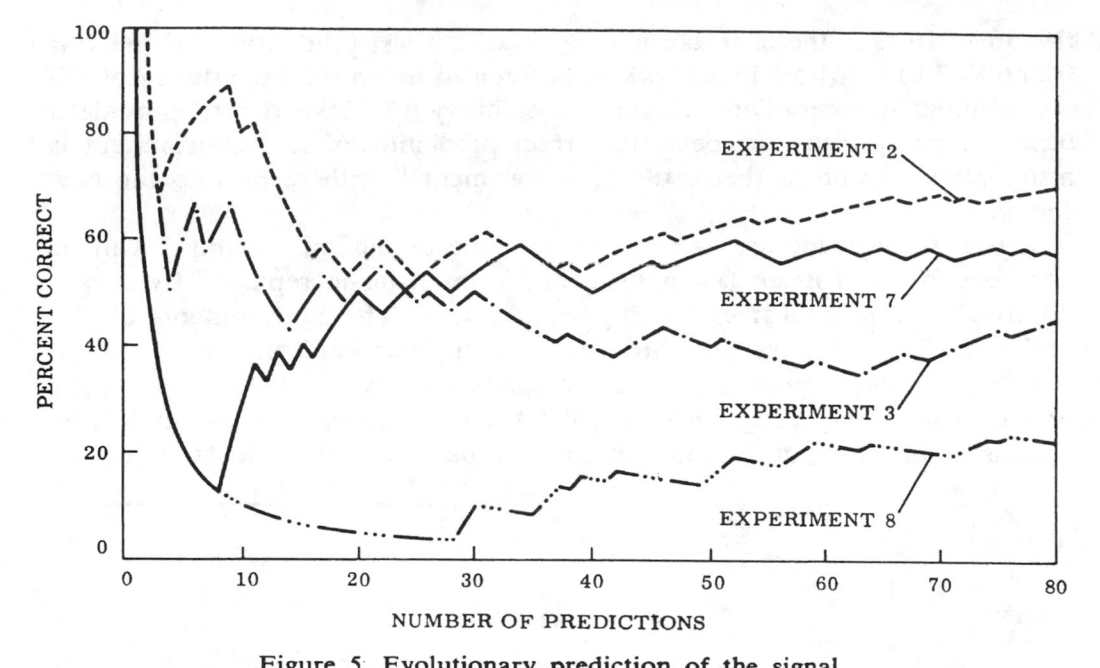

Figure 5. Evolutionary prediction of the signal.

additional times, the results being shown in Fig. 6. As expected, the variability is an inverse function of the score.

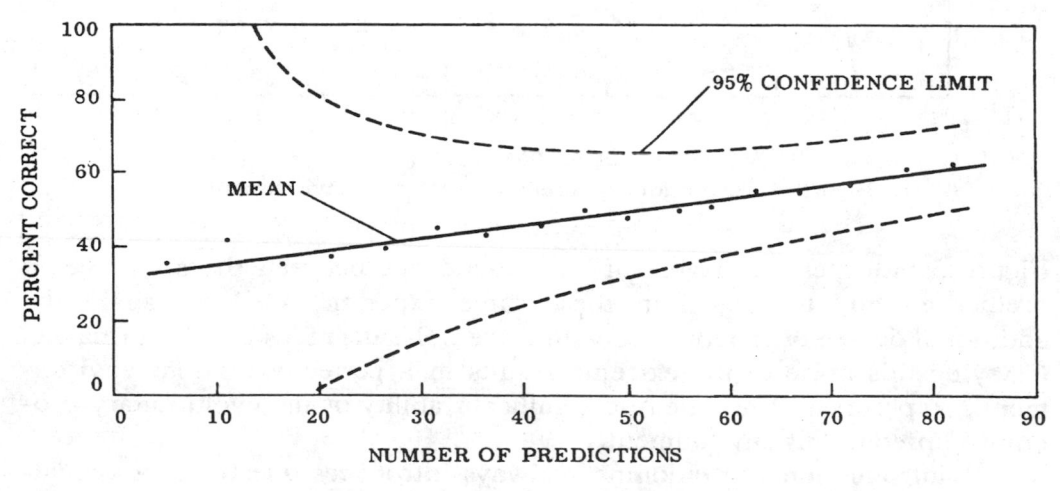

Figure 6. Statistics of Experiment 2.

The second set of experiments were concerned with the prediction of purely stochastic environments. Experiment 10 required the prediction of a zeroth-order four-symbol Markov environment, the arbitrarily chosen probabilities being 0.1, 0.2, 0.3, and 0.4. This information, as prior knowledge, would dictate the continual prediction of the most probable symbol giving the asymptotic score of 40%. At the other extreme, perfectly random prediction would have an expected score of 25%. As shown in Fig. 7

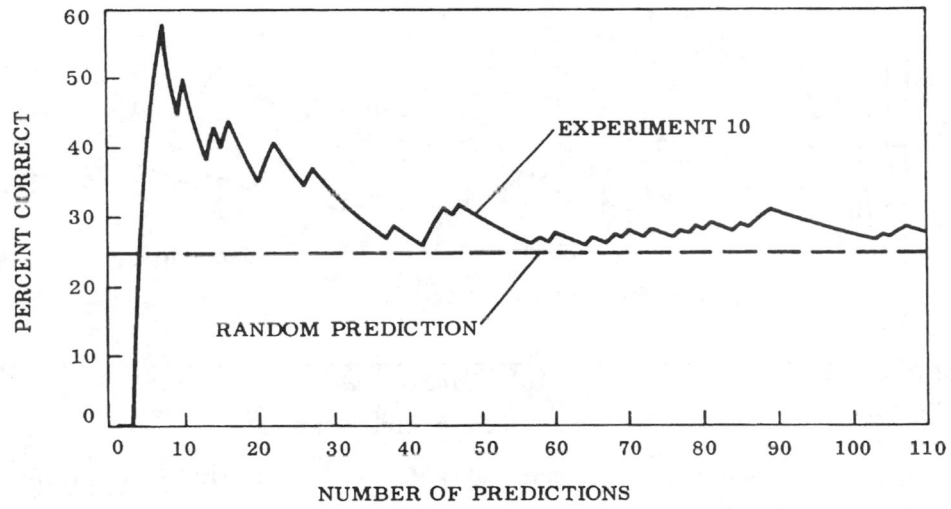

Figure 7. Evolutionary prediction of zeroth-order Markov environment.

the evolutionary score settled between these extremes thus demonstrating the purposeful extraction of information from the previous symbols.

The first order environment of Experiment 11 was generated by the arbitrary transition matrix shown in Table 1. The actual environment had the transition matrix of relative frequencies shown in Table 2.

TABLE 1

Arbitrary transition matrix

	0	1	2	3
0	0	0.8	0.1	0.1
1	0.1	0	0	0.9
2	0.9	0.1	0	0
3	0	0.1	0.8	0.1

TABLE 2

Transition matrix of relative frequencies

	0	1	2	3
0	0	0.822	0.071	0.107
1	0.035	0	0	0.965
2	0.915	0.085	0	0
3	0	0.077	0.862	0.061

The marginal frequencies of this environment were 0.236, 0.241, 0.249 and 0.274, respectively.

With prior knowledge that the process is first order it would be possible to attain the score of 89.5% on the 200th prediction in the manner shown in Fig. 8. *But even without this knowledge the evolutionary prediction*

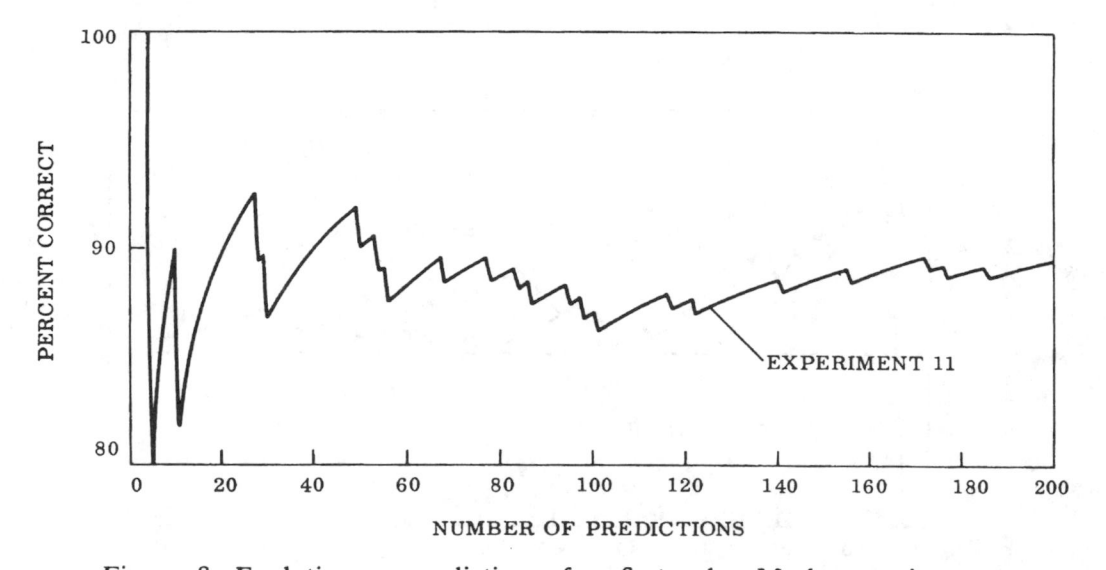

Figure 8. Evolutionary prediction of a first-order Markov environment.

technique attained this same score at each point. Analysis of the sequence of predictions revealed that at the end of this experiment the environment was properly characterized by the maximum transition probabilities of each row. Other experiments were conducted on first and second order processes with similar results.

To generate a more difficult environment, the powers of 2 and 3 were rank-ordered and reduced modulo 8, that is, 1234010300010300100-3010030100300. . . After the first 300 predictions (Experiment 12) the percent correct score reached 88.7. 1,401 different offspring were evaluated using an *all or none* error matrix (all off-diagonal elements have a value of 7 and all main-diagonal elements a value of 0). The predictor machines were generally of six states. In order to avoid the reduction to three symbols in the latter portion of the sequence, the powers of 2 and 3 were rank-ordered and reduced modulo 7 yielding 12341220414254112431122-461424512143122461. . . After the first 216 predictions (Experiment 13), the percent correct score was 56.6, this being found through the evaluation of 3,671 offspring, which were generally of about 21 states. The score for the last 50 predictions was 78%. Certainly the prediction capability was far better than chance would yield. Prediction of this sequence based on the most probable symbol up to each point in time yields a score of only 8.5%.

A set of experiments was conducted in order to evaluate the evolutionary technique as a means for detecting the existence of correlation between variables, this correlation to be used to enhance the sequential prediction of one or more of the sensed variables. In Experiment 14, a random sequence of binary symbols were presented to an arbitrarily chosen finite-state machine, M (shown in Table 3). Note that four different symbols comprise the output alphabet of this machine. The evolutionary program was required to predict each next symbol in the output sequence, this being predictable only to the extent that the structure of the machine M is super-imposed upon the otherwise random driving signal. The percent correct score, shown in Fig. 9, reveals increasing stability as the score settles around 51% after the first 152 predictions (with 20 symbols as the initial recall).

TABLE 3. Machine M

Present State	Input Symbol	Next State	Output Symbol	Present State	Input Symbol	Next State	Output Symbol
1	0	3	2	3	1	4	1
1	1	5	0	4	0	5	3
2	0	1	2	4	1	3	0
2	1	2	3	5	0	2	0
3	0	1	1	5	1	1	2

This prediction score should be considerably improved if the program were permitted to evolve finite-state representations on the basis of both the 4-symbol output sequence and the 2-symbol input sequence. In fact, with this additional information, the prediction score should asymptotically approach 100% as the recall increases in view of the deterministic nature of machine M. Experiment 15 was conducted to investigate whether or not this would take place through the evolutionary technique. At each point in time two-input single-output finite-state machines were evaluated in terms of their predictive-fit of the known recall of the 4-symbol output sequence while driven by the preceding recall of both the 4-symbol and 2-symbol sequences. The percent correct score, shown in Fig. 9, attains a value of 80%.

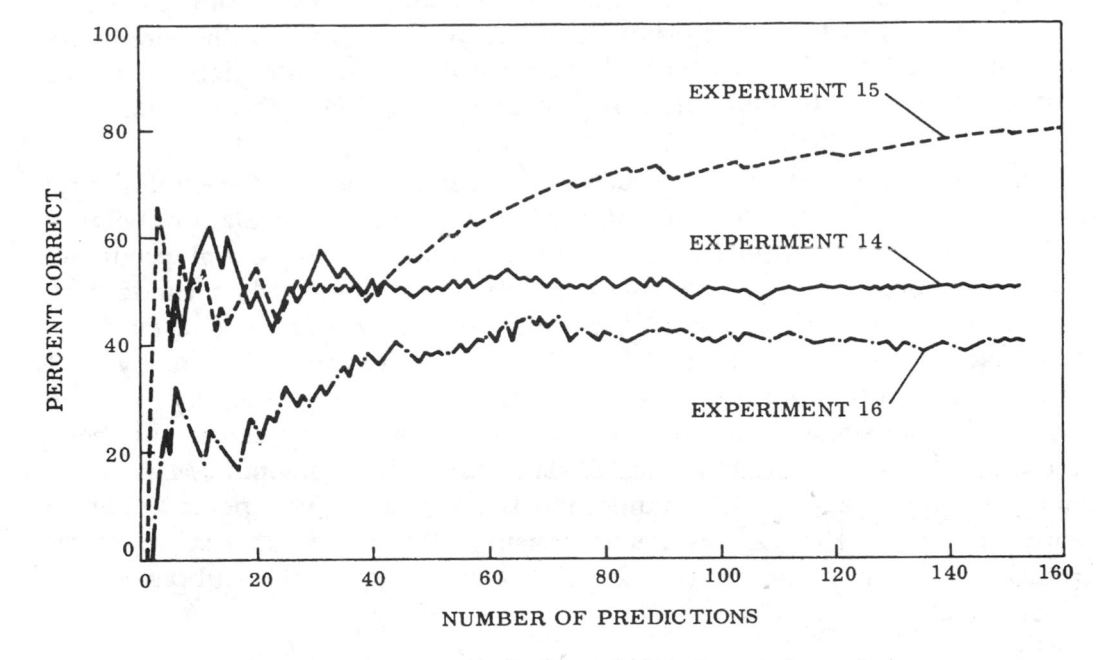

Figure 9. Evolutionary prediction of the 4-symbol environment.

In some sense, the sequence of evolved double-input single-output machines should provide an increasingly accurate representation of the transduction performed by machine *M*. Figure 10 is a graphical representation for the finite-state machine which evolved after 160 predictions. This machine, *N,* has the states designated by letters in order to avoid any confusion with the number-designated states of machine *M*. It is possible to compare machine *N* to machine *M* in the following way: examine each state of *M* in order to determine if there is some state in machine *N* which in some sense will perform the same function when given the same sequence of input symbols. (Note that the input to each state of machine *N* consists of an ordered pair of symbols, the first being the input to the state machine *M* and the second being the previous output of machine *M* before it entered that state.)

To illustrate, consider state 1 of machine *M*. Since the outputs of the transitions which enter this state are either 1 or 2, the possible corresponding inputs to machine *M* are 0, 1; 0, 2; and 1, 2. Examination of state *B* of machine *N* shows that 0, 1 and 0, 2 both yield the output symbol 2 which corresponds to the output of state 1 when it receives an input symbol of 0. Similarly 1, 1 and 1, 2 produce an output of 0 which corresponds to the output of state 1 when it receives input symbol 1. Similarly, 1, 1

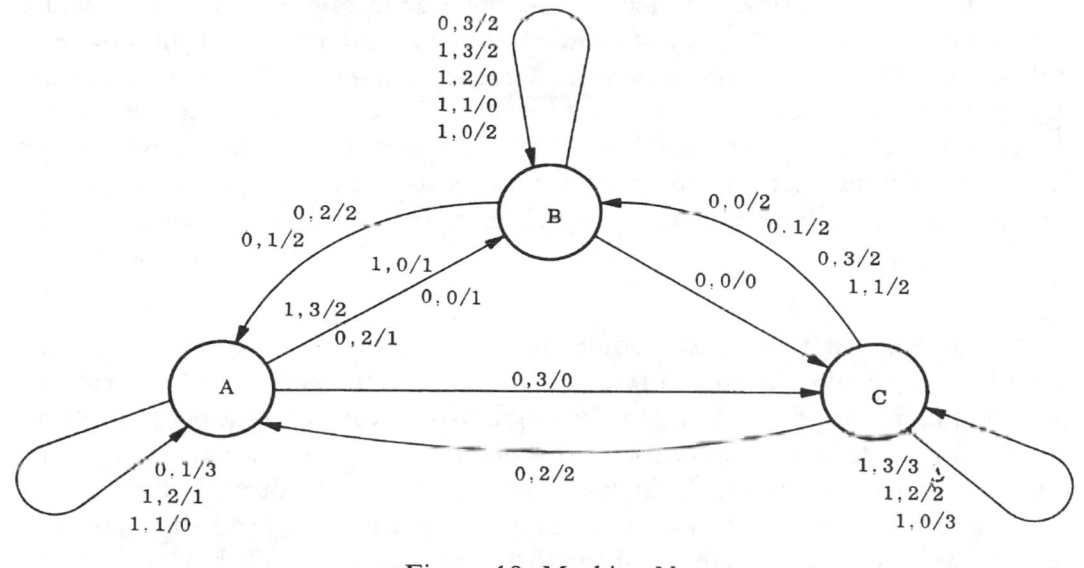

Figure 10. Machine N.

and 1, 2 produce output symbol 0 which corresponds to the output of state 1 when it receives an input symbol 1. Thus, state 1 of machine *M* may be said to be contained in state *B* of machine *N* in the sense that machine *N* will yield the same response if given a corresponding sequence of inputs. Following this logic it can be shown that state 3 is contained in state *A* and state 5 is contained in state *B*.

Viewing the problem in the large, it is often essential to erect hypotheses concerning the logical relation between two or more separately sensed variables. It would appear that the evolutionary technique offers a means for finding multiple input predictor machines which can be translated into reasonable representations for the observed logical relationship.

Experiment 16 was conducted to confirm the claim that the evolutionary technique had, indeed, extracted useful information from the correlative input variable in Experiment 15. In Experiment 16 two-input single-output machines were evolved in the very same manner as before but in this case the binary variable which was offered was uncorrelated with the output sequence taken from *M* in that it was independently generated. The percent correct score, shown in Fig. 9 remains below the corresponding scores which resulted from the previous two experiments. In other words, the additional information which was furnished in the form of an independent binary variable formed a *distraction* resulting in poorer predictive-fit and predictions. More specifically, the more complex machines which were considered made successful mutation and selection less likely.

In view of the apparent need, a formal technique was devised which permits expression of the set of finite-state machines which are functionally contained within any given machine. Thus it becomes possible to translate each predictor machine into a set of hypotheses concerning the transduction logic of the environment. The finite-state machines which are functionally contained in the set of machines used for successive prediction can be examined for consistency which may be expected to provide additional insight.

*   *   *

It is of interest to consider some of the ways in which the evolutionary technique is similar to and different from other approaches to the problem of artificial intelligence. It might be argued that the evolutionary program is to some extent like a *linear threshold device* in that each finite-state machine can be represented in terms of a network of threshold elements. True, but in general such a network requires more complex logic than that of a linear threshold device. A linear threshold device which has no internal feedback is representable as a *one state* machine. The evolutionary technique develops multi-state machines as required.

It might be argued that the evolutionary technique is a *heuristic program* on the grounds that it is in essence "reasonable." It is indeed reasonable, but it was not written in an attempt to simulate the logical behavior displayed by the human decision maker.

An essential aspect of intelligent behavior is versatility. In contrast to the statistical decision procedures which operate well only within the restricted domain of stationary environments, the evolutionary technique has been applied with considerable success over a wide range of goal functions and environments. This technique utilizes the available evidence at each point in time, making only one assumption concerning the mature of the environment: that the logic which would have best satisfied the goal within the remembered experience is that logic which should be used for the present decision. In general, the fewer the assumptions, the greater the versatility.

The evolutionary technique is more closely related to some of the iterative *function-fitting techniques* of modern mathematics. These are almost always restricted to the consideration of point functions rather than transduction functions as represented by finite-state machines. The evolutionary search technique is therefore carried out in the meaningful domain of possible decision makers which affect a causal universe.

Of course, a final choice among alternative methods to enhance or replace human decision making must rest on economic grounds. Thus far the evolutionary program remains experimental, the primary concern here being feasibility. A comparative economic evaluation will be carried out later in this program of research.

CONCLUSIONS

The key to artificial intelligence lies in automating an inductive process which will generate useful hypotheses concerning the logic which underlies the experienced environment. The creatures of natural evolution are just such hypotheses, survival being the measure of success. As described above, some fundamental aspects of this process can be replicated with finite state machines proceeding through iterative mutation and selection in order to find a *best* logic to satisfy the given goal in the light of the available memory of experience. Real-time decisions can then be based upon the extracted results of this continuing fast-time evolution.

A series of experiments were conducted in order to evaluate this concept. An 8-symbol language evolutionary program was given the goal of predicting a variety of time-series on the basis of the previous symbols. These *environments* consisted of a cyclic signal embedded in various kinds and amounts of noise, purely stochastic environments which contain the signal only in the form of the transition matrix, nonstationary samples of deterministic sequences, and multivariate sequences which were the input and output of an arbitrary transduction. In every case, the results attained compared favorably to those which would be expected from conventional techniques using only the same information. It was, in fact, possible to identify some of the cyclic and stochastic properties of the signals from the predictor machines.

The success of these experiments demonstrates that useful models of the regularity within a sequence of observation can, indeed, be found by the evolutionary technique. Although this demonstration was restricted to a variety of goals relevant to prediction there would appear to be no need to maintain this constraint. So long as the correctness of each response can be evaluated prior to the next response, the goal may be made to reflect the immediate concern of the decision maker.

The first phase of computer technology was devoted to the development of equipment which would carry out a large number of simple operations quickly and reliably. The second phase of computer technology was devoted to the development of languages which permit the detailed instruction of this equipment in an efficient manner. Computer technology is now entering a new phase in which it will no longer be necessary to specify exactly how the problem is to be solved. In contrast it will only be necessary to provide an exact statement of the problem in terms of *goal* and *costs* in order to allow the evolution of the best program possible within the available computation capability. Solution of the problem then includes a statement of the discovered algorithm. The old saw "the computer never knows more than the programmer" is simply no longer true.

The scientific method consists of *induction,* inductive inference, followed by independent verification. Hypotheses, generated so that they cover the available evidence and additional data points, are individually evaluated in terms of the validity of their inference. Those that prove worthy are modified, extended, or combined to form new hypotheses which carry on a "heredity of reasonableness." As the hypotheses correspond more and more closely with the logic of the environment they provide an "understanding" which can be used for the improvement of goal-seeking behavior in the face of that environment.

The correspondence between natural evolution and the scientific method is obvious. Individual organisms in nature serve as hypotheses concerning the logical properties of their environment. Their behavior is then an inductive inference concerning some as-yet-unknown aspects of their environment. Their health is a measure of their suitability. Their offspring include the heredity of reasonableness as well as additional information resulting from mutation and recombination.

The evolutionary technique described above is then a realization of the scientific method in which the hypotheses are restricted only in the sense that they be finite-state machines. The creation of successive hypotheses requires the introduction of randomness as well as the prevailing logic of inheritance. The versatility which is essential to intellect is a natural product of the iteration. In essence, the scientific method is an essential part of nature. It is no wonder, then, that its overt exercise has provided mankind with distinct benefits and now permits even its own automation through the artificial evolution of automata.

REFERENCES

1. L. J. Fogel, A. J. Owens and J. Walsh, "On the Evolution of Artificial Intelligence," *Proceedings of the Fifth National Symposium on Human Factors in Electronics, IEEE,* San Diego, May 5-6, pp. 63-76.
2. L. J. Fogel, A. J. Owens and J. Walsh, "An Evolutionary Prediction Technique," presented before International Conference on Microwaves, Circuit Theory, and Information Theory, *IEEE,* Tokyo, Japan, September 7-11, 1964. Summary published in Part 3 of the *Proceedings,* pp. 173-174.

COMMENTS AND REPLIES

GOLD* *(Stanford):* Fogel reports having used, as input in his *evolution simulator,* the characteristic function of the set of prime numbers, *P(t),* defined on the positive integers to be

$$P(t) = 1 \text{ if } t \text{ is prime}$$
$$0 \text{ if } t \text{ is not prime}$$

As he noted, $P(t)$ can be generated by no inputless finite automation (their output functions are necessarily ultimately periodic). However, $P(t)$ is a primitive recursive function, and it is of interest to note that an evolution simulator which tries primitive recursive functions, rather than being restricted to the functions produced by inputless finite automata, would eventually identify $P(t)$ correctly.

This is the content of the proof of Theorem 7 (actually, its application to Corollary 7) of my paper titled "Limiting Recursion." * To see this, observe that the goal of Fogel's evolution simulator, if it is presented with a nonprobabilistic function of time, is what I call *identification of a function in the limit.* In order to construct an evolution simulator which will identify any primitive recursive function in the limit, the key point to note is that the class of primitive recursive functions is effectively enumerable. That is, it is possible to order the primitive recursive functions,

$$p_1(t), p_2(t) \ldots p_a(t) \ldots$$

with repetitions, in such a way that there is a single algorithm which will compute $p_a(t)$ for any given values of a and t. It is only necessary for the evolution simulator to do the following: at time t it considers the first t primitive recursive functions of the enumeration and chooses that one which best agrees with the inputs received so far. It is important that the enumeration be effective so that the evolution simulator can have the capability computing any desired value of any specified primitive recursive function.

Fogel's evolution simulator is very similar to Friedberg's computer program which he calls *Harvey.*†

FOGEL: I appreciate the remarks of Gold concerning primitive recursive functions. *Harvey* was one of a number of programs (Herman, Sherman, Sampson, Homer, Teddy, Ramsey, et al.) written by Friedberg, et al., to automatically formulate sets of instructions which might solve certain simple problems. These programs differed from one another in their

---

* E. Mark Gold, "Limiting Recursion," to appear in the *Journal of Symbolic Logic.*
† R. M. Friedberg, "A Learning Machine," *IBM Journal of Research and Development, 2,* 2-13 (1958) and *3,* 282-287 (1959).

specific heuristics but generally relied upon the rule that *the frequency with which each instruction on record is active depends partly on how well it is performing currently and partly on its long-term record.* A review of experimental results demonstrated significant difficulty in finding worthwhile programs by such means. In his summary, Friedberg admitted, "Where we should go from here is not entirely clear." In my opinion these experiments failed because of a lack of sufficient logical *heredity.* Modification of a single instruction causes a radical change in a program whereas a single mutation of a finite-state machine does not destroy its general character.

HORMANN *(System Development Corp.):* It would be illuminating to see how Fogel's *evolving organisms* would do in the following experiment:

> Suppose that an offspring α has been produced to achieve goal *A* successfully, and that α has gone through subsequent evolutionary processes, producing its offspring β (perhaps many generations down) to achieve goal *B*. Is β likely to retain some of the characteristics of its ancestor α. Then how would the offspring β do when presented with goal *A*? Would it achieve goal *A* immediately, or would it have to go through the whole process again as if it had been presented with a new goal? How would the degree and aspects of similarity between goal *A* and goal *B* affect the process?

If we consider the *evolving organism* as one individual, and if we view the *mutation process* as an aspect of adaptation or learning, then my questions seem to imply the presence of *transfer of learning* phenomenon, that is, either positive or negative effects of previous learning on new learning.

But there are, perhaps, other inferences that might be drawn from my questions. Fogel stressed that the new offspring produced to achieve a specified goal would not necessarily be more complex (in terms of number of states) than its ancestor, unless the particular goal intrinsically required more complexity. This would be considered a good feature if the environment of the organism were always restricted to one goal or one kind of goal. However, it seems reasonable to assume that the more intelligent and adaptive an organism were, the richer in problems its environment could be. (That is, the environment could contain a greater potential for problem situations; some of these problems might never be presented to the organism.) In the light of this consideration, increasing complexity in *descendants* and retention of some of the *ancestors'* characteristics might be a desirable feature. Such complexity and trait-retention would enable new generations to cope more effectively with new environments that contained some of the problems their ancestors had faced before.

FOGEL: I appreciate Hormann's suggested experiment. Although this specific experiment has not yet been conducted, it is reasonable to expect that the machines which evolve under goal *A* will initially increase in complexity when the goal is changed to goal *B*. Gradually these machines will reduce complexity in order to more efficiently satisfy goal *B*, this at an implicit cost to goal *A*. Eventually, the inheritance of *learned aspect* of goal *A* will disappear. Of course, the more similar the goals, the greater the *transfer of training* and the less the increase in complexity. In fact, this increase in complexity may furnish a measure for the dissimilarity of successful goals. These remarks are based upon experiments in which the goal remained invariant but the environment was suddenly and radically changed.

KLEYN *(Northrop Nortronics)*: Fogel's paper at this symposium exposed me to his work for the first time and I am not sure that I fully absorbed the implications of the evolutionary approach to artificial intelligence. But it appears to me that the comments made during the discussion session at the symposium to the effect that the evolution scheme possesses many aspects of a learning scheme, are valid. Nor need this view detract from the accomplishments presented, for surely much remains to be done in constructing models of learning and Fogel's approach is distinct and novel and his experimental results invite attention.

Admittedly, the finite state machines which are replicated and mutated during the evolutionary process do not themselves learn. The learning is exhibited by the master computer program. Thus, one could argue that evolutionary *learning* which requires discarding many *organisms* is difficult to identify with learning by individual organisms which do not discard portions of themselves. But as the title of the paper indicates, Fogel fully intends to identify the evolutionary process with artificial intelligence which is a property usually associated with individuals, not generations. Furthermore, the master computer program does not discard any organisms in a physical sense, it only discards, or better said, re-organizes those parts of the computer and program which comprise the organism. Viewed in this manner the program appears an excellent and novel approach to a learning or self organizing system.

Fogel emphasizes the prediction aspect of intelligence and, unless I am mistaken, includes such functions as detection, discrimination and pattern recognition as subordinate to and serving the function of prediction.

It is interesting to cast pattern recognition proper into a prediction model. Pattern recognition poses the problem of discovering the *many-to-one* mapping which associates categories with a set of inputs. These categories are the ones into which a human perceiver maps the inputs. In the case of pattern recognition through learning the problem is to construct a machine which discovers this mapping for itself. What information does such a machine have to go on? Just the input sets and some symbols or special names which the human perceiver uses to designate the input sets. From this viewpoint, the task of the machine is to predict the response of the environment, which here is the human perceiver, to the input sets which both are exposed to. It remains to be seen if there is an advantage in viewing the problem in this manner.

FOGEL: With respect to Kleyn's commentary, it is well to note that the learning does take place by the evolutionary program and not by the individual *organisms*. Clearly the program is Darwinian and not Lamarckian. This is not to say that individual learning could not be added to the procedure. The results of research in this direction will form the basis for a future paper.

Again, let me remark that prediction was used as a means for the identification of the existence of signal in noise (detection) and for the representation of that signal in terms of reflection of the finite-state predictor-machine (discrimination). If reference signals are available, these may be predicted to yield predictor-machine representations which may be individually compared to the evolved predictor-machines in order to permit pattern recognition. But the goal need not be restricted to prediction. It may well be generalized to allow the direct evaluation of the worth of each model in terms of the worth of its sequence of previous responses. In such a case, control decision-making is accomplished with prediction becoming an identified intervening variable. Let me emphasize that prediction is but a particular goal which serves the purpose of demonstration.

OVERTON *(North American Autonetics Div.):* A ubiquitous problem regarding research concerns the translation of results into terms which other scientists and the general public can understand. Such translation is obviously necessary if the results of research are to be adequately exploited; by definition, the specialist doing the research is not intimately conversant with the many practical problems to which the results might conceivably be applied. Furthermore, it is obviously impossible for one researcher to compare his work with that of another if he does not understand the other's concepts and terminology. This situation—the existence of personal concepts which have been expressed in language made up by a researcher for his own use—exists to an appreciable extent in the cybernetic sciences.

Fogel has contributed to an amelioration of this situation with his paper, "Artificial Intelligence Through the Simulation of Evolution." By describing what some of us are doing in new terms and with different analogies, he has helped us to look at our research in a different light.

The above statements, of course, represent a personal reaction to his paper. I know Fogel feels the specific details of his work are more significant. But, given the fact that he obviously knows more than I do about what he is doing, it is my opinion that his translation is the more valuable contribution.

Among the particularly interesting analogies which one may draw from his translation are these:

- The state of a learning program after one "pass through a lesson"* is analogous to a mutated system (in this case, a machine) after one generation;

- The analysis of the nature of the lesson is analogous to the degree of prediction that the environment will permit some system to reach;

- Most basically, there is an analogy between learning and evolution.

Among the results of this way of looking at learning machine work are these ideas:

- The retaining of tentative memory of the results of the last few passes through the lesson would seem to be profitable because it is analogous to keeping the few best systems after a period of mutation;

- It may be well to add size or cost of memory to the evaluation criteria because of the analogy with the efficiency with which an animal uses its food;

- It is possible that the statistics applicable to genetic behavior might be found to include techniques which would be useful in analyzing the possible results of the combination, or *mating* of systems of learning.

FOGEL: Overton's commentary deserves clarification and correction. I am happy if the evolutionary approach to artificial intelligence provides a new

---

* This jargon is used at Autonetics

vantage point from which it is possible to review other research in this field. In general, the analogies cited are valid, but I would like to add a few discriminating remarks.

- A *generation* occurs when an offspring is identified which demonstrates a better score with respect to the given goal over the available recall than the best score of the retained machines. The learning of the evolutionary program may proceed even without the passage of generations. These just mark significant points of success. The program may be made to *pass through a lesson* more than once using the last evolved machine as a hint in that it serves as the original machine for the next review of the lesson.

- The evolutionary program may be used for prediction, interdiction or post-diction. Prediction was chosen only as a means to demonstrate the capability of the evolutionary technique. A *lesson* should only be characterized in terms of its predictability if this happens to be the goal to be achieved.

- Indeed, evolution does provide learning at the *specie* level rather than at the level of the individual.

The resulting *ideas* may be stated as follows:

- Inheritance of the best few theories is always good practice. Here the term *best* only becomes meaningful if the goal is well-defined so that each hypothesis may be tested. The mutative construction of new models should always continue (in the case of the evolutionary program, the theories, hypotheses, models, conjectures, representations—call it what you will—took the form of finite-state machines);

- Cost is always an important factor in data processing. The demand for realism requires that cost minimization be a part of the goal to be achieved;

- It is felt that worthwhile rules for the combination of offspring may arise from studies of genetic coding or from pure mathematical as well as from statistical analysis of natural evolution.

5. NATURAL AUTOMATA AND PROSTHETIC DEVICES

LAWRENCE J. FOGEL, Discussion Leader

Decision Science, Inc.
San Diego, California

WARREN S. McCULLOCH, Supporting Discussant

Research Laboratory of Electronics
Massachusetts Institute of Technology
Cambridge, Massachusetts

FOGEL: I would like to begin this session on natural automata and prosthetic devices by having Dr. Marvin Minsky present a film to you which demonstrates the performance of a computer-operated mechanical hand. [1] Dr. Minsky will narrate the film and explain the details of how the mechanical hand operates.

MINSKY: The movie about to be shown depicts the doctoral research of Heinrich Ernst which is documented in his 1961 dissertation (Ref. 38b) and in this film.

Ernst's goal was to attach a hand to a computer and to get it to behave moderately intelligently, and he did a marvelous job. I thought you would like to see what I think was a milestone in the connection between a computer and the outside world.

The computer he used is the TX-O computer at MIT. It is a small computer by today's standards. However,

---

[1] The editor regrets that the reader cannot share in viewing the film that was shown at the conference. The film is entitled, "MH-1, A Computer-Operated Mechanical Hand." Any questions concerning the contents of this film that are not covered in the discussion presented here may be addressed to Prof. Marvin Minsky, Department of Electrical Engineering, Massachusetts Institute of Technology, Cambridge, Mass. 02139. See scenario in this chapter.

Reprinted with permission from *Aids to Biological Communication: Prosthesis and Synthesis,* L. J. Fogel and W. S. McCulloch "Natural automata and prosthetic devices," Vol. 2, D. M. Ramsey-Klee (ed.), Vol. 2, pp. 221-262. © 1970 by Gordon and Breach, Science Publishers, Inc.

it was the first solid state general-purpose machine. It has a simple, straightforward order code with some "microprogramming" features. It had 4000 words of memory at that time, and was a very fast computer for its era (I think it had a 6-microsecond cycle time.)

The mechanical arm was a standard remote manipulator operated by moving a hand-held gripping device; a pantograph mechanism caused the machinery in the mechanical arm to perform the same motions. Ernst activated all the degrees of freedom by servomotors connected to the computer (Figs. 27 and 28).

The computer program is organized in terms of a hierarchy of rather complicated procedures. At the start, or when there is nothing else to do, the area into which the arm can reach is scanned in a systematic, thorough pattern. This procedure is painful to watch because it is rather slow. When the arm hits something, the program calls on a subprogram which has two or three ways of identifying the object that has been encountered. The program only knows about two or three kinds of objects and distinguishes between them by making certain length measurements. The device has no vision, in the sense of dealing with images, but it can feel when it bumps into something.

It can also tell just before it bumps into something because it has photocells which act much as proximity detectors or the bristles on a cat. When it gets near a black object, the photodetector says, "You are near a black object," and then by moving, it can measure the size of the object. In its cognitive universe the only objects it knows are the table, small blocks, and large blocks, and it can distinguish between these by their size. If it finds a small object, it assumes it is a block and calls on a subroutine which says how to pick up a block. This involves making adjustments until it is parallel to the grasp.

The hand has tactile detectors at certain points and some photocells at the fingertips and the middle of the palm. The hand does not close its grippers until the center of the palm has touched something and thus can

FIGURE 27. The modified servomanipulator.

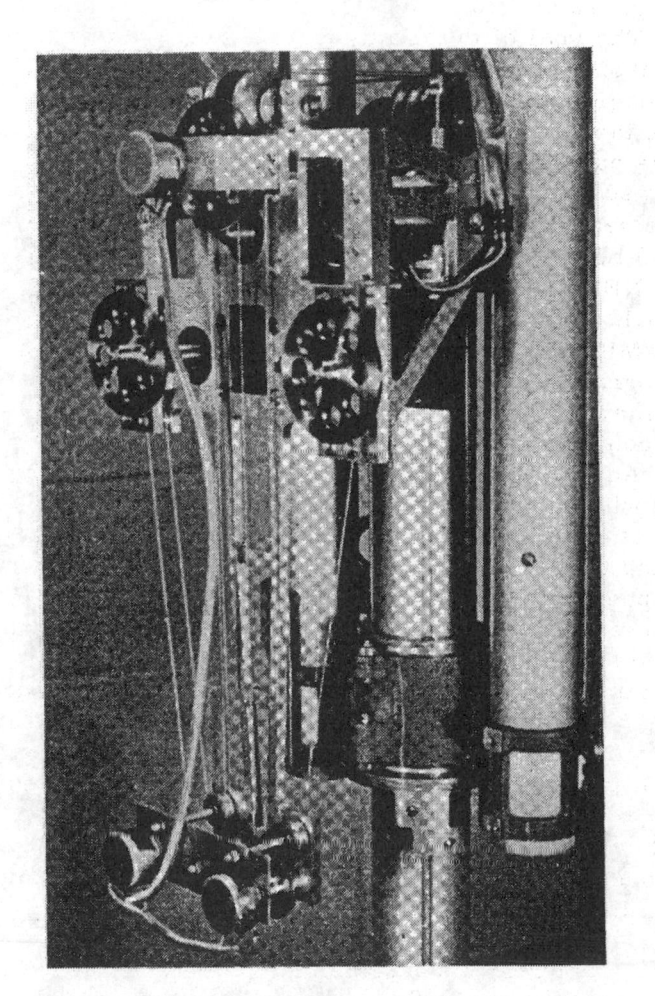

FIGURE 28. Motor and sensor organs.

assume that the object is fairly well within its grasp
(Figs. 29 and 30).

The goal of the first program prepared by Ernst is
to find a number of blocks and build them into a tower.
It begins with a search phase, in which the hand scans
the table surface until it collides with an object. When
this happens, the hand passes closely around the block
to ascertain its size and position and then goes back into
the search phase to find a second block which eventually
will be placed on the first block.

NEWELL: If the block moved even a small amount,
would it escape?

MINSKY: Because the servos are not very accurate, the
program is written to always check if an object that was left
somewhere is still there. The entire program is designed
to continually reverify things that have been remembered.

An interesting feature of the program is demonstrated
by what happens when an assistant puts the block the hand
is searching for between the fingers of the hand. The
hand takes it immediately because it realized what is
happening; it does not continue its search process. This
constitutes a rather clever piece of program organization.
The system keeps testing for various conditions. If
nothing else is happening, it goes to search mode. When
it finds something, it measures it and gets it in its hand.
But if it finds something in its hand, it simply jumps to
that part of the program which tells it what to do if it
already has something in its hand. Ernst did not have to
add any special programming for the contingency of the
block being dropped into its hand; the program automatically skips over the acquisition phase.

The goal of the second program is to put blocks into a
box. One problem in performing this task is to tell the difference between a block and a box. If the hand finds something big enough to put its hand into, then it is a box.

PAXSON: Dr. Minsky, what kind of buffering is
being used to obtain a time match between the hand and
the computer? The computer is presumably doing its computations rapidly with the results of its operations stored
for use on the time scale of the mechanical equipment.

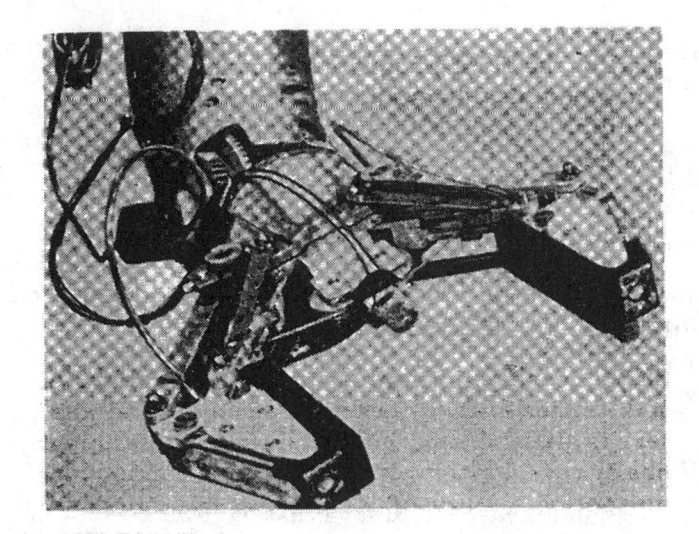

FIGURE 29. The hand.

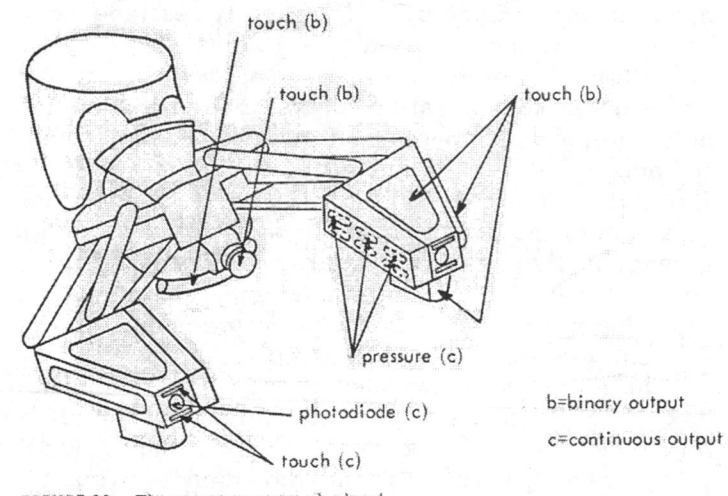

touch (b)

touch (b)

touch (b)

pressure (c)

photodiode (c)

touch (c)

b=binary output

c=continuous output

FIGURE 30. The sense organs on the hand.

260

NATURAL AUTOMATA

SCENARIO OF "MH-1, A COMPUTER OPERATED MECHANICAL HAND"*

Title of scene	Comments
PART 1: PICTURES OF EQUIPMENT AND INSTALLATION	
Title: MH-1, A COMPUTER-OPERATED MECHANICAL HAND	Hand holds board on which title is printed.
Total: manipulator	Overall view of servomanipulator.
Closeup: hand with sense organs	Pointing out different sense organs on the fingers.
Semitotal: master arm	View of motors and potentiometers for wrist and tong control.
Title: TX-O COMPUTER	Name shield of the TX-O computer.
Panorama: TX-O installation	View of arithmetic unit, memory, in-out panel, power supply.
Semicloseup: MH-1 control unit	Shows front panel of control unit.
Panorama: console of TX-O	
Closeup: Flexowriter	The computer program as typed on the flexowriter for the first half of the film.
PART 2. BUILDING OF A TOWER WITH WOODEN BLOCKS	
Total: search initiation	Hand moves into position to start search motion for block on the table. This block is to be the base of the tower.
Semicloseup: search motion	Shows how the hand ascertains the distance from the table during the search motion.
Total: search motion	Hand finishes one line of the search pattern and advances into the next line.
Closeup: search motion	Hand touches the table.
Total: approach of block	Hand approaches block and hits it.

* Editor's note: The entire scenario of Dr. Ernst's film is included for those readers desiring more information about the detailed contents of the movie.

INFORMATION AND CONTROL PROCESSES

Title of scene	Comments
Closeup: scanning of block	Hand passes closely behind block to ascertain its size and position, then starts to go back into search motion for second block which should eventually be placed on the first block.
Total: putting in second block	Assistant puts in second block while hand starts to search for it.
Total: coffee break	Assistant can be seen drinking coffee while hand searches for block.
Semitotal: get block	Hand hits block, goes behind it and takes it.
Semiclose: transport block	Hand lifts block and puts it down behind first block where tower is to be built. Hand starts to feel for base block, keeping the second block in its fingers.
Closeup: deposit block	Hand hits base block, goes behind it, lifts second block and puts it on the first block.
Total: search motion	The hand goes back into the search motion for the third block.
Semitotal: block presentation	The assistant puts the block the hand is searching for between the fingers of the hand. MH-1 takes it immediately because "it realizes what is going on."
Closeup: block deposition	Third block is deposited (not quite squarely) on the two previous blocks.
Total: search motion	Hand starts to grope for the fourth block (the awareness of the MH-1 system has been cut off for that phase).
Total: block presentation	The operator tries unsuccessfully (since there is no awareness) to present the block to the hand directly; MH-1 does not react.
Closeup: scan motion	If the hand is aware that it bumps into the table, it automatically adjusts its height. Without awareness it drags clumsily over the table in that case.

NATURAL AUTOMATA

Title of scene	Comments
Closeup:	MH-1 has found the block and tries to deposit it on the tower. However, it bumps into the tower and the tower collapses.

PART 3: PUTTING BLOCKS INTO BOX

Semitotal: discover box	Hand has gone through search motion (not shown) and now hits box.
Semiclose: box exploration	Hand feels around box to ascertain its size and position.
Total: search motion	MH-1 searches for block to be put into box.
Closeup: take block	MH-1 hits block and tries to take it. A sense element is not responding properly; the hand makes little jerks back and forth.
Total: repairs	The assistant enters the picture with a voltmeter and traces the trouble to a sense element that is stuck.
Closeup: take block	Hand is successful in picking up block in this second attempt.
Semitotal: go to box	Hand transports block into vicinity of box, hits box and goes behind box.
Closeup: deposit block	Hand lifts block and deposits it in box.
Total: search motion	Hand goes back to search for more blocks. Assistant shifts box into way of hand.
Semiclose: hit object	Hand hits box, thinks it is block, but finds out object is bigger than block when it tries to take it.
Semiclose: scanning motion	Hand goes into scanning motion to determine nature of obstacle (END). (Hand will then find out that it is a box and return to that part of the program which deals with boxes.)

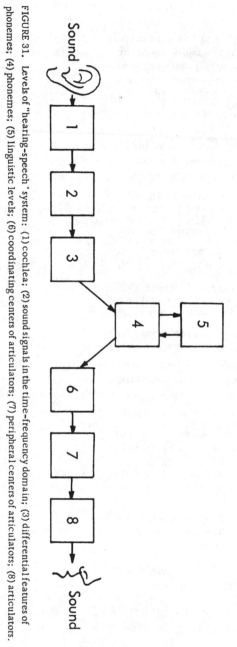

FIGURE 31. Levels of "hearing–speech" system: (1) cochlea; (2) sound signals in the time–frequency domain; (3) differential features of phonemes; (4) phonemes; (5) linguistic levels; (6) coordinating centers of articulators; (7) peripheral centers of articulators; (8) articulators.

INFORMATION AND CONTROL PROCESSES

MINSKY: If the servos take long to get anywhere, the computer simply keeps waiting for a response from them. It took about 5 minutes for the mechanical arm to grope through one of these tasks. Ernst could have speeded up the film, but he resisted this temptation.

McCARTHY: Dr. Minsky, were you on Ernst's thesis committee?

MINSKY: Yes, Professor Claude Shannon and I were both on the committee.

I would like to add one final comment. This research was done in 1961 and we assumed that the world would beat a path to this particular doorstep, but nothing happened in this field for 4 or 5 years.

McCARTHY: So we had to do it ourselves.

MINSKY: Right, we had to do it ourselves. We have recently begun new projects to develop a mechanical hand, with the goal of incorporating as much visual pattern recognition as we can possibly achieve. We also hope to develop better hands and arms than available previously.

FOGEL: Dr. Minsky, thank you very much for your commentary.

Dr. Kozhevnikov has been kind enough to volunteer to present some of the work in which he is currently engaged.

KOZHEVNIKOV: I want to present some of our views about speech perception and speech production. We consider this process as a kind of action of certain natural automatic systems. A fully automatic system receives sound waves by ear, and it speaks by means of articulatory apparatus. The result of the muscular activity is sound waves, acoustical speech.

We are physiologists and we are trying to understand what kind of processes are going on inside of the brain during perception and production of speech. We are trying to understand the processing of information by which we can perceive sounds, and the processes which control our speech mechanisms.

It is necessary to consider different levels of the treatment of information. In more simplified form I think it is possible to present it by Fig. 31. We can consider the first level — the cochlea — as some kind of frequency analyzer which has a very low Q in its filters.

INFORMATION AND CONTROL PROCESSES

It is possible to describe the following level as some kind of matrix, which receives signals from different filters of the cochlea. This matrix acts as a window in the time domain. The picture of the signal which we can see through this window is divided into as many as 100 steps, for example. This matrix acts as a kind of buffer in a digital computer. The length of every step is equal to 1 to 2 milliseconds, perhaps. Thus, we can see through this window a segment of the sound which is as long as 0.1 to 0.2 seconds.

In the case of the treatment of speech signals we have to suppose, as the next level, the level of differential features of phonemes. It is possible, again in an over-simplified way, of course, to assume the weighted summation of the signals from certain cells of the matrix. When the sum of signal energy exceeds some value, a decision is made about the presence of a particular feature of a class of phonemes. In general, it means a loss of useless information, a gain of useful information, and the ultimate detection of a definite phoneme.

The output from this level is a sequence of phonemes. This sequence serves as input to the linguistic levels. The information is treated here according to grammar rules and the dictionary of the language which we use.

Now I want to go in the opposite direction. Our articulatory apparatus consists of many physiological systems: the respiratory system, the system of muscles of the tongue, lips, larynx, and so forth. The only physiological action which is responsible for the production of sound is muscle contraction — nothing more. We have to take into consideration that during speech production there are different nervous signals that control this system. It is well known that there are special centers which control, for example, the respiratory system. The action of these centers is not connected solely with speech production. They have a general physiological task, but in the case of speech they receive special control signals from the levels placed still higher, and all these signals have to be coordinated to organize normal speech. In general, it is possible to consider the information from the linguistic levels as the input to the coordinating centers which govern the physiological structures.

In our experiments we are trying to study the processes which go on in the natural automaton described above. First, we have made special devices to measure main articulatory events. It was rather a difficult task to construct practically useful transducers.

Just recently we have developed a set of devices that register simultaneously as many as 15 processes, mainly mechanical and aerodynamical. On these records we can discriminate different phonetic complexes and measure their characteristics. The main initial emphasis of our work was an investigation of the time features of oral speech (Ref. 60).

We tried to find the smallest element which did not change in spite of changes of context. It proved to be not phonemes nor simple syllables, but rather syllables which end with a vowel sound — open syllables. We came to this conclusion from an investigation of the changes in time structure of speech during the changes of its tempo.

The patterns of speech were estimated statistically. The decrease of correlation between neighboring sounds was just at the limits of open syllables.

Experiments on artificial stuttering provoked by means of delayed acoustical feedback were performed, and the results confirmed our ideas about open syllables. Oral speech is a sequence of open syllables, at least for the Russian language. It seems to be necessary to use a set of such syllables as the main alphabet in systems for automatic recognition of speech.

Two lines of investigation are now in progress. They were both instigated by our wish to delve deeply into the system. One approach is psychophysical. The part of the nervous system which is responsible for perception of speech consists of many levels. It is possible to use certain methods of psychological scaling for the description of the signals at every level.

ROSENBLITH: What kind of scaling do you have in mind?

INFORMATION AND CONTROL PROCESSES

KOZHEVNIKOV: For example, we can try to find equal distances on the scales of loudness, pitch, and duration of the sounds. The second line of investigation is a biophysical one. We are working to obtain a sufficient amount of data about control of processes which are the basis for speech production. As an example, I can mention the action of the respiratory system. The respiratory muscles receive control signals from the centers which are responsible in the first place for vegetative breathing. Very characteristic patterns appear in the course of speech. The picture reflects both the syllabic content and the prosodic features of speech. The experimental material and some speculations have been published in Russian (Ref. 61).

Before concluding my remarks I want to emphasize that it has proven useful to treat the human speech mechanism as a kind of natural automaton. I should like to add that the leader of this work is my wife, Dr. Ludmila Chistovich.

FOGEL: Thank you very much, Dr. Kozhevnikov. I would like to ask if there are any comments relevant to this particular topic before we move on to another subject area.

NEWELL: Yes; I should like to hear somebody, perhaps Dr. David, comment on the efforts in this country to try to produce speech from phonemes.

DAVID: There are a number of efforts going on in this country not too different from what has been described by Dr. Kozhevnikov. Some years ago, people trying to automate speech recognition assumed that if you analyzed the acoustic wave precisely, it might be possible to recognize speech sounds simply from those acoustic measurements. Some early work of this kind was done at the Bell Laboratories (Refs. 62-66), and there was early work at other laboratories too.

I think now every one of us realizes that you will not be able to recognize speech in the sense that the human recognizes speech by dealing with the acoustics alone. You have to understand in a rather deep way the physiology of how that speech wave was put together. In other

words, you have to understand all the constraints in the
generator to which Dr. Kozhevnikov has alluded.

There is a great deal of work, some at MIT, some
at the Bell Laboratories, some at the Haskins Laboratories, some in Sweden at Dr. Fant's laboratory, and
some in England, aimed at trying to understand the
mechanisms by which speech is produced (Refs. 67-69).
This includes everything from the articulatory to the
phonemic, linguistic, and semantic levels.

Let me describe briefly some work that was done by
Kelly and Gerstman at Bell Laboratories some time ago
(Refs. 70, 71). They programmed a simulation of a vocal
tract analog and proceeded to try to make it speak, using
a sequence of phonemic symbols as input. They also
put in information about the duration of these phonemes.
Now, the vocal tract analog is driven directly by a number of continuously varying parameters which, in effect,
control the position of the analog articulators. If the
analog is of some other kind, there are other parameters
to be controlled. These continuous controls are generated by a program which synthesizes appropriate signals
from the phoneme symbol sequence.

In addition to specifying the phoneme sequence, the
input must specify the duration of the phonemes and
their pitch. These specifications provide the stress
and intonation pattern of the speech.

BIGELOW: Was there any phase information at all,
such as one gets from explosive sounds?

DAVID: That would all be inherent in this vocal tract.
One interesting problem which you run into is this: We
found that making this analog talk was very easy, except
for finding out what to put in for duration and pitch. It
proved very difficult to take a phoneme sequence, which
might represent a sentence, and find rules for the stress
pattern in terms of physical duration and the intonation
pattern in terms of pitch.

What we ended up doing, of course, was taking a
human utterance, analyzing it, and measuring durations
and pitches. We found very little linguistic work which
related the phoneme sequence to questions of duration and
pitch, or to put it another way, intonation and stress

pattern. So one of the problems which we have faced, and on which some progress has been made is: How do you, given the phoneme sequence, determine duration and pitch inputs according to rule?

There are some interesting questions that one can ask about systems of this general kind. You might ask the following question: If you make speech according to some set of rules using an analog of the vocal tract, is this speech recognizable by a machine — uniquely recognizable by a machine? You would think that the answer would be yes, but if you look closely, you find that the transformation from phonemic symbols to control signals is nonunique, and so the answer is no. You could come close, of course, but there would be some ambiguities.

You can also ask the second question: Is there any set of rules to put in this box which will be satisfactory for all people? Suppose you try to determine what these rules should be by examining numerous people's speech; are those rules different for various talkers? It turns out that they are. If you take two talkers from the same cultural background so that you might think their rules would be the same, you find that in order to imitate a particular person's speech, you need a different set of rules.

I think the point Professor Kozhevnikov has made, namely, that in speech recognition you have to know a great deal more about the speech than merely the acoustic wave, is one that has impressed speech researchers deeply and is one that I think is quite appropriate for this meeting. You have to know a great deal about linguistics. You also probably have to involve the semantic level in order to do a really good job of recognition.

BIGELOW: Is this also true for nonsense syllables spoken from one person and heard by another?

DAVID: I think that it is. The levels you have to go to are not as high in this case, but you find out, for example, that people make finer discriminations between phonemes of their own language than of a foreign language. So even in hearing nonsense syllables, the phonemes which arc in a person's native language have a strong influence on what he hears and how he interprets an acoustic sequence.

LETOV: Mr. Chairman, most of the remarks of Dr. Kozhevnikov and Dr. David were devoted to a description of particular automata which produce speech. I would like to know something more about the structure and activity of the feedback loop in this artificial mechanism which permits an individual to produce reasonable speech.

FREMONT-SMITH: Some years ago Dr. Alvin Barrett at Columbia University College of Physicians and Surgeons developed an apparatus, a cylinder in which a person stood, where the pressure could be changed. Oxygen and helium were the gases and he changed the pressure so that respiration became unnecessary. I have been in this apparatus, and after a minute or two the pressure goes up and down and you do not have to breathe; you just stand there and you discover that you do not have to breathe.

When you speak in this situation, the entire respiratory muscular apparatus of speech is eliminated. Because of the helium your speech is very high pitched. It occurred to me that there might be something in this approach that would simplify study of certain aspects of the speech process as well as complicate other aspects.

Dr. Barrett was not studying speech at all; he was simply trying to put the lung at rest, and he succeeded.

FOGEL: I should like to initiate a discussion concerning prosthetics for natural automata. Let me take the next few minutes to indicate a viewpoint which presents not only a philosophical attitude toward this problem, but also presents some direct practical problems which, once solved, would allow us to proceed in many situations that today are limited by the capability of the human operator.

A prosthetic is a device that aids or extends human capacity in a particular regard. Obviously, you have to be able to describe what the goal is for the prosthetic to have a meaning. There must be a purpose. The traditional prosthetics, such things as crutches, automobiles, airplanes, and space ships, are all crutches of only slightly different kinds. Each is an attempt to facilitate transportation of the human operator.

WASSERMAN: May I interrupt on a point of definition? To keep it clear in our minds (and this is in terms of the medical application of the term "prosthetic"), we think of something being prosthetic to something else if it substitutes for a missing function, one that was once present but is no longer there; whereas the crutch and the brace are referred to as orthotic to the system, supplying an augmentative function for a slightly impaired function which is not necessarily missing but weakened.

FREMONT-SMITH: What do you use, then, as a term for the extension of a human function which has not been weakened; it simply has not become that good yet?

WASSERMAN: Hyperorthotics.

FOGEL: In the hope that I will not transgress the linguistic barrier, I will use the word prosthesis for the remainder of this discussion and learn the other terms at my leisure.

I see two classes of prosthetics. In the first class are those that are intended to aid the human operator in his gathering of information. These traditionally are means of sensing, communication, and transportation. Another class of prosthetics are intended to aid the human operator in affecting his environment. These are such things as tools, hammers, and weapons. These are devices that are extensions of his ability to control his environment.

We have input-output systems surrounding the human operator. What I want to talk about is the black box system which intervenes between the human operator and the real world. He receives information through multiple sensors of the real world. His decisions are implemented through some output system which affects the real world in many ways. Because all aspects of the real world affect each other in some sense, what he sees in the future is to some degree affected by what he has done in the past.

ROSENBLITH: The human operator is not part of the real world?

NATURAL AUTOMATA

FOGEL: I am taking him out of the real world for the sake of this discussion. I want to focus attention on this black box which translates data from the sensed real world to him and the similar black box which accepts the results of his decisions and translates these to the real world. Usually there are displays which translate information from the sensors directly into a language acceptable to the human operator; and there are controls which accept the information from the man and transmit it to the equipment, which in turn translates it into the "language" of the real world through the affectors that are available. I want to discuss the problem of design for both of these black boxes.

The designer first asks the question, "What is the task of the human operator who will handle this system?" With this task defined he presumes knowledge of the nature of the human operator and the nature of the external real world. He proceeds to design the input black box that connects the available sensors to the displays (including, of course, the displays), and he proceeds to design the output black box (even recognizing that every control is also a display). Almost always this is done in such a manner as to produce deterministic mechanisms. In other words, he designs fixed transducers.

O. SCHMITT: Do you include transponders among your transducers, or do you represent them by separate loops when you mean them?

FOGEL: I will indicate them separately should I refer to them.

O. SCHMITT: Fine! You will say them separately when they have a closing loop?

FOGEL: That is correct. Early systems for driving displays were invariant transducers. As the systems under control became more versatile, the human operator was required to deal with diverse and variable environments. The designer was asked to provide input systems having some adaptive characteristics so that as the environment changes, the black box will process the data in a different manner to still display useful and relevant data. Similarly, adaptive mechanisms were designed for the controls. From 1950 through 1960

there was extensive research on different kinds of
adaptive mechanisms; some were heuristic, some were
based on algorithmic procedures, but for the moment I
simply want to consider all of these as transducers
which change their characteristics based upon the nature
of the incoming information, intending to provide the
human operator with information needed to satisfy his
given goal — to allow him to perform the task which he
has been assigned.

Now as the vehicle or system under control becomes
more versatile, the operator's immediate goals are
harder to define in the sense that versatility permits
the pursuit of higher level goals. The specifics of
personal behavior are gradually left up to his own imagination. In fact he is encouraged to become more and
more autonomous.

Under such circumstances we should like to have
systems of input transduction and output transduction
which are both controllable and adaptive. By controllable,
I mean that the human operator can instruct the system
as to his goal or subgoal. The system will decide not
only how to adapt to the variations and foibles of the
environment, but also will perform this adaptation with
respect to different criteria as to what is relevant as the
immediate goal changes. This is distinctly different
from pure adaptation.

To be both adaptive and controllable the system must
be designed to accept various statements of goal. We
must ask how we can specify a goal and what is the nature
of the goal in the sense that we can instruct the machine
as to the goal. Let me defer discussion of the details of
this subject for the moment in order to examine where
such a design philosophy leads.

Suppose, for the moment that I have, by some magic,
an ability to give the black box not only a statement of my
goal but also hints as to the expected nature of the environment. Is it not reasonable to suspect that the better
the adaption of those systems with respect to my goal,
the less there remains to be done in the decisionmaking,
the transduction that must be performed in order to

effectively couple the input display system with the output control system for adequately directing performance of the mission? In other words, the availability of controllable adaptive displays leads to the design of autonomous automata, systems which become completely automatic at the lowest level, leaving the human operator free to adjudicate among the various goals available to the system at the tactical level. In essence, the human is removed from the tactical level to the strategic level. With increased adaptive and controllable adaptive systems operating at the higher level, the human operator can turn his attention to still higher level goals. This approach to the design of man-machine systems offers a way of allowing human decisionmaking to operate at its highest level, and at the same time relegates to the machine that which it can do best — follow specific instructions. With such a system in operation the human operator can maintain cognizance over lower level operations so that, should they fail to perform adequately, he can choose to intervene and make appropriate changes of the subgoal or the relevant adaptive mechanism. On the other hand, with such controllable adaptive capability within the system the human operator can proceed to higher and higher levels of operation in order to gradually extract himself from necessary functioning at the lower and more routine levels.

Now let us consider how all this might come about. The problem that I would like to address concerns various ways in which this type of controllable adaptive transducer of information can be realized in hardware.

To be fully specified, a goal must be described in terms of all possibilities of some parameter, each of these being weighted by its relative importance at each point in future time. To illustrate, the importance weighting of the playboy is concentrated on the very near future. For him, "Tonight is the night; all of the future beyond tonight is of no value." The more serious-minded person holds a more extensive weighting function. This function might have a rise of importance at some future time where his offspring are expected to be of

particular concern to him. The only common properties
of all such relative weighting functions is that they must
fall between zero and unity at any particular time and
the entire area enclosed under the function must sum
to unity.

Note that this entire weighting function may change
with time or be a function of actual performance. Let
us say that my object is to predict a particular future
event. In general, the worth of a correct prediction
might strongly depend upon my present situation. In
fact, I might hold greatest interest in correctly predicting the event to occur in the immediate future, and this
with an all-or-none error function. Prediction of this
same event at a more distant point in time might have
lesser relative importance, with the loss function of
prediction relative to this time being described by the
mean-squared error. That is to say, if my immediate
prediction is not correct, it is too late to take corrective
action; but if my longer range prediction is in error,
there may still be some value in being nearly correct.
The problem of guiding a missile to a target is of just
this kind.

Now let us take a look at each of these more specifically. Consider the problem of predicting the immediate future.

NEWELL: You began by talking about payoff and
somehow we have shifted to prediction.

FOGEL: Prediction is simply an arbitrary choice to
embody the discussion. I said that whatever the parameter of importance may be, there is some relative
weighting of importance over the future. Do you agree
with this?

NEWELL: Right.

FOGEL: As a particular case in point, I want to
consider the problem of prediction rather than the problem of selecting a most suitable response, because it is
simpler. If you wish to discuss the problem of selecting
that present response which will yield maximum future
payoff, we can do so; but introducing such added complexity may not be necessary. The overall worth of
behavior can be viewed as a composite of an ability to

predict one's environment, followed by the suitable choice of a response in the light of the prediction. I am examining prediction only to keep the illustration both explicit and simple.

BIGELOW: Is there a value function in your example?

FOGEL: Do you mean other than the weightings on the parameter of interest over time?

BIGELOW: Yes, I do. There is some kind of space in your example that you have not described, but let us suppose that being near a target is a measurable function in this space.

FOGEL: Let us say it is a rendezvous problem.

BIGELOW: All right, but it is not quite clear that predicting where the missile will be and the value of intercepting it are identical.

FOGEL: That is exactly why I made the simplification, in order that this be so.

BIGELOW: Then the value is unity under all circumstances of success from a direct hit to a complete miss?

FOGEL: Evidently I am not making myself clear. For the sake of this presentation, let me assume that the worth of performance is measured by your ability to predict. I grant you, in the real world, this is generally not the case, but for the sake of this discussion let me make this assumption. Now note that all correct decisions (represented on the main diagonal of the error-cost matrix) need not be of equal worth. It is a common oversimplification to assume that they are all equal. Similarly, off-diagonal terms can have arbitrary assigned weights as to the relative cost of individual errors.

I expressed the loss function in terms of an error-cost matrix. To have a complete specification of goal, each point in future time must have such a matrix. I choose to examine the nature of goals in such a detailed manner to help myself answer the question, "How can I build a system which will accept even a small portion of this kind of complex expression?"

Now I would like to offer one answer to this question, a procedure that will behave with respect to an arbitrary goal (not necessarily mean-square error reduction) in the face of a wide variety of environments, a procedure that will behave in a manner that is adaptable; that is, it will modify its behavior based upon the nature of its experience, and at the same time will be controllable in that it will accept a change of goal while performing the mission. This is achieved through evolutionary programming — but before I describe this in greater detail — are there any questions?

BIGELOW: When you say "change the goal," do you mean change your strategy or change the actual value system that you are trying to predict?

FOGEL: I mean change the time weighting of importance and/or the error-cost matrix associated with each point in future time.

BIGELOW: Is the matrix the goal?

FOGEL: This became a complete expression of the goal when I made the simplifying assumption that all the importance weighting is concerned with prediction of only each next symbol to be observed. I am not asking you to predict the next few symbols; I am asking you to predict only the next one, so that all of your importance is associated with your next symbol at each point in time. The goal is measured by the worth of performance.

NEWELL: The goal is to predict this sequence according to the following cost matrix. If we talk about changing the goal, at least with the overtone of generality that you are giving it, it should be like saying, "I don't want to predict. I would like to get rich."

FOGEL: I do not deny that possibility, but for the sake of keeping this simple enough so that we can talk about it, I am saying that my interest at the present time is only to predict, and I make it even simpler, only to predict each next symbol, not the sequence of all future symbols. Thus I only have to consider a single error-cost matrix at each point in time. We could predict the next symbol and the one after that, each with some relative weighting of importance and have a different matrix for each one. There is no reason such a more complex problem is not addressable by the very same technique.

NATURAL AUTOMATA

DAVID: But you have not answered Mr. Bigelow's question, which was: Are you changing the strategy of predicting or are you just changing the value metric by which you judge progress?

FOGEL: My goal, specifically in this problem (and again I am going to make it simple) is to predict each next symbol, predict it in such a manner as to minimize the cost with respect to a given error-cost matrix.

DAVID: So, you are simply changing the value function?

FOGEL: We can change the cost of each correct and incorrect prediction. These are arbitrary weightings. Do you not agree with me?

MINSKY: Does this mean you are changing the goal each time by changing some of the numbers in the matrix?

FOGEL: That is right. The goal is given to you at a point in time and your object is to predict each next symbol so as to minimize the cost with respect to this goal. In fact, I want this goal to be an arbitrary function of time so that, as I am predicting, this goal can be changed, but of course is known to me at each point in time.

SAMUEL: It seems to me that you are using the word g o a l with two different connotations. The goal is to minimize that matrix. That matrix is not the goal.

FOGEL: Oh, but is that matrix not an essential part of the goal? If I do mean-squared tracking, is the goal to do tracking or is it to do mean-squared tracking?

ARBIB: That is the point. The goal is to track and the specification of the goal involves specifying the matrix, but the matrix is not a goal.

FOGEL: It is part of the description of the goal.

MINSKY: Dr. Arbib, you are saying he is confusing a function with a set of ordered pairs. He has a perfect right to say the goal is minimizing the matrix, and when you change the matrix it is a different goal.

FOGEL: Then we are agreed. Now you may recall that I should like to have a display system, a control system, or both, so that I can allow these weightings to be arbitrary functions of time as long as they are specified to me or are computable functions of the performance up to each point in time.

BIGELOW: Suppose you have made a prediction on the ith occasion; that means you have estimated A_i and you have adjusted the matrix in a certain fashion.

FOGEL: You have not adjusted it. It may have been adjusted for you by some outside agency, or it may stay the same. The matrix, itself, is a specification to you and need not be a function of your performance.

BIGELOW: All right. You do something with respect to your performance, do you not? There is a matrix and there is a dependent time series resulting.

FOGEL: Right.

BIGELOW: A goal is equivalent to X_i minus a given X at each particular moment; correct? Besides the pure paperwork which you are talking about, you are going to do something. You are going to take a physical action of some sort.

FOGEL: Correct; predict the next number.

BIGELOW: There are two different things happening, are there not? There is the computation to take action and there is an action. You do take action, do you not?

FOGEL: Then we are still confusing the fact that I simplified this problem of prediction and response into prediction alone by saying that the response will always be proper in the light of the prediction and, therefore, the value of response.

BIGELOW: You have presented them to us additively. You do the computation for the prediction and then you observe a response. Allow me to say it consecutively like that. Then the question I want to raise is the following one: Suppose you have made a computation which is a prediction on the ith occasion and you have observed the ith response; then for the jth occasion the matrix will have been affected by the prediction-response pair that occurred on the ith occasion.

FOGEL: It may have been affected, but it need not have been.

BIGELOW: But in general it will.

FOGEL: In general it will.

NATURAL AUTOMATA

BIGELOW: The point I am trying to raise, then, is that this being so, you cannot make the prediction for the ith occasion on the basis of one step. It has to be made on the basis of all previous steps, because the entire history of the matrix itself is built into what you mean by prediction.

FOGEL: That is true.

MINSKY: Does the matrix at the ith step depend on your prediction the time before?

FOGEL: It may or it may not. It depends on what the goal is.

BIGELOW: In general it will.

FOGEL: In general it will. In a rendezvous task, it certainly will.

MINSKY: So, it is possible under certain circumstances that it is in your interest to make a bad guess because you might get a better matrix later.

FOGEL: Exactly so. We want to keep that as a possibility so that you can start playing games with goals and subgoals.

MINSKY: So you have what is called an extensive game.

FOGEL: I would like to have a system whereby I can let these weightings be functions of time or functions of performance, and I offer you one technique which allows me that freedom, that technique being evolutionary programming. Let me assume that I am given an error-cost matrix expressed in a finite language. Let me restrict the logic of transduction to being described by finite-state machines. I start with some arbitrary finite-state machine. Now I exercise this machine on the history that is available to me, making only the assumption that this history is comprised of true data. As each output is noted, it is evaluated with respect to the given error-cost matrix, the total error cost being a measure of the inappropriateness of this machine with respect to the given goal.

This machine is going to be the progenitor in our simulated evolution. I take this machine and randomly alter one of its properties, then similarly drive this new machine by the history, and go through the same evaluation procedure on the sequence of symbols that emerge and sum the costs. Comparison of these two scores reveals the more suitable machine. This one is retained to serve as the parent of other machines, other trial offspring.

Now I proceed to mutate this machine by randomly changing it in, say, one of five possible modes of mutation — these being to randomly change an output symbol, to add a state (randomly connected), to delete a state (randomly chosen), to randomly change the initial state, or to randomly change a next state. This new machine has all the properties of its parent machine except for the single random change. In general, I will get a new sequence of predictions for that machine and thus a new score. If that score is better in the sense of lesser cost, then I keep this new machine and throw away the original one. If it is worse, I go back to the parent and create another offspring (Ref. 72).[2]

BIGELOW: How long does it take to discover that it is better?

FOGEL: Let a generation be defined to occur whenever a new machine is found that is better than all previously retained machines (there is reason to believe that you want to retain not only the best machine but rather the best few). You go through thousands of generations in minutes on a high-speed digital computer.

BIGELOW: After you make a mutation, then, is all of the history prior to that irrelevant to the new machine?

FOGEL: Not so.

MINSKY: Why do you mutate randomly?

FOGEL: Because I do not have any nonrandom procedure which is known to be useful.

MINSKY: What about first trying something rational such as the condition that has been best in the past?

FOGEL: That certainly is a reasonable thing to do and we can talk about introducing that kind of procedure as a cleanup process.

BIGELOW: If you mutate the structure, then the prior history is no longer relevant.

---

[2] Editor's note. A brief description of the evolutionary programming technique has been included as an appendix.

FOGEL: Not so. When we mutate the structure, we reevaluate the offspring, which is some finite-state machine, on all the available history for the moment. Each machine has made some sequence of predictions and has some value. It may be more or less complex than its predecessors. Note that I have not as yet made a real prediction.

MacKAY: Do you throw away the previous machine at once or do you store it for a number of moves in case you are at a local maximum?

FOGEL: We get around the local maximum problem by a slightly different procedure. Let me put your question aside for the moment and talk about it later.

So far there has not been a real prediction. We pass through an arbitrary number of machines making pseudo-predictions and scoring their worth with respect to the given arbitrary goal. Then at some point in time when an order is received to make a real prediction, for example, when you have run out of money for computation, when you have run out of real time, or when your predictive-fit score has reached some required level, you take that machine representing the best logic found up to that point in time and you predict with it (that is, how it would have responded in the light of the history). Then, right or wrong, you proceed with that logic serving as the parent for future generations.

MINSKY: In the prior session it occurred to me that the biological way of thinking may not always be good. Everything you have said in describing your technique is in terms of evolutionary process—mutation, generating offspring, and so forth. That is an old but very suggestive biological idea. I submit that there has been technical progress in the area of artificial intelligence. Everything you have suggested, it seems to me, can be put much more clearly and sharply in terms of tree searches.

FOGEL: I can suggest reasons why this is probably not true.

MINSKY: I would like to hear them, because using random mutations is the last thing you would do only if you could not think of a more rational procedure.

FOGEL: That is right, and in fact we adopt just this approach.

MINSKY: The question is: What is the justification for looking at this problem from a biological point of view?

FOGEL: First of all, I agree with you that, when you can find deterministic ways which reflect the goal, by all means use them. You fall back on randomness only as a last recourse. I agree with this. In fact, we do not change the output symbol randomly because there is, indeed, a deterministic routine which recommends the change. There is at least one way of introducing determinism into the procedure.

As to your second point, "Can you do things with this procedure that you cannot do through a tree search?" In the tree search can you accept arbitrary goals which are changing in time as you proceed iteratively through the tree structure?

MINSKY: Of course. Is there any particular difficulty in that?

NEWELL: Dr. Samuel does it in his checker program all the time. He changes the payoff function with a polynomial.

MINSKY: His payoff function when he is imagining an end game is different from that used in the beginning.

FOGEL: That is true.

MINSKY: My question really is: Is the model of evolution, which was once suggestive, appropriate to use in the light of new knowledge and technology?

FOGEL: Let us consider some specific aspects of this technique. For example, I note that if the evolutionary program is viewed as a tree search, then the tree defines a function space, and the search is to find functions which are suitable for describing the logic of an unknown environment with respect to any given goal. Let me offer a problem and ask you to find the solution by means of tree search.

MINSKY: The question is not, "Is tree search better, because this is a tree search?" The question is, "Is the language of mutation and offspring as good as the language of subgoal, and so forth?"

FOGEL: The language is trivial as far as I am concerned. It is the solution of the problem that justifies the means.

Here is a sequence of numbers taken from an infinite environment.

1 2 3 4 1 2 2 6 4 1 4 2 5 4 1 1 2 4
3 1 2 2 4 6 1 4 2 4 5 1 2 1 4 ...

I ask you to predict each next number. Let me give you a hint. This sequence is nonperiodic which, when looked upon statistically, appears nonstationary.

BIGELOW: No 7's, no 9's, and no 8's. It is very nonrandom.

FOGEL: Let me tell you how this sequence was generated and ask you if a tree search would solve the prediction problem. You rank-order the powers of 2 and the powers of 3; then reduce these numbers to the residue, module 7. Now, this is a deterministic sequence. If you know the rule, it is simple to predict each next number.

McCARTHY: What reason is there to suppose that this resembles anything that an organism can be expected to predict?

FOGEL: I am not trying to simulate organisms. I am trying to design a black box which will perform data transformation with respect to an arbitrary goal, so I can in principle couple two of these together and keep myself from having to worry about a lower level tactical problem.

McCARTHY: Is it reasonable to try to make intelligent machines or goal-seeking machines that are entirely independent of the kind of goal that you specify?

FOGEL: Versatility is the keynote here.

McCARTHY: The kinds of numerical problems that people can concoct are not a reasonable sample of the kinds of problems that people face.

MINSKY: Or that people can solve.

McCARTHY: I consider your example an extremely unrealistic problem to pose. Whoever would want to predict such a sequence?

FOGEL: For the particular example I gave you, the evolutionary program without any hint correctly predicted

56.6 percent of the first 216 predictions, and for the last 50 of these predictions, the score was 78 percent correct. In other words, it demonstrated distinct learning.

MINSKY: In fact, how was it doing it?

FOGEL: In fact, it was doing it by using a succession of different finite-state machines at each point in time (as you would expect because of the nonstationarity of the process). When you offer the program a stationary process, it characterizes the low-order statistics of the process in terms of their modal properties, and thus constructs the best deterministic bettor against that environment. If you offer a periodic process, it will characterize the cycle loops that comprise that process.

MINSKY: Let me ask the question: How did it do compared to a conditional probability machine that looks at the last four or five digits?

FOGEL: We ran experiments in which the evolutionary program was required to predict low-ordered Markov processes with considerable success.

MINSKY: What order?

FOGEL: Zero-, first-, and up to second-order Markov processes. You can address higher-order Markov processes by the evolutionary technique, but then you have to extend the memories of the finite-state machine. I will discuss this if it is of interest. But I want to get to a practical point, if I may.

If, indeed, I can predict sequences of symbols in some meaningful manner, there must be within the predictor some model of the underlying logic which generated this sequence of symbols.

Let us consider an arbitrary problem, then, of a different kind. I give you an unknown plant, P, and I ask you to make it track an arbitrary reference signal, that is, make the output follow some random sequence of symbols. The plant may be digital; it may be a man; it may indeed be anything at all. Is this not an interesting problem?

DAVID: What is supposed to come out of it?

FOGEL: I have a plant, P, which I can drive in any way I wish in terms of symbols. The object is to make it put out a sequence of symbols which will have minimum cost with respect to a given goal (expressed in the form of an error-cost matrix) as this plant tracks an arbitrary sequence of reference symbols.

PAPERT: Is some assumption being made about the nature of the sequence it is tracking?

FOGEL: No. Let me put it this way: I give you a vehicle of arbitrary unknown transfer properties and I ask you to guide it so that it will follow the white line of a highway you have never seen before in such a way as to minimize the total cost of error with respect to any given error-cost matrix. This matrix might be all or none, but it is more likely to be asymmetrical, some function of time or a function of position along the road. This is the general problem of control.

MINSKY: If your evolutionary programming technique is supposed to be an advance on, say, the Uttley machine (Refs. 32a, 32b), it is up to you to characterize the kinds of problems on which it does better and the kinds on which it does worse. If your technique performs as well as a fourth-order Markoff process, then you should clarify why it is not subject to the objections raised against Friedberg's machine (Refs. 36a, 73) or other learning machines that employ a random change of computer programs. There must be some sequences on which your technique performs badly and on which other simpler machines are better.

When one talks about a learning machine of this type, one really ought to characterize the class of problems for which it is good, and I think this is what John McCarthy was saying in another way. What class of problems is your technique good at solving? It is not enough to say it is good at [solving] all.

PAPERT: I think you can say something stronger than that. An abstract characterization of all learning situations, including this one, is that you are searching for some function in some determined space of functions (it might be finite or infinite) to satisfy some set of conditions.

FOGEL: That is right.

INFORMATION AND CONTROL PROCESSES

PAPERT: If you are looking for a function about which you know nothing except that it is in some very, very large space of functions, I contend it will take a very, very long time to find it. The only way out of the exponential blowup is by avoiding the model of a blind hunt in an arbitrary space of functions, for example, by building very specific structured knowledge into the system. But what you have described is a blind hunt.

FOGEL: Is evolution a blind hunt? Friedberg's machine failed because it was a blind hunt. Each change in a computer program gives you a new program which has no logical relation to the preceding program. If you simply rearrange the steps in a program, it is a new program. But, in contradistinction, the evolutionary program is not a blind hunt because each machine has a strong inheritance in terms of logical properties that are relevant subproperties of the environment.

MINSKY: Does the similarity between adjacent machines in the set have any relation to the function for which you are searching in the space? It is not enough that your machines be similar. Friedberg's programs are very similar when he changes them, because he makes only a small change and they inherit a tremendous amount. I am not trying to raise a philosophical question. I am asking you what class of problems you think this technique is good at solving, and I am saying, in effect, that I will not accept all as an answer.

FOGEL: Let me indicate some of the problems evolutionary programming can address; then I want to come back to the problem of gradually removing the human operator from the control loop. If you give me a zero-order or first-order Markov process, you very quickly evolve a finite-state machine which offers a deterministic characterization of the model properties of the transition matrix.

MINSKY: That sounds like an Uttley-type machine.

FOGEL: It acts like an Uttley-type machine; it becomes a probability bettor. If you now give me a higher-order Markov process, so that it has lower-order processes embedded within it, the evolutionary program will discover and use these lower-order properties. It will not find higher-order properties because finite-state machines only remember their present state. If, on the other hand, you want to search for all, say, second-order Markov properties and predict on these, then all I must do is recharacterize the experience in terms of successive overlapping couplets. Thus, I have artificially extended the memory of each finite-state machine and the evolutionary program can address second-order Markov processes. Similar extension to triplets, and so forth, permits addressing higher-order environments.

I have oversimplified the program in one sense. The worth of each machine is not simply measured by ability to attain the goal in the past, but is rather that ability degraded by some cost for its complexity, such as some constant times the number of states in the machine. By adjusting this constant I can make the evolution behave as a statistical observer or behave as a deterministic observer. In fact, the program can be made to perform a tradeoff between these two extremes. If you desire only a statistical view, I will increase the value of this constant, that is, introduce a large cost for complexity. This may reduce the evolved machine to a one-state machine so that the decision rule becomes based on the marginal probabilities exhibited by the environment.

Suppose that, as you offer a specific prediction problem, you add that you suspect the environment to be a second-order Markov process or to have some other particular property. I can obviously draw a finite-state machine that represents just that knowledge and insert that machine into the program at any point. If your hint was a good one, that machine will rise to the "top of the heap" and I will benefit from the hint; but if that hint was inappropriate, that machine will soon be discarded and there will have been some cost associated with my accepting it for trial. In short, the evolutionary program can accept hints and use these as they justify their individual worth. The process need not always start from the state of "no knowledge."

INFORMATION AND CONTROL PROCESSES

Now let me return to the control problem. The history of input symbols received by the plant is used as a basis to evolve a succession of finite-state machines having output symbols which agree with the appropriate past response of the plant with minimum error score. For the moment let me assume that the plant is time-invariant. In that case the evolutionary procedure should converge on a best (in the sense of the given goal) representation within the domain of finite-state machines permitted within the search space.

At each point in time the evolved machine can be viewed as a representation of the relevant logic of the plant, this being used to identify that input symbol which will cause the plant to output the most desirable symbol, given that it is in its present state. If the essence of the logic of the plant has been truly discovered and the plant is indeed in the indicated state, then it becomes a simple procedure to drive the plant in the appropriate manner by substituting the best input symbol for the reference symbol.

We have now run a number of successful control experiments with time-invariant plants, these being arbitrary finite-state machines. In each case a highly nonlinear plant was made to track a random sequence of digits with minimum magnitude of the error.

In summary, I suspect that if you allow me sufficient history and sampling rate, I should be able to represent an arbitrary plant reasonably well (even though it may have variable properties) and, therefore, perform some desired control of the plant.

Here is a technique, a general approach, which allows me to address the problem of how I design a transducer which will link the given sensors to the displays, and another connecting the controls to the given effectors in such a way as to reflect my goal at each point in time. This technique allows instruction of the system as to the nature of the goal structure and any available hints. As the human operator notes that the display and control signals become meaningfully reflective of his goal, he can conceivably begin to simplify his decision making and eventually couple these together, turning his attention to higher-level decisions.

ARBIB: When we talk about the formal theory of automata, we find that looking at the state graph does not give us direct insight into the types of input sequence which the automaton will accept. One of the ways we get around this is by replacing the description of the automaton by what is technically known as the language of regular events. I ask you, therefore, if regular events describe sequences better than state graphs, why are you not doing your mutation on the regular events, because this will give a completely different history?

FOGEL: Good question. In fact, I might choose to describe the machines in terms of a semigroup representation. Each of these different representations offers different modes of mutation. I could even identify each finite-state machine by a Gödel number and select to mutate these in some particular way. Selecting the description of the machine imposes a structure on the search procedure. The Gödel numbering offers a nominal scale with no obvious inheritance characteristic. State diagrams offer a means for maintaining some "physiological" properties through inheritance. Regular expressions would offer a means for maintaining some "psychological" properties through inheritance (the regular expression being an indication of the class-recognition ability of the machine).

McCARTHY: I want to elaborate that point in the following way. Depending on how you represent the behavior, then a change in the representation produces different kinds of change in the behavior.

FOGEL: That is exactly my point.

McCARTHY: Suppose we examine human intellectual behavior and we ask what kinds of changes in the representation of this behavior ordinarily occur. The answer is (at least if this meeting is to be considered as a means of changing behavior) that we change each other's behavior by telling each other sentences, and that behavior is represented by action with respect to a collection of facts. This is the representation that gets changed by meeting, not a representation in terms of state transitions in some automaton. We do not represent ourselves as automata and have other people change the states.

FOGEL: Let me restate this argument. I see three classes of finite-state machine representations. The first class uses arbitrary names. Gödel numbering is in this class. You give each machine a number, a name. Mutation of this name yields another name, one that appears to be unrelated in terms of similar behavioral properties.

In the second class each machine is represented in terms of the things that it can separately discriminate within its environment. The semigroup and the regular expression representations are examples of this technique. It is possible to describe people in terms of those properties of the environment they can discriminate. Each machine is characterized only in terms of the things it can recognize as being discriminatively different. Mutation within this class consists of an alteration of the ability of the machine to differentiate and respond to various aspects of its environment.

In the third class each machine is characterized in terms of its internal functions. For example, I would describe a machine by listing each of its possible states and the response of that machine to each of the possible input symbols while it is in each of these states. People are often described in this manner. They are in an alert state in which each input receives careful attention and immediate response, or they may be in a panic state and to some extent behave randomly. There may even be a state of behavior called the "do-nothing" state. No matter what somebody says, there is no response. In this manner I can characterize each machine in terms of its modes of behavior and their connectivity. I choose this representation for the evolutionary program in view of its intuitive reasonableness with respect to the inheritance properties of the real-world environment. This is one reason this kind of simulated evolution works and Friedberg's evolution did not.

PAPERT: A deep-rooted rationalism tells me that randomness cannot be the right way. There is a simple dilemma. If you are trying to home-in on an automaton in a large space of equiprobable automata, nothing will save you from long search, and systematic enumeration is at least as good, probably twice as good, as random variation. If the space of automata is structured in some way relevant to your search, then systematic procedures using this structure are certainly vastly better than stochastic methods.

FOGEL: Let me offer a converse theorem. The theorem says that induction itself is a process which is nondeterministic, and if the creation of new models is a useful thing to do, it can only be done by the introduction of noncorrelative data; that is, it is a nondeterministic procedure. In this case the mutation noise is the introduction of randomness in a purposeful manner. Any truly intelligent creature must have randomness in its behavior. Otherwise, it is not sufficiently versatile.

Let me come back to my original purpose. I want to ask the group whether it thinks that this problem is deserving of engineering attention, that is, the problem of designing controllable adaptive mechanisms, and if so, what techniques do we have in the offing which will allow us to design, not adaptive, but controllable adaptive transducers? I have expressed my opinion and indicated one approach. Is there anyone else who wishes to comment? Are there any alternative suggestions for allowing machines to accept goals, relative to the problem of designing controllable adaptive systems?

McCARTHY: There is the general problem-solver scheme for goal seeking (Refs. 74, 75); there is the advice-taker scheme for goal seeking (Ref. 76). All of these schemes can be described. They differ from your scheme primarily in the ways in which behavior is represented and, therefore, in the kinds of change which are appropriate.

Your third class of finite-state machine representation, where each machine is characterized in terms of its internal functions, is subject to a great deal of subdivision. For example, behaviors may be represented by computer programs in machine language; they may be represented by computer programs in ALGOL. A

INFORMATION AND CONTROL PROCESSES

small change in a program in ALGOL is different from
a small change in a machine-language computer program.

In the advice-taker model, behavior is represented by
a collection of sentences and a program which acts
relevant to these sentences. The plausibility of the
advice-taker model can be illustrated by the interaction
occurring at this conference. We are changing each
other's behavior allegedly by telling each other things.
If you ask somebody, "What did you learn from this
meeting?" he can tell you in the form of a few sentences.

The implausibility of the other approaches, at least
with regard to the learning systems with which we are
familiar, is that it is difficult to describe what a person
has learned in terms of a change in a finite-state machine
or in terms of a change in a computer program.

FOGEL: I took on a task which changed character
from the time it was first given to me until the time it
was completed, and I would like to review the history of
this change of character. I think it might be instructive
to all of us.

The task was to lead the discussion on Natural Automata and Prosthetic Devices. First, I learned that
there were a number of other people who had material
to present about this subject, and certainly their interest
was to be served because of the group. So the first
portion of the session was devoted to hearing something
of what Dr. Kozhevnikov is doing, followed by some
comments from Dr. David. Then we saw a movie and
heard a discussion of an artificial hand. These were
interesting contributions.

I learned that my idea of prosthetics was a generalization over that offered by the dictionary.

Next, I posed a problem of practical import in hopes
that I would hear some suggestions as to how to aid the
designer in mechanizing controllable adaptive transducers, because I think this is a practical way of reaching toward more and more autonomous autopilots, toward
removing the requirement that people perform direct
control of systems, thereby allowing them to turn their
attention to higher-level decision making. To stimulate

294

the discussion, I planned to mention a technique. Somehow I became seduced into talking about it in greater depth than I had intended and yet not at sufficient depth and with enough detail to avoid raising a number of questions that required attention as they were raised. So I feel that I have done injustice to research in which I have shared.

I asked for other techniques to be offered which might be used to address this particular engineering problem, and I found, to my surprise, that there are indeed many of these. They are various methods of tree search, goal-seeking techniques, heuristic programming, and other concepts; but somehow it would seem that these have not yet been brought to the attention of the designer who faces the problem of realizing controllable adaptive systems.

I should like to conclude my remarks by asking that those among us who have the designer's interest at heart, who are practical enough in the sense of being aware of the real world in terms of its needs from science, should ask ourselves where these techniques that are the outgrowth of research, heuristic and nonheuristic, can help us design specific transducers which will link available sensors to the display and the controls to given effectors in such a way that these links can gradually assume greater and greater responsibility for the task currently performed by the man — from the man. His responsibility is presently diffused to include his input and output systems. Hopefully, he will allow his responsibility to be shared to a greater and greater extent at the lower levels of performance. We need mechanisms which will allow this to become a reality.

I plead only for logic and experiment. In the words of Galileo, "In questions of science, the authority of thousands is not worth the humble reasoning of a single individual"; and in the words of Cole Porter [1933]:

Experiment, Make it your motto day and night.
Experiment, And it will lead you to the light.
The apple on the top of the tree is never
too high to achieve.

INFORMATION AND CONTROL PROCESSES

EXPERIMENT, Be curious,
Tho' interfering friends may frown,
Get furious At each attempt to hold you down.
If this advice you only employ,
The future can offer you infinite joy and merriment,
EXPERIMENT, and you'll see.

* Lines from the lyric of "Experiment," Cole Porter, 1933, quoted with the permission of Warner Bros. Seven Arts Music.

Chapter 8
Evolutionary Experimentation

I. Rechenberg (1965) "Cybernetic Solution Path of an Experimental Problem," Royal Aircraft Establishment, Library Translation 1122.

AS graduate students at the Technical University of Berlin in 1964, Ingo Rechenberg, Hans-Paul Schwefel, and Peter Bienert studied fluid mechanics with a particular interest in experimenting with devices in a wind tunnel. Their hope was to create a cybernetic "research robot" that could autonomously solve engineering problems. They took on the challenge of finding optimum shapes for three-dimensional surfaces so as to minimize drag. Initially, they attempted to optimize a contiguous series of plates connected by variable hinges (Figure 8.1a). The traditional deterministic methods of varying a single parameter (a hinge angle) at a time, or applying discrete gradient search failed to provide a solution because the series of configurations eventually stalled in local minima. Rechenberg suggested that random alterations could be applied simultaneously to all parameters in order to escape from suboptimal solutions, followed by selection applied in a manner similar to natural evolution. The initial results of this search strategy (an "evolution strategy") were presented to the members of the Institute of Hydrodynamics on June 12, 1964.

The early demonstrations of the technique were performed on a mechanical calculating machine. In these experiments, a single parent set of parameters was mutated to yield an offspring para-meter vector. If the offspring offered a superior solution it was retained as the new parent; otherwise, it was discarded and further mutation was applied to the original parent. This was later described as a $(1 + 1)$ evolution strategy, where the " $+$ " symbol indicates that the competition for survival is conducted between parents and offspring. Mutations were implemented as binomial random variables centered on the parent's position.

These experiments were then carried over to the problem of finding the optimum configuration of a hinged plate (as above). A series of six plates hinged on five protractors was subjected to an airflow. The goal was to find a shape for the overall plate that minimized the difference in measured airflow beyond the plate using a series of vertically arranged barometric instruments. Each hinge offered 51 settings, each at 4° intervals, this allowing for a variety of plate configurations. Mutations were determined by tossing dice and probabilistically adjusting the settings of the hinges. Figure 8.1b shows the rate of improvement in the design of the plate over successive iterations of the

(a)

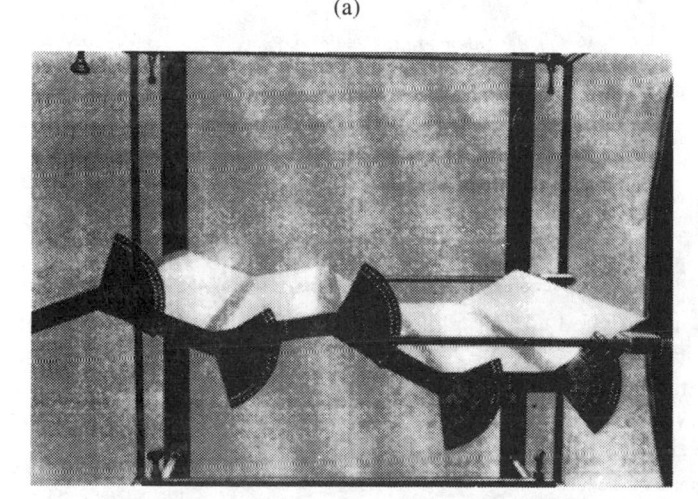

(b)

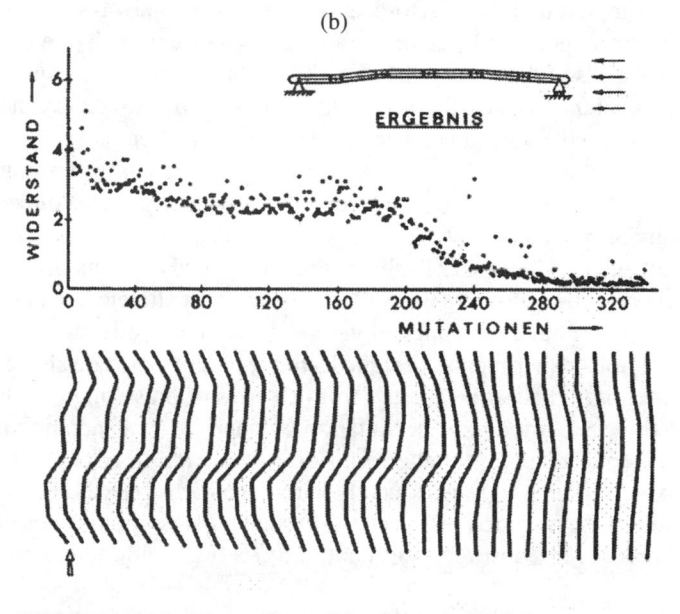

Bild 3. Verlauf der Optimierung der parallel angeströmten Gelenkplatte

Fig. 8.1. (a) Six hinged plates (five hinges) were set up in a wind tunnel. The objective was to find the settings of the hinges to obtain minimum pressure difference beyond the plate. (b) The resulting evolution of the shape of the plate and realized error as a function of the iteration (from Rechenberg, 1973, pp. 25–26).

(1 + 1) procedure, and the final convergence to the optimum configuration of a flat plat.

Concurrent to the experiments with the hinged plates, the three students constructed a plastic box measuring about six inches on a side and about three inches deep. Inside the box, they placed a landscape, a model of hills and valleys. The box was covered with a plastic lid that was perforated with a grid of small pin holes (each individually drilled). A variety of optimization techniques could be demonstrated by placing pins through the cover until they touched the surface of the landscape inside the box. For example, a hill-climbing technique consisted of initially placing two pins at random and choosing the pin that was raised above the cover at the greatest height. The other pin was removed and reinserted at a point next to the current best pin. If this new height was greater, it was retained, while if it was lower it was removed and reinserted again at another neighboring location. The box was a convenient device for contrasting the evolution strategy (new pins were placed based on random variations from the current best pin) with other techniques.

The (1 + 1) strategy served as a useful procedure that could be implemented in practice.[1] In 1965, attention was turned to evolving the best shape for a pipe elbow to maximize the vented nozzle pressure of a gas being forced through the pipe. A series of rods was placed at the pipe elbow so as to determine its shape (Figure 8.2). The positioned settings for the rods were varied at random, again using thrown dice to determine new configurations. The outcome of the experiment generated a shape that was superior to conventional engineering thought (Lichtfuß, 1965).

Shortly before the work of Lichtfuß (1965), Schwefel (1965) implemented the evolution strategy on a computer (a Zuse Z23). The use of binomial mutations was observed to lead to stagnation at points that were sometimes not even locally optimal. This eventually led to utilizing zero mean Gaussian mutations applied to continuous parameters, a method that is still in use. Normally distributed mutations afforded the possibility to always escape from any point on an error surface. This change of variation operator also led to many theoretical results regarding the adjustment of the step size of the mutation operator so as to achieve the most rapid convergence rates on particular functions.

In 1966, Schwefel experimented with a two-phase flashing nozzle (Figure 8.3). The problem was to find the three-dimensional shape of a convergent-divergent (i.e., supersonic) nozzle for a hot-water flashing flow that would offer maximum energy efficiency. Again, a (1 + 1) evolution strategy was applied, although the application required some modifications to previous experimental design. To realize a given nozzle shape, the nozzle had to be created from a series of brass rings, each having its own diameter. Mutation was applied by generating a binomial random variable from dice, at first with a relatively large variance and subsequently with a smaller variance. The result of the mutation was used to vary the diameter of the chosen brass ring. Constraints on continuity between adjacent rings

(a)

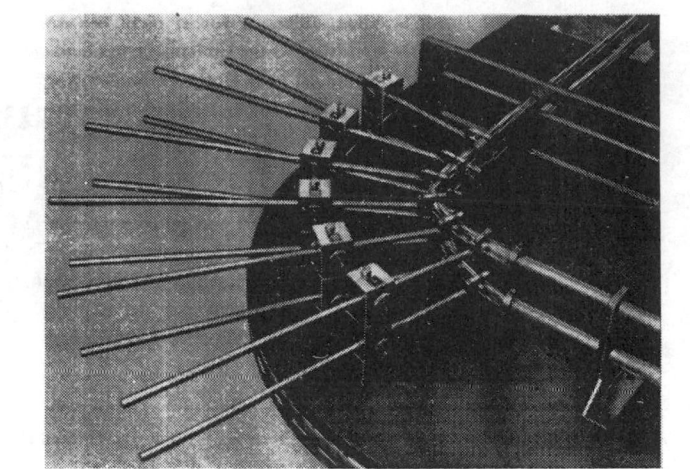

Bild 5. Versuchsaufbau – flexible Rohrumlenkung

(b)

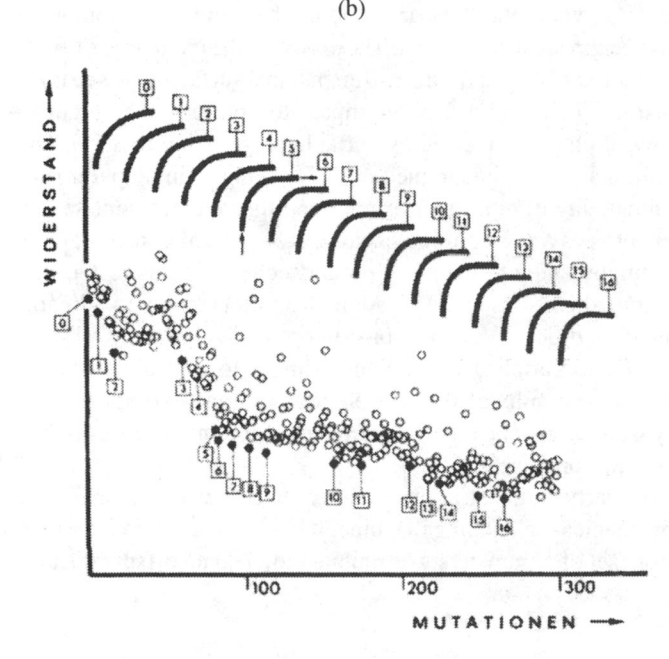

Bild 6. Verlauf der Optimierung des Rohrkrümmers

Fig. 8.2. (a) The configuration for an experiment with a flexible pipe. Rods were used to determine the shape of the bend. The settings of the rods were determined using an evolution strategy. The upper pipe was the parent configuration and the lower pipe was the offspring configuration. The difference in flow between the two pipes was measured and indicated whether or not a superior offspring had been discovered at each iteration. (b) The resulting evolution of shape and utility of the bent pipe (from Rechenberg, 1973, pp. 31–32).

necessitated adding additional rings on either side of the mutated ring. Also, the length of the optimum nozzle configuration was unknown *a priori*. Thus, the experiment involved a variable-length encoding and incorporated duplication and

---

[1]Schwefel (1996) noted that Bienert (1967) constructed a robot that could perform the actions of the (1 + 1) procedure automatically.

(a)

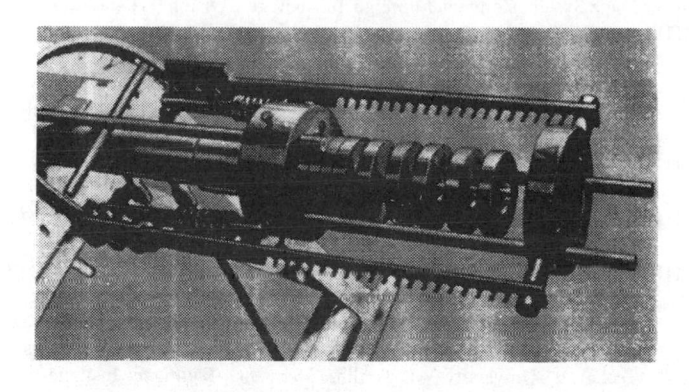

Bild 8. Versuchsaufbau – segmentierte Düse

(b)

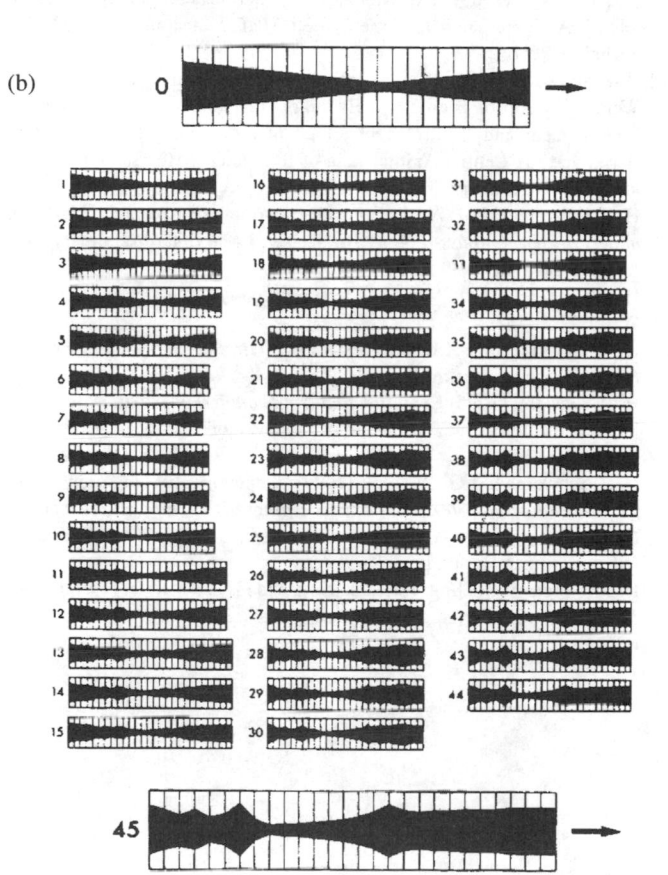

Bild 9. Entwicklung einer Zweiphasen-Überschalldüse von der Anfangsform 0 zur Optimalform 45

Fig. 8.3. (a) The setup for an experiment with the flashing nozzle. The nozzle was configured using a series of brass rings. (b) The initial configuration involved a single pinch point; however, over successive iterations, a superior design was realized that possessed multiple pinch points and greater length (from Rechenberg, 1973, pp. 34–35).

deletion operators. As with the bent elbow (above), the result of the experiment was a design that was superior to conventional engineering wisdom (Schwefel, 1968; Klockgether and Schwefel, 1970).

Subsequently, greater concentration was given to the mathematical properties of the evolution strategy.[2] Rechenberg (1973) developed convergence rate formulae for two function optimization problems (involving the so-called "corridor" and "sphere" models). These formulae led to one of the first implementations of adjusting the step size of the evolutionary search online: the 1/5 rule. This rule suggests that the (approximate) maximum rate of convergence on the two studied functions, when possessing a large number of dimensions, is obtained when the ratio of successful mutations to all mutations is 1/5. Schwefel (1977, 1981) offered an online procedure for conducting this adjustment.

Rechenberg, Bienert, and Schwefel recognized early on that procedures for adjusting the step size of the evolutionary search deserved particular attention (Schwefel, 1996). Rechenberg's notes from 1967 (Rechenberg, 1994a) indicated a comment that the mutation distribution used to search for superior designs could itself be subject to mutation and selection (i.e., a second level of evolution). In essence then, an evolutionary search for the optimum "strategy parameters" could be made simultaneously to the search for the optimum solution to the engineering problem at hand. Schwefel (1977) offered an algorithmic procedure to achieve this evolution (see Bäck and Schwefel, 1993, for details), applying it to the continuous-valued evolution strategy that used zero mean Gaussian mutation (Rechenberg, 1973). The application required the use of a population of parents and a potential surplus of generated offspring.[3] Two approaches were offered (Schwefel, 1975), denoted as $(\mu + \lambda)$ and (μ, λ), in which μ parents generate λ offspring and, in the first case, the best μ parents are selected from the parents and offspring, while in the latter case the μ parents are selected only from the λ offspring. Recombination was also applied to the strategy parameters (Rechenberg, 1973, Schwefel, 1977). Since these early developments, several other extensions have been incorporated into evolution strategies, (e.g., the use of a life span parameter for parents, Schwefel and Rudolph, 1995) and the interested reader may wish to examine Rechenberg (1994b), Schwefel (1995), and Bäck (1996) for more details.

Although most efforts in evolution strategies are currently implemented in computer simulations (Schwefel and Männer, 1991; Männer and Manderick, 1992; Davidor et al., 1994, Voigt et al., 1996), Rechenberg and a group at the Technical University of Berlin still studies the ideas of evolutionary experimentation on physical devices (e.g., winglets), particularly

[2]Earlier and independently, Rastrigin (1965) and Matyas (1965) studied convergence properties of similar stochastic optimization procedures (also see Brooks, 1958).

[3]There is some difference in recollection as to the time of the first introduction of a population within evolution strategies. Schwefel (1996) provided a historical document from 1971 indicating that a (9 + 1) evolution strategy was introduced in 1969; however, Rechenberg (1996) indicated that he earlier used the (9 + 1) strategy with the plastic box and pins as a tool for explaining the evolutionary method. The surplus of offspring was developed later to allow for successive adaptation of the "strategy" parameters associated with mutation.

with an interest to emulate naturally evolved inventions found in the living world.

The paper reprinted here (Rechenberg, 1965) is the English translation of a lecture presented in September 1964. The best available version that could be obtained was a relatively poor-quality facsimile. The paper was typeset at the publisher with original figures scanned electronically As a result, several figures may be unintelligible. Much of the original German remains in the figures as well, but the reader should be able to discern the general ideas presented without undue difficulty.

References

[1] T. Bäck (1996) *Evolutionary Algorithms in Theory and Practice*, Oxford, NY.

[2] T. Bäck and H.-P. Schwefel (1993) "An overview of evolutionary algorithms for parameter optimization," *Evolutionary Computation,* Vol. 1:1, pp. 1–23.

[3] P. Bienert (1967) "Aufbau einer optimierungsautomatik für drei parameter," Dipl.-Ing. thesis, Technical University of Berlin, Institute of Measurement and Control Technology.

[4] S. H. Brooks (1958) "A discussion of random methods for seeking maxima," *Operations Research,* Vol. 6, pp. 244–251.

[5] Y. Davidor, H.-P. Schwefel, and R. Männer (eds.) (1994) *Parallel Problem Solving from Nature 3,* Springer, Berlin.

[6] J. Klockgether and H.-P. Schwefel (1970) "Two-phase nozzle and hollow core jet experiments," *Proc. 11th Symp. on Engineering Aspects of Magnetohydrodynamics,* D. G. Elliot (ed.), California Institute of Tech., Pasadena, CA, pp. 141–148.

[7] H. J. Lichtfuß (1965) "Evolution eines rohrkrümmers," Dipl.-Ing. thesis, Technical University of Berlin, Hermann Föttinger Institute for Hydrodynamics.

[8] R. Männer and B. Manderick (eds.) (1992) *Parallel Problem Solving from Nature 2,* North-Holland, Amsterdam.

[9] J. Matyas (1965) "Random optimization," *Automation Remote Control,* Vol. 26, pp. 244–251.

[10] L. Rastrigin (1965) Random Search in Optimization Problems for Multiparameter Systems (translated from Russian: Sluchainyi poisk v zadachakh optimisatsii mnogoarametricheskikh sistem, Zinaine, Riga) Air Force System Command Foreign Technology Division FTD-HT-67-363.

[11] I. Rechenberg (1965) "Cybernetic solution path of an experimental problem," Royal Aircraft Establishment, Library Translation 1122.

[12] I. Rechenberg (1973) *Evolutionsstrategie: Optimierung technischer Systeme nach Prinzipien der biologischen Evolution,* Frommann-Holzboog, Stuttgart, Germany.

[13] I. Rechenberg (1994a) personal communication, Technical University of Berlin, Germany.

[14] I. Rechenberg (1994b) *Evolutionsstrategie '94,* Frommann-Holzboog, Stuttgart, Germany.

[15] I. Rechenberg (1996) personal communication, Technical University of Berlin, Germany.

[16] H.-P. Schwefel (1965) "Kybernetische evolution als strategie der experimentellen forschung in der strömungstechnik," Dipl.-Ing. thesis, Technical University of Berlin, Hermann Föttinger Institute for Hydrodynamics.

[17] H.-P. Schwefel (1968) "Experimentelle optimierung einer zweiphasendüse tell 1," AEG Research Institute Project MHD-Staustrahlrohr 11034/68, Technical Report 35.

[18] H.-P. Schwefel (1975) "Bindre optimierung durch somatische mutation," Working Group of Bionic and Evolution Techniques at the Institute of Measurement and Control Technology of the Technical University of Berlin and the Central Animal Lab. of the Medical Highschool of Hannover Technical Report.

[19] H.-P. Schwefel (1977) *Numerische Optimierung von Computer-Modellen mittels der Evolutionsstrategie,* Birkhäuser, Basle, Interdisciplinary Systems Research 26.

[20] H.-P. Schwefel (1981) *Numerical Optimizaton of Computer Models,* John Wiley, Chichester, U.K.

[21] H.-P. Schwefel and R. Männer (eds.) (1991) *Parallel Problem Solving from Nature,* Springer, Berlin.

[22] H.-P. Schwefel (1995) *Evolution and Optimum Seeking,* John Wiley, NY.

[23] H.-P. Schwefel (1996) personal communication, Univ. of Dortmund, Germany.

[24] H.-P. Schwefel and G. Rudolph (1995) "Contemporary evolution strategies," *Advances in Artificial Life: Proceedings of the Third European Conference on Artificial Life,* F. Morán, A. Moreno, J. J. Merelo, and P. Chacón (eds.), Springer, Berlin, pp. 893–907.

[25] H.-M. Voigt, W. Ebeling, I. Rechenberg, and H.-P. Schwefel (eds.) (1996) *Parallel Problem Solving from Nature 4,* Springer, Berlin.

Cybernetic Solution Path of an Experimental Problem (Kybernetische Lösungsansteuerung Einer Experimentellen Forschungsaufgabe)

by Ingo Rechenberg (Hermann - Föttinger - Inst. für Strömungstechnik der Technischen Universität Berlin). Translated by B. F. Toms.

SUMMARY

This is the text of a lecture given at the Annual Conference of the WGLR at Berlin in September 1964. It is concerned with the application of an evolution strategy according to Darwin to the problem of finding the body shape which gives the smallest drag within a certain class of bodies. It is shown that there are cases where the evolution strategy is superior to the method of steepest descent and to the Gauss-Seidel strategy. The method is applied to the simple case of a kinked plate.

INTRODUCTION

In cybernetics it is axiomatic that common theories can be applied to apparently widely separated fields of science. The increasingly evident points in common between biology as the theory of organisms and technology as the theory of mechanisms provide a good example of this.

In this lecture we shall report on the attempt to apply a principle of biology to technical systems, namely, the evolution of organisms, a theory which is closely linked with the name of Charles Darwin, has since then been contested, but which today is regarded by biologists as established.

The question is this: Could the efficiency of technical systems likewise be increased by

(a) continuously introducing small random changes (mutations) into the system, and

(b) by subjecting the newly arising systems to continuous selection according to certain quality principles?

Would this process lead to success in a reasonable length of time? This question can be answered in the affirmative on the basis of theoretical considerations and on account of the first practical results.

A system of interarticulated surfaces was constructed in the wind tunnel of the Hermann-Föttinger Institute in Berlin. This shutter-type structure could be altered stepwise, so that 345 million forms were possible. Each form had a certain drag force. The object was to find which shape offered the least drag. Each drag measurement takes about 30 seconds, and another 30 seconds are required to set up another form. If one were to measure the drag of all the possible shapes, one would need about 600 years. The application of the evolution strategy made it possible to obtain the desired result in one day. The solution is in this case a flat plate, that is to say, all the surface elements are lined up parallel to the airflow. We knew, of course, beforehand that this shape would have the least drag. That this shape in fact was the one which actually emerged is a proof of the applicability of the evolution method to technical systems.

LECTURE

In recent years, cybernetics has shown by numerous examples that automata can be endowed with simple human mental functions. These results have led to an ever increasing abandonment of the view held hitherto that certain abilities are possessed only by living organisms, in particular by man. As an example of this I will quote some of the titles of papers presented during today's session* under the heading of "Cybernetic problems in air and space travel":

Automatic teaching machines,

Learning systems,

Self-adapting systems and

Automatic recognition systems.

I should now like to report on an experiment the object of which was to mechanize a certain type of man's experimental

*This paper was presented at the Annual Conference of the WGLR at Berlin in September 1964.

research activity. This would save the experimenter time-consuming false moves in his activity, and the end result of such a development would be to leave the solution of experimental problems to "automatic research machines".

Man's research activities are extraordinarily complex. It would therefore be foolhardy to hope to design an automatic research machine which would provide the solution for any given problem. Instead, it is necessary at the outset of such an undertaking to limit oneself to a certain class of experimental research problems. I should therefore like to begin with a brief analysis of experimental research activity.

An experiment always postulates the interplay of two activities: An operation on the object (the action) and the observation of the resulting effects (the recording of a reaction).

Between the action and the reaction lies a structure which is not directly accessible to the experimenter because of his limited powers of recognition. To signify this, cybernetics has coined the figurative expression "the black box". (Fig. 1).

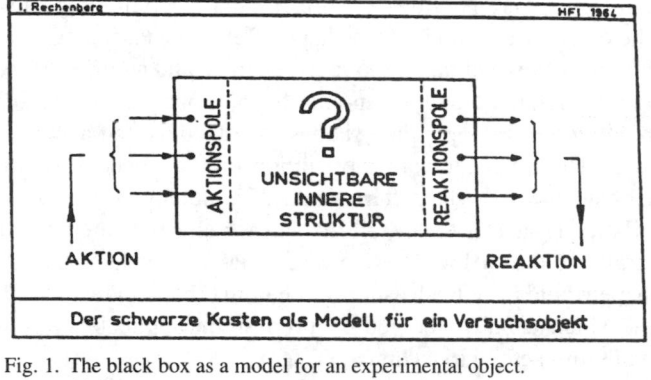

Fig. 1. The black box as a model for an experimental object.

Aktion	= action
Aktionspole	= action poles
unsichtbare innere Struktur	= invisible inner structure
Reaktionspole	= reaction poles
Reaktion	= reaction

The experimenter sees himself as facing a box which he cannot see through and the inner structure of which is inaccessible to him. He can actuate only a few elements which project from the box, for example, mechanical and electrical switches. We shall call these elements "action poles". Other elements projecting from the box allow the experimenter to observe the effects of his operations, in the given case by employing suitable measuring devices. We shall call these elements "reaction poles". The model of the black box thus suggests that three "regions" are to be distinguished in every experimental object:

The action poles on which the experimenter performs his operations;

the invisible inner structure which holds within it the logical connections between action and reaction;

the reaction poles on which the experimenter makes his observations.

The proposed division of the experimental object into three parts implies three possible questions which the experimenter may ask of the experimental object, (Fig. 2).

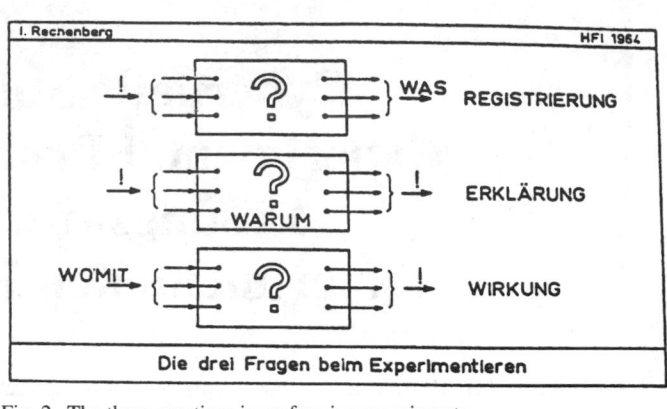

Fig. 2. The three questions in performing experiments.

womit	= how	Registrierung = recording
warum	= why	Erklärung = explanation
was	= what	Wirkung = effect

(1) WHAT is the reaction to a given action? The purpose of research here is merely the accumulation of facts about the object. The question: "What can be found out?" stands at the beginning of every empirical investigation in both the scientific and the technical fields.

(2) WHY does the reaction to an action occur in the observed way? The purpose of research here is to give an explanatory description of the action-reaction mechanism within the black box. The investigator is searching for a model which satisfactorily simulates the internal structure of the black box. The question: "Why does this or that phenomenon occur?" is the basis of scientific research.

(3) HOW (by means of what action) can a given reaction be obtained? The purpose of research in this case is to alter the experimental object so that a desired effect can be achieved. The question: "How can a certain effect be achieved?" is the fundamental problem of technological development.

To what end has this three-part scheme been devised? The answer is: I propose that, depending on the question which is posed, a distinction should be made between three types of automatic research machines.

Automatic research machines of type 1 would be machines which register the reactions of the experimental object to a given action programme. Such research machines have already been in existence a long time. A special illustrative example of this is the earth satellite. The preset action program is the trajectory of the satellite, the physical state of cosmic space is to be compared with the invisible inner structure of the black box, and the reactions registered are those of cosmic radiation, the magnetic field, meteorite collisions, electron density and so on.

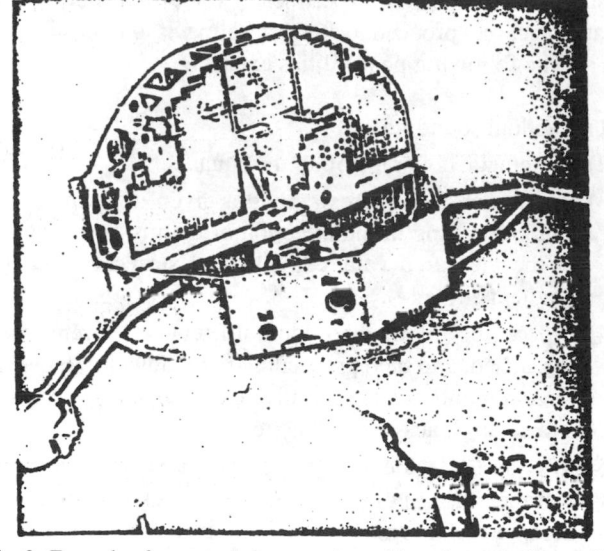

Fig. 3. Example of an automatic research machine of type 1. The satellite OSO carrying instruments for 13 experiments.

Automatic research machines of type 2 would be machines which, for example, translate the action-reaction mechanism into a mathematical formula of the best possible fit. I have as yet no knowledge of such automata and I do not want to speculate about them here.

Automatic research machines of type 3 would be machines which continue to vary the actions performed on the experimental object in accordance with a certain plan until a desired effect is achieved at the reaction poles.

Before dealing with the possibility of realising an automatic research machine of type 3, I should like to use a model to provide a visual illustration of the points made so far (Fig. 4).

An experimenter is standing on a plane surface which he has divided into equal squares. Beneath this plane surface there is a hilly surface which is invisible to him. The experimenter can

(1) record, at different points selected by him, the depth of the surface below him by means of a plumb line;

(2) try, by means of suitable soundings, to give a pictorial description of the surface below him (describe it as being conical, undulating, angular, etc.);

(3) desire to find, in the shortest possible time, the greatest elevation of the surface below him (perhaps because he might wish to climb down there).

In the next diagram I have tried to provide a scheme for the experimenter's procedure in solving a research problem (Fig. 5). A cycle can be found, which in essence can be described as follows:

Performance of an action on the experimental object,

Observation of a reaction on the part of the experimental object, in the given case, by means of appropriate measuring devices,

Processing of the observed phenomena and comparison with similar situations,

Recording the data, drawing diagrams, developing formulae, etc.,

Comparing the results with the aims of the investigation, and hence

Deciding what the next operation should usefully be.

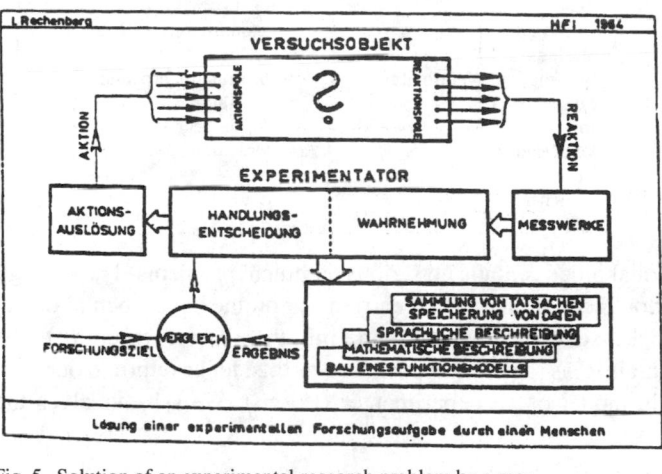

Fig. 5. Solution of an experimental research problem by a man.

Versuchsobjekt	= experimental object
Aktionspole	= action poles
Reaktionspole	= reaction poles
Aktion	— action
Experimentator	= experimenter
Reaktion	= reaction
Aktions-Auslösung	= initiation of action
Handlungs-Entscheidung	= decision on action
Wahrnehmung	= perception
Messwerke	= measuring devices
Sammlung ven Tatsachen	= accumulation of data
Speicherung von Daten	= recording of data
Sprachliche Beschreibung	= verbal description
Mathematische Beschreibung	= mathematical description
Bau eines Funktionsmodells	= construction of a functional model
Forschungsziel	= aim of research
Vergleich	= comparison
Ergebnis	= result

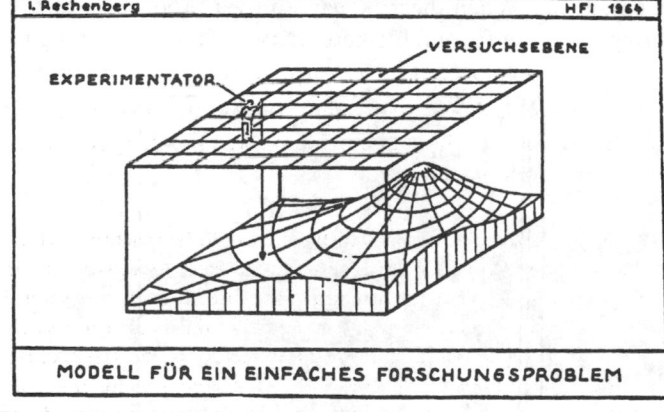

Fig. 4. Model for a simple research problem.

Experimentor	= experimenter
Versuchsebene	= experimental plane

The next diagram (Fig. 6) shows the same cycle again, but this time adapted to the possibilities of automation. It can now be seen that if one attempts to design an automatic research machine of type 3, the essential element in this block diagram is the strategy required. The realisation of the remaining func-

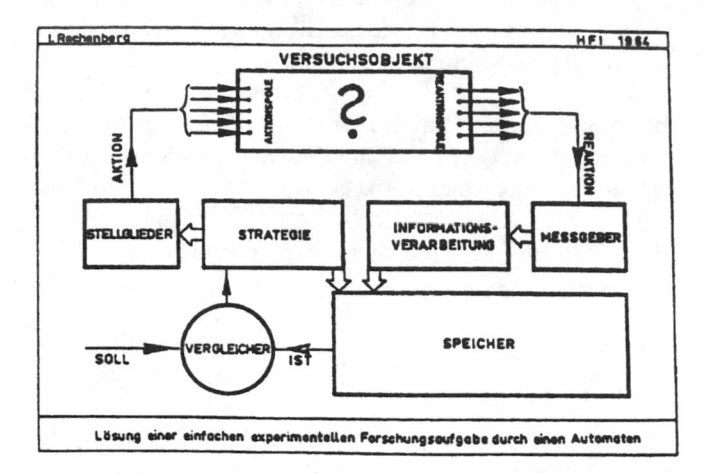

Fig. 6. Solution of a simple experimental research problem by an automatic machine.

Versuchsobjekt	= experimental object
Aktionspole	= action poles
Reaktionspole	= reaction poles
Aktion	= action
Reaktion	= reaction
Stellglieder	= actuating mechanism
Strategie	= strategy
Informations-Verarbeitung	= information processing
Messgeber	= transducer
soll	= should be
Vergleicher	= comparator
ist	= is
Speicher	= store

tional blocks entails no serious technical problems. The strategy should enable the action carried out on the experimental object to be so controlled as to permit the desired reaction to be obtained as rapidly as possible. For this, let us return to our simple model of an experimenter (Fig. 7). We ask ourselves the

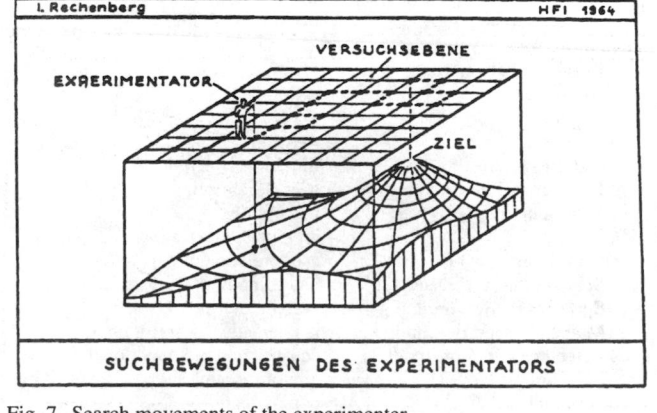

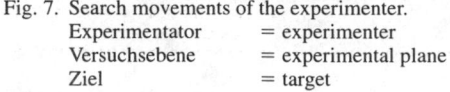

Fig. 7. Search movements of the experimenter.

Experimentator	= experimenter
Versuchsebene	= experimental plane
Ziel	= target

question: "How must the experimenter (or a corresponding automatic machine) proceed to attain his aim without the aid of vision?". The following possibilities exist:

(1) Random soundings.

(2) Systematic coverage of all the squares.

(3) Probing of the squares around a given measuring point and proceeding in the direction of the maximum change of the measured values (method of steepest descent or gradient strategy).

(4) Proceeding in one direction until relative extreme value is reached, and then proceeding perpendicular to the first direction until another relative extreme value is reached, and so on (Gauss-Seidel strategy).

(5) Toss up to decide which of the squares surrounding the last measuring point shall be used to take the next measurement. If this measurement gives a worse value, return to the starting point and toss up again. If the measurement gives a better value, however, remain in the new position and toss up for next one.

I now wish to show how the movements of the experimenter on the experimental plane surface would look if the strategies just described were followed. For this purpose a box with a base 20 × 20 cm was used. In the top 10000 holes were drilled at regular distances from each other. At these discrete points "hills" which lie underneath the top of the box can be sounded with a rod. If each of the places sounded with the rod is marked by a point, the following fields of search are obtained (Figs. 8–12). (Note: for figures 8–12 a small box with only 2600 holes was used.)

Which strategy now proves to be the most effective and should be programmed into an automatic research machine? You will be surprised when I state that the evolution strategy, the last of these described above, is the most suitable. The biological principle of "mutation and selection" is thereby transferred to a technological system, albeit in a very simplified fashion. Mutations in biology are random small changes in organisms. When there is marked over-production of off-spring the more favourable hereditary variants stand a better

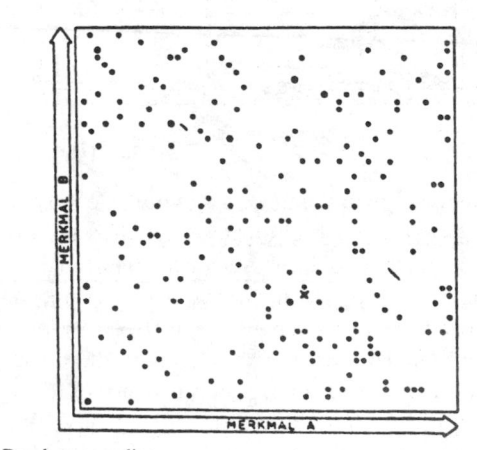

Fig. 8. Random soundings.

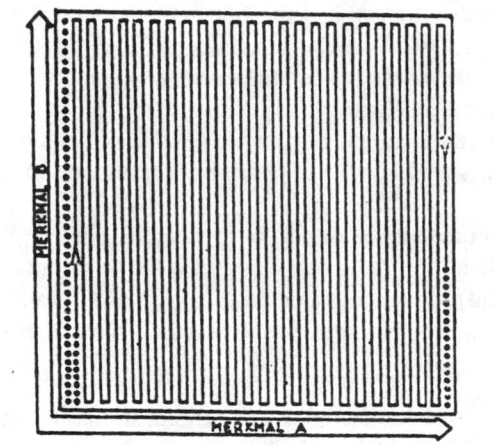

Fig. 9. Systematic scanning of the entire plane.

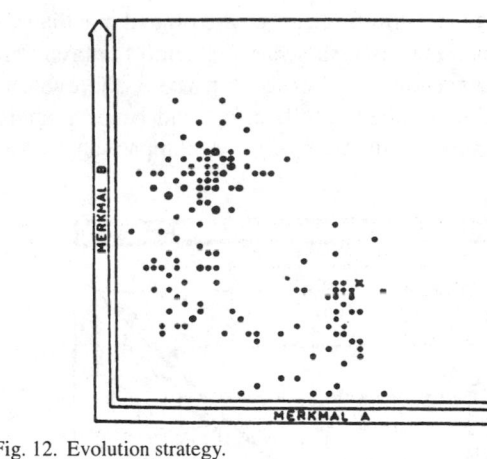

Fig. 12. Evolution strategy.

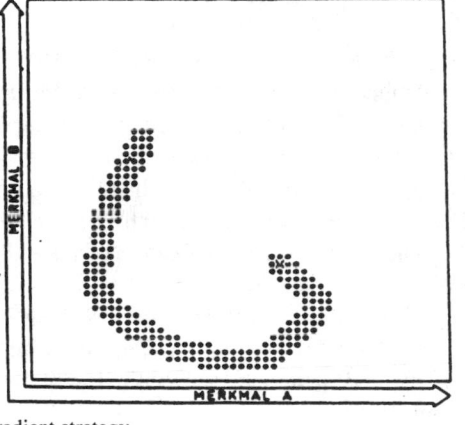

Fig. 10. Gradient strategy.

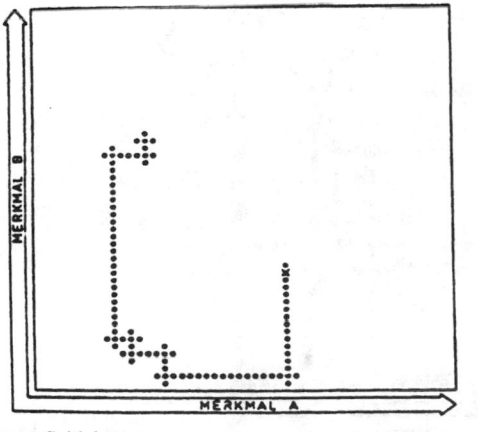

Fig. 11. Gauss-Seidel strategy.

chance of survival than the more unfavourable variants. Thus if the changes serve a useful purpose, the affected organisms will continue further to improve with retention of the mutated character; if they do not serve a useful purpose, they die out. Nowadays, it is accepted as sufficiently proven that the highly organised creatures as we know them today have developed purely in accordance with this principle. The evolution of organisms has, however, taken many millions of years. The application of an evolution strategy to technologi-

cal systems would naturally only have any point in its favour if the period of development, or search time, were considerably shorter.

Let us now estimate the search time. We consider an inclined plane (Fig. 13), (Note: Any surface can be thought of as composed of many such planes to a first approximation). I started to toss a coin at the position where the arrow starts. After 69 throws I reached the top. Altogether, I advanced 16 squares. We shall use the term "search effort" to denote the number of throws divided by the minimum number of squares to be crossed between the starting point and the end point. For the above example, as actually carried out, the search effort $\sigma = 69/16 = 4.3$. If the gradient procedure were employed, σ would be $32/16 = 2$. This treatment can now be carried out with mathematical precision and an upper and lower limit can be found for the search coefficient σ using the evolution strategy if more than two parameters or dimensions n are involved. After some algebra we find for the upper limit

$$\hat{\sigma}_{max} = 3^n \Big/ \left(\sum_{i=1}^{n/2} \frac{(n-1)!}{(n-2i+1)!\,(i-1)!\,(i-1)!} + \sum_{i=1}^{n/2} \frac{(n-1)!}{(n-2i-1)!\,i!\,(i-1)!} \right)$$

and for the lower limit $\sigma_{min} = 3$.

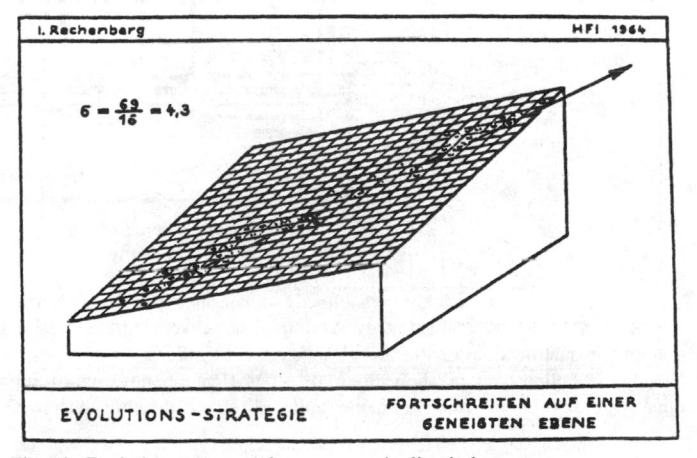

Fig. 13. Evolution strategy. Advance on an inclined plane.

In Fig. 14 these two limiting curves are plotted in a diagram.

We obtain the surprising result that, in the most unfavourable case for $n > 10$ and on the average for $n > 6$, the evolution strategy requires a smaller search effort and hence a shorter search time than the gradient strategy. A comparison between the evolution strategy and the Gauss-Seidel strategy leads one to expect a similar conclusion. Since the two remaining strategies, random soundings and the systematic sampling of all possible values, demand a very high search effort, it is to be expected that the application of the evolution strategy in problems with a large number of parameters gives the shortest search time*.

I now come to the last part of my lecture, namely, the testing of the evolution strategy by applying it to the solution of an experimental problem. In order that the practical test might be as effective as possible, the problem must have the following prerequisites:

(a) It must be as complex as possible, so that it cannot readily be solved analytically;

(b) the solution of the problem must be known so that the experiment will be conclusive.

I decided on the following experiment (Fig. 15).

A system of plane surfaces linked together at 5 hinges was put into a wind tunnel. The articulations can be altered stepwise. Each articulation has 51 steps. Thus there are $51^5 = 345$ million different shapes of this model. Each shape has a certain drag force. The value of the drag is measured by a pitot-tube traverse of the wake. The task is to find the shape of least drag. We all know that this is a flat plane.

The system was set in an initial position as far as possible removed from the plane configuration. The mutations were selected by a method which may be illustrated by a Galton pin-board (Fig. 16).

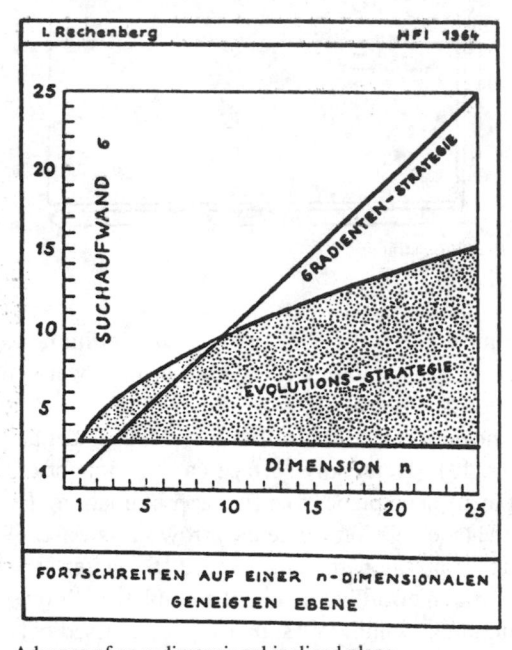

Fig. 14. Advance of an n-dimensional inclined plane.
Suchaufwand σ = search effort σ
Gradienten-Strategie   = gradient strategy
Evolutions-Strategie   = evolution strategy
Dimension n = dimension n

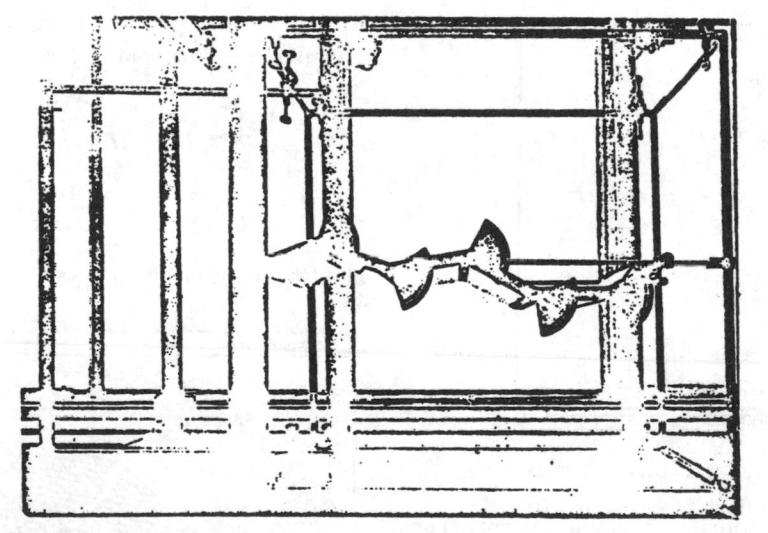

Fig. 15. Experimental set-up.

*Note: The working model of the inclined n-dimensional plane was originally devised only for evaluating the search effort for the evolution strategy. A comparison between different search strategies can only be carried out with certain qualifications on this model. In later investigations, therefore, the inclined plane is replaced by an inclined ridge falling away parabolically to the sides. Here too, however, for problems with a large number of parameters, the evolution strategy remains superior to strictly determined strategies.

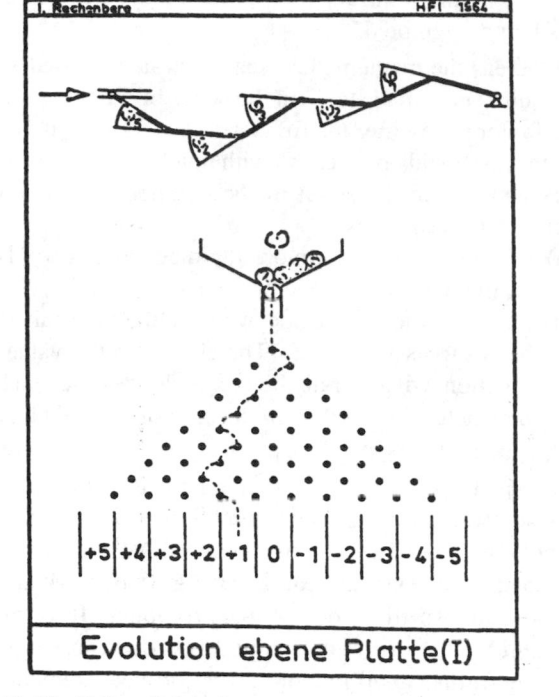

Fig. 16. The Galton pin-board.
Evolution of flat plate (I)

The points indicate a pin pyramid between two walls. Five balls inscribed φ_1, φ_2, φ_3, φ_4, φ_5 fall one after another through this pyramid and land in compartments standing underneath which are marked from +5 to −5. The balls represent the parameters of the system. The labels on the compartments indicate the changes which the parameters lying in the respective compartments have to undergo. A mutation thus occurs which

- dies out if the drag is greater than that of the initial shape;

- is retained and becomes the starting point for new mutations if the drag is smaller.

It is a property of the Galton pin-board that the inner compartments are much more frequently loaded with balls than the outer compartments. Small mutations will thus occur (e.g. +1, 0, −1) with significantly greater frequency than large muta-

tions (e.g. +5 or −5). It has intentionally been arranged that way. The small mutations are designed for the continuous further development of evolution. Very great jumps which occur from time to time, and which incidentally are mostly abortive, avert the danger of development being held up in front of a small barrier.*

Fig. 17 shows the course of the first experiment. The generally decreasing value of the drag is plotted against the number of mutations. In the lower part of this diagram, the "living" form of the articulated plate found after every ten mutations is shown. After 340 mutations, the plate has practically assumed the plane shape.

In this experiment, the plate system was arranged in such a way that the last, fixed, hinge point and the first, movable, hinge point did lie on a line parallel to the airstream. These boundary conditions were changed in a subsequent experiment by raising the leading hinge by $\frac{1}{4}$ of the length of the plate above the trailing hinge. How must the plate system be shaped under these conditions for the drag to reach a minimum value? As the initial shape of the evolution process, a plane configuration of the plate system was chosen. The flat

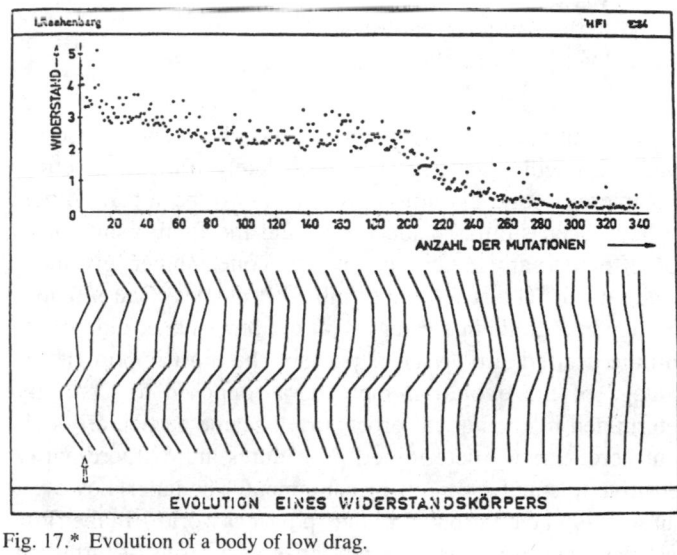

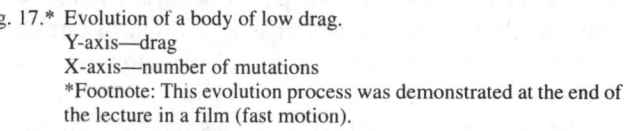

Fig. 17.* Evolution of a body of low drag.
Y-axis—drag
X-axis—number of mutations
*Footnote: This evolution process was demonstrated at the end of the lecture in a film (fast motion).

*Note: In the meantime, further theoretical investigations have shown that combinations of large and small mutation steps are of fundamental importance for the effectiveness of the evolution strategy. In this investigation, the working model of the inclined plane of Fig. 13 was replaced by a ridge falling away parabolically to the sides and rising linearly in a given direction. We select

(a) very small mutation steps. Although we shall be able to record many successes, the rate of progress will nevertheless be very slow on account of the small size of the steps;

(b) very large mutation steps. Here, admittedly, one success will permit us to make a considerable advance, but since the probability of success if now very small, the rate of progress will once again be slow.

It is evident, therefore, that a combination of (a) and (b) allows of considerably more rapid progress than would be the case if either (a) or (b) were used alone.

plate set at an angle against the airstream possesses a relatively high drag since the flow over the upper surface of the plate breaks away, leading to a marked eddy formation. The result of evolution is an S-shaped profile with greater curvature at the trailing end than at the leading end and with a relatively low drag (Fig. 18).

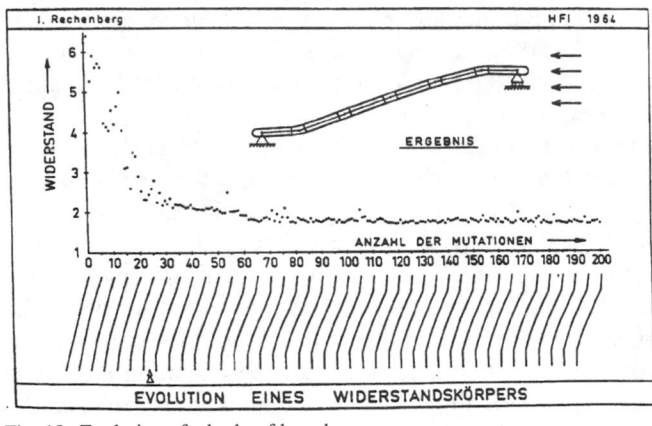

Fig. 18. Evolution of a body of low drag.
Y-axis—drag
X-axis—number of mutations
Ergebuis = final result

This result, incidentally, could not have been known beforehand. Admittedly, such a problem is not likely to arise in practice.

Numerous other evolution processes have been carried out; all have been similarly successful and the final results have not differed to any notable degree from one another (given the same set-up). The strictly determined strategies (Gauss-Seidel strategy and gradient strategy), on the other hand, did not attain the desired aim. What is perhaps the most important advantage of the evolution method, as opposed to a strictly determined mathematical procedure, became particularly evident here. Strictly determined procedures are extraordinarily sensitive to small errors of measurement. The latter, however, can scarcely be eliminated in the "physical world". In the flow past the plate, for example, they arise as a result of turbulent fluctuations in the flow. Small errors of measurement, however, cause a strictly determined strategy to be constantly diverted from its proper direction. It is thus transformed into a stochastic strategy which is particularly unsuitable for this purpose.

The advantages which I expect from an evolution strategy, as opposed to a strictly determined mathematical strategy, with regard to the solution of problems of the kind considered here, can be summarised as follows:-

(1) When a large number of parameters are involved, the evolution strategy attains the desired result more rapidly than the more familiar strictly determined search strategies, assuming the size of the search steps to be the same in both cases. So far, this could only be proved for the case of an n-dimensional hyperplane

rising in any arbitrary direction. A more general proof is being attempted.

(2) Whereas the mathematical search strategies used so far require very small steps (in the sense of the truncation of a Taylor series after the first term), the evolution method can and should operate also with much larger steps which exceed the linear region in the neighbourhood. Taking larger steps signifies.

(a) in many situations a more rapid advance towards the desired aim, and

(b) a shorter time to decide whether the step taken has been successful or not. (The change in the value of a function will generally be greater in the case of a large parameter change than in the case of a small change.)

(3) A so-called "steady signal" in the measurement of the value of a function is a mathematical fiction. Disturbances are always present, which give rise to errors of measurement. The effectiveness of the strictly determined mathematical search strategies is markedly reduced by small errors of measurement. It is of the nature of the evolution method (as a stochastic search method) that small random errors of measurement cannot have a decisive influence on the development of the process.

(4) There are cases in which the more familiar mathematical search strategies must reach deadlock. Such cases are frequently observed in physics; examples are hysteresis phenomena and locally limited extremal values. The evolution method can generally cope with such situations without difficulty.

(5) The algorithm of the evolution strategy is extraordinarily simple. This implies that the effort required for an automatisation of the search process is relatively low.

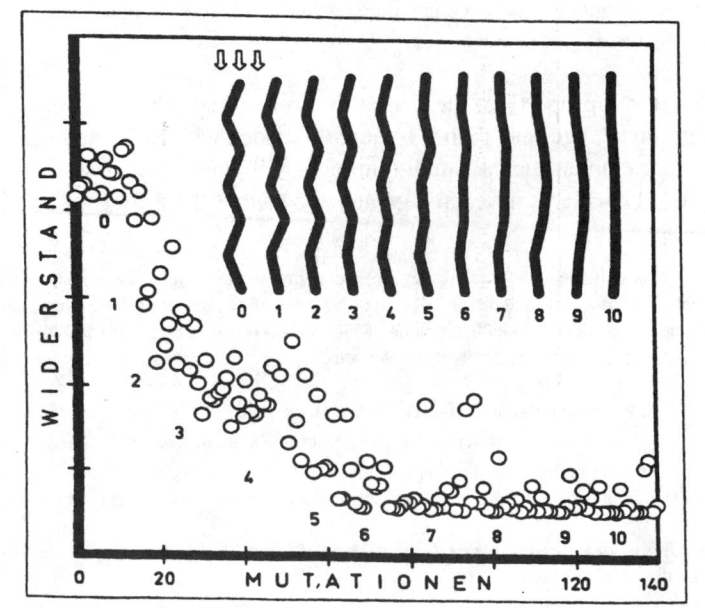

Fig. UNF-1

Fig. UNF-2

Library Translation No. 1122
Royal Aircraft Establishment

007 :
001.89 :
533.6.072

CYBERNETIC SOLUTION PATH OF AN EXPERIMENTAL
PROBLEM (KYBERNETISCHE LÖSUNGSANSTEUERUNG
EINER EXPERIMENTELLEN FORSCHUNGSAUFGABE).
Ingo Rechenberg (Hermann - Föttinger - Inst. für Strömungstechnik
der Technischen
Universität Berlin) Translated by B.F. Toms.   July 1965.

This is the text of a lecture given at the Annual Conference of the
WGLR at Berlin in September 1964. It is concerned with the application of an evolution strategy according to Darwin to the problem of
finding the body shape which gives the smallest drag within a certain
class of bodies. It is shown that there are cases where the evolution
strategy is superior to the method of steepest descent and to the
Gauss-Seidel strategy. The method is applied to the simple case of a
kinked plate.

Library Translation No. 1122
Royal Aircraft Establishment

007 :
001.89 :
533.6.072

CYBERNETIC SOLUTION PATH OF AN EXPERIMENTAL
PROBLEM (KYBERNETISCHE LÖSUNGSANSTEUERUNG
EINER EXPERIMENTELLEN FORSCHUNGSAUFGABE).
Ingo Rechenberg (Hermann - Föttinger - Inst. für Strömungstechnik
der Technischen
Universität Berlin) Translated by B.F. Toms.   July 1965.

This is the text of a lecture given at the Annual Conference of the
WGLR at Berlin in September 1964. It is concerned with the application of an evolution strategy according to Darwin to the problem of
finding the body shape which gives the smallest drag within a certain
class of bodies. It is shown that there are cases where the evolution
strategy is superior to the method of steepest descent and to the
Gauss-Seidel strategy. The method is applied to the simple case of a
kinked plate.

Library Translation No. 1122
Royal Aircraft Establishment

007 :
001.89 :
533.6.072

CYBERNETIC SOLUTION PATH OF AN EXPERIMENTAL
PROBLEM (KYBERNETISCHE LÖSUNGSANSTEUERUNG
EINER EXPERIMENTELLEN FORSCHUNGSAUFGABE).
Ingo Rechenberg (Hermann - Föttinger - Inst. für Strömungstechnik
der Technischen
Universität Berlin) Translated by B.F. Toms.   July 1965.

This is the text of a lecture given at the Annual Conference of the
WGLR at Berlin in September 1964. It is concerned with the application of an evolution strategy according to Darwin to the problem of
finding the body shape which gives the smallest drag within a certain
class of bodies. It is shown that there are cases where the evolution
strategy is superior to the method of steepest descent and to the
Gauss-Seidel strategy. The method is applied to the simple case of a
kinked plate.

Library Translation No. 1122
Royal Aircraft Establishment

007 :
001.89 :
533.6.072

CYBERNETIC SOLUTION PATH OF AN EXPERIMENTAL
PROBLEM (KYBERNETISCHE LÖSUNGSANSTEUERUNG
EINER EXPERIMENTELLEN FORSCHUNGSAUFGABE).
Ingo Rechenberg (Hermann - Föttinger - Inst. für Strömungstechnik
der Technischen
Universität Berlin) Translated by B.F. Toms.   July 1965.

This is the text of a lecture given at the Annual Conference of the
WGLR at Berlin in September 1964. It is concerned with the application of an evolution strategy according to Darwin to the problem of
finding the body shape which gives the smallest drag within a certain
class of bodies. It is shown that there are cases where the evolution
strategy is superior to the method of steepest descent and to the
Gauss-Seidel strategy. The method is applied to the simple case of a
kinked plate.

Chapter 9
Evolution and Optimization

H. J. Bremermann, M. Rogson, and S. Salaff (1966) "Global Properties of Evolution Processes,"
In *Natural Automata and Useful Simulations,* edited by H. H. Pattee, E. A. Edlsack, L. Fein, and
A. B. Callahan, Spartan Books, Washington D.C., pp. 3–41.

HANS Bremermann's contributions to evolutionary computation spanned over 35 years (e.g., Bremermann, 1958, 1994). Bremermann (1958) recognized a connection between individual learning and evolutionary learning and sought to explain the nervous system as a model of the individual's environment (similar ideas were offered in Wiener, 1961, and Atmar, 1976). A more formal description of the similarities between these forms of learning led to an idealized model of evolution: heritable genes were encoded in binary strings which were processed by reproduction (sexual or asexual), selection, and mutation. This framework was also suggested by Bledsoe (1961a, 1961b), and was essentially akin to what later became known as a *genetic algorithm* (see Chapter 15).

Bremermann (1958) recognized natural evolution as an optimization process and suggested using simulated evolution for discovering extrema of functionals. He offered some fundamental results for the *onemax* function (i.e., where a binary string receives a fitness score equal to the number of ones in the string): the most favorable mutation probability for improvement without degeneration when the number of incorrect bits is small with respect to the length of the vector is approximately equal to $1/n$, where n is the length of the string. This result was later rediscovered by Mühlenbein (1992). Bremermann (1958) also derived the generalized optimum probability of mutation as $1 - (m/n)^{1/(n-m)}$, where m is the number of bits that are correct. Bäck (1996, p. 206) independently offered a slightly different formula, and only recently have these early contributions become more widely recognized (Mühlenbein and Schlierkamp-Voosen, 1995). In the paper included here, Bremermann et al. (1966) extended the result to hold for any fitness function that is a function of n binary variables and is a monotone function of the Hamming distance of its argument vector from the solution vector.

Preliminary experiments at general function optimization were undertaken in Bremermann (1962) and extended and reviewed in Bremermann and Rogson (1964), Bremermann et al. (1965, 1966), Bremermann (1967a, 1968, 1970, 1973), and oth-

ers. The first attempts centered on evolving the solutions of linear systems of equations. Continuous variables were transformed into a binary coding and a trial vector, **x,** was evaluated in how well it minimized the norm:

$$\|Ax - b\|_2$$

where A is the coefficient matrix and b is the target vector. New candidate vectors were created either by mutation or by versions of "sexual recombination" between independently evolved trial vectors. Recombination was implemented to involve two or more parent vectors both by random combinations and linear superposition. None of these schemes proved suitable, however, particularly for systems of equations with large condition numbers. Bremermann (1962) speculated that linear programming problems might be more appropriate for these recombination operators.

Bremermann and Rogson (1964), however, offered a different approach for solving linear programming problems. Essentially, the method was to pick an initial feasible vector and determine a parent vector in the direction normal to the plane defined by the cost function. A collection of offspring vectors was created by mutating the parent vector along a cone of random directions[1] defined about the difference between itself and the initial feasible vector. The best offspring from this "buckshot" of new candidates was determined and selected as the new parent if it was superior to the old parent. The process was then iterated until a prescribed error tolerance was achieved. This algorithm was able to discover nearly optimal solutions to some linear programming problems of five dimensions. Mating operations were added for larger systems such that subsets of the population of generated offspring would be used to compute a new candidate vector. But the time constraints of the available computers limited the experimentation (e.g., 46 iterations on a 30–dimensional problem required seven minutes). Bremermann et al. (1965, 1966), and Bremermann (1968) reported results on extensions of these procedures applied to linear and convex programming problems (Bremermann et al., 1966, is reprinted

---

[1]Goguen and Goguen (1967) explored the idea of evolving in uniformly random directions in general optimization problems.

here), and Goguen (1966) offered mathematical considerations on mating, including blending search directions.

Although Bremermann (1962; and others) was motivated to simulate evolution in order to generate an efficient optimization process, he often wrote about attempts to gain a better understanding of the natural processes of evolution through simulation. Given the general acceptance of mathematical genetics, it is noteworthy that Bremermann was highly cautious of idealized models of evolution:

"One might think that the mathematics of evolution has essentially been worked out. In fact, the opposite is the case. Dobzhansky [1967], in his address to the 5th Berkeley Symposium on Statistics and Probability pointed out that most of the mathematical genetics makes assumptions that simplify the mathematics to a point where it becomes tractable, but only at the price of oversimplification. Dobzhansky writes [1967]: ' . . . The classical model of genetic population structure has until recently received the lion's share of attention. It has the advantage of simplicity but the disadvantage of misrepresenting reality. It is not entirely played out, and probably never will be, since it does contain a grain of truth, for some genes and for environments its simplifying assumptions are satisfactory as approximations. But the biological reality is different, and if I may say so, more interesting than the classical model suggests.' In the same article Dobzhansky further writes: 'The difficulty stems from the premises and the assumptions. Most exasperating is the habit of certain mathematical geneticists who make their assumptions implicit rather [than] explicit, on the grounds that to them the truth of their assumptions seems self-evident.'

Dobzhansky lists in his paper the assumptions that underlie the classical model and he criticizes in particular the assumption of a 'constant environment' as unrealistic. In the author's studies another frequently made assumption appears most restrictive: independence of individual genes. . . . Even when interaction between genes is introduced it is usually done in a way (e.g., by making specific assumptions about pairwise interactions and neglecting all 'higher order interactions') that makes the model hardly more realistic. There are many phenotypic features that depend necessarily upon entire blocks of interacting genes (cf. Bremermann [1963, 1967b])."

If a realistic model of the evolution of a species under mutation and selection is mathematically intractable, computer simulation may give some insight" (Bremermann, 1968).

Similar cautions are observed in Bremermann et al. (1966). Yet Bremermann (1994) clearly endorsed the use of simulations: "One can gain insights into the global course of evolution by *simulating* it in model systems. Such simulations should be part of every biology curriculum." The intent of Bremermann's long warning was to insist upon an explicit acknowledgment of the limitations and assumptions inherent to any specific model or simulation, and to ensure that extrapolations from such models are made with care.

Bremermann's simulations represent some of the first efforts to incorporate the notion of a population into simulated evolution, and he offered several conjectures on possible uses of evolutionary simulations, including the potential for studying coevolving systems (Bremermann, 1958), incorporating heuristics into random search operators (Bremermann, 1962), and using evolutionary optimization to train neural networks (Bremermann, 1968).[2] Bremermann's efforts consistently anticipated developments in evolutionary computation.

References

[1] J. W. Atmar (1976) "Speculation on the evolution of intelligence and its possible realization in machine form," Ph.D. diss., New Mexico State University, Las Cruces, NM.

[2] T. Bäck (1996) *Evolutionary Algorithms in Theory and Practice,* Oxford, NY.

[3] W. W. Bledsoe (1961a) "The use of biological concepts in the analytical study of systems," technical report of talk presented to ORSA-TIMS national meeting, San Francisco, CA, Nov.

[3] W. W. Bledsoe (1961b) "Lethally dependent genes using instant selection," technical report, Panoramic Research, Inc., Palo Alto, CA, Dec.

[4] H. J. Bremermann (1958) "The evolution of intelligence. The nervous system as a model of its environment," technical report, No. 1, Contract No. 477(17), Dept. of Mathematics, University of Washington, Seattle, July.

[5] H. J. Bremermann (1962) "Optimization through evolution and recombination," *Self-Organizing Systems-1962,* M. C. Yovits, G. T. Jacobi, and G. D. Goldstein (eds.), Spartan Books, Washington D.C., pp. 93–106.

[6] H. J. Bremermann (1963) "Limits of genetic control," *IEEE Trans. Military Electronics,* Vol. MIL-7, pp. 200–205.

[7] H. J. Bremermann (1967a) "Quantitative aspects of goal-seeking self-organizing systems," *Progress in Theoretical Biology,* Vol. 1, Academic Press, NY, pp. 59–77.

[8] H. J. Bremermann (1967b) "Quantum noise and information," *Proc. of the 5th Berkeley Symp. on Mathematical Statistics and Probability,* Vol. 4, University of California Press, Berkeley, CA, pp. 15–22.

[9] H. J. Bremermann (1968) "Numerical optimization procedures derived from biological evolution processes," *Cybernetic Problems in Bionics,* H. L. Oestreicher and D. R. Moore (eds.), Bionics Symposium 1966, Dayton, OH, Gordon and Breach, NY, pp. 543–561.

[10] H. J. Bremermann (1970) "A method of unconstrained global optimization," *Math. Biosciences,* Vol. 9, pp. 1–15.

[11] H. J. Bremermann (1973) "On the dynamics and trajectories of evolution processes," *Biogenesis, Evolution, Homeostasis,* A. Locker (ed.), Springer-Verlag, NY, pp. 29–37.

[12] H. J. Bremermann (1994) "Self-organization in evolution, immune systems, economics, neural nets, and brains," *On Self-Organization,* Springer Series in Synergetics, Vol. 61, R. K. Mishra, D. Maaß, and E. Zwierlein (eds.), Springer, Berlin, pp. 5–34.

[13] H. J. Bremermann (1996) personal communication, Berkeley, CA.

[14] H. J. Bremermann, and M. Rogson (1964) "An evolution-type search method for convex sets," ONR Tech. Report, Contracts 222(85) and 3656(58), Berkeley, CA, May.

[15] H. J. Bremermann, M. Rogson, and S. Salaff (1965) "Search by evolution," *Biophysics and Cybernetic Systems,* M. Maxfield, A. Callahan, and L. J. Fogel (eds.), Spartan Books, Washington, D.C., pp. 157–167.

[16] H. J. Bremermann, M. Rogson, and S. Salaff (1966) "Global properties of evolution processes," *Natural Automata and Useful Simulations,* H. H.

[2] Bremermann (1996) indicated that he discussed the potential for using evolutionary optimization to train perceptrons with Frank Rosenblatt at Cornell University about 1961 or 1962. Dress (1997) offered that Bremermann's intuitions on evolving neural networks may have been influenced by interactions with Iben Browning, who later offered a randomized approach to such network design in Browning (1964), which includes an appendix by Bremermann.

Pattee, E. A. Edlsack, L. Fein, and A. B. Callahan (eds.), Spartan Books, Washington, D.C., pp. 3–41.

[17] I. Browning (1964) "A self-organizing system called 'Mickey Mouse,'" annual report under support of Sandia Corp., No. 52–2850.

[18] T. Dobzhansky (1967) "Genetic diversity of environments," *Proc. of the 5th Berkeley Symp. on Mathematical Statistics and Probability,* Vol. 4, pp. 295–304.

[19] W. B. Dress (1997) personal communication, Oak Ridge National Laboratory, Oak Ridge, TN.

[20] J. A. Goguen (1966) "Some considerations on evolutionary algorithms," ONR Tech. Report, Contracts 3656(08) and 222(85), Berkeley, CA, Feb.

[21] J. A. Goguen and N. H. Goguen (1967) "Contributions to the mathematical theory of evolutionary algorithms," ONR Tech. Report, Contracts 3656(08) and 222(85), Berkeley, CA, Sept.

[22] H. Mühlenbein (1992) "How genetic algorithms really work: Mutation and hill-climbing," *Parallel Problem Solving from Nature 2,* R. Männer and B. Manderick (eds.), North-Holland, The Netherlands, pp. 15–26.

[23] H. Mühlenbein and D. Schlierkamp-Voosen (1995) "Analysis of selection, mutation and recombination in genetic algorithms," *Evolution and Biocomputation,* W. Banzhaf (ed.), Springer, Berlin, pp. 142–168.

[24] N. Wiener (1961) *Cybernetics,* Part 2, MIT Press, Cambridge, MA.

313

GLOBAL PROPERTIES OF EVOLUTION PROCESSES

H. J. Bremermann, M. Rogson, and S. Salaff

University of California, Berkeley

PART I. INTRODUCTION

Over 100 years ago Darwin conceived the theory of evolution through mutation of hereditary traits and survival of the fittest. His theory was supported through Mendelian genetics, by later experiments with populations of organisms, and recently through the discoveries of the DNA structure and of the genetic coding process.

Biology today would be unthinkable without the theory of evolution. Nevertheless there are vast areas of ignorance. While there is a fairly clear picture of how protein synthesis is controlled by DNA, the manifold interactions of proteins with each other and with the cell environment are not well understood. Many feedback loops are involved; there are even chemical switches that can act on the genes and enzymes themselves and block or activate them selectively. A system of such complexity is beyond the reach of an explicit mathematical analysis. Knowledge is less complete when it comes to the interaction and control of cells in multicellular organisms. The details of the process by which anatomy and behavior of macroorganisms arise from their DNA are largely unknown.

Genetic studies leave out the unknown particulars of the link between the gene and the phenotypical characteristics. Thus, Mendelian genetics is a model which abstracts the many intermediate steps between the nucleotide

Reprinted with permission from *Natural Automata and Useful Simulations*, H. J. Bremermann, M. Rogson, and S. Salaff, "Global properties of evolution processes," edited by H. H. Pattee, E. A. Edlsack, L. Fein, and A. B. Callahan, pp. 3-41. © 1966 by H. J. Bremermann and Spartan Books.

sequence and the final phenotypical observable to a direct causal relationship. Newton's mechanics abstracts the whole earth to a mass point and yet computes its orbit around the sun. Similarly, genetics reduces a segment of DNA to a gene and the many intermediate steps between a gene and phenotypic expression to a causal connection, and still derives important results.

Conventional mathematical genetics goes a step further and ascribes a "selection" advantage or disadvantage to individual genes. A typical result is the celebrated Fisher Theorem which predicts the fate of homologous alleles in a population. The assumption that individual genes possess an intrinsic selection advantage or disadvantage is a very strong one but is not satisfied in many cases. In other words, the model oversimplifies.

We may describe the "fitness" of a genotype numerically (compare Moran[22]) by a *fitness function*. We write $f(g_1, \ldots g_n)$, where $g_1, \ldots g_n$ are the gene "variables"; they run over the alleles of the individual genes. The assumption that individual genes confer a selection advantage (or disadvantage) implies that the fitness function $f(g_1, \ldots g_n)$ is a *linear* function (compare Moran[22]).

Note that a linear function is determined by its values at n points. If each gene has k alleles, then a linear fitness function is determined by $k \times n$ genotypes. If we drop the linearity assumption, then the function has k^n possible values. Even for $k = 2$, k^n becomes a "transastronomical number" for $n > 300$ (larger than the estimated number of particles in the universe). Compare Ashby[1] and Bremermann[10]. The number $k \times n$ in comparison is "almost negligibly small."

Any genetic theory that assumes independent gene selection (linear fitness function) has limited validity. On the other hand, if we give up the assumption of linearity of the fitness function, then we run into severe mathematical difficulties (compare Moran[22]).

Evolution is an optimization process: The goal of a species is optimal adaption to its environment. The properties of the fitness function determine how difficult it is to find its optimum. If the function is subject to no constraints whatsoever, then knowledge of its values on a subset of its domain yields no information about its values on the remaining set.

If the domain consists of N points and if K points are sampled, then the probability of finding the maximum of a function in the sampled points is simply $\frac{K}{N}$.

Extrapolating from Drosophila studies, Curt Stern[26] has estimated the possible number of genes in humans as being from 2,000 to 50,000, with

10,000 being the average number. More recent estimates, derived from DNA studies, tend to put the number of genes at 10^6 or higher (Nirenberg[23]). The number of alleles is, by definition, at least 2; however, it is probably higher for most genes. If we take the minimal estimate of 2 alleles and 2,000 genes, we have $2^{2,000} \approx 10^{600}$ possible genotypes, and thus the fitness function has a range of $N = 10^{600}$ points. No biological or physical process can sample that many points. Hence, in practice, the problem to optimize a general function of 2,000 binary variables is unsolvable.

If, however, the function to be optimized is known to have certain properties, then the task becomes simpler: Linearity is an extreme assumption, which makes the problem almost trivial. Larger classes of functions are, for example: Piecewise linear functions, quadratic functions, and convex functions. The difficulty of finding the maximum of a function is reduced for these classes by utilizing their properties.

The general properties of biological fitness functions are not known. However, some of the mechanics of biological optimization are known: Species evolve and thus optimize their fitness through many types of mutation, selection, and recombination (as well as some less obvious processes, such as transduction).

The major purpose of the work which is described in this paper is the study of the effects of mutation, mating, and selection on the evolution of genotypes in the case of non-linear fitness functions. In view of the mathematical difficulties involved, computer experimentation has been utilized in combination with theoretical analysis. A second objective, which is reported elsewhere, is the study of the evolutionary processes which are useful as a numerical method for optimization in general.

Optimization problems are among the most outstanding problems of numerical analysis. Theoretical methods (e.g., finding the critical points of the function to be optimized) are often of little practical use. (Cf. Gelfand and Tsetlin[18])

Such problems as optimization of profit, minimization of cost, optimal routing of aircraft, optimal scheduling of a production line, optimal strategies in games, minimal circuits, optimal heuristics for problem solving, optimal assignment of weights in a neural net, etc., more often than not lead to numerical optimization problems that tax the capacity of even the largest and fastest computers.

Finally, a better insight into biological evolution processes is necessary to understand transgenetic evolution. Artifacts, cultures, and technologies change and evolve. There are no molecular "genes," but there is change and evolution. Research and development efforts are deliberate attempts

316

to develop methods and technologies which bring about a speedy evolution of products. Any basic insight into the biological evolution process should be useful.

One of the results of the experiments was unexpected: The evolution process may stagnate far from the optimum, even in the case of a smooth convex fitness function. This phenomenon occurred with great persistency. It can be traced to a bias that is introduced into the sampling of directions by essentially mutating one gene at a time.

One may tend to think that mating would offset this bias; however, in many experiments, mating did little to improve convergence of the process. Many modifications of the basic scheme were tried, but the "trap phenomenon," as we came to call stagnation at points other than the optimum, persisted with great tenacity. Also, in a new series of experiments, we found evolutionary schemes that converge much better, but with no known biological counterpart.

In view of these experiences, the question arises: Are biological species really optimally, or nearly optimally, adapted to their environments? Or, is an "ecological niche" in most cases merely a stagnation point? (A stagnation point in the case that the species in question could do much better if it would mutate several genes simultaneously.)

According to Stone,[27] the mutation rate for Drosophila genes is about 10^{-4} to 10^{-5}. Accordingly, the probability that n particular genes mutate simultaneously is about 10^{-4n} to 10^{-5n}. Obviously, already for $n = 5$, such a multiple mutation is extremely unlikely. (Stone's figures imply that the number of fruit flies that ever lived since the appearance of the genus is from 10^{18} to 2×10^{25} flies.) In our experiments, stagnation points at which many genes had to be changed simultaneously were quite common.

The work on evolution reported here has developed several by-products; namely, studies on the problem of optimization in general, concern about physical limitations of data processing, and the theory of limitation of genetic control.

Biological evolution is a form of *data processing*. Data processing, biological and otherwise, is subject to limitations as a result of the uncertainty principle of quantum mechanics and the difficulties in storing and transmitting information without being affected by thermal noise. The essence of these considerations has appeared in ref. 9. Ashby[1] has pointed out some of the fundamental consequences. See also Novotny,[24] Bledsoe,[2] and Bremermann.[10]

Finally, the work with evolutionary models has led to a concern with optimization problems in general and with "self-organizing" systems. The findings have been reported in an address by Bremermann to the Biophysical Society and an amplified version of the talk is to be published.[10] The problem of optimizing pattern recognition schemes led Dr. W. W. Bledsoe to an interest in evolutionary procedures. In this case, the nature of the "fitness function" is not known and no competing optimization procedures are available. Some of the earlier experiments of the work reported in this paper were carried out by Bledsoe. Bledsoe also did theoretical studies of evolving populations.[3, 4] Moreover, Bremermann has received much stimulation through many discussions with Dr. Bledsoe.

The work reported in parts II and III was jointly planned and done by Bremermann and S. Salaff. Salaff did the programming and much of the theoretical analysis and write-up. In a similar way Bremermann and Rogsow cooperated on part IV.

PART II. CONVEX QUADRATIC AND PIECEWISE LINEAR FITNESS FUNCTIONS GENERATED BY SYSTEMS OF LINEAR EQUATIONS

The first "fitness functions" that we considered were generated by systems of linear equations. Numerical methods to compute the solutions of such systems are readily available. Thus, we can readily compute the optimum. Moreover, the properties of the fitness function are known: The function generated is either quadratic (and differentiable) or piecewise linear (and not differentiable at the "corners"). Authors who have worked with fitness functions of unknown properties have often encountered unpleasant surprises. Friedberg[17] found that some of his evolution processes designed to let a computer evolve a simple program did a 1000 times worse than pure random search.

We could have generated quadratic functions in a different way; however, we found linear systems convenient to work with. By using different norms, we generate both quadratic and piecewise linear functions at the same time.

Studies with quadratic fitness functions have general significance. Any twice differentiable function at a local maximum can be approximated by a quadratic function of the kind that we are considering in the following. Thus, as one would expect in the case of an arbitrary twice differentiable fitness function, an evolution process converges in a similar way (at a local maximum), as reported in the following. The problem of finding the largest of several local maxima is a problem of a different kind which we, therefore, have not investigated. Repeated search for local maxima could find it.

We found that the evaluation process can stagnate at a point far from the optimum, without having fitness functions with more than one local optimum. Some of the reasons are analyzed in the following.

Instead of maximizing a function we have been minimizing a function. By multiplying with −1 we would have had a maximization problem. Since the minimization and maximization are equivalent, we worked with minimization rather than maximization.

LINEAR SYSTEMS

A linear system is a set of n equations in n unknowns:

$$a_{11}x_1 + a_{12}x_2 + \ldots + a_{1n}x_n = b_1$$
$$a_{21}x_1 + a_{22}x_2 + \ldots + a_{2n}x_n = b_2$$
$$a_{n1}x_1 + a_{n2}x_2 + \ldots + a_{nn}x_n = b_n$$

We will assume throughout this paper that the a_{ij} and b_i are real.

If we put $A = \begin{bmatrix} a_{11} \ldots a_{1n} \\ \cdot \quad \cdot \\ \cdot \quad \cdot \\ \cdot \quad \cdot \\ a_{n1} \quad a_{nn} \end{bmatrix}, \qquad b = \begin{bmatrix} b_1 \\ \cdot \\ \cdot \\ \cdot \\ b_n \end{bmatrix},$

$x = \begin{bmatrix} x_1 \\ \cdot \\ \cdot \\ \cdot \\ x_n \end{bmatrix}$, we have the matrix form of the problem: $Ax = b$.

RESIDUE, NORM

If x is any vector, we write $r = b - Ax$ and we call r the residue vector. We measure the "length" of r by two different norms: the *euclidean* norm

$$\| r \|_E = \sum_{i=1}^{n} r_i^2$$

and the *absolute* norm

$$\| r \|_A = \sum_{i=1}^{n} | r_i |$$

We use both $\| r \|_E$ and $\| r \|_A$ as "unfitness function."

Of these two norms, the *absolute* is easier to compute, involving no multiplications. The *euclidean* is preferable for analysis since it is a differentiable function of the components of r. It also represents the euclidean length of a vector in n-space, and is thus more intuitive. Both norms have been used in evolution experiments.

In the immediately following theoretical analysis we will use the *euclidean* norm.

IMPROVEMENT

The vector y is called an improvement over x if $\| b - Ay \| < \| b - Ax \|$. The method we are going to employ starts with an arbitrary initial $x^{(0)}$ and varies it (according to different schemes explained in the following) until an improvement $x^{(1)}$ is found. This process is then repeated a number of times, generating a sequence $x^{(i)}$ of trial solutions and the associated residues $r^{(i)}$. Each $x^{(i)}$ is an improvement over its predecessor.

MODES OF VARIATION

In part V it is shown that when the number of genes that has not reached its correct (solution) value is small compared to n, then the optimal mutation probability is $\frac{1}{n}$. Thus optimal improvement is obtained when, on the average, one gene is mutated on each attempt.

This analysis is strictly valid only under the assumption that the fitness function is a monotone function of the distance between solution and try. However, early experiments conducted by W. W. Bledsoe showed that in the case of linear systems, maximal convergence was obtained for a mutation probability equal to $\frac{1}{n}$, where n is the number of variables.

STEP SIZE

The amount δ by which x_j is altered is determined as follows.

1. δ is directly proportional to $\| r \|$, *where* $\| r \| = \| b - Ax \|$.
2. δ is inversely proportional to n, the number of variables.
3. δ is directly proportional to a random number z between zero and one. We have worked with:
   a. the uniform distribution between -1 and 1.
   b. the normal distribution with mean 0, standard deviation 1.

Early experiments indicated a slight advantage in using distribution *a*. In the following, unless otherwise noted, we use the uniform distribution (case *a*).

The change δ would vary $x = (x_1, \ldots, x_n)$ to $\tilde{x} = (x_1, x_2, \ldots, x_j + \delta, \ldots, x_n) = x + \delta e^j$, where $e^j = (0, 0, \ldots, 1, \ldots, 0)$ (1 in the jth component). If $\tilde{x}$ is an improvement, the change is accepted; otherwise a new random component of x is chosen and we determine a new step size by which to vary it, test for improvement again, and so on. Let us call the operation of selecting a component of a trial solution x at random and computing the step size according to 1 -- 3 a "try."

The improvement process is halted when either $|| r^{(m)} || < \epsilon$ for an arbitrary preassigned $\epsilon > 0$ and some integer m, or the number of tries exceeds a fixed maximum number L.

We wish to motivate the particular method of improvement chosen. Let us consider the problem $Ax = b$, where $x_1, \ldots, x_n$ are restricted to integer values. We assume a solution in integers exists. If x_i is an integer between 0 and 9 its binary representation contains at most four binary digits. For example, $6 = 0110$, $9 = 1001$, and so forth.

Thus the integer problem can be transformed into one in which the x_i are binary digits.

Regarding the x_i as binary valued "genes" of the "organism" x, we seek the most efficient mode of mutating the genes x_i of x towards the solution $\overline{x} = (\overline{x}_1, \ldots, \overline{x}_n)$. By discretizing the x_i we have made possible this simple analogy with the genetic concept of mutation.

CONTINUOUS VARIABLES

In the case of continuously varying x_i we decided to follow a scheme that changes *exactly* one variable at a time (instead of *in the average* one variable).

Each try at improvement consists in a change in one component (randomly selected) of the trial solution vector.

This rule has the advantage of relative ease in computation as compared with a method which changes each component x_i with a certain probability.

At a given stage of the improvement process, we have a current trial solution vector, say $x = x^{(m)}$. We seek an improvement in x by changing a randomly chosen component x_j. Listed on page 9 are three factors which determine the proposed change δ in x_j. From these,

$$\delta = \frac{|| r || z}{n},$$

where z is a random number, uniformly distributed in $[-1, 1]$.

If $|| r ||$ is large, x is not close to the solution x, and so we need to take relatively large steps to approach it. Hence the factor $\| r \|$ in δ. The factor $\dfrac{1}{n}$ was inserted so that as the number of unknowns increases, and the problem gets more difficult, smaller steps will be taken.

Finally, the random number z allows for the minus sign, as well as for small steps.

COMPARISON OF RESIDUES

We next compare the error

$$\| r \| = \| b - Ax \|$$

of x with the error

$$\| \widetilde{r} \| = \| b - A\widetilde{x} \|$$

associated with the attempted improvement x. We have

$$\widetilde{x} = x + \delta \cdot e^j$$

So $\widetilde{r} = b - A(x + e^j) = r - \delta A e^j$ or $\widetilde{r}_i = r_i - \delta \cdot a_{ij}$, since

$$Ae^j = \begin{bmatrix} a_{11} \ldots a_{1n} \\ \cdot \quad \cdot \\ \cdot \quad \cdot \\ \cdot \quad \cdot \\ a_{n1} \ldots a_{nn} \end{bmatrix} \begin{bmatrix} 0 \\ \cdot \\ 1 \\ \cdot \\ 0 \end{bmatrix} = \begin{bmatrix} a_{1j} \\ \cdot \\ \cdot \\ \cdot \\ a_{nj} \end{bmatrix}$$

Hence

$$(\widetilde{r}_i)^2 - (r_i)^2 = -2\delta r_i a_{ij} + \delta^2 a_{ij}^2$$

and

$$\sigma = \| \widetilde{r} \|_E - \| r \|_E = -2\delta \sum_{i=1}^{n} r_i a_{ij} + \delta^2 \sum_{i=1}^{n} a_{ij}^2$$

The column sums $\sum_{i=1}^{n} a_{ij}^2$ can be precomputed.

If the sign of σ is negative, then $\widetilde{x}$ is an improvement. If it is positive, $\widetilde{x}$ is not an improvement, and another try is made, with a new randomly selected component, and a new value of δ given by $\dfrac{\| r^{(m)} \| \cdot z'}{n}$ where z' is a new random number in $[-1, 1]$.

We agree to call the case $\sigma = 0$ an improvement.

The only exception to the above is the case when the total number of tries is about to exceed L, the upper limit on this number, in which case the method stops as $x^{(m)}$.

Figure 1 illustrates a flow diagram of our method. It was programmed in FORTRAN for the 7090 Computer at the University of California Computer Center, Berkeley.

Many variations of the scheme described here were tried. For as little as three and four variables "traps" of many kinds occurred that had not been anticipated. We even left purely evolutionary schemes and incorporated "steepest descent" steps, and so on. No matter what we did, the evolution process performed poorly.

CONDITION NUMBER

Before listing our results we shall comment on the mathematical aspects of the problem. We first used non-singular matrices A with integer entries chosen more or less arbitrarily, and found the behavior of the method hard to control. Closer examination showed that non-convergence was intimately associated with condition number.

The condition number P of a real symmetric matrix is the ratio

$$|\lambda| / |u|$$

where $\lambda =$ largest eigenvalue of A, in absolute value

$u =$ smallest eigenvalue of A, in absolute value.

The condition number H of a real non-symmetric matrix B is

$$\sqrt{|\theta| / |\phi|}$$

where $\theta =$ largest eigenvalue of $B^T B$, in absolute value

$\phi =$ smallest eigenvalue of $B^T B$, in absolute value.

The P and H condition numbers coincide for symmetric matrices. A discussion of these and other measures of condition may be found in von Neumann and Goldstine.[28]

MATRICES USED

We chose the matrix

$$A = \alpha I + J,$$

where I is the $n \times n$ identity matrix

J is the $n \times n$ matrix each of whose entries is 1, and

α is a positive scalar.

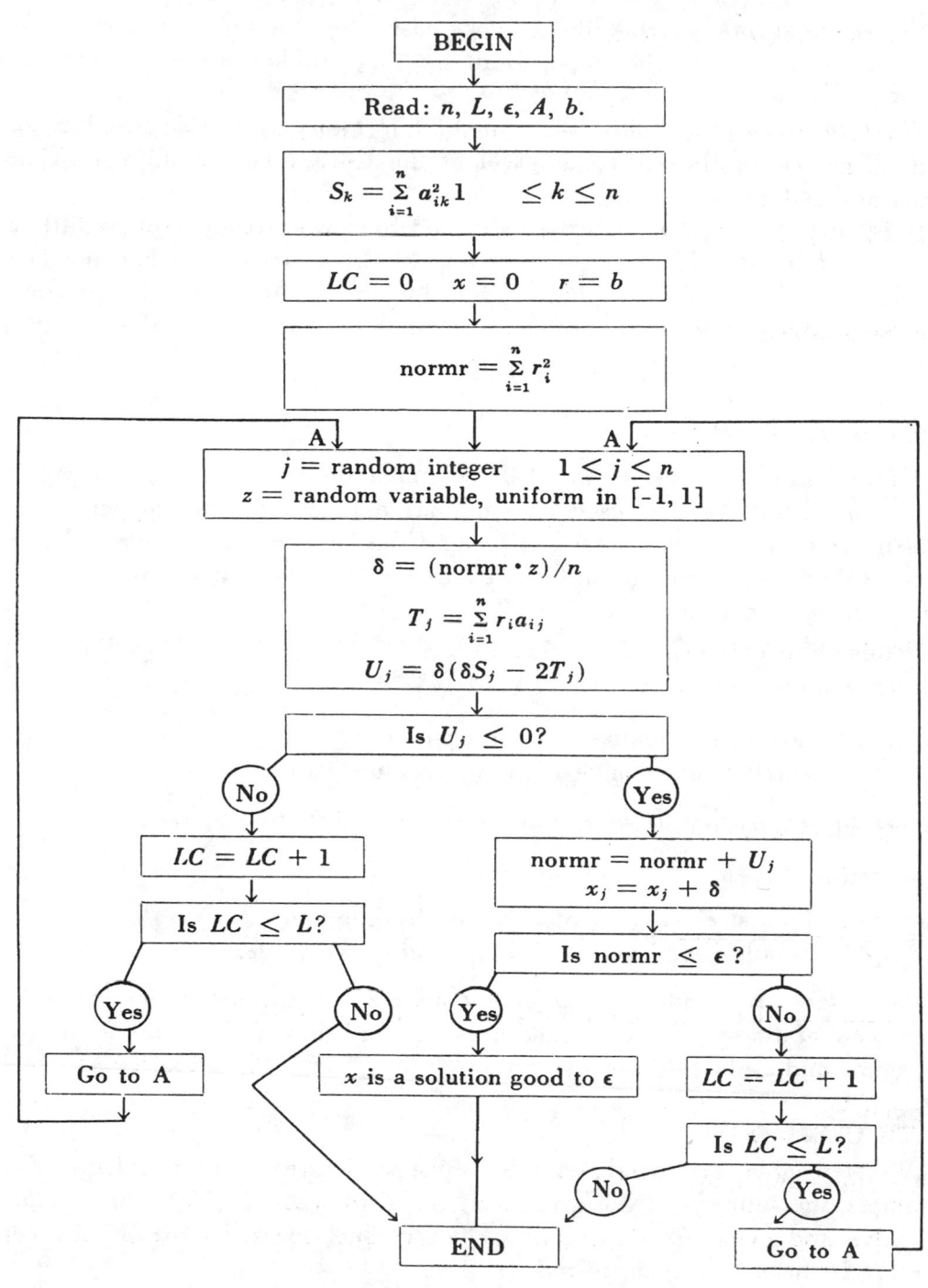

Figure 1. Flow diagram.

For the vector b we took

$$b = (1, 3, 5, \ldots, (2n-1)).$$

This choice was suggested in a paper by Hooke and Jeeves.[19] (These authors are also concerned with search techniques for the solution of linear equations and other problems.)

If $D_n = |(\alpha - \lambda)I + J|$ where I and J are $n \times n$, the recursion relation

$$D_n = (\alpha - \lambda)D_{n-1} + (\alpha - \lambda)^{n-1}$$

can be established readily, leading to

$$D_n = (\alpha - \lambda)^{n-1}(\alpha - \lambda + n)$$

The eigenvalues of A are thus

$$\alpha \qquad n-1 \text{ times}$$
$$\alpha + n \quad \text{once, so}$$
$$P = \frac{\alpha + n}{\alpha}$$

and this ratio can easily be varied with α.

Furthermore, the exact inverse of A is

$$\frac{1}{\alpha} I - \frac{1}{\alpha(n+\alpha)} J$$

so the exact solution of $Ax = b$ is

$$x = A^{-1}b$$
$$= \frac{1}{\alpha} b - \frac{1}{\alpha(n+\alpha)} Jb$$
$$= \frac{1}{\alpha(n+\alpha)} \Big(n + \alpha - n^2, \; 3(n+\alpha) - n^2, \; 5(n+\alpha) - n^2, \ldots$$
$$(2n-1)(n+\alpha) - n^2 \Big)$$

RESULTS

We were able, first of all, to solve some moderately sized problems. For example, the following table gives the time T in seconds (7090 time) taken to solve and print out the solution. The column marked "normr" denotes the value of the absolute error norm $\sum_{i=1}^{n} |r_i|$ at the termination of the method.

n	α	P	T	normr
12	11	2.09	10.9	.000075
20	19	2.05	20.0	.000092
30	30	2.0	61.8	.000098

In order to conserve machine time we did not try to push n higher, except to note that considerably more than two minutes was required to approach the solution when $n = 50$.

As previously mentioned (page 12) the success of the method depends heavily on the condition number P of the matrix A. The following results for the case $n = 5$ show this strikingly. Here we return to the use of the euclidean norm $\sum\limits_{i=1}^{n} r_i^2$.

The first column contains the condition number and the second contains the value of normr achieved after 1.2 seconds.

Dependence of performance on condition is thus clearly exhibited. We now discuss explanations for this dependence and offer several suggestions to improve performance in the case of large condition number.

P	normr
3	.0014
5	.0106
7	.0143
7.5	.0231
8	.0976
8.5	.1436
9	9.7708
11	5.8215

SURFACES OF CONSTANT ERROR

The euclidean norm of the error vector r can be written

$$r^2 = (b - Ax)^T (b - Ax)$$

where T denotes transpose. In solving $Ax = b$, we choose a sequence $x^{(i)}$, with $x^{(0)} = 0$, so that r^2 decreases with each choice. But the corresponding sequence $r^{(i)}$ of norms does not necessarily decrease to zero, and this has been the difficulty with the method. We noted that whenever the condition number of A exceeded 5 or so, the sequence $(r^i)^2$ decreased monotonically to a non-zero limit.

Now $r^2 = (b^T - x^T A^T)(b - Ax)$
$$= b^T b - 2 b^T A x + x^T A^2 x$$

since A is symmetric.

The surfaces $r^2 = constant = k > 0$ are the hyperellipsoids

$$x^T A^2 x - 2 b^T A x + b^T b = k$$

This is so because α has been chosen positive, making $A^2 = \alpha^2 I + (n + 2\alpha) J$ positive definite.

Let C be an orthogonal matrix which diagonalizes A:

$$C^T A C = \Lambda = \begin{vmatrix} \alpha & & & & \\ & \alpha & & & \\ & & \ddots & & \\ & & & \alpha & \\ & & & & \alpha + n \end{vmatrix}$$

If we make the transformation

$$y = C^T x - \Lambda^{-1} C^T b$$

or

$$x = Cy + C\Lambda^{-1} C^T b = Cy + A^{-1} b$$

the hyperellipsoid $r^2 = k$ becomes

$$y^T \Lambda^2 y = k$$

or

$$\alpha^2 y_1^2 + \alpha^2 y_2^2 + \ldots + \alpha^2 y_{n-1}^2 + (\alpha + n)^2 y_n^2 = k$$

$$\frac{y_1^2}{\left(\dfrac{k}{\alpha^2}\right)} + \frac{y_2^2}{\left(\dfrac{k}{\alpha^2}\right)} + \ldots + \frac{y_{n-1}^2}{\left(\dfrac{k}{\alpha^2}\right)} + \frac{y_n^2}{\left(\dfrac{k}{(\alpha + n)^2}\right)} = 1$$

From this we can read off that the semi-axes of $r^2 = k$ are of lengths

$$\frac{\sqrt{k}}{\alpha}, \quad \frac{\sqrt{k}}{\alpha}, \quad \ldots, \quad \frac{\sqrt{k}}{\alpha}, \quad \frac{\sqrt{k}}{\alpha+n}$$

Clearly, the common center is the solution $x = A^{-1}b$.

For ease in illustration we shall now restrict attention to the case $n = 2$, where we have the ellipses

$$\frac{y_1^2}{\left(\dfrac{k}{\alpha^2}\right)} + \frac{y_2^2}{\left(\dfrac{k}{(\alpha+2)^2}\right)} = 1$$

The matrix $C = \dfrac{1}{\sqrt{2}}\begin{bmatrix} 1 & 1 \\ -1 & 1 \end{bmatrix}$, and so the major and minor axes of

the ellipses $r^2 = k$ are $x_1 + x_2 = \dfrac{4}{\alpha+2}$, $x_2 - x_1 \dfrac{4}{\alpha(2+\alpha)}$ respectively.

Figure 2 illustrates the case $\alpha = \dfrac{1}{2}$.

It can be seen geometrically that improvement in a coordinate direction will be difficult from points on the major axis $x_1 + x_2 = \dfrac{4}{\alpha(2+\alpha)}$, especially when α is near zero. In the figure the condition number is 5 and it is evident that improvement from point P^0 can be made in small steps only.

MAXIMUM IMPROVEMENT POSSIBLE

More precisely, we calculated the maximum improvement possible, with α and k arbitrary, from a point P^0 located at an extremity of the major axis. If d denotes this quantity, we have

$$d = \frac{\sqrt{2k}}{\alpha+2+\dfrac{2}{\alpha}}$$

As is known, a small positive value of α makes $P = 1 + \dfrac{2}{\alpha}$ large. The ellipses are then very elongated and the quantity d will be less than it will be for a larger value of α. Further, the direct dependence of d on $\sqrt{k}$ indicates the following fact, which is rather evident geometrically. If $k_1 < k_2$ are two positive numbers, and if P_1^0 P_2^0 are two points located at the extremities of

328

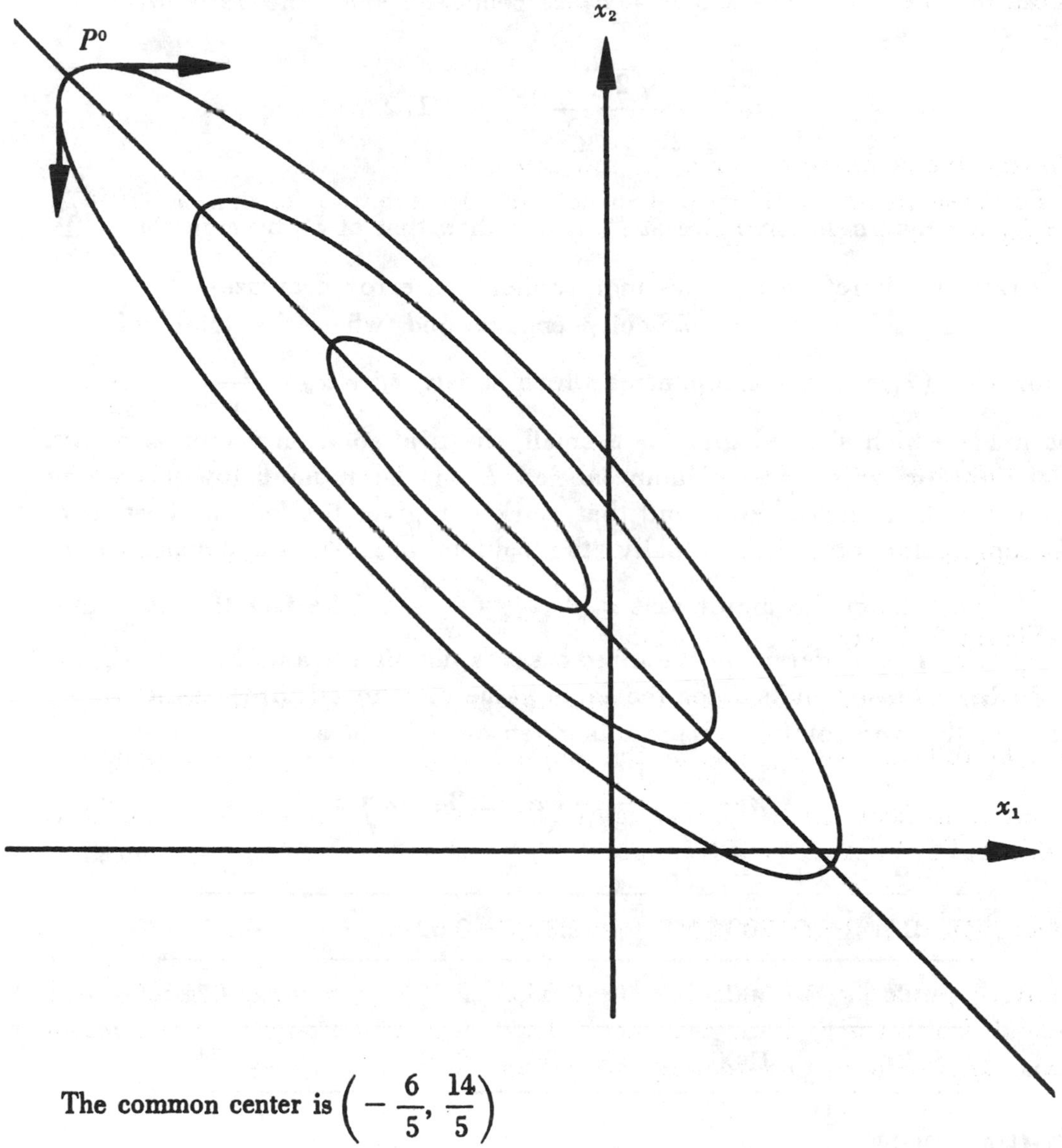

The common center is $\left(-\dfrac{6}{5}, \dfrac{14}{5} \right)$

Major axis is of length $\dfrac{2\sqrt{k}}{5}$

Minor axis is of length $\dfrac{2\sqrt{k}}{5}$

Figure 2. Error ellipses for $\alpha = \dfrac{1}{2}$.

the major axes of $r^2 = k_1$, $r^2 = k_2$ respectively, then the ratio $d_1/d_2 = \sqrt{k_1}/\sqrt{k_2}$, where

$$d_i = \frac{\sqrt{2k_i}}{\alpha + 2 + \dfrac{2}{\alpha}} \qquad i = 1, 2.$$

that is, the maximum step size at P_1 is less than that at P_2 by a factor $\sqrt{\dfrac{k_1}{k_2}}$. Improvement therefore becomes more difficult as error decreases.

The analysis shows the difficulty encountered when the trial solution vector $x = (x_1, x_2)$ has components which satisfy $x_1 + x_2 = \dfrac{4}{\alpha + 2}$. A test was made which showed that for α small, the trial solution vector is drawn to the major axis. The column labeled L in the table below gives the number of tries for fixed α, and that marked x gives the last, or best, trial solution vector obtained. Finally, the column ρ gives the distance ρ of $x = (x_1, x_2)$ from the major axis $x_1 + x_2 = \dfrac{4}{\alpha + 2}$. The fact that the number of tries L is different in the three cases is not significant. Similar results would have been obtained for the same value of L in all three cases.

Recall that the solution vector x is given on page 14 as

$$x = \frac{1}{\alpha(2 + \alpha)}\left(\alpha - 2, 3\alpha + 2\right)$$

α	L	x	ρ
.2	200	(2.31, −0.62)	.0891
.02	500	(−0.05, 2.03)	.0007
.002	1000	(−0.04, 2.04)	.0004

COMPARISON

For the purpose of comparison a standard method, elimination with maximum pivot, was applied to the original linear problem $Ax = b$. The method is coded in FORTRAN, and is known as ISIMEQ in the SHARE library. The time T taken to solve n equations was approximately

$$T = 32.5n^3 + 175n^2 \ microseconds$$

The solution times given on page 15 for our method are very much greater, indicating its non-competitiveness.

CASE OF n DIMENSIONS

The restriction to the case $n = 2$ does not give a sufficiently true picture of the error surfaces $r^2 = k$. The n dimensional hyperellipsoid

$$\frac{y_1^2}{\dfrac{k}{\alpha^2}} + \frac{y_2^2}{\dfrac{k}{\alpha^2}} + \ldots + \frac{y_{n-1}^2}{\dfrac{k}{\alpha^2}} + \frac{y_n^2}{\dfrac{k}{(\alpha+n)^2}} = 1$$

may be described as an n-sphere

$$\sum_{i=1}^{n} y_i^2 = \frac{k}{\alpha^2}$$

flattened along the y_n axis. The "surface of elongation," corresponding to the line $y_2 = 0$ in two dimensions and the plane $y_0 = 0$ in the three dimensions, is the hyperplane $y_n = 0$. Improvement will be difficult from this hyperplane, and the effect will be magnified as the number of dimensions of the original problem is increased.

ABSOLUTE NORM

We defined

$$\| r \|_A = \sum_{i=1}^{n} | r_i |$$

Since no multiplications are required in computing the absolute norm, a great deal less time is taken than for the euclidean norm. Furthermore, the condition $\| r \|_A < \epsilon$ is more restrictive on the error vector r than the corresponding $\| r \|_E < \epsilon$ in the sense that if all $| r_i | < \dfrac{1}{n}, \| E < \| r \|_A$.

The absolute norm creates a special kind of "trap." Consider the linear problem

$$x_1 + x_2 + x_3 = 6$$

$$\frac{x_2}{10} + x_3 = 5$$

$$x_3 = 3.$$

The solution is $x_1 = -17, x_2 = 20, x_3 = 3$. We use the absolute norm of the error vector r.

Suppose we arrive at the trial solution $x_1 = -23 - 9k$, $x_2 = 20 + 10k$, $x_3 = 3 - k$ where $k > 0$. Then for this vector, $r_1 = r_2 = 0$ and $r_3 = k > 0$. So $\| r \|_A = k$.

Attempted improvement in x_1 or x_2 will force r_1 or r_2 positive without affecting r_3, thus increasing $\| r \|_A$. The attempted improvement from x_3 to $x_3 + \delta$, for any δ, leads to

$$\| r \|_A = 2 \mid \delta \mid + \mid k - \delta \mid.$$

But

$$2 \mid \delta \mid + \mid k - \delta \mid \geq 2 \mid \delta \mid + k - \mid \delta \mid = \mid \delta \mid + k \geq k$$

Hence no improvement is possible, when a single variable is changed, no matter how small the change.

This is a new kind of trap which we did not encounter when the euclidean norm was applied.

THE ERROR SURFACES ($\| r \|_A =$ constant)

In two dimensions we have the squares $\mid y_1 \mid + \mid y_2 \mid =$ constant

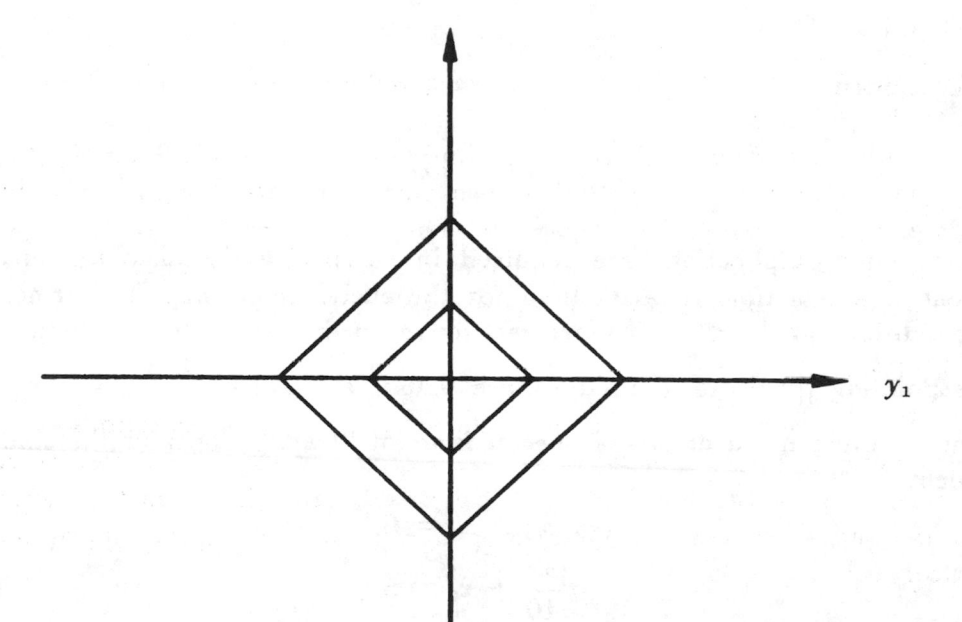

In higher dimensions any two dimensional subset is of this form. The set $\mid y_1 \mid + \mid y_2 \mid \ldots + \mid y_n \mid \leq$ constant is a convex polyhedron, and thus the fitness function is piecewise linear.

In order to circumvent the trap it may be possible to use a norm of the form

$$\| r \| = \sum_{i=1}^{n} \gamma_i \, | r_i |$$

where the γ_i are randomly chosen coefficients such that $\sum_{i=1}^{n} | \gamma_i | = n$. One could change these coefficients in different cycles. This is a topic for further study.

EQUATIONS IN INTEGERS

Previously we briefly considered the case where the x_i are restricted to integer values, and the solution is in integers. A variant of the method we have been describing is available for this situation. In it, the random changes are chosen from a continuous distribution. Thus starting from the zero vector, we make integer changes in one component at a time so as to decrease $\| r \|_A$. Another interesting form of trap occurs here.

Suppose we are trying to solve the trivial system

$$x_1 + x_2 = 1$$
$$x_2 = 1$$

Starting with $x = (0, 0)$, $r = (1, 1)$ we could improve to $x = (0, 1)$, $r = (0, 0)$, the solution.

We could also decrease $\| r \|_A$ by improving to $x = (1, 0)$, $r = (0, 1)$. From here, however, no further improvement in one component only is possible, since $\| r \|_A \geq 1$ if $\| r \| \neq 0$. This is the simplest case of "dependence," a phenomenon discussed by Bledsoe.[4] Two or more genes have to be changed simultaneously in order to solve this sort of problem.

DEPENDENCE

W. W. Bledsoe constructed a system of ten linear equations in ten unknowns with solution $x_1 = 1$, $x_2 = 2, \ldots, x_{10} = 10$. The system turned out to be very badly conditioned. Bledsoe used an evolution method similar to ours, and after many cycles the program arrived at the approximate solution

$$(0, 8, 10, 1, 9, 10, 8, 9, 10, 5)$$

(Changes were restricted to integers between 0 and 10).

Bledsoe found that in order to reduce the residue further, six of the 10 components had to be changed simultaneously. If each component is changed with a probability p, then a particular sixtuple is changed with probability p^6. If one works, for example, with $p = 10^{-1}$ (which was found to be optimal), then $p^6 = 10^{-6}$. Thus, in this method, one must expect something like a million futile tries before a single step of improvement is achieved.

If there are more than 10 variables, the problem gets exponentially worse. The number of different k-tuples of n components is $\binom{n}{k}$. For n large $\binom{n}{k} \approx 2^n$.

For example, if $n = 100$, then $\binom{n}{k} \approx 2^{100}$, thus there are about 2^{100} k-tuples. Any straightforward search through all possible k-tuples is out of the question, and relying on chance is even worse.

Thus the problem of *dependence* presents serious obstacles. Dependence means that the increments in which improvement can be obtained, lie in a lower dimensional submanifold.

It is clear that additional techniques have to be employed in order to circumvent the difficulties arising from poorly conditioned systems. Such techniques transform the method from a strictly evolutionary one into something else. In the spirit of the term "self-organizing," one would try to develop a program, such that the past performance in solving a given problem is taken into account. The program should "learn" that improvement is very difficult in certain directions, and be provided with a means of avoiding such directions.

LINEAR TREND IN TIME TAKEN

Two papers on the solution of linear equations have reported successful search methods for solving linear equations in which the time taken for solving a system of n equations increases linearly with n (Hooke and Jeeves[19] and Curtis[14]).

Curtis found that in theory the time required to correct one component of the solution vector by a *Monte Carlo* method was equal to n plus a fixed large number b. He predicted that there would be a certain value of n, depending on b, beyond which it would be more profitable to employ a Monte Carlo method rather than a standard elimination or iteration method.

We did not pursue the direction of the very interesting work of Hooke and Jeeves, and Curtis, but keeping closer to the example of biological evolution, we experimented with patterns of sexual-type evolution.

Biological evolution uses reshuffling of genotypes through mating on a grand scale. We thus turned to experiments with schemes of recombination and superposition of approximate solutions.

MATING

The methods discussed so far correspond to *asexual evolution*. We will now discuss methods resembling *sexual evolution*. The central feature of sexual evolution is mating: two or more individuals pool their genetic material. Each individual has either a double set of instructions (diploid) which is reduced to a single set of instructions (meiosis) before mating, or there is a single set of instructions (haploid). In this other case genetic material may be exchanged between two or several individuals, leading to a scrambling of the genotypes.

There is the question as to how a double set of instructions asserts itself in the case where some genes disagree. One possible mode is the following: One of the two states of a gene supersedes the other; for example—AA, aA and Aa all act like A, only aa acts like a. Another possible mode is that the result is intermediate. We will now describe various patterns of data processing that more or less resemble biological patterns of mating.

m-TUPLE MATING

Suppose $x^{(1)}, x^{(2)}, \ldots, x^{(m)}$ constitute an *m*-tuple of vectors, each of which has the binary components 0 or 1. There are a number of ways to mate them to form an "offspring vector" z. Note that for $m = 2$ we have pairwise mating.

One way to combine the $x^{(j)}$ is the following: choose

$$z_i = \begin{cases} 0 \text{ if half or more than half of the } x_i^{(j)} = 0 \\ 1 \text{ otherwise} \end{cases}$$

This could be termed "majority" mating since if the majority (or half) of the $x_i^{(j)} = 0$ for $j = 1, \ldots, m$, then the corresponding component z_i of the offspring is also zero. Obviously this method gives a majority decision in all cases only if m is odd.

MATING BY CROSSING OVER

Another type of *m*-tuple mating of binary component vectors is the following:

Choose m integers n_k, $k = 1, \ldots, m$, such that $0 \leq n_k \leq m$ and $\sum\limits_{k=1}^{m} n_k = n$.
Then choose n_k components from $x^{(k)}$, $k = 1, \ldots, m$, subject to the restriction that exactly one of the $x_i^{(k)}$ is chosen, for $i = 1, \ldots, n$.

For example, with $m = 3$ we could select $n_1 = n_2 = 1$, $n_3 = n - 2$, and take the first component of $x^{(1)}$, the last component of $x^{(2)}$, and the remaining ones from $x^{(3)}$.

$$
\begin{array}{ll}
x^{(1)} & \underbrace{x_1^{(1)}}, x_2^{(1)}, \ldots \qquad , x_n^{(1)} \\[4pt]
x^{(2)} & x_1^{(2)}, x_2^{(2)}, \ldots \qquad , \underbrace{x_n^{(2)}} \\[4pt]
x^{(3)} & x_1^{(3)}, \underbrace{x_2^{(3)}, \ldots, x_{n-1}^{(3)}}, x_n^{(3)} \\[4pt]
z & x_1^{(1)}, x_2^{(3)}, \ldots, x_{n-1}^{(3)} \; x_n^{(2)}
\end{array}
$$

We will call this latter method "mating by crossing over," because of the analogy to a biological pattern.

In our work with linear equations in integer variables, we performed several experiments, using majority mating and mating by crossing over, on approximate solutions. The results were somewhat inconclusive. No definite benefit was obtained.

CONTINUOUSLY VARYING COMPONENTS

If the components $x_i^{(j)}$ of trial solutions $x^{(j)}$ are not discrete, but vary continuously, mating by averaging is indicated. Quite simply, the average z is given by

$$
z_i = \frac{1}{m} \sum_{j=1}^{m} x_i^{(j)} \qquad i = 1, \ldots, n
$$

Weighted averages can also be formed, according to

$$
z_i = \sum_{j=1}^{m} \lambda_j x_i^{(j)} \qquad i = 1, \ldots n
$$

where

$$
\sum_{j=1}^{m} \lambda_j = 1
$$

A case in which the average or weighted average of several trial solutions yields real improvement has already been briefly mentioned.

Suppose we are trying to solve

$$
Ax = b,
$$

where $A = \alpha I + J, b = (1, 3, \ldots (2n-1))$ as previously, and suppose the condition number $\dfrac{\alpha + n}{\alpha} > 5$. Then we have seen that, for $n = 2$, the method we described for solving linear equations is likely to become stuck, or trapped, at a point on the major axis of the error ellipse $r^2 = k$. Let $x^{(1)}$ and $x^{(2)}$ be two distinct such points, obtained on separate runs of the method, with $x^{(1)}$ on the ellipse $r^2 = k_1$, $x^{(2)}$ on $r^2 = k_2$. Then the weighted average $z = \lambda x^{(1)} + (1 - \lambda) x^{(2)}$, will also be a point on the major axis, lying between the two. So if $x^{(1)}$ and $x^{(2)}$ happen to lie on opposite sides of the common center, there will be a value of λ, $0 \le \lambda \le 1$, for which z represents the solution; i.e., for which $r = 0$. For example, if $x^{(1)}$, $x^{(2)}$ are at opposite extremities of the *same* ellipse, $z = \dfrac{x^{(1)} + x^{(2)}}{2}$, their average will be the solution.

Even if $x^{(1)}$, $x^{(2)}$ are not on opposite sides, we have obtained the location of the long axis as the line joining them.

In general, for $n > 2$, and large condition number, knowledge of m stagnation points $x^{(1)}, \ldots, x^{(m)}$ which have occurred on the hyperplane $y_n = 0$ will enable us to gain a better approximation by averaging, in the case where these points are randomly distributed.

No large scale experiments were conducted. The inconclusiveness of those experiments that were conducted may be traced to a failure to achieve stagnation points that were distributed sufficiently at random, even though they were arrived at by "random walks." The starting points of these random walks were not sufficiently random.

Rather than pursuing this question further we turned instead to linear programming where the effect of mating by averaging can easily be seen to have spectacular effects.

PART III. FITNESS FUNCTIONS DERIVED FROM LINEAR PROGRAMMING PROBLEMS

The problem is to minimize

$$z = \sum_{i=1}^{n} c_i x_i + c_0 \tag{1}$$

subject to the linear inequalities

$$\sum_{j=1}^{n} a_{ij} x_j \le b_i \qquad i = 1, \ldots, m \tag{2}$$

where $m \ge n$, and the a_{ij}, b_i, c_i are real. z is called the *cost* or *cost function*.

The vector $x = (x_1, \ldots, x_n)$ is said to be a *feasible solution* if the x_i satisfy (2). The set of feasible solutions is convex. That is, if $x^{(1)}$ and $x^{(2)}$ are feasible, so is $\lambda x^{(1)} + (1 - \lambda) x^{(2)}$ for $0 \le \lambda \le 1$.

Hence the average of two or more feasible solutions is itself feasible. This fact forms the basis for the remainder of this paper.

We will call the hyperplane $\sum_{i=1}^{n} c_i x_i + c_0 = 0$ the *cost plane*. We will call the $n-1$ dimensional hyperplanes

$$\sum_{j=1}^{n} a_{ij} x_i = b_i \qquad i = 1, \ldots, m \tag{3}$$

faces. A *vertex* is then defined as the intersection of n faces. (We shall assume that no two faces are parallel, and that the intersection of any set of n hyperplanes is a well defined point.) Similarly, an edge is given by the intersection of $n-1$ faces. (Again we assume that any set of $n-1$ hyperplanes defines a line.)

Unless the cost plane is parallel to an edge, the optimal feasible solution is at a vertex. If two or more vertices are optimal, then the whole line segment connecting them is optimal.

SIMPLEX METHOD

The *simplex* method proceeds from a vertex $x^{(0)}$ to another vertex $x^{(1)}$ on the same edge, so as to decrease the cost. $x^{(1)}$ is said to be neighboring to $x^{(0)}$. A sequence $x^{(i)}$ of vertices is selected, each neighboring the former, and with a lower cost value (1). The simplex method is an algorithm. It converges to the solution in a finite number of steps (if the solution is unique).

As in the preceding discussion of linear systems of equations, we do not attempt to compete with the well established simplex algorithm. It is intended rather to test out the evolutionary idea and in particular mating, on a theoretically well understood problem.

BLIND SEARCH FOR OPTIMAL SOLUTION

It is of interest to formulate linear programming as a search problem.*
The m hyperplanes (3) determine $\binom{m}{n}$ vertices of which one (or on occasion several neighboring ones) is the required optimum. Obviously, one way to solve the problem is to search through all vertices and to see which one has lowest cost. This method, however, is not practical due to the large number of vertices.

---

* For a general discussion of search, compare Gelfand and Tsetlin.[18]

For sufficiently large n we can easily estimate $\binom{m}{n}$ by the approximation $\binom{m}{n} \approx 2^m$ valid for $m = 2n$. For $m = 100$, $n = 50$, $\binom{m}{n} \approx 2^{100} \approx 10^{30}$. Assuming that one microsecond* of computer time was necessary to locate, and then test each vertex for optimality, the total time for searching through all vertices would be approximately 3×10^{16} years! $m = 100$, $n = 50$, however, is a rather modest problem as linear programming goes.

DESCRIPTION OF EVOLUTION METHOD

To illustrate the method consider a problem in two variables, four equations, $m = 4$, $n = 2$ (Fig. 3). l_1, l_2, l_3 and l_4 denote the faces. We assume $c_0 = 0$ in the cost function. The line $z = c_1 x_1 + c_2 x_2 = 0$ passes through the origin, and the shaded area is the set of feasible solutions. Most linear programming problems include the requirement $x_j \geq 0$ for the variables. For simplicity's sake this requirement is omitted here.

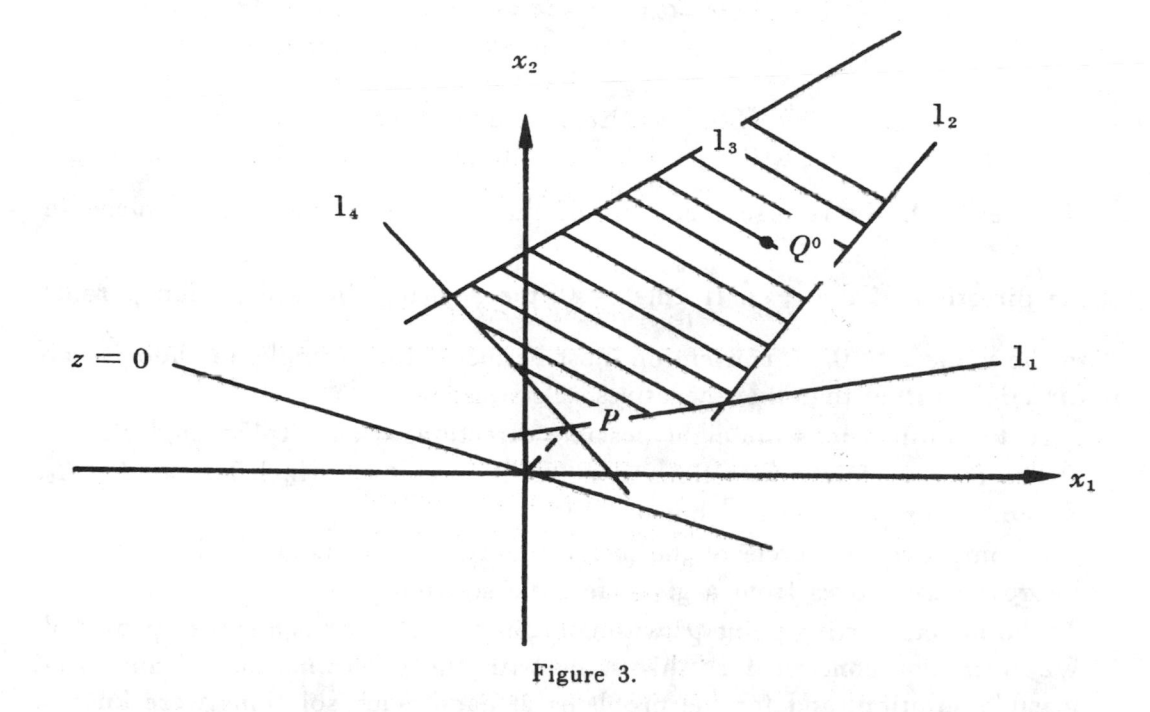

Figure 3.

The optimal solution in this case is the point P, the intersection of l_1 and l_4. Consider the feasible solution Q^0. Our method attempts to determine a

---

* Which is an extremely low estimate.

sequence of such improved feasible solutions starting at Q^0 that converges to P.

DESCRIPTION OF THE METHOD

In this way, e_i, $i = 1, \ldots, n$ is chosen randomly from -1, 0 or 1, and a direction in n space whose components are $e_i, \ldots, e_n$ is established. If $\sum_{i=1}^{n} c_i e_i < 0$, then a move from $x = (x_1, \ldots, x_n)$ to $x = (x_1 + \lambda e^1, \ldots, x_n + \lambda e_n)$ with $\lambda > 0$ is made with λ chosen as large as possible consistent with the condition that x' must be a feasible solution. Then z' corresponding to x' is given by

$$z' = \sum_{i=1}^{n} c_i x_i' = \sum_{i=1}^{n} c_i (x_i + \lambda e_i)$$

$$= \sum_{i=1}^{n} c_i x_i + \lambda \sum_{i=1}^{n} c_i e_i$$

$$= z + \lambda \sum_{i=1}^{n} c_i e_i < z.$$

If $\sum_{i=1}^{n} c_i e_i > 0$, we reverse the signs of the e_i; improvement takes place in the direction of the $-e_i$. If $\sum_{i=1}^{n} c_i e_i = 0$, the e_i determine a direction parallel to the plane $z = 0$. A convention must be made, for example, to choose such directions rather than their negatives.

In two dimensions the eight possible directions occur at $45°$ angles.

The four preferred directions at a feasible point Q^0 which lead to R_1, R_2, R_3, and R_4 on the faces are shown in Fig. 4.

An improvement cycle of the method consists of a fixed number, $3n$,* of improvement moves from a given feasible starting point.

The initial starting point Q^0 with coordinate vector X^0 is assumed provided. We were not concerned at this stage with the determination of an initial feasible solution, and for the problems at hand, such solutions were known.

---

* This number was chosen because a smaller one, say n, appears insufficient to provide enough direction for landing on most of the faces, and a larger one would take too much time.

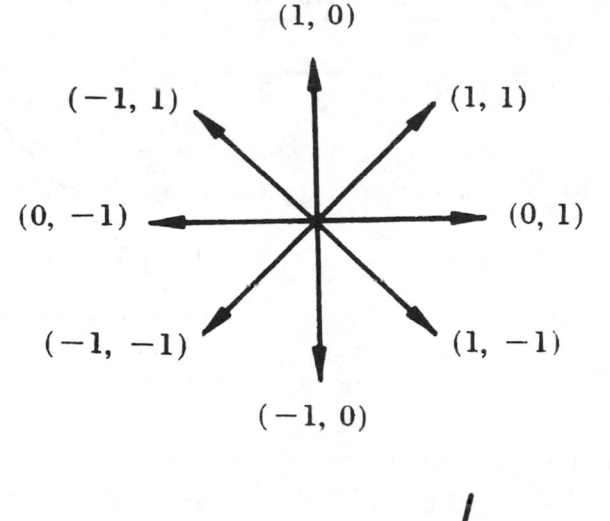

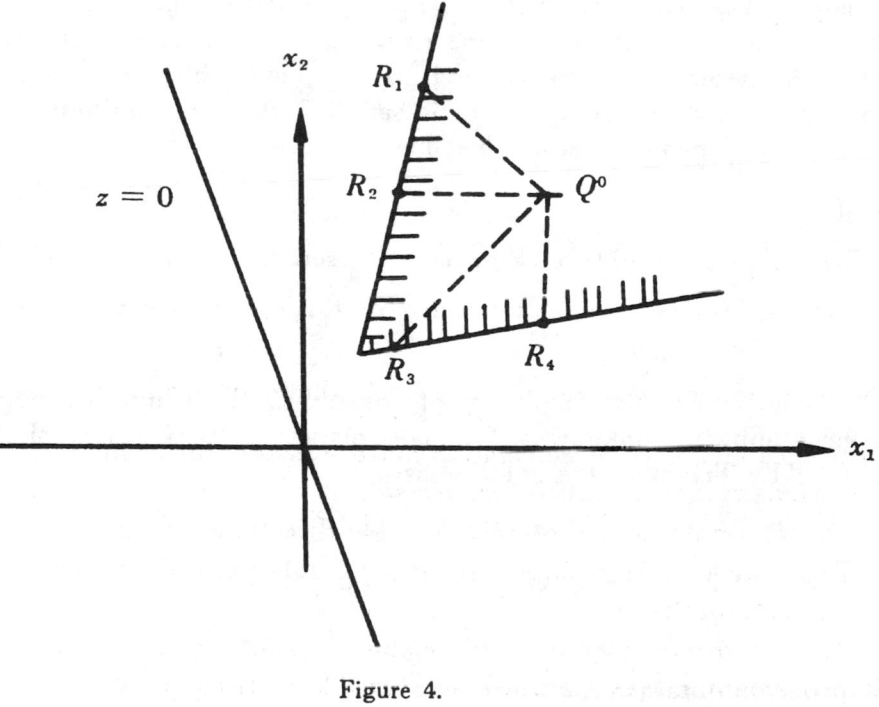

Figure 4.

After the first cycle of improvements there are in general slightly less than $3n$ new vectors (due to possible repetitions), each of which lies on a certain *face*. If more than one vector is on a given face, then we select that vector for which the associated cost is minimal. This set of m or fewer feasible solutions is then averaged. Due to convexity, this average furnishes a starting vector for the next cycle.

The Sequence of Operations

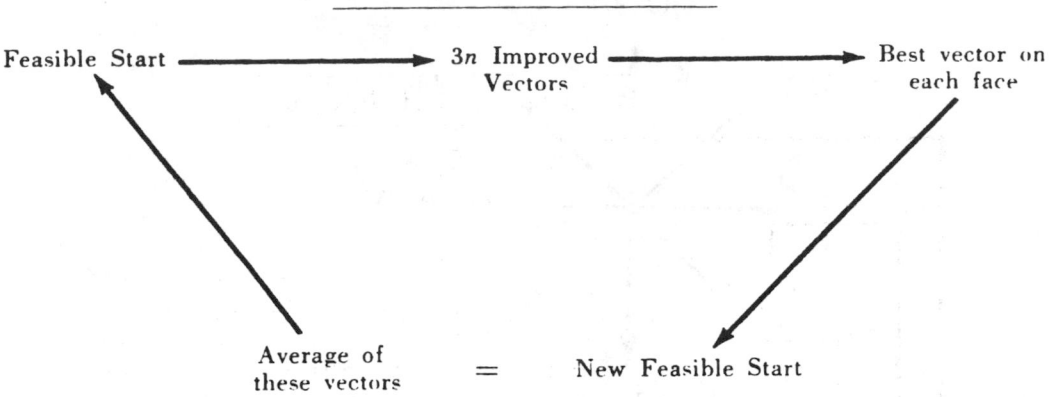

We formulate a criterion for termination of the method as follows. If the cost at the start of a cycle has not decreased by more than a fixed small $\epsilon > 0$ relative to the cost at the start of the preceding cycle, the method stops. A theory of convergence is not yet available, and there are many examples where this requirement is satisfied at a non-solution. Nevertheless the criterion seems a reasonable one to employ.

RESULTS

The following problem (Fig. 5) was used to test the method:

minimize $\qquad z = x_1 + x_2 + \ldots + x_{n-1} + 1.1x_n$

subject to $\qquad x_i \geq 1, \qquad 1 \leq i \leq n$

This is an especially simple set of constants, whose number n is the same as the number of unknowns. The cost plane $z = 0$ is skewed slightly by the factor 1.1. Trivially, the solution is

$$x_i = 1, \qquad 1 \leq i \leq n$$

With $n = 8$ from the given initial feasible vector

$$(1.51, 5.51, 9.51, 1.11, 6.01, 9.31, 1.11, 8.51)$$

the program obtained the approximation

$$(1.0003, 1.0004, 1.0005, 1.0004, 1.0001, 1.0003, 1.0003, 1.0003),$$

requiring an average of twenty cycles to reduce the error by one order of magnitude.*

---

* This result was actually obtained using a very slightly improved method developed later.

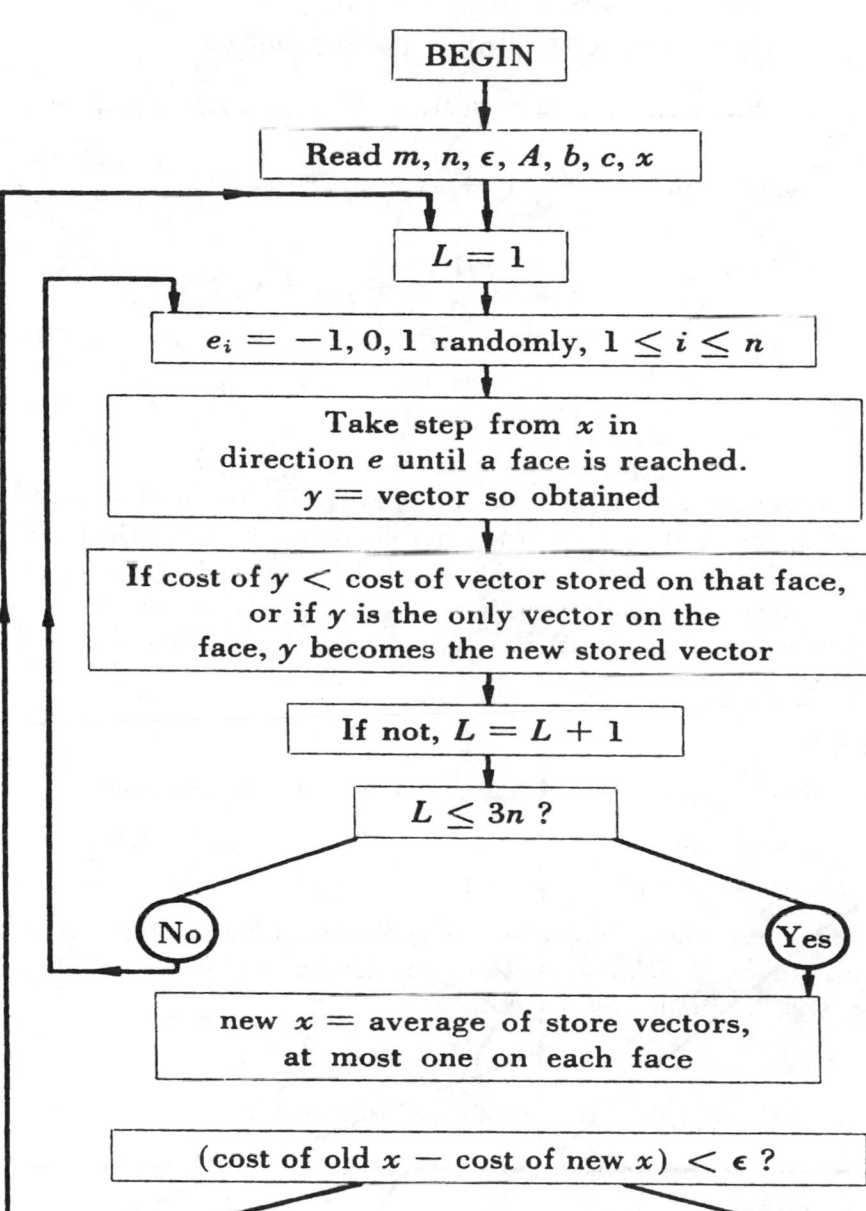

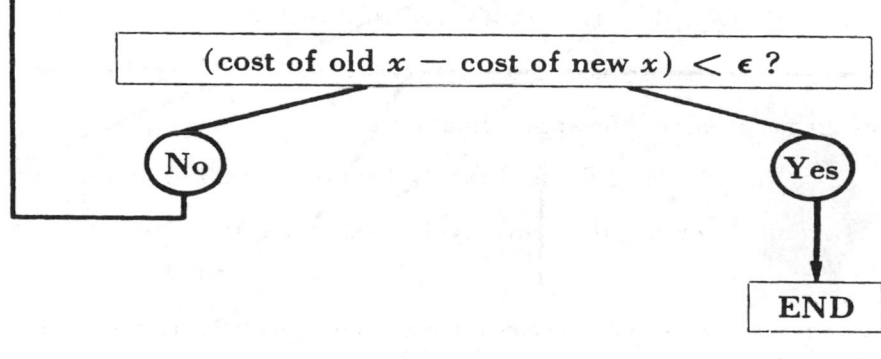

Figure 5.

A less trivial problem (Fig. 6) is the following:

minimize $\qquad z = x_1 + x_2 + \ldots + x_n$

subject to $\qquad \dfrac{1}{2}x_1 + x_2 + \ldots + x_n \geq 1$

$$x_1 + \frac{1}{2}x_2 + \ldots + x_n \geq 1$$

$$\cdots$$

$$x_1 + x_2 + \ldots + \frac{1}{2^z n} \geq 1$$

whose solution is

$$x_i = \frac{2}{2n-1}, \qquad 1 \leq i \leq n$$

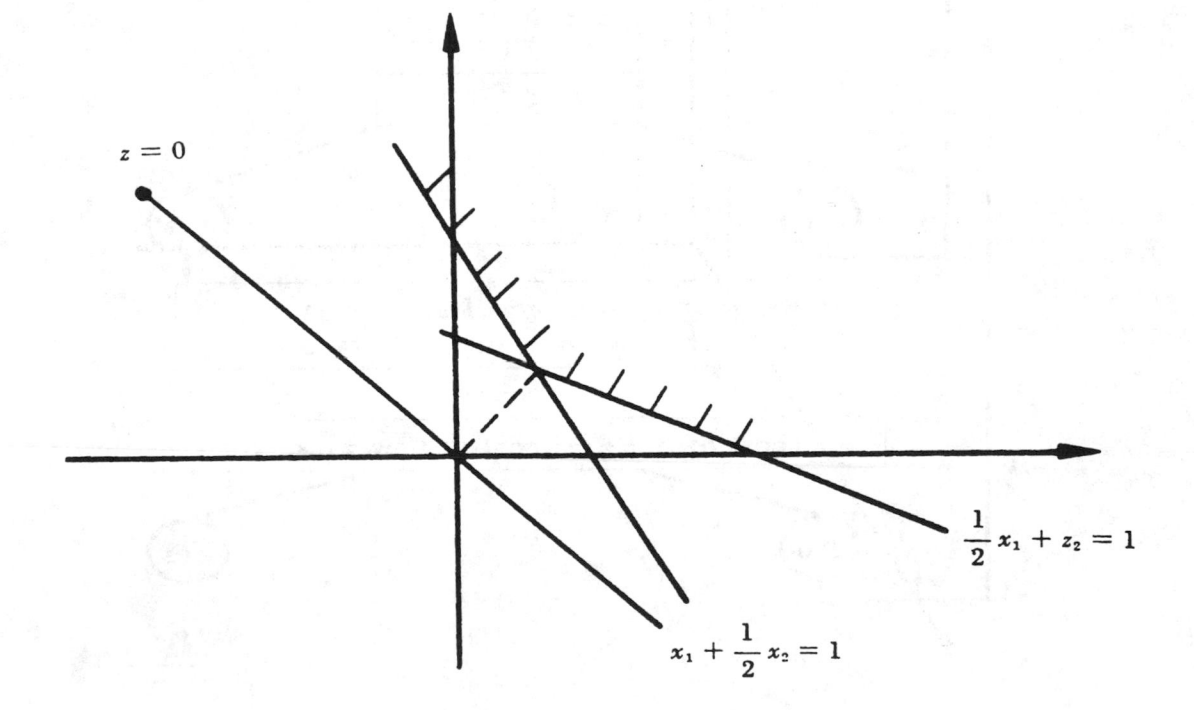

Figure 6.

With $n = 4$ the start

$$(1, 2, 1, 2, 1, 2, 1, 2, 1)$$

was provided, which is relatively close and symmetrically placed with respect to the solution $x_i = \dfrac{2}{17}$. Yet this solution could not be reached; instead the

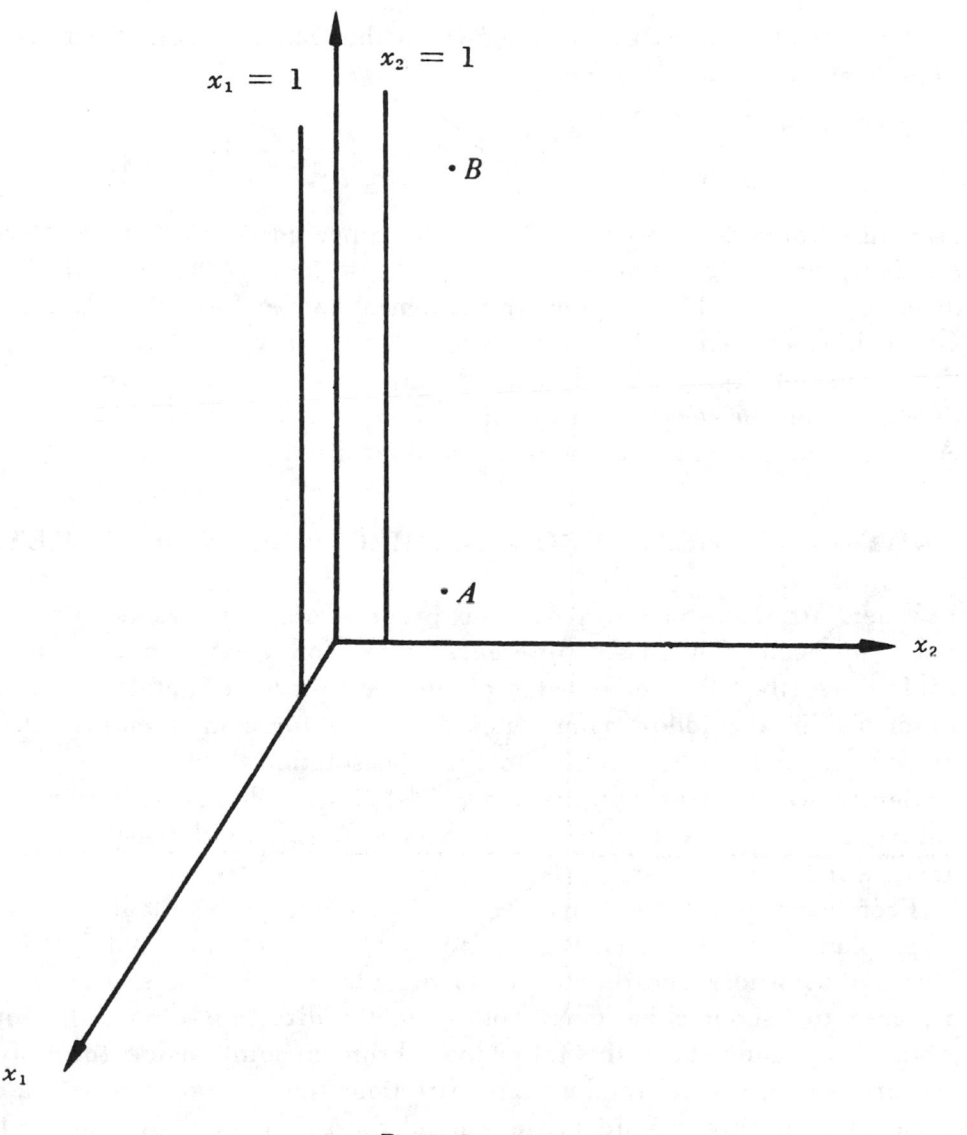

Figure 7.

average vector, after many attempts at improvement, was located close to the ninth face $x_1 + x_2 + \ldots + \dfrac{1}{2} x_9 = 1$. In examples like this one (where the faces are nearly parallel), we found that instead of approaching the true solution, the average vectors approached one of the faces at a point other than the solution.

EXAMPLE

The following is an example in which this can be clearly seen (Fig. 7). The three dimensional problem

minimize $x_1 + x_2 + 1.1x_3$

subject to $x_i \geq 1, \quad 1 \leq i \leq 3.$

Starting from A: $x_1 = x_2 = 2, x_3 = 3$, convergence to the solution P is relatively rapid. Starting from B: $x_1 = x_2 = 2, x_3 = 20$, there is only one direction $(0, 0, -1)$ in which improvement to the face $x_3 = 1$ is possible. Other improvements will confined to $x_1 = 1$ or $x_2 = 1$ and averaging of these will not decrease the x_3 component sufficiently. So convergence is very dependent on the selection of certain unique direction numbers from $-1, 0, 1$. An unsymmetric start compounds the difficulties.

PART IV. LINEAR PROGRAMMING, "BUCKSHOT METHOD"

When Mr. Rogson took over the programming, it was decided to try a new approach. The basic difference is the following: In the experiments so far described the components of the vector x were mutated. The method described in the following mutates directions for components of the vector from a fixed feasible solution to the trial solution.

Before we describe our method in detail we will briefly outline it in an intuitive manner. Our method assumes a given initial feasible solution Y_0 from which we can start.

From this point we travel in the direction perpendicular to the cost hyperplane until we reach the boundary of the constraint polyhedron. After this point. whose coordinate vector we denote by x, has been found. we proceed to "shoot a buckshot volley" in the direction $x - y_0$. By "buckshot volley" we understand the following: From a point inside the polyhedron we proceed in several random trial directions that are contained in a circular cone of some given solid angle around the given fixed direction. The distribution of the random directions is analogous to the trajectories traced by individual grains of a blast of buckshot. We then determine the intersection points of the trajectories with the polyhedron. The coordinate vectors of these intersection points are feasible vectors since they satisfy $Ax \leq b$ (with the equal sign occurring in at least one component). Among these vectors there is, we hope, one vector x' having a cost that is less than the cost of x and also such that the norm of the difference between x and x'; i.e., $\sum_{i=1}^{n} (x'_i - x_i)^2$ is greater than a certain ϵ which we will accept as the smallest change in vectors.

If the norm of the difference is greater than ϵ and we have not exhausted a given maximum number of loops we then exchange x' and x and shoot another "volley" around the new direction determined by the new x and y_0. This method is continued until we either are not improving our approximation fast enough; that is, $\sum(x'_i - x_i)^2 \leq \epsilon$, or we exceed the maximum number we allowed for iterations.

In what follows we shall try to describe the above mentioned "buckshot" technique. To simplify the description we want to introduce some further notation: by e_i we will understand the vector that has a 1 in the ith component and 0's elsewhere. In other words:

$$e_i = (\delta_{i1}, \delta_{i2}, \ldots, \delta_{ij}, \ldots, \delta_{in}) \qquad \text{where } \delta_{ij} = \begin{cases} 1 \text{ if } i = j \\ 0 \text{ if } i \neq j \end{cases}$$

is Kronecker's *delta*. For the "buckshot" effect we need to have some dimensions for the cone which will determine how "wide" or "open" the buckshot will be. For this we read in two factors h, l, a high and a low factor, respectively. The factor h will be multiplied by a random number, uniformly distributed between 0 and 1, during the course of each buckshot. The factor l will be set to $\frac{1}{2} \sum_{i=1}^{n} (x'_i - x_1)^2$ if $l < \sum_{i=1}^{n} (x'_i - x_i)^2$, to reduce the small cone and increase our rate of convergence as the approximation comes closer to a limit point. The process starts as follows: We generate a set of feasible vectors by modifying the direction given by y_0 and x, our initial feasible vector and our current best approximation, respectively, by adding to it:

$$d = y_0 - x + fe_i \qquad \text{for } i = 1, \ldots, n$$

where f is h or l, the high and low factors, respectively. This gives us for each of the n-dimensions 4 new vectors and since we include x in the family we have $4n + 1$ feasible vectors. These vectors, which we shall call the "parents," are then "pairwise mated." If we denote by x^j the members of the parents set, $j = 1, \ldots, 4n+1$, then by pairwise mating we mean the follow-

347

ing: We determine the coordinate vector z of the intersection of the lines through y_0 in the direction

$$d = y_0 - 0.5(x^j + x^k) \qquad \text{where } 1 \leq j \leq 4n+1$$
$$1 \leq k \leq 4n+1 \text{ and } j \neq k$$

with the boundary of the constraint polyhedron. By pairwise mating we thus obtain $\binom{4n+1}{2} = 2n(4n+1)$ vectors. We note that for large values of n this method implies a prohibitively large number of operations. To somewhat counteract this problem we have introduced a modification which will enable us to skip over a number of parents in the "mating" process. We have thus been able to analyze some higher dimensional cases. By ranking this family of vectors according to the cost we can proceed to find the one with lowest cost function value and proceed with the iteration method described above. We shall now give a description of how the method works in a step by step manner:

1. Given the initial feasible vector y_0, the constraints $Ax \geq b$ and the cost function $cx = z$, our first step is to find the vector x in the direction normal to the plane defined by the cost function. This vector x now becomes the first vector in our scheme and the iteration cycle begins.

2. We take the direction determined by y_0, our initial feasible vector, and our current best approximation vector x:

$$d = y_0 - x = (y_{0_1} - x_1, \ldots, y_{0_n} - x_n)$$

and effect our buckshot effect around this direction.

3. From the family generated in step 2 we pick the vector x' which has the least cost and we evaluate

$$t = |x' - x| = \sqrt{\sum_{j=1}^{n} (x'_j - x_j)^2}$$

If $t > \epsilon$ then we proceed to exchange x and x' and check whether this was the last permissible iteration. We next proceed to check if our running low factor for the buckshot effect, l, is less than t; if it is then we set $l = (1/2)t$, otherwise we leave it the same. Then, if we have not exceeded the maximum number of iterations we return to step 2. If not we finish this run by saving the information gathered so far. On the other hand, if $t < \epsilon$ then we assume that our improvement is too slow and we also proceed to save the information and exit.

We have so far repeatedly mentioned that we travel along a given direction through the initial feasible vector until we hit a face of the polyhedron. For a detailed description of an implementation of this task, see the flowchart or Rogson.[25] For the initialization of the technique we give as a direction vector the normal to the cost plane; i.e., the plane defined by the linear cost function c. We do this in order to get an "optimal" start to the method since further directions are going to be defined by the succeeding approximations together with the given initial feasible vector y_0. Before reporting the experimental data obtained from the method discussed above we will describe a few changes which were introduced during the course of experimentation:

1. Instead of travelling into a "mutant direction," until we hit a face of the polyhedron, we minimized the cost in the subspace defined by d and the previous best vector. This modification did not affect the rate of convergence at all among the analyzed problems.

2. We recall that the "buckshot" effect was accomplished by generating a set of "parents" by (asexual) mutations on our last best vector. These parents are then mated in a pairwise manner. We introduced the following change: If we find, while producing a family of mutants for mating, a mutant whose cost is smaller than the cost of the last best vector minus ϵ, then we take this mutant as the new best vector, and continue with a new iteration cycle. This change, seemingly minor in character, allowed for a tremendous upsurge in the time rate of improvement—that is, timewise the convergence was greatly accelerated. Because of this we were able to handle problems with almost complete success which previously were not feasible. Furthermore, in those cases where the unmodified version worked well there was also an increase in the accuracy of convergence.

3. A third change that was suggested for the buckshot effect was not to "mate" parents which lie on the same hyperplane. That is, if we let x and z denote two arbitrary parents we would not mate them if for some i

$$b_i - A_i x = 0 = b_i - A_i z$$

   This change does not seem to affect the experiments in any measurable way.

4. The last suggested change was to move the original vector y_0 from its static location to a more centralized location. Several different methods were attempted, none of which showed any improvement on preliminary tests.

PART V. APPENDIX: OPTIMAL MUTATION PROBABILITY

Suppose we have a fitness function that is a function of n binary variables and is a monotone function of the Hamming distance of its argument vector from the solution vector: $f(x) = g(\| x - x^{(0)} \|)$, where g is a monotone function defined on the integers $0, \ldots n$, $x^{(0)}$ is the solution vector,

$$\| x - x^{(0)} \| = \sum_{j=1}^{n} | x_j - x_j{}^{(0)} |$$

Suppose each component of x mutates with a probability p. Let $I(p)$ be the probability that the resulting mutant constitutes an improvement over its "parent." For what mutation probability p does $I(p)$ assume a maximum? This question can be answered easily.

Suppose ζ of the n components of x disagree with $x^{(0)}$ while the others agree. A mutation among the components that agree will produce a degeneration while a mutation among the ζ "bad" components leads to an improvement.

The probability for at least one improvement in the "bad" part and no degeneration in the "good part" is:

$$I_0(p) = (1 - (1 - p)\zeta)(1 - p)^{n-\zeta} = (1 - p)^{n-\zeta} - (1 - p)^n$$

$I(p)$ is composed of this part plus $I_1(p) + I_2(p) + \ldots + I_{\zeta-1}(p)$, the probability of having exactly one degeneration combined with at least two improvements, etc.,

$$I_1(p) = \left[1 - (1 - (1-p)^2)^{(n-\zeta)(n-\zeta-1)/2} \right] \left[1 - (1-p^2)\frac{\zeta-1}{2} \right],$$

etc. developing $I_0(p)$ into a binomial series we find $I_0(p) = \zeta p + \zeta \left(\frac{\zeta-2n+1}{2} \right) p^2 + \ldots$. Developing $I_1(p)$ we observe that the first factor is less than 1 while the second gives $\frac{(\zeta-1)}{2} p^2 + \ldots$.

We have $\dfrac{dI_0(p)}{dp} = (n-\zeta)(1-p)^{n-\zeta-1} - n(1-p)^{n-1} = 0$. Solving for p we have $1 - \dfrac{\zeta}{n} = (1-p)\zeta$ and $p = 1 - \sqrt[\zeta]{1 - \dfrac{\zeta}{n}}$

Assuming that ζ is small compared with n we have $p \approx \dfrac{1}{n}$.

For this value, however, the contribution of $I_2(p)$ is small as compared with $I_0(p)$. The development of $I_3(p)$ begins with p^3, etc. Neglecting these contributions we find: If ζ is small as compared with n, then $I(p)$ takes its maximum for $p = \dfrac{1}{n}$, that means that in the average, one gene per individual per generation is mutated.

In Drosophila, surprisingly enough this condition is nearly satisfied. The estimated number of genes is 10^4. The observed mutation rate (W. S. Stone[27]) per locus is 10^{-4} to 10^{-5}. Taking into consideration that some genes have more than two alleles, this constitutes excellent agreement.

REFERENCES

1. W. Ross Ashby, "Constraint Analysis of Many Dimensional Relations," Technical Report No. 2, Grant AF-AFOSR7-63, Urbana, Ill., May 1964.

2. W. W. Bledsoe, "A Quantum-Theoretical Limitation of the Speed of Digital Computers," *IRE Trans. Elec. Comp.*, vol. EC-10, no. 3, Sept. 1961.

3. W. W. Bledsoe, "The Use of Biological Concepts in the Analytical Study of Systems," Technical Report, Panoramic Research Inc., Palo Alto, Calif., 1961.

4. W. W. Bledsoe, "Lethally Dependent Genes Using Instant Selection," Technical Report, Panoramic Research Inc., Palo Alto, Calif., 1961.

5. H. J. Bremermann, "The Evolution of Intelligence," ONR Technical Report No. 1, Contract Nonr 477 (17), University of Washington, Seattle, 1958.

6. ———, "Limits of Genetic Control," *IEEE Transactions on Mil. Electronics*, vol. MIL-7, April-July 1963, pp. 200-205.

7. ———, "The Principle of Genetic Cost," Technical Report, Contracts Nonr 22(85) and Nonr 3656(08), 1963.

8. ———, "Non-Genetic Control in Biological Systems," Technical Report, Contracts Nonr 222(85) and Nonr 3656(08), 1963.

9. ———, "Optimization through Evolution and Recombination," *Self-Organizing Systems 1962*, Yovits, Jacobi, and Goldstein, eds., Spartan Books, Wash., D. C., 1962.

10. ———, "Quantitative Aspects of Goal-Seeking Self-Organizing Systems," to appear in vol. 1, no. 1, *Progress in Theoretical Biology*, 1966.

11. H. J. Bremermann, M. Rogson and S. Salaff, "Search by Evolution," *Biophysics and Cybernetic Systems*, M. Maxfield, A. Callahan, L. J. Fogel, eds., Spartan Books, Wash., D. C., 1965.

12. H. J. Bremermann and S. Salaff, "Experiments with Patterns of Evolution," Technical Report, Contracts Nonr 22(85) and 3656(08), Berkeley, 1963.

13. H. J. Bremermann and M. Rogson, "An Evolution-Type Search Method for Convex Sets," Technical Report, Contracts Nonr 222(85) and 3656(08), Berkeley, 1964.

14. J. Curtis, "A Theoretical Comparison of the Efficiencies of Two Classical and a Monte Carlo Method for Computing One Component of the Solution of a Set of Linear Algebraic Equations," *Symposium on Monte Carlo Methods*, Wiley, New York, 1954.

15. Ferguson and G. Dantzig, "The Allocation of Aircraft to Routes: An Example of Linear Programming Under Uncertain Demand," *Manag. Sci.*, vol. 6, pp. 45-73.

16. R. A. Fisher, *The Theory of Inbreeding*, Oliver and Boyd, London, 1949.

17. R. M. Friedberg, "A Learning Machine," *IBM J. Res. and Dev.*, part I, vol. 2, pp. 2-13, Jan 1958, part II, vol. 5, pp. 183-191, July 1959.

18. I. M. Gelfand and M. L. Tsetlin, "Some Methods of Control for Complex Systems," *Russian Math. Surveys*, vol. 17, no. 1, 1961.

19. R. Hooke and T. Jeeves, "Direct Search—Solution of Numerical and Statistical Problems," *Journal of the A.C.M.*, April 1961.

20. Albert Madanski, "Inequalities for Stochastic Programming Problems," *Manag. Sci..* vol. 6, no. 2, pp. 197-205.

21. P. A. P. Moran, *The Statistical Processes of Evolution Theory*, Clarendon Press, Oxford, 1962.

22. ———, "Unsolved Problems in Evolutionary Theory," to appear in *Proceedings of the Fifth Berkeley Symposium on Mathematical Statistics and Probability*.

23. M. W. Nirenberg, "The Genetic Code II," *Scientific American*, vol. 20, no. 3, Mar. 1963, pp. 80-94.

24. G. V. Novotny, "Machines That Think Like People," *Electronics*, July 13, 1964.

25. M. Rogson, "A Search Method in Convex Programming," Technical Report, Contract Nonr 222(85), Berkeley, Feb. 1965.

26. C. Stern, *Principles of Human Genetics*, 2d Ed., Freeman, San Francisco, 1960.

27. W. S. Stone, "The Dominance of Natural Selection and the Reality of Superspecies (Species Groups) in the Evolution of Drosophila," *Studies in Genetics*, Univ. of Texas, publ. No. 6205, Mar. 1962.

28. J. von Neumann and H. H. Goldstine, "Numerical Inverting of Matrices of High Order," *Bulletin of the AMS*, Nov. 1947.

29. Wilde, *Optimum Seeking Methods*, Prentice-Hall Inc., Englewood Cliffs, N. J., 1963.

Chapter 10
Evolutionary Algorithms
for System Identification

H. Kaufman (1967) "An Experimental Investigation of Process Identification by Competitive Evolution," *IEEE Trans. Systems Science and Cybernetics,* Vol. SSC-3, No. 1, June, pp. 11–16.

TO effectively control a system, one must first know something about it. Mathematical descriptions of dynamic systems are often sought in order to achieve control of a plant. System identification concerns the problem of designing and optimizing these mathematical models. Successful identification relies on determining an appropriate model structure and associated parameters given observed data and the known physics of the plant. The typical procedure is for the engineer to postulate a particular model form and apply a numerical technique to optimize the parameters of the model in light of a performance index (often mean squared error).

Caines (1988) regarded system identification as the invention and evaluation of scientific theories—that is, identification is performed using the scientific method. Suppose that some unknown aspects of a system are to be estimated. Data are collected in the form of previous observations or known results, and combined with newly acquired measurements. After examining the data for erroneous components, a class of models of the system that is consistent with the data is generalized. This is an inductive process. The class is then reduced and more formally described with fixed parameters which are typically deduced. This specific hypothesized model (or models) is then tested in its ability to predict the behavior of the system, this process iterating until a sufficient level of credibility is attained. Observation of the attributes of new data, taken together with an estimate of the confidence with which this observation appears to confirm the predicted behavior of the system, constitutes a test of the hypothesized model. If the confidence exceeds some appropriate level, the model has been "confirmed" (Ornstein, 1965). The recognition that there are similarities between the processes of evolution and the scientific method suggests that simulated evolution could be used for system identification, generating models of arbitrary input-output devices (Fogel, 1964; Fogel et al., 1966, pp. 108–112).

The paper reprinted here, Kaufman (1967), extended the use of simulated evolution in Fogel et al. (1966) by applying it to the problem of process identification in the spectral domain. The task was to devise a model that would generate a best fit to a chosen observable process in light of a performance index. The proposed evolutionary algorithm differed from the earlier evolutionary programming (Fogel et al., 1966) by operating on transfer functions and providing the possibility of randomly recombining functional blocks of different models (possibly more than two). The effort incorporated the concept of a population of contending models and was shown to be effective at modeling a simple linear plant and a more complicated, nonanalytical plant. Kaufman (1967) also represents one of the earliest efforts to incorporate a hybrid method that coupled an evolutionary search for model structure with a deterministic procedure to adjust model parameters (sequential estimation, Detchmendy and Sridhar, 1965). This general approach of using evolution to provide an initial framework to then be fine tuned by local optimization techniques has been incorporated in several more recent efforts (Harp et al., 1989; Miller et al., 1989; Belew et al., 1992; Yao and Liu, 1996; and others).

Subsequent research in evolutionary algorithms for system identification focused mainly in the time domain. Early attempts by Lutter and Huntsinger (1969) and Burgin (1973) evolved finite state machines to model temperature data from a water cooling tower and characteristics of a flight test of an X-15, respectively. The use of finite automata as models naturally suggested the development of n-step-ahead predictors, essentially offering the potential to generate difference equations that capture the underlying logic of the considered plants. It is somewhat remarkable, however, that subsequent research in evolutionary identification using more generalized model structures has also been concentrated primarily in the time domain rather than the frequency domain. Perhaps this is in part due to the attention that was given to chaotic systems beginning in the late 1980s (Gleick, 1987; Devaney, 1989; Casdagli and Eubank, 1991; and others); spectral analysis of chaotic systems averages out the dynamics of interest.

Fogel (1990, 1991, 1992; and others) used evolutionary programming to simultaneously adapt the structure and parameters of auto-regressive moving-average models (ARMAX) of signals, particularly with concern to impulsive signals of short duration. Information criteria (Akaike, 1974; Risannen, 1978) were applied to trade off the goodness-of-fit provided by a particular model with the number of degrees of freedom. The results indicated that evolutionary optimization could yield suitable models where gradient searches for optimal parameter sets would fail. This approach has also been pursued in Yang et al. (1996) and others.

Efforts at evolving parameter estimates for fixed models from presumed and real-world mechanical systems include Johnson and Husbands (1991), Kim et al. (1996), and others. Also of related interest are efforts to use evolutionary algorithms to design infinite impulse response (IIR) and finite impulse response (FIR) filters (e.g., Kristinsson and Dumont, 1992; Flockton and White, 1993; and others) as well as the evolution of a variety of data structures for predicting chaotic times series (Kargupta and Smith, 1991; Iba et al., 1993; McDonnell and Waagen, 1994; Mulloy et al., 1996; and others). The general success of evolving models for predicting or identifying temporal systems, particularly those displaying chaotic behavior, should be considered well established.

References

[1] H. Akaike (1974) "A new look at the statistical model identification," *IEEE Trans. Auto. Control,* Vol. 19:6, pp. 716–723.

[2] R. K. Belew, J. McInerney, and N. N. Schraudolph (1992) "Evolving networks: Using the genetic algorithm with connectionist learning," *Artificial Life II,* C. G. Langton, C. Taylor, J. D. Farmer, and S. Rasmussen (eds.), Addison-Wesley, Reading, MA, pp. 511–547.

[3] G. H. Burgin (1973) "Systems identification by quasilinearization and by evolutionary programming," *J. Cybernetics,* Vol. 3:2, pp. 56–75.

[4] P. E. Caines (1988) *Linear Stochastic Systems,* John Wiley & Sons, NY.

[5] M. Casdagli and S. Eubank (eds.) (1991) *Nonlinear Modeling and Forecasting,* SFI Studies in the Sciences of Complexity, Proc. Vol. XII, Addison-Wesley, Reading, MA.

[6] P. M. Detchmendy and R. Sridhar (1965) "Sequential estimation of states and parameters in noisy nonlinear dynamical systems," *JACC preprints* (cited in Kaufman, 1967).

[7] R. L. Devaney (1989) *An Introduction to Chaotic Dynamical Systems,* 2nd ed., Addison-Wesley, Redwood City, CA.

[8] S. J. Flockton and M. S. White (1993) "Pole-zero system identification using genetic algorithms," *Proc. of the 5th Intern. Conf. on Genetic Algorithms,* S. Forrest (ed.), Morgan Kaufmann, San Mateo, CA, pp. 531–535.

[9] D. B. Fogel (1990) "System identification through simulated evolution," master's thesis, UCSD.

[10] D. B. Fogel (1991) *System Identification through Simulated Evolution: A Machine Learning Approach to Modeling,* Ginn Press, Needham, MA.

[11] D. B. Fogel (1992) "Using evolutionary programming for modeling: An ocean acoustic example," *IEEE J. Oceanic Eng.,* Vol. 17:4, pp. 333–340.

[12] L. J. Fogel (1964) "On the organization of intellect," Ph.D. diss., UCLA.

[13] L. J. Fogel, A. J. Owens, and M. J. Walsh (1966) *Artificial Intelligence through Simulated Evolution,* John Wiley, NY.

[14] J. Gleick (1987) *Chaos: Making a New Science,* Viking, NY.

[15] S. A. Harp, T. Samad, and A. Guha (1989) "Towards the genetic synthesis of neural networks," *Proc. of the 3rd Intern. Conf. on Genetic Algorithms,* J. D. Schaffer (ed.), Morgan Kaufmann, San Mateo, CA, pp. 360–369.

[16] K. Kristinsson and G. A. Dumont (1992) "System identification and control using genetic algorithms," *IEEE Trans. Systems, Man and Cybernetics,* Vol. 22:5, pp. 1033–1046.

[17] H. Iba, T. Kurita, H. de Garis, and T. Sato (1993) "System identification using structured genetic algorithms," *Proc. of the 5th Intern. Conf. on Genetic Algorithms,* S. Forrest (ed.), Morgan Kaufmann, San Mateo, CA, pp. 279–286.

[18] T. Johnson and P. Husbands (1991) "System identification using genetic algorithms," *Parallel Problem Solving from Nature,* H.-P. Schwefel and R. Männer (eds.), Springer, Berlin, pp. 85–89.

[19] H. Kargupta and R. E. Smith (1991) "System identification with evolving polynomial networks," *Proc. of the 4th Intern. Conf. on Genetic Algorithms,* R. K. Belew and L. B. Booker (eds.), Morgan Kaufmann, San Mateo, CA, pp. 370–376.

[20] H. Kaufman (1967) "An experimental investigation of process identification by competitive evolution," *IEEE Trans. Systems Science and Cybernetics,* Vol. SSC-3, No. 1, June, pp. 11–16.

[21] J.-H. Kim, H.-K. Chae, J.-Y. Jeon, and S.-W. Lee (1996) "Identification and control of systems with friction using accelerated evolutionary programming," *IEEE Control Systems Magazine,* Vol. 16:4, pp. 38–47.

[22] K. Kristinsson and G. A. Dumont (1992) "System identification and control using genetic algorithms," *IEEE Trans. Systems, Man and Cybernetics,* Vol. 22:5, pp. 1033–1046.

[23] B.E. Lutter and R. C. Huntsinger (1969) "Engineering applications of finite automata," *Simulation,* Vol. 13, pp. 5–11.

[24] J. R. McDonnell and D. Waagen (1994) "Evolving recurrent perceptrons for time-series modeling," *IEEE Trans. Neural Networks,* Vol. 5:1, pp. 24–38.

[25] G. F. Miller, P. M. Todd, and S. U. Hegde (1989) "Designing neural networks using genetic algorithms," *Proc. of the 3rd Intern. Conf. on Genetic Algorithms,* J. D. Schaffer (ed.), Morgan Kaufmann, San Mateo, CA, pp. 379–384.

[26] B. S. Mulloy, R. L. Riolo, and R. S. Savit (1996) "Dynamics of genetic programming and chaotic time series prediction," *Genetic Programming 1996: Proc. of the 1st Annual Conference,* J. R. Koza, D. E. Goldberg, D. B. Fogel, and R. L. Riolo (eds.), MIT Press, Cambridge, MA, pp. 166–174.

[27] L. Ornstein (1965) "Computer learning and the scientific method: A proposed solution to the information theoretical problem of meaning," *J. of the Mount Sinai Hospital,* Vol. 32:4, pp. 437–494.

[28] J. Rissanen (1978) "Modeling by shortest data description," *Automatica,* Vol. 14, pp. 465–471.

[29] H.-T. Yang, C.-M. Huang, and C.-L. Huang (1996) "Identification of ARMAX model for short term load forecasting: An evolutionary programming approach," *IEEE Trans. Power Systems,* Vol. 11:1, pp. 403–408.

[30] X. Yao and Y. Liu (1996) "Evolutionary artificial neural networks that learn and generalise well," *Proc. of 1996 Intern. Conf. on Neural Networks: Plenary, Panel and Special Sessions,* IEEE Press, NY, pp. 159–164.

IEEE TRANSACTIONS ON SYSTEMS SCIENCE AND CYBERNETICS   VOL. SSC-3, NO. 1   JUNE 1967

An Experimental Investigation of Process Identification by Competitive Evolution

HOWARD KAUFMAN, MEMBER, IEEE

Abstract—The feasibility of using evolutionary techniques to construct a model for a given process was investigated. Programs were written to synthesize six competing models and to adjust, combine, and eliminate these according to their performance with respect to the actual process. Results show that although the use of evolutionary techniques is promising for identifying processing having nonanalytic properties, it is also, at the same time, very costly with respect to computer time.

INTRODUCTION

ATTENTION has recently been given to the application of the principle of biological evolution towards the solution of decision-oriented problems.[1] To apply this principle to decision-making problems, a competing *colony* of solutions, along with a test for survival and a means for reproduction, are necessary. Initially, a colony of solutions can be randomly created and, if necessary, adjusted to perform a certain function. After this stage, the solutions are compared according to some performance index and unsatisfactory ones eliminated. Finally, new solutions must be produced from mutations and from selective combinations of the remaining solutions.

Process identification is one such decision-oriented problem suitable for an evolutionary means of solution. Contemporary practical identification schemes include cross-correlation techniques, input-output sampling, matched filter identification, and pulse-transfer techniques.[2] These methods yield satisfactory approximations to most linear and to several nonlinear plants, but do not in general give good models for plants containing highly nonlinear or nonanalytic blocks such as saturation and dead zones. Thus, one may conclude that trial and error techniques, which consider several of the possible plant analyticities, might be successful. One such technique involves the use of evolutionary principles.

The distinguishing characteristic of an evolutionary procedure for process identification is the dynamic competition among many models to optimize a performance index which relates their conformance with the process. Those models which do not satisfactorily duplicate the process characteristics are *exterminated* while those that perform satisfactorily are allowed to remain and reproduce. Thus, beginning with a randomly *created* colony of models, a statistically steady-state colony might be reached in which most members closely approximate the given process.

Manuscript received November 17, 1966. This paper is based on work performed under an internal research project at Cornell Aeronautical Laboratory, Buffalo, N. Y.
The author is with Cornell Aeronautical Laboratory, Buffalo, N. Y.

THEORY

To establish a specific evolutionary procedure for plant identification, let the problem be posed as follows: Given an unknown single-input, single-output plant with measurable input and output quantities, establish a model, whose dynamics are well defined, and which yields the same output as the plant in response to the same input applied to the plant.

The proposed evolutionary method of solving this problem utilizes a collection or colony of several models in which the performance of each model is competitively rated. Initially this colony is to be statistically created from a stockpile of transfer function blocks as shown in Fig. 1.

Each transfer function block is intuitively assigned a probability corresponding to the expected complexity of the plant. Then each of the N models in the original colony is constructed from a random number of randomly chosen transfer function blocks in accordance with the assigned probabilities.

Upon establishment of this initial colony, each model must go through a learning or adjustment phase during which each of its parameters (time constants, gains, dead zone cutoffs, saturation limits, etc.) is adjusted towards an optimal value. One such method of adjustment is sequential estimation, described by Detchmendy and Sridhar.[3] Briefly, if the model states $X_m(t)$'are described by the vector equation

$$\dot{X}_m = g(X, t)$$

where x_{m_i} is the model's output and the plant observations $Y(t)$ are described by

$$Y(t) = h(t, X)$$

where $h(t,X)$ relates the states $X(t)$ to the observable process outputs, it is desired to estimate in the least squares sense, the current state $X(T)$ based upon the measurements $Y(t)$ in the interval $0 \leq t \leq T$. For the particular problem considered, $Y(t) = x_p(t)$ the plant output, and the functional to be minimized is

$$\int_0^t [x_{m_i}(t) - x_p(t)]^2 dt.$$

From Detchmendy and Sridhar, the equations to be solved to satisfy the last equation are:

$$\dot{X} = g(X, t) + Q(t)(\nabla h^T)(Y - h(X))$$

$$\dot{Q} = (\nabla g^T)^T Q + Q(\nabla g^T) + Q(Z - \nabla h^T P(\nabla h^T)^T)Q$$

Reprinted from *IEEE Transactions on Systems Science and Cybernetics*, Vol. SSC-3, No. 1, pp. 11-16, June 1967.

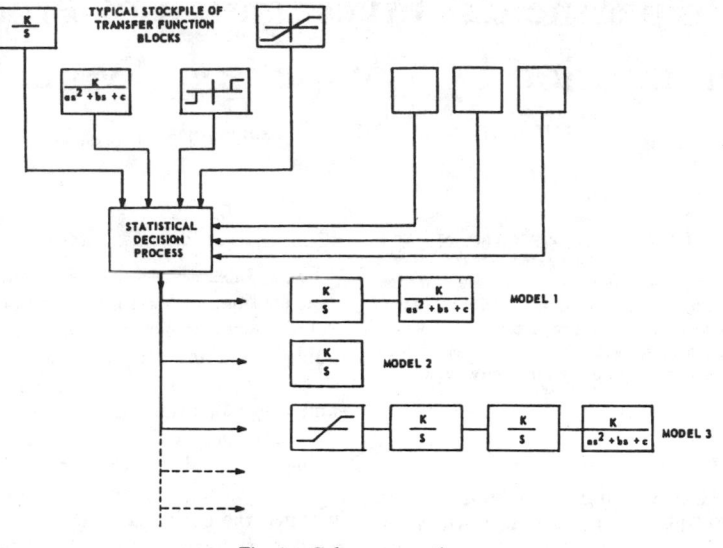

Fig. 1.   Colony generation.

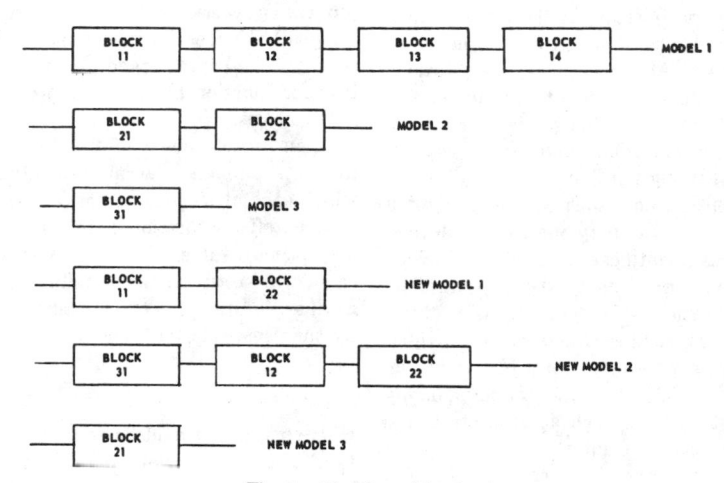

Fig. 2.   Model combination.

where

$$Z_{ij} = \left\{ \frac{\partial^2}{\partial x_i \partial x_j} [h^T] \right\} (Y - h)$$

$$h = x_1 = x_{m_1}$$

$$Y = \text{measured plant output } x_p$$

and

$$\begin{bmatrix} \partial/\partial x_1 \\ \cdot \\ \cdot \\ \cdot \\ \partial/\partial x_n \end{bmatrix}.$$

Initial values of X are to be chosen so as to reflect the physical situation, while the diagonal terms in $Q(0)$ reflect in some manner the confidence in the selected $X(0)$.

After a sufficiently long adjustment period, each of the N model's performance is rated according to a performance index such as

$$J = \int_{t_A}^{T_{max}} |x_{m_i}(t) - x_p(t)| dt$$

or

$$J = \int_{t_A}^{T_{max}} |x_{m_i}(t) - x_p(t)|^2 dt$$

where t_A is the time at which the adjustment phase ends,

and T_{max} is the selected lifetime of the model colony being considered. A set of satisfactory models is then chosen to consist of either those models, each of whose performance index lies below a given threshold, or the $K < N$ models with smallest J's. The remaining unsatisfactory models are deleted from the colony.

It is then postulated that the remaining set of models contains those transfer function blocks, the proper combination of which will yield a satisfactory plant approximation. Thus, to fill the vacancies in the colony caused by deletion of the unsatisfactory models, new models are generated from the random interchange of transfer function blocks of the remaining or parent models as shown in Fig. 2. Another means of generating a new model is the actual mutation or change in some transfer function block of an existing model. Clearly it is very possible that some of these new models may be identical in form to the parent models.

These new models (except for those identical to some parent model) must next go through the adjustment phase prior to their evaluation. The colony is then re-evaluated as a whole, resulting in the deletion and generation of another generation of models. This process is repeated until the colony has reached a steady state in which all models are identical.

EXPERIMENTAL RESULTS

Background

To further investigate evolutionary techniques, a set of digital programs was written to identify a given process using up to three transfer function blocks in the model configuration shown in Fig. 3. Six individual models were used and their performance assessed according to the criterion

$$J = \sum_{t_i = t_A}^{T_{max}} \left| x_{m_i}(t_i) - x_p(t_i) \right|$$

where t_A and T_{max} were read into the program.

Initially, the number of blocks in the ith model configuration Ni was randomly chosen as follows:

$$P(Ni = 1) = 0.2$$

$$P(Ni = 2) = 0.4$$

$$P(Ni = 3) = 0.4.$$

These probabilities were intuitively selected in accordance with the assumed form of the process to be identified. Upon selection of the number of blocks, the configurations were also randomly chosen from the blocks shown in Fig. 4. Again these were chosen to conform with the expected configurations of the given processes. The rules for selection were as follows:

Case I: Ni = 1

  Permissible block types—1, 2
  Probability of each—$P(1) = P(2) = 1/2$

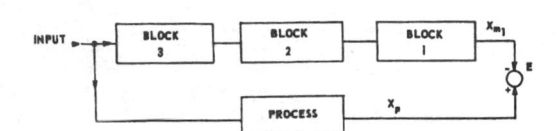

Fig. 3. Experimental block diagram.

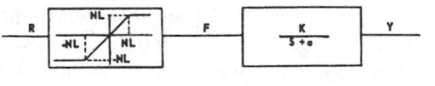

Fig. 4. Transfer function block stock pile.

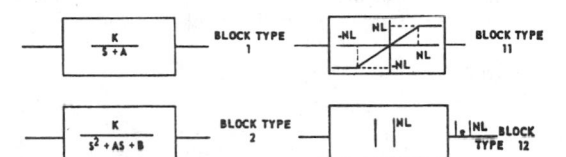

Fig. 5. Typical model.

Case II: Ni = 2

  a) *Block 1*

  Permissible block types—1, 2
  Probability of each—$P(1) = P(2) = 1/2$

  b) *Block 2*

  Permissible block types—1, 2, 11, 12
  Probability of each—$P(1) = P(2) = P(11) = P(12) = 1/4$

Case III: Ni = 3

  a) *Block 1 and Block 2*

  Permissible block types—1, 2
  Probability of each—$P(1) = P(2) = 1/2$

  b) *Block 3*

  Permissible block types—11, 12
  Probability of each—$P(11) = P(12) = 1/2.$

Upon selection of the six model configurations, the training or parameter adjustment phase was initiated. This was accomplished using the method of sequential estimation described earlier.

As an example, the state equations for the particular model shown in Fig. 5 will be developed.

Let

$$x_1 = y$$
$$x_2 = a$$
$$x_3 = K$$
$$x_4 = NL.$$

Then

$$\dot{x}_1 = x_3 F - x_2 x_1$$
$$\dot{x}_2 = 0$$

$$\dot{x}_3 = 0$$

$$\dot{x}_4 = 0$$

where

$$F = R \text{ if } |R| < x_4$$

$$F = x_4 \text{ if } |R| \geq x_4.$$

The states for all possible model configurations are now defined in the following. In all cases the last state variable takes on the value of the nonlinear variable term NL, if present in a preceding block. Thus, if, in the previous example, the saturation block were not present, x_4 would always appear as 0, and F would equal R.

Model $K/(s + a)$

$$x_1 = y$$
$$x_2 = a$$
$$x_3 = K$$
$$x_4 = NL$$

Model $K/(s^2 + as + b)$

$$x_1 = y$$
$$x_2 = \dot{y}$$
$$x_3 = a$$
$$x_4 = b$$
$$x_5 = K$$
$$x_6 = NL$$

Model $K/(s + a)(s + b)$

$$x_1 = y$$
$$x_2 = \dot{y}$$
$$x_3 = a$$
$$x_4 = b$$
$$x_5 = K$$
$$x_6 = NL$$

Model $K(s + c)(s^2 + as + b)$

$$x_1 = y$$
$$x_2 = \dot{y}$$
$$x_3 = \ddot{y}$$
$$x_4 = K$$
$$x_5 = a$$
$$x_6 = b$$
$$x_7 = C$$
$$x_8 = NL$$

Model $K/(s^2 + as + b)(s^2 + cs + d)$

$$x_1 = y$$
$$x_2 = \dot{y}$$
$$x_3 = \ddot{y}$$
$$x_4 = \dddot{y}$$
$$x_5 = K$$
$$x_6 = a$$
$$x_7 = b$$
$$x_8 = c$$
$$x_9 = d$$
$$x_{10} = NL.$$

At the end of the adjustment or training phase, the models were rated in accordance with the previously stated performance criteria. In the actual program, this training period lasted 10 seconds, and the performance was measured only during the latter 5-second portion of the interval, so that large initial errors caused by poor selection of initial conditions on x and Q might be eliminated by proper training. Otherwise, a model with proper configuration but erroneous initial states might be exterminated because of poor initial performance.

Those four models having the best performance were allowed to remain and recombine for the purpose of producing two new configurations. The rules for combination were as follows:

1) Number of blocks in new model is equal to the number of blocks in any one of the remaining four models with probability 1/4.

2) First block is equal to the first block of any one of the remaining four models with probability 1/4.

3) If model is to have two blocks, the second block is chosen with equal probability of being the second block of one of the remaining four models which have two or three blocks.

4) If model is to have three blocks, the second block is chosen with equal probability of being the second block of one of the remaining four models which have linear second blocks. The third block is chosen with equal probability of being the third block of one of the remaining four models which have three blocks.

Upon the creation of these two new models, their parameters are adjusted, their performance evaluated, and the process repeated.

Experimental Results

To demonstrate the applicability of evolutionary techniques, the previously described procedure was used to identify two processes. In both cases, the training period was 10 seconds, the performance J was evaluated during the latter five seconds, and the integration time increment was 0.5 seconds. The initial Q matrix consisted of all *ones*

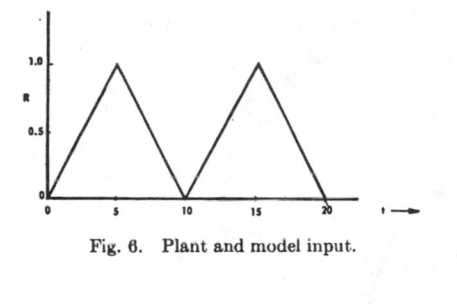

Fig. 6.   Plant and model input.

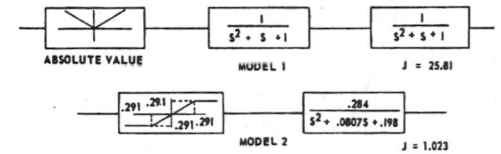

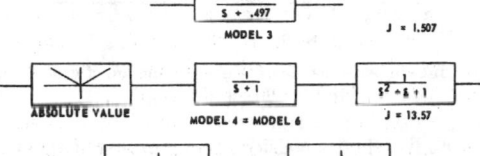

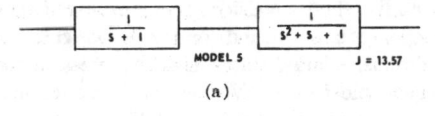

(a)

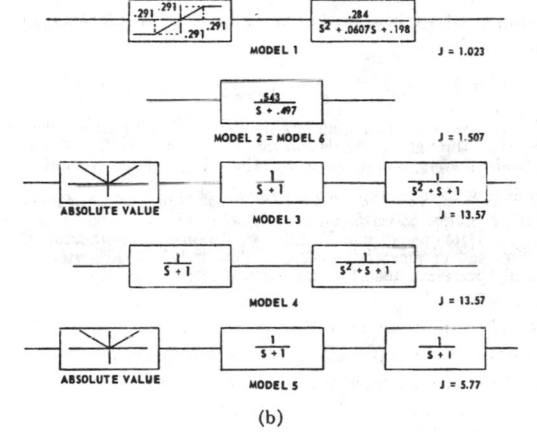

(b)

Fig. 7.   (a) Generation 1 colony for the plant $1/(s + 1)$. (b) Generation 3 colony for the plant $1/(s + 1)$. (In generation 2, models 1–4 were the same as in generation 3. Model 5 was identical to model 4, and model 6 was identical to model 1.)

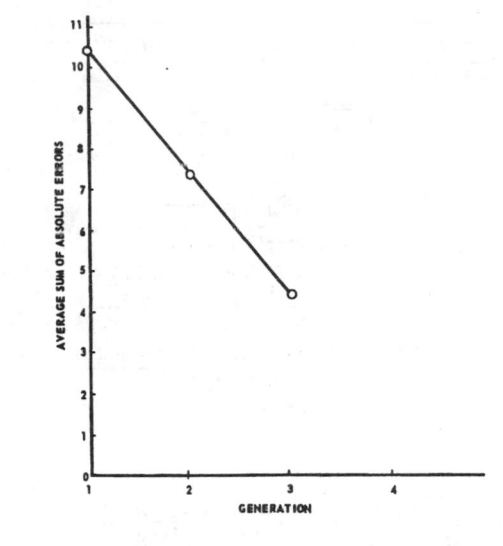

Fig. 8.   Change in performance criteria as a function of generation for the plant $1/(s + 1)$.

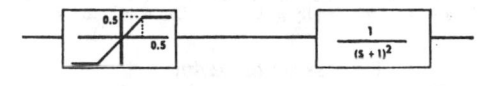

Fig. 9.   Plant to be identified.

except for the diagonal terms which were *threes*, and the initial values of all states were 0.1 except for the exponential nonlinearity with tag 12 which had an initial state of 1. The input to both plant and model was the triangular wave shown in Fig. 6.

The first process to be identified was the simple linear plant described by the transfer function $1/(s + 1)$. Re-sults, which appear in Fig. 7, show that after three generations, two of the six models (numbers 2 and 6) have this configuration (*but only the second smallest error criterion J*). Model 1 with an error criteria of 1.023 is better by 0.484. This is due to the fact that models 2 and 6 approximate the plant by $0.543/(s + 0.497)$ which is not a perfect identification. Apparently the configuration of model 1 is a more satisfactory identification. Figure 8, which clearly illustrates the decrease in average absolute error with successive generations, shows that a trend towards a steady-state colony whose members all closely approximate the given process does definitely exist.

The next process to be identified was the plant, shown in Fig. 9, which contains a nonanalytic saturation element. Figure 10 shows the results for the first and sixth generations. An acceptable configuration can be either a non-linear block of type 11 and a linear block of type 2 or the same nonlinear block with two linear blocks of type 1. These results along with Fig. 11 show that beginning with a colony with an average performance index of 11.3 and containing only one proper configuration (model number 2) a steady state with an average index of 0.25 was reached in six generations. Although this model [as shown in Fig. 10(b)] does not exactly duplicate the plant because of parameter differences, a sufficiently long adjustment period after this generation would probably result in a satisfactory approximation.

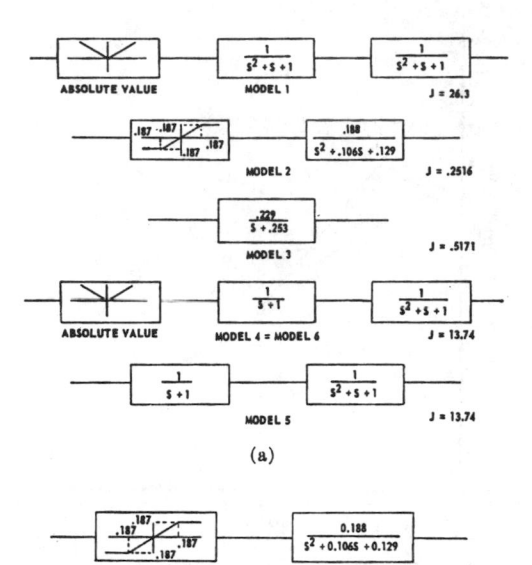

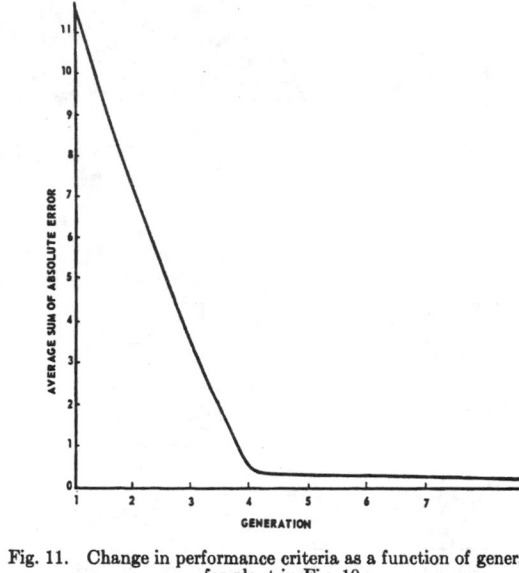

Fig. 11.   Change in performance criteria as a function of generation for plant in Fig. 10.

Fig. 10.   (a) Generation 1 colony for the plant $1/(s + 1)^2$ preceded by a 0.5 saturation nonlinearity. (b) Generation 6 for the plant $1/(s + 1)^2$ preceded by a 0.5 saturation nonlinearity.

DISCUSSION OF RESULTS

Although the preceding results demonstrated that evolutionary techniques can be used to obtain an acceptable configuration for an unknown process, the computation time required for each generation was approximately 0.03 hour on the IBM 7044 computer. To further improve the results, it may be necessary to increase the training time to approximately 30 seconds and to evaluate performance over the last 10 seconds. This could possibly increase the computation time to approximately 0.1 hour per generation. In general, it appears that the techniques described may be useful for evolving a model for a plant having nonlinear and/or nonanalytic properties.

Future studies should consider faster means of solving Detchmendy's equations and optimum assignment of initial conditions. A larger colony of many types of models is also a desirable feature. Various performance indices should be used with means for penalizing complicated model structure. Finally, a wider class of representative and practical processes should be identified using these evolutionary techniques.

REFERENCES

[1] L. J. Fogel, A. J. Ownes, and M. J. Walsh, "Intelligent decision-making through a simulation of evolution," *IEEE Trans. Human Factors in Electronics*, vol. HFE-6, pp. 13–23, September 1965.
[2] J. E. Gibson, *Nonlinear Automatic Control*. New York: McGraw-Hill, 1963, pp. 500–518.
[3] P. M. Detchmendy and R. Sridhar, "Sequential estimation of states and parameters in noisy nonlinear dynamical systems," *JACC preprints*, 1965.

Chapter 11
Co-Evolution, Self-Adaptation,
and Crossover

J. Reed, R. Toombs, and N. A. Barricelli (1967) "Simulation of Biological Evolution and Machine Learning: I. Selection of Self-Reproducing Numeric Patterns by Data Processing Machines, Effects of Hereditary Control, Mutation Type and Crossing," *J. Theoret. Biol.*, Vol. 17, pp. 319–342.

R EED et al. (1967) is a truly remarkable contribution. This paper offers some of the earliest experiments in three distinct areas: co-evolution in gaming, self-adaptation, and comparing the effectiveness of different operators. In retrospect, the paper could have had a profound effect on the course of experimentation in evolutionary computation had it received greater attention. The framework included the evolution of probabilistic strategies for playing a simplified game of poker using parameters to affect the probabilities of crossover and mutation. The recombination operators included the one-point crossover popularized in Holland (1975), as well as the uniform crossover[1] popularized 22 years later in Syswerda (1989) (also see Ackley, 1987). Across a series of different environments with continuous and discrete parameters, Reed et al. (1967) offered the first consistent evidence that recombination could be used to advantage on problems that were highly separable (decomposable into independent subproblems) and were coded in a low cardinality alphabet. Their results also indicated that crossover provided no advantage in the cases of continuous parameters or strong interfering interactions between parameters.[2]

The poker game centered on hands comprising a single card, which could either be high or low. Three betting options were provided: pass, low bet, and high bet, these being of two, three, or seven units, respectively. The strategies for playing the game were composed of eight parameters, four of which defined probabilities for making various types of wagers, with the other four determining the mutation probabilities and the associated effective step size, as well as the potential for a strategy to crossover

with another strategy. These "self-adaptive" or so-called "strategy" parameters were also subject to random variation.[3] In a particular setting for which an optimum strategy could be determined via the game theory of Von Neumann, the evolutionary simulation converged on nearly optimal plans, confirming the utility of the self-adaptive method.

This possibility to allow mating and mutation rates to evolve along with the solutions to the posed problem was suggested as early as Bledsoe (1961), and was later developed independently within evolution strategies (Rechenberg, 1973; Schwefel, 1977), evolutionary programming (Fogel et al., 1991), and genetic algorithms (Rosenberg, 1967; also 1970). This is a case of convergent evolution on a fundamental idea: the possibility to have the evolutionary algorithm evolve itself. Some of the experiments in Reed et al. (1967) used a tag bit to determine whether or not a strategy was suitable for crossover, a procedure later offered in Fogel and Atmar (1990), Schaffer and Eshelman (1991), Spears (1995), and others. The use of self-adaptation to adjust both recombination and mutation parameters in Reed et al. (1967) showed significant foresight.

Another basic contribution of the paper was the extension of evaluating solutions based on how well they compete with other members of the population, as opposed to a fixed fitness function that was provided *a priori*. This had been explored in an earlier paper by Barricelli (1963) (see Chapter 6), as well as later in Fogel and Burgin (1969), but did not gain real attention again until Axelrod (1987) (see Chapter 19). Reed et al. (1967) demonstrated a precursor to the use of co-evolution that has

[1]One-point crossover exchanges one contiguous length of coding between two solutions whereas uniform crossover provides the possibility of choosing each component of a new offspring from either parent.

[2]This was also later described as *epistasis* in Holland (1975). It is noteworthy that Holland (1975) consistently indicated that crossover should be particularly useful when facing epistatic interactions.

[3]Toombs (1997) indicated that the idea to use self-adaptation grew from Barricelli's earlier work in patterns emerging from rules governing local interactions of numbers in a grid (see Chapter 6). In particular, different mutation operations were observed to affect the type of patterns that could emerge, leading to the suggestion that the computer could be used to evolve appropriate variation operators simultaneously with the search for a solution to a particular problem.

been applied in many recent efforts (Hillis, 1991; Angeline and Pollack, 1993; Sebald and Schlenzig, 1994; Rosin and Belew, 1995; and others). Other notable contributions of Reed et al. (1967) include the use of a population of as many as 10,000 individuals, which is large even by current standards, and a form of tournament selection to determine which strategies would become parents for the next generation.

Several sentences in Reed et al. (1967) convey the surprise of the result that crossing over might provide no benefit to the speed of adaptation (or perhaps even a negative effect),[4] and separability (i.e., independent parameters) was in part identified as a key to judging the effectiveness of discrete recombination methods. Only as recently as Salomon (1996) was it demonstrated that the design of evolutionary algorithms relying on crossover and small mutation rates (e.g., genetic algorithms) is best suited for problems posing independent parameters, and that the optimization performance of such techniques can be greatly reduced under coordinate rotations that induce parametric interdependencies. In a somewhat discouraging result, Salomon (1996) also indicated that these separable problems can be solved by a class of procedures that are much more efficient than genetic algorithms, being of order $O(n)$ rather than $O(n \log n)$ for n parameters. Moreover, Jones (1995a) demonstrated a mutation-based hill climbing procedure that outperformed genetic algorithms on the so-called Royal Road functions, which were specifically designed to possess "building blocks" that would be suitable for recombination in a genetic algorithm (Mitchell et al., 1992; Jones, 1995b).[5] The problem of discovering which circumstances are particularly appropriate for specific operators remains open (Eshelman and Schaffer, 1993; Wolpert and Macready, 1997). Nevertheless, the results of Reed et al. (1967) could have served to temper some of the hyperbole concerning recombination in evolutionary algorithms (e.g., Bridges and Goldberg, 1987; Goldberg, 1989, pp. 106; Whitley and Hanson, 1989; and others).

References

[1] D. H. Ackley (1987) "An empirical study of bit vector function optimization," *Genetic Algorithms and Simulated Annealing*, L. Davis (ed.), Pitman, London, pp. 170–204.

[2] P. J. Angeline and J. B. Pollack (1993) "Competitive environments evolve better solutions for complex tasks," *Proc. of the 5th Intern. Conf. on Genetic Algorithms*, S. Forrest (ed.), Morgan Kaufmann, San Mateo, CA, pp. 264–270.

[3] R. Axelrod (1987) "The evolution of strategies in the iterated prisoner's dilemma," *Genetic Algorithms and Simulated Annealing*, L. Davis (ed.), Pitman, London, pp. 32–41.

[4] N. A. Barricelli (1963) "Numerical testing of evolution theories: Part II: Preliminary tests of performance, symbiogenesis and terrestrial life," *Acta Biotheoretica*, Vol. 16, No. 3–4, pp. 99–126.

[5] W. W. Bledsoe (1961) "The use of biological concepts in the analytical study of systems," technical report, Panoramic Research, Inc., Palo Alto, CA.

[6] H. J. Bremermann (1962) "Optimization through evolution and recombination," *Self-Organizing Systems*, M. C. Yovits, G. T. Jacobi, and G. D. Goldstine (eds.), Spartan Books, Washington D.C., pp. 93–106.

[7] C. L. Bridges and D. E. Goldberg (1987) "An analysis of reproduction and crossover in a binary-coded genetic algorithm," *Genetic Algorithms and Their Applications: Proc. of the 2nd Intern. Conf. on Genetic Algorithms*, J. J. Grefenstette (ed.), Lawrence Erlbaum, Hillsdale, NJ, pp. 9–13.

[8] L. Davis (1991) "Bit-climbing, representational bias, and test suite design," *Proc. of the 4th Intern. Conf. on Genetic Algorithms*, R. K. Belew and L. B. Booker (eds.), Morgan Kaufmann, San Mateo, CA, pp. 18–23.

[9] K. A. De Jong (1975) "An analysis of the behavior of a class of genetic adaptive systems," Ph.D. diss., Univ. of Michigan, Ann Arbor, MI.

[10] T. Dobzhansky (1955) *Evolution, Genetics, & Man*, John Wiley, NY.

[11] L. J. Eshelman and J. D. Schaffer (1993) "Crossover's niche," *Proc. of the 5th Intern. Conf. on Genetic Algorithms*, S. Forrest (ed.), Morgan Kaufmann, San Mateo, CA, pp. 9–14.

[12] D. B. Fogel and J. W. Atmar (1990) "Comparing genetic operators with Gaussian mutations in simulated evolutionary processes using linear systems," *Biological Cybernetics*, Vol. 63, pp. 111–114..

[13] D. B. Fogel, L. J. Fogel, and J. W. Atmar (1991) "Meta-evolutionary programming," *Proc. of the 25th Asilomar Conf. on Signals, Systems and Computers*, R. R. Chen (ed.), Maple Press, San Jose, CA, pp. 540–545.

[14] L. J. Fogel and G. H. Burgin (1969) "Competitive goal-seeking through evolutionary programming," final report under Contract No. AF 19(628)-5927, Air Force Cambridge Research Labs.

[15] A. S. Fraser (1957) "Simulation of genetic systems by automatic digital computers. I. Introduction," *Australian J. Biological Sciences*, Vol. 10, pp. 484–491.

[16] D. E. Goldberg (1989) *Genetic Algorithms in Search, Optimization and Machine Learning*, Addison-Wesley, Reading, MA.

[17] W. D. Hillis (1991) "Co-evolving parasites improve simulated evolution as an optimization procedure," *Emergent Computation*, S. Forrest (ed.), MIT Press, Cambridge, MA, pp. 228–234.

[18] J. H. Holland (1975) *Adaptation in Natural and Artificial Systems*, Univ. of Michigan Press, Ann Arbor, MI.

---

[4]Crossover was routinely viewed as the primary source of important variation in prior work by Barricelli (see Chapter 6), Fraser (see Chapter 3), and many others at the time. Holland (1992) offered, however, that the first attempts in simulated evolution "fared poorly because they followed the emphasis in biological texts of the time and relied on mutation rather than mating. . . . " In retrospect, this would appear to be a misjudgment not only because many early attempts in simulated evolution did suggest and explore crossover (e.g., Fraser, 1957; Bledsoe, 1961; Bremermann, 1962; and others), but there was already evidence of the conditions that would favor or disfavor its effectiveness (Reed et al., 1967). With regard to the emphasis in biological texts of the time, Mayr (1963, pp. 175–176) noted "it is held by contemporary geneticists that mutation pressure as such is of small immediate evolutionary consequence in sexual organisms, in view of the relatively far greater contribution of recombination and gene flow to the production of new genotypes. . . . " Dobzhansky (1955, p. 253) wrote "Mendel's work showed that gene recombination in sexually produced progenies creates an immense amount of genetic variability and of raw materials for evolution. . . . Sex arose in organic evolution as a master adaptation which makes all other evolutionary adaptations more readily accessible." Thus it is not clear that Holland (1992) fairly represented the biological thought on the relative importance of mutation and mating in the 1950s and 1960s.

[5]Similar results were offered in Davis (1991) where a randomized hill-climbing procedure outperformed a genetic algorithm across a series of test functions including those from De Jong (1975).

[19] J. H. Holland (1992) "Genetic algorithms," *Scientific American,* July, pp. 66–72.

[20] T. Jones (1995a) "Crossover, macromutation, and population-based search," *Proc. of the 6th Intern. Conf. on Genetic Algorithms,* L. Eshelman (ed.), Morgan Kaufmann, San Mateo, CA, pp. 73–80.

[21] T. Jones (1995b) "A description of Holland's royal road function," *Evolutionary Computation,* Vol. 2:4, pp. 411–417.

[22] E. Mayr (1963) *Animal Species and Evolution,* Belknap Press, Cambridge, MA.

[23] M. Mitchell, S. Forrest, and J. H. Holland (1992) "The royal road for genetic algorithms: Fitness landscapes and GA performance," *Toward a Practice of Autonomous Systems: Proc. of the 1st Europ. Conf. on Artificial Life,* F. J. Varela and P. Bourgine (eds.), MIT Press, Cambridge, MA, pp. 245–254.

[24] I. Rechenberg (1973) *Evolutionsstrategie: Optimierung technischer Systeme nach Prinzipien der biologischen Evolution,* Frommann-Holzboog, Stuttgart, Germany.

[25] J. Reed, R. Toombs, and N. A. Barricelli (1967) "Simulation of biological evolution and machine learning: I. Selection of self-reproducing numeric patterns by data processing machines, effects of hereditary control, mutation type and crossing," *J. Theoret. Biol.,* Vol. 17, pp. 319–342.

[26] R. S. Rosenberg (1967) "Simulation of genetic populations with biochemical properties," Ph.D. diss., Univ. of Michigan, Ann Arbor, MI.

[27] R. S. Rosenberg (1970) "Simulation of genetic populations with biochemical properties: II. Selection of crossover probabilities," *Mathematical Biosciences,* Vol. 8, pp. 1–37.

[28] C. D. Rosin and R. K. Belew (1995) "Methods for competitive co-evolution: Finding opponents worth beating," *Proc. of the 6th Intern. Conf. on Genetic Algorithms,* L. Eshelman (ed.), Morgan Kaufmann, San Mateo, CA, pp. 373–380.

[29] R. Salomon (1996) "Re-evaluating genetic algorithm performance under coordinate rotation of benchmark functions. A survey of some theoretical and practical aspects of genetic algorithms," *BioSystems,* Vol. 39, pp. 263–278.

[30] J. D. Schaffer and L. Eshelman (1991) "On crossover as an evolutionarily viable strategy," *Proc. of the 4th Intern. Conf. on Genetic Algorithms,* R. K. Belew and L. B. Booker (eds.), Morgan Kaufmann, San Mateo, CA, pp. 61–68.

[31] H.-P. Schwefel (1977) Numerische Optimierung von Computer-Modellen mittels der Evolutionsstrategie, Birkhäuser, Basle, Interdisciplinary Systems Research 26.

[32] A. V. Sebald and J. Schlenzig (1994) "Minimax design of neural net controllers for highly uncertain plants," *IEEE Trans. Neural Networks,* Vol. 5:1, pp. 73–82.

[33] W. M. Spears (1995) "Adapting crossover in evolutionary algorithms," *Evolutionary Programming IV: Proc. of the 4th Ann. Conf. on Evolutionary Programming,* J. R. McDonnell, R. G. Reynolds, and D. B. Fogel (eds.), MIT Press, Cambridge, MA, pp. 367–384.

[34] G. Syswerda (1989) "Uniform crossover in genetic algorithms," *Proc. of the 3rd Intern. Conf. on Genetic Algorithms,* J. D. Schaffer (ed.), Morgan Kaufmann, San Mateo, CA, pp. 2–9.

[35] R. I. Toombs (1997) personal communication, The Boeing Corporation, WA.

[36] D. Whitley and T. Hanson (1989) "Optimizing neural networks using faster, more accurate genetic search," *Proc. of the 3rd Intern. Conf. on Genetic Algorithms,* J. D. Schaffer (ed.), Morgan Kaufmann, San Mateo, CA, pp. 391–396.

[37] D. H. Wolpert and W. G. Macready (1997) "No free lunch theorems for optimization," *IEEE Trans. on Evolutionary Computation,* Vol. 1:1, pp. 67–82.

J. Theoret. Biol. (1967) **17**, 319–342

Simulation of Biological Evolution and Machine Learning

I. Selection of Self-reproducing Numeric Patterns by Data Processing Machines, Effects of Hereditary Control, Mutation Type and Crossing

. Jon Reed

Matematisk Institut, Oslo Universitet, Norway

Robert Toombs and Nils Aall Barricelli

*Department of Genetics, University of Washington,
Seattle, Washington, U.S.A.*

(*Received* 14 *February* 1967)

The effect of crossing and of different types of mutations (or genetic control) on the speed of selective adaptation has been investigated by data processing machines. The procedure is based on the use of self-reproducing numeric patterns, or arrays of numbers and/or single bits of information. Each number or bit was used to identify a property of the pattern, which could be subject to selection. Some of the properties represented crossing, mutation or reproductive characteristics of the pattern. Some other properties represented other characteristics essential for the pattern's ability to survive and reproduce, for example, by identifying the pattern's strategy in a game of poker used as a means to select the patterns to be reproduced. A measure of the speed of improvement as a function of generation number has been defined and used to compare the improvement under various conditions.

The results obtained can be summarized as follows: under conditions simulating regular Mendelian (non-polygenic nor quantitative) inheritance, crossing greatly enhances the speed of selective adaptation, particularly if interaction between the expressions of different hereditary factors is avoided; under conditions simulating polygenic control of quantitative characters, crossing does not enhance the speed of selective adaptation. Furthermore, in a period of rapid selective adaptation the ability to interbreed spreads rapidly in the population as a positively selected characteristic when conditions simulating regular Mendelian inheritance are used. On the other hand, the breeding characteristic spreads less rapidly or not at all under conditions simulating quantitative characters under polygenic control.

In the simple game arrangements used in our adaptive selection experiments, the patterns were very often able to develop an optimum game strategy. However, in some instances a "selective instability" characterized by statistical fluctuations preventing the achievement of an optimum game strategy was developed. Selective instability, if not properly controlled, is likely to assume larger proportions in experiments of a more complicated nature, and may constitute a serious problem for the practical applications of adaptive selection methods by data processing machines.

1. Introduction

Experiments designed to investigate, by the use of high speed computers, the development of an evolutionary process based on hereditary changes and selection have been undertaken by various researchers (Lewontin, personal communication; Bremermann, 1962; Bremermann & Salaff, 1963; Bossert, 1967).† A common approach is to use self-reproducing numerical patterns presenting hereditary properties designed to simulate the genetic properties of living organisms. Observations relevant to the problem of estimating the speed of evolutionary adaptation and of evaluating the possibility of obtaining optimal adaptations by these selection procedures have been presented by Bossert and by Bremermann (quoted above).

The aims of the adaptive selection experiments performed by data-processing computers to be presented in this paper are of two general categories:

(i) to acquire experience in the use of evolutionary processes based on mutation, crossing and selection of numerical patterns for the development of short programs or number sequences useful for specific tasks;

(ii) to test the effects of various methods of controlling the variability of the patterns by mutation and the effect of crossing for the speed of improvement or learning.

The experiments presented here differ from the evolution experiments published earlier by Barricelli (1957, 1962, 1963) in that a fixed pattern of numbers, each one controlling a specific set of reciprocally exclusive characteristics, is used instead of a pattern developed by successive evolutionary steps. The fixed pattern method was first used by one of the authors (J. Reed) in a series of adaptive selection experiments performed on the FREDERIC Computer at Lilleström, near Oslo, Norway. We shall briefly outline these experiments.

2. The FREDERIC Adaptive Selection Experiments

The experiments with the fixed pattern method were originally carried out in order to develop by an evolutionary procedure a set of good von Neumann parameters for a poker game strategy. As a test for the method, an extremely

† Related research designed to use evolutionary processes in problem solving has been done by various authors (Friedberg, 1958, 1959; Samuel, 1963; Fogel, Owens & Walsh, 1964; Bremermann & Rogson, 1964; Fogel, 1963; and Bremermann, Rogson & Salaff, 1965).

simplified version of poker was used. Each pattern consisted of eight positive numbers of which four were used as parameters affecting the mutation and crossing properties of the pattern. The other four numbers were used as parameters (actually representing betting probabilities under different circumstances) to define a betting strategy in the poker game. Each pattern was permitted to play 20 poker games against an opponent as a standard procedure to select the patterns to be reproduced.

2.1. RULES OF THE GAME

Each player (pattern) received only one card (hand). There were only two types of cards: high cards (high hand) and low cards (low hand). The probability of receiving high hand is called h, and $1-h$ is therefore the probability of receiving low hand. In Reed's original experiments, h was equal to one-half. Only three bets were permitted: Pass, low bet and high bet. In order to bet pass, a player would have to pay two pennies; to bet low, it would cost three pennies; to bet high would cost seven pennies. If the two opponents bet differently, the highest bet always won. If the bets were the same, the higher hand won. If, also, the hands were the same, each player would get his money back.

Four numbers (or betting probabilities) are sufficient to define the betting strategy in this kind of poker game:

L_P = probability of betting pass with low hand;
H_P = probability of betting pass with high hand;
L_L = probability of betting low with low hand;
H_L = probability of betting low with high hand.

When these four numbers are given the remaining two probabilities, L_H (betting high with low hand) and H_H (betting high with high hand) are determined by the relations

$$\left.\begin{array}{l} L_H = 1 - L_P - L_L, \\ H_H = 1 - H_P - H_L. \end{array}\right\} \tag{1}$$

The optimum game strategy can be determined by von Neumann's game theory. In the particular case in which the probability h of receiving high hand is 1/2 one finds:

Optimum strategy for $h = 1/2$

$$\left.\begin{array}{lll} H_P = 0, & H_L = 0, & H_H = 1, \\ L_P = 0, & L_L = 1, & L_H = 0. \end{array}\right\} \tag{2}$$

For different h values the optimum betting probabilities are different (see Appendix 3).

As mentioned above, each numeric pattern consisted of eight numbers not larger than 1, namely a, b, c, d, L_P, H_P, L_L, H_L, of which four were the betting probabilities L_P, H_P, L_L, and H_L used to define the betting strategy. The remaining four numbers, a, b, c and d were used to define the mutation and crossing properties of the pattern. One hundred patterns like this were listed in the machine memory. To begin with (at generation 0), each pattern consisted of random numbers. A program written for the FREDERIC computer allowed each pattern to play 20 games of poker against an opponent (meaning another pattern selected in the same set of 100 patterns by a procedure which is irrelevant in this connection). In each game the two opponents received a card (or hand, high or low) and would bet with probabilities determined according to their betting parameters. After 20 games, the loser would be erased from the machine memory, and the winner would be reproduced. As a result, in each generation 50 patterns (losers) would be erased from the machine memory, and the remaining 50 patterns (winners) would be reproduced. During reproduction, mutations and crossing could take place in a manner determined by the remaining four parameters a, b, c, d according to the following rules:

2.2. MUTATION RULES

These were specified by the definitions of the following three parameters:

$1-a$ = random mutation probability (or the probability that one of the eight numbers, randomly selected, will be replaced by a random number in the duplicated pattern);

$1-c$ = systematic mutation probability (each one of the four betting probabilities has a probability $1-c$ of receiving an increment d —see next parameter—in the duplicated pattern);

d = increment for systematic mutations.

2.3. CROSSING RULES

In one of the evolution experiments performed, only patterns with $b > 0.5$ are crossed. Crossing takes place before mutation between two duplicating patterns, and is carried out in the following way: each number of the four (two parental and two progeny) patterns, is taken with 50% probability from one parent and 50% probability from the other, with the only reservation that c and d are always taken from the same parent, and otherwise completely at random.

In another evolution experiment, *crossing was obligatory irrespective of the b value*. In a third experiment, *no crossing was allowed*. The three experiments will be respectively designated: "free crossing", "mandatory crossing" and "inhibited crossing" experiment.

In all three experiments, the high hand betting probabilities were nearly optimized in less than 200 generations and approached the optimum values $H_P = 0$, $H_L = 0$ and $H_H = 1$ calculated by von Neumann's game theory. The low hand betting probabilities, which are less important for the quality of the game, were not optimized yet and still presented considerable differences in different patterns. The quality of the game was fully competitive with average human players uninformed about game theory.

The improvement in game strategy was not very different in the three experiments. The fastest improvement was observed in the free crossing experiment followed by the inhibited crossing experiment. The slowest improvement was observed in the mandatory crossing experiment, a result of considerable biologic interest which will be discussed in the next section.

These results have been confirmed in similar experiments performed later on an IBM 7094 machine which are described in a more detailed fashion in sections 4 and 5.

3. Significance of the FREDERIC Experiments

For many biologists the most startling result of the FREDERIC experiments is the small effect crossing has on the speed of improvement in betting strategy. Particularly, the fact that crossing inhibition did not drastically reduce the speed of improvement may be surprising to many. A reduction of this speed (or an increase of the time required for each improvement) proportional to the number of hereditary factors controlling the game strategy would have been expected according to Fisher's law (see Appendix 1). However, if we try to apply Fisher's law in this particular case, we run immediately into serious problems. What is the number of genes or the number of spreading hereditary factors which control the game strategy? The game strategy is controlled by four parameters H_P, H_L, L_P, L_L, of which the closest analogy in living organisms would be four quantitative characters probably controlled by polygenes. No rule has been given for counting the number of polygenes in a manner useful for the application of Fisher's law. We are not better off if we try to count the number of hereditary characters spreading in the population without interacting with each other. Any one of the four parameters H_P, H_L, L_P, L_L can have 100 different values in the 100 pattern population. Many of these 100 values could be spreading useful characters which, however, interact with each other both genetically (since they exclude each other and are therefore alleles) and epistotically (since the selective value of each one of these depends on the frequency of the others). Evidently, Fisher's law was never meant to be applied to a case like this (see section 4).

The mathematical theory for the evolution by quantitative (or polygenic) characters has been developed along different lines and has yielded results which, although less well known, are fully applicable to the determination of the evolutionary improvement in this particular area. The result is in agreement with the FREDERIC experiment result. No appreciable difference in the speed of evolutionary improvement is expected between crossing and non-crossing organisms (or patterns) if only quantitative characters are involved (see Appendix 2).

Another significant result of the FREDERIC experiments is the possibility they disclosed to optimize or nearly optimize a game strategy by this kind of adaptive selection methods. The question is posed whether the results could be improved and extended to a much larger range of possible applications by taking into account population genetical (particularly evolution mechanical) results and theorems.

4. The Scope of the IBM 7094 Adaptive Selection Experiments

The FREDERIC experiments can easily be modified in a manner which would make the selective adaptation more close to an evolution process based on the machine equivalent of independent Mendelian characters, a case in which Fisher's law would be applicable. A repetition of the FREDERIC experiments both in their original form and in the modified (or Mendelized) form seemed therefore justified as a means to test the theoretic predictions concerning the speed of evolutionary improvement in the two cases. The desired modification (or Mendelization) of the FREDERIC experiments was made particularly simple by the fact that in the case considered above in which the probability of receiving a high hand is 50%, the correct strategy is to bet always high with high hand, always low with low hand and never bet pass. In other words it is possible to optimize the betting strategy without using any quantitative characters (such as betting probabilities) at all. The numeric patterns could be allowed to acquire by mutation or crossing a finite and rather small number of allelic (hereafter designated as Mendelian) characters† such as (i) the property (A_H) of always betting high with high hand, or the opposite (allelic) property (a_H) of never betting high with high hand; (ii) the property (B_H) of always betting low with high hand if (a_H) is present, or the allelic property (b_H) of always betting pass with high hand if (a_H) is present; (iii) the property (F_H) of allowing Mendelian characters

† The term "Mendelian character" shall hereafter be used to designate one out of a finite and usually very small number of allelic properties specified by one bit or a few bits of the numeric pattern. On the other hand, the term "Darwinian character" (see Appendix 2) is used to designate a quantitative property (such as, for example, a betting probability) which ideally could be subject to continuous variation.

like (A_H), (a_H), (B_H), (b_H) to control the betting strategy with high hand, and the allelic property (f_H) of allowing quantitative characters (like betting probabilities, hereafter designated as "Darwinian characters") to control the betting strategy. Similar Mendelian properties A_L, a_L, B_L, b_L can be defined for the low hand betting strategy, and another pair of allelic properties (F_L) and (f_L) are allowed to decide whether these Mendelian properties or the corresponding Darwinian (quantitative) properties should control the low hand betting strategy.

The 7094 selection experiments have been based on the methods outlined above and on a method of machine representation similar to the one used in the FREDERIC experiments, but with a few minor modifications in the mutation and crossing rules designed to facilitate a faster optimization of the betting strategy.

5. Darwinian Selection Experiments

A list of 500 patterns is recorded in 500 consecutive memory locations of the IBM 7094. Each pattern consists of a set of eight numbers which comprises all together 36 bits and can therefore be recorded in a single memory location of the 7094 computer. The eight numbers, with their designations and bit contents are recorded in Table 1 in the same order in which they appear in a machine memory location. Each binary number referred to in Table 1 is normalized by dividing it by a number K ($K = 7$ if it is a three bit number, and $K = 31$, if it is a five bit number) so that the variability range of each number is contained between 0 and 1.

TABLE 1

List of the eight numbers in each pattern

Sequence order	Symbolic designation	Verbal designation	No. of bits
1	C	Crossover parameter	3
2	H_H	High bet on high hand probability	5
3	$H_L(H)$	Low bet on high hand probability if no high bet takes place	5
4	F	Bits free for other usages	5
5	M	Mutation probability	3
6	L_H	High bet on low hand probability	5
7	$L_L(H)$	Low bet on low hand probability if no high bet takes place	5
8	m	Mutation size (upper limit)	5
		Total no. of bits	36

The meaning of the four betting probabilities H_H, $H_L(H)$, L_H and $L_L(H)$ in Table 1 is obvious. Notice that $H_L(H)$ and $L_L(H)$ are probabilities *a posteriori*. The probabilities *a priori*, H_L (low bet on high hand) and L_L (low bet on low hand), are given by the formulas

$$\left. \begin{aligned} H_L &= (1-H_H)H_L(H), \\ L_L &= (1-L_H)L_L(H). \end{aligned} \right\} \tag{3}$$

The mutation probability M gives the probability that a mutation will take place in the pattern during its reproduction. The mutation is determined in the following manner. By a random process it is decided which one of the eight numbers shall be mutated. The mutation is performed by adding to it or subtracting from it (with equal probability) a random number smaller or equal to m (which is the upper limit of the mutation size listed in the table above). However, if the result of this operation was a negative number N, the number to be mutated was replaced by $N+1+1/K$. Similarly, if the result after normalization was a number N larger than 1, the result would be substituted by $N-1-1/K$, where K is defined above.

The crossing parameter C in the table above specifies the crossing probability in the free crossing experiments. However, experiments without crossing and experiments with obligatory crossing were also performed and in these cases the parameter C was ignored. In the free crossing experiments the following crossing rules have been applied. During reproduction each pattern occupying an odd position in the list has a probability R to cross with the following pattern occupying an even position. The crossing probability R is equal to the modulus of the difference between the C parameters of the two patterns. In this way the crossing probability is determined by the C-parameters of the two patterns. This crossing procedure has the advantage of making possible the development of two or more meeting types with different C-values. When crossing takes place, the two patterns themselves are not changed, but their progeny will be hybrids resulting from a single crossover in their bit sequence. For example, if we call a_1, a_2, a_3, ..., a_{36}, the 36 bits of one of the parental patterns, and b_1, b_2, b_3, b_4, ..., b_{36}, the 36 bits of the other parental pattern, the two progeny patterns would look as follows: $a_1, ..., a_n, b_{n+1}, ..., b_{36}$ and $b_1, ..., b_n, a_{n+1}, ..., a_{36}$, if the crossover has occurred between the nth and the $(n+1)$th bit. The location of the crossover is, of course, determined by a random number.

The selection procedure is the same as the one used in the FREDERIC experiments, and the same kind of simplified poker game was used. Each pattern plays 20 games against an opponent. The loser was removed from the machine memory. The winner was reproduced, and during reproduction, crossing and mutation could take place according to the rules specified above. Experiments with, and without crossing have been performed, and in the first case both free crossing and obligatory crossing have been used. The result was basically similar to the one obtained in the FREDERIC experiments, but betting probabilities closer to the optimum values have been obtained by this procedure. In Fig. 1 the result of the experiments performed with a 50% probability of receiving high hand is represented by various diagrams which show how the various parameters (such as betting probabilities, mutation probabilities and crossover frequencies) changed during three experiments as a function of generation number. Each experiment was repeated several times (see legend of Fig. 1) and the average results are plotted in the figure.

In one of the experiments (dotted lines) no crossing was allowed, in the other two experiments respectively free crossing (solid lines) and obligatory crossing (dashed lines) have been used. The figure gives the four betting probabilities H_H, H_L, L_H, L_L, the mutation probability M, the upper mutation size m, and a measure of the crossover frequency.

In all three experiments, the four betting probabilities listed above had a value 0 to begin with. We observe in Fig. 1 that the probability H_H of betting high with high hand rapidly grows to 1 (its optimum value) in all three experiments. The probability L_L of betting low with low hand grows much more slowly to high values approaching the optimum value which is 1. The other two betting probabilities H_L and L_H remain close to 0, which is their optimum value at the end of the experiment, but not at the beginning, where some fluctuations can be observed particularly in the low hand high bet probability L_H. We notice that the diagrams are very similar in the three experiments, and there is no good evidence that crossing has any influence on the speed of improvement.

The mutation probability M decreases in all three experiments, slightly more rapidly in the "100% crossing" case than in the other two. In all experiments, the mutation probability was not allowed to decrease below 0·05 according to specific instructions given to the computer. The mutation size m was set very low (less than 0·1) at the beginning of the experiments in order to simulate polygenic (invisible) mutations (cf. Mather, 1941, 1943). Only in the inhibited crossing experiments (dotted line in Fig. 1) did it show a slight tendency to increase. In the other experiments, it remained nearly unchanged.

In Fig. 2 the percentage of organisms which developed optimum game strategy (identified by $H_H = 1$, $H_L = 0$, $L_H = 0$, $L_L = 1$) is plotted as a function of generation number for all three experiments. It is evident from this figure that crossing does not increase the speed of adaptation in our Darwinian selection experiments.

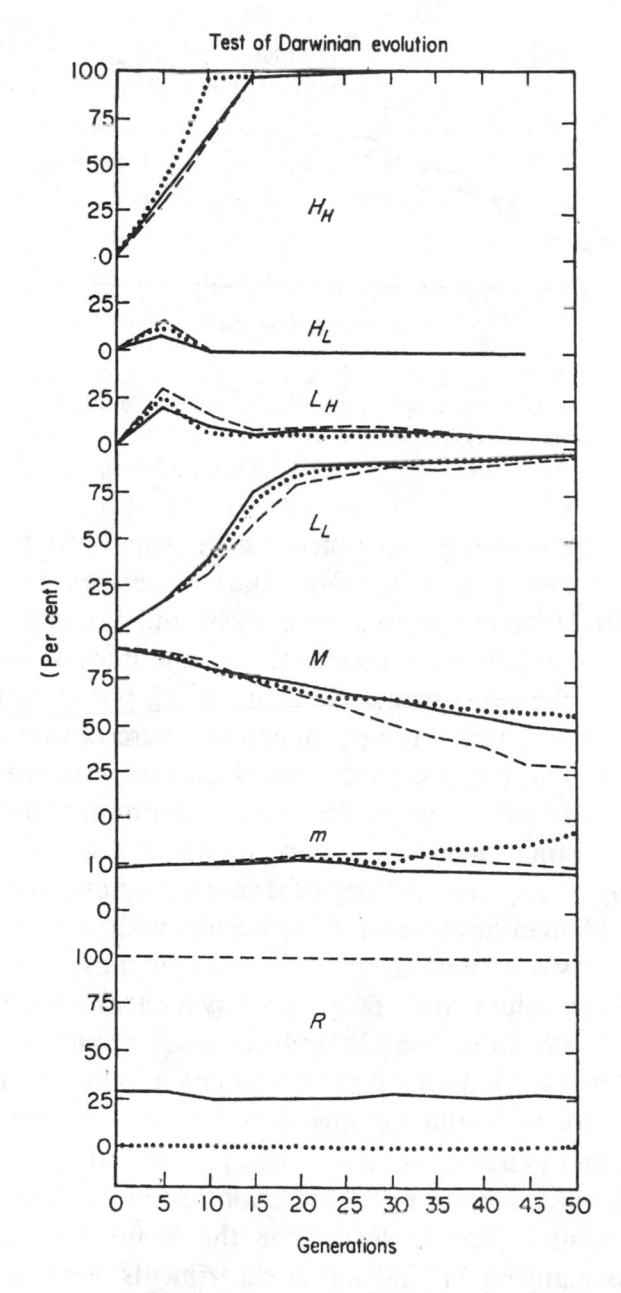

FIG. 1. Average values of the parameters H_H, H_L, L_H, L_L, M and m [see Table 1 and formula (3)] and the crossover probability R calculated from the distribution of C-values in the population, are given in per cent as a function of generation number. All experiments were performed using a probability of receiving high hand $h = 50\%$. ($\cdots$) no crossing (average of three experiments); (——) free crossing (average of three experiments); (– – –) mandatory crossing (average of three experiments).

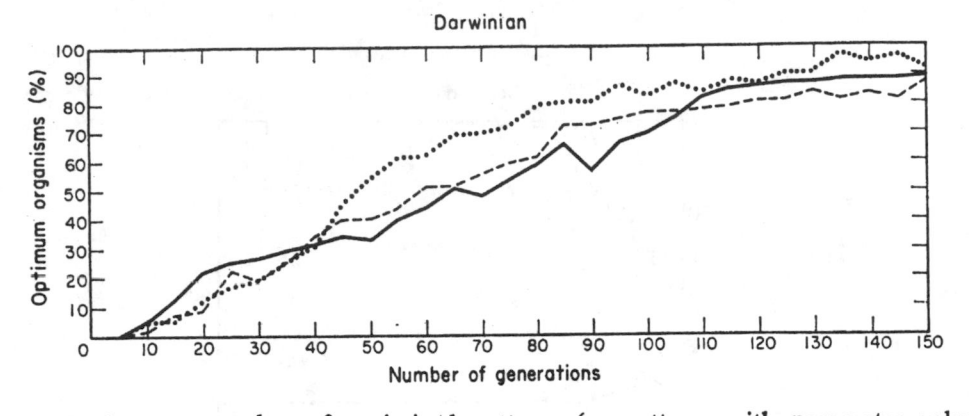

FIG. 2. Average number of optimized patterns (or patterns with parameter values $H_H = 1$, $H_L = 0$, $L_H = 0$, $L_L = 1$) given as a function of generation number for the same experiments used in Fig. 1.

A series of free crossing experiments have also been performed using different probabilities of receiving high hand. In these experiments, only 100 (instead of 500) numeric patterns were used, and the initial 100 numeric patterns were random 36 bit numbers. In all experiments, free crossing was used. In many of the experiments the final betting probabilities were nearly optimized. The high hand betting probabilities were always 1 for high bet, and 0 for low bet and for pass at the end of each experiment. The low hand betting probabilities were usually close to the optimum values calculated by von Neumann's game theory, which are shown in Fig. 3 (see Appendix 3). The dots in Fig. 3 are final betting probabilities or average values of such probabilities obtained in several experiments with the same high hand probability. It is worth noting, however, that in those cases in which the predicted optimal values are not 1 nor 0 (which are the cases $h = 0.30$, $h = 0.35$ and $h = 0.40$ in Fig. 3), individual experiments usually did not identify the optimum values of the low hand betting probabilities. Considerable fluctuations would be observed during each experiment as the generation number increased, and as long as the mutation rate was sufficient to maintain some variability no stabilization of the low hand betting probabilities was obtained. Nevertheless, when the arithmetic means of several (in one case as many as 11) individual experiments were calculated a good agreement with theory was obtained as shown in Fig. 3. A similar failure to reach optimum values was observed also by Friedberg (1958, 1959), Bremermann (1958) and Bossert (1967) in their selection experiments. This phenomenon, which will hereafter be designated as "selective instability", may give rise to considerable difficulties in the utilization of these kinds of evolutionary processes if a method to overcome this problem is not found. An

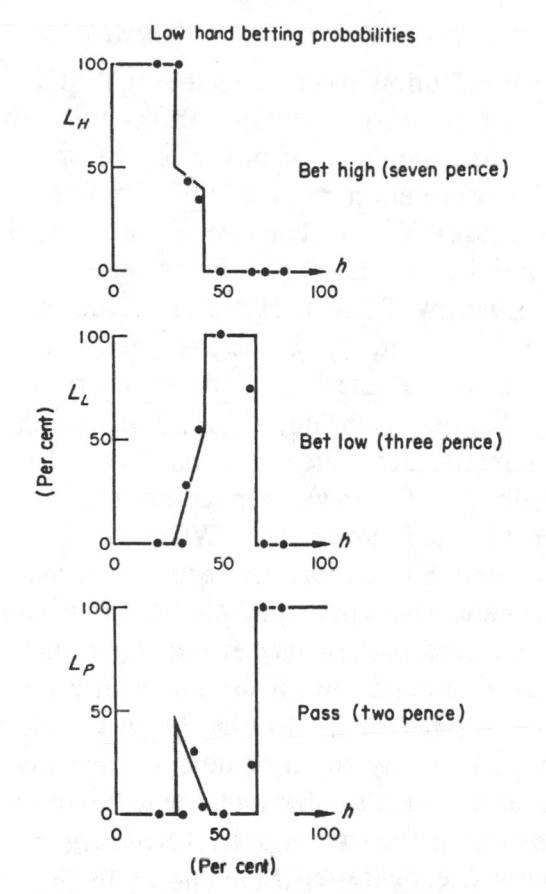

Low hand betting probabilities

FIG. 3. Theoretic and observed low hand betting probabilities L_H (= high betting probability), L_L (= low betting probability) and L_P (= passing probability) as a function of h (= the probability of receiving high hand). Each diagram presents the optimum betting probability calculated as a function of h by Von Neumann's game theory. The dots represent the final values of the same probabilities obtained in a series of evolution experiments lasting from 150 to 300 generations. Most final values are averages calculated for several evolution experiments. For example, the dots in position $h = 35\%$ represent the average value found in 11 experiments. In this particular region (namely for $h = 30$, $h = 35$ and $h = 40$), individual experiments often presented considerable fluctuations (cf. text: selective instability).

investigation is being undertaken in order to clarify the problem and find appropriate countermeasures. In our experiments one of the causes of selective instability is that optimum game strategy does not confer an advantage against all competitors. It only ensures that a pattern following an optimum game strategy will not be at a disadvantage when playing against any particular opponent. However, we do not know whether this is the only explanation of selective instability. The results of this investigation will be presented in a separate paper.

6. Mendelian Evolution Experiments†

In the Mendelian evolution experiments the eight numbers in each pattern are the same as in Darwinian evolution (Table 1). However, the interpretation of several of the numbers (or of the bits by which they are formed) can be modified by information inserted in the five free bits left in the fourth number for later usages. When these five bits are 0's, the meaning of the remaining seven numbers in the pattern is the same as in Darwinian evolution, and is still given by Table 1. However, if the first of these five bits, hereafter designated as L-bit, is a 1, the low hand betting strategy is Mendelized† (which means that the low hand betting parameters are replaced by an all or none betting arrangement—see footnote). Similarly, if the last of the five bits, hereafter designated as H-bit, is a 1, the high hand betting strategy is Mendelized (see below). Mendelization of the high hand betting strategy is obtained in the following way. When the H-bit is a 1, the machine ignores all but the first bits of the numbers H_H and $H_L(H)$. These two first bits shall hereafter be designated H_H-bit (or high-hand-bet-high bit) and H_L-bit (or high-hand-bet-low bit). If the H_H-bit is a 1, the pattern will always bet high on high hand while if the H_H-bet is a 0, the pattern will not. If the pattern does not bet high, it will bet low if the H_L-bit is a 1 and pass if the H_L-bit is a 0. This way an all or none (Mendelized) betting behavior on high hand is defined. In a similar way, the low hand betting strategy can be Mendelized by using the first bits L_H and L_L of the low hand betting probabilities. These Mendelization rules are summarized in Table 2.

With this arrangement, experiments have been done using mandatory crossing, free crossing and inhibited crossing.

In Fig. 4, the results of three experiments, each one repeated ten times, are presented using the mean parameter values as in Fig. 1. When Fig. 4 is compared with Fig. 1, we observe the effect of Mendelization on the speed of improvement in each parameter. The differences between the selection diagrams obtained with inhibited crossing, and the same diagrams obtained with free crossing or obligatory crossing present a result of particular interest in this connection. As already mentioned, in quantitative evolution, Fig. 1, no advantage resulting from crossing is observed. If anything, the "no crossing" diagrams seem to be slightly ahead of the others. In Fig. 4 we find, on the other hand, that there is a noticeable difference between the three cases when Mendelian evolution is used; obligatory crossing gives consistently the fastest improvement while the absence of crossing appears

---

† The terms Mendelian or Darwinian evolution, Mendelization or Darwinization of a betting strategy are only used for convenience and do not necessarily imply that quantitative (Darwinian) characters follow hereditary rules other than Mendelian.

TABLE 2

List of numbers and controlling bits in each pattern

Sequence order	Symbolic designation	Mendelian bits used	Bit designation	Verbal bit designations†	No. of bits in each parameter
1	C	—	—	—	3
2	H_H	First	H_H	High-hand-bet-high bit	5
3	$H_L(H)$	First	H_L	High-hand-bet-low bit	5
4	—	First	L	Low hand Mendelization-bit	5
		Fifth	H	High hand Mendelization-bit	
5	M	—	—	—	3
6	L_H	First	L_H	Low-hand-bet-high bit	5
7	$L_L(H)$	First	L_L	Low-hand-bet-low bit	5
8	m	—	—	—	5
				Total No. of bits	36

† For verbal designation of parameters see Table 1.

to give the slowest improvement and/or the largest fluctuations in the rate of progress. The same can be observed in Fig. 5, particularly when compared with Fig. 2.

The Fig. 5 shows the per cent of optimized patterns as a function of generation number. We see that optimization is slowest in the "no crossing" experiments and fastest in the mandatory crossing experiments. The effect of crossing is, however, not as strong as a superficial evaluation based on Fisher's law may have suggested. The following reasons may give part of the explanation.

(1) The selective values of the Mendelian characters H_H, L_L, H and L are drastically different since the property H_H of betting high with high hand is by far more important than the others—a fact of considerable importance in a case in which only four Mendelian characters are considered.

(2) The selective value of the L_L property depends strongly on the abundance of H_H individuals and is actually lower than the L_H property (of betting high or bluffing with low hand) at the beginning of the experiments when there are few H_H patterns.

(3) Also, the selective value of the Mendelizing (H and L) bits depends on the presence of the other bits needing Mendelization.

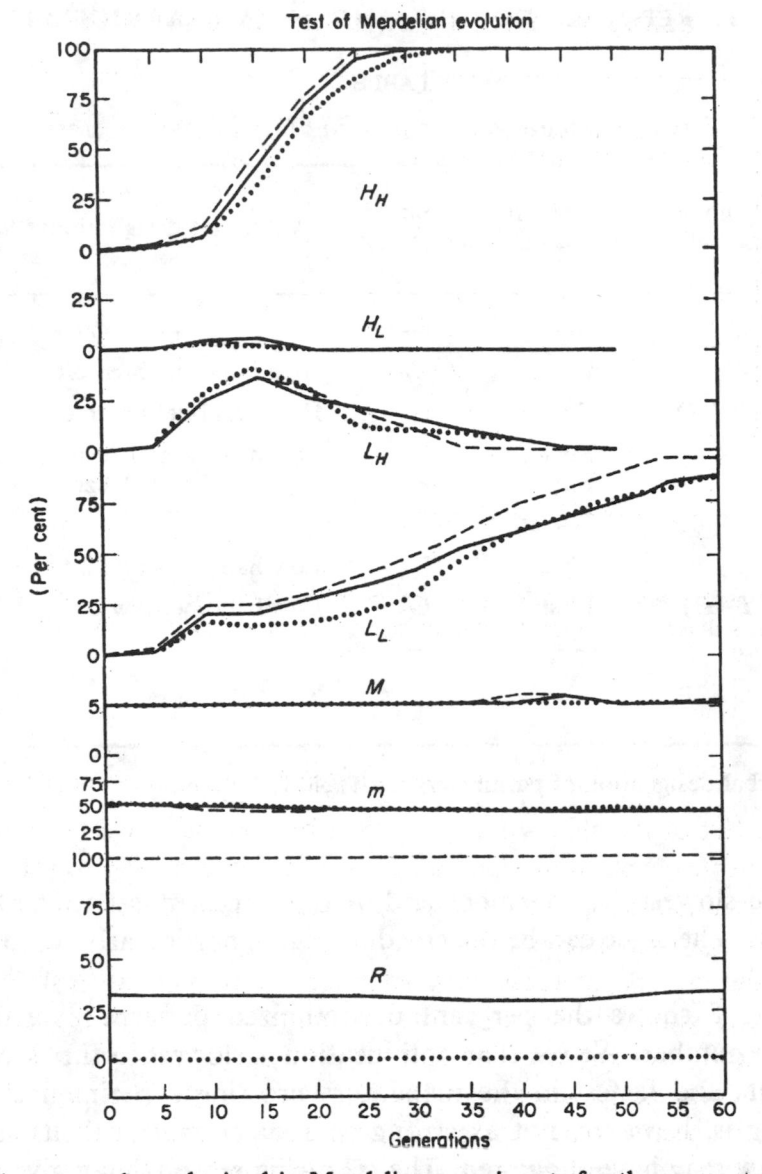

FIG. 4. Average values obtained in Mendelian experiments for the parameters H_H, H_L, L_H, L_L, M, m [see Table 2 and formula (3)] and the crossover probability (R) calculated from the distribution of C values in the population, are given as a function of the generation number. All experiments were performed using a probability of receiving high hand $h = 50\%$. ($\cdots$) no crossing (average of ten experiments); (——) free crossing (average of ten experiments); (– – –) mandatory crossing (average of ten experiments). The initial mutation probability M of each pattern in these experiments was 0·05. The initial mutation size m was given by the five-bit binary number 10000 (which when normalized is equivalent to the decimal number 0·516, see Fig. 4). This value of m makes it possible to change by mutation only the first bit (the only bit which is different from zero in the binary expression 10000 of m) in each game parameter. Therefore, as long as m is unchanged, each pattern had the possibility to optimize its game strategy only by Mendelian mutations. To optimize the game strategy by quantitative (or Darwinian) heredity would require changes not only in the first bit but in all the five bits of the H_H and L_L parameters. Mutations in the H-bit (see Table 2) were also made possible, at the same rate as in the other Mendelian bits, by special instructions inserted in the program.

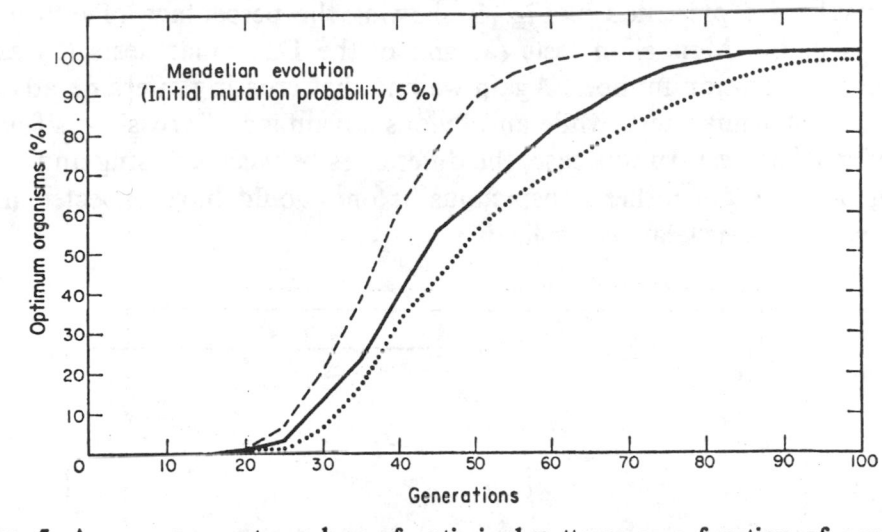

FIG. 5. Average per cent numbers of optimized patterns as a function of generation number. Each diagram represents average values for the same ten experiments used in Fig. 4.

7. A Simplified Test of Darwinian and Mendelian Selection

A better test of the effect of crossing can be obtained by a modification of the experiments designed to eliminate most of the interaction and the selective value differences among the various characters. Instead of four poker-game parameters, we used five parameters, x_1, x_2, x_3, x_4, x_5 for the Darwinian test, and the first bit of each parameter for the Mendelian test. Instead of using a poker game, the selection of winning patterns to be reproduced was made in the following way. The sum of the five parameters was compared in the two competing patterns in the Darwinian test. The sum of the first bits of the respective parameters was compared in the Mendelian test. The organism with the highest sum was the winner. If the two competing organisms had the same sum, highest priority was given to the parameter (or bit) x_1, next highest to x_2, and so forth. If also the priorities were the same, the winning competitor was randomly determined. At the beginning of the experiment all five numbers (or bits) were zeroes. To optimize their survival strategy, the numerical patterns would have to increase the five numbers to the highest possible values in Darwinian evolution, or replace the five zero bits by 1's in Mendelian evolution. The mutation, crossing and reproduction rules were the same as in the preceding experiments. However, we were able to operate with a population of 10,000 patterns (instead of 500 patterns) because of the shorter machine time required to determine the winning numerical patterns with this arrangement.

The results are presented in Fig. 6, showing the percentage of optimum organisms in the Mendelian tests (a) and in the Darwinian tests (b) as a function of generation number. Again we find that crossing offers no advantage in the Darwinian test, while an obvious advantage of crossing is found in the Mendelian test. In this case, the differences between crossing and non-crossing patterns are rather conspicuous as one would have expected in a case in which Fisher's law is applicable.

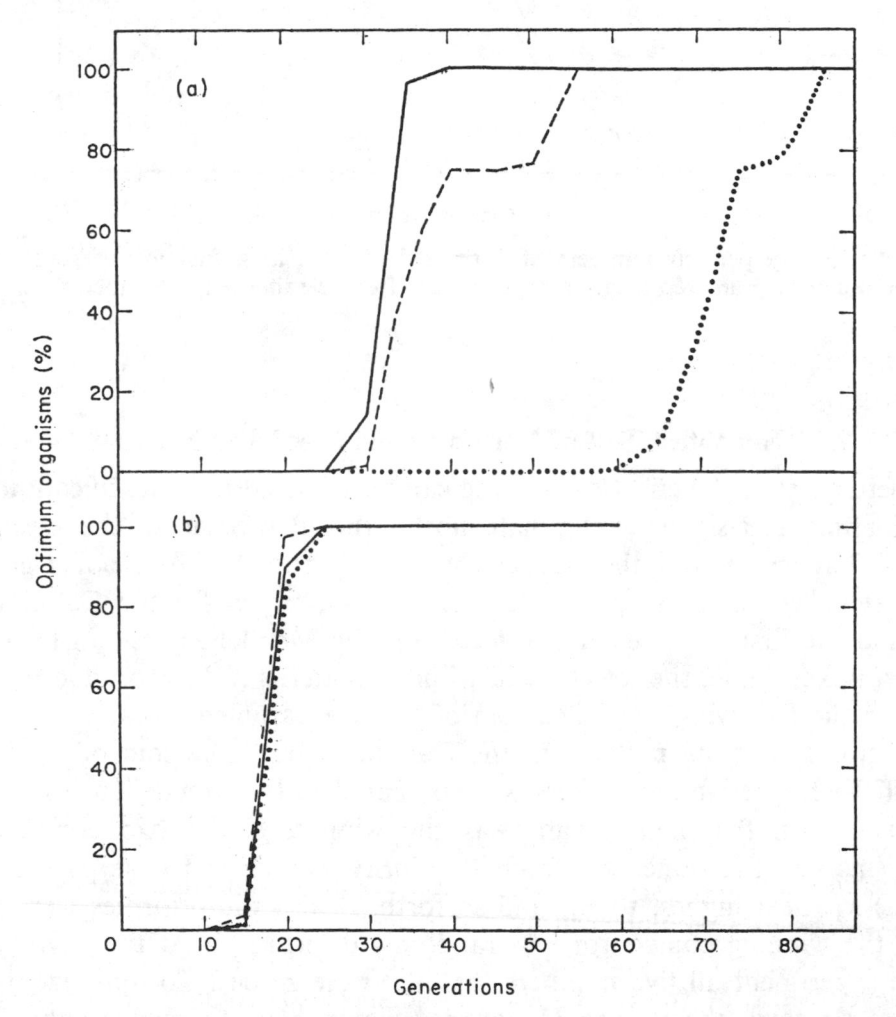

FIG. 6. Average number of optimized patterns in the simplified tests of (a) Mendelian and (b) Darwinian selection, as a function of generation number. 10,000 patterns were used in these experiments. Each diagram plots the average of four experiments. The initial mutation frequency was 0·0003 in the Mendelian experiments and 1·00 in the Darwinian experiments. The initial mutation size was given by the five-bit binary number 10000 in the Mendelian case (as in Fig. 4) while in the Darwinian experiments a low m (00001) was used. In the free crossing experiments (both Darwinian and Mendelian) the crossing parameter C was initially given by random three-bit numbers.

T.B.

23

8. Selection for Crossing Ability

A question which arises at this point is whether the ability to interbreed has a positive selective value. The fact that crossing increases the speed of adaptation of the whole species does not necessarily imply that the ability to interbreed presents a selective advantage for the individual or its progeny (Crow & Kimura, 1965). If there is a selective advantage, theoretical considerations indicate that such advantage may be present under conditions in which selective adaptation is taking place. The following experiment has been designed to test this point. The two free crossing experiments, one Mendelian and one Darwinian, of the type described in section 7, and Fig. 6, were repeated with the following modification. As in a usual free crossing experiment, we started with random crossing parameters: however, the crossing rules used to decide whether two consecutive patterns (winners) designated as possible mates would cross were changed. If both patterns have 1's in the first bit (crossing bit) of their respective crossing parameters, they would cross, while they would not if one or both of them had a 0 in the crossing bit. Under these conditions about 50% of the patterns would have no crossing possibilities at the beginning of the experiment. If crossing presents a selective advantage, we would expect the crossing patterns (with crossing bit 1) to increase in number from about 50% to nearly 100% of the population.

Figure 7 shows the result of this experiment, both in the Mendelian and the Darwinian case. In four Mendelian experiments, the average percentage of crossing patterns in the population rapidly increases to 100%, while the average in four Darwinian experiments fluctuates randomly. Crossing is definitely a hereditary characteristic presenting a significant selective advantage in Mendelian experiments but not in Darwinian experiments. The Mendelian case in Fig. 7 is a good illustration of the following point made by Fisher (1958): ". . . an organism sexually reproduced can respond so much more rapidly to whatever selection is in action, that if placed in competition on equal terms with an asexual organism similar in all other respects, the latter would certainly be replaced by the former". The question whether crossing is a characteristic without selective advantage in itself and developed only for the benefit of the species (Fisher, 1958, chapter 3; Crow & Kimura, 1965) is thus answered negatively.

9. Conclusion

The adaptive selection experiments performed by the simple arrangement used in these investigations, have been particularly useful as a means to test some of the basic evolution mechanical results and theorems. The main

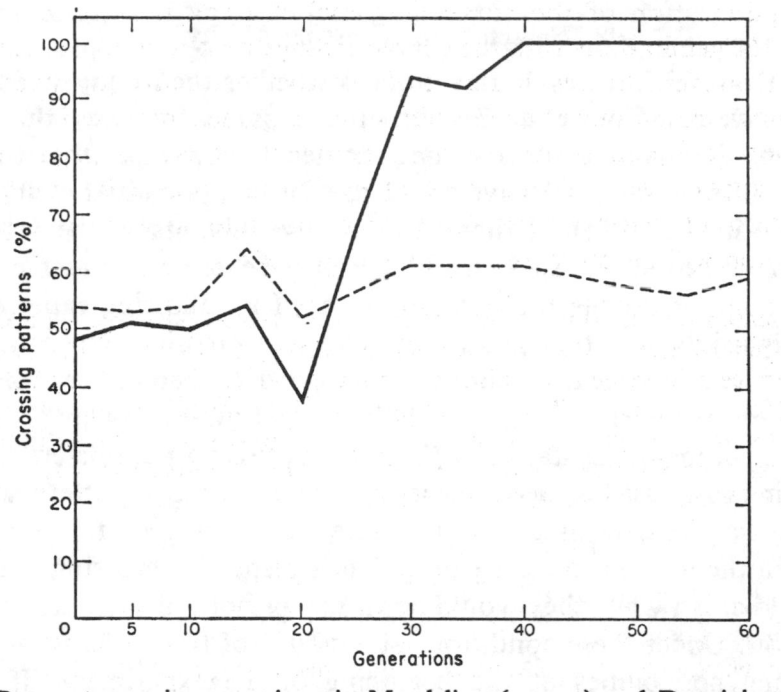

FIG. 7. Per cent crossing organisms in Mendelian (———) and Darwinian (-----)
selective adaptation experiments of the type used in Fig. 6, with a modified crossing rule
(see section 8). Each diagram represents the average of four experiments.

results can be summarized as follows. Crossing enhances the speed of selective
adaptation when the survival strategy (or in our case, game strategy) is
controlled by hereditary characters organized in a Mendelian arrangement
with few alleles in each locus. Crossing has little or no effect on the speed
of selective adaptation when the survival (game) strategy is controlled by
quantitative characters organized in a fashion designed to simulate polygenic
inheritance with frequent but small accidental changes (invisible mutations)
of the hereditary characters involved. These results are in agreement with
theoretic expectation. It has also been verified that the effect of crossing
in a Mendelian arrangement can be more or less inhibited when the Mendelian characters interfere or interact with each others expression (compare
Figs 5 and 6).

The question whether crossing presents a selective advantage for a variety
of individuals able to cross, can be answered in the following way. In a
situation of rapid adaptive selection a variety of individuals able to interbreed
with each other (but not with the rest of the population) rapidly increases
from 50 to 100% of the population with a Mendelian arrangement, but
not with a polygenic arrangement, in which case only random fluctuations
are observed (Fig. 7).

The optimization of the survival (game) strategy is often completed in less than 150 generations with the simple arrangement used in our experiments (Fig. 3). However, the development of selective instability (or major random fluctuations), maintaining various non-optimal game strategies in the majority of the population in some of the experiments, raises a problem for the practical applications of this kind of evolution process by computer. An investigation of selective instability and possible means to control this phenomenon will be presented in a separate paper.

This research was supported by research grant GM–12581 from the Division of General Medical Sciences of the National Institutes of Health, United States Public Health Service. The same grant also supported about 20% of the IBM 7094 machine time used for this research. The remaining 80% of the machine time was courteously supplied by the computing facility of the University of Washington, Seattle, Washington. The FREDERIC machine time used by Jon Reed was supplied by the Norwegian Defence Research Establishment.

REFERENCES

BARRICELLI, N. A. (1957). Methodos. Milano IX, n. 35, 36, p. 1–40.

BARRICELLI, N. A. (1962). Acta Biotheoretica. Leiden XVI, pp. 69–98.

BARRICELLI, N. A. (1963). Acta Biotheoretica. Leiden XVI, pp. 99–126.

BARRICELLI, N. A. (1964). "Elements of Evolution Mechanics". Mimeographed series of lectures given at AEC Computing Center, NYU, N.Y., 1959; Vanderbilt University, Nashville, Tenn., 1961; University of Washington, Seattle, Wash., 1964.

BOSSERT, W. (1967). "Mathematical Optimization: Are there Abstract Limits on Natural Selection?" Mathematical Challenges to the Neo-Darwinian Interpretation of Evolution, The Wistar Institute Symposium, Monograph No. 5.

BREMERMANN, H. J. (1958). ONR Technical Report No. 1, Contract NONR 477 (17).

BREMERMANN, H. J. (1962). "Optimization Through Evolution and Recombination Self-Organizing Systems", (Yovits, Goldstine, Jacobi, eds). Washington, D.C.: Spartan Books.

BREMERMANN, H. J. & ROGSON, M. (1964). ONR Technical Report, Contracts 222 (85) and 3658 (58), Berkeley, California.

BREMERMANN, H. J., ROGSON, M. & SALAFF, S. (1965). Proceedings of the Second Cybernetic Sciences Symposium, October 13, 1964, p. 157.

BREMERMANN, H. J. & SALAFF, S. (1963). ONR Technical Report, Contracts 222 (85) and 3656 (08), Berkeley, California.

CROW, J. F. & KIMURA, M. (1965). *American Naturalist*, 99 (909). pp. 439–450.

FISHER, R. A. (1958). "The Genetical Theory of Natural Selection". New York: Dover Publications, Inc.

FOGEL, L. J. (1963). "Biotechnology: Concepts and Applications", chapter 10. New York: Prentice-Hall.

FOGEL, L. J., OWENS, A. J. & WALSH, M. J. (1965). Biophysics and Cybernetics Sciences Symposium, October 13, 1964.

FRIEDBERG, R. M. (1958). IBM Journal, Research and Development, Part I, 2, pp. 2–13.

FRIEDBERG, R. M. (1959). IBM Journal, Research and Development, Part II, 3, pp. 183–191.

LEWONTIN, R. C. (1961). *J. Theoret. Biol.* 1, 382.

MATHER, K. (1941). *J. Genet.* 41, 159.

MATHER, K. (1943). *Proc. R. Soc.* B. 132, 308.

ROGSON, M. (1964). ONR Technical Report, Contracts 222 (85) and 3656 (08), Berkeley, California.

SAMUEL, A. L. (1963). "Some Studies in Machine Learning Using the Game of Checkers", (E. A. Feigenbaum and Julian Feldmann, eds), pp. 71–105. New York: McGraw-Hill.

VON NEUMANN, J. & MORGENSTERN, O. (1944). "Theory of Games and Economic Behavior". Princeton University Press.

Appendix 1

Fisher's law (cf. Fisher, 1958; Barricelli, 1964) is known for the elegant simplicity with which it expresses by a quantitative measure, designated as "speed of evolution", the competitive advantage conferred to a species by the ability of its components to interbreed. The "speed of evolution" can roughly be measured† by the number of positively selected mutations able to spread to a fixed large portion of the population (for example, 50% of the population) in a fixed period of time (which must be large compared with the spreading time of each individual mutation). Fisher's law can be stated as follows. *Fisher's law: The maximum speed of evolution a species is capable of (or potential speed of evolution) is proportional to the number of genes. A species unable to interbreed is to be considered in this connection as a species with only one gene.*

The second statement follows from the following definition of gene to be used in the application of Fisher's law.

(1) Neighboring genes must be efficiently separable by crossovers. This means that several crossovers must occur between any pair of neigh-

---

† The rigorous measure of the speed of evolution is defined as follows:

$$E = \sum_{r=1}^{n} S_r \frac{di_r}{dt},$$

where t = time (usually measured in number of generations), n is the number of positively selected mutations $(M_1, M_2, \ldots, M_n)$ expanding in the population, $S_1, S_2, \ldots, S_n$ are the selective values of the n mutations, i_r is the fraction of the gene population inheriting the mutation M_r. The mean $\bar{E}[T]$ of the speed of evolution E over a period of time T very long compared with the time each individual mutation M_r needs in order to increase its frequency from $i_r = 0$ to $i_r = 1$ (meaning the time needed for the mutation M_r to expand to the entire population) is:

$$\bar{E}[T] = \sum_{r=1}^{n} \frac{1}{T} \int_0^T S_r \frac{di_r}{dt} \, dt \approx \sum_{r=1}^{n} S_r[T] = n\bar{S},$$

where $\bar{S}$ is an average of the selective values S_r of the n mutations for the period of time T. If $\bar{S}$ is taken as a unit for the speed of evolution over the period T, the last formula expresses that the mean speed of evolution in that period can roughly be represented by the number n of mutations expanding to the entire population in that period.

An objection to the above consideration is that, for various reasons, not all positively selected mutations succeed in expanding to the entire population. A better measure of the mean speed of evolution $\bar{E}[T]$ may therefore be the number of mutations which succeed in expanding to at least 50% of the gene population (or some other large fraction of the gene population) during the time period T.

boring genes during the time required for a positively selected mutation to spread to any appreciable fraction of the population. Unless this condition is fulfilled, the two neighboring genes will, under certain circumstances, behave as a single gene with a larger number of alleles, and not as two separate genes, as far as the application of Fisher's law is concerned.

(2) To achieve the maximum speed of evolution at any particular moment, mutations expanding in different genes must not interact with one another phenotypically (or epistatically) to the point of removing or reverting the selective value of one of the mutations. Strong epistatic interactions between two genes may justify disregarding one of them in the application of Fisher's law.

The importance of crossing in biological evolution can be expressed by the following startling example of an application of Fisher's law: In a breeding species with, for example, 5000 genes fulfilling the above conditions, the potential speed of evolution is 5000 times greater than it would be if the species did not have the ability to interbreed. For such a species the loss of the ability to interbreed would be equivalent to a reduction by a factor 5000 of the maximum speed of evolution the species would be able to attain whenever required.

A modified version of Fisher's law which does not depend on the gene-notion and expresses directly the speed of evolution instead of the potential speed of evolution, is the following:

The speed of evolution at any moment is proportional to the number of positively selected mutations which are able to spread without interfering with one another either epistatically or as a result of insufficient crossovers. In a species without crossing only one mutation at the time is able to spread without interfering with other spreading mutations because of lack of crossovers.

In this form Fisher's law is almost tautological, since the speed of evolution can be measured by the number of spreading mutations.

Appendix 2

In quantitative (Darwinian) evolution changes in the population can be expressed by the time variations of the mean values $\bar{x}_1, \bar{x}_2, \ldots, \bar{x}_p$ of a certain number of variables $x_1, x_2, \ldots, x_p$ (such as, for example, body length, body weight or weight ratios, amount of a certain pigment, etc.) which can be measured† in each individual of the population. The following formula, which is nearly a literal mathematical translation of Darwin's

† The measures must be made in a representative sample of individuals kept under standard environmental conditions or by other means must be corrected for environmental influences in order to obtain measures of the variable $x_1, x_2, \ldots, x_p$ which reflect hereditary characteristics of the various individuals rather than environmental influences.

selection law, expresses the time derivatives of $\bar{x}_1, \bar{x}_2, \ldots, \bar{x}_p$ in terms of the covariances V^{ij}, with $i, j = 1, 2, \ldots, p$, of the $x_1, x_2, \ldots, x_p$ variables and the partial first derivatives of the selective value $s(x_1, x_2, \ldots, x_p)$ given as a function of the x_r variables (cf. Barricelli, 1964):

$$\frac{d\bar{x}_r}{dt} = \sum_{i=1}^{p} V^{ir} \frac{\partial s(\bar{x}_1, \bar{x}_2, \ldots, \bar{x}_p)}{\partial \bar{x}_i}. \qquad (A.2.1)$$

From this formula the following expression for the speed of evolution E (see Appendix 1) can be obtained in first approximation:

$$E = \sum_{i,j=1}^{p} V^{ij} \frac{\partial s}{\partial \bar{x}_i} \frac{\partial s}{\partial \bar{x}_j}. \qquad (A.2.2)$$

This formula can also be derived as a first approximation from Fisher's fundamental theorem, assuming absence of epistatic interactions, and time constant selective value s in each point of the $x_1, x_2, \ldots, x_p$ space. The derivation of formula (A.2.1) involves little more than a literal application of Darwin's principle of natural selection. The formula could have been derived either by Darwin himself, or more likely, by one of the mathematicians of his time, even if completely ignorant of Mendel's heredity laws.

A main property of these formulas to which we wish to attract the reader's attention is that neither the number of genes nor the contribution of individual genes to the variances and covariances V^{ij} appear in the expressions of E and $d\bar{x}_r/dt$. This means that it makes no difference for the speed of evolution E nor for the time variation of the means $\bar{x}_r$ whether the quantitative characters $x_1, x_2, \ldots, x_p$ are controlled by several genes or by a single gene. In other words, the question whether the species does or does not have a breeding and crossover mechanism has no effect for the speed of quantitative (Darwinian) evolution, at least in the first approximation range in which the above formulas are valid.

In contrast to this result we may remind the reader that in Mendelian evolution the speed of evolution E (see footnote to Appendix 1 on page 339) is directly proportional to the number of genes n and is therefore strongly enhanced by crossing.[†] The numeric evolution experiments presented in this paper were partly designed to test this theoretically predicted difference of the effect of crossing in Mendelian and Darwinian evolution.

[†] Also in Darwinian evolution it is found that crossing has a strong—in this case inhibitory—effect on the time variation of the variances and covariances V^{ij}. If we calculate the time derivatives of the V^{ij} we find that they are strongly dependent on the contributions of individual genes and are therefore strongly influenced by the number of genes separable by crossovers.

Appendix 3

The simplified poker game described in section 2 can be studied theoretically by standard methods in game theory. One finds an optimal strategy given by

$$H_P = 0, \qquad H_L = 0, \qquad H_H = 1$$

and (if h designates the probability of receiving high hand):

for $0 \leq h < \dfrac{2}{7}$: $\quad L_P = 0, \quad L_L = 0, \quad L_H = 1;$

for $h = \dfrac{2}{7}$: $\qquad L_P \leq \dfrac{7}{15}, \quad L_L = 0, \quad L_H \geq \dfrac{8}{15};$

for $\dfrac{2}{7} < h < \dfrac{3}{7}$: $\quad L_P = \dfrac{3-7h}{3-3h}, \quad L_L = \dfrac{7h-2}{3-3h}, \quad L_H = \dfrac{2-3h}{3-3h};$

for $h = \dfrac{3}{7}$: $\qquad L_P = 0, \quad L_L \geq \dfrac{7}{12}, \quad L_H \leq \dfrac{5}{12};$

for $\dfrac{3}{7} < h < \dfrac{2}{3}$: $\quad L_P = 0, \quad L_L = 1, \quad L_H = 0;$

for $h = \dfrac{2}{3}$: $\qquad L_P, L_L$ arbitrary, $\quad L_H = 0;$

for $\dfrac{2}{3} < h < 1$: $\quad L_P = 1, \quad L_L = 0, \quad L_H = 0;$

for $h = 1$: $\qquad L_P, L_L, L_H$ arbitrary.

Chapter 12
Evolving Populations

W. Bossert (1967) "Mathematical Optimization: Are There Abstract Limits on Natural Selection?" *Mathematical Challenges to the Neo-Darwinian Interpretation of Evolution,* P. S. Moorhead and M. M. Kaplan (eds.), The Wistar Institute Press, Philadelphia, PA, pp. 35–46.

A symposium was held on April 25 and 26, 1966, at the Wistar Institute of Anatomy and Biology with the title: "Mathematical Challenges to the Neo-Darwinian Interpretation of Evolution." The symposium's purpose was to bring together mathematicians and evolutionary biologists in order to clarify issues and remove potential misunderstandings (Mayr, 1976, p. 54). The participants included Chairman Peter Medawar, Murray Eden, Stanislaw Ulam, Ernst Mayr, Richard Lewontin, Sewall Wright, C. H. Waddington, and others. Of the many papers presented, Bossert (1967) (reprinted here) concerned the use of simulated evolution for optimization.

Bossert (1967) remarked on the stagnation evidenced in previous efforts by Hans Bremermann. There was an acknowledgment of the need for multiple genetic changes when the isoclines of the fitness surface aligned themselves across rather than along the genetic axes. As a potential method to overcome this stagnation, Bossert (1967) proposed possibly the earliest instance of the idea to use multiple competing populations,[1] as well as a notion to introduce variation (heterogeneity) in the fitness surface. The idea of competing populations was further enhanced by the potential for individuals to migrate between populations. The results of a simple experiment demonstrated success in overcoming a previous instance of stagnation.

The discussion following the paper concentrated mainly on the use of the term fitness; however, key insights that arose later in evolutionary computation appear in the discourse, including the intuition that strong (epistatic) interactions between genetic loci will result in a rough adaptive landscape (offered by R. Levins), although the difficulty that this presents to an evolutionary optimization met with some contention. Nils Barricelli concluded the discussion by relating the efforts described in Reed et al. (1967) restating that mating schemes (crossing over) may not lead to increased rates of optimization (see Chapter 11). The participants in the recorded discussion are (in order of appearance): Peter Medawar, William Bossert, Alex Fraser, R. Levins, C. H. Waddington, Stanislaw Ulam, Ernst Mayr, John

C. Fentress, Murray Eden, Walter Howard, George Wald, J. L. Crosby, and Nils Barricelli.

It may be of some historical interest that in the first presentation of the workshop, Eden (1967) remarked on the simulation of evolution: " . . . every attempt to provide for 'computer' learning by random variation in some aspect of the program and by selection has been spectacularly unsuccessful, even though the number of variants a computer can try can easily run into the billions. Of course, the simple explanation may be that the computer programmers weren't smart enough to set up the problem right. It seems to me that an adequate theory of adaptive evolution would supply the computer programmer with the correct set of ground rules and perhaps some day it will." Perhaps the comment spoke more to the state of knowledge of what experiments had been performed to date than to Eden's judgment of every previous attempt in evolutionary computation. But the statement did go unchallenged by those in attendance who might have had knowledge to the contrary (e.g., Crosby, Fraser, Barricelli).

References

[1] T. Bäck (1996) *Evolutionary Algorithms in Theory and Practice,* Oxford, NY.
[2] W. Bossert (1967) "Mathematical optimization: Are there abstract limits on natural selection," *Mathematical Challenges to the Neo-Darwinian Interpretation of Evolution,* P. S. Moorhead and M. M. Kaplan (eds.), The Wistar Institute Press, Philadelphia, PA, pp. 35–46.
[3] M. Eden (1967) "Inadequacies of neo-Darwinian evolution as a scientific theory," *Mathematical Challenges to the Neo-Darwinian Interpretation of Evolution,* P. S. Moorhead and M. M. Kaplan (eds.), The Wistar Institute Press, Philadelphia, PA, pp. 5–19.
[4] R. A. Jarvis (1970) "Adaptive global search in a time-variant environment using a probabilistic automaton with pattern recognition supervision," *IEEE Trans. Systems Science and Cybernetics,* Vol. SSC-6:3, pp. 209–217.
[5] R. A. Jarvis (1975) "Adaptive global search by the process of competitive evolution," *IEEE Trans. Systems, Man, and Cybernetics,* Vol. SMC-5:3, pp. 297–311.
[6] S. W. Mahfoud (1994) "Crossover interactions among niches," *Proc. of the 1st IEEE Conf. on Evolutionary Computation,* IEEE Press, NY, pp. 188–193.

---

[1]The use of competing populations (and subpopulations) was later taken up by Jarvis (1975) following Jarvis (1970), and has received considerable recent attention (Tanese, 1987, 1989; Manderick and Spiessens, 1989; Mühlenbein, 1991; Mahfoud, 1994; Bäck, 1996; and many others).

[7] B. Manderick and P. Spiessens (1989) "Fine-grained parallel genetic algorithms," *Proc. of the 3rd International Conference on Genetic Algorithms,* J. D. Schaffer (ed.), Morgan Kaufmann, San Mateo, CA, pp. 428–433.

[8] E. Mayr (1976) *Evolution and the Diversity of Life,* Belknap, Harvard.

[9] H. Mühlenbein (1991) "Evolution in time and space—The parallel genetic algorithm," *Foundations of Genetic Algorithms,* G. J. E. Rawlins (ed.), Morgan Kaufmann, San Mateo, CA, pp. 316–337.

[10] J. Reed, R. Toombs, and N. A. Barricelli (1967) "Simulation of biological evolution and machine learning. I. Selection of self-reproducing numeric patterns by data processing machines, effects of hereditary control, mutation type and crossing," *J. Theoret. Biol.,* Vol. 17, pp. 319–342.

[11] R. Tanese (1987) "Parallel genetic algorithm for a hypercube," *Genetic Algorithms and Their Applications: Proc. of the 2nd Intern. Conf. on Genetic Algorithms,* J. J. Grefenstette (ed.), Lawrence Erlbaum, Hillsdale, NJ, pp. 177–183.

[12] R. Tanese (1989) "Distributed genetic algorithms for function optimization," Ph.D. diss., Univ. of Michigan, Ann Arbor, MI.

Mathematical Optimization: Are There Abstract Limits on Natural Selection?

DR. WILLIAM BOSSERT
Harvard University
Cambridge, Massachusetts

This will be a change of pace from the previous papers. My presentation will be on a much less general level. It does follow, however, a suggestion of Dr. Eden's that we might experiment with evolutionary operators such as mutation and natural selection on a level somewhat between that of the biochemist and the natural historian.

These operators have been examined to a degree in connection with the problem of finding maxima of mathematical functions defined on a number of variables, in particular the optimization problems of operations research. At first glance the process of evolution seems quite appropriate for solution of this abstract problem. Points in the space of variables can be removed by selection inversely according to the value of the function to be optimized at the point, the analogue of natural selection. New points can be generated by chance displacements from existing points, the analogue of asexual reproduction and mutation. Several applications of "evolution" optimization have actually been carried out using just these basic operators.

There are strong arguments against the practical use of the technique, however. In general it is less efficient, in terms of number of calculations, than alternative analytic methods. This is no surprise to the biologist. The rate of evolution in asexual organisms is relatively slow. Another difficulty is of more interest to us here. The evolutionary process, at least in the simplest abstract analogue, is akin to the general class of steepest ascent optimization procedures. Although such procedures in theory assure eventual convergence to the optimum, in the case of unimodal functions, in practical application on such functions, they often terminate in an *apparent* convergence to a non-optimum point.

We can examine some experiments which demonstrate this stagnation. Bremermann and Saloff (Experiments with Patterns of Evolution, Tech. Report, Univ. of Calif., 1963) at Berkeley applied the operators, mutation and selection, to the problem of solving a set of simultaneous linear equations, a problem which yields a unimodal fitness function. Each unknown of the equations was associated with a character under genetic control, assuming the operation of the genes was independent. Mutation rates for displacements in the two characters were set according to various underlying genetic models of control. The process of solution was then iterated—asexual reproduction, mutation, and selection—according to the fitness function, from an arbitrary initial point or genotype in the space of the two characters. Sometimes, rarely, the process led to the movement of the point or population of points to the position in the character space known to be the solution of the system of equations, hence the optimum genotype.

More often the result of the process was similar to that shown in figure 1. This particular case is not from Bremermann and Saloff but is something I have computed using the process mentioned. In this figure the two characters (unknowns) x_1 and x_2 can take on values linearly arrayed over their respective axes. The optimum genotype in this particular problem is for $x_1=4$, $x_2=5$. A typical evolutionary trajectory for a point, the surviving offspring in successive generations, is shown in the

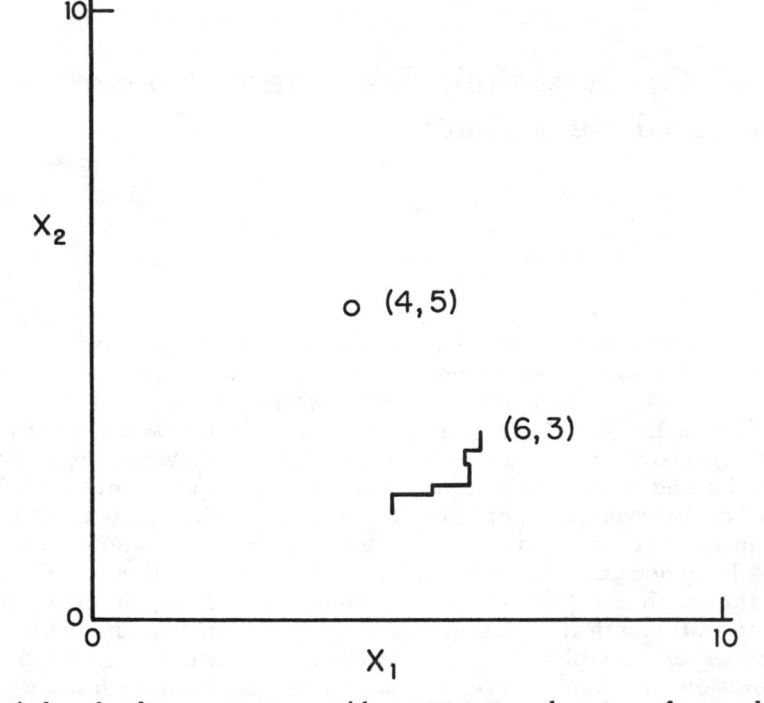

Fig. 1. Typical path of mean genotype, with respect to two characters, of a population evolving asexually under the fitness surface shown in figure 2. The population stagnates at $x_1=6$, $x_2=3$ instead of approaching the optimum $x_1=4$, $x_2=5$.

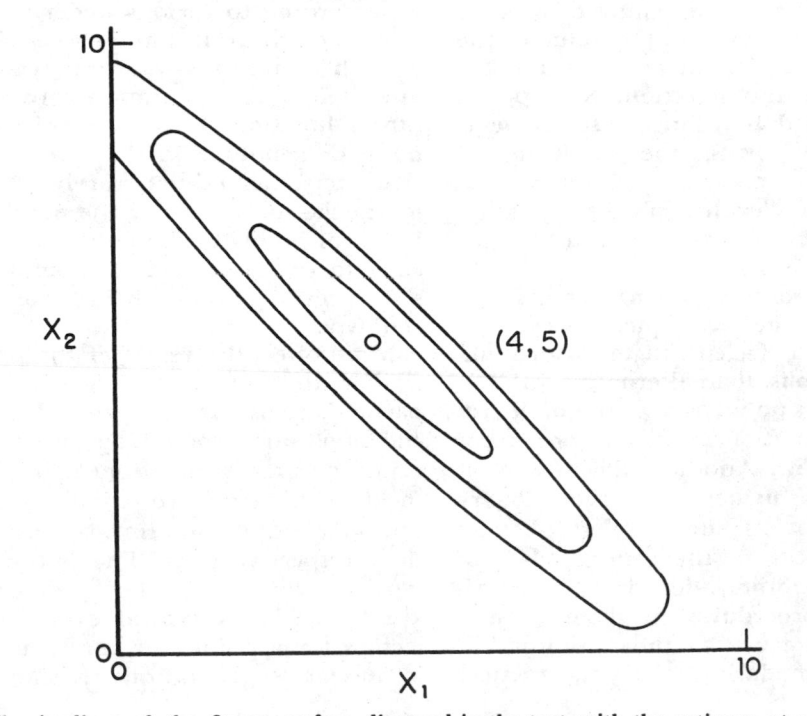

Fig. 2 The isoclines of the fitness surface discussed in the text, with the optimum at $x_1=4$, $x_2=5$.

figure beginning at about $x_1=4.5$ and $x_2=2$ and terminating not at the optimum but at $x_1=6$, $x_2=3$. The time course of this evolution was such that the movement from the initial to final point required on the order of 10 iterations of the mutation-selection process, while there was no further movement toward the optimum in 1000 succeeding iterations. The process had evolved a non-optimum genotype and stagnated. On examining the fitness function over which the process was operating, figure 2, we can come up with a ready explanation for this behavior. In this figure isoclines of fitness are shown about the optimum point. The actual values of the fitnesses are not of particular concern since selection operated in the process by removing less fit points, or genotypes, relative to others in the population. Picture this fitness surface in three dimensions. It is a long, thin ridge oriented along the long axes of the ellipses of iso-fitness. The point in figure 1 moved up the surface to a position directly on the edge of the ridge. From this position further improvement required a simultaneous change in x_1 and x_2, a simultaneous mutation at the two loci, in the negative direction for x_1, and the positive direction for x_2, of similar magnitudes. Whatever the underlying genetic model, such a joint mutation should occur with very low frequency, hence the stagnation. This simple model of evolution just does not contain any mechanism for the storage of genetic variability. Each mutation must occur at exactly the right time or be lost through selection. This explanation, as we shall see, is only partly correct.

A step which overcomes this particular difficulty is the inclusion of sexual reproduction in the process. Bremermann and Saloff attempted this and found that the difficulties of stagnation were not "automatically overcome." I found this to be true in calculations similar to theirs. The operation of mating and reproduction was defined explicitly as the production of offspring whose genotype was the arithmetic mean, with respect to all characters, of two randomly selected individuals in the parent population. Selection, again, was the removal of the least fit individuals from the population

according to the function to be optimized, and mutation, the random modification of individual characters. The results of these trials in which the initial population was distributed uniformly over $0 < x_1, x_2 < 10$, are summarized in figure 3. Here R^2, a holdover of the terminology of Bremermann and Saloff, is inversely related to the square of the fitness. Notice that on the average there was a rapid increase in fitness (approach to the solution) in the first few generations with much less change over the succeeding 1000 generations. I think it is fair to say that there is virtual stagnation here. Since, with mating, simultaneous movement in all characters is common, it is clear that our first explanation of the stagnation must be extended somewhat.

Consider the appearance of the fitness surface, figure 2, from a position on the edge of the ridge, that is, at the end of one of the long axes of an ellipse of iso-fitness. In any given circle on the character plane about this position, only a small fraction of the area is associated with an increased fitness. In the evolutionary process as defined in this simple model, even with mating, the population samples the fitness surface within some circle of variation and discards those individuals lying downhill in favor of those in the region of increase. The frequency of more favorable individuals generated will be small compared to less favorable individuals generated, when the population is on the ridge; hence the increase in population fitness must be small. By this explanation, which seems to be complete, the observed stagnation is due simply to the fact that on the edge of the ridge only a narrow angle of directions (assuming finite step sizes) leads up the ridge, as opposed to the flank of the ridge on which nearly half of the possible directions leads upward. This particular phenomenon could be of some biological interest since it contains within it the "curse of dimensionality"—the more characters involved, the stronger the effect. For example, if the ratio of favorable to unfavorable regions is $1/4$ in two dimensions, it would be $1/8$ for a comparably ridged function over three dimensions and less than .001 in ten dimensions. Fitness in natural populations is probably

393

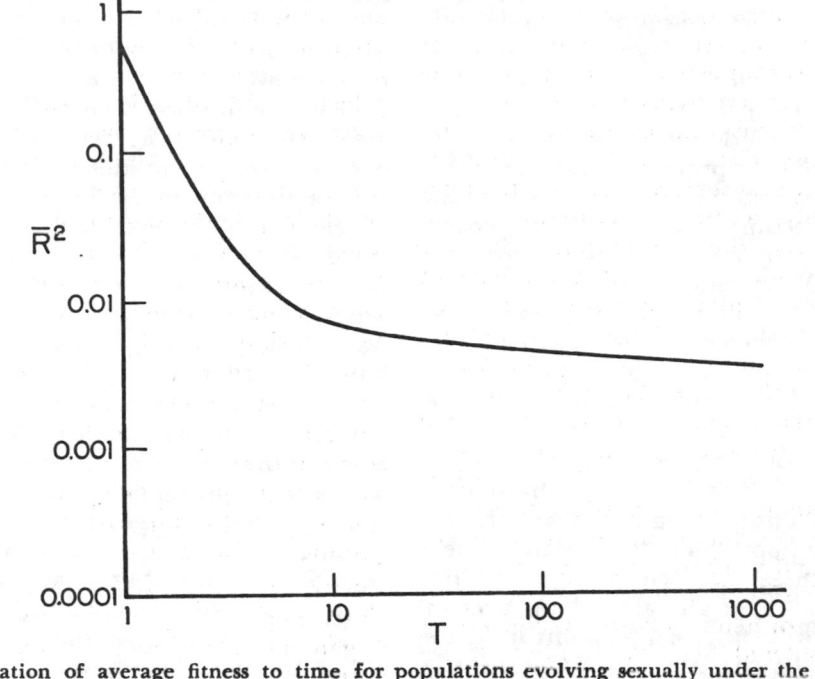

Fig. 3 Relation of average fitness to time for populations evolving sexually under the fitness surface of figure B-2. Time is in generations and $\bar{R}^2$ is inversely related to fitness. The curve is the average of nine populations.

dependent on a large number of characters, and stagnation could therefore perhaps be common in natural evolution. Beneath the conclusion, of course, is the assumption that mating, selection, and mutation in a population independent from all others, in a fixed environment, is a complete statement of the evolutionary process. Surely this could be accepted only in the rarest of situations. The phenomenon was interesting enough, however, to stimulate the consideration of additional operations in the evolutionary process in this particular abstract context.

A crude examination of the populations in the previous case seemed to indicate that lack of genetic variability went hand in hand with stagnation. The population, due to the possibility of rapid movement to the ridge but difficult movement thereafter, was usually compressed into a small region along the ridge with a low effective variability. There seemed to be two general, and recognized, evolutionary operators either of which might overcome this reduc-

tion in variability along the ridge. One would be to apply selection at a higher level, that of the population, as well as at the individual level. Explicitly this could be accomplished by introducing a number of populations, instead of one, into a single environment uniformly over the character space, letting them evolve independently as before, except that at random intervals the population with the least average fitness at the time is removed. In this form the process would merely select that population whose initial position in the space led to the most favorable position on the ridge, giving the stagnation previously found, although at a somewhat greater fitness level on the average. Only by the arbitrary insertion of populations after the beginning of the experiment or the addition of evolutionary interpopulation interactions to the process can stagnation be overcome. Both of these modifications are of biological interest; they will be discussed briefly later.

An alternative to selection by population in maintaining variability along the ridge is, of course, varying the ridge itself; that is, through the introduction of heterogeneity in the fitness surface either with time or over locations connected by migration. The former provides an interesting problem in automatic control and, by the way, evolutionary theory. Clearly, very small translations of the fitness surface over the character space possibly, with time, will not significantly displace the population from the ridge to provide any valuable increase in the frequency of more favorable genotypes. On the other hand, too large or too rapid a translation might keep the population always far from the ridge. There is then some range of translation rates at which the population can track the ridge, so to speak, but still not be caught on it. The relationships surrounding this situation are probably well-known for a deterministic ascent procedure and perhaps could be obtained for our evolutionary process. It was the introduction of heterogeneity over locations that I considered here. A number of environments with different fitness functions over the two-character space of genotypes was defined. The optimum points, orientation of the axes of the iso-fitness ellipses,

and the eccentricity, or "ridgedness," of the surface were all selected at random for each environment. A number of populations equal to the number of environments was selected and placed in each. At random intervals migration was permitted by the transfer of individuals between environments, introducing them into the instance of their population in the new environment. A modification involving the selection of populations and speciation as well as migration produced results quite similar to the simplest migration system. In this case the original populations were placed in each of the environments as before. At random intervals the least fit population was removed from a randomly selected environment and replaced by a copy of a second population selected at random over all populations and environments. This introduced population recognized no kinship to other populations in the environment and began an independent evolution. The results of a number of trials of this type are summarized in figure 4. A comparison of this with figure 3 demonstrates that the stagnation has been overcome.

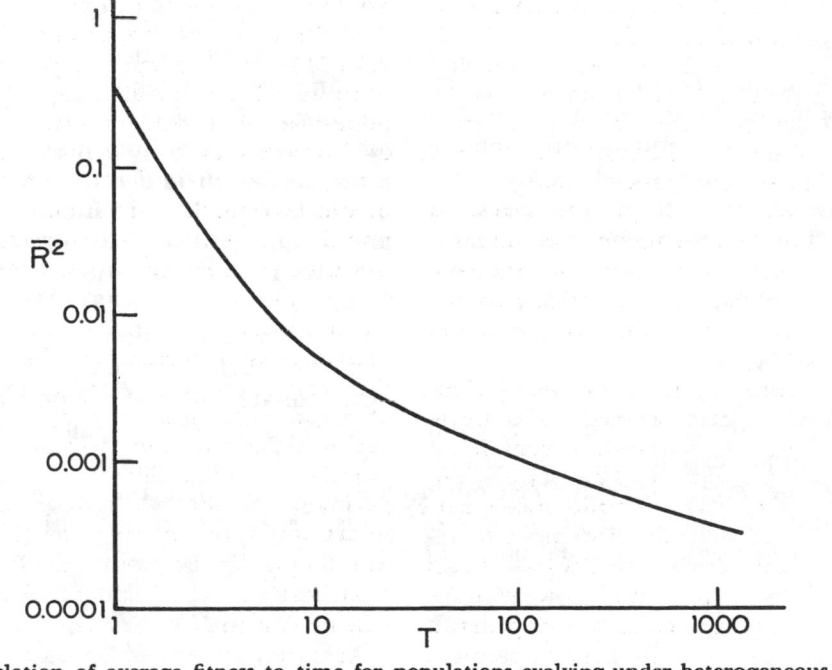

Fig. 4. Relation of average fitness to time for populations evolving under heterogeneous fitness surfaces with migration.

DR. WADDINGTON: What is the relative magnitude of the standard deviation in terms of your distance from the optimum?

DR. BOSSERT: The range in the populations was on the order of three-quarters of a unit by the scale of figures 1 and 2, although it was somewhat lower, I can't say exactly how much, for populations on the ridge of the fitness function.

DR. WADDINGTON: The variation is about 0.75 and you come to 0.001 away from the optimum in your earlier figure?

DR. BOSSERT: I must apologize for having been so vague on the nature of R^2 which appears in the figures. To explain it in detail requires a more complete definition of the abstract problem which led to the fitness function used. I had avoided doing this because it was of no biological interest except that it generated unimodal fitness functions over a character space such as the one shown in figure 2. I agree with you now, that it is important to complete the discussion of this point.

Recall that the abstract problem set was the solution of a set of simultaneous linear equations, e.g., in two dimensions,

$$a_1 X_1 + a_2 X_2 = Y_1$$
$$a_1 X_1 + b_2 X_2 = Y_2$$

where a_1, a_2, b_1, b_2, Y_1, and Y_2 are all given. X_1 and X_2 are to be determined. An arbitrary pair of values x_1 and x_2, corresponding to a genotype in our model, yields values y_1 and y_2 when substituted into the set of equations. The fitness of the trial values of x is not measured by their approximation to the solution but by the difference between the given values of Y and the y values they produce. Precisely,

$$R^2 = (Y_1 - y_1)^2 + (Y_2 - y_2)^2$$

This relation mapped on the x_1, x_2 plane gives the unimodal surface, with elliptic isoclines whose eccentricity is determined by the condition number of the matrix of coefficients in the system of equations. The R^2 measures distance in the Y plane, while population variation is interesting relative to distances in the X plane. The two are not directly comparable.

DR. ULAM: Isn't it true that there is no absolute standard of excellence or fitness; it is only a relative question, always?

DR. BOSSERT: Yes, and it was this thought that led me to neglect a definition of R^2 earlier. Between two individuals only the sign and not the magnitude of the fitness is of importance when selection is carried out in the deterministic manner as in the models discussed. Fitness can be tied, of course, to the probability of survival in which case the absolute magnitude is certainly of importance. One general criticism of these models and, in fact, of current evolution theory is that the concept of fitness needs to be broadened.

To summarize briefly, the basic evolutionary operators of mutation and natural selection do not perform well, in the sense of translating a population to which they are applied over a space of genotypes to that genotype which is optimum. The various groups of applied mathematicians who recognized this quickly discarded the operators as abstract mathematical tools and gave passing concern about the state of our understanding of evolution. The finding is no surprise and of little interest to biologists since mutation and selection do not complete the "neo-Darwinistic interpretation of evolution." Additional operators such as migration and selection of populations easily overcome the difficulty of stagnation at non-optimum fitness levels, which turned these particular mathematicians away. It is unfortunate that communication ended on both sides at such an elementary level. From the brief exploratory continuation I have made I feel that there are developments possible in this context which could be important in both disciplines.

Discussion

PAPER BY DR. WILLIAM BOSSERT

The Chairman, DR. MEDAWAR: Thank you. Exactly what do you mean by a broader view of fitness? In what way is it broader than the one we are working with?

DR. BOSSERT: I was disturbed, in our morning's discussion, that no one brought up the fact that fitness might be closely related to the adaptive form already achieved.

The Chairman, DR. MEDAWAR: How could it be otherwise?

DR. FRASER: Would you put that algebraically? Are you saying that the fitness is a dependent as a frequency-dependent function?

DR. BOSSERT: The fact is that the fitness is not determined solely by the environment; it is not a competition between individuals and the environment, but it is a competition of one against the other in the structure of the community of population so far defined. I hope we are going to get into this.

DR. LEVINS: Population mass curves your fitness space.

DR. BOSSERT: Yes, that puts it very concisely. This is the consideration I would hope to add to the discussion of fitness. Let's follow the displacement for several populations in a two-component space. In figure 5a two populations are introduced to the environment. The letters A and B stand for mean phenotypes of populations A and B. The deviations are not shown, but the variation in each population is on the order of one-tenth of the maximum range in each X and Y.

After a few generations, the populations displace as shown in figure 5b. A third population, C, has been introduced at this time. As we follow the displacement of these populations, we see they approach a configuration in the spaces which is quite independent of the order and exact values of the mean phenotypes in which populations are introduced into the environment.

Figure 5c shows the triangular pattern one would achieve in a couple of hundred of these selective steps, no matter in what

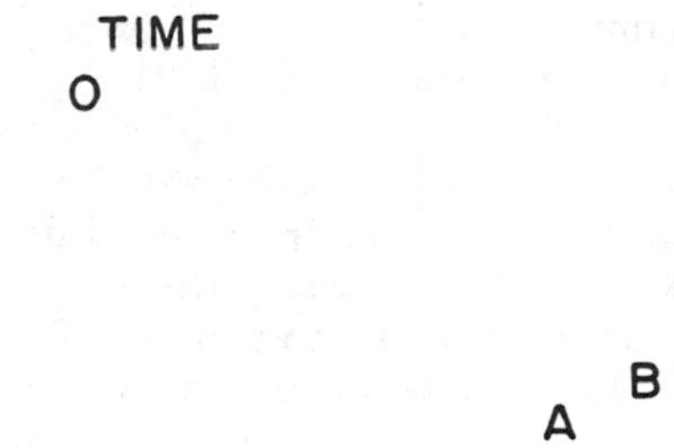

TIME 0

B
A

Figure 5 (a)

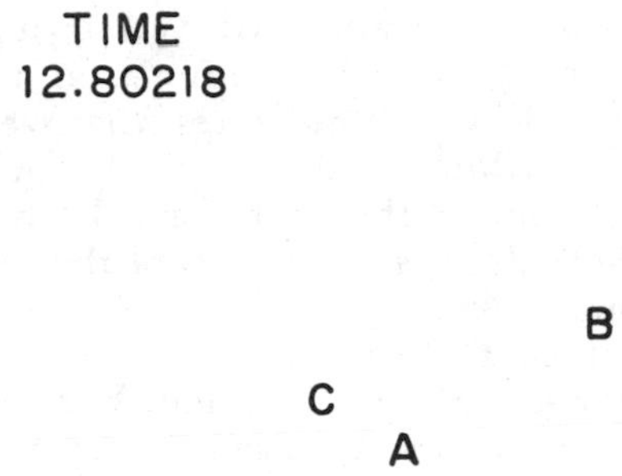

TIME 12.80218

B
C
A

Figure 5 (b)

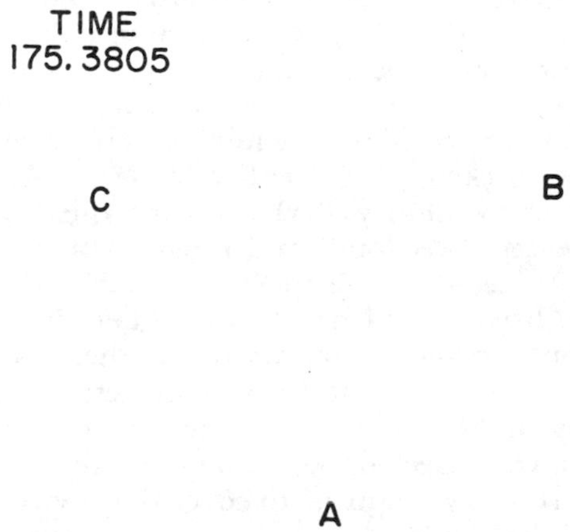

TIME 175.3805

C
B

A

Figure 5 (c)

Fig. 5 Mean phenotypes of three populations with respect to two components of mating behavior. The standard deviation about these means is approximately one-tenth the total range shown in each case.

The time is given in generations:
(a) Initial contact of two populations; (b) Insertion of third population; (c) Pattern achieved after a period of displacement.

order the population had been introduced, and no matter what initial values of the components were used for each of the introduced populations.

DR. WADDINGTON: Under what auspices do they come into this pattern? Is this with exclusion of intermating or is this productivity?

DR. BOSSERT: This is due to the exclusion of intermating. There is no difference in fitness correlated with the two components assumed. Here the fitness function over this space has been warped by the populations themselves from a constant plane to a figure with three hills.

DR. ULAM: I do not quite understand some of your diagrams. Are you trying to say that there is no absolute function, i.e., a given function which measures what we call fitness but which really depends on the relative numbers; that is to say, the fitness of one individual is a function of the fitness of his competitors?

DR. BOSSERT: Yes.

DR. ULAM: Well, the game that nature seems to be playing is difficult to formulate. When different species compete, one knows how to define a loss: when one species dies out altogether, it loses, obviously. The defining win, however, is much more difficult because many coexist and will presumably for an infinite time; and yet the humans in some sense consider themselves far ahead of the chicken, which will also be allowed to go on to infinity. Perhaps it is a question of greater "freedom" and range of actions and choices that one species can make.

DR. BOSSERT: I don't care to take it too far, but, in fact, what we do see here is a presentation of a strategy with respect to the use of the environment that one might not have expected. For example, an optimum strategy might be to cover the environment uniformly.

You see, in this last case the populations have arrayed themselves in a uniform pattern. Had we had six populations here, you would have seen them evolve a cluster in which the six populations achieved a uniform coverage of the subspace that they occupied in these two figures.

DR. MAYR: I think you ought to emphasize that is is a very specific case.

This is reinforcement of isolating mechanisms.

DR. BOSSERT: Yes, I hope I began with that.

DR. LEVINS: Atrophic competition produces something quite similar.

DR. FENTRESS: I am still confused by the definition of fitness itself. This was the point raised earlier. It sounds as if you are speaking of fitness in an absolute sense. As a biologist, I am having trouble understanding what you are getting at.

The Chairman, DR. MEDAWAR: Dr. Bossert, could you possibly begin a sentence with the words "Fitness is," and proceed from there?

DR. BOSSERT: All right. Fitness is the relative probability of an individual giving offspring which will take part in the next reproductive population, or the relative numbers that an individual will add to the succeeding reproductive population.

DR. FENTRESS: How can you approach a certain level? This, I don't understand. You say that the animal is approaching a certain level of fitness, as if this were a fixed thing, but you could increase the number of offspring.

DR. BOSSERT: No, no, it is a relative thing.

DR. FENTRESS: You had the stagnation. The animal has been approaching something which I still don't understand.

DR. FRASER: It seems to me that he has W and S confounded in such a way that what he is really talking about is intrinsic rate of increase; and he is comparing them between points on a fitness space.

DR. EDEN: Dr. Bossert, let me see if I understand your model. You are saying that if you define two populations on one parameter with a distribution function on each, the region in which they overlap corresponds to hybridization in some sense. That is to say, where the populations overlap in a property, you don't know whether an element is a member of population A or population B.

DR. BOSSERT: Yes, except by the test of mating. The offspring will have viability depending on whether or not they are from a hybrid mating.

DR. EDEN: But where they don't overlap, they breed true, so to speak, either A or B, with whatever probabilities exist.

DR. BOSSERT: Yes.

DR. EDEN: Let me ask the question, then: Presumably, there must be some minimal distance between, let us say, the mean values of the populations which will push them apart through the selection mechanism which you have proposed; but there must be some threshold below which these two populations will fuse and become a single population; is that correct?

DR. BOSSERT: No, not in the examples I have given; because I have assumed that the offspring of these hybrid matings have negligible viability, so in fact they can never fuse. The other case is quite interesting but I don't think we should take time to go into it.

DR. WALTER HOWARD: Does your fitness have to do with successful breeding?

DR. BOSSERT: That is one component of it. All I am saying is that in this particular illustration, this component might have been considered separate from the additional components of the environment.

DR. HOWARD: But your figures are based entirely on that, your curves?

DR. BOSSERT: No, the curves are based on a combination of the two.

DR. LEVINS: First, when we talk about fitness it is always subject to certain constraints which are not always made explicit. Usually in population genetics we talk about optimization of fitness in the sense of finding a gene frequency from among the set of available alternatives which maximizes the fitness function. If you introduce new alleles from the outside, you change it completely.

In the same way, when you talk about phenotypic components of fitness, you can say that a heavy shell on an organism will make it resist certain kinds of predators better but a light shell will help it run away better; so, depending on the kind of situations that it faces, you can sort out adaptations and conflicts.

In the matter of protective coloration, then you have a new component of fitness in which the kinds of escape are not conflicting anymore; the whole surface is changed.

But what I would like to touch on is what I think is the real significance of Dr. Bossert's presentation. First of all, that selection can be extraordinarily slow and that, therefore, we are not justified in looking only at equilibrium phenomena when we talk about the maintenance of genetic variability. In fact, there must be a large number of situations in which there is quasi-equilibrium, populations stuck around the ridges or at saddle points. Secondly, the behavior of such a population, when it is displaced from a quasi equilibrium, will not be the same as that of a population displaced from a true equilibrium. At a true or stable equilibrium the population will return. When displaced from a quasi equilibrium, it could go on higher or do some other things that we don't really know; so this is a problem for the mathematician—the behavior around points of quasi equilibrium.

Third, the shape of this fitness space in which the points are moving depends very much on the way the phenotype is integrated in the development system. Roughly, strong interactions between genetic loci will give a very rough surface with lots of ridges. The simplest situations of additive effects just give you a simple sloping surface in which points can move quite easily. Therefore, high degrees of genetic epistasis between loci will produce systems remarkably resistant to natural selection, systems which will have very long memories and, therefore, would be important in higher order adaptive mechanisms.

Finally, I suggest that this may be the kind of situation that accounts for the anomalous observations of the Australian grasshopper, which is persistently hanging around on a point which is practically a minimum of its fitness surface.

DR. BOSSERT: May I first say that I think Dr. Levins was much too kind to attribute both of those points to my presentation. The second certainly was his.

DR. WADDINGTON: This business of displacement from a quasi equilibrium and then coming back not exactly to where you started from—this is surely the basis of

Sewall Wright's ideas. He displaces from a quasi equilibrium or sub-peak. He displaces by a small population sample, whereas here you are displacing by some determinate form of displacement.

DR. BOSSERT: I think it is important that this is not a sub-peak.

DR. WADDINGTON: This is a ridge; when there is an increase upwards it is exceedingly slow. You have quite steep fitness gradients going downward, but a very slow fitness gradient going up. The thing can't go fast until you go down; and then you get another rush at the hill and you come up to a different place. But surely this kind of thing is exactly what Sewall Wright was talking about in saying that if you got small enough population samples, you would be wandering around for purely statistical reasons and, therefore, would tend to come down a bit and then go up again a bit further.

DR. BOSSERT: In the examples given, the gradients on the slopes were several times the gradient up the ridge, but the differences in rate of movement of the population were much greater than this. The stagnation is due to the sharpness of the ridge rather than to the difference in selective gradients directly.

DR. LEVINS: This is different in the sense that in the average Sewall Wright case, the point returns toward the peak; selection is pulling it back toward the peak until it has gotten very far away. In that situation, even a very small displacement from the ridge will send it off in a different direction. Furthermore, because of the shape of the ridge, if a population has a large phenotypic variation, most of the selection is stabilizing, keeping it on the ridge, rather than sending it up the ridge.

DR. WADDINGTON: Yes, but you have invented the ridges. The ridges are the particular fitness surface you happen to have invented. You might have invented a different one.

There is a second point I want to make. This idea that there is a great deal of interaction and epistasis, in fact what I call canalization, gives you a rough fitness surface in which you can't wander about at liberty but have to follow the peaks and hills. This is the whole point, I think, of canalization; but Alex is shaking his head.

DR. FRASER: The idea that if you have got epistasis, then selection will be ineffective, has not been confirmed experimentally. When somebody decides to have a look in an experimental animal, he selects and finds that he is selecting on the basis of epistatic variants. This happens; it has happened too many times to be disregarded. You can simulate a genetic system in a machine where, by selecting for an optimum, you have put in a tremendous amount of epistasis, and obtain significant effects of selection. If there is fluctuation of population size—and I was quite intrigued by the deduction of the necessity for this by the second speaker this morning—your selection can be highly effective. We have selected in Drosophila, and had the whole of our advance being a first by third chromosome interaction, not an additive component change at all. To make statements that you can't play around with epistatic variants is not convincing. The statement that you have a highly flat fitness surface with a lot of little ridges in it is convincing.

DR. WALD: I would like a chance to ask a stupid question. With this as a definition of fitness, can any meaning be assigned to too great a reproductive success in one population over another?

DR. CROSBY: The point I would like to raise is, in a sense, complicating the issue of fitness. I have been working for some time on this question of hybridization, in a rather different way from the way in which Dr. Bossert has. The situation which is particularly interesting from the point of view of this discussion is the one in which the initial selection of the isolating factors is slow, and which has a maximum degree of overlap. Under those circumstances, the fitness is determined not by anything to do with the environment or with the genotype, but solely by the question of numbers; because under these circumstances the less frequent species is at a disadvantage.

Let's take species A and B; if A is less numerous, it will in fact produce proportionally more hybrids among its descendants than B. This means that unless the isolating factors are produced and established by selection rather quickly, one of the two species is going to be wiped out.

We then have a rather pretty situation. Suppose for the moment B is the majority one; what you are saying is that its fitness will be increased if it develops isolating factors. This is not necessarily true; because its fittest position may be that in which it is able to wipe the other one out through hybridization.

DR. BOSSERT: You have brought up a very difficult thing. You have pointed out that there might be varying fitness over different overall complex strategies. For example, the strategy of resisting displacement might prove superior in some contexts. In the work I have done, the genetic mechanisms underlying the mating behavior were always identical in the various populations. So, what happens is that you find the smaller population being forced to move, or having selections for rapid movement. In that case, one finds further reduction in numbers in this rapid displacement to the point that even though the rapid displacement seems to be advantageous in that it decreases the relative number of hybrid matings, one finds it disadvantageous because one is selecting for individuals which have a very small probability of finding a mate at all, either A or B.

DR. CROSBY: In fact, the decrease in numbers is exponential. There is no question about it, it just shoots away, and selection cannot keep pace with it.

DR. BOSSERT: Yes, and I would predict that in most instances of displacement, one would find, in fact, the most rapidly displacing population becoming extinct.

DR. CROSBY: There is, in fact, a selective advantage to one species in being able to outhybridize the other, and I think this may well have happened a lot of times in the course of evolution.

DR. BARRICELLI: I have done some similar experiments together with Robert Toombs in Seattle, which in part are a repetition of experiments done by Jon Reed in Norway. They are based on the following: The method is basically the same as described by Dr. Bossert. Its purpose is to optimize self-reproducing numeric patterns using a high-speed computer. Each numeric pattern was composed of eight numbers comprising altogether 36 bits and which could therefore be stored in a single memory location of the IBM 7094 machine we were using. Four of the numbers are betting probabilities in a poker game, actually a very simplified version of a poker game in which each player received only one card, which could be either a high card (high hand) or a low card (low hand). There were 100 organisms or patterns and they played 20 games of poker, 50 organisms against the other 50. The losers were eliminated, and the winning organisms reproduced, so at the end of each generation we ended up with 100 organisms again.

The first number in each pattern (organism) was a crossing parameter; then we had a number representing the probability of betting high with high hand. Then we had the probability of betting low with low hand followed by the mutation probability. Then we had the probability of betting high with a low hand, and low with a low hand, and then we had a parameter regulating the mutation size. All of these numbers could be modified by mutations.

We did three experiments in this quantitative fashion. Incidentally, the optimum game strategy was calculated by Von Neumann's method, and we found that if the probability of receiving a high hand was 50 percent, the optimum game strategy would be always to bet high with high hands (probability 1 for betting high, 0 for betting low and 0 for passing), and always to bet low with low hands (probability 0 for betting high, 1 for betting low, and 0 for passing with a low hand).

In this particular case, when the probabilities of receiving high or low hands are 50% each, only the two extreme values of the betting probabilities—1 and 0—are compatible with optimum game strategy. We succeeded actually by the procedure described, in obtaining betting probabilities close to the optimum values within two or three hundred generations.

We have also used other probabilities for receiving a high hand. In some of these cases the optimal values of the betting probabilities were not 1 or 0, but intermediate. This situation is more similar to the case presented by Dr. Bossert. We also observed the phenomenon Bossert has described, namely that the organism improved up to a certain point, which was not optimal. With different experiments we get different results which usually differ from the optimum betting probabilities.

The strange thing about all these experiments, which was observed first by Jon Reed, was that when we used crossing (either obligatory crossing or free crossing, meaning crossing regulated by the crossing parameter in each pattern) or no crossing at all, we got approximately the same speed of evolution. Crossing made no difference; Jon Reed was very surprised about that. Fortunately, we had already made some theory on a population genetic basis about evolution by quantitative characters. We expected that crossing would make no difference as long as one operates with quantitative characters only. We were therefore, able to explain Reed's result; however, this posed a challenge. What if we could repeat the experiments without using quantitative characters? If that were possible, one should expect a considerable enhancement of the speed of evolution produced by crossing, according to Fisher's law.

By a modification of the experimental arrangement we have been able to do just that. Instead of using betting probabilities which could have a continuous value, we used only the first bit in each betting parameter. If that bit was 1 in the parameter for betting high, the numeric organism would always bet high. If it was 0, then the machine would look at the first bit in the parameter for betting low; if that bit was 1, the organism would bet low; if it was 0, the organism would pass. This way we had defined a kind of Mendelian system, based on a few well-defined properties (such as the property of always betting high with high hand, or always betting low), rather than using intermediate probability values or quantitative characters. With this arrangement, crossing proved to enhance the speed of evolution in agreement with Fisher's Law. In other words, when we have four genes, or four bits controlling the betting probabilities as in our case, we get a speed of improvement about three-and-a-half times higher by allowing crossing than without crossing, which is fairly close to the result expected according to Fisher's Law.

Chapter 13
Artificial Ecosystems

M. Conrad and H. H. Pattee (1970) "Evolution Experiments with an Artificial Ecosystem," *J. Theoret. Biol.*, Vol. 28, pp. 393–409.

NATURAL evolution is characterized by levels of hierarchy: the gene, the individual, the population, the ecosystem. The interactions across these levels of hierarchy are important in determining the course of evolution. Genetic interactions affect individual behaviors, as do populational interactions. Selection at the level of the individual affects gene frequencies, while simultaneously affecting ecosystems by modifying, among many factors, food webs. Each level of the hierarchy can affect all others either directly or indirectly. Simulations of evolution that are offered in the hope of gaining a greater general understanding of evolutionary processes should allow for the systematic study of their hierarchical character.

The paper reprinted here (Conrad and Pattee, 1970) offered one of the earliest simulations of a hierarchical artificial ecosystem (details on the programming can be found in Conrad, 1969). A population of cell-like individual organisms was subjected to a strict materials conservation law which induced competition for survival. The organisms were capable of mutual cooperation, as well as executing biological strategies that included genetic recombination and the modification of the expression of their genome. No fitness criteria were introduced explicitly as part of the program. Instead, the simulation was viewed as an ecosystem in which genetic, individual, and populational interactions would occur and behavior patterns would emerge.

The organisms were composed of what were essentially genetic subroutines. Individual phenotypes were determined by the manner in which these routines were used by the organisms. The fixed set of routines that they provided were somewhat limiting, but Conrad and Pattee (1970) accepted this limitation so long as the "process of modifying these features produces a system whose behavior converges to that of a natural ecosystem." In particular, they were concerned that (1) behavior that is characteristic of ecological succession processes should potentially emerge, (2) the processes of the evolutionary search must agree with biological fact (i.e., nonbiological operators that would make search more "efficient" were precluded), and (3) the simulation should be as simple as possible to allow for studying both the fundamental features of ecosystems as well as the minimum necessary conditions for natural evolution.

The events in the evolution program took place in an environment termed the *world*. The world was a one-dimensional string of *places*. The first element in this array was linked to the last, forming a loop, to avoid end effects. Each place in the world was characterized by a state (A or B) and a certain number of material parts called *chips*. The organisms determined the control of the flow of the chips in the world. Chips were essential to persistence (i.e., survival) and reproduction, but no explicit behavioral strategies were introduced to acquire these chips. Any such behaviors emerged from the initial conditions of the simulation as the conservation of matter (a fixed number of total chips) induced a competition for chips among the organisms.

Each organism in the simulation was associated with a place in the world. Organisms operated over a range of contiguous places (i.e., a territory). Events occurred in discrete periods with the lifespan of an organism extending over a number of these periods. Organisms interacted in a two-phase system: In the first phase they interacted with the environment and the other organisms in their local areas, while in the second phase chips were collected for the various behaviors, reproduction occurred (if applicable), and chips from "dead" organisms were deutilized (i.e., returned to a "matter pool"). The effects of birth and decay determined the composition of the next iteration, and the process was then repeated.

As indicated above, the organisms were modeled with an explicit genotype-phenotype. An organism's genome was mapped to a phenotype according to a coding function. Sixteen possible pairs of genomic symbols were associated with six types of phenome symbols. The organism's immediate behavior depended on an "internal state" and an "input state." The sequence of internal states, combined with the input states and the other organisms in the population, determined the chip-collecting behavior of the organism. Organisms could allocate chips for self-repair, and reproduction was allowed when a sufficient number of chips had been stored. The genetic program of the organisms was subjected to both point and size mutations, as well as a recombination operator which consisted of breaking and splicing the genomes at random places. The location of daughter cells created by reproduction was restricted, but was also under genetic control.

The population size ranged from 200 to 400 organisms, depending on the number of chips made available at the start of the simulation. The distribution of the chips was affected only by the organisms' behavior and therefore the fitness criteria varied during the simulation as organisms interacted (noise was also added to the environment in some experiments). Attention was focused on features such as chip utilization (ratio of chips used to chips not used), matching ratio (successful matches to those attempted), and changes in size of population (survival curve).

Some general observations of these experiments included: (1) the matching ratio of phenotypic characters to environmental conditions increased when the environment contained no noise, (2) adding noise led to a diversity of phenotypic types with no indication of impending homogeneity, (3) in general, the probability of recombination tended to decrease, (4) utilization of the environment tended to increase, and (5) the predominant organisms carried phenome sequences that were not executed and therefore of no selective value.

The description of this model (termed EVOLVE) and results were extended in Conrad (1981) and led to a new version of the simulation (EVOLVE II in Conrad and Strizich, 1985; also see Strizich, 1982). In this new implementation, the environment was modeled with parameters for light, temperature, and available food (denoted by "mass," which could be passed back to the environment in the form of organism waste and "dead bodies"). Total mass was again conserved throughout the simulation. Organisms possessed three genes and their phenotypic traits were determined from their genotypes. A historical record was also maintained for each organism with 11 entries including, for example: (1) mass required to reproduce, (2) mass collected, (3) age, (4) number of reproductions, (5) match to temperature, and other features. Food was allocated to organisms on the basis of how well they matched the environment relative to other organisms. An "aging number" determined the likelihood for surviving to another cycle and those with sufficient mass were reproduced.

During reproduction, each gene was given a 1/3 chance of mutation. Each gene was composed of two values: the first coded for the phenotypic trait the gene determined and the second coded for that gene's amenability to evolution (i.e., the likelihood of mutation to a gene that codes for a similar function). Phenotypic development was accomplished in a simple one-gene/one-trait manner (i.e., each gene separately encoded a trait). The environment was updated according to a user-selected schedule of varying the light and temperature, as well as updating the decaying mass left over from other organisms, and the cycle was then repeated.

A series of experiments examined the relative effects of varying different parameters, such as decay rate on mass and the cost of amenability. Under constant environments the temperature and light genes of the organism converged toward the environment, but in more variable environments the genes often matched only the lower bounds of the environmental parameters. Organisms were observed to carry traits that were advantageous to the population even though they led to the production of offspring that had traits which were disadvantageous to the individual. For example, high aging values with low amenability led to a higher probability of individual death but a greater chance of creating offspring with low aging values and thus greater probability of survival.

The model was again extended in EVOLVE III (Rizki and Conrad, 1985; see also Rizki, 1985; Rizki and Conrad, 1986), which comprised three levels of organization: populations of organisms, individual organisms, and gene structures. Fifteen phenotypic traits were determined: (1) temperature optimum (i.e., the ideal temperature for the organism), (2) temperature tolerance, (3) light intensity optimum, (4) light intensity tolerance, (5) rate of energy intake, (6) rate of energy outflow, (7) the minimum energy that must be maintained for self-repair (i.e., below this level the organism dies), (8) maximum energy capacity, (9) the amount of energy required prior to reproduction, (10) protection mechanisms, (11) aggressive mechanisms, (12) developmental period, (13) adult period, (14) rate of reproduction, and (15) mutation rate. It is noteworthy that the behavioral characteristics explicitly included parameters regarding the manner of interaction with other organisms, as well as a self-adaptive parameter regarding mutation rate.

Each trait was encoded by a maximum of 15 genes, and the set of genes used for each trait did not have to be unique (i.e., overlapping sets of genes were possible). Actions were triggered by environmental states. For example, migration was triggered if an organism failed to gather sufficient energy or was displaced by another organism. Migration moved the organism to a new location, with an associated energy cost for the movement. A sophisticated encoding scheme was implemented to model genetic structure. A single strand of nucleic acid was coded as a string of binary digits using two bits of information for each base. The linear sequence of nucleic acids was partitioned into critical and modifier sections. Critical sections were at fixed locations and were of primary importance in determining the "major level of gene function." Codons were extracted from the critical sections and used as arguments to a function which performed a dot product with a weighting vector to trade off the importance of different codon values. A similar procedure was implemented for the modifier values except that these were limited to effect only small changes.

At the highest level of organization, the ecosystem contained a set of populations (each composed of individual organisms) and an abiotic environment. The primary distinction between populations was their food source. In the simplest case this took the form of two populations: producers and decomposers. Decomposers consumed the waste products of producers and returned mass that became nutrients for producers after a certain time delay. This formed a food cycle. No restrictions were placed on the number of populations; additional types of populations could be introduced to utilize alternative food sources, creating a food web. Even predator organisms were also possibly included and their success was determined by the phenotypic traits of protection and aggression.

The environment was modeled to a two-dimensional space divided into regions. Each region had a number of attributes including its temperature, light intensity, piles of food of various types, and adjacency to other regions. Organisms could move between regions at a cost. The number of organisms in a region was potentially unlimited, but only one organism of each type of population could be actively feeding, thus resulting in competition. Conflicts between similar types were decided on the basis of phenotypic appropriateness (i.e., match to environment).

A review of the series of simulations (EVOLVE I, II, and III) appeared in Conrad and Rizki (1989). Symbiosis between organisms was later introduced in O'Callaghan and Conrad (1992) and resulted in "arms races"—escalations in values of phenotypic traits that were counteracting. The symbiosis was not forced into the simulation, but rather only the capability for such symbiosis was provided.

Conrad and Pattee (1970) framed their investigation around the question of which processes are essential to produce evolution, and offered, "The only way to approach this problem is to design an artificial ecosystem which we can test and modify." This presages the theme of artificial life, offered 17 years later in Langton (1987): "life as it could be, rather than life as it is." The Tierra simulations of Ray (1992) (Chapter 23) were also similar in design to EVOLVE and its successors, with CPU time allocated to competing programs taking the place of chips, and fitness criteria to measure the effectiveness of the programs in the simulation emerging from the competition for these resources.

Michael Conrad's contributions to evolutionary computation are not limited to the study of simulated ecosystems. In addition to his study of evolutionary models of brain behavior (Conrad, 1974) (see Chapter 17), Conrad (1977, 1978, 1979, 1982, 1983, 1990; Conrad and Rizki, 1980) has long argued that the evolvability of a system is itself subject to selection: selection will tend to eliminate organisms (genetic systems) that are less evolvable. Evolvability is facilitated by increasing the number of amino acids or by greater reliance on polygenic and pleiotropic inheritance (Conrad and Rizki, 1980). Although there is a cost in terms of energy expended by the individual, amenability to evolution as a trait is proposed to be able to persist by genetic hitchhiking along with other advantageous traits whose appearance makes amenability possible. Conrad and Rizki (1980) therefore described this effect as "amenability pulls itself up by its own bootstraps," or more simply as the "bootstrap effect." Models of this bootstrapping effect were developed in Conrad and Rizki (1980).

Conrad (1988) has also speculated that advances in evolving programs will come from a bifurcation in the development of computer hardware, that is, the design of a family of architectures associated with evolutionary machines (see also Atmar, 1976). Akingbehin and Conrad (1989) suggested that cellular automata systems may be quite usefully implemented as such evolutionary machines, a notion that has been subsequently advocated by de Garis (1994, 1996, and others). ATR in Japan is actively pursuing the possible design and construction of such machines.

References

[1] K. Akingbehin and M. Conrad (1989) "A hybrid architecture for programmable computing and evolutionary learning," *J. Parallel and Distributed Computing,* Vol. 6, pp. 245–263.

[2] J. W. Atmar (1976) "Speculation on the evolution of intelligence and its possible realization in machine form," Ph.D. diss., New Mexico State University, Las Cruces, NM.

[3] M. Conrad (1969) "Computer experiments on the evolution of coadaptation in a primitive ecosystem," Ph.D. diss., Biophysics program, Stanford, CA.

[4] M. Conrad (1974) "Evolutionary learning circuits," *J. Theoret. Biol.,* Vol. 46, pp. 167–188.

[5] M. Conrad (1977) "Evolutionary adaptability of biological macromolecules," *J. Mol. Evol.,* Vol. 10, pp. 87–91.

[6] M. Conrad (1978) "Evolution of the adaptive landscape," *Theoretical Approaches to Complex Systems,* R. Heim and G. Palm (eds.), Springer Lecture Notes on Biomathematics No. 21, Springer, Heidelberg, pp. 147–169.

[7] M. Conrad (1979) "Bootstrapping on the adaptive landscape," *BioSystems,* Vol. 11, pp. 167–182.

[8] M. Conrad (1981) "Algorithmic specification as a technique for computing with informal biological models," *BioSystems,* Vol. 13, pp. 303–320.

[9] M. Conrad (1982) "Natural selection and the evolution of neutralism," *BioSystems,* Vol. 15, pp. 83–85.

[10] M. Conrad (1983) *Adaptability: The Significance of Variability from Molecule to Ecosystem,* Plenum Press, NY.

[11] M. Conrad (1988) "Prolegomena to evolutionary programming," *Advances in Cognitive Science: Steps Toward Convergence,* M. Kochen and H. M. Hastings (eds.), AAAS Selected Symposium 104, American Association for the Advancement of Science, NY, pp. 150–168.

[12] M. Conrad (1990) "The geometry of evolution," *BioSystems,* Vol. 24, pp. 61–81.

[13] M. Conrad and H. H. Pattee (1970) "Evolution experiments with an artificial ecosystem," *J. Theoret. Biol.,* Vol. 28, pp. 393–409.

[14] M. Conrad and M. M. Rizki (1980) "Computational illustration of the bootstrap effect," *BioSystems,* Vol. 13, pp. 57–64.

[15] M. Conrad and M. M. Rizki (1989) "The artificial worlds approach to emergent evolution," *BioSystems,* Vol. 23, pp. 247–260.

[16] M. Conrad and M. Strizich (1985) "Evolve II: A computer model of an evolving ecosystem," *BioSystems,* Vol. 17, pp. 245–258.

[17] H. de Garis (1994) "Growing an artificial brain: The genetic programming of million-neural-net-module artificial brains with trillion cell cellular automata machines," *Proc. of the 3rd Ann. Conf. on Evolutionary Programming,* A. V. Sebald and L. J. Fogel (eds.), World Scientific, River Edge, NJ, pp. 335–343.

[18] H. de Garis (1996) "'Cam-Brain': ATR's billion neuron artificial brain project. A three year progress report," *Proc. of 1996 IEEE Conf. on Evolutionary Computation,* IEEE Press, NY, pp. 886–891.

[19] C. G. Langton (1987) "Artificial life," *Artificial Life: The Proc. of an Interdisciplinary Workshop on the Synthesis and Simulation of Living Systems,* C. G. Langton (ed.), Addison-Wesley, Reading, MA, pp. 1–47.

[20] J. O'Callaghan and M. Conrad (1992) "Symbiotic interactions in the EVOLVE III ecosystem model," *BioSystems,* Vol. 26, pp. 199–209.

[21] T. Ray (1992) "An approach to the synthesis of life," *Artificial Life II,* C. G. Langton, C. Taylor, J. D. Farmer, and S. Rasmussen (eds.), Addison-Wesley, Reading, MA, pp. 371–408.

[22] M. M. Rizki (1985) "A discrete event simulation of an evolutionary ecosystem," Ph.D. diss., Wayne State University, Detroit, MI.

[23] M. M. Rizki and M. Conrad (1985) "Evolve III: A discrete events model of an evolutionary ecosystem," *BioSystems,* Vol. 18, pp. 121–133.

[24] M. M. Rizki and M. Conrad (1986) "Computing the theory of evolution," *Physica 22D,* pp. 83–99.

[25] M. Strizich (1982) "Ecosys: A model of an evolving ecosystem," master's thesis, Wayne State University, Detroit, MI.

J. theor. Biol. (1970) **28**, 393–409 Reproduction in whole or in part is permitted by the publisher for any purposes of the United States Government

Evolution Experiments with an Artificial Ecosystem

MICHAEL CONRAD† AND H. H. PATTEE

W. W. Hansen Laboratories of Physics,
Stanford University, Stanford, Calif. 94304, U.S.A.

(*Received 5 February* 1970)

The technique of ecosystem reconstruction provides a mechanism for examining assumptions about natural ecosystems and their evolution. A hierarchical computer program has been developed with genetic, organismic and population levels embedded in an ecosystem. The program consists of a population of cell-like organisms subject to a strict materials conservation law. The conservation law induces a competitive interaction among organisms. Organisms can also participate in co-operative interactions and are capable of executing representative biological strategies, such as genetic recombination and modulation of the expression of the genome. Preliminary experiments with the system show that the efficiency of individual organisms may be restricted, but with a concomitant increase in the utilization of the environment. The nature of the restrictions probably differs from those that occur in natural biology, due to specializing assumptions in the model, but the mechanism of their appearance may be a general ecosystem process.

1. Introduction

The present theory of evolution is concerned primarily with the statistics of populations of genes. The statistical behavior of these populations is dependent on what are called *fitness* criteria. Unfortunately, fitness functions are not a part of the theory itself, but must be inferred from observation, experience, and intuition. In order to improve the predictive capacity of the theory of evolution, many attempts have been made to define general aspects of fitness, but this has always proved exceptionally difficult. We believe that this difficulty is a reflection of our lack of understanding of the origin of hierarchical organizations—that is, collections of units which exhibit short term *deterministic* mechanisms in the individual which are to a large degree the long term result of the *statistical* behavior of large groups of these individuals. In evolutionary theory this is often expressed by the statement that evolution operates on populations of phenotypes, not on the genotype or the individual, but at the same time it is the gene which constructs the individual

† Present address: Apartment 56, 421 Leavenworth Street, San Francisco, California, U.S.A.

according to deterministic mechanisms which are not dependent at the time of construction on any of the statistical behavior of populations. We know that in nature the behavior of living systems involves many hierarchical levels of interaction, from the genetic level up through the individual, the population, and ultimately to the entire ecosystem. A theory of evolution which does not reflect this hierarchical level structure cannot be expected to give a clear picture or to give useful predictions.

In this paper we describe preliminary evolutionary experiments using a computer program in which three hierarchical levels of biological organization are distinguishable. We call these the *genetic*, the *organismic* and the *population* levels. Furthermore, we do not introduce any fitness criteria as an explicit part of the program. Instead, we interpret the computer in this experiment as an ecosystem in which the various interactions within and between levels must take place. This means that we must introduce certain constraints which correspond in some sense to the real ecological conditions in the physical world. The two most important which we consider are: (1) the way the world is spatially divided, i.e. how the individual organisms can occupy the world; and (2) the closure of this world to matter, i.e. the conservation of matter. Notice that these are not fitness criteria, but are the most general conditions which collections of organisms must satisfy.

A number of programs have already been developed which use some features of natural evolution. In particular, Bremermann (1962, 1967) has applied evolution techniques to optimization and search processes. The work of Reed, Toombs & Barricelli (1967) uses a fixed pattern of numbers, determining certain strategies that develop in the context of a game situation. Programs or designs have also been published which incorporate the detailed dynamics of self-reproduction. For example, see Stahl (1967) for a simulation of the bacterial cell and Arbib (1967) for the design of a more general self-reproducing system. However, the complexity of such programs precludes the possibility of extending them to the population dynamics level on present day computers.

To our knowledge the program which will be described in the present paper is the first which allows individual organisms with a genotype and phenotype to evolve within the general constraints of an ecosystem. Our approach is intermediate between those in which organisms evolve in relation to arbitrary fitness criteria imposed by the programmer and those in which the dynamics at the level of the individual organisms are so detailed that evolution processes are precluded on present day computers. Hopefully this intermediate approach will provide a tool for examining many problems in evolutionary biology which are intractable from an analytical or natural experimental point of view.

While it may be clear that a multilevel hierarchical structure is a characteristic of living systems, it is by no means clear how these levels should be represented in a computer program. As a practical matter we cannot represent all the molecular details of the genetic level and still have time or room in the computer for the direct multiplication of these detailed individuals up to the population level. To avoid this problem the present program uses a common, fixed set of features or routines which can be used by all "organisms". The phenotype (or phenome) of an organism determines the way in which these routines are used. Of course this places restrictions on possible modes of evolution, restrictions not present in nature, where new functions or hierarchical levels of organization can always emerge. However, it is just these features or conditions for evolution which are the important unknowns. The use of common routines allows these unknowns to be manipulated directly. This would not be possible if each organism carried its own private program.

The features which are included in the common set of routines are to a certain extent arbitrary. This is not necessarily a defect as long as the process of modifying these features produces a system whose behavior converges to that of a natural ecosystem. It is difficult to define the characteristics of an evolution process precisely. However, we consider it a step in the right direction if the artificial ecosystem exhibits certain behavior characteristic of the natural system—for example, behavior characteristic of ecological succession processes, with the provision that the origin of new types, of crucial importance in evolution, is possible (see Margalef, 1958; Odum, 1969). We have also adopted two other ground rules. First, the processes of evolutionary search must agree with biological fact, even if unbiological search techniques are more effective. This is important, since breakdowns in the analogy between computers and biological systems might lead to such a divergence. Second, the artificial ecosystem must be as simple or primitive as possible. We are working for the minimum or necessary conditions for evolution. We hope that this is a useful strategy for distinguishing the essential properties of ecosystems from the frozen accidents. In fact, the present project grew out of an attempt to determine what observed conditions should be considered as significant in an abiogenic experiment, if the origin of life is identified with the origin of systems capable of supporting evolution (Pattee, 1966, 1969).

2. Biological Fundamentals

In order for a lineage to persist the organisms of this lineage must, on the average, produce at least one offspring before they die. Individual organisms have various strategies for achieving this. Also, lineages or species have various survival strategies. Organisms, for example, may be well adapted to a given environment, or they may be adaptable, in the sense that they can respond flexibly to a variety of environments. This is also true at the level of the species, and at other levels of biological organization. In general adaptability and adaptedness interfere with each other at a given level of organization, in the sense that an increase in adaptability results in a decrease in adaptedness. This interference can also occur between levels of organization. For example, the adaptability of a species depends on the variability of its gene pool. If the organisms of the species have complicated adaptations, however, the variability of the gene pool will be restricted. This follows because the genes of more complicated organisms must be more integrated, or adapted to one another, and therefore rearrangements of these genes are more likely to produce nonfunctional forms.

Thus the set of strategies characterizing a biological system cannot be arbitrary—only certain sets of strategies will be consistent. This is important because it means that the evolutionary search strategy of a system depends on the characteristics of organisms. At the same time, the evolution of these characteristics depend on the search strategies. This is one statement of the apparent paradox of the origin of hierarchical control.

These search strategies cannot be described without describing the spatial organization of the population. This organization, together with species specificity, influences the pattern of gene flow, and therefore the search strategy. Extensive gene flow within a population tends to keep the population homogeneous. In order for the population to differentiate, gene flow between subpopulations must be prevented. Such blockage usually, and perhaps always, follows isolation of the subpopulations. In any case it is evident that isolation would facilitate the speciation process.

The formation of species is restricted by the niches, or spheres of existence which are available. These, in turn, are modified by the formation of new species. Such niche multiplication is most evident in relation to the appearance of trophic levels. Of course species with certain types of strategies may be able to colonize new niches more rapidly than other types of species, or migrate into similar niches in different locations. The sequential change of the ecosystem will decrease only as niches are filled and as mechanisms evolve to reset the environment. The development of such stability in an ecosystem is often associated with the development of a high degree of symbiosis.

Most of the mechanisms which have been discussed are present in bacteria. Here there are three important levels of organization: the genetic, the cellular (or organismic) and the population levels. Bacteria have a variety of homeostatic or adaptability mechanisms at the organism level—for example induced enzyme synthesis. In general the presence of an unnecessary inducible enzyme places a bacterial strain at a disadvantage. The bacterial population can also exhibit a greater or lesser degree of adaptability. Since many bacteria have a high growth rate, mutation is of course an important mechanism of search. Bacteria are also capable of exchanging genetic material through conjugation and transduction. Ravin (1960) has suggested that such processes may play an important role in the life of micro-organisms, and that bacterial populations could support a wide variety of breeding structures. Of course the main processes of present day microbial ecology may involve the culturing of already existing types in different environments. It is likely, however, that sexuality played an important role in relation to the origin of these types.

Bacteria participate in symbiotic relationships in the sense that certain types may require substances produced by other types. There are no trophic levels in the sense in which they occur in metazoan populations, but a number of species may be required to cycle materials back to their original form. The organization of the bacterial population, as a whole, may be controlled by chemical messages between units. For example, bacterial populations often reach a certain size and stop growing, despite the fact that nutrients are not limiting and poisons have not accumulated. This fact turns out to be pertinent when the behavior of the model ecosystem is analyzed.

3. Representing the Fundamentals

The significance of the facts reviewed in the previous section may not be universally agreed upon, but that is not crucial for our purposes. Most of the features discussed are present in micro-organisms. Presumably these are most similar to primitive life forms and it is possible that any systems capable of supporting evolution would rapidly develop a spectrum of processes including those discussed. The question we are asking is whether or not these processes, or some subset of them, can actually produce evolution. It is possible that not all of these processes are essential. It is also possible that some important processes in nature have been overlooked. The only way to approach this problem is to design an artificial ecosystem which we can test and modify.

We shall only describe this evolution program at a functional level since a listing of the program and a detailed description of its operation is given elsewhere (Conrad, 1969). It is written in LISP, a list processing language, and in the LISP assembly language. The program and its variations have been run for about three hours on the IBM 360/67 computer at the Stanford Computation Center.

The events in the evolution program take place in what we call the *world*. The world is a one-dimensional string of *places*. The string is closed, or forms a loop, to avoid end effects. Each place in the world is characterized by a state (A or B) and a certain number of material parts called *chips*. Organisms (or cells) may be viewed as selective catalytic shunts in a cycle of these parts, i.e. cells control the rate of flow of matter. The world, as a whole, is subject to a strict materials conservation law. This induces the fundamental competitive interaction among the organisms.

Each organism is attached to some place in the world. It should be emphasized that the organism is not located only at this place, and the place does not correspond to an ecological niche. (Certainly such niches must not be artificially imposed by the programming.) Rather, the organism is allowed to operate over a certain number of contiguous places (its territory), but is associated with the place of attachment for certain of its activities and at the end of every period or cycle of the program.

The temporal organization of the model is not in real physical time since events occur at discrete periods. The periods are not generation times, and the lifetime of an organism can extend over any number of these periods. However, the consequences of an organism's behavior must depend on the behavior of other organisms—that is, it must appear as if organisms are functioning simultaneously. This is achieved, inside the computer, by using the device of a *two-pass* system, allowing interaction among processes arbitrarily separated by the sequential operations of the computer. In the first pass organisms interact with the environment and with other organisms in local sequence time. The consequences of an organism's behavior in what corresponds to global, physical time are determined during the second pass. Here chips are collected, organisms reproduce, and chips from decaying organisms (detritus chips) are returned to the matter pool. The net effect of birth and decay determine the composition of the new biota, and the process is repeated. In the actual program the second pass consists of a number of subpasses. The overall flow of information in the program is illustrated in Fig. 1.

The organisms have a genome and a phenome. The genome is mapped into the phenome according to a doublet code. This mapping is not designed to represent the known processes of protein synthesis, or embody any particular logic of self-reproduction. However the genotype-phenotype distinction serves as a basis for efficiently describing various representative strategies of construction and interaction, which is the same function it performs in real cells.

There are 16 possible codons or pairs of genome symbols. These are usually mapped into six types of phenome symbols. The sequence of phenome symbols determines how an organism uses common routines. The immediate behavior of an organism depends on its *internal state* and on its *input state*. The first input state is the state associated with the organism's attachment

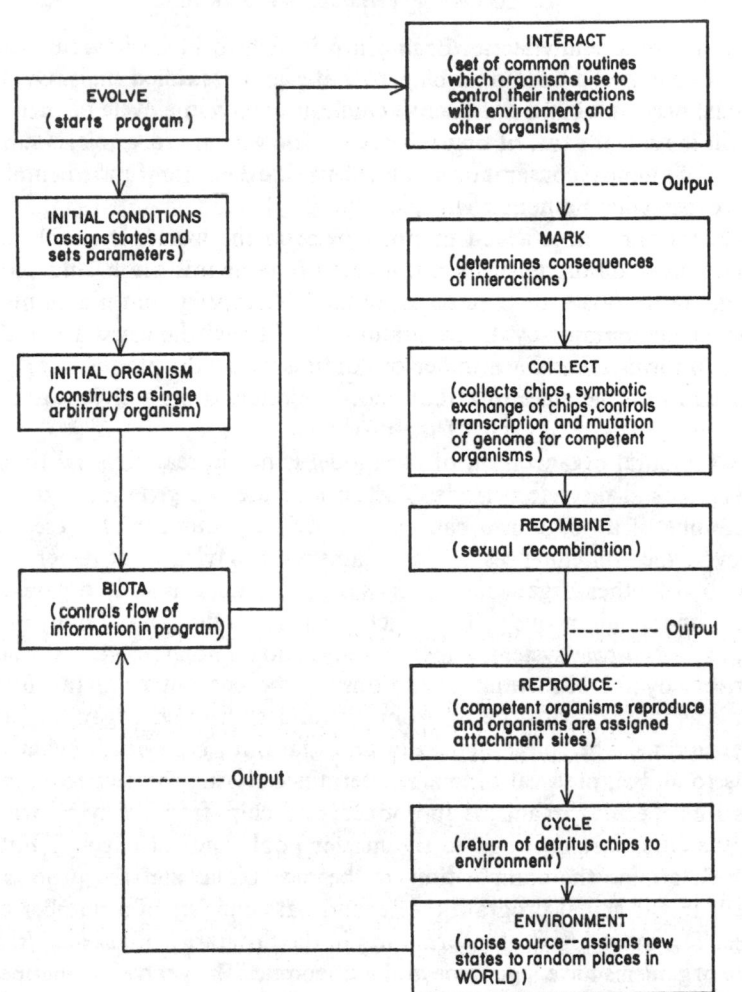

FIG. 1. Flow of control in evolution program. The diagram is highly simplified. A new period is initiated when BIOTA is re-entered after ENVIRONMENT is executed. The second pass begins with MARK.

site. The organism's first internal state is the first symbol on its phenome. The second input state is the state immediately to the right of the first input state, and the second internal state is the next symbol on the organism's phenome, and so forth.

Six routines are possible. When the organism is in state A or B it compares itself to the input state, or tries to match the environment. When it is in state

C it looks for a conjugate, or organism with which it can recombine. When it is in state D it allocates a chip to a repair process—that is, to its self-maintenance—if it in fact succeeds in collecting this chip. When it is in state F it looks for a symbiont, or organism with which it can share chips. When it is in state E it jumps into a mode in which the rule determining the next internal state is somewhat more complicated. We call this the parametric mode. Here the system can select the next phenome symbol from a pair of possible phenome symbols on the basis of the state of the environment. This represents a modulation of the genome or an inducible property, in the sense that the genome can express itself differently in different environments. The organism must pay for the extra genetic material, but only for the part of the phenome which it uses. The operation of an organism is illustrated in Fig. 2.

WORLD	P1	P2	P3	P4	P5	P6	P7	P8	P9	P10	P11	P12
INPUT STATE	A	A	B	A	B	B	A	B	A	A	A	
INTERNAL STATE		A	C	B	D	F	E	AB	AA	E	D	

—————————ORGANISM—————————

FIG. 2. Representative organism. The organism is attached at place P2. The first internal state is A and the first input state is A. The organism records this fact, and will collect a certain number of chips during the second pass. The second internal state is C. The organism will later look for a conjugate at position P3. The third internal state is B and the input state is A. These do not match and the organism goes to its next internal state. This is D, and the organism can allocate a chip to repair processes, assuming that it collects a chip. The fifth internal state is F, and the organism will later look for a symbiont at place P6. The next internal state is E and the organism goes into a parametric mode. The seventh internal state is one of the pair A or B. Thus the organism will always match place P8. The internal state which follows is also one of a pair, but in this case both members of the pair are A's. This is wasteful and, in fact, the organism does not match place P9. The organism now goes into an E state and jumps out of the parametric mode. Thus the next state is D and the organism is entitled to another repair chip. In general, if the second alternative in the parametric mode is a non-matching type symbol it cannot be used. Also it is wasteful if the E symbol signalling the parametric mode occurs in an odd position, since the immediately following symbol will not be used.

Each organism carries a record of its own activities. The record is an accounting device and not part of the structure of an organism. During the first pass each organism marks all the places which it utilizes and stores the names of these places on its record. It also stores the location of possible symbionts and conjugates, as well as some other accounting information. After each organism in the system has interacted with the environment, the number of chips returned at each place is determined. This is the quotient of the number of chips and the number of marks associated with the place. Only a whole number of chips can be withdrawn and therefore a certain number of chips, associated with an unusable remainder, may be unavailable. In the second pass each organism collects chips from the places it has matched. The symbiotic exchange of chips with other organisms takes place at this time. Organisms can develop recognition codes enabling them to exert some selective control over their symbiotic interactions. A certain number of chips which are collected are used for repair and arc immediately returned to the matter pool. Repair chips represent energy which an organism uses for its maintenance. The probability that an organism will decay depends on the number of repair chips which it collects.

Chips which are not used for repair are allocated to reproduction. When an organism doubles its size, in terms of chips, it becomes competent to reproduce or fission. In this case the genome is copied and mutated. Both point and size (increment or decrement) mutations are possible. Redundancy in the code allows some control over the effective mutation rates of phenome symbols. If the organism has the sexual property it looks for a conjugate in the appropriate places. Recombination consists of breaking and splicing genomes at random places. Organisms can develop recognition codes which enable them to exert some selective or specific control over the flow of genetic information.

The transcribed genome is translated into a phenome and a daughter cell is produced. The daughter cell may be attached at the same place as the parent, or one place to the right or left. In most versions of the program the probability of dispersing or migrating in this sense is under genetic control. If the parent organism decays, chips must be returned to the environment. This is also true if an organism collects more chips than it can use. This replacement is achieved by dividing the number of chips to be returned by the length of the phenome. The quotient (rounded to the lower integer) is returned to each place corresponding to the position of a phenome symbol. The remainder is also returned to the environment, one chip at a time starting at the attachment site of the organism. Chips are strictly conserved, but the pattern of replacement is, in general, not the same as the pattern of withdrawal. This is important because it means that the environment is not automatically reset by the program. If resetting the environment is important for survival, then organisms must evolve mechanisms for achieving this themselves.

Thus, the redistribution of chips in the environment is affected only by the activities of organisms. The distribution of A and B states, on the other hand, may be affected by noise. A and B states can be interpreted as representing sets of environmental conditions, such as temperature or salinity, which are reasonably independent of the structure of the matter cycle. Organisms can withdraw chips if they are in a state which allows them to function under the given conditions.

4. Some Comments on the Model

The present model establishes a one-dimensional biology. There is a discrete one-dimensional world with one-dimensional organisms. The ecology, physiology and genetics of these organisms are therefore as simple as we can imagine.

The discreteness and relatively rigid form of this model at both the organism and population levels is clearly an imperfect representation of biological systems. Within this framework, however, there is a remarkable degree of internal freedom and behavioral flexibility. The reproductive success of organisms, or groups of organisms, depends on both the biological and physical environments. These change in the course of evolution, and consequently the criteria of fitness are subject to evolutionary change. Likewise, the processes of speciation and niche selection are not imposed by the programming. For example, the sexual recognition codes control the flow of genetic information in the population. Speciation occurs (in the biological sense) when subpopulations are genetically isolated. Such isolation cannot arise from a single change in a sexual recognition code; in fact, the population could maintain a range of such codes. Thus the possibility of speciation or development of disjoint sets of codes is present in the model, but whether or not this possibility will be used is not known in advance. It should be remarked that the present program does not use any external mechanisms for spatially isolating populations. In fact, this may be an unrealistic freedom in the model.

The amount of information which can be processed in the environment of the computer is much smaller than can be processed by natural systems. For example, in the experiments to be described the average population ranged from 200 to 400 organisms, depending on the number of chips initially assigned to the world. This point should be kept in mind when interpreting the statistical behavior of the system. It should also be remembered that the organisms are operating in the information-rich constraints of the computer environment, and can therefore evolve free of many of the internal constraints necessary for the self-reproduction of organisms in the present day environment. Furthermore, the types of organisms which can evolve are limited by the fixed nature of the common routines. For example, there are severe restrictions on the trophic structure of the system. In fact only a single trophic level is possible, along with symbiotic interactions at this level. The system could be modified to accommodate an arbitrary number of trophic levels, but we felt that this complication might obscure the minimum initial constraints required for evolution.

These minimum conditions can be studied by further simplifying the system. For example, any particular organism strategy can be removed by altering the code. The evolution process can begin with a single arbitrary organism. The program is also capable of accepting populations of organisms as inputs, and most of the parameters in the system are under operator control.

5. Experimental Results

(A) GENERAL COMMENTS

Several variations of the evolution program have been prepared. The behavior of the program is very complex, and it is neither practical nor instructive to follow the dynamics in detail. Rather, a mode of description must be adopted which gives a simple, but not overly distorted, picture of the system. The data analysis associated with the evolution program has proved to be somewhat difficult, since it is hard to know in advance what features of the system will be significant.

The program reports vital statistics of organisms, as well as statistics regarding the matter pool, the size of organisms, and the fraction of organisms executing various strategies. The *utilization* of the matter pool is expressed by the ratio of chips bound in organisms to chips free in the environment. The *matching ratio* is the quotient of the number of successful matches to the number of attempted matches. The relative changes in size of the population can be expressed as a *survival curve*.

The system has also been examined relative to certain classification schemes. For example, organisms can be classified in terms of the types of strategies which they execute, in terms of size, in terms of the functionality of the parametric strategy, and so forth. It should be emphasized that these classification schemes are arbitrary, and the accuracy of the picture which they provide cannot exceed the extent to which they express important elements of the system. The data provided by the various population quantities and classification schemes must always be considered in conjunction with one another.

The only way of understanding the causal relations of the system with certainty is by controlling its dynamics through alterations in the programming. In practice, experiments focused on limited aspects of the system's behavior, and many variations on the program were run relative to these.

The behavior of the program will be summarized in the remainder of this section. A detailed presentation of results and data analysis, as well as a description of all alterations in the programming, can be found elsewhere (Conrad, 1969).

(B) TYPICAL BEHAVIOR

The first evolution program (system I) was run for only 41 periods. The interpretation of its behavior, however, is clear cut and establishes a good perspective.

The system I environment was noiseless, and organisms were allowed to persist indefinitely if they could allocate a sufficient number of chips to repair processes, a condition rapidly discovered by the population. The matching ratio increased as the run progressed, and the organisms became increasingly efficient with respect to the relative number of matching type symbols on their phenomes (see Fig. 3). However, the behavior of system I was in many ways the antithesis of evolution. Both birth and death rates became very small. Organisms maintained themselves by withdrawing

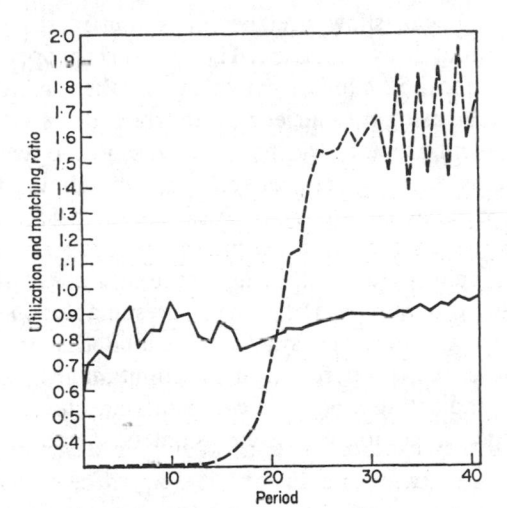

FIG. 3. Utilization and matching ratio curves for system I. The growth phase ended in the neighborhood of period 25. – – – –, Utilization curve; ———, matching ratio curve.

repair chips; only a small number of chips were used for self-reproduction. Inefficient organisms were eliminated (most markedly) when the number of free chips fell below a critical level. As a consequence the population became homogeneous, one type tended to predominate, and chips in certain places were never utilized. This was associated with the fact that the dispersion of the population was retarded, due to the low reproduction rate. Essentially, the collectively stable behavior of the population precluded its further evolution.

Subsequent versions of the program were therefore modified in a variety of ways. In particular, a noise level was established and repair strategies were designed so that the probability for an organism to decay is never zero. These are certainly more realistic conditions. One version (system III) was run for 251 periods. This was the longest run, and is the only one which will be described here.

The composition of the system III biota showed a succession of patterns. The diversity of types, as with system I, increased to a maximum at the end of the growth phase. Aside from this, and in contrast to system I, there is no indication that the composition of the population will cease to change. The size of organisms, as well as the diversity of sizes, also tended to increase, allowing organisms to interact with a greater number of places. The changing composition of the population is associated with this size increase, since the probability distribution for different types of organisms is related to average size (in the absence of selection).

The various strategies show stretches of overall increase or decrease; in general trends cannot be extrapolated. The symbiosis property often occurred with a frequency above its equilibrium value, and the symbiosis codes tended to concentrate, increasing the efficiency with which this strategy was utilized. In general the sexual property was not favored, and the sexual codes tended to spread out, thus reducing the probability of recombination. However, the sexual property exhibited stretches of consistent increases. These were associated with consistent increases in the parametric property. This pattern was always observed immediately following perturbation experiments in which the logic of the program was modified, presumably changing optimum strategies. In one case, however, system III exhibited a stretch of consistent decreases in the sexual property, with concomitant increases in the parametric property, indicating a possible competitive exclusion of types.

On the average, utilization of the environment increased as the run progressed.

(C) CO-ADAPTIVE PROPERTIES

Some of the organisms in system III were efficient, and well-organized parametric types did occur. However, the predominant types of organisms were definitely inefficient. Many organisms carried phenome sequences of no apparent selective value, and the parametric property was by and large non-functional, in the sense that it was rarely used to increase the organism's ability to match the environment. Organisms with efficiently placed parametric symbols exhibited no clear advantage over those with inefficiently placed symbols. Also, organisms with a conservative dispersal strategy (whose offspring have a better chance of retaining correlations with the environment) did not predominate over dispersive types. The matching ratio showed con-

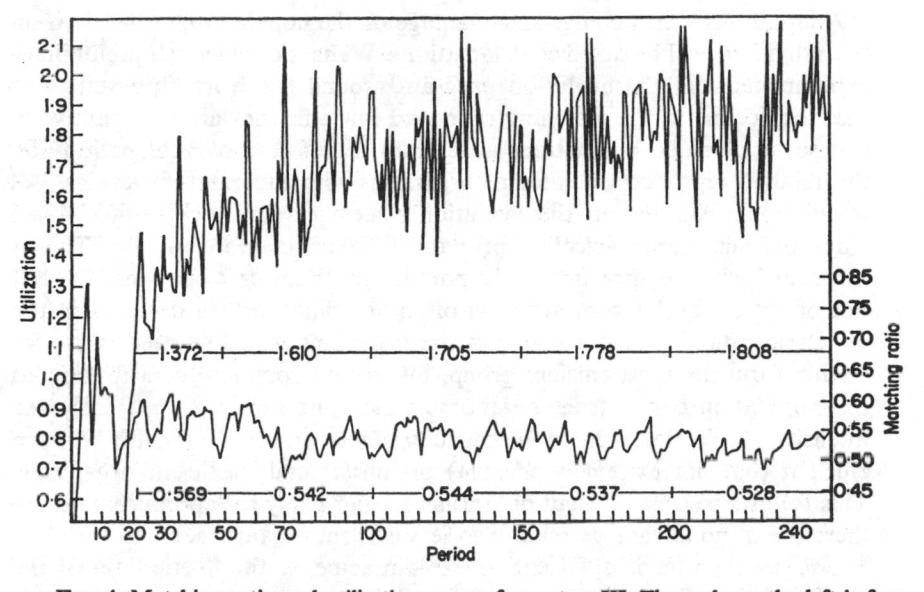

FIG. 4. Matching ratio and utilization curves for system III. The scale on the left is for the utilization (upper curve). The scale on the right is for the matching ratio (lower curve). Average values for representative blocks of the run are shown under each curve. Note decreasing average matching ratio and increasing average utilization. The growth phase ended sharply at period 20.

siderable fluctuation, but declined slightly, on the average, as the run progressed (see Fig. 4).

The low efficiency of organisms and the decline in the matching ratio are somewhat surprising since improvement in these properties would often be regarded as a criterion for evolution. Yet system I, which had a steadily increasing matching ratio, did not exhibit behavior at all characteristic of evolution.

The low efficiency and decline in the matching ratio were not artifacts, since these were exhibited by many arbitrary variations of the system. The decline is associated with the fact that the number of chips returned to an organism for matching a particular place is the quotient (rounded to the lowest integer) of the number of chips and the number of matches for that place. Thus a decrease in the number of matches may increase the utilization of the environment, and conversely. This is a feature of the programming, but it is not unrealistic, since overly severe competition usually reduces the efficiency of populations in nature. In fact, the matching ratio can be induced to increase by cycling chips more rapidly through the system.

Adaptations which confer an advantage on the population, rather than on the individual, will be called co–adaptations. We have oriented our preliminary experiments with the evolution program around the working assumption that the decline in the matching ratio and low efficiency are co–adaptive in this sense. Such co–adaptations arise from the effect of overcompetition on the relative reproduction rates of organisms with different efficiencies. All population quantities in the evolution system exhibit oscillations. When chips become scarce, selection presumably becomes more intense. This is associated with a contraction in the population. In the case of system III (but not of system I) this contraction is often associated with a decrease in the matching ratio. In such a situation the organisms which succeed in reproducing form the most efficient group, but the predominant contribution to the population comes from organisms which just survive. Apparently the intensity of selection reduces the capacity of selection to distinguish between efficient (but not extremely efficient) organisms and inefficient organisms. This follows because efficient organisms are not likely to reproduce, and are therefore at no advantage relative to less efficient organisms.

This mechanism is sufficient to explain some of the fluctuations of the matching ratio, but it does not account for the overall decrease. Organisms in the evolution program may be grouped together on the basis of the places they match. In fact, associations of less efficient organisms have been found which make a greater contribution to the subsequent population than comparable associations of more efficient organisms. This raises the possibility that group or interdeme selection is operating, but the relative importance of this mechanism is not yet clear. It might be remarked that removal of all unnecessary strategies (by an alteration of the code) increases the fluctuation of the matching ratio, suggesting that these strategies helped to retain co-adaptations.

6. How Useful is this Model?

This simple evolution program shows many analogies to natural evolution, but the departures from natural evolution are also striking. Different types of organisms may develop, but there is no evidence that speciation, in any biological sense, is taking place. Mutual dependencies may exist among some of these types, but there does not seem to be any niche multiplication in the sense of the developing biota creating roles for widely different organisms. The system applies some of the built-in routines in unexpected ways, but of course there is no possibility for new routines to appear.

It is evident that the richness of possible interactions among organisms and the realism of the environment must be increased if the model is to be improved. For example, the system exhibited some integration or coadaptive behavior at the population level. However, this integration was slight by comparison to that which develops in the course of natural evolution, and quite different in character. Natural biological systems are not known to control competition by developing disorganized phenotypes. Ordinarily such systems prevent overutilization of the environment through mechanisms of inter- or intraspecies communication, e.g. control of predator-prey relationships, regulation of growth rates, territoriality, self-thinning, etc. Natural systems may also increase stability by rapid cycling of materials back to the environment through the detritus pathway. The evolution program did not allow good internal control in this respect, and, in fact, the elastic growth characteristics of the population would have reduced the importance of such control. However, oscillations in population quantities are decreased by imposing an increase in the rate at which chips are cycled through the system.

One point is clear, that the processes of variation and natural selection alone, even when embedded in the context of an ecosystem, are not necessarily sufficient to produce an evolution process. This was evident in the case of system I. This would imply that variation and selection processes associated only with replicating nucleic acid molecules (as distinct from systems capable of using such a store of instructions to control specific reactions) would probably not evolve in any effective way (see Mills, Peterson & Spiegelman, 1967).

7. Conclusions

It is, of course, not yet clear what necessary conditions are to be associated with evolution processes. However, we have shown how assumptions about such conditions can be examined through the technique of ecosystem reconstruction. We feel that the present study indicates the feasibility and usefulness of this synthetic approach as a source of new ideas and information about fundamental problems in evolutionary biology.

In addition to their use as models of biological evolution, the behavior of such programs, especially their shortcomings, appear to us as potentially valuable for understanding the general theory of the origin of hierarchical control at all levels, including the design of computers with more creative or life-like problem-solving abilities. Most current efforts at problem solving are still algorithm-oriented rather than evolution-oriented; that is, we tend to concentrate on procedures which can be definitely specified rather than on general search and selection strategies.

Experience with the present model re-enforces our feelings that the most profound and significant processes of evolution—the innovations, the origins of new hierarchical levels of organization—are still outside the scope of this type of program and remain to be discovered. Experience with other computer programs intended to model higher mental processes, such as general problem solving and artificial intelligence studies, have shown similar weaknesses. As one might expect, true acts of creation at all levels remain difficult to imitate.

We are indebted to Dr Michael Arbib and Dr A. K. Christensen for helpful suggestions. In part the paper was prepared at the AIBS Colloquium on Theoretical Biology at Fort Collins, Colorado, and the Center for Theoretical Studies, University of Miami, Coral Gables, Florida. The work was supported by Office of Naval Research Contract Nonr-225-90 and National Science Foundation Grant GB 6932.

REFERENCES

ARBIB, M. (1967). *J. theor. Biol.* **14**, 131.

BARRICELLI, N. A. (1962). *Acta biotheor.* **16**, 69.

BREMERMANN, H. J. (1962). In *Self-Organizing System.* (Yovits, M. C., Jacobi, G. T., & Goldstein, G. D., eds.), p. 93. Washington, D.C.: Spartan Books, Inc.

BREMERMANN, H. J. (1967). *Progr. theor. Biol.* **1**, 59.

CONRAD, M. (1969). Thesis, Biophysics Program, Stanford University.

MARGALEF, R. (1958). *Gen. Syst.* **3**, 36.

MILLS, D. R., PETERSON, R. L. & SPIEGELMAN, S. (1967). *Proc. natn. Acad. Sci. U.S.A.* **58**, 217.

ODUM, E. P. (1969). *Science, N.Y.* **164**, 262.

PATTEE, H. H. (1966). In *Natural Automata and Useful Simulations.* (Pattee, H. H., Edelsack, E. A., Fein, L., & Callahan, A. B., eds.) p. 73. Washington, D.C.: Spartan Books, Inc.

PATTEE, H. H. (1969). In *Towards a Theoretical Biology*, Vol. 2. (Waddington, C. H., ed.) p. 269. Edinburgh: Edinburgh University Press.

RAVIN, A. W. (1960). *Bact. Rev.* **24**, 201.

REED, J., TOOMBS, R. & BARRICELLI, N. A. (1967). *J. theor. Biol.* **17**, 319.

STAHL, W. R. (1967). *J. theor. Biol.* **14**, 187.

Chapter 14
Soft Selection

M. Pincus (1970) "An Evolutionary Strategy," *J. Theoret. Biol.,* Vol. 28, pp. 483–488.

R. Galar (1985) "Handicapped Individua in Evolutionary Processes," *Biological Cybernetics,* Vol. 53, pp. 1–9.

SELECTION in evolutionary algorithms serves to stochastically cull the least-fit individuals from the population. The probability that a particular solution will survive into the next generation is typically a function of its fitness relative to other solutions in the population. The stringency of selection can be measured by the likelihood of retaining solutions with lesser fitness into future generations.[1] "Hard" selection methods devote attention solely to a collection of best individuals. The $(\mu + \lambda)$ selection of evolution strategies (Schwefel, 1981), where μ parents create λ offspring and the best μ of the $\mu + \lambda$ individuals are selected to be parents of the next generation, is one example. "Soft" selection methods assign nonzero probabilities of survival to solutions that would not be retained under hard selection. One such procedure is the fitness-proportional selection of genetic algorithms (Holland, 1975), where each individual is assigned a probability of replication that is equal to the ratio of its fitness to the mean fitness of all individuals in the population.

Pincus (1970) (reprinted here) was one of the first papers to identify that accepting individuals with lower fitness can lead to more rapid rates of escaping local minima. In particular, Pincus (1970) identified the hard selection found in Bremermann et al. (1966) and Bossert (1967) as being a source for stagnation at ridges on the fitness landscape. A Markov chain analysis indicated that higher probabilities of escaping local optima could be attained by temporarily accepting individuals of lower fitness.

Later, Galar (1985) (reprinted here) offered that not only are points of evolutionary stagnation overcome by soft selection, but innovations in evolutionary processes may be much more likely to occur from "mutations of mutants" than from the variations of superior individuals that would survive hard selection. This followed the seminal works of Manfred Eigen and colleagues (Eigen and Schuster 1977, 1978a, 1978b; Eigen 1985; and others, also later Fontana and Schuster, 1987; Wang, 1987) who proposed the concept of a "quasi-species" in molecular evolution (i.e., a distribution of macromolecular types with closely interrelated sequences) and the recognition that neutral mutations in these quasi-species may be of particular significance in facilitating evolutionary optimization.

Galar (1985) devoted attention to asexual populations undergoing fitness-proportional selection and concluded that evolutionary inventions are more likely to occur when selection does not act simply to eliminate the least-fit individuals in a population. Galar (1987, 1989) explicated the analysis in more detail to consider the time required for a population to cross a gap between maxima in a fitness landscape defined by multimodal Gaussian hills. The expected number of trials required to cross such a gap increased with the number of dimensions both for hard and soft selection; however, for a particular studied case, soft selection was on average 30 times faster than hard selection (Galar, 1997, remarked that the order of magnitude can approach infinity depending on the conditions). Galar (1989) suggested that evolutionary optimization algorithms might do well to switch between hard and soft selection in an attempt to alternatively optimize the rate of convergence or the rate of saddle crossing.

Galar (1991, 1994a) went further to explore the effectiveness of asexual or sexual reproduction at crossing saddles on the fitness landscape. Sexual recombination was implemented as an averaging (blending) of continuous varying traits, following Bremermann et al. (1966)[2] and others. In simulations, recombination by blending showed accelerated local search at the expense of longer waiting times to cross saddles.[3] Galar (1994b) showed, particularly under conditions of random recombinations, that large populations are less innovative than small

---

[1] Alternatively, the stringency of selection can be measured by the expected "takeover time," the number of generations required for a single solution to dominate a population. Bäck (1994) provides a mathematical treatise on the relative stringency of diverse selection methods with respect to this measure.

[2] Recall that Bremermann used a variety of recombination methods applied to two or more parents, including averaging, crossing over, and majority logic (see Chapter 9).

[3] Beyer (1995; and others) has proved that rates of local convergence are improved for intermediate recombination operators, particularly when applied simultaneously to multiple parents.

populations, and speculated that there may be a reasonable analogy to a 1988 report from the U.S. National Science Foundation that 98% of all radical improvements come from small enterprises. Perhaps soft selection should be encouraged in industry, despite bringing no immediate rewards and only "vague promises of undefined gains in the future" (Galar, 1994b).

An independent and notable contribution of both Pincus (1970) and Galar (1985) was the recognition that evolutionary algorithms could be modeled as Markov chains, where the next state of a population is stochastically dependent only on the current state, with probabilities that are time-homogeneous. This recognition was also explicit in other early efforts (e.g., De Jong, 1975). Markov chains have been used to describe the asymptotic convergence properties of a variety of evolutionary algorithms (Eiben et al., 1991; Fogel, 1994; Rudolph, 1994; and others), as well as the transient dynamics of small or infinite populations (e.g., De Jong et al., 1995; and Nix and Vose, 1991, respectively).

Unfortunately, exact probability models of evolutionary algorithms become computationally intractable for even small population sizes, n, when individuals can assume a modest number of alternative configurations, k. The number of entries in the probability transition matrix for the associated Markov chain increases as k^{2n}. For only 10 individuals, each of say 10 bits in a binary coded genetic algorithm, there are 1024^{20} probabilities to consider. In practice, the number is much bigger than this, and by consequence this method of analysis is somewhat limited.

References

[1] T. Bäck (1994) "Selective pressure in evolutionary algorithms: A characterization of selection mechanisms," *Proc. of the 1st IEEE Conf. on Evolutionary Computation,* IEEE Press, NY, pp. 57–62.

[2] H.-G. Beyer (1995) "Toward a theory of evolution strategies: On the benefits of sex—the $(\mu/\mu,\lambda)$ theory," *Evolutionary Computation,* Vol. 3:1, pp. 81–111.

[3] W. Bossert (1967) "Mathematical optimization: Are there abstract limits on natural selection?" *Mathematical Challenges to the Neo-Darwinian Interpretation of Evolution,* P. S. Moorhead and M. M. Kaplan (eds.), The Wistar Institute Press, Philadelphia, PA, pp. 35–46.

[4] H. J. Bremermann, M. Rogson, and S. Salaff (1966) "Global properties of evolution processes," *Natural Automata and Useful Simulations,* H. H. Pattee, E. A. Edelsack, L. Fein, and A. B. Callahan (eds.), Spartan Books, Washington D.C., pp. 3–41.

[5] K. A. De Jong (1975) "An analysis of the behavior of a class of genetic adaptive systems," Ph.D. diss., Univ. of Michigan, Ann Arbor, MI.

[6] K. A. De Jong, W. M. Spears, and D. F. Gordon (1995) "Using Markov chains to analyze GAFOs," *Foundations of Genetic Algorithms 3,* L. D. Whitley and M. D. Vose (eds.), Morgan Kaufmann, San Mateo, CA, pp. 115–137.

[7] A. E. Eiben, E. H. L. Aarts, and K. M. Van Hee (1991) "Global convergence of genetic algorithms: A Markov chain analysis," *Parallel Problem Solving from Nature,* H.-P. Schwefel and R. Männer (eds.), Springer, Berlin, pp. 4–12.

[8] M. Eigen (1985) "Macromolecular evolution: Dynamical ordering in sequence space," *Ber. Bunsenges Phys. Chem.,* Vol. 89, pp. 658–667.

[9] M. Eigen and P. Schuster (1977) "The hypercycle, a principle of natural self-organization: Part A: Emergence of the hypercycle," *Naturwissenschaften,* Vol. 64, pp. 541–565.

[10] M. Eigen and P. Schuster (1978a) "The hypercycle, a principle of natural self-organization: Part B: The abstract hypercycle," *Naturwissenschaften,* Vol. 65, pp. 7–41.

[11] M. Eigen and P. Schuster (1978b) "The hypercycle, a principle of natural self-organization: Part C: The realistic hypercycle," *Naturwissenschaften,* Vol. 65, pp. 341–369.

[12] D. B. Fogel (1994) "Asymptotic convergence properties of genetic algorithms and evolutionary programming," *Cybernetics and Systems,* Vol. 25:3, pp. 389–407.

[13] W. Fontana and P. Schuster (1987) "A computer model of evolutionary optimization," *Biophys. Chem.,* Vol. 26, pp. 123–147.

[14] R. Galar (1985) "Handicapped individua in evolutionary processes," *Biological Cybernetics,* Vol. 53, pp. 1–9.

[15] R. Galar (1987) "Global random search with soft selection," *Cybernetics and Systems: Present and Future,* J. Rose (ed.), Lytham St. Annes, Thales, pp. 817–820.

[16] R. Galar (1989) "Evolutionary search with soft selection," *Biological Cybernetics,* Vol. 60, pp. 357–364.

[17] R. Galar (1991) "Simulation of local evolutionary dynamics of small populations," *Biological Cybernetics,* Vol. 65, pp. 37–45.

[18] R. Galar (1994a) "Soft selection in global optimisation," *Proc. of the 17th National Conf. on Circuit Theory and Electronic Networks,* Wroclaw, Poland, pp. 9–18.

[19] R. Galar (1994b) "Evolutionary simulations and insights into progress," *Proc. of the 3rd Annual Conference on Evolutionary Programming,* A. V. Sebald and L. J. Fogel (eds.), World Scientific, River Edge, NJ, pp. 344–352.

[20] R. Galar (1997) personal communication, Univ. of Wroclaw, Poland.

[21] J. H. Holland (1975) *Adaptation in Natural and Artificial Systems,* Univ. of Mich. Press, Ann Arbor, MI.

[22] A. Nix and M. D. Vose (1991) "Modeling genetic algorithms with Markov chains," *Annals of Math. and Art. Intell.,* Vol. 5, pp. 79–88.

[23] M. Pincus (1970) "An evolutionary strategy," *J. Theoret. Biol.,* Vol. 28, pp. 483–488.

[24] G. Rudolph (1994) "Convergence analysis of canonical genetic algorithms," *IEEE Trans. Neural Networks,* Vol. 5:1, pp. 96–101.

[25] H.-P. Schwefel (1981) *Numerical Optimization of Computer Models,* John Wiley & Sons, Chichester, U.K.

[26] Q. Wang (1987) "Optimization by simulating molecular evolution," *Biological Cybernetics,* Vol. 57, pp. 95–101.

An Evolutionary Strategy

MARTIN PINCUS

Department of Mathematics, Polytechnic Institute of Brooklyn,
Brooklyn, New York, U.S.A.

(*Received 2 February* 1970)

Evolution has been described as an optimization process. This paper presents a stochastic optimization procedure which contains an arbitrary probability distribution. Certain choices of the probability distribution lead to optimization methods that are analogous to some evolutionary operators representing asexual reproduction, sexual reproduction, mutation and selection, which have been introduced by other authors. The selection process differs from previous methods in that there is a positive probability of evolving to a lower fitness level. However, this may overcome certain previously encountered difficulties in optimization, such as the phenomenon of stagnation points, and may also have some biological significance.

1. Introduction

Various methods of solving optimization problems have been interpreted as possible evolutionary strategies by Bremermann, Rogson & Salaff (1966) and Bossert (1967). In this note I present a stochastic optimization procedure that contains arbitrary probability distributions. Certain choices of these distributions will be shown to resemble (with high probability) the evolutionary operators introduced by Bremermann *et al.* (1966). However, the selection method is different from that employed by Bremermann *et al*. This selection method should give a higher probability of escaping from the stagnation points encountered by the latter and may also be of some biological significance.

2. The Optimization Problem

Mathematically, the optimization problem to be investigated can be simply stated. Let $F(x_1, \ldots, x_n)$ be a continuous function over a closed, bounded region S of n-dimensional space. Under the assumption that there exists a unique point $z = (z_1, \ldots, z_n)$ in S at which F attains its global minimum, the problem is to find the co-ordinates of the minimizing point z. (Whether one seeks the minimizing point or maximizing point is irrelevant, since minimizing F is equivalent to maximizing $-F$.)

This problem can present formidable mathematical difficulties. The classical approach to this problem, setting the partial derivatives of F equal to zero and solving the resulting equations, is not often of much value for the following reasons. First, F may not have partial derivatives. Even when F has derivatives the minimum may be attained on the boundary and therefore the partials may not vanish at the point where the global minimum is attained. Furthermore, even if F attains its global minimum on the interior of S, one would then have to solve n simultaneous, non-linear algebraic equations, for which there is no general method of solution. Even if they could be solved, among the solutions would be points of local extrema. Each would have to be tested in turn to see which is the global minimum. What is wanted is a step-by-step method of generating a sequence of points in S that will converge, at least approximately, to the required minimizing point z.

From the evolutionary point of view, F is a measure of the fitness of the evolving population and different methods of generating the sequence of points (which represent genotypes) approximating the minimizing point are interpreted as asexual reproduction, mutation, selection, etc. This is discussed below.

The optimization procedure presented here also illustrates a point about the relation between deterministic and stochastic models of real phenomena, a relationship that was much discussed in Dr H. Pattee's group at the Colloquium on Theoretical Biology held in Fort Collins, July and August 1969. The point is that among the most important aspects of probabilistic models are those properties that behave in a localized manner with probability very near one, i.e. for practical purposes, deterministically. The various laws of large numbers are mathematical examples of this type of behavior. Also, in a study describing the behavior of certain types of random nets, Kauffman (1969) found that in nets of low connectivity the cycle lengths were surprisingly short and stable. It was this localized behavior of the random nets that was felt to be significant and might be widely used by Nature in the construction of living systems.

The stochastic optimization procedure discussed in this note is another example. The mathematical problem stated above contains no stochastic elements. I propose to solve it by utilizing a highly localized property of a stochastic process, the law of large numbers for Markov chains. Actually, this is the idea behind many of the Monte Carlo methods of numerical analysis (Hammersley & Handscomb, 1964) of which the method used here is one example.

The Monte Carlo method used was developed by Metropolis, Rosenbluth, Rosenbluth, Teller & Teller (1953) to approximate certain expressions occurring in statistical mechanics. The application of the method to optimization problems is based on a theorem in Pincus (1968) and is explained more fully in Pincus (manuscript in preparation) where there is also some discussion of the errors made in the approximations used. I outline the method here.

To each minimization problem of the type described above one associates a Markov chain $\{X_i\}$ whose state space and transition matrix is constructed in the following way. Partition S into N non-overlapping subsets $\{S_1, S_2, \ldots, S_N\}$. For simplicity I assume the subsets $\{S_i\}$ have equal volumes. Choose a point $y^j = (y_1^j, \ldots, y_n^j)$ in each set $S_j, j = 1, \ldots, N$. Let $P^* = (p_{ij}^*)$, $1 \le i, j \le N$, be any matrix satisfying $p_{ij}^* = p_{ji}^*$, $p_{ji}^* > 0$ and

$$\sum_{j=1}^{N} p_{ij}^* = 1$$

for all i and j, and let $\Pi_j = \exp(-\lambda F(y^j))$ where λ is a fixed, positive real number. (The larger λ is, the better the approximation given by the method will be.) The transition matrix $P = (p_{ij})$ that one wants is defined for $1 \le i$, $j \le N$ by

$$p_{ij} = \begin{cases} p_{ij}^* \Pi_j / \Pi_i & \text{if } \Pi_j/\Pi_i < 1 & i \ne j \\ p_{ij}^* & \text{if } \Pi_j/\Pi_i \ge 1 & i \ne j \\ p_{ii}^* + \sum_{j: \Pi_j < \Pi_i} p_{ij}^* (1 - \Pi_j/\Pi_i) & & i = j \end{cases}$$

A Markov chain with the above transition matrix has an invariant distribution given by $\{\Pi_i\}$. By the strong law of large numbers for Markov chains (Chung, 1967) it follows that with probability one the sequence of sample averages

$$\frac{1}{m} \sum_{j=1}^{m} X_j$$

converges to the n-dimensional vector whose ith component is

$$\sum_{j=1}^{N} y_i^j \exp(-\lambda F(y^j)) / \sum_{j=1}^{N} \exp(-\lambda F(y^j)).$$

The co-ordinates of this vector are approximately equal to the co-ordinates z_i of the minimizing point. The larger N and λ are, the better the approximation should be (Pincus, 1968, 1970).

Although the transition matrix appears to be complicated, a Markov chain with this transition matrix can be simulated rather easily as follows: Given that the chain is in state y^i at time k, i.e. ($X_k = y^i$), the state at time $k+1$ is determined by choosing another state according to the distribution

$\{p_{ij}^*, j = 1, \ldots, N\}$. If the state chosen is y^j, one then calculates the ratio Π_j/Π_i. If $\Pi_j/\Pi_i \geq 1$ (i.e. the fitness is increased), one accepts y^j as the new state. If $\Pi_j/\Pi_i < 1$ (i.e. the fitness is decreased), one takes y^j as the new state at time $k+1$ with probability Π_j/Π_i, and y^i as the new state at time $k+1$ with probability $1 - \Pi_j/\Pi_i$. It should be noted that there are no difficult calculations to carry out in the above procedure as it only involves the ratios $\Pi_j/\Pi_i = \exp[\lambda(F(y^i) - F(y^j))]$. The common denominators of the Π_j cancel out.

Some of the specific evolutionary operators introduced by Bremermann *et al.* (1966) can be approximately realized by various specializations of the choice of state space $\{y^i\}$ and transition matrix P^* which is arbitrary except for the conditions given above.

For instance, in asexual evolution, mutation is represented by changing a single component (chosen at random) of the present state. Geometrically the process moves only in directions along the co-ordinate axes. This can be represented by choosing for S a rectangular parallelopiped and partitioning S by a rectangular grid where the planes of subdivision are parallel to the co-ordinate planes. By choosing for the y^j the centers of the rectangles, adjacent states differ only in one co-ordinate. For fixed i choose the distribution $\{p_{ij}^*, j = 1, \ldots, N\}$ so that most of the probability falls on those states y^j that differ only in one co-ordinate from the state y^i.

Through sexual evolution the offspring receives a more varied genotype upon which selection may act than is possible through asexual reproduction and mutation of one gene at a time. Geometrically, the mating schemes in Bremermann *et al.* allow the optimization method to search for improvements in directions other than those parallel to one of the co-ordinate axes, although the directions are still limited. For instance, Bremermann and his co-workers' averaging method generates a new point (offspring) on the line between points represented by the parents. From this point of view sexual evolution can be represented in our method in many ways. For example, for each state y^i we can assign a few preferred directions by choosing the probability distribution $\{p_{ij}^*, j = 1, \ldots, N\}$ so that most of the weight is assigned to those states y^j in the preferred directions from the fixed state y^i. Furthermore, the final estimate of the minimizing point used in the present scheme is the average

$$\left(\frac{1}{m} \sum_{j=1}^{m} X_j \right)$$

of all the states sampled. This is directly analogous to Bremermann and his co-workers' averaging method to represent sexual reproduction. Mutations are represented in all these schemes due to the requirement that $p_{ij}^* > 0$ for all i and j.

In the optimization procedures used by Bremermann *et al.*, they found that the process may stagnate at a point far from the optimum. This occurs at ridges and, according to Bossert (1967), is due to the fact that most of the volume around points lying on the ridge contains only points of lower fitness, while the only new genotypes accepted by Bremermann *et al.* are those that give higher fitness.

In contrast, from the description given above for simulating the Markov chain, it can be seen that there is a positive probability of accepting a genotype y^j of lower fitness. The distribution $\{p_{jk}^*, k = 1, \ldots, N\}$ may be weighted so that it leads, with high probability, to states nearer to the minimum than the genotypes on the ridge, or it may lead to a path that entirely avoids the ridge. This would give the process a higher probability of escaping from stagnation points. In biological terms, this suggests that it might be advantageous at times to accept the risk of evolving to a lower fitness level temporarily, in order to proceed to optimal fitness more rapidly.

It can also be seen from the description of the Markov chain that, in the long run, the chain will spend most of the time visiting states near the minimizing point. This is in fact the idea behind Metropolis' Monte Carlo method and is somewhat reminiscent of the statement: "The idea that evolution comes about from the interaction of a stochastic and a directed process was the essence of Darwin's theory." (Wright, 1967).

In order to take into account a changing environment, it may be of interest to let the fitness function F depend on time in a smooth way and to see how well the Markov chain (which becomes a non-stationary process since the transition probabilities now depend on time through the Π_i) would follow the now time-dependent minimizing point.

To model more drastic changes occurring at random times, we could choose a sequence of random times (T_i), e.g. the jump times of a Poisson process, and a sequence of auxiliary fitness functions $\{F_i(x_1, \ldots, x_n, t)\}$ to construct a new fitness function F defined by $F(x_1, \ldots, x_n, t) = F_i(x_1, \ldots, x_n, t)$ for $T_i \leq t < T_{i+1}$, $i = 1, 2, \ldots$. That is, the fitness function F is a well behaved function between the random times T_i, T_{i+1} but may have discontinuities at the endpoints of this time interval.

In particular, it would be of interest to note if certain choices of the P^* matrix interpretable as evolutionary operators would follow the minimizing point more efficiently than others. It should be noted that in the time-dependent cases, the mathematical analysis of the original problem no longer applies.

These suggestions could be carried out in many different ways. However, further discussion of the foregoing ideas would be more profitable after testing a few of these models on a computer.

This work was supported by the National Aeronautics and Space Administration and by the American Institute of Biological Sciences at the Colloquium on Theoretical Biology held in Fort Collins, Colorado in July and August 1969.

REFERENCES

BOSSERT, W. (1967). In *Mathematical Challenges to the Neo-Darwinian Interpretation of Evolution*. (P. S. Moorhead and M. M. Kaplan, eds.) p. 35. The Wistar Institute Symposium Monograph No. 5. Philadelphia: Wistar Institute Press.

BREMERMANN, H. J., ROGSON, M. & SALAFF, S. (1966). In *Natural Automata and Useful Simulations*. (H. H. Pattee, E. A. Edelsack, L. Fein & A. A. Callahan, eds.) p. 3. Washington, D.C.: Spartan Books.

CHUNG, K. L. (1967). *Markov Chains with Stationary Transition Probabilities*, 2nd Ed. Berlin: Springer.

HAMMERSLEY, J. M. & HANDSCOMB, D. C. (1964). *Monte Carlo Methods*. London: Methuen's Monographs on Applied Probability and Statistics.

KAUFFMAN, S. A. (1969). *J. theor. Biol.* **22**, 437.

METROPOLIS, N., ROSENBLUTH, A. W., ROSENBLUTH, M. N., TELLER, A. H. & TELLER, E. (1953). *J. chem. Phys.* **21**, 1087.

PINCUS, M. (1968). *Ops Res.* **16**, 690.

WRIGHT, S. (1967). In *Mathematical Challenges to the Neo-Darwinian Interpretation of Evolution*. (P. S. Moorhead and M. M. Kaplan, eds.) p. 117. The Wistar Institute Symposium Monograph No. 5. Philadelphia: Wistar Institute Press.

Handicapped Individua in Evolutionary Processes

R. Galar

Institute of Engineering Cybernetics, Technical University of Wroclaw, Wybrzeze Wyspianskiego 27, PL-50-370 Wroclaw, Poland

Abstract. Usually, when considering an evolutionary development, the fittest individua and their succesfull mutations recive most attention. The less fit ones, existing due to selection-mutation equilibria, are regarded, as an inevitable waste. It appears that these handicaped individua play an important role in evolution, as they make escapes from evolutionary traps possible. It is concluded that crucial improvements come from mutations of mutants. The implications seem to account to some extent for the observed speed and mode of evolution. An appropriate stochastic model is introduced, numerical experiments reported and results discussed.

1 Introduction

Considerations concern the problem of improvement of a fine tuned multi-parameter system. If the values of parameters fit so that only a specific change of several of them at once can bring an improvement while any lesser change results in deterioration, it can be extremely difficult to achieve an improvement.

An evolving population shows a tendency to move uphill in the fitness landscape. The movement is relatively smooth and fast as long as improvements can result from mutations of single traits of a type which currently dominates in the population. Once such possibilities are exhausted, and some "local summit" or "narrow ridge" in the fitness landscape attained, the population finds itself in an evolutionary trap. Then, there is no real progress but a kind of quasi-equilibrium (metastability) between selection and mutations. It can last long before some successful mutation "breakes the trap" and triggers off the next stage of development. Traps can occupy only a very minute fraction of the fitness landscape, nevertheless, as they act as attractors, it is conceivable that most of the time populations spend in traps. This can be one of reasons of the phenomenon of punctuated equilibria (e.g. Stanley 1982).

The problem of "escape from evolutionary trap" is well known. Estimations of improvement probability in cases corresponding to evolutionary traps (e.g. Bremermann 1966; Conrad 1979) have in common a more or less implicit assumption of hard selection. The hard selection allows only for a "survival of the fittest" and, in consequence, an improvement can come only from a mutation of the fittest. Such approach seems to be consistent with a way of dealing with population dynamics apart from mutation processes (Ginzburg 1983). Estimates thus obtained give the probabilities of improvement so small as to arouse doubts about optimization powers of evolution (Cohen 1973).

To deal with escapes from evolutionary traps, it is necessary to study mutation-selection equilibria resulting from soft selection; i.e. survival of handicapped mutants. A deterministic theory of such processes was given by Eigen (1971) and developed by Thompson and McBride (1974) and Jones et al. (1976). Lately, McCaskill (1984) gave a stochastic version of the Eigen's model. A model applied here belongs to the same class and was developed for simulation needs (Galar et al. 1980; Kwasnicka et al. 1983). Similar problems are also investigated using stochastic models with highly aggregated parameters of population (Wagner 1984).

A thesis is that escapes from evolutionary traps, if happen, happen almost entirely because of mutations of individua which are handicapped with regard to currently the most fit. Section 2 presents the model and a concept of the diversity induced gain in improvements. Section 3 deals with deterministic estimates of escape probabilities, especially for cases with large numbers of traits. Section 4 describes a method and the results of an immediate simulation. Section 5 gives concluding remarks. The probabilities of escape, however small, are orders of magnitude greater than those estimated when assuming hard selection.

The author's interest in the subject come from optimisation research. The author regrets that references he is able to give are far from being adequate.

2 Model

A simple, stochastic mutation-selection model of an asexually reproducing haploid population with non-overlapping generations in an isolated, constant, uniform and saturated environment is postulated. An evolving system defined by a set of functionally constrained but genetically independent traits is considered. An estimate of the diversity induced gain in improvements is given.

2.1 Law of Evolution

The population evolves in a manner defined by probabilities of reproduction and mutation of particular individua:

$$P[e'He''|e' \in G(i), e'' \in G(i+1)] = q(h(e'))\Big/ \sum_{e \in G(i)} q(h(e)), \tag{1}$$

$$P[h(e'')=x''|e'He'', h(e')=x'] = m(x'',x'), \tag{2}$$

where

$e, e', e'' \in E$	– individua of species E
$x, x', x'' \in T$	– types from space T
$G(i)$	– i-th generation; $i=0,1,2,\dots$
$e'He''$	– relation: e'' is a descendant of e'
$h : E \to T$	– classification function defining types of individua
$q : T \to R^+$	– quality function defining fitness of types
$m : T \times T \to [0,1]$	– modification function defining probabilities of mutation from type to type

Let us consider the environment as an agency providing each generation with a number of compartments. As the environment is saturated and isolated all compartments have to be occupied by descendants from the previous generation. Formula (1) gives the probability that the occupant of the compartment denoted by ″ is a descendant of the individuum from the compartment denoted by ′. For any compartment, chances of implanting one's own descendant are distributed among individua proportionally to their fitness.

Formula (2) states that the probability of the mutation from one type to another depends only on these types. The replication is regarded as the mutation to the same type.

States of the population can be defined by listing numbers of individua of particular types or, as implied above, by listing types of individua in particular compartments. The second approach, however unusual, is suitable for simulation when types are more numerous than individua. Since (1) and (2) suffice to determine transition probabilities between states, evolution can be considered as a Markov chain.

The mean number of individua of type x in the next generation can be derived considering the expected numbers of descendants: of one type x individuum; of all type x individua; of all type x individua which will turn to be of type x', for all types x:

$$E(N(i+1, x'))$$
$$= c(i+1)/c(i) \sum_{x \in T} N(i,x)\, m(x',x)\, q(x)/q_A(i), \tag{3}$$

where

$N(i,x)$	– number of type x individua in generation $G(i)$
$c(i)$	– capacity of environment at time i
$q_A(i)$	– average quality at time i:

$$q_A(i) = 1/c(i) \sum_{x \in T} N(i,x)\, q(x)$$

For large populations (3) can be used as a deterministic model:

$$v(i+1, x') = 1/q_A(i) \sum_{x \in T} v(i,x)\, m(x',x) \tag{4}$$

where

$v(i,x) = N(i,x)/c(i)$ – type x fraction of population at time i

For a finite number of adequately numbered types (4) can be rewritten in a matrix notation:

$$v(i+1) = 1/q_A(i)\, Mv(i) \tag{5}$$

where

$v(i) = (v_1(i), \dots, v_r(i))$	– population distribution on types
$M = (m_{jk}q_k)$	– transfer matrix; $j,k = 1,\dots,r$
q_k	– fitness of k-type
m_{jk}	– probability of k to j-type mutation

Formula (5) is a discrete time version of the Eigen's model; (Eigen 1971) of the evolution of a set of macromolecular information cariers. Fitness q_k is proportional to Eigen's rate parameter A_k, probability of exact replication m_{kk} equals quality factor Q_k, product $q_k m_{jk}$ equals quantity φ_{jk}.

432

2.2 Space of Types

To use model (1) and (2); space of types T, quality function $q(\cdot)$, and modification function $m(\cdot, \cdot)$ should be defined. Let types, i.e. evolutionary characteristics of considered species, be determined by some w traits: $x = (x_1, ..., x_w)$. Let us assume a discrete, w-dimensional space T with the distance given by the absolute norm:

$$x \in T \subset R^w; \quad T = \{..., -1, 0, 1, ...\}^w;$$

$$\|x, x'\| = \sum_{k=1}^{w} |x_k - x_k'|. \tag{6}$$

Let the fitness change gradually with the distance and depend on a degree of tuning of traits. Let all traits be vital, i.e. a great enough change of any trait can make the fitness arbitrarily close to zero. The basic function considered is:

$$q(x) = A \exp(-a\|x, x'\|^2); \quad x, x' \in T; \quad a, A \in R. \tag{7}$$

Let mutations result from statistically independent changes of particular traits and let mutation probability depend only on a distance between types. There exists the unique probability mass function $f(\cdot)$ assigning probabilities to mutation distances:

$$f(d) = m(x', x) = ((1-p)/(1+p))^w p^d;$$

$$x, x' \in T; \quad p \in (0, 1), \tag{8}$$

where

$d = \|x, x'\|$ – mutation distance

$(1-p)/(1+p)$ – probability that particular trait stays the same.

The mean mutation distance is equal $2np/(1-p^2)$. When small, the value of parameter p is close to the probability that a value of a particular trait will increase (decrease) by one.

2.3 Advantage of Diversity

Type x^0 will be called suboptimal of order n if the nearest better type lies at distance n:

$$n = \min\{\|x, x^0\| : q(x) > q(x^0)\}. \tag{9}$$

If the order is high enough with regard to the number of traits, mutations probabilities and the local fitness, the region of the suboptimal type constitutes an evolutionary trap. An improvement happens when some mutation produces an individuum better than x^0. Let B denote the set of types better then x^0; $B = \{x : q(x) > q(x^0)\}$. The expected number of improvements in generation $G(i+1)$ is:

$$u(i) = c(i+1)/c(i) \sum_{x \in B, e \in G(i)} m(x, h(e)) q(h(e))/q_A(i). \tag{10}$$

The aim is to compare expected numbers of improvements for a homogenous population composed of suboptimal individua and for a diversified population in statistical quasi-equilibrium. For the homogenous population of the size c the expected number is:

$$u_h = c \sum_{x \in B} m(x, x^0). \tag{11}$$

For the diversified, equilibrium population the number is:

$$u_d = c/q_A \sum_{x \in B, x' \in T'} v(x') m(x, x') q(x'), \tag{12}$$

where

$T' = T \backslash B$ – set of types no better then x

$q_A, v(x)$ – equilibrium values of $q_A(i), v(i, x)$.

The ratio of expected number of improvements (12) and (11) will be a measure of an evolutionary advantage which results from the population diversity.

3 Estimates by Deterministic Model

Assuming population large enough, deterministic model (5) is employed to get the quasi-equilibrium distribution and the gain in improvements. At first, a trap involving two traits only is considered. This illustrates the problem and provides data for Sect. 4. A large number of traits gives a multitude of types what prevents straightforward estimation of quasi-equilibria. To gain some insight into multi-dimensional cases, equilibria in a symmetrical fitness landscape are considered. Similar types are aggregated, an aggregated model developed and distributions at equilibria estimated. Preserving the original shape near optimum the fitness landscape is modified to create a trap. The quasi-equilibrium is computed as a perturbance of the genuine one.

3.1 Introductory Example

There are two traits; $x = (x_1, x_2)$. The fitness is either zero or is given by:

$$q(x) = \exp[-0.3\sqrt{(x_1-4)^2 + (x_2-4)^2}]$$
$$+ 2\exp[-0.3\sqrt{(x_1-9)^2 + (x_2-9)^2}]. \tag{13}$$

All 69 types with the non-zero fitness are shown in Table 1. The types better then type (4, 4) are denoted by "+". The remaining types have their fitness given in precents of $q(4, 4)$. As (7, 8) is the nearest better type to (4, 4), (4, 4) is the suboptimal type of order 7 and there is an evolutionary trap around (4, 4).

433

Table 1. Introductory example: fitness of types [*100%/$q(4,4)$]

x_2 \ x_1	0	1	2	3	4	5	6	7	8	9
0	0	0	0	0	31.6	31.8	0	0	0	0
1	0	0	0	39.3	42.3	42.3	39.8	0	0	0
2	0	0	42.8	51.4	56.5	55.6	50.9	45.4	0	0
3	0	39.3	51.4	65.4	75.3	71.3	62.8	55.4	49.4	0
4	31.6	42.3	56.5	75.3	100.0	83.4	72.3	64.9	59.2	54.0
5	31.8	42.3	55.6	71.3	83.4	82.3	77.2	73.4	70.2	66.1
6	0	39.8	50.9	62.8	72.3	77.2	79.7	82.0	83.6	81.6
7	0	0	45.5	55.4	64.9	73.4	82.0	91.6	+	+
8	0	0	0	49.4	59.2	70.2	83.6	+	+	+
9	0	0	0	0	54.0	66.1	81.6	+	+	+

Table 2. Introductory example: population mass distribution at quasi-equilibrium [*100%]

x_2 \ x_1	0	1	2	3	4	5	6	7	8	9
0	–	–	–	–	0.005	0.002	–	–	–	–
1	–	–	–	0.02	0.07	0.03	0.007	–	–	–
2	–	–	0.03	0.21	0.82	0.34	0.07	0.01	–	–
3	–	0.02	0.21	1.70	7.19	2.65	0.53	0.08	0.01	–
4	0.005	0.07	0.82	7.19	36.83	11.03	2.06	0.31	0.04	0.005
5	0.002	0.03	0.34	2.65	11.03	5.25	1.43	0.32	0.06	0.01
6	–	0.007	0.07	0.53	2.06	1.43	0.63	0.26	0.08	0.02
7	–	–	0.01	0.08	0.31	0.32	0.26	0.19	*	*
8	–	–	–	0.01	0.04	0.06	0.08	*	*	*
9	–	–	–	–	0.005	0.01	0.02	*	*	*

The mutation probability is given by (8). The expected number of improvements in a population of c suboptimal individua is:

$$u_h = c((1-p)/(1+p))^2 \, p^7(2+3p+2p^2+p^3). \qquad (14)$$

Parameter p is adjusted to get the mean distance of mutation 0.2 (or 0.5). For a population of 50 individua an improvement can be expected once for about 15 millions (30,000) generations.

To compute the quasi-equilibrium distribution one must assume absence of population mass flow (mutations) to types better than the suboptimal (McCaskill 1984). The distribution was computed using (5) repeatedly. System (5) has one stable solution as shown by Jones et al. (1976). Table 2 gives the results obtained after 100 runs for the initially suboptimal population. Number of improvements expected in quasi-equilibrium (11) greately exceeds the number given by (14). Improvements in the population of 50 can be expected once for 69 (2.2) generations. The gain in improvements is more than 200,000 (10,000).

The illustrated direct application of model (5) can not go much farther. Transfer matrix M already has 3721 elements and one more trait would push this number to over 100,000.

3.2 Multidimensional Equilibrium Distribution

The evolving system is defined by (6), (7), (8). The aim is to estimate the equilibrium distribution around the optimum.

Let S_r^k denote the set of types on distance r from type x^k:

$$S_r^k = \{x : \|x, x^k\| = r\} \subset T \subset R^w. \qquad (15)$$

The number of types on absolute metric sphere S is given by:

$$s(r) = \sum_{j=1}^{\min(r,w)} 2^j \binom{w}{j}\binom{r-1}{j-1}. \qquad (16)$$

Number $s(r)$ grows rapidly with r and w; e.g. for $r=10$ there is more than 4 millions types if $w=10$ and more than 10^{26} types if $w=1000$.

Let (7) be modified to get the zero fitness if the distance to optimum x^0 exceeds r_m. Radius r_m has to be large enough not to disturb the original equilibrium distribution. Now, model (5) applies but is transcomputable owing to the number of types in sphere $S_{r_m}^0$. To make the problem computable, types are assembled so that the fitness and the mutation probability depend only on categories the types belong to.

434

All types in the same category can be aggregated into one "lumped type" and equations (5) can be reformulated to describe the dynamics of lumped types.

Because of the symmetry of (6), (7), and (8), types with the same set of absolute values of traits (irrespective of sign or order) belong to the same category. An adequate aggregation gives no errors but the resulting number of lumped types is still too large. Let x^r denote the "standard type" on sphere S_r^0:

$$x^r = (x_1^r, \ldots, x_r^r, \ldots, x_w^r); \quad x_j^r = 1 \text{ if } j \leq r;$$
$$x_j^r = 0 \text{ if } j > r. \tag{17}$$

It is convenient to aggregate whole spheres, assuming that each type on sphere S_r^0 has mutation probabilities the same as standard type x^r. If $w > r_m$, thus inflicted distortion of the original model is of little consequences; it slightly prefers inward mutations. The model has a form:

$$v_r(i+1) = 1/q_A(i) \sum_{s=0}^{r_m} v_s(i) q_s m_{rs}; \quad r = 0, \ldots, r_m \tag{18}$$

where

$v_r(i)$ – fraction of population on sphere S_r^0 at time i

q_r – fitness on distance r from x^0: $q_r = q(x^r)$

m_{rs} – transfer probability from sphere S_s^0 to S_r^0:

$$m_{rs} = P[\|h(e^r), x^0\| = r \mid eHe^r, \|h(e), x^0\| = s].$$

The probabilities of transfer from sphere to sphere depend on parameter p and number of traits w. Let us consider mutations of descendants of types on sphere S_r^0 in terms of the population mass transfer. Exact replications make some mass stay on sphere S_r^0; mutations at distance 1 transfer mass to spheres S_{r+1}^0 and S_{r-1}^0; mutations at distance 2 transfer mass to spheres S_{r+2}^0, S_r^0, and S_{r-2}; etc. All mutations at odd distances transfer mass from sphere S_r^0 to S_{r+1}^0, while mutations at even distances transfer mass to sphere S_{r+2}^0; etc. The transfer probability is given by:

$$m_{rs} = ((1-p)/(1+p))^w \sum_{j=0}^{s} p^j b_{rs}; \quad t = |s-r| + 2j, \tag{19}$$

where b_{rs} denotes the number of types common to S_r^0 and S_s^0:

$$b_{rs} = \binom{r}{k} \sum_{n=0}^{t-k} \binom{t-1}{t-k-n} \sum_{m=0}^{w} 2^m \binom{w-s}{m} \binom{r-k}{n-m};$$
$$k = (t-r+s)/2. \tag{20}$$

Model (18) depends on parameters w, a, p defining in turn number of traits, fitness hill slope, mutation probability. The standard values: $w = 1000$, $a = 0.1$, $p = 0.0005$ give 1 for the mean distance of mutations (Bremermann et al. 1966) and 10% drop in fitness at distance 1 from optimum. The equilibrium distribution

was computed using (18) repeatedly till $\sum_r [v_r(i+1) - v_r(i)]$ got less then 10^{-6}. In examined cases; for $v_0(0) = 1$, it was sufficient to execute $20 \div 200$ iterations, adopt radius of distribution $r_m = 25$ and allow mutations to distance 10. The equilibrium distribution for the standard set of w, a, p with some data on mutations and fitness are given in Table 3. One can note the known effects:

– In spite of the small probability of any particular mutation, most of descendants undergo some mutation. This large population mass transfer goes both to inner and outer spheres. The amounts are comparable if radius r of original sphere S_r^0 exceeds number w of traits. In our case $r \ll w$ and the bulk of mass goes outside.

– As the population mass influx to center $S_0^0 = x^0$ is very small, the mass attributed to the optimal type come from replication. Mean equilibrium fitness q_A relatively to optimal fitness q_0 is close to replication probability $f(0)$. The mean equilibrium fitness depends little on the shape of the fitness landscape.

– The finite population statistical equilibrium distribution in the space of types forms a kind of a nebula. This centred on the optimal type nebula is very sparse. Of all types encompassed a tiny fraction only can ever come into being and even the optimal type might well not be present within a generation. Almost all individua are unique in type, still, for any selected trait majority have it fixed at the value of the optimal type.

Figures 1–3 show impacts of parameters on the equilibrium distribution. Fixed parameters assume standard values. For large w and small p the distribution depends on product wp; Fig. 4.

3.3 Gain in Improvements

It is convenient to consider an evolutionary trap given by:

$$q(x) = \exp\left\{ a\left[\langle x, x^n \rangle / n - \left(\min_{t \in R} \{\|x - tx^n\|\} \right)^2 \right] \right\};$$
$$x, x^n \in T. \tag{21}$$

Nearby suboptimal type x^0, function (21) gives almost the same fitness as (7). Standard type x^n is the closest better type. If types x are regarded as elements of w-dimensional real space, (21) gives a "narrow ridge" diagonal to the first n coordinates. The fitness grows along and drops across a line $x = tx^n$. As types belong to space T, at each type kx^n exists a trap of order n. For types which the first n traits have values 0 or 1, one get:

$$q(x) = q_{rs} = \exp[a(r/n - (n/2 + s - |n/2 - r|)^2)], \tag{22}$$

where

$$r = \sum_{k=1}^{n} x_k; \quad s = \sum_{k=n+1}^{w} |x_k|; \quad (x_1, \ldots, x_n) \in \{0, 1\}^n.$$

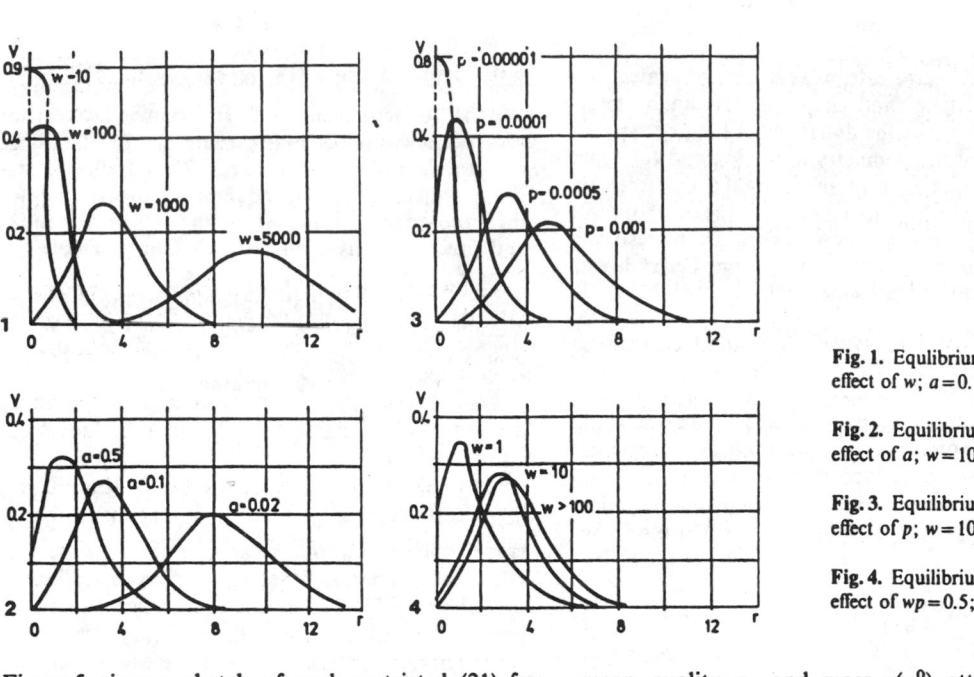

Fig. 1. Equilibrium distribution – effect of w; $a=0.1$, $p=0.0005$

Fig. 2. Equilibrium distribution – effect of a; $w=1000$, $p=0.0005$

Fig. 3. Equilibrium distribution – effect of p; $w=1000$, $a=0.1$

Fig. 4. Equilibrium distribution – effect of $wp=0.5$; $a=0.1$

Figure 5 gives a sketch of such restricted (21) for $a=0.1$, $n=8$. The distinct fitness gap along axis r and the similarity of (7) and (21) in the region of type x^0 are visible.

A symmetry of the evolutionary trap is more complex than of the optimum, what hinders estimating of the quasi-equilibrium distribution. To simplify, the distribution for (21) is obtained as a perturbance of the distribution for (7). We assume that, at equilibrium

mean quality q_A and mass $v(x^0)$ attached to the optimum are practically the same for both distributions.

Type x is intermediary between x^a and x^b if values of x_k are from a range x_k^a to x_k^b. Let set T^n consist of types intermediary between x^0 and x^n except type x^n. Traits of types from T^n are equal 0; only first n traits can be equal 0 or 1. The symmetry of (21) makes all $\binom{n}{r}$

Table 3. Multi-dimensional example: data relevant to the state of statistical equilibrium $w=1000$, $a=0.1$, $p=0.0005$, $q_A=0.36959$

r	$s(r)$	$f(r)$	$s(r)f(r)$	q_r/q_0	v_r	g_r	g_r'	g_r''
0	1	0.36788	0.36788	1.00000	0.0066	–	–	–
1	$2.0e3$	0.00018	0.36788	0.90484	0.0681	–	–	–
2	$2.0e6$	$9.2e{-}8$	0.18394	0.67032	0.1950	–	–	–
3	$1.3e9$	$4.6e{-}11$	0.06131	0.40657	0.2722	$4e1$	$1e2$	$1e3$
4	$6.7e11$	$2.3e{-}14$	0.01532	0.20190	0.2323	$5e2$	$4e3$	$2e4$
5	$2.7e14$	$1.1e{-}19$	0.00307	0.08208	0.1370	$8e3$	$1e5$	$2e7$
6	$8.9e16$	$5.7e{-}21$	0.00051	0.02732	0.0603	$7e4$	$2e6$	$2e9$
7	$2.5e19$	$2.9e{-}24$	0.00007	0.00745	0.0208	$6e5$	$5e7$	$2e11$
8	$6.4e21$	$1.4e{-}27$	0.00001	0.00166	0.0059	$3e6$	$7e8$	$2e13$
9	$1.4e24$	$7.2e{-}31$	$1.0e{-}6$	0.00030	0.0014	$2e7$	$2e10$	$2e15$
10	$2.8e26$	$3.6e{-}34$	$1.0e{-}7$	0.00005	0.0003	$1e8$	$3e11$	$2e17$

$$XeY = X \cdot 10^Y$$

$s(r)$ — number of types on sphere S_r^0
$f(r)$ — probability of mutation to the type at distance r
$s(r)f(r)$ — probability of mutation at distance r
q_r/q_0 — fitness on sphere S_r^0 relative to fitness at optimum
v_r — fraction of population on sphere S_r^0 at equilibrium
g, g', g'' — gain in improvements; $b=0$, 0.5, 0.9

436

intermediary types on sphere S_r^0 equivalent to standard type x^r. At the equilibrium, any type loses as much mass by mutations of descendants as it gains from mutations of other types. For w large, mutations to x^r from outside of set T^r can be neglected. By (4), considering mutations from T^r only, the x^r type fraction of population at equilibrium is:

$$v(x^r) = 1/(q_A - q(x^r)f(0)) \sum_{s=0}^{r-1} \binom{r}{s} v(x^s) q(x^s) f(r-s).$$ (23)

Knowing $v(x^0)$ and q_A, values $v(x^r)$ can be computed successively.

Let set B of types better then x^0 be restricted to x^n and let set T' of types no better then x^n be restricted to T^n. Formula (12) for the expected number of improvements assumes a form:

$$u_d = c/q_A \sum_{r=1}^{n-1} v(x^r) q(x^r) f(n-r).$$ (24)

Formula (11), which gives the number of improvements expected in the suboptimal populations of size c, assumes the form $u_h = c f(n)$. Ratio u_d/u_h gives the gain in improvements:

$$g_n = 1/q_A \sum_{r=1}^{n-1} v(x^r) q(x^r)/p^r.$$ (25)

As a whole, the simplifying assumptions underestimate the gain in improvements. The accuracy of estimates grows with w, a, and n growing and p diminishing.

Table 3 gives the gain in improvements for standard values of parameters w, a, p; and for orders of traps n from 3 to 10. The improvements are evidently due to mutations of intermediary types. The probability of an improvement in the homogeneously suboptimal population is relatively negligible.

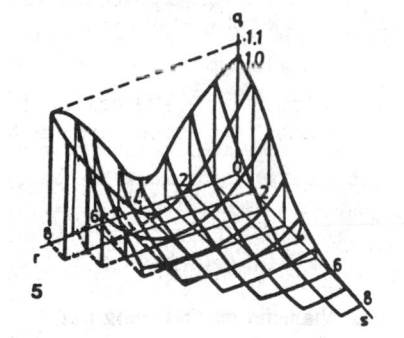

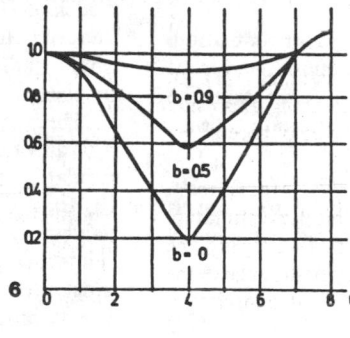

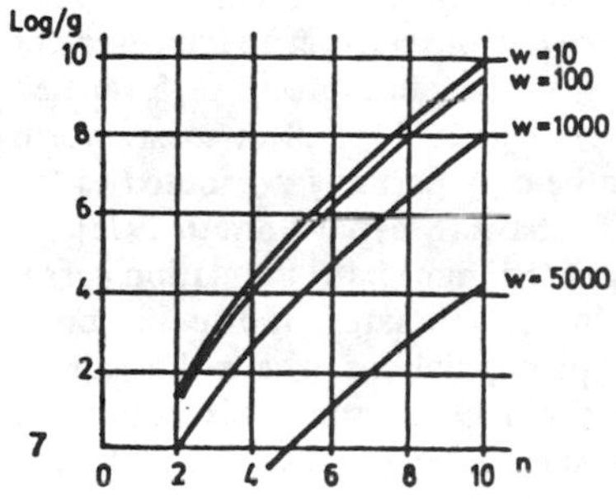

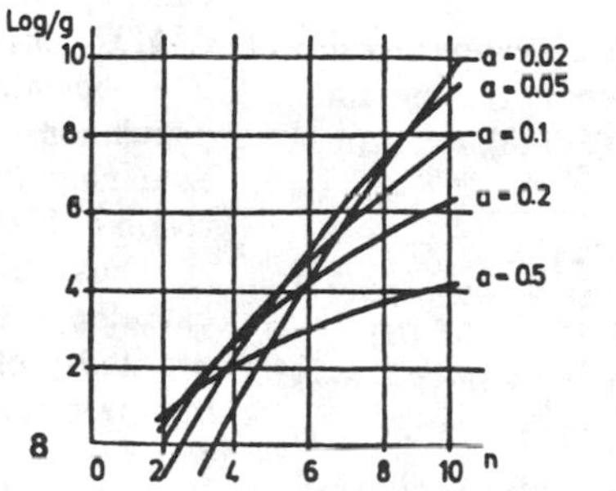

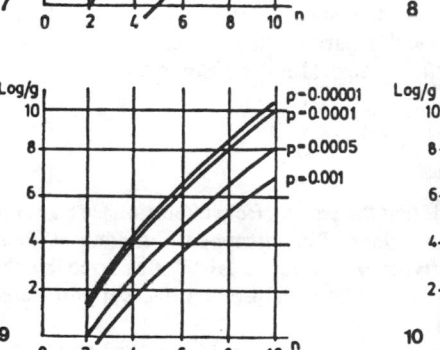

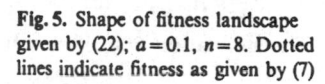

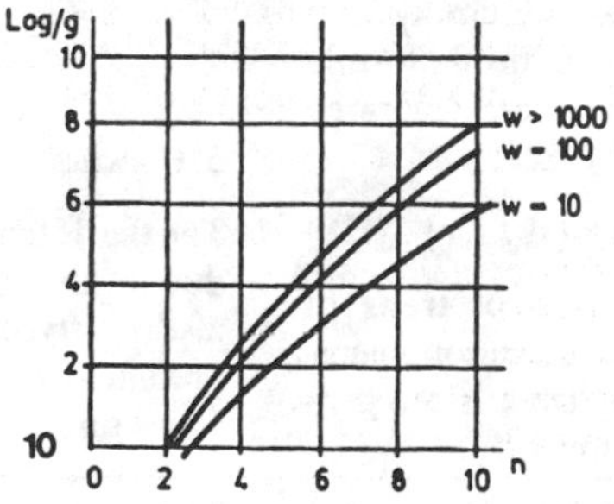

Fig. 5. Shape of fitness landscape given by (22); $a=0.1$, $n=8$. Dotted lines indicate fitness as given by (7)

Fig. 6. Fitness gap profile

Fig. 7. Gain in improvements – effect of w; $a=0.1$, $p=0.0005$

Fig. 8. Gain in improvements – effect of a; $w=1000$, $p=0.0005$

Fig. 9. Gain in improvements – effect of p; $w=1000$, $a=0.1$

Fig. 10. Gain in improvements – effect of $wp=0.5$; $a=0.1$

Table 3 gives also gains for fitness gaps less distinct than the one given by (21). Gains g' and g''; see Fig. 6, correspond to fitness: $q(r,b) = q(x^r) + b(1 - q(x^r))$; $b' = 0.5$, $b'' = 0.9$. These are less exact estimates, as it can not be hold that the original optimal distribution is only slightly perturbed.

Figures 7–9 show the impact of parameters w, a, p on the gain in improvements. As the gain and mutation probability (8) both drop with the growth of w, the probability of improvement depends on w more than it is implied by (11). The inverse is with p as the gain drops and the mutation probability grows with growing p. A growth of a makes escape from traps of low order more and from traps of high order less probable. Figure 10 shows that for w large and p small the gain depends on product wp.

4 Simulation Estimates

Model (1), (2) can be conveniently used for simulation. The state of the next generation is a multidimensional random variable which distribution depends on the state of the present one. The population state at time i is stored as an ordered set of w-tuples $(x_1^k, \ldots, x_w^k)$; $k = 1, \ldots, c(i)$, representing types of individua. According to (1), the probability that the descendant is produced by one of the first k ordered individua is given by:

$$D(k) = 1/[c(i) q_A(i)] \sum_{j=1}^{k} q(x^j). \tag{26}$$

According to (2), the mutation probability depends only on types involved. Let h denote the mutation increment of k-trait value. Because of (8), the probability distribution of h is:

$$F(h) = \begin{cases} p^{-h}/(1-p) & \text{for} \quad h = 0, -1, -2, \ldots \\ 1 - p^{h+1}/(1-p) & \text{for} \quad h = 1, 2, \ldots. \end{cases} \tag{27}$$

The simulation algorithm sets types for the next generation. To get the type of an individuum its progenitor has to be fixed. Let members of each generation be numbered in some way. Let individuum j from $G(i+1)$ be a descendant of individuum k_j from $G(i)$. Let r denote a realization of an uniformly distributed random variable. Numbers k are generated from the inverse of distribution $D(k)$:

$$k_j = \min\{k : r < D(k)\}; \quad r \in [0, 1); \quad k \in \{1, \ldots, c\}. \tag{28}$$

Traits of individuum j assume values of traits of progenitor k_j modified by random mutation increments $h_k : x_k^j = x_k^{kj} + h_k$. Mutation increments are generated from the inverse of distribution $F(h)$:

$$h_k = \min\{h : r < F(h)\}; \quad r \in [0, 1); \quad k \in \{1, \ldots, c\}. \tag{29}$$

To simulate a population of c individua with w traits the memory of size $2c(w+1)$ is necessary. The space of types needs to be stored what allow to examine continuous spaces too. Reckoning of the next generation requires $c(w+1)$ random numbers; formulae (28) and (29) must be used c and cw times respectively.

The above procedure was applied to the problem described in Sect. 3.1. For the initially suboptimal population of 50 and mean mutation distance 0.2 (0.5), the simulation was carried out until the first improvement was found. For 100 trials the average waiting time to an improvement was 157 (16) generations, the standard deviation was 117 (9) and the average distance of the improving mutation was 1.5 (2.0). The deterministic model gives the waiting time 69 (2.2) generations, but this estimate applies to quasi-equilibrium. The transition from suboptimal to quasi-equilibrium population takes, for the deterministic model, 65 (18) generations. Altogether, the deterministic estimate of the waiting time is 134 (20) generations what is rather close to the value 157 (16) given by the simulation.

The population size of 50 can be considered large regarding the number of types involved. An analogous simulation was made for a small population of 10. In 100 trials the average waiting time was 245 (47), the standard deviation was 211 (37), the average distance of improving mutation was 1.2 (1.7). The waiting time is definitely shorter than the one resulting from the inverse proportion of population size and expected number of improvements. Moreover, with probability 0.74 (0.6), one has to wait longer to get an improvement in a population of 50 than in a group of 5 isolated populations of size 10 each. This effect seems worth examining. It can be only partially attributed to the action of Muller's ratchet (Maynard Smith 1978).

Above results of the immediate simulation corroborate the deterministic estimates and point on a peculiarity of small populations where the gain in improvements can still be greater. This peculiarity seems to be important, as practically all populations are small with regard to the number of types involved (if only the number of traits exceeds few).

5 Conclusions

The thesis that the escapes from evolutionary traps are due to mutations of handicapped individua of intermediary types seems to be fairly supported by the presented estimations. This, in turn, support some views on evolution:

– For evolution, if crucial improvements come from outsiders, the maintenance appears as important as the generation of diversity. Evolution proceeds because of leniency of natural selection.

- External attempts to improve the population by admitting only the best individua results in a pseudo-development; it gives immediate increase of the average fitness but hinders a further progress.

- Evolution, if crucial improvements are concerned, is basically unpredictable. At a quasi-equilibrium, there can be no available data on what an improvement would be and when it would happen.

The same concerns history. Intermediary types responsible for escapes constituted minute fractions of bygone generations. With random evidence only, missing links are not likely to be found.

The conclusions depend on the models validity. One can point here that the accepted assumptions agree with prominent views on evolution (Mayr 1983); the deterministic version coincide with the renowned model of Eigen (1971); model dynamics resemble in a way real life processes of development (Galar et al. 1980).

The notions of traits and mutations should not be referred to the gene loci events directly. We deal rather with effects of mutations on higher levels where complex traits recognisable by environment (like control loop parameters) are affected. If so, numbers of traits about a hundred can be regarded as large.

There is an interesting problem how recombinations due to sexual reproduction distort obtained results. The answer is not easy and exceeds the scope of this paper. The author's guess is that recombinations are not predominantly important, as long as the particular problem of the evolution of an isolated, freely mixing population is considered. It seems reasonable to look for the advantage of recombinations in the spread of innovations between generally isolated populations. The simulations of small population point this way.

Usually, an escape from an evolutionary trap is explained by some external influences like environmental changes or increases in mutation rates (Kirkpatrick 1982). Leaving aside the relative importance of these factors, it seems worth stressing that the ability to escape and to resulting step-wise development are the basic properties of spontaneously evolving populations, even in all constant and no interaction cases.

Acknowledgement. The author wishes to thank Prof. A. Rybarski for criticism and encouragement.

References

Bremermann HJ, Rogson M, Salaff S (1966) Global properties of evolution processes. In: Patecc HH, Edelsack EA, Fein L, Callahan AB eds) Natural automata and useful simulations. Spartan Books, Washington, pp 3–41

Cohen D (1973) The limits of optimization in evolution. In: Locker A (ed) Biogenesis evolution homeostasis. Springer, Berlin Heidelberg New York, pp 39–40

Conrad M (1979) Bootstraping on the adaptative landscape. Bio Systems 11:167–182

Eigen M (1971) Self-organization of matter and the evolution of biological macromolecules. Naturwissenschaften 58:465–523

Galar R, Kwasnicka H, Kwasnicki W (1980) Simulation of some processes of development. In: Dekker J, Savastano G, Vansteenkiste GC (eds) Simulation of systems. North-Holland Publishing Company, Amsterdam Oxford New York, pp 133–142

Ginzburg LR (1983) Theory of natural selection and population growth. Benjamin/Cummings Publishing Company Inc, Menlo Park

Jones BJ, Enns RH, Rangnekar S (1976) On the theory of selection of coupled macromolecular systems. Bull Math Biol 38:15–28

Kirkpatrick M (1982) Quantum evolution and punctuated equilibria in continuous genetic characters. Am Nat 119:833–848

Kwasnicka H, Galar R, Kwasnicki W (1983) Technological substitution forecasting with a model based on biological analogy. Technol Forecast Soc Change 23:41–58

Maynard Smith J (1978) The evolution of sex. Cambridge University Press, Cambridge

Mayr E (1983) How to carry out the adaptationist program? Am Nat 121:324–334

McCaskill JS (1984) A stochastic theory of macromolecular evolution. Biol Cybern 50:63–73

Stanley SM (1982) Macroevolution and the fossil record. Evolution 36:460–473

Thompson CJ, McBride JL (1974) On Eigens theory of the self-organization of matter and the evolution of biological molecules. Math Biosci 21:127–142

Wagner GP (1984) Coevolution of functionally constrained characters: prerequisities for adaptive versality. Biol. Syst 17:51–55

Received: November 6, 1984

R. Galar
Institute of Engineering Cybernetics
Technical University of Wroclaw
Wybrzeze Wyspiankiego 27
PL-50-370 Wroclaw
Poland

Chapter 15
Schema Processing
and the K-Armed Bandit

J. H. Holland (1973) "Genetic Algorithms and the Optimal Allocation of Trials," *SIAM J. Computing*, Vol. 2:2, pp. 88–105.

JOHN Holland's interest in simulating evolution grew in the 1960s from a desire to understand the logical processes involved in adaptation. In particular, his interest stemmed from two separate lines of inquiry. First, owing to Fisher (1930), there was a recognition that much of the process of adaptation could be formalized mathematically. Second, contemporary research in growing cellular automata (e.g., Burks, 1960) as well as neural and logic networks (Selfridge, 1958; Holland, 1959) demonstrated that simple rules could generate flexible behaviors. Combining these notions appeared to provide the potential for a mathematical framework to study the evolution of behaviors in complex systems.

The first written descriptions of such a proposal were offered in Holland (1962a, 1962b). These papers proposed that a study of adaptation should recognize that (1) adaptation occurs in the face of an environment, (2) adaptation is a populational process, (3) individual behaviors can be represented by programs, (4) new behaviors can be generated by the random variation of programs by methods including recombination, minor changes, insertions, etc., and (5) the outputs of two programs are often related if their structures are related (e.g., if two programs hold subroutines [building blocks] in common, it is often the case that their behaviors have common features). Adaptation was viewed in terms of a formalism in which programs of a population interact in an environment and gain "activation releases" from the environment determined by the appropriateness of their behavior. The combination of random variations and selection acting in light of this measured appropriateness would be expected to lead to a general adaptive system.

Holland (1967) extended this framework of adaptation to more formal questions of mathematics: (1) how robust is a particular adaptive strategy? and (2) does a particular strategy make better use of available information than another? Effort was fo-cused on obtaining results about the rate of adaptation of particular strategies in various classes of linear or nonlinear environments. Adaptation was viewed as a search trajectory through a state space of possible solutions to a particular problem.[1] Performance had to be measured by some criterion, and Holland (1967) noted that there were many possible such criteria (e.g., maximum accumulated utility, attainment of target state in minimum time, survival, etc.) but chose primarily to examine functions of the form of accumulated payoffs. Attention was devoted to strategies that were either linear or nonlinear functions of the environment, and environments that were either linear or nonlinear in their structure.

The paper reprinted here then, Holland (1973), represents the culmination of almost a decade of thought on measuring the performance of an adaptive strategy in light of payoff information gained by sampling an environment (possibly nonlinear). The analysis centers on how to best allocate some number of trials to two (or more) random variables with well-defined means and variances so as to minimize expected losses, this allocation being based on limited sampling of each random variable.[2]

The problem of allocating trials to random variables grew out of a perspective that any particular sample in a state space of solutions can be viewed as a schema (a template) that provides partial information regarding all other partial solutions that match that particular schema. For example, under a binary encoding of length four bits, the value assigned to solution [1001] gives partial information about the values that may be expected from sampling in the hyperplanes [#001], [1#01], [10#1], and so forth, where the # symbol matches either symbol in {0,1}. Each hyperplane was viewed as analogous to a random variable and thus the analogous problem of sampling optimally from various hyperplanes was related to sampling optimally from a set of mutually independent random variables.

---

[1]Interestingly, the first formal problem Holland (1967) described was that of a trajectory through a space of finite automata. The output of a particular automaton would be input to another automaton that represented the environment, and would elicit the next environmental symbol. This framework has clear similarities to an approach offered earlier in Fogel (1964; Fogel et al., 1966) to evolve finite state machines.

[2] This analysis formed the basis of Chapter 5, "On the Optimal Allocation of Trials," of the seminal publication Holland (1975).

This general framework was subsequently incorporated into a "genetic reproductive plan" (which later became known more conventionally as a *genetic algorithm*). A search through a space of candidate structures (solutions) was conducted by recombining building blocks of current structures, along with occasional random mutation, coupled with reproducing structures in proportion to their relative fitness. This procedure was studied and extended in the 1960s and 1970s by several of Holland's graduate students (Bagley, 1967; Rosenberg, 1967; Cavicchio, 1970; Hollstien, 1971; Frantz, 1972; Martin, 1973; De Jong, 1975), both in terms of the effects of varying representations and operators, as well as mathematical properties of convergence and function optimization.

Although the current use of genetic algorithms often focuses on optimization problems, Holland (1996) noted that he never put emphasis on optimization per se. Indeed, the criterion of minimizing expected losses from a series of samples does not equate with finding a single maximal sample. Thus the optimal allocation of trials offered in Holland (1973) does not provide justification for or against the use of genetic algorithms in function optimization applications (see De Jong 1992, 1993 for related discussions). Instead, it offers a mathematical description of a quantified approach to assessing the trade-offs between exploration (sampling in new regions in hopes of novel discoveries) and exploitation (sampling in known regions to take advantage of already acquired knowledge) in adaptive systems (see also Box, 1954; Box and Youle, 1955).

References

[1] J. D. Bagley (1967) "The behavior of adaptive systems which employ genetic and correlation algorithms," Ph.D. diss., Univ. of Michigan, Ann Arbor, MI.

[2] G. E. P. Box (1954) "The exploration and exploitation of response surfaces: Some general considerations and examples," *Biometrics*, Vol. 10, pp. 16–60.

[3] G. E. P. Box and P. V. Youle (1955) "The exploration and exploitation of response surfaces: An example of the link between the fitted surface and the basic mechanism of the system," *Biometrics*, Vol. 11, pp. 287–323.

[4] A. W. Burks (1960) "Computation, behavior and structure in fixed and growing automata," *Self-Organizing Systems*, M.C. Yovits and S. Cameron (eds.), Pergamon Press, NY, pp. 282–309.

[5] D. J. Cavicchio (1970) "Adaptive search using simulated evolution," Ph.D. diss., Univ. of Michigan, Ann Arbor, MI.

[6] K. A. De Jong (1975) "Analysis of the behavior of a class of genetic adaptive systems," Ph.D. diss., Univ. of Michigan, Ann Arbor, MI.

[7] K. A. De Jong (1992) "Are genetic algorithms function optimizers?" *Parallel Problem Solving from Nature, 2*, R. Männer and B. Manderick (eds.), North-Holland, Amsterdam, pp. 3–13.

[8] K. A. De Jong (1993) "Genetic algorithms are NOT function optimizers," *Foundations of Genetic Algorithms 2*, L. D. Whitley (ed.), Morgan Kaufmann, San Mateo, CA, pp. 5–17.

[9] R. A. Fisher (1930) *The Genetical Theory of Natural Selection*, Clarendon Press, Oxford.

[10] L. J. Fogel (1964) "On the organization of intellect," Ph.D. diss., UCLA.

[11] L. J. Fogel, A. J. Owens, and M. J. Walsh (1966) *Artificial Intelligence through Simulated Evolution*, John Wiley, NY.

[12] D. R. Frantz (1972) "Non-linearities in genetic adaptive search," Ph.D. diss., Univ. of Michigan, Ann Arbor, MI.

[13] J. H. Holland (1959) "Cycles in logical nets," Ph.D. diss., Univ. of Michigan, Ann Arbor, MI.

[14] J. H. Holland (1962a) "Outline for a logical theory of adaptive systems," *J. Assoc. Comp. Mach.*, Vol. 9, pp. 297–314.

[15] J. H. Holland (1962b) "Concerning efficient adaptive systems," *Self-Organizing Systems—1962*, M. C. Yovits, G. T. Jacobi, and G. D. Goldstein (eds.), Spartan Books, Washington, D.C., pp. 215–230.

[16] J. H. Holland (1967) "Nonlinear environments permitting efficient adaptation," *Computer and Information Sciences—II*, J. T. Tou (ed.), Academic Press, NY, pp. 147–164.

[17] J. H. Holland (1973) "Genetic algorithms and the optimal allocation of trials," *SIAM J. Comput.*, Vol. 2, pp. 88–105.

[18] J. H. Holland (1975) *Adaptation in Natural and Artificial Systems*, Univ. of Michigan Press, Ann Arbor, MI.

[19] J. H. Holland (1996) personal communication, University of Michigan.

[20] R. B. Hollstien (1971) "Artificial genetic adaptation in computer control systems," Ph.D. diss., Univ. of Michigan, Ann Arbor, MI.

[21] N. Martin (1973) "Convergence properties of a class of probabilistic adaptive schemes called sequential reproductive plans," Ph.D. diss., Univ. of Michigan, Ann Arbor, MI.

[22] R. Rosenberg (1967) "Simulation of genetic populations with biochemical properties," Ph.D. diss., Univ. of Michigan, Ann Arbor, MI.

[23] O. G. Selfridge (1958) "Pandemonium: A paradigm for learning," *Proc. of the Symp. on Mechanization of Thought Processes*, Teddington, England, pp. 511–529.

GENETIC ALGORITHMS
AND THE OPTIMAL ALLOCATION OF TRIALS*

JOHN H. HOLLAND†

Abstract. This study gives a formal setting to the difficult optimization problems characterized by the conjunction of (1) substantial complexity and initial uncertainty, (2) the necessity of acquiring new information rapidly to reduce the uncertainty, and (3) a requirement that the new information be exploited as acquired so that average performance increases at a rate consistent with the rate of acquisition of information. The setting has as its basis a set $\mathscr{A}$ of structures to be searched or tried and a performance function $\mu: \mathscr{A} \to$ real numbers. Within this setting it is determined how to allocate trials to a set of random variables so as to maximize expected performance. This result is then transformed into a criterion against which to measure the performance of a robust and easily implemented set of algorithms called *reproductive plans*. It is shown that reproductive plans can in fact surpass the criterion because of a phenomenon called *intrinsic parallelism*—a single trial (individual $A \in \mathscr{A}$) simultaneously tests and exploits many random variables.

1. Introduction. There is an extensive and difficult class of optimization problems characterized by:

(1) substantial complexity and initial uncertainty;

(2) the necessity of acquiring new information rapidly to reduce the uncertainty;

(3) a requirement that the new information be exploited as acquired so that average performance increases at a rate consistent with the rate of acquisition of information.

These problems derive from a whole range of long-standing questions, such as the following.

How is the productivity of a plant or process to be improved, while it is operating, when many of the interactions between its variables are unknown?

How does one improve performance in successive plays of a complex game (such as Chess or Go or a management game) when the solution of the game is unknown (and probably too complex to implement even if it were known)?

How does evolution produce increasingly fit organisms under environmental conditions which perforce involve a great deal of uncertainty vis-à-vis individual organisms?

How can the performance of an economy be upgraded when its mechanisms are only partially known and relevant data is incomplete?

In each case rapid improvement in performance is highly desirable (or essential), though the combination of complexity and uncertainty makes a direct approach to optimization unfeasible. Often the complexity and uncertainty are great enough that the optimum will be attained, if at all, only after an extensive

---

* Received by the editors August 3, 1972.

† Department of Computer and Communication Sciences, University of Michigan, Ann Arbor, Michigan 48104. This research was supported in part by the National Science Foundation under Grant GJ-29989X.

period of trial and calculation. Problems of this kind will be referred to here as "problems of adaptation," a usage similar to that of Tsypkin [8], though broader.

Problems of adaptation can be given a more precise formulation along the following lines. Let $\mathscr{A}$ be the set of objects or *structures* (control policies, game strategies, chromosomes, mixes of goods, etc.) to be searched or tried. Generally $\mathscr{A}$ will be so large that it cannot be tried one at a time over any feasible time period. Let $\mu : \mathscr{A} \to [r_0, r_1]$, where $[r_0, r_1]$ is an interval of real numbers, be a "performance measure" (error, payoff, fitness, utility, etc.) which assigns a level of *performance* $\mu(A)$ to each structure $A \in \mathscr{A}$. Conditions (2) and (3) above then reduce to:

(2′) Obtain new information about μ by trying previously untried structures in $\mathscr{A}$ (it being assumed that the outcome of trying $A \in \mathscr{A}$ is the information $\mu(A)$).

(3′) Assure that $\bar{\mu}_t$, the average of the outcomes of the first t trials, increases rapidly whenever the search of $\mathscr{A}$ reveals an A with $\mu(A) > \bar{\mu}_t$.

With no more formalization than this, a dilemma comes into sharp focus. The rate of search is maximized when each successive trial is of a previously untried $A \in \mathscr{A}$. On the other hand, repeated trials of any A' for which $\mu(A') > \bar{\mu}_t$ will increase $\bar{\mu}_t$ more rapidly. In other words, if the rate of search is maximized, the information cannot be exploited, whereas if information obtained is maximally exploited, no new information is acquired.

The basic problem then is to find a resolution of this dilemma. The simplest precise version of the dilemma arises when we restrict our attention to two random variables ξ_1, ξ_2, defined so that the outcome of a trial of ξ_i is a performance $\mu(\xi_i, t)$. The object then is to discover a procedure for distributing some arbitrary number of trials, N, between ξ_1 and ξ_2 so as to maximize the expected payoff over the N trials. If for each ξ_i we know the mean and variance (μ_i, σ_i) of its distribution, the problem has a trivial solution (namely, allocate all trials to the random variable with maximal mean). The dilemma asserts itself, however, if we inject just a bit more uncertainty. Thus we can know the mean-variance pairs but not which variable is described by which pair; i.e., we know pairs (μ, σ) and (μ', σ') but not which pair describes ξ_1. (This is a version of the much studied 2-armed bandit problem, a prototype of important decision problems. See, for example, Bellman [2] and Hellman and Cover [4].) If it could be determined through observation which of ξ_1 and ξ_2 has the higher mean, then from that point on, all trials could be allocated to that random variable. Unfortunately, unless the distributions are nonoverlapping, no finite number of observations will establish *with certainty* which random variable has the higher mean. Here the tradeoff between gathering information and exploiting it appears in its simplest terms. Gathering information requires trials of *both* random variables, with a consequent decrement in average payoff (because the average performance of one of the random variables is less than maximal). On the other hand, premature exploitation of the apparent best random variable, by allocation of most or all trials thereto, runs the risk of a large loss (because there is a nonzero probability that the apparent best is really second best).

A procedure for allocating trials between ξ_1 and ξ_2 will be said to optimally satisfy the conditions (2′) and (3′) if it maximizes the expected performance over N trials. It should be noted that any increase in the uncertainty, such as not knowing some of the means or variances, can only result in a lower expected performance.

Thus a procedure which solves the given problem yields an upper bound on expected performance under increased uncertainty—a criterion against which to measure the performance of various feasible algorithms.

In these terms the objective of this paper is two-fold. First, making the natural extension to an arbitrary number r of random variables determine an upper bound on expected performance under uncertainty. Second, compare the expected performance of the general class of algorithms known as *reproductive plans* (see Holland [5] and later in this paper) to this criterion. It will be shown that, even under conditions of maximum uncertainty, reproductive plans closely follow the criterion. Moreover, reproductive plans can use a single trial to test many random variables simultaneously, a property designated *intrinsic parallelism*. Thus, reproductive plans can exceed the optimum for one-at-a-time testing of random variables. The latter part of the paper will show how this advantage is attained and that it increases in direct proportion to the number of random variables r.

2. Definition of the problem. Several definitions will have to be added to give precise mathematical form to the problem of searching $\mathscr{A}$ under conditions (1), (2′) and (3′). (This is only a partial formalization of problems of adaptation, sufficient for present purposes—a more complete formulation is given in Holland [5].) First of all, let the elements of $\mathscr{A}$ be represented by strings of length l over a set of symbols $\Sigma = \{\sigma_1, \cdots, \sigma_k\}$; i.e., each $A \in \mathscr{A}$ is represented by (or designates) a string of symbols (alleles, weights, etc.) $\sigma_{i_1}\sigma_{i_2} \cdots \sigma_{i_l}$, where $\sigma_{i_h} \in \Sigma$. For simplicity in what follows, $\mathscr{A}$ will simply be taken to be the set of strings (rather than the abstract elements represented by the strings). Let $\sigma_1 \square\square \cdots \square$ designate the set of all elements of $\mathscr{A}$ beginning with the symbol σ_1. (For example, $\sigma_1\sigma_1\sigma_4$, $\sigma_1\sigma_2\sigma_1$, and $\sigma_1\sigma_3\sigma_3$ would belong to $\sigma_1\square\square$, but $\sigma_2\sigma_1\sigma_1$ would not.) More generally, let any string ξ of length l over the augmented set $\Sigma \cup \{\square\} = \{\sigma_1, \cdots, \sigma_k, \square\}$ designate a subset of $\mathscr{A}$ as follows: $A \in \mathscr{A}$ belongs to the subset designated by $\xi = \delta_{i_1}\delta_{i_2} \cdots \delta_{i_l}$ if and only if (i) whenever $\delta_{i_j} \in \Sigma$ the string A has the symbol δ_{i_j} at the jth position, and (ii) whenever $\delta_{i_j} = \square$ any symbol from Σ may occur at the jth position in A. (For example, the strings $\sigma_1\sigma_1\sigma_1\sigma_3$ and $\sigma_3\sigma_1\sigma_2\sigma_3$ belong to $\square\sigma_1\square\sigma_3$ but $\sigma_1\sigma_1\sigma_1\sigma_2$ does not.) The set of $(k + 1)^l$ strings defined over $\Sigma \cup \{\square\}$ will be called the set Ξ of *schemata*; they amount to a decomposition of $\mathscr{A}$ into a large number of subsets based on the representation in terms of Σ.

If now there is a probability distribution P over $\mathscr{A}$, say the probability $P(A)$ that $A \in \mathscr{A}$ will be selected for trial, $\mathscr{A}$ can be treated as a sample space and each schema ξ designates an event on $\mathscr{A}$. Accordingly, the performance measure μ becomes a random variable, the elementary event A occurring with probability $P(A)$ and yielding payoff $\mu(A)$. Moreover, the restriction $\mu|\xi$ of μ to a particular subset ξ is also a random variable, $A \in \xi$ being chosen with probability $(P(A))/(\sum_{A \in \xi} P(A))$ and yielding payoff $\mu(A)$. In what follows, ξ will be of interest only in its role of designating the random variable $\mu|\xi$; therefore ξ will be used to designate both an element of Ξ and the corresponding random variable with sample space ξ, distribution $(P(A))/(\sum_{A \in \xi} P(A))$, and values $\mu(A)$. As a random variable, ξ has a well-defined average μ_ξ and variance σ_ξ^2; intuitively, μ_ξ is the payoff expected when an element of ξ is randomly selected (under the distribution P).

Using the decomposition of $\mathscr{A}$ into random variables gives by P and Ξ, it is possible to formalize the earlier discussion concerning optimal allocation of trials.

For the 2-schemata case, let $n_{(2)}$ be the number of trials allocated to the schema with the lowest *observed* payoff rate at the end of N trials. Let $q(n_{(2)})$ be the probability that the schema with the highest observed payoff rate is actually second best. (In detail $q(n_{(2)})$ is actually a function of $n_{(2)}$, N, (μ, σ), and (μ', σ'); hence the necessity of knowing the mean-variance pairs.) If μ_1 is the mean of ξ_1 and μ_2 the mean of ξ_2, then the expected payoff rate for the allocation $(N - n_{(2)}, n_{(2)})$ is

$$\max(\mu_1, \mu_2) - [(N - n_{(2)})q(n_{(2)}) + n_{(2)}(1 - q(n_{(2)}))] \cdot |\mu_1 - \mu_2| \cdot \frac{1}{N}.$$

An optimum value for $n_{(2)}$ can be obtained from this expression by standard techniques. (The development below is somewhat more complicated since, in general, one cannot guarantee a priori that the random variable with the highest observed payoff rate at the *end* of the N trials will have received a predetermined number of trials by that time.)

Though the derivations are much more intricate, the extension from the 2-schemata case to the r-schemata case is conceptually straightforward. It is possible to obtain useful bounds on the payoff rate as a function of the total number of trials, N, together with bounds on the total number of trials which should be allocated to the schema with the highest observed payoff rate. Because the upper and lower bounds so obtained are close to one another (relative to N), the action of the optimal procedure is pretty clearly defined. Using this information it is possible to define a realizable procedure which approaches the optimum as N increases.

If the r-schemata must be tested one at a time, it is clear that one can do no better than the procedure just outlined. If, on the other hand, information about *several* schemata can be obtained from trial of a single individual, $A \in \mathscr{A}$, the rate of improvement could exceed the optimal rate for the one-schema-at-a-time procedure. It should be remarked at once that, for this improvement to take place, the information must not only be obtained but *used* to generate subsequent individuals $(A \in \mathscr{A})$ for trial—each of which will reveal further information about a variety of schemata. The second part of this paper studies a specific set of reproductive plans, the *genetic plans*, which can do just this. It will be shown that genetic plans follow the general course of the optimal procedure when artificially constrained to one-schema-at-a-time searches, but advance much more rapidly when not so constrained.

3. Optimal allocation of trials. For notational convenience in the 2-schemata case, let ξ_1 be the schema with highest mean, ξ_2 the schema with lowest mean. (The observer, of course, does not know this.) Let $\xi_{(1)}(\tau, N)$ be the schema with the highest *observed* payoff rate (average per trial) after an allocation of N trials according to plan τ; let $\xi_{(2)}(\tau, N)$ designate the schema with lowest *observed* rate. Note that for any number of trials n, $0 \le n \le N$, allocated to $\xi_{(2)}(\tau, N)$, there is a positive probability, $q(N - n, n)$, that $\xi_{(2)}(\tau, N) \ne \xi_2$ (assuming overlapping distributions). Equivalently, $q(N - n, n)$ is the probability that the observed best is actually second best.

THEOREM 1. *Given N trials to be allocated to two random variables ξ_1 and ξ_2, with means $\mu_1 > \mu_2$ and variances σ_1, σ_2 respectively, the minimum expected loss results when the number of trials allocated ξ_2 is*

$$n^* \sim \left(\frac{\sigma_2}{\mu_1 - \mu_2}\right)^2 \ln\left[\left[\left(\frac{\mu_1 - \mu_2}{\sigma_2}\right)^4 \left(\frac{N^2}{8\pi \ln N^2}\right)\right]\right].$$

The corresponding expected loss per trial is

$$l^*(N) \sim \frac{\sigma_2^2}{(\mu_1 - \mu_2)N}\left[2 + \ln\left[\left[\left(\frac{\mu_1 - \mu_2}{\sigma_2}\right)^4 \left(\frac{N^2}{8\pi \ln N^2}\right)\right]\right]\right].$$

Proof. (Given two arbitrary functions, $Y(t)$ and $Z(t)$, of the same variable t, "$Y(t) \sim Z(t)$" will be used to mean $\lim_{t \to \infty} (Y(t)/Z(t)) = 1$ while "$Y(t) \cong Z(t)$" means that under stated conditions the difference $Y(t) - Z(t)$ is negligible).

In determining the expected payoff rate of a plan τ over N trials, two possible sources of loss must be taken into account: (1) The *observed* best $\xi_{(1)}(\tau, N)$ is really second best, whence the $N - n$ trials given $\xi_{(1)}(\tau, N)$ incur an (expected) cumulative loss $|\mu_1 - \mu_2| \cdot (N - n)$; this occurs with probability $q(N - n, n)$. (2) The observed best is in fact the best, whence the n trials given $\xi_{(2)}(\tau, N)$ incur a loss $|\mu_1 - \mu_2| \cdot n$; this occurs with probability $q(N - n, n)$. The expected loss $l(N)$ over N trials is thus

$$|\mu_1 - \mu_2| \cdot [(N - n)q(N - n, n) + n(1 - q(N - n, n))].$$

In order to select an n which minimizes the expected loss, it is necessary first to write $q(N - n, n)$ as an explicit function of n. To derive this function let S_2 be the sum of the outcomes (payoffs) of n trials of ξ_2 and let S_1 be the corresponding sum for the $N - n$ trials of ξ_1. Then $q(N - n, n)$ is just the probability that $S_2/n < S_1/(N - n)$ or, equivalently, the probability that $S_1/(N - n) - S_2/n < 0$. By the central limit theorem, S_2/n approaches a normal distribution with mean μ_2 and variance σ_2^2/n; similarly, $S_1/(N - n)$ has mean μ_1 and variance $\sigma_1^2/(N - n)$. The distribution of $S_1/(N - n) - S_2/n$ is by definition the sum (convolution) of the distributions of $S_1/(N - n)$ and $-(S_2/n)$; by an elementary theorem (on the convolution of normal distributions), this is a normal distribution with mean $\mu_1 - \mu_2$ and variance $\sigma_1^2/(N - n) + \sigma_2^2/n$. Thus the probability $\Pr\{S_1/(N - n) - S_2/n < 0\}$ is the tail $1 - \Phi(x_0)$ of a normal distribution $\Phi(x)$ in standard form so that

$$x = \frac{y - (\mu_1 - \mu_2)}{\sqrt{\sigma_1^2/(N - n) + \sigma_2^2/n}}$$

and $-x_0$ is the value of x when $y = 0$.

The tail of a normal distribution is well approximated by

$$\Phi(-x) = 1 - \Phi(x) \lesssim \frac{1}{\sqrt{2\pi}} \cdot \frac{e^{-x^2/2}}{x}.$$

Thus

$$q(N - n, n) \lesssim \frac{1}{\sqrt{2\pi}} \cdot \frac{e^{-x_0^2/2}}{x_0}$$

$$= \frac{1}{\sqrt{2\pi}} \frac{\sqrt{\sigma_1^2/(N - n) + \sigma_2^2/n}}{\mu_1 - \mu_2} \exp \frac{1}{2}\left[\frac{-(\mu_1 - \mu_2)^2}{\sigma_1^2/(N - n) + \sigma_2^2/n}\right]$$

(from which we see that q is a function of the variances and means as well as the total number of trials, N, and the number of trials, n, given ξ_2). Upon noting that q decreases exponentially as a function of n, it becomes clear that, to minimize loss as N increases, the number of trials allocated the observed best, $N - n$, should be increased dramatically relative to n. This observation (which will be verified in detail shortly) enables us to simplify the expression for x_0. Whatever the value of σ_1, there will be an N_0 such that, for any $N > N_0$, $\sigma_1^2/(N - n) \ll \sigma_2^2/n$, for n close to its optimal value. (In most cases of interest this occurs even for small numbers of trials since, usually, σ_2 is at worst an order of magnitude or two larger than σ_1.) Using this we see that, for n close to its optimal value,

$$x_0 \lesssim \frac{(\mu_1 - \mu_2)\sqrt{n}}{\sigma_2}, \qquad N > N_0.$$

We can now proceed to determine what value of n will minimize the loss $l(n)$ by taking the derivative of l with respect to n:

$$\frac{dl}{dn} = |\mu_1 - \mu_2| \cdot \left[-q + (N - n)\frac{dq}{dn} + 1 - q - n\frac{dq}{dn}\right]$$

$$= |\mu_1 - \mu_2| \cdot \left[(1 - 2q) + (N - 2n)\frac{dq}{dn}\right],$$

where

$$\frac{dq}{dn} \lesssim \frac{1}{\sqrt{2n}}\left[-\frac{e^{-x_0^2/2}}{x_0^2} - e^{-x_0^2/2}\right]\frac{dx_0}{dn} = -\left[\frac{q}{x_0} + x_0 q\right]\frac{dx_0}{dn}$$

and

$$\frac{dx_0}{dn} \lesssim \frac{\mu_1 - \mu_2}{2\sigma_2\sqrt{n}} = \frac{x_0}{2n}.$$

Thus

$$\frac{dl}{dn} \lesssim |\mu_1 - \mu_2| \cdot \left[(1 - 2q) - (N - 2n) \cdot q \cdot \frac{x_0^2 + 1}{2n}\right].$$

n^*, the optimal value of n, satisfies $dl/dn = 0$, whence we obtain a bound on n^* as follows:

$$0 \lesssim (1 - 2q) - \left(\frac{N}{2n^*} - 1\right) \cdot q \cdot (x_0^2 + 1)$$

or

$$\frac{N}{2n^*} - 1 \lesssim \frac{1 - 2q}{q \cdot (x_0^2 + 1)}.$$

Noting that $1/(x_0^2 + 1) \lesssim 1/x_0^2$ and that $1 - 2q$ rapidly approaches 1 because q decreases exponentially with n, we see that $(N - 2n^*)/n^* \lesssim 2/(x_0^2 q)$, where the error rapidly approaches zero as N increases. Thus the observation of the preceding paragraph is verified, the ratio of trials of the observed best to trials of second-best growing exponentially.

Finally, to obtain n^* as an explicit function of N, q must be written in terms of n^*:

$$\frac{N - 2n^*}{n^*} \lesssim \frac{2\sqrt{2\pi}\sigma_2}{\mu_1 - \mu_2} \cdot \frac{1}{\sqrt{n^*}} \cdot \exp\left[\frac{(\mu_1 - \mu_2)^2 n^*}{2\sigma_2^2}\right].$$

Introducing $b = (\mu_1 - \mu_2)/\sigma_2$ and $N_1 = N - n^*$ for simplification, we obtain

$$N_1 \lesssim \frac{\sqrt{8\pi}}{b} \cdot \exp\left[\frac{b^2 n^* + \ln n^*}{2}\right]$$

or

$$n^* + \frac{\ln n^*}{b^2} \gtrsim \frac{2}{b^2} \cdot \ln\left(\frac{b}{\sqrt{8\pi}} \cdot N_1\right),$$

where the fact that $(N - 2n^*) \sim (N - n^*)$ has been used, with the inequality generally holding as soon as N_1 exceeds n^* by a small integer. (To get numerical bounds on σ_2 when it is not explicity known, note that for bounded payoff (all $A \in \mathscr{A}$, $r_0 \leqq \mu(A) \leqq r_1$) the maximum variance occurs when all payoff is concentrated at the two extremes. That is,

$$P(r_0) = P(r_1) = \tfrac{1}{2} \quad \text{and} \quad \sigma_2^2 \leqq \sigma_{\max}^2 = (\tfrac{1}{2}r_1^2 + \tfrac{1}{2}r_0^2) - (\tfrac{1}{2}r_1 + \tfrac{1}{2}r_0)^2 = \left(\frac{r_1 - r_0}{2}\right)^2.)$$

We obtain a recursion for an ever better approximation to n^* as a function of N_1 by rewriting this as

$$n^* \gtrsim b^{-2} \ln\left[\frac{(bN_1)^2}{8\pi n^*}\right].$$

Thus

$$n^* \gtrsim b^{-2} \ln\left[\frac{(bN_1)^2}{8\pi(b^{-2}\ln((bN_1)^2/8\pi n^*))}\right]$$

$$\gtrsim b^{-2} \ln\left[\frac{b^4 N_1^2}{8\pi} \cdot \frac{1}{\ln((bN_1)^2/8\pi) - \ln n^*}\right]$$

$$\gtrsim b^{-2} \ln\left[\frac{b^4 N_1^2}{8\pi(\ln N_1^2 - \ln(b^2/8\pi))}\right]$$

$$\gtrsim b^{-2} \ln\left[\frac{b^4 N_1^2}{8\pi \ln N_1^2}\right],$$

where, again, the error rapidly approaches zero as N increases. Finally, where it is desirable to have n^* approximated by an explicit function of N, the steps here can be redone in terms of N/n^*, noting that N_1/n^* rapidly approaches N/n^* as N increases. Then

$$n^* \sim b^{-2} \ln \left[\frac{b^4 N^2}{8\pi \ln N^2} \right],$$

where, still, the error rapidly approaches zero as N increases.

The expected loss per trial $l^*(N)$ when n^* trials have been allocated to $\xi_{(2)}(\tau, N)$ is

$$l^*(N) = \frac{1}{N}|\mu_1 - \mu_2| \cdot [(N - n^*)q(N - n, n^*) + n^*(1 - q(N - n^*, n^*))]$$

$$= |\mu_1 - \mu_2| \cdot \left[\frac{N - 2n^*}{N} q(N - n^*, n^*) + \frac{n^*}{N} \right]$$

$$\gtrsim |\mu_1 - \mu_2| \cdot \left[\frac{2n^*}{N x_0^2} + \frac{n^*}{N} \right]$$

$$\gtrsim \frac{|\mu_1 - \mu_2|}{b^2 N} \cdot \left[2 + \ln \left(\frac{b^4 N^2}{8\pi \ln N^2} \right) \right]. \qquad\qquad \text{Q.E.D.}$$

The expression for n^* (and hence the one for $l^*(N)$) was obtained on the assumption that the n^* trials were allocated to $\xi_{(2)}(\tau, N)$. However, there is *no* realizable sequential algorithm which can "foresee" in all cases which of the two schemata will be $\xi_{(2)}(\tau, N)$. There will always be observational sequences wherein *each* schema has a positive probability of being $\xi_{(2)}(\tau, N)$ even after τ has allocated $n > n^*$ trials to one. (For example, τ may have allocated exactly n^* trials to each and must decide where to allocate the next trial even though each schema has a positive probability of being $\xi_{(2)}(\tau, N)$.) Thus, no matter what the plan τ, it will in some cases allocate $n > n^*$ trials to a schema ξ (on the assumption that ξ will turn out to be $\xi_{(1)}(\tau, N)$) only to be confronted with the fact that $\xi = \xi_{(2)}(\tau, N)$. For these sequences the loss will perforce exceed the optimum. Hence $l^*(N)$ is not attainable by any realizable sequential algorithm τ—there will always be outcome sequences which lead τ to allocate too many trials to $\xi_{(2)}(\tau, N)$.

There is, however, a realizable plan τ_0 for which the expected loss per trial $l(\tau_0, N)$ quickly approaches $l^*(N)$; i.e.,

$$\lim_{N \to \infty} \frac{l(\tau_0, N)}{l^*(N)} = 1.$$

τ_0 initially allocates n^* trials to each schema (in any order) and then allocates the remaining $N - 2n^*$ trials to the schema with the highest observed payoff rate at the end of the $2n^*$ trials.

COROLLARY 1.1. *Given N trials, τ_0's expected loss, $l(\tau_0, N)$, approaches the optimum $l^*(N)$. That is, $l(\tau_0, N) \sim l^*(N)$.*

Proof. The expected loss per trial $l(\tau_0, N)$ for τ_0 is determined by applying the earlier discussion of sources of loss to the present case:

$$l(\tau_0, N) = \frac{1}{N} \cdot |\mu_1 - \mu_2| \cdot [(N - n^*)q(n^*, n^*) + n^*(1 - q(n^*, n^*))],$$

where q is the same function as before, but here the probability of error is irrevocably determined after only n^* trials have been allocated to *each* schema; i.e.,

$$q(n^*, n^*) \sim \frac{1}{\sqrt{2\pi}} \frac{\sqrt{\sigma_1^2/n^* + \sigma_2^2/n^*}}{\mu_1 - \mu_2} \exp\left[\frac{-(\mu_1 - \mu_2)^2}{\sigma_1^2/n^* + \sigma_2^2/n^*}\right].$$

(Note that n^* is *not* being redetermined for τ_0; n^* is the number of trials determined above.) Rewriting $l(\tau_0, N)$ we have

$$l(\tau_0, N) = |\mu_1 - \mu_2| \cdot \left[\frac{N - 2n^*}{N} q(n^*, n^*) + \frac{n^*}{N}\right].$$

Since, asymptotically, q decreases as rapidly as N^{-1}, it is clear that the second term in the brackets will dominate as N grows. Inspecting the earlier expression for $l^*(N)$ we see the same holds there. Thus, since the two second terms are identical,

$$\lim_{N \to \infty} \frac{l(\tau_0, N)}{l^*(N)} = 1. \qquad\qquad \text{Q.E.D.}$$

To treat the case of r schemata we need a new determination of the probability that the observed best is not the schema with the highest mean. To proceed to this determination let the r schemata be $\xi_1, \xi_2, \cdots, \xi_r$ and let $\mu_1 > \mu_2 > \cdots > \mu_r$ (again, without the observer knowing that this ordering holds).

THEOREM 2. *Given N trials to be allocated to r random variables $\{\xi_1, \xi_2, \cdots, \xi_r\}$, with means $\mu_1 > \mu_2 > \cdots > \mu_r$ and variances $\sigma_1, \sigma_2, \cdots, \sigma_r$ respectively, the minimum expected loss per trial $l_r^*(N)$ is bounded above and below by $l_{N,r}''$ and $l_{N,r}'$, respectively, where*

$$l_{N,r}' \sim \frac{(r - 1)\sigma_2^2}{(\mu_1 - \mu_2)N}\left[2 + \ln\left[\left(\frac{\mu_1 - \mu_2}{\sigma_2}\right)^4 \left(\frac{N^2}{8\pi(r - 1)^2 \ln N^2}\right)\right]\right]$$

and

$$l_{N,r}'' \sim \frac{(r - 1)\mu_1\sigma_2^2}{(\mu_1 - \mu_2)^2 N}\left[2 + \ln\left[\left(\frac{\mu_1 - \mu_2}{\sigma_2}\right)^4 \left(\frac{N^2}{8\pi \ln N^2}\right)\right]\right].$$

Proof. We are interested in the probability q_r that the average of the observations of any $\xi_j, j > 1$, exceeds the average for ξ_1; that is, the probability of error

$$q_r(n_1, \cdots, n_r) = \Pr\left\{\left(\frac{S_2}{n_2} > \frac{S_1}{n_1}\right) \text{ or } \left(\frac{S_3}{n_3} > \frac{S_1}{n_1}\right) \text{ or } \cdots \text{ or } \left(\frac{S_r}{n_r} > \frac{S_1}{n_1}\right)\right\}.$$

When a given number of trials $n_0 = \sum_{i=2}^r n_i$ has been allocated to $\xi_2, \xi_3, \cdots, \xi_r$ to minimize the probability of error q_r, that error will clearly be largest if $\mu_2 = \mu_3 = \cdots = \mu_r$. (In other words, when $\mu_j \lesssim \mu_2$, an allocation of $n_j < n_2$ trials to ξ_j will yield

$$\Pr\left\{\left(\frac{S_j}{n_j} > \frac{S_1}{n_1}\right)\right\} < \Pr\left\{\left(\frac{S_2}{n_2} > \frac{S_1}{n_1}\right)\right\};$$

hence for a given number of trials, a greater reduction in q_r can be achieved if the means μ_j are not all equal.) Moreover, for those cases where $\mu_2 = \mu_3 = \cdots = \mu_r$, the worst case occurs when the largest of the variances $\sigma_2, \sigma_3, \cdots, \sigma_r$ is in fact the common variance of each of $\xi_2, \xi_3, \cdots, \xi_r$. Given this worst case, $(\mu_2, \sigma_2) = (\mu_3, \sigma_3) = \cdots = (\mu_r, \sigma_r)$, q_r will be minimized for an allocation of n_0 trials to $\xi_2, \cdots, \xi_r$ if (as nearly as possible) equal numbers of trials are given each schema (since each schema contributes equally to the probability of error).

From these observations we can obtain bounds on q_r. As before, let

$$q(n_1, n_2) = \Pr\left\{\frac{S_2}{n_2} > \frac{S_1}{n_1}\right\}.$$

When $\sum_{i=2}^{r} n_i = (r-1)m$ trials are allocated to $\xi_2, \cdots, \xi_r$ so as to minimize the probability of error, we have

$$\Pr\left\{\left(\frac{S_2}{n_2} > \frac{S_1}{N-(r-1)m}\right) \text{ or } \cdots \text{ or } \left(\frac{S_r}{n_r} > \frac{S_1}{N-(r-1)m}\right)\right\}$$

$$< \binom{r-1}{1} q(N-(r-1)m, m) - \binom{r-1}{1} 2! q(N-(r-1)m, m) + \cdots,$$

where the right-hand side is obtained by noting that the events $\{(S_i/n_i > S_1/(N-(r-1)m)), i = 2, \cdots, r\}$ are independent and, under the best allocation in the worst case, $n_i = m$ for $i = 2, \cdots, r$, so that

$$\Pr\left\{\frac{S_i}{m} > \frac{S_1}{N-(r-1)m}\right\} = q(N-(r-1)m, m).$$

Thus, when $(r-1)m$ trials are allocated to $\xi_2, \cdots, \xi_r$ to minimize the probability of error, we have the following bounds:

$$q(N-(r-1)m, m) < q_r(n_1, \cdots, n_r)$$

$$= \Pr\left\{\left(\frac{S_2}{n_2} > \frac{S_1}{N-(r-1)m}\right) \text{ or } \cdots \text{ or } \left(\frac{S_r}{n_r} > \frac{S_1}{N-(r-1)m}\right)\right\}$$

$$< (r-1) q(N-(r-1)m, m).$$

Using the upper and lower bounds on q_r thus obtained, we can proceed to upper and lower bounds, $l_{N,r}^{II}$ and $l_{N,r}^{I}$ respectively, on the expected loss $l_N(n_2, \cdots, n_r)$ for N trials:

$$l_{N,r}^{I}(m) = (\mu_1 - \mu_2)[(N-(r-1)m)q + (r-1)m(1-q)] < l_N(n_2, \cdots, n_r) < l_{N,r}^{II}(m)$$

$$= \mu_1[(N-(r-1)m)(r-1)q + (r-1)m(1-(r-1)q)],$$

using the earlier discussion of sources of loss and the fact that $(\mu_1 - \mu_2) < (\mu_1 - \mu_j) < \mu_1$ for all j. As before we can determine the optimal value of m for these bounds, m^{**} and m^{*} respectively, by setting $dl/dm = 0$. Letting $q'' = (r-1)q$ and $q' = q$, we have

$$\frac{dl^{(i)}}{dm} = \mu^{(i)}\left[(N-2(r-1)m)\frac{dq^{(i)}}{dm} - 2(r-1)q^{(i)} + (r-1)\right] = 0.$$

Or, noting that $1 - 2q^{(i)}$ rapidly approaches 1, we have

$$m^{(i)} \sim \frac{N}{2(r-1)} + \frac{1}{2}\left(\frac{dq^{(i)}}{dm}\right)^{-1}.$$

$q^{(i)}$ decreases with m, so dq/dm is a negative quantity. Since $dq''/dm = (r-1)dq'/dm$, we have $dq''/dm < dq_r/dm < dq'/dm$ and

$$m^{**} \sim \frac{N}{2(r-1)} + \frac{1}{2}\left(\frac{dq''}{dm}\right)^{-1} = \frac{N}{2(r-1)} + \frac{1}{2(r-1)}\left(\frac{dq'}{dm}\right)^{-1}$$

$$> \frac{N}{2(r-1)} + \frac{1}{2}\left(\frac{dq'}{dm}\right)^{-1} \sim m^*.$$

That is, $(r-1)m^{**}(N) > n_{opt} = \sum_{i=2}^{r} n_{i,opt} > (r-1)m^*(N)$. Thus by determining the optimal m for each of the bounds we obtain bounds on n_{opt}, the number of trials which should be allocated to schemata other than ζ_1 in order to minimize expected loss.

m^* is directly obtained from the previous 2-schemata derivation by using $(r-1)m$ in place of n and taking the derivative of q with respect to m instead of n. The result is

$$m^* \sim b^{-2}\ln\left(\frac{b^4 N_1^2}{8\pi(r-1)^2 \ln N_1^2}\right) \sim b^{-2}\ln\left(\frac{b^4 N^2}{8\pi(r-1)^2 \ln N^2}\right),$$

where $N_1 = N - (r-1)m$ now.

m^{**} is similarly obtained using $(r-1)q$ for q throughout. The result is

$$m^{**} \sim b^{-2}\ln\left(\frac{b^4 N^2}{8\pi \ln N^2}\right) \sim m^* + \frac{2\ln(r-1)}{b^2}.$$

The corresponding upper and lower bounds on the expected loss per trial are

$$l''_{N,r}(m^{**}) \sim \mu_1 \cdot \frac{r-1}{b^2 N}\left[2 + \ln\left(\frac{b^4 N^2}{8\pi \ln N^2}\right)\right]$$

and

$$l'_{N,r}(m^*) \sim (\mu_1 - \mu_2) \cdot \frac{r-1}{b^2 N}\left[2 + \ln\left(\frac{b^4 N^2}{8\pi(r-1)^2 \ln N^2}\right)\right]. \qquad \text{Q.E.D.}$$

4. Allocation of trials by genetic plans. We now have bounds on the best possible performance (in terms of minimizing expected loss) of any plan which tests one random variable (schema) at a time. The objective now is to obtain a measure of the performance of genetic plans in similar circumstances so that a comparison can be made with this criterion. This comparison will reveal two things: (1) Even when the genetic plan is constrained to test one schema at a time, losses decrease at a rate proportional to that decreed by the criterion (though, initially, the plan does not have information about the means and variances required to calculate an optimal allocation of trials). (2) Intrinsic parallelism (tests of many schemata with a single trial) is used to advantage by genetic plans, enabling them to greatly surpass the one-schema-at-a-time criterion. Because both of these points came through convincingly under approximations less severe than "$\sim$", weaker approximations will be used wherever they substantially simplify the derivation.

Specifically, let us consider reproductive plans using genetic operators on a nonincreasing population (i.e., for all t, the average effective payoff rate of the population, $\bar{\mu}_t'$, is 1). Such plans can be diagrammed as in Diagram 1. The genetic operators, $\omega \in \Omega$, of step 7 are either of the form

$$\omega : \mathcal{A} \times \mathcal{A} \rightarrow \mathcal{A} \times \mathcal{A}$$

or else

$$\omega : \mathcal{A} \rightarrow \mathcal{A}.$$

DIAGRAM 1

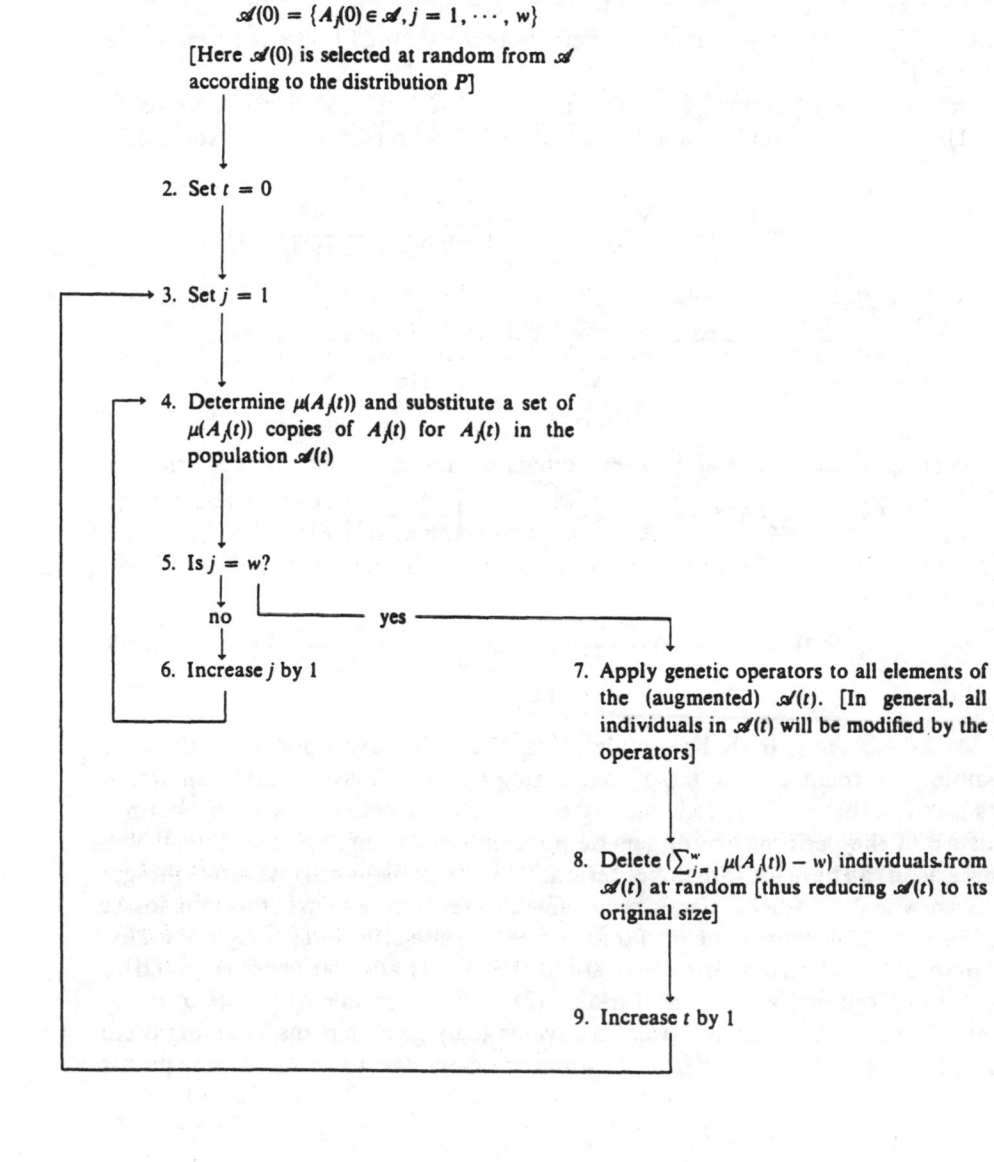

1. Select an initial population

$$\mathcal{A}(0) = \{A_j(0) \in \mathcal{A}, j = 1, \cdots, w\}$$

[Here $\mathcal{A}(0)$ is selected at random from $\mathcal{A}$ according to the distribution P]

2. Set $t = 0$

3. Set $j = 1$

4. Determine $\mu(A_j(t))$ and substitute a set of $\mu(A_j(t))$ copies of $A_j(t)$ for $A_j(t)$ in the population $\mathcal{A}(t)$

5. Is $j = w$?

   no   yes

6. Increase j by 1

7. Apply genetic operators to all elements of the (augmented) $\mathcal{A}(t)$. [In general, all individuals in $\mathcal{A}(t)$ will be modified by the operators]

8. Delete $(\sum_{j=1}^{w} \mu(A_j(t)) - w)$ individuals from $\mathcal{A}(t)$ at random [thus reducing $\mathcal{A}(t)$ to its original size]

9. Increase t by 1

The intended interpretation is that (pairs of) individuals selected from the population $\mathscr{A}(t)$ are transformed by the operator into new (pairs of) individuals. The operators are conservative in the sense that they do not alter the size of the population. Formal definitions of various genetic operators can be found in Holland [6], but for the analysis below, it is necessary to know only that (i) arguments for the operators are chosen at random from $\mathscr{A}(t)$, and (ii) the conditional probability $o_{\xi t}$ that $A \in \xi$, once selected, will be transformed to some $A' \notin \xi$, is generally small and decreases to a value negligibly different from zero as the proportion of ξ in $\mathscr{A}(t)$ approaches 1.

It should be noted that the reproductive plan modifies the distribution P (over $\mathscr{A}$) as the number of trials increases. As a consequence the marginal distributions for the schemata of interest $\{\xi\}$, hence the means $\{\mu_\xi\}$, may change as N grows large. However, the central limit theorem holds for sequences of independent random variables with variable distribution as long as they are uniformly bounded. This condition holds for all schemata when there is an upper bound on performance (l.u.b.$_{A \in \mathscr{A}} \{\mu(A)\} < \infty$), and we shall proceed accordingly.

THEOREM 3. *Given N trials, a reproductive plan with genetic operators can be expected to allocate $N_{\xi_{(1)}}$ trials to the schema (random variable) $\xi_{(1)}$ with the best observed average payoff, where*

$$N_{\xi_{(1)}} \cong (>) \frac{N_{\xi_{(1)}}(0)}{\hat{z}_{(1)}} \exp\left[\frac{\hat{z}_{(1)} n_\rho}{n_\rho(0)}\right],$$

with $\hat{z}_{(1)}$ being the average of the logarithms of the observed payoffs for $\xi_{(1)}$, $N_{\xi_{(1)}}(0)$ being the trials allocated to $\xi_{(1)}$ at the outset, and $n_\rho = N - N_{\xi_{(1)}}$.

Proof. The increase in the number of instances of schema ξ during step 4 of the plan is given by

$$\mathscr{N}'_\xi(t) = \sum_{A \in \xi(t)} \mu(A),$$

where $\xi(t)$ is the set of instances of ξ in the population $\mathscr{A}(t)$. The instances of ξ in $\mathscr{A}(t)$ constitute a sample of ξ under the modified distribution P_t holding at time t. The value of $\mathscr{N}'_\xi(t)$ can be written in terms of $\mathscr{N}_\xi(t)$, the expected number of individuals in $\xi(t)$, by using the average

$$\hat{\mu}_{\xi t} \stackrel{\text{def}}{=} \frac{1}{\mathscr{N}_\xi(t)} \sum_{A \in \xi(t)} \mu(A)$$

of the observations of ξ at time t. In these terms

$$\mathscr{N}'_\xi(t) = \hat{\mu}_{\xi t} \mathscr{N}_\xi(t).$$

We will concentrate here on (typical) reproductive plans wherein the operators in step 7 are applied to elements of $\mathscr{A}(t)$ independently of their identity (representation). As mentioned earlier, the common characteristic of these operators is such that the conditional probability $o_{\xi t}$ of $A \in \mathscr{A}(t)$ being transformed to $A' \notin \xi(t)$ decreases to a negligibly small value as the size of $\xi(t)$ approaches that of $\mathscr{A}(t)$. Hence at the end of step 7 we can expect

$$\mathscr{N}''_\xi(t) \cong (1 - o_{\xi t}) \hat{\mu}_{\xi t} \mathscr{N}_\xi(t),$$

where the factor $(1 - o_{\zeta t}) \to 1$ as $\mathcal{N}_\zeta(t) \to w$. (This ignores new instances of ζ formed by the operators from other $A \notin \zeta(t)$.) Finally, after the deletions of step 8 (which are again uniform over $\mathcal{A}(t)$), we expect

$$\mathcal{N}_\zeta(t + 1) \cong \frac{1}{\bar{\mu}_t}(1 - o_{\zeta t})\hat{\mu}_{\zeta t}\mathcal{N}_\zeta(t),$$

since the expected value of the increase, $\sum_{j=1}^w \mu(A_j(t))$, is just $\bar{\mu}_t = \sum_{A \in \mathcal{A}} \mu(A)P_t(A)$, where P_t is the (modified) distribution over $\mathcal{A}$ at time t. Putting this recursion in explicit form we get

$$\mathcal{N}_\zeta(t + 1) \cong \left(\prod_{t'=0}^t \hat{\mu}'_{\zeta t'}\right)\mathcal{N}_\zeta(0) = \mathcal{N}_\zeta(0) \exp\left[\ln\left(\prod_{t'=0}^t \hat{\mu}'_{\zeta t'}\right)\right]$$

$$= \mathcal{N}_\zeta(0) \exp\left[\sum_{t'}^t \ln(\hat{\mu}'_{\zeta t'})\right]$$

$$= \mathcal{N}_\zeta(0) e^{\hat{z}_t t},$$

where $\hat{\mu}'_{\zeta t} = \hat{\mu}_{\zeta t}(1 - o_{\zeta t})/\bar{\mu}_t$ and $\hat{z}_t = \sum^t \ln(\hat{\mu}'_{\zeta t'})/t$. Using the fact that

$$\sum_{t'=1}^t f(t') \geq \int_{t'=0}^t f(t')\,dt'$$

for monotone increasing functions, the *total* number of trials of ζ to time t, $N_\zeta(t)$, can be approximated by

$$N_\zeta(t) \cong \mathcal{N}_\zeta(0) + \mathcal{N}_\zeta(0) \sum_{t'=1}^t e^{\hat{z}_t t'} \geq \mathcal{N}_\zeta(0) + \mathcal{N}_\zeta(0) \int_{t'=0}^t e^{\hat{z}_t t'}\,dt',$$

assuming that at each time t, $\hat{\mu}'_{\zeta t}$ (the rate of increase of $\zeta(t)$) is in excess of 1. (It should be noted that this approximation to $N_\zeta(t)$ assumes $\hat{z}_1 \cong \hat{z}_2 \cong \cdots \cong \hat{z}_t$. For small t, there may be a considerable error if $\hat{z}_{t'}, t' = 1, \cdots, t$, swings over a wide range, though at worst the error (as a fraction of the true value) will be considerably less than $e^{-\hat{z}_t}$. Moreover, as t increases, $\hat{z}_t$ changes more and more slowly because it is an average, while earlier terms in the sum being approximated are swamped by the larger later terms. Thus the probability of a given error steadily decreases as t increases.)

$$N_\zeta(t) \cong \mathcal{N}_\zeta(0) + \frac{\mathcal{N}_\zeta(0)}{\hat{z}_t}[e^{\hat{z}_t t} - 1]$$

$$\cong \mathcal{N}_\zeta(0)\left[\frac{e^{\hat{z}_t t}}{\hat{z}_t} + \left(1 - \frac{1}{\hat{z}_t}\right)\right].$$

Therefore the total number of trials of a schema ζ increases exponentially as a function of time (assuming the performance of ζ is consistently better than the average).

Let $\xi_{(1)}$ be the schema receiving the greatest number of trials over the interval t, and let $\xi_{(1)}(t')$ designate the set of instances of $\xi_{(1)}$ present in the population $\mathcal{A}(t')$ at time t'. Let $n_p(t)$ be the total trials allocated to all other *individuals* $\{\mathcal{A}(t') - \xi_{(1)}(t')\}$ from $t' = 0$ through $t' = t$. Since for all t' the number of individuals in

$\mathscr{A}(t')$ remains constant, the total number of trials $N(t) = N(0) \cdot t$. It follows that

$$\frac{n_\rho(t)}{n_\rho(0)} = \frac{N(t) - N_{\xi_{(1)}}(t)}{N(0) - N_{\xi_{(1)}}(0)} \leq \frac{N(t)}{N(0)} = t.$$

Hence

$$N_{\xi_{(1)}} \cong (>)\mathscr{N}_{\xi_{(1)}}(0)\left[\frac{1}{\hat{z}_{(1)}} \cdot \exp\left(\frac{\hat{z}_{(1)}n_\rho(t)}{n_\rho(0)}\right)\right],$$

where $\hat{z}_{(1)}$ is the observed $\hat{z}_t$ for $\xi_{(1)}$. Or

$$n_\rho \cong (<)(n_\rho(0)/\hat{z}_{(1)})\ln\left[\hat{z}_{(1)}N_{\xi_{(1)}}/\mathscr{N}_{\xi_{(1)}}(0)\right]. \qquad \text{Q.E.D.}$$

The following correspondence allows comparison of this result to the one obtained earlier for optimal allocation.

	"Optimal" [*]	"Reproductive" [ρ]
$N_{\cdot}$, trials allocated to $\xi_{(1)}$	$N_{1\cdot} = N_1$	$N_{1\rho} = N_{\xi_{(1)}}$
n, trials allocated to other schemata	$n_\cdot = (r-1)m^*$	$n_\rho = n_\rho$

Thus we have

$$N_{1\cdot} \sim \frac{(r-1)\sqrt{8\pi}}{b}\exp\left[\frac{1}{2}\left(\frac{b^2 n_\cdot}{r-1}\right) + \frac{1}{2}\ln\left(\frac{n_\cdot}{r-1}\right)\right],$$

where $b = (\mu_1 - \mu_2)/\sigma_2$, vs.

$$N_{1\rho} \cong (>)\frac{N_{1\rho}(0)}{\hat{z}_{(1)}}\exp\left[\frac{\hat{z}_{(1)}n_\rho}{n_\rho(0)}\right],$$

where $\hat{z}_{(1)} = \sum^t \ln(\hat{\mu}'_{\xi_{(1)}t'})/t$.

Clearly the two plans behave in roughly the same way, the number of trials allocated to the "best" in each case increasing exponentially as a function of the total number of trials allocated to all other schemata. However, a comparison of expected loss per trial yields much more interesting information. For the reproductive plan the expected loss per trial is bounded above by

$$l''_\rho = \frac{\mu_1}{N}[N_{1\rho}r'q(N_{1\rho}, n'_\rho) + (1 - r'q(N_{1\rho}, n'_\rho))n_\rho],$$

where r' is the number of schemata which have received n'_ρ (or more) trials under the reproductive plan. It is critical to what follows that $r' \cdot n'_\rho$ need *not* be equal to n_ρ. Each $A \in \mathscr{A}$ is a trial of 2^l distinct schemata. As $\mathscr{A}(t)$ is transformed into $\mathscr{A}(t+1)$ by the reproductive plan, *each* schema ξ having instances in $\mathscr{A}(t)$ can be expected to have $(1 - o_\omega)\mu_\xi/\bar{\mu}_t$ instances in $\mathscr{A}(t+1)$. Thus, over the course of several time steps, the number of schemata r' receiving n'_ρ trials will be much, much greater than the number of trials n_ρ allocated to *individuals* A (where $A \in \mathscr{A}$ but $A \notin \xi_1$) even when n'_ρ approaches or exceeds n_ρ. (As a simple example consider a set of three trials $\{\sigma_1\sigma_1\sigma_1\sigma_2, \sigma_2\sigma_1\sigma_1\sigma_1, \sigma_2\sigma_2\sigma_2\sigma_1\}$. Each of the 6 schemata $\{\sigma_2\square\square\square,$ $\square\sigma_1\square\square, \square\square\sigma_1\square, \square\square\square\sigma_1, \sigma_2\square\square\sigma_1, \square\sigma_1\sigma_1\square\}$ receives 2 trials, so that for $n'_\rho = 2$ we have $r' = 6$ and $n'_\rho \cdot r' = 12$ though n_ρ is clearly 3 (or less). (See below.) This observation, that generally $r'n'_\rho \gg n_\rho$, is an explicit consequence of the reproductive plan's *intrinsic parallelism* (each trial of an individual $A \in \mathscr{A}$ is a useful trial of a great many schemata).

THEOREM 4. *The ratio of the upper bound on the expected loss per trial for a reproductive plan, l''_ρ, to the corresponding lower bound for optimal allocation, l'_*, varies inversely as the number, r', of schemata being tried. Specifically,*

$$\frac{l''_\rho}{l'_*} \to \frac{(\mu_1 - \mu_2)^2 n_\rho(0)}{2\sigma_2^2 \mathscr{Q}_{(1)}} \left(\frac{1}{r' - 1}\right),$$

where the parameters are as defined in the statements of the previous theorems.

Proof. Substituting the earlier expressions for $N_{1\rho}$ and $q(N_{1\rho}, n'_\rho)$ in l''_ρ, and noting that $(1 - r'q(N_{1\rho}, n'_\rho))n_\rho < n_\rho$, gives

$$l''_\rho \lesssim \frac{\mu_1}{N} \left[\frac{r'N_{1\rho}(0)}{\mathscr{Q}_{(1)}b\sqrt{2\pi}} \exp\left[\frac{\mathscr{Q}_{(1)}n_\rho}{n_\rho(0)} - \frac{b^2 n'_\rho + \ln n'_\rho}{2}\right] + n_\rho \right].$$

If $b^2 n'_\rho/2 \geqq \mathscr{Q}_{(1)} n_\rho/n_\rho(0)$, it is clear that the first term decreases as n_ρ increases, but the second term, n_ρ, increases. In other words, if n'_ρ increases at a rate proportional to the rate of increase of n_ρ, the expected loss per trial will soon depend almost entirely on the second term, as was the case for optimal allocation. Thus, for n'_ρ so specified, we can compare losses by taking the ratio of the respective second terms:

$$\frac{l''_\rho}{l'_*} = \frac{n_\rho}{(r - 1)m^*}.$$

(A quick comparison of the first terms of l''_ρ and l'_* also shows that the above condition on n'_ρ is sufficient to assure that the first term of l''_ρ is always less than the first term of l'_*.) This comparison is conservative in the sense that the *upper* bound on the reproductive plan's losses is compared to the *lower* bound on the optimal allocation's losses.

To proceed, let the reproductive plan's loss per trial over N trials be compared to that of an optimal allocation of N trials to the r'-schemata which received n'_ρ or more trials under the reproductive plan. (It should be noted that the above condition on n'_ρ can be made as weak as desired by simply choosing $n_\rho(0)$ large enough.) Substituting the explicit expressions derived earlier for n_ρ and m^* as a function of N gives

$$\frac{l''_\rho}{l'_*} \lesssim \frac{b^2 n_\rho(0) \ln\left[\mathscr{Q}_{(1)}(N - n_\rho)/N_{1\rho}(0)\right]}{(r' - 1)(\ln\left[b^4 N^2/8\pi(r' - 1)^2 \ln N^2\right])\mathscr{Q}_{(1)}}.$$

Simplifying and deleting terms which do not affect the direction of the inequality, we get

$$\frac{l''_\rho}{l'_*} \lesssim \frac{b^2 n_\rho(0) \ln(\mathscr{Q}_{(1)}N)}{(r' - 1)\mathscr{Q}_{(1)}[2 \ln b^2 N - \ln(8\pi(r' - 1)^2 \ln N^2)]}.$$

458

Or, as N grows,

$$\frac{l''_\rho}{l'_*} \to \frac{b^2 n_\rho(0)}{2\hat{2}_{(1)}}\left(\frac{1}{r'-1}\right).$$ Q.E.D.

Thus the reproductive plan effectively exploits its intrinsic parallelism—its losses for a given number of trials N, in relation to an optimal (one-schema-at-a-time) allocation, are reduced by the factor r'. We can get some idea of how large this reduction is by looking more closely at the relation between N, n'_ρ and r'. This relation in turn is more easily approached if we first look more closely at schemata. A schema will be said to be *defined on* the set of positions $\{j_1, \cdots, j_h\}$ at which $\delta_{i_j} \neq \square$. Given Σ with k symbols, there are k^h distinct schemata defined on any given set of $h \leq l$ positions; moreover, no matter what set of positions is chosen, *every* $A \in \mathscr{A}$ is an instance of one of these k^h schemata. That is, the set of schemata so defined partitions $\mathscr{A}$, and any distinct set of positions gives rise to a different partition of $\mathscr{A}$. (For example, given the alphabet $\Sigma = \{\sigma_1, \sigma_2\}$ and strings of length $l = 4$, the set of schemata defined on position 1 is $\{\sigma_1\square\square\square, \sigma_2\square\square\square\}$. It is clear that every string in $\mathscr{A}$ begins either with the symbol σ_1 or else the symbol σ_2, hence the given set partitions $\mathscr{A}$. Similarly the set defined on position 2, $\{\square\sigma_1\square\square, \square\sigma_2\square\square\}$, partitions $\mathscr{A}$, and the set defined on positions 2 and 4, $\{\square\sigma_1\square\sigma_1, \square\sigma_1\square\sigma_2, \square\sigma_2\square\sigma_1, \square\sigma_2\square\sigma_2\}$, is still a different partition of $\mathscr{A}$, a refinement of the one just previous.) There are $\binom{l}{h}$ distinct ways of choosing h positions $\{1 \leq j_1 < j_2 < \cdots < j_h \leq l\}$ along a string of length l, and h can be any number between 1 and l. Thus there are $\sum_{h=1}^{l} \binom{l}{h} = 2^l$ distinct partitions induced on $\mathscr{A}$ by these sets of schemata. It follows that when the reproductive plan generates N trials, they will be simultaneously distributed over each of these partitions. That is, *each* of the 2^l *sets* of schemata (defined on the 2^l distinct choices of positions) receives N trials.

We can get a *rough* estimate of the number, r', of schemata receiving n'_ρ or more trials by assuming the N trials are distributed uniformly and independently over each partition. Two factors perturb the estimate: (1) Given a uniform initial distribution P, the reproductive plan will make the distribution increasingly non-uniform as n_ρ increases. However, until N gets fairly large relative to n_ρ the departure is small enough to make the estimate useful. (2) When a given schema defined on h positions receives n'_ρ or more trials, then so must every schemata of which it is a subset. (For example, let $\square\sigma_2\square\sigma_1$ receive 2 trials, say $\{\sigma_2\sigma_2\sigma_2\sigma_1, \sigma_1\sigma_2\sigma_1\sigma_1\}$. These are at the same time trials of $\square\sigma_2\square\square$, and also of $\square\square\square\sigma_1$. Hence $\square\sigma_2\square\square$ and $\square\square\square\sigma_1$ also receive at least 2 trials.) Similarly, if a given schema receives less than n'_ρ trials, then so will every schema of which it is a superset. These are clearly violations of the assumption of independence. Nevertheless, when N is small relative to k^l (so that only a small fraction of schemata have been tried), departures from independence are small enough to allow a useful estimate. Some thought about the number of dependencies relative to the total number of schemata tried, or a small Monte Carlo simulation, are convincing in this respect. Though the estimate is rough, the value of r' obtained for typical values of N, b, $\mathfrak{L}_{(1)}$, etc. is clearly of the right order of magnitude.

The average number of trials per schema for a set of schemata defined on h positions is N/k^h. Under the assumption of uniform, independent trials, the Poisson distribution gives the number of schemata receiving n'_ρ or more trials:

$$\sum_{n=n\rho}^{\infty} \left(\frac{N}{k^h}\right)^n \left(\frac{1}{n!}\right) e^{-N/k^h}.$$

There are $\binom{l}{h}$ distinct sets of k^h schemata defined on h positions, so that the number r' of schemata in Ξ, $h = 1, \cdots, l$, receiving at least n'_ρ trials is then

$$\sum_{h=1}^{l} \binom{l}{k} k^h \sum_{n=n'_\rho}^{\infty} \left(\frac{N}{k^h}\right)^n \left(\frac{1}{n!}\right) e^{-N/k^h}.$$

This is a very large number as long as n'_ρ is smaller than $N/2$, as it always would be in practice. Even when N is quite small (so that the estimate is good), the number is substantial. For example, if the representations are of length $l = 32$ with two symbols in Σ (so that $\mathscr{A}$ contains $2^{32} \cong 4 \times 10^9$ elements) and if $N = 16$ with $n'_\rho = 8$, then $r' > 700$ schemata can be expected to receive in excess of n'_ρ trials. The numbers chosen here are clearly very conservative—if $N = 32$, $r' > 9000$ for l and n'_ρ as given; any increase in l produces even more dramatic increases in r'.

The advantages implied by this analysis have been observed in a variety of computer tests (Bagley [1], Cavicchio [3], Hollstien [7]).

5. Conclusion. Intrinsic parallelism in the search of schemata offers a tremendous advantage to any optimization procedure which can exploit it. Reproductive plans with genetic operators (genetic algorithms) are the only procedures so far studied which exhibit this phenomenon. They have the additional desirable properties of easy implementation, compact storage and automatic use of the large amounts of relevant information encountered during operation, and robustness (efficient operation under maximal uncertainty). For these reasons it is recommended that genetic algorithms be given serious consideration whenever a problem of natural or artificial adaptation arises.

REFERENCES

[1] J. D. BAGLEY, *The behavior of adaptive systems which employ genetic and correlation algorithms*, Ph.D. dissertation, University of Michigan, Ann Arbor, 1967.

[2] R. BELLMAN, *Adaptive Control Processes*, Princeton University Press, Princeton, N.J., 1961.

[3] D. J. CAVICCHIO, *Adaptive search using simulated evolution*, Ph.D. dissertation, University of Michigan, Ann Arbor, 1970.

[4] M. E. HELLMAN AND J. M. COVER, *Learning with finite memory*, Ann. Math. Statist., 41 (1970), pp. 765–782.

[5] J. H. HOLLAND, *A new kind of turnpike theorem*, Bull. Amer. Math. Soc., 75 (1969), pp. 1311–1317.

[6] ———, *Processing and processors for schemata*, Associative Information Techniques, E. L. Jacks, ed., Elsevier, New York, 1971, pp. 127–146.

[7] R. B. HOLLSTIEN, *Artificial genetic adaptation in computer control systems*, Ph.D. dissertation, University of Michigan, Ann Arbor, 1971.

[8] YA Z. TSYPKIN, *Adaptation and Learning in Automatic Systems*, Academic Press, New York, 1971.

Chapter 16
Classifier Systems

J. H. Holland and J. S. Reitman (1978) "Cognitive Systems Based on Adaptive Algorithms,"
Pattern-Directed Inference Systems, D. A. Waterman and F. Hayes-Roth (eds.), Academic Press,
NY, pp. 313–329.

EFFORTS in artificial intelligence in the late 1960s and early 1970s centered mainly on developing "heuristic programs" (Feigenbaum and Feldman, 1963, p. 6; Feigenbaum et al., 1971; Shortliffe, 1974; and others). Humans with particular expertise in a specific domain provided approximate rules of their behavior. These were often codified in a series of if-then, condition-action instructions (i.e., production systems) and executed on a computer. Such "expert systems" were hoped to be completely reliable, never fatigued, and able to offer the same decisions as the human experts they emulated. But there were several drawbacks to this approach: (1) the rules obtained might not be correct, (2) the rules might not be self-consistent, (3) there might be disagreement between different experts, (4) the rules might be incomplete, and (5) rule-based systems were inherently brittle, not robust. This last shortcoming was perhaps the most severe. Expert systems might perform well in a limited domain, and even then only on limited data, but would offer poor or even no performance at all in another setting. Yet true intelligence requires adaptive behavior in the face of changing circumstance.

Holland (1975, pp. 141–158) suggested that such adaptive behavior could be facilitated by manipulating strings in a language of detectors and messages using a genetic algorithm (see Chapter 15). This "broadcast language" contained 10 symbols: {0, 1, *, :, ◊, ∇, ▼, ∆, p, '}. The symbols 0 and 1 were used as the essential descriptors of an environment in order to maximize the number of schemata that could be generated. Other symbols determined the input-output behavior of the system. For example, the symbol * indicated that the following string of symbols was an active broadcast unit, while : served as a punctuation. Thus the string *1100:11 would indicate that when a signal 1100

is detected, the rule would respond with the message 11. The symbol ◊ served as a wild card or "don't care" term, so that the string *1◊00:11 would broadcast a 11 if either 1100 or 1000 were detected. For specific definitions of the other symbols, see Holland (1975, pp. 145–148). Essentially, each string acted as a classifier which responded with a particular output when stimulated by any string from a specified set of inputs. With this framework, it was possible to describe a variety of production systems and apply genetic algorithms to optimize the associations between detectors and messages that defined system behavior (see also Holland, 1976).

Holland and Reitman (1978) (reprinted here)[1] constructed a simple "classifier system"[2] (termed CS-1 for "cognitive system") to operate on a one-dimensional array of nodes. Each node was tagged by a binary feature vector (k bits). The ends of the array designated specific rewards of food or water, respectively, and the challenge was to move left or right on the array so as to maintain sufficient levels of nourishment, which diminished regularly over time. A collection of 100 individual classifiers (detector:message pairs) competed for activation at each node. Scoring was made using a reinforcement learning method based on the product of the sum of the number of features matched by the string and the predicted payoff of the string (the amount and type of need satisfied in past usage of the string, adjusted by exogenous factors). The top ten classifiers were given a probability of being activated in proportion to their performance measure. Crossover was used to generate new classifiers. The results indicated that this classifier system could learn to reach a reward state at either end of the array much faster than a simple

---

[1]Holland and Reitman (1978) appears in Waterman and Hayes-Roth (1978) where references for all of the publications were collected in a single listing. Readers who wish to verify a particular citation are referred to Waterman and Hayes-Roth (1978).

[2] A classifier system is a restricted broadcast system in that individual classifiers are described solely by the symbols {0, 1, #} (# was routinely substituted for ◊ as a wild card symbol) and cannot create other classifiers (Holland, 1992, pp. 171–172). Wilson (1994) commented that efforts to realize the full potential of the broadcast system have met with "mixed success" primarily because of difficulties in comprehending the complexity of the system. Thus the simpler system of classifiers over {0, 1, #} has received greater attention and generated more useful results.

random walk or a reinforcement learning procedure in the absence of crossover.[3]

Rather than compete individual rules, Smith (1980, 1983) proposed a classifier system in which entire rule-sets were measured based on their performance as a functional unit and competed with other complete rule sets. A system of recombination, inversion, and mutation was applied to randomly vary collections of rules. Smith (1980) evolved a classifier system to play draw poker against a hand-crafted rule base developed in Waterman (1970), which was calibrated to perform at about the same level as an experienced human player. The environmental conditions were described in terms of the value of the player's hand, the amount of money in the pot, the most recent wager, a measure of the probability that a bluff would be successful, the number of cards the opponent drew, and so forth. A "correct" decision was defined by a set of logical arguments about poker and the values of the environmental variables at each stage (Waterman, 1968). Over a long evolution (40,000 rounds of poker) the classifier system was able to dominate the hand-crafted poker player of Waterman (1970). Analysis showed that Waterman's program was not designed for such extended play, which left it susceptible over time to a simple bluffing strategy. After the design flaw was corrected, Waterman's poker player proved a more formidable opponent, and against this opponent, a rule base evolved that agreed with the logical arguments about 82% of the time.

These two approaches to engineering classifier systems, one offering credit to individual rules, the other scoring complete rule sets in their entirety,[4] have engendered a voluminous literature on the application of genetic algorithms to evolving rules of behavior (e.g., Booker, 1982; Goldberg, 1983; Holland, 1986; Holland et al., 1986; Grefenstette, 1988; Smith, 1989; and others; for a recent review see Wilson, 1994, 1995). Many of these efforts were directed to gaining a better understanding of how to apportion credit to individual rules (e.g., Holland, 1985; Westerdale, 1985; Riolo, 1987; and others),[5] or structuring crossover to effectively recombine building blocks of classifiers (e.g., Antonisse and Keller, 1987; Bickel and Bickel, 1987; and others; for early reviews, see Holland, 1987; De Jong, 1987). Others examined the equivalence/mapping of classifier systems to other computational frameworks (e.g., Forrest, 1985; Davis, 1989). The early work of Wilson (1985) using classifiers to describe the behavior of an artificial animal (i.e., an "animat") has led to a broader framework for studying adaptive behavior using simulation (e.g., Wilson, 1987; Meyer and Wilson, 1991; Meyer et al., 1993; and others; similar suggestions were made in Pask, 1962, 1969). Classifier systems have also been used to model economies and other complex adaptive systems without using explicit fitness criteria (e.g., using the *Echo* system, Holland, 1992, pp. 186–194; Holland, 1995, pp. 93–160) with an interest for examining their emergent properties.

References

[1] H. J. Antonisse and K. S. Keller (1987) "Genetic operators for high-level knowledge representations," *Genetic Algorithms and Their Applications: Proc. of the 2nd Intern. Conf. on Genetic Algorithms,* J. J. Grefenstette (ed.), Lawrence Erlbaum, Hillsdale, NJ, pp. 69–76.

[2] A. S. Bickel and R. W. Bickel (1987) "Tree structured rules in genetic algorithms," *Genetic Algorithms and Their Applications: Proc. of the 2nd Intern. Conf. on Genetic Algorithms,* J. J. Grefenstette (ed.), Lawrence Erlbaum, Hillsdale, NJ, pp. 77–81.

[3] L. B. Booker (1982) "Intelligent behavior as an adaptation to the task environment," Ph.D. diss., Univ. of Michigan, Ann Arbor, MI.

[4] H. B. Cribbs and R. E. Smith (1996) "Classifier system renaissance: New analogies, new directions," *Genetic Programming 1996: Proc. of the 1st Annual Conference,* J. R. Koza, D. E. Goldberg, D. B. Fogel, and R. L. Riolo, MIT Press, Cambridge, MA, pp. 547–552.

[5] L. Davis (1989) "Mapping classifier systems into neural networks," *Advances in Neural Processing Systems 1,* D. S. Touretzky (ed.), Morgan Kaufmann, San Mateo, CA, pp. 49–56.

[6] K. De Jong (1987) "On using genetic algorithms to search program spaces," *Genetic Algorithms and Their Applications: Proc. of the 2nd Intern. Conf. on Genetic Algorithms,* J. J. Grefenstette (ed.), Lawrence Erlbaum, Hillsdale, NJ, pp. 210–216.

[7] E. A. Feigenbaum, B. G. Buchanan, and J. Lederberg (1971) "On generality and problem solving: A case study involving the DENDRAL program," *Machine Intelligence 6,* B. Meltzer and D. Michie (eds.), American Elsevier, NY, pp. 165–190.

[8] E. A. Feigenbaum and J. Feldman (1963) "Artificial intelligence," *Computers and Thought,* E. A. Feigenbaum and J. Feldman (eds.), McGraw-Hill, NY.

[9] S. Forrest (1985) "Implementing semantic network structures using the classifier system," *Proc. of an Intern. Conf. on Genetic Algorithms and Their Applications,* J. J. Grefenstette (ed.), Lawrence Erlbaum, Hillsdale, NJ, pp. 24–44.

[10] D. E. Goldberg (1983) "Computer-aided gas pipeline operation using genetic algorithms and rule learning," Ph.D. diss., Univ. of Michigan, Ann Arbor, MI.

[11] D. P. Greene and S. F. Smith (1994) "Using coverage as a model building constraint in learning classifier systems," *Evolutionary Computation,* Vol. 2:1, pp. 67–91.

[12] J. J. Grefenstette (1988) "Credit assignment in rule discovery systems based on genetic algorithms," *Machine Learning,* Vol. 3, pp. 225–245.

[13] J. H. Holland (1975) *Adaptation in Natural and Artificial Systems,* Univ. of Michigan, Ann Arbor, MI.

---

[3] Holland and Reitman (1978) indicated the "power of the genetic algorithm" to optimize the performance of the system. The classifier system that used both recombination and reinforcement outperformed the system that used only reinforcement and could not generate new classifiers. However, no results with a control procedure offering an alternative means of generating new classifiers (e.g., mutation) were reported. Thus the degree to which this improvement depended on the effective recombination of useful classifiers remains uncertain. Recent efforts have been made to directly compare the effectiveness of various variation operators (e.g., Greene and Smith, 1994).

[4] These two approaches of scoring individual classifiers or complete sets of rules were termed the *Michigan* and *Pitt* approaches, respectively, owing to the universities where they were originally proposed.

[5] Cribbs and Smith (1996) remarked that the common credit assignment method in the Michigan approach, termed the *bucket brigade* (Holland et al., 1986, pp. 71–75), is very similar to another reinforcement technique termed *Q-learning* (Watkins, 1989).

[14] J. H. Holland (1976) "Adaptation," *Progress in Theoretical Biology*, R. Rosen and F. M. Snell (eds.), Vol. 4, Plenum, NY, pp. 263–293.

[15] J. H. Holland (1985) "Properties of the bucket brigade," *Proc. of an Intern. Conf. on Genetic Algorithms and Their Applications*, J. J. Grefenstette (ed.), Lawrence Erlbaum, Hillsdale, NJ, pp. 1–7.

[16] J. H. Holland (1986) "Escaping brittleness: The possibilities of general purpose machine learning algorithms applied to parallel rule-based systems," *Machine Learning: An Artificial Intelligence Approach*, Vol. 2, R. S. Michalski, J. G. Carbonell, and T. M. Mitchell (eds.), Morgan Kaufmann, Los Altos, CA, pp. 593–623.

[17] J. H. Holland (1987) "Genetic algorithms and classifier systems: Foundations and future directions," *Genetic Algorithms and Their Applications: Proc. of the 2nd Intern. Conf. on Genetic Algorithms*, J. J. Grefenstette (ed.), Lawrence Erlbaum, Hillsdale, NJ, pp. 82–89.

[18] J. H. Holland (1992) *Adaptation in Natural and Artificial Systems*, 2nd ed., MIT Press, Cambridge, MA.

[19] J. H. Holland (1995) *Hidden Order: How Adaptation Builds Complexity*, Addison-Wesley, Reading, MA.

[20] J. H. Holland, K. J. Holyoak, R. E. Nisbett, and P. G. Thagard (1986) *Induction: Processes of Inference, Learning, and Discovery*, MIT Press, Cambridge, MA.

[21] J. H. Holland and J. S. Reitman (1978) "Cognitive systems based on adaptive algorithms," *Pattern-Directed Inference Systems*, D. A. Waterman and F. Hayes-Roth (eds.), Academic Press, NY, pp. 313–329.

[22] J.-A. Meyer, H. L. Roitblat, and S. W. Wilson (eds.) (1993) *From Animals to Animats 2: Proc. of the 2nd Intern. Conf. on Simulation of Adaptive Behavior*, MIT Press, Cambridge, MA.

[23] J.-A. Meyer and S. W. Wilson (eds.) (1991) *From Animals to Animats: Proc. of the 1st Intern. Conf. on Simulation of Adaptive Behavior*, MIT Press, Cambridge, MA.

[24] G. Pask (1962) "The simulation of learning and decision-making behavior," *Aspects of the Theory of Artificial Intelligence: The Proc. of the 1st Intern. Symp. on Biosimulation*, C.A Muses (ed.), Plenum, NY, pp. 165–215.

[25] G. Pask (1969) "The computer-simulated development of populations of automata," *Mathematical Biosciences*, Vol. 4, pp. 101–127.

[26] R. L. Riolo (1987) "Bucket brigade performance. I. Long sequences of classifiers," *Genetic Algorithms and Their Applications: Proc. of the 2nd Intern. Conf. on Genetic Algorithms*, J. J. Grefenstette (ed.), Lawrence Erlbaum, Hillsdale, NJ, pp. 184–195.

[27] E. H. Shortliffe (1974) "MYCIN: A rule-based computer program for advising physicians regarding antimicrobial therapy selection," Ph.D. diss., Stanford Univ., Stanford, CA.

[28] R. E. Smith (1989) "Default hierarchy formation and memory exploitation in learning classifer systems," Ph.D. diss., Univ. of Alabama, Tuscaloosa, AL.

[29] S. F. Smith (1980) "A learning system based on genetic adaptive algorithms," Univ. of Pittsburgh, Pittsburgh, PA.

[30] S. F. Smith (1983) "Flexible learning of problem solving heuristics through adaptive search," *Proc. of the 8th Intern. J. Conf. on Artificial Intelligence*, A. Bundy (ed.), William Kaufmann, Inc., Los Altos, CA, pp. 422–425.

[31] D. A. Waterman (1968) "Machine learning of heuristics," Ph.D. diss., Stanford Univ., Stanford, CA.

[32] D. A. Waterman (1970) "Generalization learning techniques for automating the learning of heuristics," *Artificial Intelligence*, Vol. 1, pp. 121–170.

[33] D. A. Waterman and F. Hayes-Roth (1978) *Pattern-Directed Inference Systems*, Academic Press, NY.

[34] J. C. H. Watkins (1989) "Learning with delayed rewards," Ph.D. diss., King's College, London.

[35] T. H. Westerdale (1985) "The bucket brigade is not genetic," *Proc. of an Intern. Conf. on Genetic Algorithms and Their Applications*, J. J. Grefenstette (ed.), Lawrence Erlbaum, Hillsdale, NJ, pp. 45–59.

[36] S. W. Wilson (1985) "Knowledge growth in an artificial animal," *Proc. of an Intern. Conf. on Genetic Algorithms and Their Applications*, J. J. Grefenstette (ed.), Lawrence Erlbaum, Hillsdale, NJ, pp. 16–23.

[37] S. W. Wilson (1987) "Classifier systems and the animat problem," *Machine Learning*, Vol. 2, pp. 199–228.

[38] S. W. Wilson (1994) "ZCS: A zeroth level classifier system," *Evolutionary Computation*, Vol. 2:1, pp. 1–18.

[39] S. W. Wilson (1995) "Classifier fitness based on accuracy," *Evolutionary Computation*, Vol. 3:2, pp. 149–175.

COGNITIVE SYSTEMS BASED ON ADAPTIVE ALGORITHMS[1]

John H. Holland and Judith S. Reitman
The University of Michigan

The type of cognitive system (CS) studied here has four basic parts: (1) a set of interacting elementary productions, called lassifiers, (2) a performance algorithm that directs the action of the system in the environment, (3) a simple learning algorithm that keeps a record of each classifier's success in bringing about rewards, and (4) a more complex learning algorithm, called the genetic algorithm, that modifies the set of classifiers so that variants of good classifiers persist and new, potentially better ones are created in a provably efficient manner.

Two "proof-of-principle" experiments are reported. One experiment shows CS's performance in a maze when it has only the ability to adjust the predictions about ensuing rewards of classifiers (similar to adjusting the "weight" of each classifier) vs. when the power of the genetic algorithm is added. Criterion was achieved an order of magnitude more rapidly when the genetic algorithm was operative. A second experiment examines transfer of learning. Placed in a more difficult maze, CS with experience in the simpler maze reaches criterion an order of magnitude more rapidly than CS without prior experience.

1.0 OVERVIEW

In broadest terms, the cognitive systems described here infer environmental patterns from experience and associate "appropriate" response sequences with them. To determine appropriate response sequences the systems keep track of selected performance measures

[1] Research reported in this paper was supported in part by the National Science Foundation under grant DCR 71-01997 and by the Horace H. Rackham School of Graduate Studies under grant 387156.

such as the rate of accumulation of certain resources or the rate of reduction of predetermined needs. These performance measures enter directly into determining what patterns and associations are inferred and preserved. Because behaviors (patterns and associations) required for good performance in a given environment may be difficult to determine a priori, the systems of interest to us must be capable of learning most behaviors falling within the limits set by the system's basic capacities (inputs, outputs, memory capacity, and procedures given a priori). In particular, the systems must be capable of organizing these behaviors into "cognitive maps" appropriate to the environment and usable for prediction and lookahead.

It follows that the core of such a cognitive system (CS) is the learning algorithm—its way of inferring patterns, associations, and predictions. This emphasis on learning is reinforced by the observations that intelligent systems inevitably require a steady succession of changes as experience is gained. CS makes these changes itself, rather than requiring the designer to supply them. The problems encountered are familiar: learning of new productions, generalization, focus of attention and conflict resolution between competing productions, access and use of knowledge already acquired and the use of information not directly associated with the performance measures. (Discussion of these problems occur, for example, in Hayes-Roth and Burge [F. Hayes-Roth76e], McDermott and Forgy [J. McDermott77b], Samuel [Samuel63], Soloway and Riseman [Soloway77], and Waterman [Waterman70]. The systems we study approach these problems by generating and combining elementary productions (simple condition-action rules) to generate behavior. Learning is accomplished through a powerful and flexible algorithm belonging to the class of provably efficient genetic algorithms [Holland75].

The designer of an intelligent system should give possible changes careful consideration from the outset, even if he intends to make all the changes himself. Without such consideration it is unlikely that the changes dictated by experience will be easily implemented. Since it is impossible to determine, at the outset, which changes experience will require, the designer should consider the full range A of systems that could result from various combinations of changes. If the designer concentrates only on his initial design, a single variant in A, he gains little insight into A as a whole. Not knowing much about A, he wanders through it blindly as experience accumulates, making "local improvements" but searching A very inefficiently. In the literature of artificial intelligence, Samuel's [Samuel63] approach to game-playing is a particularly good example of such prior consideration of A. He characterizes the range of strategies A by specifying allowable changes, then searches A by making progressive changes in the strategy initially tried.

It is vital that the changes allowed for, i.e., the range A, be rich enough to give a reasonable chance of correcting faults in the initial design. In other words, the system should not only learn, but some guarantee should be given that it can adapt to a wide range of situations. Given any program of behavior falling within CS's basic limits (storage capacity, etc.), A should contain a variant that can carry it out. One way of assuring sufficient richness is to find a set of allowable changes that makes A computationally complete within CS's basic limits. (A is computationally complete if each procedure executable by a general-purpose computer can be effected by some system in A. For CS this means that any complex of behaviors that can be defined precisely can be attained by some sequence of changes. The learning algorithm now becomes a procedure for searching A, making changes in accord with the patterns it finds in the environment.

Because learning has a central role, all parts of CS have been designed with a view to their suitability as part of the domain of the learning algorithm. The basic elements of the domain are elementary productions called classifiers. Each classifier is defined by specifying a set of conditions to which it is sensitive—collectively called a *taxon*—and a signal it broadcasts when these conditions are fulfilled. A classifier can be sensitive to signals broadcast by other classifiers as well as to signals from the environment. That is, the taxon specifies conditions on both kinds of signals.

In more detail: CS encodes all signals by using an array of *detectors*, δ_i, $i = 1, \ldots, \beta$, each detector being sensitive to some attribute of the environment or some attribute of the (set of) signals from other classifiers. Using I_1 to denote the set of environmental conditions, and I_2 to denote all possible sets of broadcast signals, the i-th detector can be thought of as a function

$$\delta_i : I_j \to \{0,1\} \quad i = 1, \ldots, \beta \quad j = 1 \text{ or } 2$$

where δ_i takes the value 1 ("on") whenever detector i detects the appropriate attribute in a signal; and otherwise takes value 0 ("off"). Each combination of environmental and internal signals s yields an β-digit binary number $(\delta_1[s], \ldots, \delta_\beta[s])$ via the detectors. More formally, the set

$$\{\delta_1(s), \delta_2(s), \ldots, \delta_\beta(s) : s \in I_1 \times I_2\}$$

constitutes a set of representations of all combinations of signals $I_1 \times I_2$.

The taxon of a classifier specifies a subset of the β-tuples, any one being sufficient to cause the classifier to broadcast its signal. In the simpler versions a taxon specifies, for each detector, whether or not the classifier attends to that detector and, if it does, what value (0 or 1) it requires. In this case, the taxon is defined by an β-tuple

John H. Holland and Judith S. Reitman

$$(\lambda_1, \lambda_2, \ldots, \lambda_\beta) \; \lambda_i \; \epsilon \; \Lambda = \{\#,0,1\} \; i = 1, \ldots, \beta$$

where $\lambda_i = \#$ indicates that the classifier does not attend to detector i (it "doesn't care"), and $\lambda_i = 0$ (or 1) indicates that detector i must be off (or on) for the overall detector configuration to satisfy the taxon. The task of the learning algorithm is to manipulate these taxa so that the appropriate signals are emitted to control CS under the various environmental conditions it faces.

Two broad principles guide the learning algorithm in the manipulation of taxa:

The first principle concerns the level of generality of a taxon. A taxon that is too general (too many #s) will give rise to inconsistent performance by signaling in inappropriate situations. A taxon that is too specific (too few #s) wastes memory capacity by specifying a situation that recurs infrequently or not at all. The learning algorithm must put a steady selective pressure upon the population of classifiers, replacing taxa that are too general or too specific by more appropriate taxa.

The second principle concerns prediction as a means of using the (usually extensive) information acquired on the way to a goal (attainment of resources, reduction of needs, etc.). Generally, when a goal is attained after a long sequence of responses, many of the responses along the way are inconsistent in the sense that they delay attainment of the goal. (As an example, consider a board game such as Go; cf. W. Reitman [W.R. Reitman77].) The learning algorithm must rid the system of such inconsistencies. Stated the other way around, the algorithm should select classifiers that produce consistent goal-directed sequences of responses. This selection depends much more on information acquired on the way than on information received at the time the goal is attained. Inconsistencies along the way can be discovered if each classifier predicts the anticipated long-term effect of its activation. Then, if a subsequently activated classifier makes a different prediction, an inconsistency is revealed. The procedure here was foreshadowed by Samuel's [Samuel63] use of a position evaluator as a predictor. When Samuel's evaluator gives a different value at a subsequent position, thereby revealing an inconsistency in expectations, weights are revised to bring the earlier prediction more in line with the subsequent evaluation. Here, inconsistencies are reconciled by having the learning algorithm punish (select against) activated classifiers making predictions that are subsequently contravened. Chains of consistent goal-oriented responses are rewarded (selected for) with the eventual result that the first element of the chain predicts the ultimate outcome.

Using these principles as guides, a genetic algorithm was modified to serve as a learning algorithm. Briefly, the genetic algorithm selects classifiers based on their predicted performance, and operates on their taxa to build classifiers that potentially respond to important regularities in the environment. It does this by making intensive use of particular combinations of 0s, 1s, and #s consistently associated with above-average performance (see below). A primary reason for this choice is the provable efficiency of *genetic algorithms* (see Chapter 7 of Holland [Holland75]). This efficiency is most easily understood by considering special classes of taxa called *schemata*. (More formally, schemata are hyperplanes in A). Schemata name or identify particular combinations of attributes and don't cares (0s, 1s, and #s) that aid in producing goal-oriented classifiers. The object of the learning algorithm then is to discover good schemata and use them in building taxa.

Formally, schemata are defined in terms of a β-tuple representation of A's elements by using an additional symbol, $\square$. $\square$ is a schema-level "don't care." For example the schema $(1, \square, \#, \square, \square, \ldots, \square)$ names the *set* of all β-tuples in A that have a 1 at the first position, a # at the third position, and any values whatsoever from $\Lambda = \{\#,0,1\}$ at the other β-2 positions. In full generality, the set Ξ of schemata is the set of all β-tuples formed by substituting $\square$s at various positions in β-tuples from A. (Formally, $\Xi = \{\Lambda \cup \{\square\}\}^{\beta}$). A schema $\xi = (\nu_1, \ldots, \nu_\beta)$ from Ξ names a subset of A as follows: $A = (\lambda_1, \ldots, \lambda_\beta) \in A$ belongs to $\xi = (\nu_1, \ldots, \nu_\beta)$ if and only if, for each position j, $\lambda_j = \nu_j$ if $\nu_j \in \Lambda = \{\#,0,1\}$; otherwise, when $\nu_j = \square$, λ_j can be any element of Λ.

The following are important, provable properties of genetic algorithms regarding schemata:

(1) If A is regarded as a sample space then *each* taxon is a legitimate sample point for each of the 2^β schemata (sets) of which it is an instance. The genetic algorithm uses this information to alter its estimates of the value of a large proportion of the schemata involved. In fact, each time the genetic algorithm is applied to the set of M taxa being used by the cognitive system, the estimated values of about $M^2 \times 2^{\beta/2}$ implicitly associated schemata are adjusted. This occurs even though only M taxa are modified—a phenomenon called *intrinsic parallelism*. It cannot be emphasized too strongly that this tremendous "speed-up" plays a critical role in the performance of this system. In effect $M^2 \times 2^{\beta/2}$ different combinations of 0s, 1s, and #s are tried, evaluated, and used to generate taxa each time the genetic algorithm is applied.

A more formal statement gives a more precise idea of what is actually occurring. Let $B = \{A_1, \ldots, A_M\}$ be the set of taxa maintained by the cognitive system at some time t, and let $\{\mu_1, \ldots, \mu_M\}$ be the associated performance predictions. That is, μ_j is a *prediction* of the value of the goal that will be attained if the classifier with taxon A_j is activated. Consider a schema ξ having instances in B. (A taxon A_j is an instance of a schema ξ if it belongs to the subset of A named by ξ, i.e., if it has the particular attributes named by ξ). Let μ_ξ be the average of the predicted performances of the instances of ξ. (Formally, $\mu_\xi = [\Sigma_{jA_j\epsilon\xi}\mu_j]/[\text{number of } A_j\epsilon\xi]$). It can be proved that each time the genetic algorithm is applied to B, about $M^2 \times 2^{\beta/2}$ schemata will have the number of their instances in B changed by amount proportional to μ_ξ. The exact statement is

$$M_\xi(t+1) = (\mu_\xi/\mu) \times (1\text{-}\epsilon) \times M_\xi(t)$$

where $M_\xi(t)$ is the number of instances of ξ in B at time t, μ is the average predicted performance of all the taxa in B, and $\epsilon << 1$ for the $M^2 \times 2^{\beta/2}$ schemata of interest. (See Lemma 7.2 of Holland [Holland75].)

(2) Under the selective pressure on schemata described in (1), the number of instances in B of a given schema comes to reflect its usefulness in generating new taxa. That is, the cumulative effect of the genetic algorithm is the *ranking* (and use) of a great many schemata. The *rank* of a schema ξ is simply the number M_ξ of taxa in B at time t that are instances of ξ. This rank reflects the system's experience with ξ prior to time t; in general, the higher the rank, the more useful the schema in generating new taxa, the higher the estimated value of that particular combination of 1s, 0s, and #s. Recall that each *taxon* is an instance of 2^β distinct schemata; here, several taxa serve to determine the rank of a given schema. It follows that a given taxon can participate in the ranking (count) of many schemata. Thus the rank of a great many schemata can be stored with the help of a few dozen selected taxa. The net result is the compact storage of a great deal of information gleaned from experience. The genetic algorithm automatically accesses and uses this information to generate new taxa. The higher the rank of a schema, the more frequently the corresponding combination of 0s, 1s, and #s is used.

(3) Because the set of schemata is such a rich cover of the space of possibilities A, almost any interactions (or correlations) between attributes (detectors) will be "discovered" by some schema. Combinations of attributes that correspond to useful regularities in the environment will give rise to corresponding high-ranking schemata in B. Because of the way in which the genetic algorithm processes schemata, high dimensionality and complex interactions pose no difficulties; the genetic algorithm actually exploits them to make an efficient search of A. (In another context, genetic algorithms have been used to find the global optimum of high-dimensional, nonlinear functions. See

DeJong [DeJong75]. Hayes-Roth's [Burge76] SLIM makes a similar use of schemata, but without the intrinsic parallelism.)

These properties, taken together, augur well for genetic algorithms as learning algorithms. By using the predicted performances $\{\mu_j\}$ the algorithm directly incorporates the two principles discussed earlier. A taxon that is too general cannot yield consistent performances or predictions. Hence it will be associated with a low predicted performance μ. Similarly a taxon that is associated with an invalid prediction will have its μ lowered (see the next section for details). In both cases, the genetic algorithm selects against such taxa (relative to taxa at more appropriate levels of generality and prediction). At the same time, the algorithm displaces too-specific taxa (because they "age" more rapidly—see the next section) to make room for newly generated taxa. Only taxa that are above average in their predictions and levels of generality will be favored. More importantly, the schemata they instance will be favored in the generation of new taxa. The resulting speed-up, a consequence of the algorithm's intrinsic parallelism, provides rapid and flexible learning.

2.0 DESCRIPTION OF COGNITIVE SYSTEM LEVEL 1 (CS-1)

CS-1 consists of a general memory store containing classifiers and three important processes that act on information in memory. The first process directs CS-1's performance by coding the situation and finding in memory actions that are appropriate to both the specifics of the situation and CS-1's goals. The second process is a form of simple learning; after a series of actions, it stores in memory information about the consequences of these actions. The third process is a more complex learning process that changes the contents of memory to allow good productions (classifiers) to endure, bad ones to be deleted, and new, potentially good productions to be created for subsequent trial. The novelty of the model is not so much in the performance or simple learning processes, but rather in the process that changes memory.

Structure. Figure 1 illustrates the structural parts of the system. Included are a *detector array* encoding the current environmental and internal states, *resource reservoirs* keeping track of the needs and goals of the system, a set of *classifiers* constituting the procedural content of memory, a *message list* keeping track of some of the system's most recent internal state, and an *effector array* specifying the action to be taken.

For CS-1 the detector array consists of a set of individual sensors, each of which is either turned on (1) or off (0) at each time-step by the

John H. Holland and Judith S. Reitman

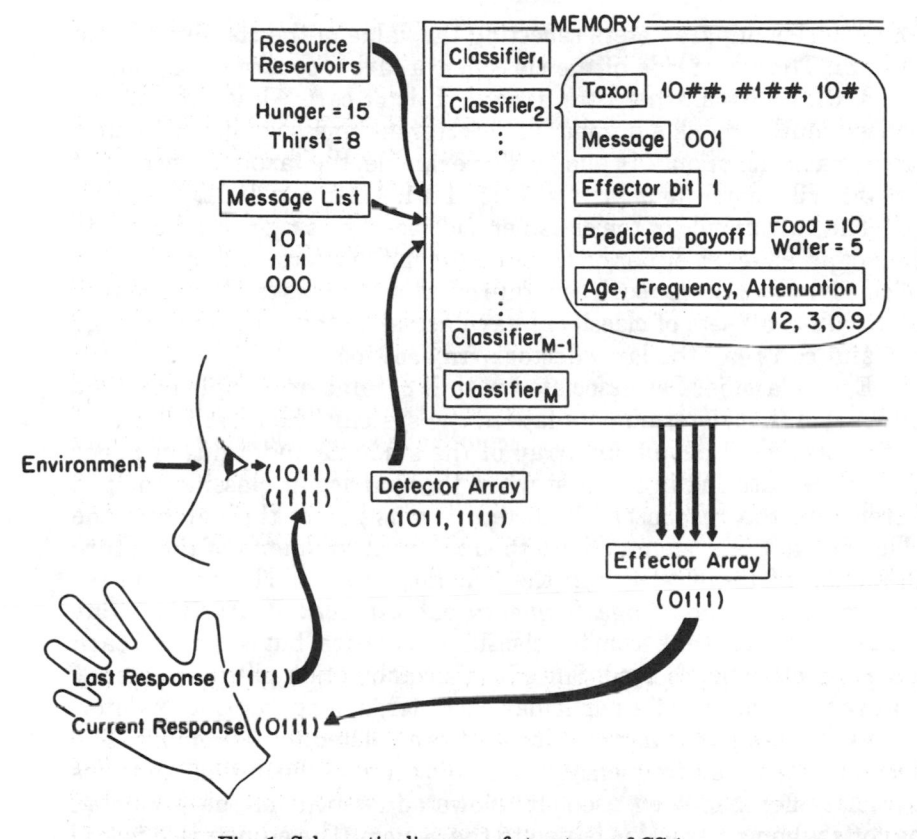

Fig. 1. Schematic diagram of structure of CS-1.

overall situation obtaining at that time. There are two sets of sensors, one set triggered by the external environment (such as retinal detectors or higher-order feature detectors), and one set by the state of memory and the effectors in the last response. Each classifier used in producing an actual response broadcasts a signal or message, also an array of 1s and 0s, stored on the message list. In the next time-step responses are chosen on the basis of both the message list and detector array, allowing the system to make an action conditional on past "thoughts" and actions. The resource reservoirs reflect simple biological needs, such as thirst and hunger, assumed in CS-1 to deplete regularly in time. Cognitive goals could be incorporated easily as needs that can be filled without introduction of externally presented rewards and that do not necessarily deplete continuously in time.

The memory of the system consists of a fixed large number of classifiers. Each classifier has a part that is sensitive to system input and messages, a signal that is emitted when the classifier is activated, and a series of parameters reflecting the value of the classifier to the system. The input side of the classifier, a taxon, consists of an array of 1s, 0s, and #s, where #s denote "I don't care." Each taxon will consequently match a number of detector and message list configurations, patterns of only 1s and 0s. For example, the taxon consisting of 1##1 will match four arrays, 1011, 1101, 1111, and 1001.

The output side of the classifier specifies a message and the condition of an effector, either on (1) or off (0), in the effector array. One set of classifiers exists for each separate effector. Actions can be coordinated because all sets of classifiers have access to the common message list and can read the last effector array setting.

Each classifier has associated with it parameters (single numbers of arrays) that determine its fate in the system. The most important is the predicted payoff, an array of the amounts and kinds of needs that were satisfied in the past when that particular classifier helped determine the responses. Predicted payoffs reflect the value of the classifier to the system and are the major determinants of the future influence of the classifier in the learning process. The three lesser parameters are called *age, frequency*, and *attenuation*. Age is a value that increases in time from the classifier's creation, but is reduced each time the classifier is used. Since it is a combination of the frequency and recency of use of a particular classifier, age appropriately determines when a particular classifier is to be deleted from memory. The second parameter, frequency, is a simple count of the number of times the classifier has been used. It allows adjustment of the predicted payoffs when a reward is issued to the system. The adjustment is less if the classifier has had heavier use and its predicted payoff can consequently be assumed to be more reliable. The third parameter, attenuation, figures in the adjustment of the predicted payoff when a reward is issued to the system, allowing appropriate allocation of this single reward value of the various classifiers that led to the reward. Between rewards, attenuation is incremented at every time t wherein an activated classifier has a successor (at time t+1) with a lower predicted payoff. This increment in attenuation is assigned to all classifiers activated between the last reward and time t+1. This gives early overpredicting classifiers less credit (an attenuated reward) than an accurately predicting one. This parameter is highly correlated with the delay between a response and the reward. On average, the classifier that is one step away from the reward will be less attenuated than one ten steps away; during the nine intervening steps, the early classifier is more likely to experience *attenuation*.

The performance process. The system performs in two cycles, one embedded in the other. In the smaller cycle, called a *stimulus-response cycle*, the input changes and a response (external or internal) is constructed. Many stimulus-response cycles can occur within each *epoch*, a larger cycle that is bounded by the issuance of a reward.

In the stimulus-response cycle, the set of classifiers is searched for those matching the particular current detector array and message list. Because many classifiers match each array (e.g., *taxa* 1###, 10##, 101#, and 1011 match array 1011), a competition is held for the one classifier whose taxon fits the specific details of the overall array (input and messages) *and* whose predicted payoff fits the current needs and goals of the system. Within each effector set, only classifiers having no direct mismatches with the detector array are considered for competition (e.g., taxa #11##, 01###, and ####1 will all fit *detector array* 01101, but *taxon* 11110# will not). A score is calculated for each matching classifier: the sum of the number of features specified in the taxon (giving weight to the more specific taxon) multiplied by the amount of the current needs fulfilled by this classifier's predicted payoff. Only the ten highest-scoring classifiers are assigned a probability of being activated; the probability is directly proportional to the classifier's relative advantage over the others in the high-scoring subset. The winner is then activated (turned on), its output condition is specified in the effector array, its message is left on the message list, the classifier's age is reduced by half, its frequency count is increased, and the attenuation parameters of other recently activated classifiers are adjusted. Within each stimulus-response cycle, many sets of classifiers hold simultaneous competitions, each set's winner producing an independent instruction for the effector array and a message for the common message list.

Briefly then, the system's performance is dependent in an interactive way on the input information (the detector array), an interpretation of that information dependent on context (the classifiers), recent actions and thoughts (the message list), the needs or goals (resource reservoir levels), and past long-term experience (the predicted payoffs).

The simple learning process. When a reward enters the system to replenish the depleting reservoirs, the epoch ends. All the predicted payoffs of the currently activated classifiers are then modified to reflect their accuracy in anticipating this reward. Those predicted payoffs that were consistent with (not greater than) this reward are maintained or increased; those that overpredicted are significantly reduced, all according to their attenuations. When the predicted payoffs of all the activated classifiers have been adjusted, the classifiers are inactivated and their attenuation parameters are returned to zero.

The adaptation of the contents of memory. The heart of the system's learning arises in continuous, autonomous changes occurring in the classifiers. The specific algorithm is a tailored genetic algorithm.

Genetic algorithms are defined and investigated in a general context in Holland [Holland75] and in a behavioral context in Holland [Holland76]. The algorithm combines variants of genetic operators (such as cross-over) with a "survival of the fittest" principle, to discover and exploit above-average feature combinations. At regular intervals, the contents of memory (the classifiers) are changed. The "survival of the fittest" aspect of the genetic algorithm is reflected in the fact that those classifiers that have performed better than average (have above-average predicted payoffs) are more likely to produce variants for trial in the next generation of memory. The variants in the next generation are produced by genetic operations such as a "crossing-over" applied to the taxa associated with good classifiers.

Specifically, the genetic algorithm proceeds as follows. From each set of classifiers controlling a single effector, two parent classifiers are chosen. Each candidate has a probability of being chosen proportional to its predicted payoff. The taxa and the signals (messages) of the two parent classifiers are crossed, respectively, at a random point, creating new taxa and signals. For example, if one "parent" taxon is #1111, the other is 00###, and the cross-over point is after the second position, the "children" *taxa* are #1### and 00111. One of the offspring is selected at random, and an entirely new classifier built around it. Its predicted payoff and effector setting are those of one of its parents, its age is an average of its parents' ages, and its frequency and attenuation parameters are set to zero.

Creation of a new classifier in a memory of fixed size requires that one be deleted; ideally, this should be a poor one. Recall that a classifier with a poor predicted payoff rarely wins competitions; without a win, its age increases steadily. Age therefore, reflects the classifier's quality as well as its frequency of use. To make room for the new classifier therefore, one with an old age is deleted. In particular, from the set of oldest classifiers, the one chosen for deletion is the one that most resembles the new classifier, resemblance being a simple count of the number of feature matches, a 1 for a 1, a # for a #.

This process, the selection of parent classifiers on the basis of good predicted payoffs, the creation of a new classifier, and the deletion of an old but similar classifier, continually revises memory. Useful productions that work well with others in memory persist; useless ones are replaced by new "trial" productions.

3.0 CS-1 IN OPERATION

CS-1 has been programmed in FORTRAN-IV on an IBM-1800 with 32K memory by Ted Wright (now at Bell Labs, Murray Hill, New Jersey) and Leslie Forman (a graduate student in Computer Science at The University of Michigan). All components of the proposed system are included in this version. Its simplicity comes from the following restrictions:

(1) Each taxon has 25 features, 8 bits in the part that attends to the environment, 1 bit for the last effector setting, and 16 bits that attend to the internal signals encoded as messages on the message list.

(2) One effector is included with two possible settings, 0 and 1.

(3) Two needs are included and each grows one unit per time-step.

(4) The environment consists of a set of positions or nodes in a graph, each node defined by a simple 8-bit array.

(5) The memory consists of 100 classifiers: 50 directing the effector to be set to 1, 50 to 0.

The initial set of classifiers are built at random such that the probability that a specific taxon bit is on or off is 0.10. (This amounts to initializing with a fair amount of generality.) The initial predicted payoffs are all equally low (no biases built in).

Figure 2a illustrates the environment of the initial test, a one-dimensional array of nodes with a reward for one of the needs at one end and a reward for the other need at the other. Each position is labeled by an eight-tuple of 1s and 0s; the system can move left or right, depending on the effector setting chosen. At the beginning of each trial, the system is assumed to be in the middle, position 7. By attending to different features of the positions as well as keeping track of messages and the last effector settings, the system should learn to move to the two rewards, alternating in proportion to the amount of reward received and the depletion rate of its two reservoirs.

In this environment, the expected time to reach a reward on the first trial can be calculated from a simple random-walk model. For this environment, with choices of moves made randomly, the system should reach the goal in nine time-steps. Figure 3 shows the number of time-steps required by the system (an average of four runs of the system, each with a different random number seed). The rate of traversal to the reward quickly becomes far better than random, and improves regularly until CS-1 moves to the rewards in the correct two-to-one proportion, each trip requiring the minimum three steps.

To illustrate the power of the genetic algorithm, the dashed line in Fig. 3 shows the performance of the system *without* the genetic algorithm operating. This performance is dependent on updating the predicted payoffs after each reward is received, but is lacking experience with newly created and tested classifiers. Without the genetic algorithm, performance is better than random (as might be expected from the success of algorithms that change weights of features) but clearly not as good as that with the genetic algorithm operating. With-

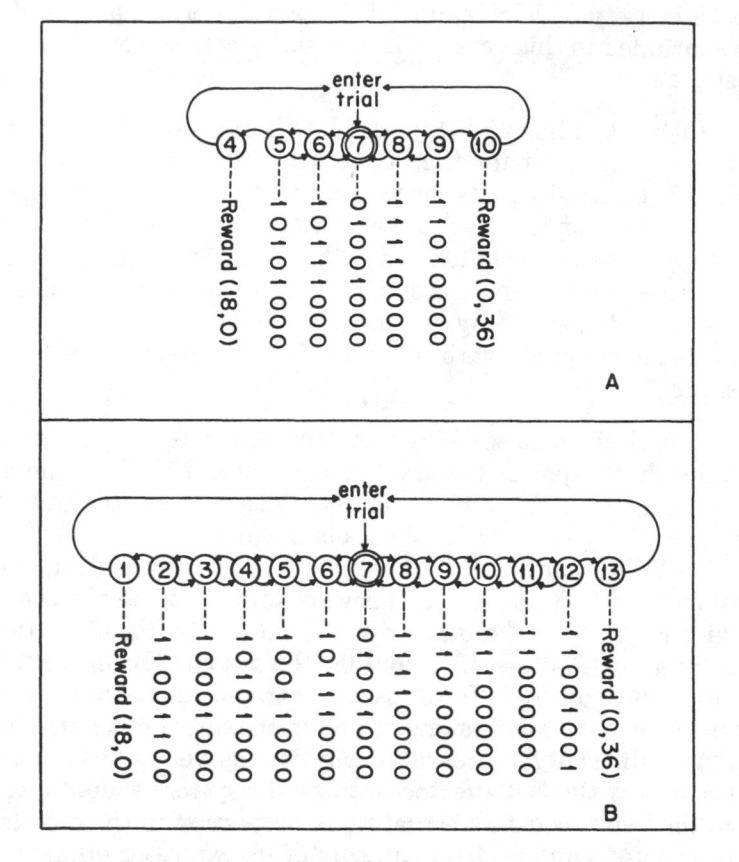

Fig. 2. Initial test (a) and transfer (b) environments for CS-1.

out the genetic algorithm, the system only reaches criterion (10 successive, three-step trials in the two-to-one proportion of alternating goal attainment) in 2161 time steps; with the learning algorithm, it reaches the criterion in 212 steps.

To demonstrate CS-1's ability to benefit from past experience, we tested it next in a simple transfer situation. First, the system experienced the environment in Fig. 2a to the criterion above. Then with memory intact, the system experienced the transfer environment illustrated in Fig. 2b, a larger one-dimensional array of nodes in which the first environment is embedded as the center seven nodes. The reward values remained the same but were positioned at the ends of this longer array. Random choices of moves in this environment result in an average of 36 time steps to reach a reward. Since the amount of the largest reward is 36 units and the reservoirs are assumed to deplete one unit per time-step, the system must do better than chance or perish!

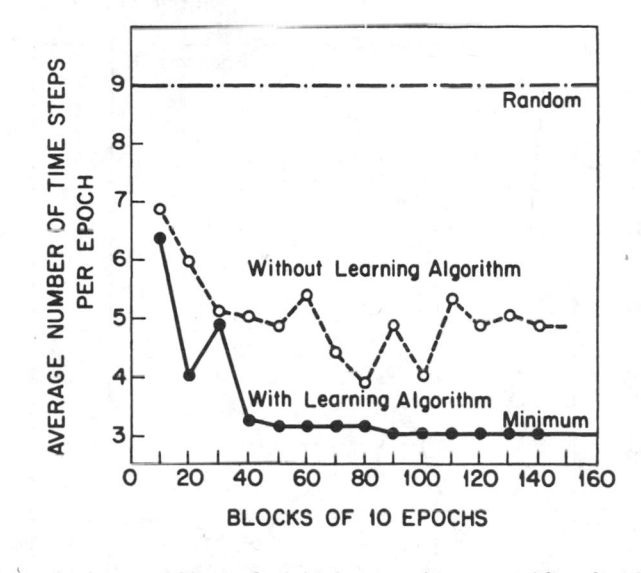

Fig. 3. Performance of CS-1 in the initial test environment with and without learning algorithm operating.

Figure 4 shows the performance of the experienced system compared to a naive system (traversing the same larger environment without prior experience). Having had even this limited experience with the seven center positions gave the system a 1865 time-step advantage over the naive system. To underscore the power of the genetic algorithm, we also tested the experienced system in the 13 position environment without the genetic algorithm. Even with the "tuned" classifiers provided by the prior learning, it failed to reach criterion in 10,000 time-steps.

4.0 EXTENSIONS

Cognitive maps. Our experiments to this point are of the "proof of principle" type. Consequently the environments used have only required the simplest cognitive maps (i.e., internal models of the environment). Nevertheless, extensive, learned cognitive maps are a major objective of these studies. Cognitive maps allow the system to use lookahead to explore, without overt acts, the consequences of various courses of action. Thus, procedures that use experience to generate good cognitive maps are of critical importance.

To see how the genetic algorithm generates coherent cognitive maps, attention must be focused on the "successful" (general and predictive) classifiers. The taxa of such classifiers generally achieve

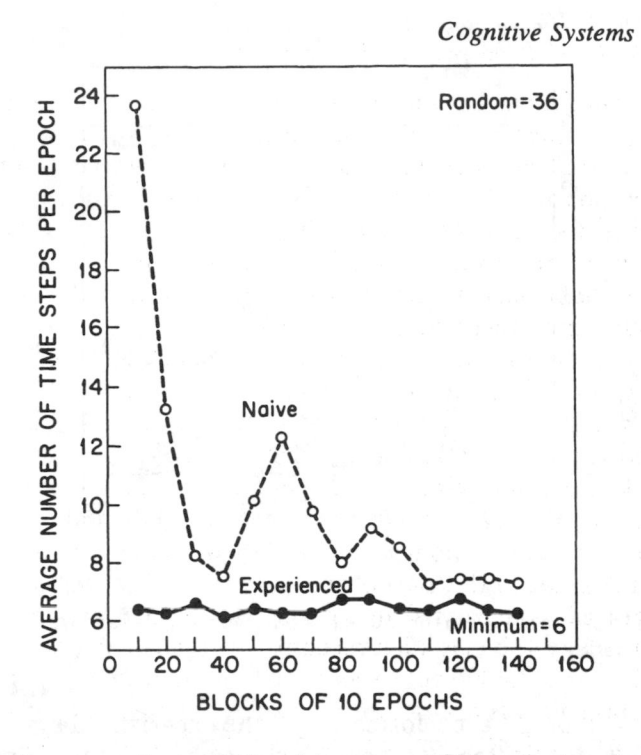

Fig. 4. Performances of naive CS-1 and experienced CS-1 in the transfer environment.

above-average match scores by attending to messages in the message list as well as environmental signals. (Recall that the messages are signals broadcast by the classifiers activated on the previous time-step.) To see this, consider two classifiers, one attending only to attributes of the environmental signals and another attending to these same attributes as well as to some correlated attributes of the message list. Clearly the latter classifier will have a higher match score than the former and will thereby tend to win out in competitions for activation. Thus the classifiers activated at time t-1 will, by the messages they place on the message list, increasingly influence which classifiers are activated at time t. Now consider a mode of operation where, at some point in time, the classifiers cease attending to environmental signals. At that point the internal signals, via the message list, would directly determine the set of classifiers activated next. These classifiers in turn would broadcast new signals activating a new set of classifiers, etc. This linking of classifiers via the message list (with environmental signals ignored) constitutes the system's cognitive map, as shown below. It is important that the cognitive map is produced automatically with no changes in the genetic algorithm.

Each classifier can be considered to name an environmental condition or pattern that has been significant in the system's experience. Each signal emitted is correlated with, and in fact represents, the system's response to that pattern. Overall, then, classifier-signal sequences constitute a kind of model of the environment; by defining the environment more carefully, this idea can be made more precise.

Assume that each response of the cognitive system alters the state of the environment according to a well-defined set of rules or laws. Technically, then, the environment can be defined (or approximated) by an automaton. The states and transition rules of the environment are initially unknown to the cognitive system; its task is to use experience to construct a model of them. Each state s of the environment gives rise to a set of environmental signals that elicit (via detectors, classifiers, etc.) a reponse from the cognitive system. The response produces a new environmental state s'. The system has constructed a good cognitive map of this part of the environment when the classifier that recognizes (is activated by) s' also has a good match with the signal emitted by the classifier recognizing s. Then the activation of the classifier for s is sufficient to activate the classifier for s' even if the environmental signals are ignored. In effect the cognitive system has enough internal information to predict the pattern (elicited by s') that will occur after the system makes its response to s. (Recall that the classifier activated by s determines the response to s.) Clearly this kind of linkage extends with no change to a whole chain s', s'', s''', . . .

The genetic algorithm, by favoring high match scores, enforces attention to internal signals. Thus each classifier will be sensitive to the signals originating from its immediate predecessors. At the same time it must be sensitive to the *current* environmental signals. This enforced combination automatically generates the cognitive map. Technically, the transition graph of the environment has been represented by the linked classifiers. Nodes of the graph, i.e., environmental states, are represented by taxa; edges, i.e., actions taken by the cognitive system, are represented by signals.

Once the system has enough experience with a part of the environment to have generated the relevant classifiers, lookahead is easily achieved. Given a classifier (or set of them) γ activated by the current environmental state s, the system simply traces out a sequence of classifiers γ', γ'', . . ., $\gamma^{(n)}$ by using the message list alone. $\gamma^{(n)}$ amounts to the system's prediction of the condition or environmental pattern that will result from the sequence of reponses elicited by γ', γ'', . . ., $\gamma^{(n-1)}$.

Using expert knowledge. There is no reason to require the cognitive system to start with a "clean slate," if the designer has substantial knowledge of the task environment(s). It is easy to supply the system with an initial cognitive map, incorporating the designer's knowledge and conjectures. This requires writing an initial program for the system in the (computationally complete) "classifier language." The system itself corrects errors, misconceptions, etc., as experience accumulates. In addition, minor changes in the system's rules allow it to accept advice from experts unfamiliar with the cognitive system. The procedure is much like Samuel's [Samuel63] book-move technique.

One of us (Holland) has already produced the relevant theory (it was actually possible to prove some theorems!) and algorithms for these extensions (lookahead and advice-taking). The next year should see a substantial body of experiments along these lines.

Further extensions. There are also extensive sketches of algorithms (with supporting theorems) that allow the system to

(1) produce hierarchies of classifiers wherein a "higher-level" classifier attends to activation of "lower-level" classifiers, yielding hierarchical generalizations and searches employing analogy;

(2) develop and use key classifiers, "landmarks," in its cognitive map, so that lookahead can skip from landmark to landmark, thus permitting lookahead depth to depend on the detail required;

(3) fill in lacunae in its cognitive map by searching out unfamiliar parts of its environment when its "resource" needs are not pressing, thus providing a kind of "curiosity";

(4) develop a simple symbolic "language" (in the manner of the man-made chimpanzee languages), permitting one to induce associations, or pieces of the cognitive map, by presenting appropriate concatenations of abstract patterns ("symbols").

Each of these additions can be achieved parsimonously. They use the same basic mechanism to activate different sequences of classifiers via different sets of attributes. Exploring the effects of these additions, and discovering their limitations, should occupy a good deal of our time, and our students', over the next several years.

Chapter 17
Evolving Neural Networks

R. R. Kampfner and M. Conrad (1983) "Computational Modeling of Evolutionary Learning Processes in the Brain," *Bulletin of Mathematical Biology,* Vol. 45:6, pp. 931–968.

ARTIFICIAL neural networks (or simply neural networks) are computer algorithms loosely based on modeling the neuronal structure of natural organisms. They are stimulus-response transfer functions that accept some input and yield some output, and are typically used to generalize an input-output mapping over a set of examples. Among the first such neural network designs was the *perceptron* (Rosenblatt 1957, 1958, 1960, 1962). A perceptron (Figure 17.1) consists of three types of units: (1) sensory, (2) associator, and (3) response. A stimulus will activate some sensory units. These sensory units in turn activate, with varying time delays and connection strengths, the associator units. These activations may be positive (excitatory) or negative (inhibitory). If the weighted sum of the activations at an associator unit exceeds a given threshold, the associator unit activates and sends a pulse, again weighted by a connection strength, onto the response units. There is an obvious analogous behavior of units and neurons, of connections and axons and dendrites. The characteristics of the stimulus-response (input-output) of the network define its behavior.

Earlier work by Hebb (1949) offered that neural networks could learn to recognize patterns by weakening and strengthening the connections between neurons. Rosenblatt (1957) and others (e.g., Keller, 1961; Block et al., 1962) studied the effects of changing the connection strengths in a perceptron by various rules. Block (1962) indicated that when the perceptron was employed on some simple pattern recognition problems, the behavior of the machine degraded gradually with the removal of associator units (i.e., the perceptrons were robust, not brittle). Rosenblatt (1962, p. 28) admitted that his perceptrons were "extreme simplifications of a central nervous system . . . " but noted that a strength of the approach was the ability to analyze the model.

The task of analysis was taken up by Minsky and Papert (1969). In particular, the computational limits of perceptrons with a single layer of modifiable connections were explicated. Minsky and Papert (1969) demonstrated that such devices were not able to calculate mathematical functions such as parity or the topological function of connectedness without an absurdly large number of predicates (Rumelhart and McClelland, 1986, p. 111).

But these limitations did not apply to networks of perceptrons consisting of multiple layers,[1] nor did their analysis address networks with feedback connections. Nevertheless, Minsky and Papert (1969, pp. 231–232) speculated that the study of multi-layered perceptrons would be "sterile" in the absence of an algorithm to usefully adjust the connections of such architectures. The chilling effect of Minsky and Papert (1969) on research in neural networks from 1970–1985 is almost a legend in the artificial intelligence community. Only a few researchers (e.g., Grossberg, 1976; Amari, 1971, Widrow et al., 1973; Kohonen, 1984; and others) continued focused investigations on neural networks during this period of time.

It is now well known that multiple layers of perceptrons with variable connection strengths, bias terms, and nonlinear sigmoid functions can approximate arbitrary measurable mappings (in fact, universal approximators can be constructed with a single hidden layer of "squashing" units, [Cybenko, 1989; Hornik et al., 1989; Barron, 1993]; also see Poggio and Girosi, 1990, for similar arguments covering radial basis function networks). Several algorithms, such as *back propagation,* have been

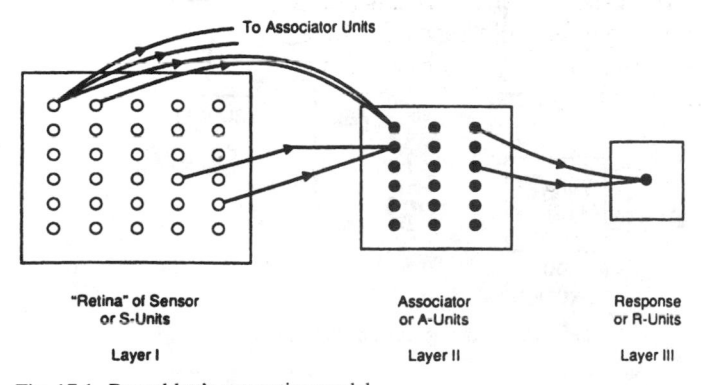

"Retina" of Sensor   Associator   Response
or S-Units   or A-Units   or R-Units

Layer I   Layer II   Layer III

Fig. 17.1. Rosenblatt's perception model.

[1] Block (1970), in a reply to Minsky and Papert (1969), wrote "Work on the four-layer Perceptrons has been difficult, but the results suggest that such systems may be rich in behavioral possibilities."

developed for training such networks[2] to generate mappings over a set of previously classified patterns (e.g., Werbos, 1974; Rumelhart and McClelland, 1986). The general procedure is to view training as a search over an n-dimensional parameter space of weights in light of an error function (i.e., response surface) and rely on calculating the gradient of this surface to guide the search. Unfortunately for the practitioner, this procedure is dependent on initial conditions and can only guarantee convergence to locally optimal weights, which may not be sufficient.

Adding more degrees of freedom to the network will eventually allow back propagation to demonstrate adequate performance on the training set, provided sufficient nodes and/or layers are available. Yet this too presents problems to the designer of the network, for a function can be constructed to map any measurable domain to its corresponding range if given sufficient degrees of freedom. Lamentably, such overfit functions generally provide very poor performance during validation on independently acquired data. Such anomalies are commonly encountered in regression analysis, statistical model building, and system identification. Assessing the proper trade-off between the goodness-of-fit to the data and the required degrees of freedom requires information criteria (e.g., Akaike's information criterion [Akaike, 1974], minimum description length principle [Rissanen, 1978], predicted squared error [Barron, 1984], or others). By relying on the back propagation method, the designer almost inevitably accepts that the resulting network will not satisfy the maxim of parsimony, simply because of the limited nature of the training procedure itself. The problems of local convergence with the back propagation algorithm indicate the desirability of training with stochastic optimization methods such as simulated evolution, which can provide convergence to globally optimal solutions (Fogel, 1992; Bäck and Schwefel, 1993; Rudolph, 1994; and others), as well as permit nondifferentiable objective functions.

Although the use of evolutionary algorithms to train and design neural networks of arbitrary configurations is now commonplace, the first insights into evolving such networks date back to some of the earliest efforts in evolutionary computation. Friedman (1956) (Chapter 2) proposed, but did not implement, an evolutionary search for a suitable connection of small functional elements that when networked appropriately could control a robot in an environment. In 1966, Hans Bremermann, who had been studying evolutionary algorithms since at least 1958 (Chapter 9) and had collaborated with Frank Rosenblatt at Cornell University in the early 1960s, was heartened by results of simulated evolution on linear equations and wrote "If [an evolutionary] search procedure can beat algorithms on such a classical and simple problem as linear equations, then we should be encouraged to try procedures on more complex problems, where no efficient algorithms are known (e.g., searching for strategies,

optimizing 'weights' in a multilayer neural net, etc.)" (Bremermann, 1968).[3]

Mucciardi and Gose (1966) suggested an "evolutionary" technique that operated on a single network and varied the weights of a network of threshold elements. The application was alphabetic character recognition, where the examples were put in six separate groups. Sixty-four weights were allowed. At the end of ten training cycles, the 32 low-weighted terms were discarded and replaced with zeroed values and subsequently retrained, but Mucciardi and Gose (1966) wrote explicitly of the possibility of randomly selecting new weight values. The results indicated significantly better performance than training a single static set of 64 weights both in training and generalization tests. Klopf and Gose (1969), following Klopf (1965), extended this work by applying different criteria for choosing which functions to replace in a single hidden layer system (Figure 17.2). The criteria involved size of weights associated with the functions, size of the products of the weights with the functions' outputs, and size of the cross correlation between functions' outputs and the desired output vector. Klopf and Gose (1969) also experimented with different replacement schedules (essentially a step size for search) where more functions were replaced early in the iterative procedure and fewer were replaced later. Gose (1969) provided a broader review of biological and automatic pattern recognition, including references in evolutionary algorithms.

The dearth of neural network research from 1970 to 1985 by consequence negatively impacted investigations into evolving such networks. An exception was Barron (1971), who suggested that guided random search could be used to evolve polynomial networks (following Brooks, 1958; Rastrigin, 1963; and others).

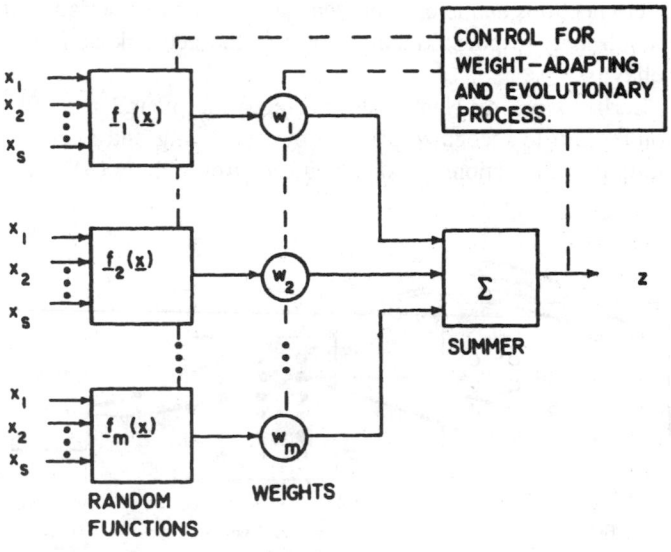

Fig. 17.2. The neural model of Klopf and Gose (1969).

---

[2]There is a diverse variety of neural architectures and various training algorithms. The discussion is limited to multilayer perceptrons only for the sake of argument. Readers interested in a more detailed study of neural networks should see Haykin (1994), Anderson (1995), and others.

[3]The citation of 1968 is correct. The symposium was held in 1966; however, the proceedings were published two years later.

Another notable exception was the effort of Conrad and colleagues (Conrad, 1974; Conrad, 1981; Kampfer, 1981; Kampfer and Conrad, 1983; Kirby and Conrad, 1986; Conrad et al., 1989; and others). Conrad (1974) proposed that the central nervous system comprises a variety of unit regions for which there are interchangeable replicas. Each region consists of neurons with a firing rate that is determined by excitase enzymes that recognize specific input patterns. These neurons are coded by genes which are heritable or culturable. Individual regions are tested and evaluated via selection circuits that control the production of culturable excitase genes on the basis of this evaluation. Thus, trial and error learning similar to natural evolution can be accomplished within the central nervous system (also see Edelman, 1987). Kampfner and Conrad (1983) (reprinted here) detailed the experimental behavior of the procedures (also see Kampfner, 1981) and suggested that "evolutionary mechanisms should be seriously considered as candidate mechanisms for neuronal adaptation." The procedures were later extended in Kirby and Conrad (1986) and Conrad et al. (1988) for a simple robot navigation task.

Interest in evolving neural networks followed increasing interest in evolutionary algorithms in the mid- to late-1980s. Ackley (1985) provided an early attempt to use evolutionary principles to design a connectionist system for optimization based on a metaphor to voting in a democracy, but a more canonical approach for optimizing weights in a network was adopted by Dress (1987; Dress and Knisley, 1987; Dress, 1989), where a single-parent, single-offspring evolutionary search was implemented based on mutating a single weight at a time (Bergman and Kerszberg, 1987, implemented a similar idea using a population). Similar efforts were made using genetic algorithms on binary-encoded weights in Caudell and Dolan (1989) and Whitley and Hanson (1989). Dolan and Dyer (1987a, 1987b) suggested an evolutionary system for optimizing neural architectures by encoding the structure in a set of production rules that were essentially a blueprint for design. Miller and Todd (1989) and Harp et al. (1989) also evolved the connectivity of neural networks, which were then subsequently trained using back propagation.

Montana and Davis (1989) offered an important departure from traditional practice in genetic algorithms by evolving weights encoded in real values, rather than binary strings, with specific genetic operators that were tailored to the application. Fogel et al. (1990) offered a similar approach using multivariate Gaussian mutations applied to all the weights of a neural network, and extended this to evolve both the weights and structure (i.e., number of hidden nodes) of a network simultaneously in Fogel (1992) (also see Collins and Jefferson, 1991; Hancock and Smith, 1991; Austin, 1992). The literature on evolving neural networks has become quite vast from 1992 to 1997, including the training and design of recurrent networks (particularly the choice of suitable variation operators, e.g., Angeline et al., 1994, following Wieland, 1991, and others), the evolution of encoding rules for network generation and development (e.g., Gruau, 1993, following Kitano, 1990, and others), input feature subset selection (e.g., Brotherton and Simpson, 1995), and so forth. Early reviews were presented in Schaffer et al. (1992) and Yao (1993).

References

[1] D. H. Ackley (1985) "A connectionist algorithm for genetic search," *Proc. of an Intern. Conf. on Genetic Algorithms and Their Applications,* J. J. Grefenstette (ed.), Lawrence Erlbaum, Hillsdale, NJ, pp. 121–135.

[2] H. Akaike (1974) "A new look at the statistical model identification," *IEEE Trans. Auto. Control,* Vol. 19:6, pp. 716–723.

[3] S.-I. Amari (1971) "Characteristics of randomly connected threshold element networks and network systems," *Proc. IEEE,* Vol. 59:1, pp. 35–47.

[4] J. A. Anderson (1995) *An Introduction to Neural Networks,* MIT Press, Cambridge, MA.

[5] P. J. Angeline, G. M. Saunders, and J. B. Pollack (1994) "An evolutionary algorithm that constructs recurrent neural networks," *IEEE Trans. Neural Networks,* Vol. 5:1, pp. 54–65.

[6] A. S. Austin (1992) "Structural level evolution of neural networks," *Proc. of the 1st Ann. Conf. on Evolutionary Programming,* D. B. Fogel and W. Atmar (eds.), Evolutionary Programming Society, La Jolla, CA, pp. 80–89.

[7] T. Bäck and H.-P. Schwefel (1993) "An overview of evolutionary algorithms for parameter optimization," *Evolutionary Computation,* Vol. 1:1, pp. 1–23.

[8] A. R. Barron (1984) "Predicted squared error: A criterion of automatic model selection," *Self-Organizing Methods in Modeling,* S. J. Farlow (ed.), Marcel Dekker, NY.

[9] A. R. Barron (1993) "Universal approximation bounds for superpositions of a sigmoidal function," *IEEE Trans. Info. Theory,* Vol. 39:3, pp. 930–945.

[10] R. L. Barron (1971) "Adaptive transformation networks for modeling, prediction, and control," *1971 IEEE Systems, Man, and Cybernetics Ann. Symp. Record,* G. L. Hollander (chair), IEEE, Anaheim, CA, pp. 254–263.

[11] A. Bergman and M. Kerszberg (1987) "Breeding intelligent automata," *Proc. of the 1st IEEE Intern. Conf. on Neural Networks,* Vol. II, IEEE Press, NY, pp. 63–69.

[12] H. D. Block (1962) "The perceptron: A model for brain functioning," *Rev. Mod. Phys.,* Vol. 34, pp. 123–125.

[13] H. D. Block (1970) "A review of 'Perceptrons: An Introduction to Computational Geometry,'" *Information and Control,* Vol. 17, No. 5, pp. 501–522.

[14] H. D. Block, B. W. Knight, and F. Rosenblatt (1962) "Analysis of a four-layer series coupled perceptron," *Rev. Mod. Phys.,* Vol. 34, pp. 135–142.

[15] H. J. Bremermann (1968) "Numerical optimization procedures derived from biological evolution processes," *Cybernetic Problems in Bionics,* H. L. Oestreicher and D. R. Moore (ed.), Gordon and Breach, NY, pp. 597–616.

[16] S. H. Brooks (1958) "A discussion of random methods for seeking maxima," *Operations Research,* Vol. 6, pp. 244–251.

[17] T. W. Brotherton and P. K. Simpson (1995) "Dynamic feature set training of neural nets for classification," *Evolutionary Programming IV: Proc. of the 4th Ann. Conf. on Evolutionary Programming,* J. R. McDonnell, R. G. Reynolds, and D. B. Fogel (eds.), MIT Press, Cambridge, MA, pp. 83–94.

[18] T. P. Caudell and C. P. Dolan (1989) "Parametric connectivity: Training of constrained networks using genetic algorithms," *Proc. of the 3rd Intern. Conf. on Genetic Algorithms,* J. D. Schaffer (ed.), Morgan Kaufmann, San Mateo, CA, pp. 370–374.

[19] R. J. Collins and D. R. Jefferson (1991) "An artificial neural network representation for artificial organisms," *Parallel Problem Solving from Nature,* H.-P. Schwefel and R. Männer (eds.), Springer, Berlin, pp. 259–263.

[20] M. Conrad (1974) "Evolutionary learning circuits," *J. Theoret. Biol.,* Vol. 46, 167–188.

[21] M. Conrad (1981) "Algorithmic specification as a technique for computing with informal biological models," *BioSystems,* Vol. 13, pp. 303–320.

[22] M. Conrad, R. R. Kampfer, and K. G. Kirby (1988) "Neuronal dynamics and evolutionary learning," *Advances in Cognitive Science: Steps Toward Convergence,* M. Kochen and H. M. Hastings (eds.), AAAS Selected Symposium 104, American Association for the Advancement of Science, NY, pp. 169–189.

[23] M. Conrad, R. R. Kampfer, K. G. Kirby, E. N. Rizki, G. Schleis, R. Smalz, and R. Trenary (1989) "Toward an artificial brain," *BioSystems,* Vol. 23, pp. 175–218.

[24] G. Cybenko (1989) "Approximations by superpositions of a sigmoidal function," *Math. Contr. Signals, Syst.*, Vol. 2, pp. 303–314.

[25] C. P. Dolan and M. G. Dyer (1987a) "Symbolic schemata, role binding, and the evolution of structure in connectionist memories," *Proc. of the 1st IEEE Intern. Conf. on Neural Networks*, Vol. II, IEEE Press, NY, pp. 287–297.

[26] C. P. Dolan and M. G. Dyer (1987b) "Toward the evolution of symbols," *Genetic Algorithms and Their Applications: Proc. of the 2nd Intern. Conf. on Genetic Algorithms*, J. J. Grefenstette (ed.), Lawrence Erlbaum, Hillsdale, NJ, pp. 123–131.

[27] W. B. Dress (1987) "Darwinian optimization of synthetic neural systems," *Proc. of the 1st IEEE Intern. Conf. on Neural Networks*, Vol. III, IEEE Press, NY, pp. 769–775.

[28] W. B. Dress (1989) "Synthetic organisms and self-designing systems," *Telematics and Informatics*, Vol. 6:3/4, pp. 351–363.

[29] W. B. Dress and J. R. Knisley (1987) "A Darwinian approach to artificial neural synthesis," *Proc. of the 1987 Intern. Conf. on Systems, Man, and Cybernetics*, Vol. 2, IEEE Press, NY, pp. 572–577.

[30] G. Edelman (1987) *Neural Darwinism: The Theory of Neuronal Group Selection*, Basic Books, NY.

[31] D. B. Fogel (1992) "Evolving artificial intelligence," Ph.D. diss., UCSD.

[32] D. B. Fogel, L. J. Fogel, and V. W. Porto (1990) "Evolving neural networks," *Biological Cybernetics*, Vol. 63:6, pp. 487–493.

[33] G. J. Friedman (1956) "Selective feedback computers for engineering synthesis and nervous system analogy," master's thesis, UCLA.

[34] E. E. Gose (1969) "Introduction to biological and mechanical pattern recognition," *Methodologies of Pattern Recognition*, Academic Press, NY, pp. 203–252.

[35] S. Grossberg (1976) "Adaptive pattern classification and universal recoding: Part I. Parallel development and coding of neural feature detectors," *Biological Cybernetics*, Vol. 23, pp. 121–134.

[36] F. Gruau (1993) "Genetic synthesis of modular neural networks," *Proc. of the 5th Intern. Conf. on Genetic Algorithms*, S. Forrest (ed.), Morgan Kaufmann, San Mateo, CA, pp. 318–325.

[37] P. J. B. Hancock and L. S. Smith (1991) "GANNET: Genetic design of a neural net for face recognition," *Parallel Problem Solving from Nature*, H.-P. Schwefel and R. Männer (eds.), Springer, Berlin, pp. 292–296.

[38] S. A. Harp, T. Samad, and A. Guha (1989) "Towards the genetic synthesis of neural networks," *Proc. of the 3rd Intern. Conf. on Genetic Algorithms*, J. D. Schaffer (ed.), Morgan Kaufmann, San Mateo, CA, pp. 360–369.

[39] S. Haykin (1994) *Neural Networks*, Macmillan, NY.

[40] D. O. Hebb (1949) *The Organization of Behavior*, John Wiley, NY.

[41] K. Hornik, M. Stinchcombe, and H. White (1989) "Multilayer feedforward neural networks are universal approximators," *Neural Networks*, Vol. 2, pp. 359–366.

[42] R. R. Kampfner (1981) "Computational modeling of evolutionary learning," Ph.D. diss., Univ. of Michigan, Ann Arbor, MI.

[43] R. R. Kampfner and M. Conrad (1983) "Computational modeling of evolutionary learning processes in the brain," *Bulletin of Mathematical Biology*, Vol. 45:6, pp. 931–968.

[44] H. B. Keller (1961) "Finite automata, pattern recognition and perceptrons," *J. Assoc. Comput. Mach.*, Vol. 8, pp. 1–20.

[45] K. G. Kirby and M. Conrad (1986) "Intraneuronal dynamics as a substrate for evolutionary learning," *Physica 22D*, pp. 205–215.

[46] H. Kitano (1990) "Designing neural networks using genetic algorithms with graph generation system," *Complex Systems*, Vol. 4, pp. 461–476.

[47] A. H. Klopf (1965) "Evolutionary pattern recognition systems," Tech. Report on Contract AF-AFOSR-978–65, Univ. Illinois, Chicago, IL.

[48] A. H. Klopf and E. Gose (1969) "An evolutionary pattern recognition network," *IEEE Trans. Systems Science and Cybernetics*, Vol. SSC-5:3, pp. 247–250.

[49] T. Kohonen (1984) *Self-Organization and Associative Memory*, Springer, Berlin.

[50] D. J. Montana and L. Davis (1989) "Training feedforward neural networks using genetic algorithms," *Proc. of the 11th Intern. Joint Conf. on Art. Int.*, Vol. 1, N. S. Sridharan (ed.), Morgan Kaufmann, San Mateo, CA, pp. 762–767.

[51] G. F. Miller and P. M. Todd (1989) "Designing neural networks using genetic algorithms," *Proc. of the 3rd Intern. Conf. on Genetic Algorithms*, J. D. Schaffer (ed.), Morgan Kaufmann, San Mateo, CA, pp. 379–384.

[52] M. L. Minsky and S. Papert (1969) *Perceptrons*, MIT Press, Cambridge, MA.

[53] A. N. Mucciardi and E. E. Gose (1966) "Evolutionary pattern recognition in incomplete nonlinear multithreshold networks," *IEEE Trans. Electronic Computers*, Vol. EC-15, pp. 257–261.

[54] T. Poggio and F. Girosi (1990) "Networks for approximation and learning," *Proc. IEEE*, Vol. 78:9, pp. 1481–1497.

[55] L. A. Rastrigin (1963) "The convergence of the random search method in the extremal control of a many-parameter system," *Automation and Remote Control*, Vol. 24, pp. 1467–1473.

[56] J. Rissanen (1978) "Modelling by shortest data description," *Automatica*, Vol. 14, pp. 465–471.

[57] F. Rosenblatt (1957) "The perceptron, a perceiving and recognizing automaton," Project PARA, Cornell Aeronautical Lab. Rep., No. 85:640–1, Buffalo, NY.

[58] F. Rosenblatt (1958) "The perceptron: A probabilistic model for information storage and organization in the brain," *Psychol. Rev.*, Vol. 65, p. 386.

[59] F. Rosenblatt (1960) "Perceptron simulation experiments," *Proc. IRE*, Vol. 48, pp. 301–309.

[60] F. Rosenblatt (1962) *Principles of Neurodynamics: Perceptrons and the Theory of Brain Mechanisms*, Spartan Books, Washington, D.C.

[61] G. Rudolph (1994) "Convergence analysis of canonical genetic algorithms," *IEEE Trans. Neural Networks*, Vol. 5:1, pp. 96–101.

[62] D. E. Rumelhart and J. L. McClelland (1986) *Parallel Distributed Processing: Explorations in the Microstructures of Cognition*, Vol. 1, MIT Press, Cambridge, MA.

[63] J. D. Schaffer, D. Whitley, and L. J. Eshelman (1992) "Combinations of genetic algorithms and neural networks: A survey of the state of the art," *COGANN-92, Intern. Workshop on Combinations of Genetic Algorithms and Neural Networks*, L. D. Whitley and J. D. Schaffer (eds.), IEEE Computer Society, Los Alamitos, CA, pp. 1–37.

[64] P. Werbos (1974) "Beyond regression: New tools for prediction and analysis in the behavioral sciences," Ph.D. diss., Harvard.

[65] D. Whitley and T. Hanson (1989) "Optimizing neural networks using faster, more accurate genetic search," *Proc. of the 3rd Intern. Conf. on Genetic Algorithms*, J. D. Schaffer (ed.), Morgan Kaufmann, San Mateo, CA, pp. 391–396.

[66] B. Widrow, N. Gupta, and S. Maitra (1973) "Punish/reward: Learning with a critic in adaptive threshold systems," *IEEE Trans. Systems, Man and Cybernetics*, Vol. SMC-5, pp. 455–465.

[67] A. Wieland (1991) "Evolving controls for unstable systems," *Connectionist Models: Proc. of the 1990 Summer School*, D. S. Touretzky, J. L. Elman, T. J. Sejnowski, and G. E. Hinton (eds.), Morgan Kaufmann, San Mateo, CA, pp. 91–102.

[68] X. Yao (1993) "A review of evolutionary artificial neural networks," *Intern. J. Intel. Syst.*, Vol. 8:4, pp. 539–567.

COMPUTATIONAL MODELING OF EVOLUTIONARY LEARNING PROCESSES IN THE BRAIN

■ ROBERTO R. KAMPFNER and MICHAEL CONRAD*
Department of Computer Science,
Wayne State University,
Detroit, MI 48202, U.S.A.

The evolutionary selection circuits model of learning has been specified algorithmically. The basic structural components of the selection circuits model are enzymatic neurons, that is, neurons whose firing behavior is controlled by membrane-bound macromolecules called excitases. Learning involves changes in the excitase contents of neurons through a process of variation and selection. In this paper we report on the behavior of a basic version of the learning algorithm which has been developed through extensive interactive experiments with the model. This algorithm is effective in that it enables single neurons or networks of neurons to learn simple pattern classification tasks in a number of time steps which appears experimentally to be a linear function of problem size, as measured by the number of patterns of presynaptic input. The experimental behavior of the algorithm establishes that evolutionary mechanisms of learning are competent to serve as major mechanisms of neuronal adaptation. As an example, we show how the evolutionary learning algorithm can contribute to adaptive motor control processes in which the learning system develops the ability to reach a target in the presence of randomly imposed disturbances.

1. Introduction. Theories or hypotheses about the principles and mechanisms underlying the function of biological systems can be analyzed using simulation. In this paper we describe the use of such a method to analyze a conceptual model of learning in the brain. By a conceptual model we mean a model which, due to the complexity of the processes considered, is formulated with the aid of natural language rather than in a closed-form mathematical framework. The objective of computational modeling is to algorithmically specify such informally stated theories precisely using the tools of computer science and to use the computer to calculate these theories (Conrad, 1981). The conceptual model studied is the evolutionary selection circuits model (Conrad, 1974a, 1976). This is a learning model concerned principally with adaptive pattern recognition and adaptive control processes in the brain. Aspects of the model have previously been analyzed from an automaton and stochastic points of view (Conrad, 1974b; Schwabauer, 1976), but it requires computational techniques to explore the full dynamics of the model.

One idea in the selection circuits model is that certain forms of ontogenetic learning involve mechanisms analogous to the variation and selection

*Also Department of Biological Sciences.

Reprinted with permission from *Bulletin of Mathematical Biology*, R. R. Kampfner and M. Conrad, "Computational modeling of evolutionary learning processes in the brain," Vol. 45, No. 6, pp. 931-968. © 1983 by Society for Mathematical Biology.

mechanisms responsible for phylogenesis. Evolutionary algorithms have previously been developed in different contexts. The initial important work in this area is due to Bremermann, who has applied the procedure to problems such as those which arise in parameter fitting (Bremermann, 1967; see also Rechenberg, 1973). An alternative genetic-plans framework which emphasizes recombination and cross-over rather than mutation has been developed by Holland (1975) and has also been applied to a variety of problem areas. A number of studies have also been made in the context of the origin of life and ecosystem evolution (Conrad 1969; Conrad and Pattee 1970; Martinez, 1979). Aside from its connection to a brain model, the selection circuits algorithm has two features which distinguish it from these earlier studies. One is that the search procedure is analogous to evolution through artificial selection rather than to evolution through natural selection. The more fundamental feature is that in analogy to natural evolution the system structure is organized in a way which is effective for search through evolution, that is, in such a way that gradual changes in structure are associated with gradual changes in function. In the absence of this gradualness property the rate of evolution would scale as a product of mutation probabilities, and therefore would be unfeasible as a source of adaptation in historical time (Conrad, 1972, 1978).

In natural genetics and in the brain the gradualism is built into the physiochemical structure of the learning system and therefore appears distinct from the search procedure itself (Conrad, 1974c, 1979). This cannot be true in the computer model since features which contribute to the gradualism as well as the search procedure must both be included in the algorithmic specification. Thus the computational model to be described here can be viewed as consisting of two parts. The first part consists of the evolutionary search procedure *per se* and the second part consists of representations of the structural and organizational features which contribute to the system's amenability to evolution. In the body of the paper we frequently refer to the search procedure as the learning algorithm since many of the experiments involve the development of suitable search procedures. But it should always be remembered that the data structures on which these procedures act are an integral part of the algorithm and are in fact more responsible for the effectiveness of the algorithm developed than the search procedure itself. The major possibilities for further improvement of evolutionary learning algorithms lie in the improvement of these representations, which only partially reflect the mechanisms considered in the conceptual model. It might be noted that a similar distinction appears to be implicit in Martinez's computational model of the evolution of cellular organization since a polynomial representation of the genome is used to achieve gradualism (Martinez, 1979).

We first describe the algorithmic specification process as it applies to the

486

evolutionary learning circuits model and report on the computational experiments which have been performed on important parameters of the search algorithm with respect to its effectiveness in adaptive pattern recognition. In Section 7 we describe computational experiments which have been performed in the context of adaptive control mechanisms in the brain. In a separate paper we will report the results which have been obtained on the sequential behavior and stability characteristics of networks of enzymatic neurons which are allowed to use feedback. Although the evolutionary selection circuits model originates in brain theory, the learning algorithm could possibly be applied to the construction of practical learning devices. Such possibilities will be discussed elsewhere.

2. *Review of the Selection Circuits Model and its Biological Correlates.*
The selection circuits model has been described in Conrad (1974a, b), which should be consulted for details. The model involves two basic ideas. The first is that the brain consists of enzymatic neurons, that is, neurons whose firing properties are controlled by postulated enzymes originally called excitases. The second idea is that important forms of learning, such as adaptive pattern recognition and adaptive motor control, are mediated by changes in the excitase composition of enzymatic neurons. The hypothesis of the selection circuits model is that processes analogous to evolution by variation and selection provide a major mechanism for guiding this change.

A formalized version of an enzymatic neuron is illustrated in Figure 1. Different types of excitases are pictured as potentially binding to different sites on the neuron membrane, represented by subcells in a tesselation structure. The degree of excitation produced by any given pattern of presynaptic input in any such subcell is determined by a *fixed* weight vector. The neuron fires if the input pattern produces a suitable degree of excitation at any site at which excitases are located. The choice of this model neuron was motivated by three important features. The first is that with n excitases it is possible to distinguish 2^n possible input patterns. The second is that by choosing different excitase types it is possible to construct neurons which can implement any of the $(2^2)^n$ dichotomies of n inputs, that is, which can implement any of the logical functions of n arguments. The third feature is that of *adding or deleting single patterns from the set of input patterns* recognized. Thus the enzymatic neuron has the gradualism property, all important for evolution, as well as powerful pattern recognition capabilities. The choice of protein enzymes as the fundamental switching elements controlling nerve impulse activity was also motivated by the amenability of proteins to graceful change in function with single variations in primary structure.

While the choice of this model neuron was originally motivated by its

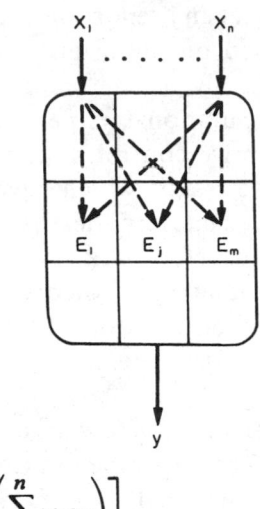

$$y = \varphi\left[\sum_{j=1}^{m} \psi\left(\sum_{i=1}^{n} w_{ij}x_i\right)\right]$$

$$\psi(u) = \begin{cases} 1 \text{ if } u \geqslant T_j \\ 0 \text{ otherwise; } T_j = \text{excitation level corresponding to } E_j \end{cases}$$

$$\varphi(v) = \begin{cases} 1 \text{ if } v \geqslant 1 \\ 0 \text{ otherwise} \end{cases}$$

Figure 1. Formalized version of an enzymatic neuron. The neuron fires only if the pattern of dendritic inputs produces a suitable degree of excitation in at least one of the loci occupied by an excitase. x_i = dendritic input i; w_{ij} = weighting coefficient of input x_i at location j; E_j = excitase molecule in location j; y = axonic output.

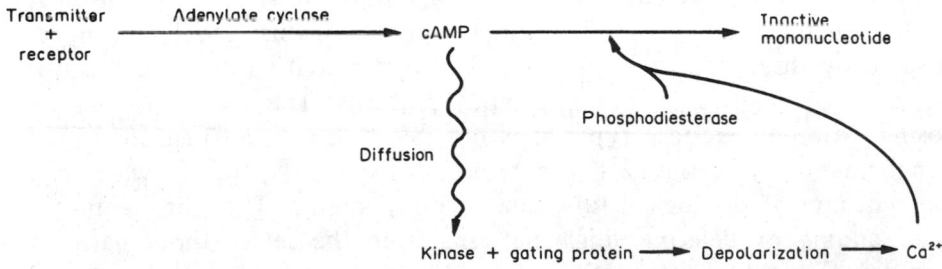

Figure 2. Biochemical correlates of the enzymatic neuron. The weight coefficients of the enzymatic neuron may be roughly interpreted as the amount of cAMP delivered to kinases located at the site j as a result of activation of dendrite i by the neurotransmitter. The excitases of the enzymatic neuron correspond to the cAMP-dependent kinases and permeability-controlling gating proteins which they activate. [See Greengard (1976) for a discussion of the biochemistry and Liberman *et al.* (1983) for a discussion of the depolarizing effects of cAMP].

posession of these features, recent investigations show that it could be based on purely experimental results. It is now known that cyclic $3',5'$-cAMP and $3',5'$-cGMP specifically control nerve impulse activity. The processes which are currently believed to occur (cf. Greengard, 1976) are illustrated in Figure 2. Neurotransmitters impinging on the presynaptic membrane activate receptors which in turn activate adenylate cyclase. Adenylate cyclase catalyzes the production of cAMP molecules which activate protein kinases. These kinases are believed to phosphorylate gating proteins which control membrane permeability. The effects of permeability which have been observed by different investigators are not uniform. However, micro-injection studies have shown that in at least some neurons cAMP causes depolarization, while cGMP causes either hyperpolarization or transient depolarization followed by hyperpolarization (Liberman et al., 1983).

A number of other proteins play an important role in this system, including calcium-activated phosphodiesterases which convert cyclic nucleotides into the inactive mononucleotides and binding protein which alter the diffusion rates of cAMP and cGMP. Alterations involving any of these proteins could contribute to evolutionary learning. However, the protein kinases appear to have properties which correspond more closely to the properties of the excitases postulated in the evolutionary learning circuits model (Conrad, 1981). The weighting coefficients (w_{ij}) in the enzymatic neuron are replaced by the amount of cAMP which diffuses to site j on the neuron due to neurotransmitter stimulation of dendrite i. Different patterns of presynaptic input give rise to different concentrations of cAMP at different sites. How the neuron interprets these concentration changes depends on the distribution of kinases. The working hypothesis is that the neuron responds to an input pattern if an above threshold concentration develops at any site at which a kinase is located. It is evident that this is an oversimplification and that the enzymatic neuron only partially reflects the rich dynamics which are now known to be possible in principle. However, the enzymatic neuron formalization does reflect these dynamics more adequately than previous formal neuron models. To the extent that it is less powerful than the real neuron, it may be reasonably assumed that the effectiveness of the learning algorithm to be described would be stronger rather than weaker than the results reported.

The learning algorithm is the procedure responsible for adaptively changing the excitases present in the enzymatic neuron. The algorithm exploits the high redundancy of brain tissue. According to Lashley's classic ablation studies this can be as high as eighty or ninety percent (Lashley, 1929). The redundant networks are supposed to be equivalent as regards connectivity and weighting coefficients of neurons. The assumption of complete interchangeability is, of course, an idealization. Different types of excitases may be added to or deleted from enzymatic neurons in these networks by various types of random processes. The behavior of the organism can be controlled by any one or any combination of these networks. The functional value of each of the networks is evaluated on the basis of the organism's performance. The excitase contents of the best performing networks are retained and transferred to the less fit networks. Many variations of this procedure have been investigated, both with a view to discovering effective procedures and with a view to discovering constraints on the process which might have significant biological correlates.

There are two possible mechanisms of transfer which were considered in the original description of the model. One involves the export of excitase RNA from the fittest to less fit neurons through glial channels. This mechanism has some analogies to the generation of variety in the immune system and to recombinant DNA processes. The second mechanism involves the random production of inducers precorrelated to specific excitase genes and the transport of these from the more fit to the less fit neurons. This mechanism has some analogies to processes involved in gene regulation. The algorithms described here are compatible with both mechanisms.

3. Major Simplifications of the Conceptual Model. To describe the process by which the conceptual model is translated into computational models which can be used for experiment, it is important to begin by considering simplifications which are made. Unless otherwise stated, these simplifications could be removed without difficulty by introducing more options into the program structure.

(i) Each excitase enzyme is regarded as responding to a unique (binary) input pattern out of those impinging upon the neuron that contains it. In the conceptual model the assumption is that each excitase is capable of recognizing a single set of similar input patterns.

(ii) The number of excitase types available to a given enzymatic neuron corresponds to that required to respond to each of the possible input patterns to that neuron. The effects of redundancy of excitase types in small neurons or limits on their numbers in larger neurons have not been considered.

(iii) Excitases are arbitrarily assigned to input patterns. In reality an excitase is specified by its sequence of amino acids and the family of input patterns it recognizes is determined by its folded structure. Folding is one of the major mechanisms of gradualism in the conceptual model, but it was felt that incorporating features which capture the essence of folding would be computationally difficult and would not contribute to working out the search procedure.

(iv) In an enzymatic neuron the specific sites to which excitases may bind are represented as locations on the cell membrane. The strength of the effect of each dendritic input on each location is represented as a weight coefficient, so that the level of excitation produced by a given input pattern at a specific location can be computed as a distinctively weighted sum. The enzymatic neuron fires when at least one of the locations which is sufficiently excited has an excitase bound to it. The number of excitases in the location should in reality have an effect on the output produced, but this is not considered in the present models.

The association of excitases with locations allows for a step-by-step modification of the properties of the neurons. Thus the representation captures one of the major gradualism-conferring features which motivated the original model. It is evident that the simplifications which have been made all reduce the realism of the model. On the other hand, by reducing the computing power and gradualistic features of the neuron they increase the likelihood that procedures found to work effectively in the computational experiments would work even more effectively in the real system.

Another important simplification is the following.

(v) Enzymatic neurons are thought of as acting on a discrete time scale and synchronously responding to each stimulus after some time delay, say τ. In this way the state of each neuron and hence of the network can be determined at times $\tau, 2\tau, 3\tau, \ldots$.

4. *Basic Organization of the Computational Models.* All the computational models developed were organized around a single basic scheme. This scheme provides a basis for incorporating specific features associated with a family of possible variations of the model. The features which are common to all the computational models are specified consistently throughout all of them. The basic organization, as well as the parameters and criteria used to define the different versions, are described below. The program is written in LISP, a function-oriented list processing language (McCarthy *et al.*, 1965), but uses FORTRAN subroutines for most numerical computations. A complete listing of one of the versions is given in Kampfner (1981).

(a) Representation of network structures. This is the aspect of the model which captures the structural and organizational features of the networks relevant to the evolution process. The data structures corresponding to each

level of the representation are illustrated in Figures 3 and 4. These diagrams can be interpreted hierarchically. Each name (e.g. NET 1) has as its value a list of properties, one of which may be another named list at the next lower level. Dashed arrows point to the list of properties. Solid arrows point to list elements at the same level if horizontal and at the next lower level if vertical. Half squares with diagonal lines terminate lists. Four levels are represented: (1) an ensemble of similar networks; (2) a network level; (3) an enzymatic neuron level; and (4) a location level. An important aspect of the representation at the network level is the specification of inputs to, outputs from and

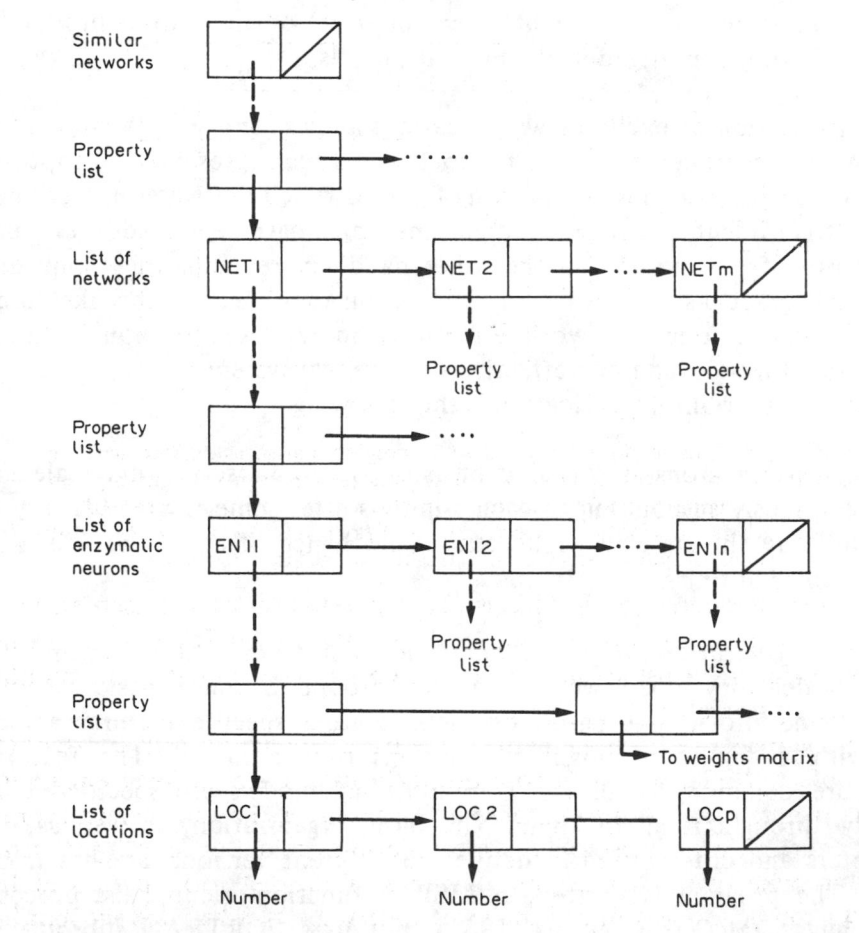

Figure 3. LISP representation of network structures. This diagram can be interpreted hierarchically. Each name (e.g. NET 1) has as its value a list of properties, one of which may be another named list at the next lower level. Dashed arrows point to the list of properties. Solid arrows point to list elements at the same level if horizontal and at the next lower level if vertical. Half squares with diagonal lines terminate lists.

interconnectivity among the elements of a network. At the neuron level the representation captures the effect of certain morphological features of the cell membrane on the firing properties of enzymatic neurons. Here specific sites to which excitases may bind are represented as locations, and the strength of the effect of each possible input to the neuron on each location is given in a weights matrix associated with each neuron. In some implementations the weights matrix was replaced by an equivalent but computationally more efficient procedure in which input patterns are mapped into numbers which represent the associated excitase type.

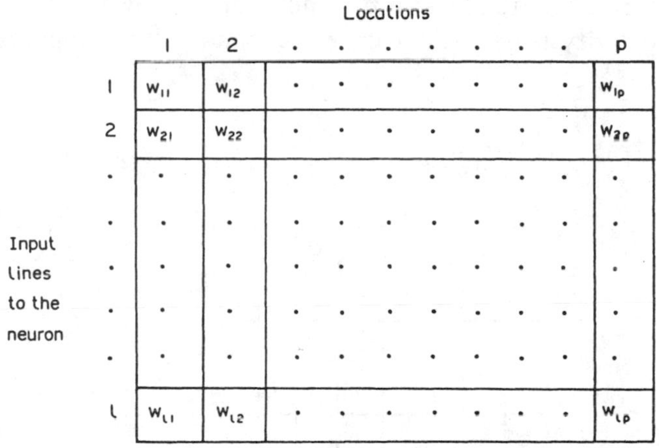

Locations

	1	2	.	.	.	.	.	.	.	p
1	w_{11}	w_{12}	.	.	.	.	.	.	.	w_{1p}
2	w_{21}	w_{22}	.	.	.	.	.	.	.	w_{2p}
.	.	.	.	.	.	.	.	.	.	.
.	.	.	.	.	.	.	.	.	.	.
.	.	.	.	.	.	.	.	.	.	.
.	.	.	.	.	.	.	.	.	.	.
.	.	.	.	.	.	.	.	.	.	.
l	w_{l1}	w_{l2}	.	.	.	.	.	.	.	w_{lp}

Input lines to the neuron

Figure 4. Schematic representation of a weights matrix associated with an enzymatic neuron. Each element w_{ij} of the matrix represents the strength of input i, impinging upon the neuron, on the excitase type represented by location j. The enzymatic neuron fires if $\Sigma_1 x_i \cdot w_{ij} \geq T_j$, where T_j is the excitation level associated with the excitase in location j.

(i) Similar networks level. This is the highest level of representation of network structures. It consists of a single object, NETS, one of whose property values is a list of m similar networks, where $m \geq 1$ is a number entered as a parameter.

(ii) Network level. Each element of the list of similar networks is denoted by an object, NET$_i$, one of whose property values is a list of n enzymatic neurons, where $n \geq 1$ is also entered as a parameter. To each particular network configuration corresponds a specification of inputs to, outputs from and interconnectivity among the elements of the network. Inputs to the networks can be thought of as l-tuples of binary digits, where l is the number of input lines impinging upon each network. A 1 in any position of the l-tuple means that the corresponding line is activated, where as a 0 means that it is not. Thus if we have n neurons in a network, each of them has $l + n$ possible inputs (l corresponding to external inputs to the network and n corresponding to inputs from other neurons and from itself). In the present computational models the inputs to and interconnectivity among the elements of a network are specified by an n by $l + n$ matrix, where the rows correspond to neurons in the network and columns correspond to possible inputs. This array is called CONNECTIVITY, and CONNECTIVITY$(i, j) = 1$ means that the jth input is connected to the ith neuron, whereas CONNECTIVITY$(i, j) = 0$ means that it is not connected. l, the number of input lines impinging upon the network, is also specified as a parameter. Outputs from the networks are specified by a linear array called NETWORK-OUTPUTS, whose ith element indicates the neuron being regarded as the ith output line from the network. s, the number of output lines, is also entered as a parameter at this level of the representation. The connectivity matrix is illustrated schematically in Figure 5.

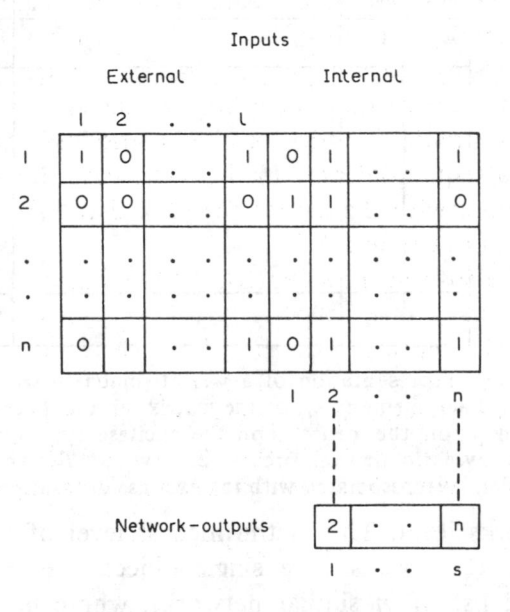

Figure 5. Some aspects of the specification of network configurations. In the matrix CONNECTIVITY the rows represent neurons and the columns represent possible inputs to each neuron. CONNECTIVITY$(i, j) = 1$ means that input j is connected to neuron i, whereas a 0 means it is not connected. The array NETWORKS-OUTPUTS specifies which neurons are emitting each of the s outputs from the network.

(iii) Enzymatic neuron level. Each element of the list of enzymatic neurons is an object, EN_{ij}, one of whose property values is a list of locations occupied by one or more excitases (i.e. a list of excitase types existing in that neuron). The firing properties of an enzymatic neuron are determined by the excitases it contains. The basic process for the learning algorithm is that of modifying the number and types of excitases existing in each neuron of a network. Each time this is done the list of locations is updated accordingly.

Another property value associated with each neuron is the name of its weights matrix, which describes the strength of each input on each location of the neuron. The list of locations and the weights matrix associated with a given neuron determines its firing properties. The list of locations specifies which excitase types are present in the neuron and the weights matrix specifies the input patterns to which each excitase type can respond.

(iv) Location level. At this level each location is associated with an excitase type. It contains the number of excitases of that type present in that neuron at any given time.

(b) Specification of the performance process. The performance process simulates the firing of networks of enzymatic neurons. It must reflect the effect of all the features represented in the network structures on the firing properties of the networks. The fixed features are the weighting coefficients and the interconnectivity of the neurons. The variable features are the patterns of excitase distribution, which change through time under the action of the learning algorithm. According to the task being learned, the performance process simulates either a combinational network (with one unit time delay) or a sequential one. In the latter case the state of the network at time t, together with the input (if any) at that time step, determines the state of the network at time $t + 1$.

More precisely, the subroutine FIRING, a LISP function which executes the performance process, acts as follows.

Let INPUT-VECTOR represent the $L = l + n$ values of the possible input lines to any neuron of the network at time t. Let WEIGHT-MATRIX (I, J) represent the weight coefficient of input x at location j (see Figure 1).

Then at time $t + 1$, for each neuron, say N_k, $k = 1, 2, \ldots, n$, of the network and for each j, where j names a location in N having excitases

$$\psi(N_k) = 1,$$

if

$$\sum_{I=1}^{L} (\text{INPUT-VECTOR}(I) * \text{WEIGHT-MATRIX}(I, J) * \text{CONNECTIVITY}(K, I))$$

$$\geqslant \text{THRESHOLD}(J) \text{ for at least one } J,$$

$$\psi(N_k) = 0, \text{ otherwise.}$$

It is easy to verify that FIRING realizes the mathematical functions described in Figure 1 (with THRESHOLD(J) represented by T_j). When the response of a network is simulated, the FIRING function is computed for all the neurons of the network.

(c) Specification of the learning algorithm. In the computational models, the learning algorithm is realized by a set of functions (or subroutines) acting upon the data structures representing the networks. Due to the complexity of the processes considered, a complete specification of the learning algorithm cannot be given *a priori*. Its basic parameters and suitable conditions for its operation have been identified and adjusted through the simulations.

The basic organization of the learning algorithm is illustrated in Figure 6.

(1) INITIALIZE. This function sets the network structures to be used in a given simulation experiment. The characteristics of the network structures are entered as parameters to this function. These parameters are: m, the number of similar networks; n, the number of neurons in each network; l, the number of input lines to each network; and s, the number of output lines from each network. Further specifications include the matrix CONNECTIVITY, described above, which defines the interconnectivity among the elements of a network, the array of NETWORK-OUTPUTS, which defines which enzymatic neurons are regarded as emitting outputs, and the weights matrix associated with each neuron. This function also provides an initial (random) assignment of excitases to each neuron of the network.

(2) EVALUATE. This function evaluates the performance of the networks, that is, applies inputs to the networks and compares the outputs obtained with the corresponding 'desired outputs'. The performance process realized by the FIRING function previously described is used by the EVALUATE function. Different levels of performance are assigned to the networks, depending on how close the actual output is to the desired one. Clearly, the criteria used in the evaluation of performance is problem dependent. If, for example, the task consists of classifying a set of input patterns into two categories, a useful performance measure is the average over all input patterns of the Hamming distances between each actual output and its corresponding desired output. For other types of tasks the evaluation criteria may be different. The learning process is terminated either when the highest possible performance associated with a task is achieved or when a predetermined number of iterations (or evaluation–modification cycles) is reached.

(3) SELECT. This function partitions the set of similar networks into two subsets, one called MOST-FIT networks and the other NOT-MOST-FIT networks. MOST-FIT networks are represented as a list whose elements are the names of those networks showing the highest performance in the last evaluation–modification cycle. The number of networks which may qualify

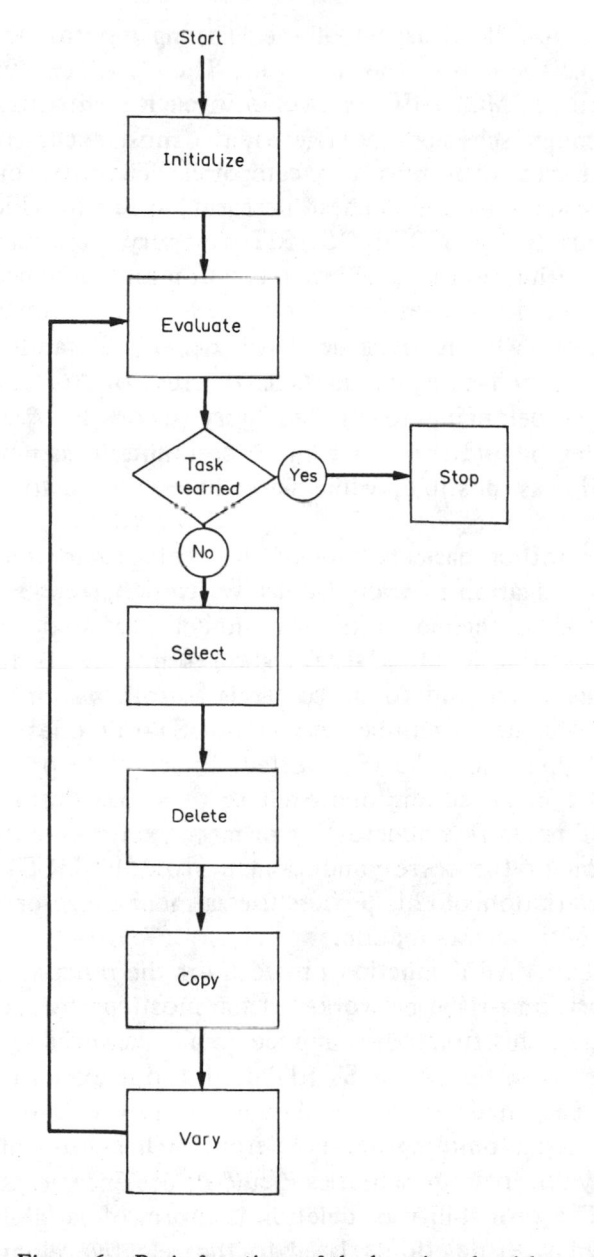

Figure 6. Basic functions of the learning algorithm.

as MOST-FIT has an upper limit, which we call P2. This is a parameter to the SELECT function and plays an important role in the performance of the learning algorithm. NOT-MOST-FIT networks, on the other hand, are a list whose elements are the names of those networks that do not qualify as MOST-FIT.

(4) DELETE. A critical aspect of the learning algorithm is the elimination of 'bad' excitases from the networks. The DELETE function deletes excitases from NOT-MOST-FIT networks in each evaluation–modification cycle. Of the many schemes investigated, the most successful one divides NOT-MOST-FIT networks into two categories. The first one consists of those networks showing the highest performance (but which are, nevertheless, grouped in the NOT-MOST-FIT category because P2 would be exceeded). Half the excitases in each location of each neuron of these networks are deleted. In cases in which there is only one excitase of a given type in a location, whether it is deleted is decided at random with probability 0.5. The second category contains the rest of NOT-MOST-FIT networks. Networks belonging to this category have all of their excitases deleted. The idea behind this procedure is to maintain as much potentially useful variability as possible, while nevertheless eliminating the harmful variations.

(5) COPY. Another basic feature of the evolutionary selection circuits model is the propagation of traits from the fittest networks to structurally similar networks. In the computational models this is done by 'copying' excitase types existing in MOST-FIT networks into NOT-MOST-FIT ones. The most succesful method found so far is as follows. First, each MOST-FIT network is assigned a number of NOT-MOST-FIT ones to which copies of its excitase types may be transported. By copying or transporting an excitase type we mean adding one excitase to a location of a neuron in a NOT-MOST-FIT network whenever one or more excitases exist in the corresponding location of the corresponding neuron of the MOST-FIT network. An important variation of this process uses a mechanism of recombination analogous to genetic recombination.

(6) VARY. The VARY function provides for the random variation in the excitase composition of the networks. In the most successful version of the learning algorithm this function is applied to all networks other than those which have been classified as MOST-FIT. It adds one excitase of a randomly chosen type to each neuron of each network with probability P_a and deletes one excitase of a randomly chosen type from each neuron of each network with probability P_d. The probabilities P_a and P_d are entered as parameters to this function. The probability of deletion is a form of variability and should not be confused with deletion related to the selective elimination of bad excitases in the DELETE function.

5. Critical Parameters of the Algorithm. In order to identify an efficient version of the learning algorithm, a large number of possibilities were investigated in a highly interactive way. In the course of this development four critical parameters of the learning algorithm were identified. These are:

(i) P1: number of similar networks; (ii) P2: maximum number of MOST-FIT networks; (iii) P3: expected number of (randomly chosen) excitase types to be added to each network in an evaluation–modification cycle; and (iv) P4: expected number of (randomly chosen) excitase types to be deleted from each network in an evaluation–modification cycle.

The experiments to be reported below all involve pattern classification. We will report on the results of varying the parameters P1–P4, varying the number of neurons in networks and varying problem size. Unless otherwise specified, these parameters have been set as follows in the pattern classification experiments described below. Parameter P1, the number of similar networks, is set to 10. Parameter P2, the maximum number of MOST-FIT networks, is set to 3.

In all these experiments the following procedure for performance evaluation was used. All input patterns were applied successively to each network and the performance with respect to each pattern was recorded. After all inputs were processed by a network its average performance was computed and an evaluation–modification cycle performed. In this way the average distance most equitably reflects the quality of the responses to all possible inputs. Evaluation after each input gives more weight to the last input and was found to be less effective.

The performance measure used in the evaluation of individual responses is a biased Hamming distance between the actual network output and the desired one. Such a distance is biased in the sense that excess firing is 'punished' more than lack of firing. For example, if the desired output is 0 0 1, then the biased Hamming distance corresponding to the actual output 0 1 0 is 3, whereas if the actual output is 0 0 0, the biased Hamming distance is 1. It was found that such a bias was necessary to prevent the accumulation of junk excitases, suggesting that selection against unfavorable genes is more difficult than selection for favorable ones.

The conditions for the selection of MOST-FIT networks are defined as follows. Let H-DISTANCE$_{i,j}$ denote the biased Hamming distance of network i with respect to input j, $i = 1, 2, \ldots, m$ and $j = 1, 2, \ldots, l$. Then the average biased Hamming distance of network i over the last l inputs is defined as

$$\text{AVERAGE-DISTANCE}_i = \sum_{j=1}^{l} \text{H-DISTANCE}_{i,j}/l.$$

Also, let us define

$$\text{MIN-AVERAGE-DISTANCE} = \min \{\text{AVERAGE-DISTANCE}_i\}.$$

The networks which qualify as MOST-FIT ones are the first three, or fewer, networks satisfying the condition

$$\text{AVERAGE-DISTANCE}_i = \text{MIN-AVERAGE-DISTANCE}.$$

The version just described was used to analyze the effect of the four main parameters of the learning algorithm on the learning process, as described below.

6. Experimental Results.

(a) Effect of varying the parameters. The analysis was carried out with respect to a task that consists of dichotomizing patterns of 4 binary digits using single-neuron networks. A typical definition of such a task is given in Table I. The corresponding network configuration is illustrated in Figure 7.

TABLE I

Desired Outputs Corresponding to Each
Input in a Task Consisting of Dichotomizing Patterns of 4 Binary Digits

No.	Input	Desired output
1	0 0 0 1	0
2	0 0 1 0	1
3	0 0 1 1	1
4	0 1 0 0	1
5	0 1 0 1	0
6	0 1 1 0	0
7	0 1 1 1	0
8	1 0 0 0	1
9	1 0 0 1	0
10	1 0 1 0	0
11	1 0 1 1	1
12	1 1 0 0	0
13	1 1 0 1	1
14	1 1 1 0	1
15	1 1 1 1	1

Figure 8 shows the result of varying parameter P1, the number of similar networks. Two series of experiments were performed, one with a maximum of 3 MOST-FIT networks (i.e. P2 = 3) and another with P2 = 5. In both series the rate of addition of excitases, P3, was fixed at 1.666 and the rate of deletion of excitases, P4, was fixed at 1.0. As shown in Figure 8, the average number of runs required to learn the task is greater when the value of P1 is relatively low. As P1 increases learning proceeds faster, until the number of runs required to learn the task reaches a minimum (e.g. an average of 7.0 for P2 = 3 and an average of 5.5 for P2 = 5, for values of P1 of 15 and 16 respectively). Further increases of P1, however, increase rather than decrease the number of runs required to learn the task. The results shown in Figure 8

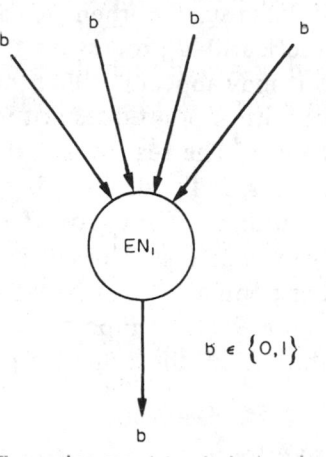

Figure 7. Network configuration used to dichotomize patterns for 4 binary digits. One single neuron is used. The neuron fires in response to an input pattern if it contains an excitase which is activated by such a pattern.

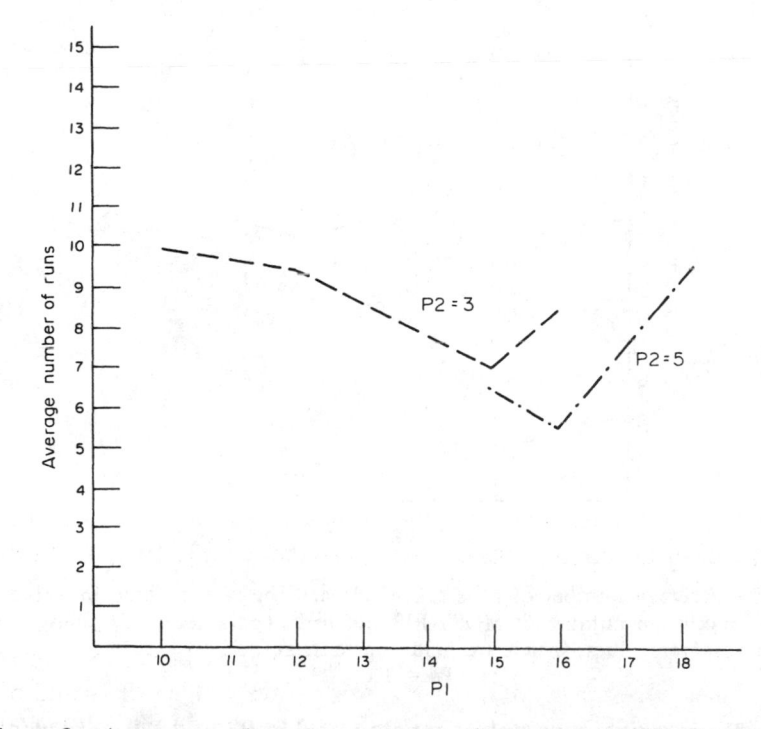

Figure 8. Average number of runs taken to learn the task defined in Table I for different numbers of similar networks (parameter P1). Parameter P3, the expected number of excitase types added to each neuron in each run, is fixed at 1.666. Parameter P4, the expected number of excitase types deleted from each neuron in each run, is fixed at 1.0. Two values of parameter P2, the maximum number of MOST-FIT networks, are used, as indicated in the figure.

suggest the following relationship between P1 and P2. If we regard P1 as a measure of the redundancy of the system, then we can say that such a redundancy tends to accelerate the learning process up to a maximum rate which depends on the value of P2. It may appear peculiar that increasing the number of similar networks beyond a critical point does not increase the rate of learning and, in fact, may decrease it. The reason is that when P2, the maximum number of MOST-FIT networks, is fixed, there is a limit to the amount of variability which the system utilizes effectively. If P1 is too large relative to P2, some good excitases may even be lost. This suggests that there are circumstances under which a smaller population can evolve faster than a large one.

The results shown in Figure 9 also support the existence of a relationship between P1 and P2, but from the opposite point of view. Here P2 is

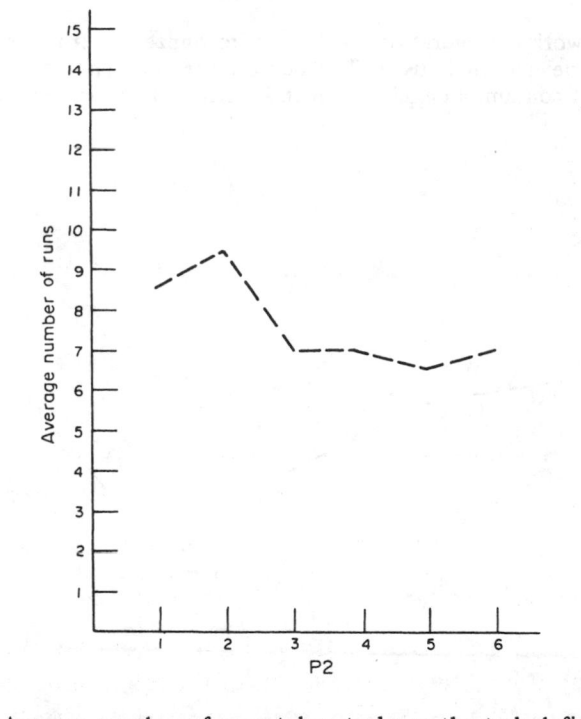

Figure 9. Average number of runs taken to learn the task defined in Table I as the maximum number of MOST-FIT networks (parameter P2) changes. Other parameter values, which are held constant, are: P1 = 15, P3 = 1.666 and P4 = 1.0.

varied while P1 remains constant at a value of 15. P3 and P4 were also held constant with the same values as for Figure 8. As can be seen in Figure 9, for fixed P1 the rate of learning tends, in general, to increase as P2 increases. However, it reaches a maximum at P2 = 5. Subsequent increases of P2 slow down the learning process. The interpretation of this result is that for a fixed number of similar networks an increase in the number of MOST-FIT networks results in the retention of potential solutions. But beyond a critical point further increase leads to a counterproductive decrease in the number of networks available for variation.

Figure 10 shows the result of varying P3, the rate of addition of excitases, on the rate of learning. In this series of simulations the values of other parameters were P1 = 15, P2 = 5 and P4 = 1.0. Because of the complementary

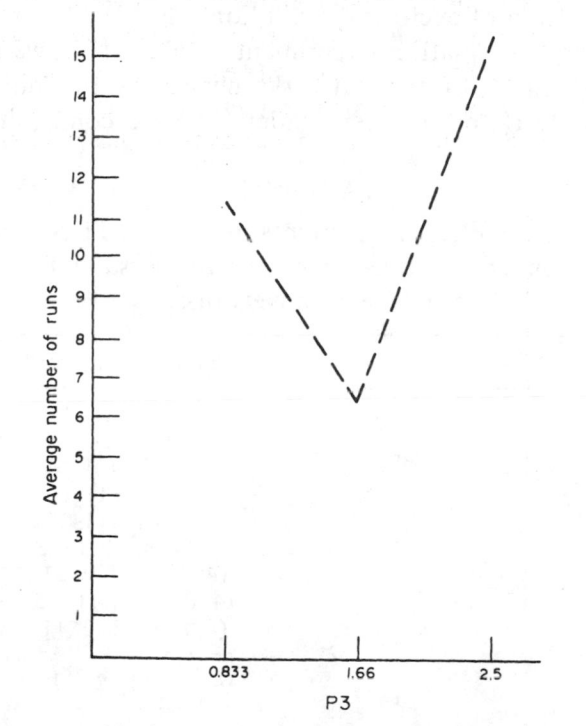

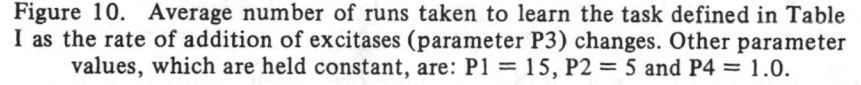

Figure 10. Average number of runs taken to learn the task defined in Table I as the rate of addition of excitases (parameter P3) changes. Other parameter values, which are held constant, are: P1 = 15, P2 = 5 and P4 = 1.0.

role played by P3 and P4 in changing the excitase contents of the networks, it seems natural to interpret this result in terms of the relationship between these two parameters. As can be seen in Figure 10 for fixed P4, a relatively low rate of addition of excitases slows down the learning process. However, a relatively high rate also slows it down. Thus it seems that there is an optimum rate of addition of excitases with respect to a given rate of deletion. In the case of Figure 13 this optimum rate corresponds to a value of P3 of 1.666. When P3 is too low (e.g. P3 = 0.8333) a plausible explanation of the slowness of learning is that not only the rate of addition of excitases is itself slow, but also the rate of deletion of excitases is relatively high. In this case too many excitases are deleted from the system in each evaluation–modification cycle.

The opposite case, in which the rate of addition of excitases is too high (e.g. P3 = 2.5), is far more interesting, for it shows the importance of gradualism in evolutionary processes. Recall that the learning algorithm builds a required pattern of excitase distribution using, in each evaluation-modification cycle, the set of MOST-FIT networks as a reservoir of potential solutions. Too high a rate of addition of excitases does not allow for a step-by-step growth of these potential solutions. This makes it less likely to reach a desired pattern of excitase distribution. This conclusion is supported by data drawn from simulation experiments. Table II shows how patterns of excitase composition of the networks change as the number of runs progresses. In Table II the numbers under 'excitase composition' indicate

TABLE II

Changes in the Excitase Contents of MOST-FIT Networks as the Number of Runs Increases; the Results Refer to Single-neuron Networks

Run	Excitase composition
1	(3)
	(4)
	(8)
	(11)
2	(4 8 11)
3	(4 8 11 13 3)
4	(4 8 13 3 2 11 15)
5	(4 8 13 2 11 15 14)
	(4 8 13 3 2 11 15)
. .	. . . .
20	(4 8 13 2 11 15 14)
. .	. . . .
25	(4 8 13 2 11 15 14 3)

excitase types existing in a MOST-FIT network. Such numbers were asigned to excitase types in such a way that they also identify the input pattern, as numbered in Table I, which activates such an excitase type. Thus when the changes occur in large jumps the number of runs required to learn the task increases. In general it has been found that versions of the learning algorithm which allow rapid initial learning usually have more difficulty completing the last steps in the learning process. This is reminiscent of colonizing populations in nature which are capable of exploiting a new environment rapidly, but which are eventually displaced by slower-evolving populations.

The data in Table II is included in the results shown in Figure 9. The corresponding value of P3 is 2.5. The excitase contents of the networks change very fast in the first four runs. However, it is not until run 25 that the desired pattern of excitase distribution is reached. When the rate of addition of excitases is lower (e.g. $P3 = 1.666$) the desired pattern is reached, in a typical simulation, in only 6 runs. This is shown in Table III.

TABLE III

Changes in the Excitase Contents of MOST-FIT Networks as the Number of Runs Increases

Run	Excitase composition
1	(2)
	(8)
	(11)
2	(2 3)
	(2 8)
3	(2 8 3 14)
4	(2 8 3 14 13 11)
	(2 8 3 14 4 3)
5	(2 8 3 14 4 13 15)
6	(2 8 3 14 4 13 15 11)

The results of Table III can be interpreted in the same way as those reported in Table II. In this case, however, parameter P3 has a value of 1.666.

The results shown in Figure 11 illustrate how the rate of deletion of excitases (parameter P4) affects the rate of learning. In this case $P1 = 15$ and $P3 = 1.666$. Such results can also be interpreted in terms of the relationship between P3 and P4, but from a complementary point of view. Here, for fixed P3 a relatively low rate of deletion (e.g. $P4 = 0.5$) slows down the learning process. A relatively high one (e.g. $P4 = 2$) slows it down considerably. The optimum rate of deletion is around $P4 = 1$. The interpretation is that for a given value of P3 a relatively low rate of deletion does not eliminate 'bad' excitases quickly enough, whereas if the deletion rate is too high 'good' excitases are deleted too fast.

(b) The effect of redundancy at the network level on the learning process. By redundancy at the network level we mean the use of networks with more than one neuron to perform tasks that can also be performed using single-neuron networks. Our study of the effect of this type of redundancy aims at investigating differences between single-neuron and redundant networks as they pertain to learning performance. The basic version of the learning algorithm is used in this analysis. In the simulations reported here 3-neuron networks were used. The network configuration is illustrated in Figure 12. In order to facilitate the comparison with single-neuron networks the task performed in the simulation with redundant networks is the same as that

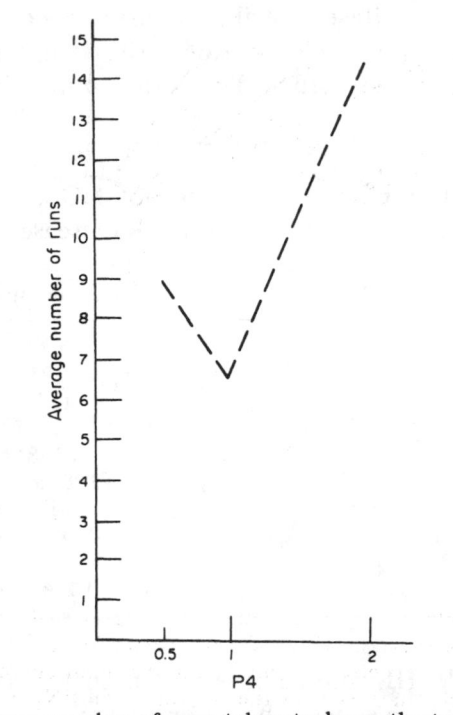

Figure 11. Average number of runs taken to learn the task defined in Table I as the rate of deletion of excitases (parameter P4) changes. Other parameter values, which are held constant, are: P1 = 15, P2 = 5 and P3 = 1.666.

used in the experiment with single-neuron networks reported above. This task is defined in Table V.

Two series of experiments are reported, both related to experiments carried out with single-neuron networks. The results of the first series, displayed in Figure 13, show the rates of learning obtained with different values of P1, the number of similar networks. These results can be compared with those shown in Figure 8 for single-neuron networks since the values of P2, P3, and P4 are the same in both cases. The results shown in Figure 13 indicate that, for comparable values of P1, learning proceeds faster in the case of single-neuron networks. As with single neurons, the learning process can be speeded up in redundant networks by increasing the number of similar networks.

In the second series of simulations the rate of addition of excitases, P3, is varied, while the values of the other parameters are held constant, with P1 = 18, P2 = 5 and P4 = 1.0. The results are displayed in Figure 14. A comparison with similar results using single-neuron networks (see Figure 10) indicates that although the value of P1 is different, the optimum rate of learning occurs when P3 and P4 have exactly the same values (1.666 and 1.0 respectively) as in the case of single-neuron networks. In the case of redundant

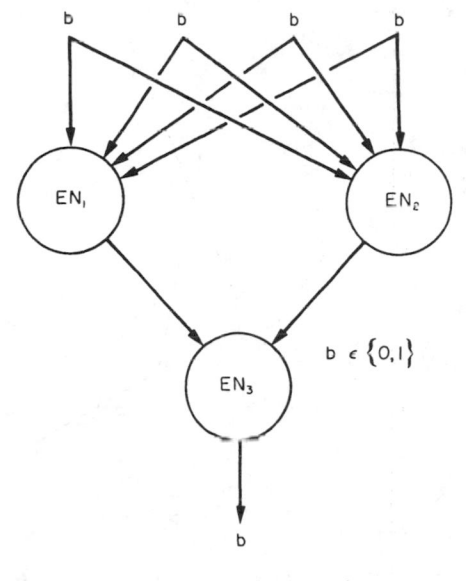

Figure 12. Network configuration used to classify patterns of 4 binary digits into two categories. Three-neuron networks are used. Enzymatic neurons EN_1 and EN_2 receive the input patterns simultaneously but process them separately. EN_3 then receives the outputs from EN_1 and EN_2 and emits the network output.

networks the optimum rate of learning is only 7.5 runs, on average, as compared with 6.5 for single neurons.

It may seem surprising that single-neuron networks are more efficient for learning than redundant ones since the greater redundancy would appear to increase the probability of discovering a solution and allows different parts of a solution to be executed by different neurons. Our result suggests that the occurrence of multiple copies of neurons in each of the similar networks aggravates the problem of deleting 'junk' excitases. It becomes more likely that undesirable excitases will appear in any given network for any given mutation rate and less likely that the selection will be able to distinguish which neuron is responsible. Also, the possibility of dividing the task among different neurons appears to make it more difficult to coordinate the learning process. But it should be noted that these conclusions apply to networks in which all neurons receive the same input.

(c) Effect of increasing problem size. A series of experiments were conducted to investigate how the performance of the learning algorithm varies as the size of the patterns increases. The results are shown in Figure 15. As expected, the number of runs required to learn such tasks increases as the number of patterns to be classified becomes larger. This number increases exponentially with the number of inputs since there are 2^n patterns of

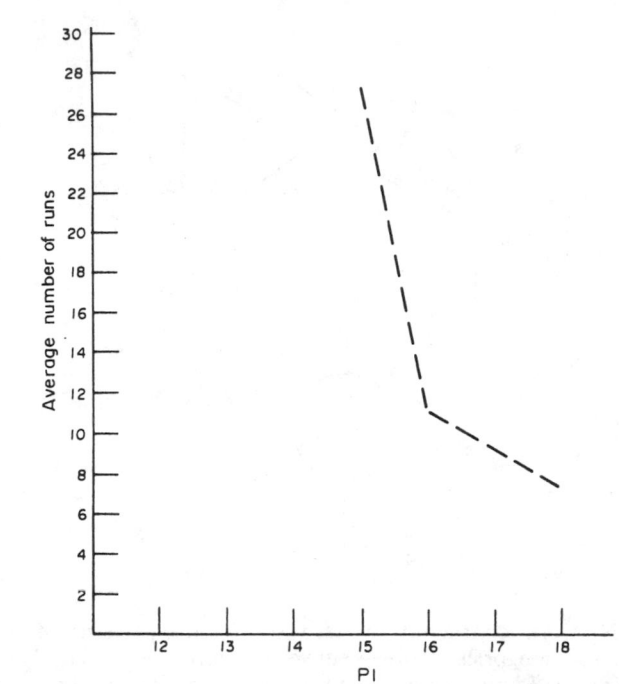

Figure 13. Average number of runs taken to learn to classify patterns of 4 binary digits with 3-neuron networks. The number of similar networks (parameter P1) is varied. Other parameter values, which are held constant, are: P2 = 5, P3 = 1.666 and P4 = 1.0. The task is that defined in Table I. A comparison of these results with those shown in Figure 8 for the same values of P2, P3 and P4 indicates that the learning process is faster for single-neuron networks. In the case of 3-neuron networks the learning process can be speeded up, but at the expense of redundancy at the system level (i.e. using more similar networks).

binary digits. In the computation models the number of excitase types also increases as 2^n (cf. Section 3, simplification (ii)). The redundancy of the systems (i.e. the number of similar networks) increases the efficiency of the system when the patterns are not too big. For relatively big patterns, however, the complexity of the tasks is necessarily reflected in the rate of learning. The important point to note is that experimentally this value appears to be a polynomial function of problem size.

   (d) Effect of recombination of traits on the learning process. In the basic. version of the learning algorithm the COPY function copies most excitases to each NOT-MOST-FIT network from only one MOST-FIT network. In some cases, however, there may be several MOST-FIT networks showing the highest performance in a given run, but each having a different excitase composition. Experiments were conducted with the aim of investigating the effect of combining the traits of all MOST-FIT networks in each of the NOT-MOST-FIT networks receiving those traits. This idea was incorporated

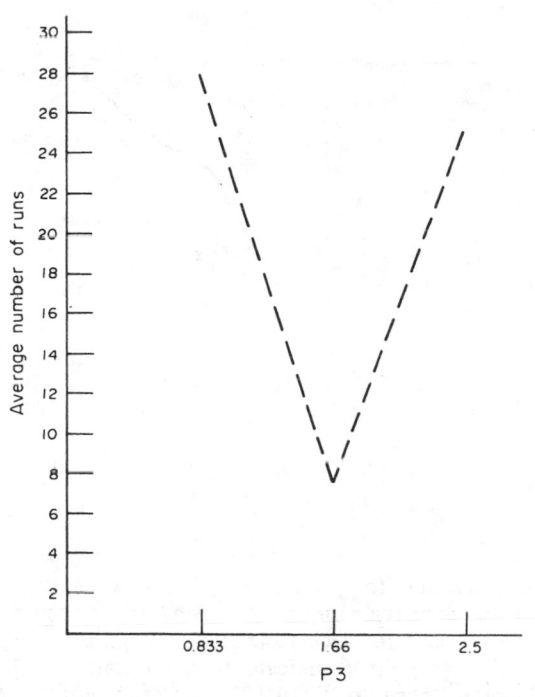

Figure 14. Average number of runs taken to learn to classify patterns of 4 binary digits with 3-neuron networks. The rate of addition of excitases (parameter P3) is varied. Other parameter values, which are held constant, are P1 = 18, P2 = 5 and P4 = 1.0. The task is that defined in Table I. A comparison of these results with those of Figure 10 indicates that although the value of P1 is different, the optimum rate of learning occurs when P3 and P4 have exactly the same values (1.666 and 1.0 respectively) as when it occurs for single-neuron networks.

by modifying the COPY function. In the modified version each excitase type existing in any MOST-FIT network is copied into each NOT-MOST-FIT network.

A series of simulations was conducted in order to compare the performance shown by the original and the modified version of the learning algorithm. The task used in these simulations involved the classification of patterns of different sizes. The data show that recombination speeds the learning process by a constant factor. Thus recombination improves the rate of learning but not significantly.

These results are shown in Figure 15. A surprising feature which should be noted in Figure 15 is the following. For small patterns the number of runs required to learn a task actually decreases as pattern size increases. This suggests that the optimum values of the parameters depends on the number of excitase types in the neuron. As the size of the problem increases we always increase the number of excitase types, leading to the unintuitive

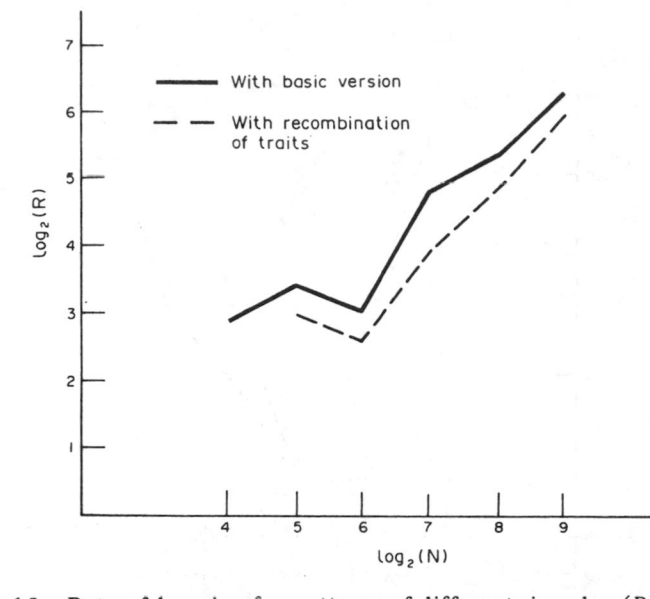

Figure 15. Rate of learning for patterns of different sizes. $\log_2(R)$ is plotted against $\log_2(N)$, where N is the number of possible distinct patterns and R is the average number of runs required to learn to recognize binary patterns of the sizes indicated. Heavy lines indicate results obtained using the basic version of the learning algorithm. Dotted lines refer to results obtained using a modified version which incorporates a form of recombination of traits (i.e. excitases) from all MOST-FIT networks when they are copied into NOT-MOST-FIT ones. In the case involving recombination each excitase type existing in any MOST-FIT network is copied into each NOT-MOST-FIT one. In all cases a training set of 15 patterns was used and the networks learned to give the desired response to each pattern of the set. In all cases the values of the main parameters of the learning algorithm were P1 = 15, P2 = 5, P3 = 1.66 and P4 = 1.0. The rate of learning appears to be a polynomial function of the number of possible patterns of each size.

effect that larger problems could be solved faster than smaller ones for some parameter settings.

(e) Variability of data. The experiments reported were performed in a highly interactive fashion under a wide variety of conditions. The trends described were discovered in the course of this interaction with the system. However, it should be recognized that the curves illustrating these trends were obtained by connecting points representing the average of either three or four experiments. The typical range of behavior is illustrated in Figure 16. We have not reported on peculiarities of particular experiments which cannot be generalized.

7. Modeling a Motor Control System. In the simulation of motor control learning it is assumed that the inputs to the networks are patterns of binary digits encoded by receptor organs which receive and process environmental

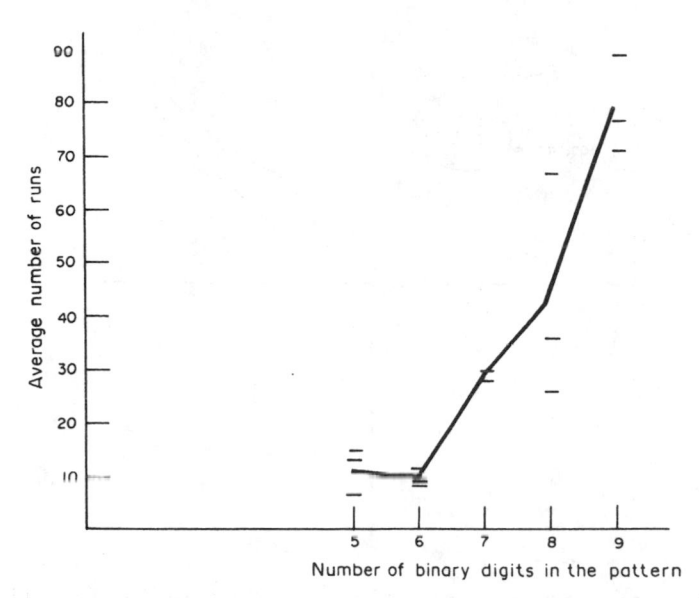

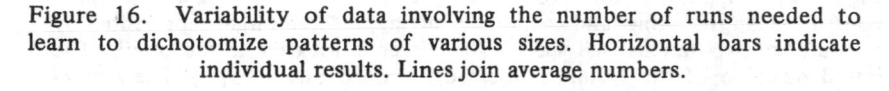

Figure 16. Variability of data involving the number of runs needed to learn to dichotomize patterns of various sizes. Horizontal bars indicate individual results. Lines join average numbers.

stimuli. The response of the networks to those binary input patterns are patterns of firing which are interpreted as signals controlling the action of effectors. Learning is accomplished through the adjustment of the patterns of firing in the networks so that appropriate, specific responses are given to the various input patterns representing environmental stimuli.

(a) Definition of the tasks. The learning goal is to approach, and eventually hit, a fixed target represented by a point in two-dimensional space. The learning system can only occupy points in the plane having integer coordinates and can only move to neighboring points through successive displacements, each of one unit distance, in one of four possible directions (i.e. towards the positive or negative side of any of the two coordinate axes). For convenience we will refer to these possible directions as North, South, East and West, following the usual convention. We will also assume that the target is always located at the origin. In a particular task the system starts at a given initial position in the plane, specified by its coordinates, and must hit the target in a specified number of time steps or stimulus–response cycles. This is illustrated in Figure 17.

Also, the system interacts with an environment which acts upon it and from which the system receives some information. The action exerted by the environment on the system can be interpreted in terms of mechanical forces which at each time step displace the system in one of the four possible directions (chosen randomly). The information that the system receives from

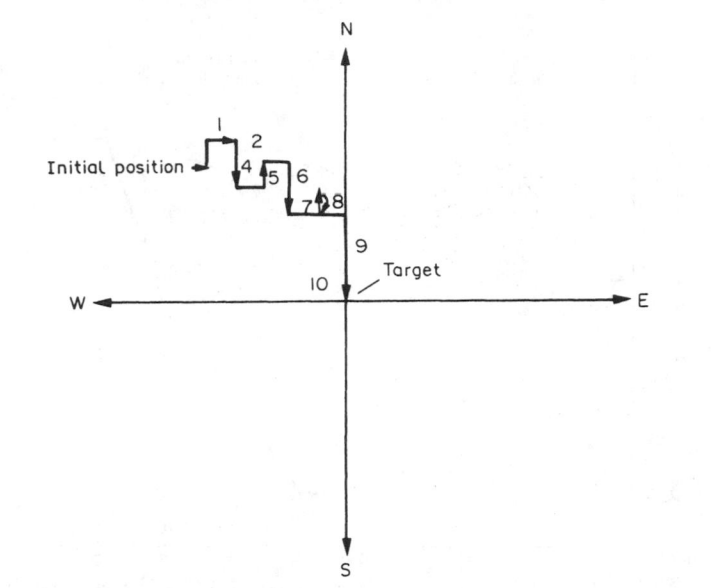

Figure 17. A possible route followed by a system performing a task which consists of approaching, and eventually hitting, a fixed target represented by a point in two-dimensional space. The target is located at the origin. The initial position of the system is located at coordinates $(-5, 5)$. The system can only occupy points in the plane having integer coordinates and at each time step it can only move to other points through one or more successive steps of one unit distance, each in one of four possible directions. For convenience those possible directions are referred to as North, South, East and West, following the usual convention. In this example the system hit the target in 10 time steps or stimulus–response cycles. The arrows indicate displacements of the system. The numbers by the arrows indicate the time step at which the corresponding displacement took place.

the environment is of two types. One type is the direction of the forces acting on the system. The other type of information is the relative position of the system with respect to the target at the present time step, say t. Thus at time $t + 1$ the system can take an action which, taking into account the action of the mechanical forces, may place it closer to the target.

Before we explain how each response of the system (i.e. its action) can be translated into movement in some direction it is necessary to consider the network configuration used. As shown in Figure 18, a network receives input patterns of 6 binary digits at time t. At time $t + 1$ it emits a response, also a firing pattern. Notice that the three left-most binary digits of an input pattern received at time t encode the direction of the mechanical forces which will be in effect at time $t + 1$, whereas the 3 right-most binary digits encode the relative position with respect to the target at time t. The response of the network, a pattern of 4 binary digits, is interpreted as an action taken by the system. Let a 1 in a given position of the output pattern indicate,

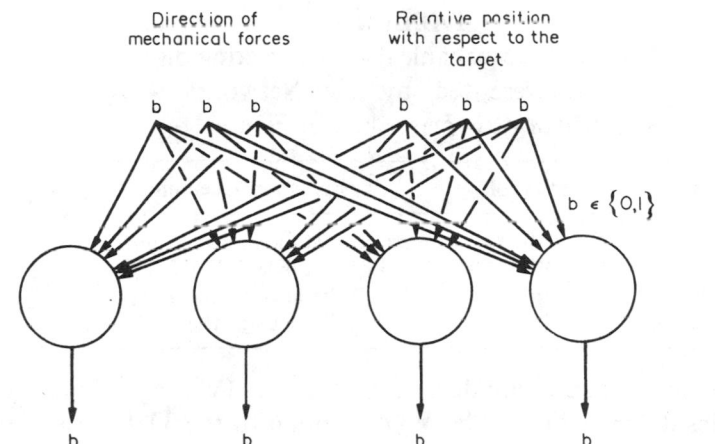

Figure 18. Network configuration used to perform taks related to motor control learning (see Figure 17). At time t the network receives an input of 6 binary digits. At time $t + 1$ the network emits as a response a pattern of 4 binary digits which correspond to this firing pattern. The 3 left-most binary digits of an input pattern encode the direction of the mechanical forces which will be in effect at time $t + 1$. The 3 right-most binary digits encode the relative position of the system with respect to the target at time t. The output from the system is a pattern of 4 binary digits which encodes the action taken by the system as follows. A 1 in a given position of the output pattern indicates, from left to right, that the system moves one unit of distance to the South, North, East and West respectively. A 0 in any of those positions indicates no movement in the corresponding direction. The total displacement of the system at each time step is computed as the composition of the system's actions with the mechanical forces. The codes for the mechanical forces and for the relative position of the system with respect to the target are described in Tables IV and V respectively. The movement of the system represented by each output pattern and the total displacements are described in Table V.

from left to right, that the system moves 1 unit of distance to the South, North, East or West respectively. Let a 0 in any of those positions indicate no movement in the corresponding direction. A 1 in a given position means that the corresponding neuron fired and a 0 means that it did not fire. Thus there are 16 possible network outputs, one corresponding to each possible firing pattern of the networks. The action of the system represented by each of these outputs is the composition of all the individual 'moves' corresponding to each position of the output pattern. For example, the output pattern 0 0 0 1 represents a movement of one unit distance to the West. As another example, the output pattern 1 1 1 1 represents no movement since the individual moves cancel each other. Furthermore, the total displacement of the system at each time step is computed as the composition of the system's action with that of the mechanical forces acting at that time step. The codes of the mechanical forces, and those for the relative positions of the system

TABLE IV

Codes for the Mechanical Forces Acting on the
System Represented by the Networks with
Configuration Described in Figure 18

Code	Symbol	Systems displacement
0 0 1	↓	1 unit distance to the South
0 1 0	↑	1 unit distance to the North
0 1 1	←	1 unit distance to the West
1 0 0	→	1 unit distance to the East

with respect to its target, are described in Tables IV and V respectively.

The codes described in Table IV correspond to the 3 left-most binary digits of the 6-digit pattern input to the networks. If a given mechanical force is detected by the system at time t, its effect will take place at time $t + 1$.

The codes of Table V correspond to the 3 right-most binary digits of the 6-digit pattern input to the networks. If a network receives any of those inputs at time t, the relative position of the system with respect to the target is, at that time step, as indicated in the table.

Table VI shows the actions represented by each network output. However, the total displacement of the system is the result of the composition of these outputs with the mechanical forces in effect at the same time step. The symbols ↓, ↑, ← and → have the same meaning as described in Table IV. Sequences of those symbols are self-explanatory; '—' means no action or displacement of the system.

TABLE V

Codes for the Relative Positions with
Respect to the Target for the System
Represented by the Network Configura-
tion Described in Figure 18

Code	Relative position of the system
0 0 0	North of the target
0 0 1	Northeast of the target
0 1 0	Southeast of the target
0 1 1	Southwest of the target
1 0 0	Northwest of the target
1 0 1	East of the target
1 1 0	South of the target
1 1 1	West of the target

(b) Simulation experiments. We report two series of simulation experiments. In both we evaluate the performance of the learning algorithm by

TABLE VI

Interpretation of Network Outputs in
Tasks Related to Motor Control Learning

Network output	Action represented
0 0 0 0	—
0 0 0 1	↑
0 0 1 0	↓
0 0 1 1	—
0 1 0 0	←
0 1 0 1	↳←
0 1 1 0	⌐←
0 1 1 1	←
1 0 0 0	→
1 0 0 1	→↑
1 0 1 0	→↓
1 0 1 1	→
1 1 0 0	—
1 1 0 1	↑
1 1 1 0	↓
1 1 1 1	—

using random search as a baseline. Tasks involving different degrees of difficulty are considered in this evaluation.

The performance of random search was also established using simulation. For this purpose we developed a version of the simulation program which tries the networks, evaluates their performance and selects the MOST-FIT networks in the same way as it is done when the learning algorithm is used. However, new structures to be tried (i.e. versions of the learning system with different patterns of excitase distribution) are selected at random from the 2^{256} structures which are possible (many are functionally equivalent). This ensures that both random search and the learning algorithm operate under comparable conditions. The degree of difficulty of the tasks may be varied, in a non-trivial way, by varying the maximum number of stimulus–response cycles allowed for hitting the target. We used the number of runs required to learn the task when random search is used as a measure of its degree of difficulty (from the point of view of learning). The smaller the number of stimulus–response cycles allowed, the more difficult it is to learn the task.

The performance of the learning algorithm is shown in Figures 19 and 20. The results shown in Figure 19 refer to tasks in which the initial position of the system is located at coordinates (5, 5). Different learning tasks are then defined by varying the number of stimulus–response cycles within which the system must hit the target. Figure 19 shows the number of runs taken, on average, to learn the task using both random search and the evolutionary learning algorithm. As can be seen, the most difficult of the tasks considered is that in which the system must hit the target in at most 6 stimulus–response cycles (the smallest number of cycles considered). In this case the task is learned in an average of 19 runs (i.e. learning cycles) using the learning algorithm. When random search is used, however, it takes 123.6 runs, on average, to learn the same task. It can also be seen in Figure

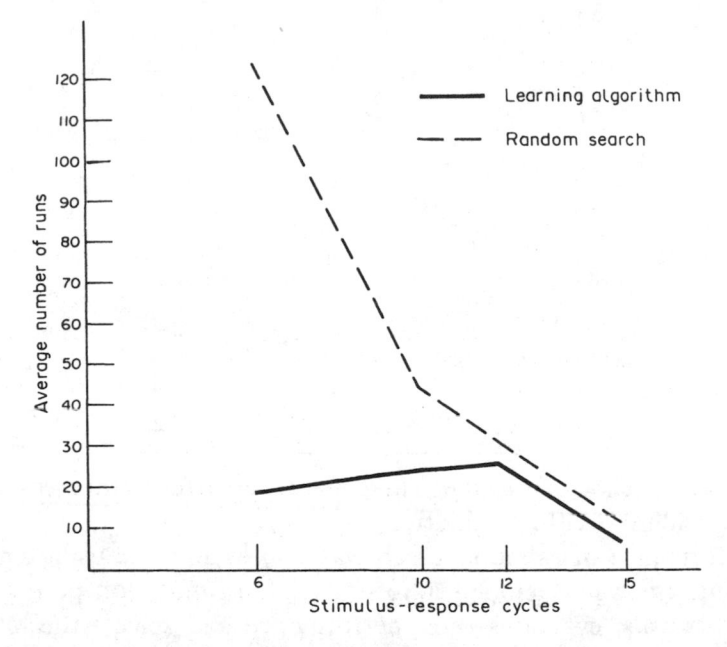

Figure 19. Average number of runs required to learn various tasks. All tasks consist of hitting a target from an initial distance of 10 units [the coordinates of the initial position are (5, 5) and the target is located at the origin]. The difference between those tasks is the maximum number of stimulus–response cycles within which the system must hit the target. The number of runs required to learn each task is shown for both random search and the learning algorithm. The number of runs taken by random search is regarded as an indication of the degree of difficulty of each task. This degree of difficulty increases as the number of stimulus–response cycles allowed decreases.

19 that the tasks become less difficult as the number of stimulus–response cycles increases. This is consistent with intuition since it becomes more likely for the system to hit the target as the number of stimulus–response cycles increases. It is important to note that the difference between random search and the learning algorithm decreases as the degree of difficulty of the task decreases.

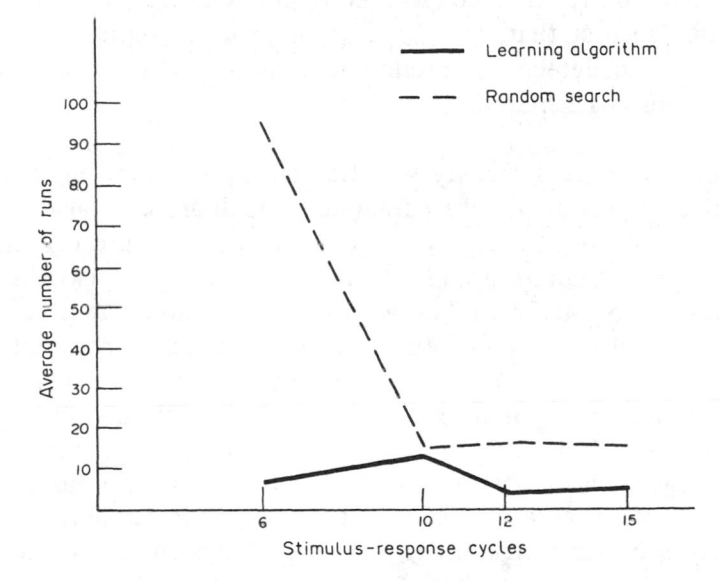

Figure 20. Average number of runs required to learn various tasks. All tasks consist of hitting a target from an initial distance of 5 units [the coordinates of the initial point are (5, 0) and the target is located at the origin]. The difference between those tasks is the maximum number of stimulus–response cycles within which the system must hit the target. The number of runs required to learn each task is shown for both random search and the learning algorithm. The number of runs taken by random search is regarded as an indication of the degree of difficulty of each task.

Figure 20 shows similar results, but with respect to a different set of tasks. These results refer to tasks in which the initial position of the systems is located at coordinates (5, 0). In this case the target is 5 units of distance away from the initial position rather than the 10 units used in the previous experiment. When the initial position is at (5, 0), however, it lies on one of the coordinate axes. This makes it more likely for a system performing a task of this type to fall into any of the quadrants adjacent to the horizontal axis on its way to the target. This means that the systems are subject to a wider variety of environmental inputs. This is to be reflected in the degrees of difficulty assigned (by random search) to each of these tasks. As before, the tasks have 6, 10, 12 and 15 stimulus–response cycles as the maximum allowed. Again, the greatest degree of difficulty corresponds to the task with the minimum number of stimulus–response cycles allowed. The most difficult task is learned with the learning algorithm in only 6.3 runs on average, whereas with random search it is learned in 95.6 runs on average. This shows the high selectivity of the learning algorithm with respect to advantageous traits ('good' responses to specific stimuli). This feature of the learning algorithm becomes most apparent precisely when the selective pressure is higher, that is, when the task to be learned is more difficult.

These results suggest that, for the type of process studied, the learning algorithm is able to detect and retain useful traits of the system until the desired performance is achieved.

8. Conclusions. The evolutionary selection circuits learning algorithm may be characterized in terms of four parameters which are critical to performance and which may be expected to be critical to any future variant of the algorithm. These parameters are: (1) the number of similar networks; (2) the maximum number of MOST-FIT networks; (3) the rate of addition of new excitases; and (4) the rate of deletion of excitases. These parameters have been examined experimentally using well-defined pattern classification tasks which are naturally posed to the neuron. These tasks involve the recognition of different patterns of presynaptic input. In the experiments we have chosen the range of values within which excitases can respond to be sufficiently narrow so that each excitase type responds to only one input pattern (Section 3, simplification (i)). This excludes single excitases from recognizing families of inputs, but makes it easier to evaluate the performance of the algorithm. By widening this band of values it would be possible to work with a smaller training set.

The main results of the computational experiments are:

(i) The algorithm is effective. The number of steps required to learn the task appeared to increase as a polynomial function of the size of the task as measured by the number of presynaptic inputs to the neuron. The polynomial character of the data suggests that the time and space requirements of alternative algorithms cannot be significantly less than those of the evolutionary algorithm. These data imply that evolutionary mechanisms are feasible mechanisms of neuronal adaptation and that from the purely computational point of view they have the same order of feasibility as other possible mechanisms.

(ii) The parameters of the algorithm have critical relationships, some of which are unexpected. The rate of learning increases with the number of similar networks, but only up to a point. Beyond this point further increase in this parameter may even result in a decrease in the rate of learning. Similarly, increasing the maximum number of MOST-FIT networks increases the rate of learning, but only up to a critical point beyond which a decrease occurs. Both effects are connected to the potential variability of the evolving system. The maximum potential variability is determined by the ratio between the number of similar networks and the number of MOST-FIT networks. If the number of similar networks grows too large relative to the number of MOST-FIT networks the variety of solutions carried by the system decreases.

(iii) There is a trade-off between the speed of learning and its accuracy. This is connected with the critical relationship between the rates of addition and deletion of excitases. If the rate of addition is increased beyond a critical point the gradualism required for evolutionary learning is lost since the possibility for small changes in the excitase composition of neurons is lost. As a consequence, increasing the rate of addition beyond this point can produce a system which learns fast, but which stagnates before it finds the optimum.

(iv) The optimal values of the parameters depend on problem size. As problem size increases performance may be improved by increasing the number of similar networks. But such enlarged systems may be less effective for smaller problems.

(v) Selection against undesirable excitases should be stronger than selection for desirable ones. If this is not the case undesirable excitases accumulate. The occurrence of many non-coding regions in DNA is possibly connected to this feature. In natural evolution the punishment of genes which detract from fitness cannot be greater than the reward for those which enhance it by an equal amount. In a selection circuits system it is possible to alter the balance between reward and punishment since measures of performance can be arbitrarily distorted by the evolutionary mechanisms. But it should be noted that in the implementation as presently constituted the behavior of the neuron does not depend on the number of excitases in a location. In reality this number might control the probability of the neuron's firing or the frequency with which it fires. Some versions of the algorithm which proved slower, such as versions which do not clear out the NOT-MOST-FIT networks, might become the versions of choice under these circumstances.

(vi) Recombination of traits from different MOST-FIT networks speeds up the learning process. But the effect of recombination and of different choices of parameter settings is not as important for the rate of evolutionary learning as is the representation of the neuronal structure on which the mechanism of variation and selection acts.

(vii) Networks of neurons learned more slowly than single neurons capable of performing the same task. To maximize the rate of adaptation it is therefore advantageous to consolidate information processing in single neurons. This observation is consistent with experiments on cyclic nucleotide control of nerve impulse activity which suggest that the intraneuronal mechanisms of information processing are more important than previously thought.

(viii) Motor control processes were learned effectively using the evolutionary selection circuits algorithm. In this case the task to be learned was interpreted as one in which the controlled system learned to approach, and eventually hit, a fixed target in two-dimensional space. Several tests were performed in which the degree of difficulty of the task was varied. The results obtained with the learning algorithm were compared with those obtained, for the same tasks, using pure random search. The learning algorithm performs much better than random, with the difference with random search growing as the difficulty of the task increased.

To put these results in perspective it is useful to consider how the features of evolutionary learning systems in nature compare to those of the model we have implemented and to other mechanisms which might be considered. There are four major points:

(1) Evolutionary learning has the advantage that new machinery, in the present case, new enzymes, is developed only when needed. This is not true for rote memorization, a mechanism easily implemented on digital computers for the tasks considered, but cumbersome to implement in single neurons. In order for a single neuron to use rote memorization it would have to be capable of systematically generating all possible outputs in response to each possible input and capable of altering its internal state so that an output evaluated as desirable would be paired to that input in the future. The neuron would have to carry this pre-existing pairing machinery for all possible input patterns. The requirements for resources and pre-existing correlation grow exponentially in this type of neuronal adaptation. Clearly direct memorization and knowledge-based processes involving memory storage and retrieval play a fundamental role in brain processes. But for problems requiring adaptation of the control characteristics of individual neurons (as in the motor control model) evolutionary mechanisms are more economical.

(2) Evolutionary learning has the advantage that the desired outputs need not be known in advance. This is a feature of the selection circuits algorithm since punishment only occurs when the network makes a false response. In the actual experiments we did specify the desired responses in advance, but only as a convenient means of evaluating the algorithm's performance rather than as a logical requirement for the algorithm's operation. It is interesting to compare this to perceptron type learning (Rosenblatt, 1961; Minsky and Papert, 1969). This relies on formal neural nets in which the weighting of the outputs of the neurons may be altered. If the network responds to its inputs when it should not, the weights of all neurons which fire are decremented; if it does not respond when it should, the weights of all

neurons which fire are incremented. For the perceptron learning algorithm to work the evaluating system must be able to recognize a failure, a requirement characteristic of statistical pattern recognition techniques. While such techniques may be effective for suitably defined problems, the requirement for labeling the patterns in advance in terms of the desirability for a particular response limits their scope as compared to evolutionary learning processes.

(3) *De novo* capabilities can always arise in evolutionary systems. In the selection circuits model new types of excitases can always appear. The delimitation of this capability in the present implementation is due to the practical limitations on computing the dynamical effects of *de novo* amino acid sequences, not on any limitation connected with the search mechanisms modeled. The possibility for novelty in evolutionary learning systems is clearly related to the fact that the desirability of the outputs need not be known in advance. Learning techniques subject to this requirement (such as perceptron type learning) cannot exhibit novelty.

(4) It is possible for enzyme-controlled mechanisms of a much more sensitive nature than those represented in our formal model to develop under the influence of the learning algorithm described, with concomitant enhancement of the information processing capabilities of neurons and neural networks. Thus the pattern recognition and response capabilities of the networks which have been used for the purpose of investigating the feasibility of evolutionary learning should not be taken as defining the information processing power of evolutionary learning systems.

These considerations, along with the experimental results reported, suggest that evolutionary mechanisms should be seriously considered as candidate mechanisms for neuronal adaptation. The evolutionary mechanism does not replace memory manipulation, statistical pattern recognition or knowledge-based procedures of the type represented in some artificial intelligence research. The potential contribution which it can make is better viewed as complementing these alternative mechanisms. Some of the contributions to higher level functions which can occur are illustrated by the experiments on the adaptation of motor control behavior described in this paper.

This work was supported by the Division of Information Systems, Office of Naval Research (Contract N00014-80-C-0365 to Michael Conrad). We are indebted to J. H. Holland, A. W. Burks, M. Savageau and R. Rada for valuable discussions.

LITERATURE

Bremermann, H. J. 1967. "Quantitative Aspects of Goal-seeking Self-organizing Systems." In *Progress in Theoretical Biology*, Ed. M. Snell, Vol. I. New York: Academic Press.

Conrad, M. 1969. "Computer Experiments on the Evolution of Coadaptation in a Primitive Ecosystem. PhD Thesis, Stanford University.

——. 1972. "Information Processing in Molecular Systems." *Currents mod. Biol.* 5, 1–14.

——. 1974a. "Evolutionary Learning Circuits." *J. theor. Biol.* 46, 167–188.

——. 1974b. "Molecular Information Processing in the CNS, Parts I and II." In *Physics and Mathematics of the Nervous System*, Eds. M. Conrad, W. Güttinger and M. Dal Cin. Heidelberg: Springer.

——. 1974c. "Molecular Automata." In *Physics and Mathematics of the Nervous System*, Eds. M. Conrad, W. Güttinger and M. Dal Cin. Heidelberg: Springer.

——. 1976. "Complementary Molecular Models of Learning and Memory." *BioSystems* 8, 119–138.

——. 1978. "Evolution of the Adaptive Landscape." In *Springer Lecture Notes in Biomathematics, No. 21. Theoretical Approaches to Complex Systems*, Eds. R. Heim and G. Palm. Heidelberg: Springer.

——. 1979. "Bootstrapping on the Adaptive Landscape." *BioSystems* 11, 167–180.

——. 1981. "Algorithmic Specification as a Technique for Computing with Informal Biological Models." *BioSystems* 13, 303–320.

——. and H. H. Pattee. 1970. "Evolution Experiments with an Artificial Ecosystem." *J. theor. Biol.* 28, 393–409.

Greengard, P. 1976. "Possible Role for Cyclic Nucleotides and Phosphorylated Membrane Proteins in Postsynaptic Actions of Neurotransmitters." *Nature, Lond.* 260, 101–108.

Holland, J. 1975. *Adaptation in Natural and Artificial Systems*. Ann Arbor: University of Michigan Press.

Kampfner, R. 1981. "Computational Modeling of Evolutionary Learning." PhD Thesis, University of Michigan, Ann Arbor.

Lashley, K. S. 1929. *Brain Mechanisms and Intelligence*. Chicago: Chicago University Press (reprinted in 1963 by Dover, New York).

Liberman, E. A., S. V. Minina, N. E. Shklovsky-Kordy and M. Conrad. 1983. "Microinjection of Cyclic Nucleotides Provides Evidence for a Diffusional Mechanism of Intraneuronal Control." *BioSystems* 15, 127–132.

Martinez, H. 1979. "An Automaton Analogue of Unicellularity." *BioSystems* 11, 133–162.

McCarthy, J., P. W. Abrahams, D. J. Edwards, T. P. Hart and M. I. Levin. 1965. *LISP 1.5 Programmers Manual*. Cambridge, MA: MIT Press.

Minsky, M. and S. Papert. 1969. *Perceptrons*. Cambridge, MA: MIT Press.

Rechenberg, I. 1973. *Evolutionsstrategie: Optimierung technischer Systeme nach Prinzipien der biologischen Evolution*. Stuttgart: Frommann-Holzboog.

Rosenblatt, F. 1961. *Principles of Neurodynamics: Perceptrons and the Theory of Brain Mechanisms*. Washington, DC: Spartan Books.

Schwabauer, R. 1976. "Enzymatic Neurons." *J. theor. Biol.* 59, 223–230.

RECEIVED 2-24-82

Chapter 18
Evolutionary Computation
and the Traveling Salesman Problem

D. E. Goldberg and R. Lingle (1985) "Alleles, Loci, and the Traveling Salesman Problem,"
Proc. of an Intern. Conf. on Genetic Algorithms and Their Applications, J. J. Grefenstette (ed.),
Lawrence Erlbaum, Hillsdale, NJ, pp. 154–159.

J. Grefenstette, R. Gopal, B. Rosmaita, and D. Van Gucht (1985) "Genetic Algorithms for the
Traveling Salesman Problem," *Proc. of an Intern. Conf. on Genetic Algorithms and Their Applications,* J. J. Grefenstette (ed.), Lawrence Erlbaum, Hillsdale, NJ, pp. 160–168.

EARLY efforts in combinatorial optimization within the field of evolutionary computation centered on the traveling salesman problem (TSP). The task is to arrange a tour of n cities such that each city is visited only once and the length of the tour (or some other cost function) is minimized. For an exact solution, the only known algorithms require the number of steps to grow at least exponentially with the number of elements in the problem. The total number of alternative tours increases as $(n - 1)!/2$, a number that grows faster than any finite power of n. The task quickly becomes unmanageable; for even $n = 50$ there are about 10^{62} different candidate tours. The two papers reprinted here (Goldberg and Lingle, 1985; Grefenstette et al., 1985) were among the first attempts to use simulated evolutionary optimization to address the TSP.

Although neither paper offered a procedure that would be deemed effective for solving the TSP given current standards, each is important in the history of evolutionary computation for other reasons. The TSP poses a representation problem, and the solutions taken in these two efforts are instructive. Both papers offered nonstandard crossover procedures that operate on specialized representations. Goldberg and Lingle (1985) reported encouraging results with a version of crossover (partially-mapped crossover or PMX) on a 10–city TSP.[1] Grefenstette et al. (1985) examined larger problems (up to 200 cities) but reported disappointing results with several combinations of representations and operators, and thereby presented some of the first evidence that the utility of specific genetic operators is representation dependent. Grefenstette et al. (1985) also offered and demonstrated some success with one of the first genetic operators combining crossover with local search (i.e., a greedy nearest-neighbor procedure). The authors suggested the potential for combining genetic algorithms with other heuristics both in initialization and in fine tuning the solutions in a final population.

Both papers highlight philosophical perspectives of the genetic algorithm community in 1985. Goldberg and Lingle (1985) offered a strong belief in the utility of crossover in evolutionary search. Consider the excerpt: "With only a unary operator to search for better string orderings, we have little hope of finding the best ordering, or even very good orderings, in strings of any substantial length. Just as mutation cannot be expected to find very good allele schemata in reasonable time, inversion cannot be expected to find good orderings in substantial problems" (Goldberg and Lingle, 1985). Grefenstette et al. (1985) did not appear to take such a strong position, as they specifically acknowledged the success of simulated annealing methods on the TSP (e.g., Bonomi and Lutton, 1984). Such methods were limited to single point-to-point searches based on unary operators. Both papers devoted space to analyzing the effect of their proposed operators on the schemata of the candidate tours and hoped to identify good subtours as building blocks for creating improved solutions.[2]

Goldberg and Lingle (1985) and Grefenstette et al. (1985) served as foundational papers for subsequent efforts on applying simulated evolution to the TSP. Oliver et al. (1987) followed

---

[1]Goldberg (1997) related that larger problems involving 20 and 33 cities were presented orally at the 1985 conference.

[2]Subsequently, Grefenstette (1987) focused on incorporating problem-specific knowledge into genetic algorithms, both in terms of seeding initial populations and tailoring search operators. In contrast, Goldberg (1989) continued to devote attention to order-based schemata and efforts to have a genetic algorithm learn how to link building blocks in a solution such that recombination could serve to reliably transmit these substrings to other solutions (see Goldberg et al., 1989; Goldberg and Bridges, 1990; Harik and Goldberg, 1997; and others).

the permutation representation of Goldberg and Lingle (1985) and compared three different tailored crossover operators both in terms of their order-based schema processing and empirical trials. Suh and Van Gucht (1987) gave further consideration to specialized genetic operators for the TSP, including a variation of 2–opt (Lin and Kernighan, 1972). The inclusion of local improvement operators was seen to improve the efficiency of the genetic algorithm. Similar efforts were included in Liepens et al. (1987) and Mühlenbein et al. (1987, 1988). Whitley et al. (1989) proposed a recombination operator acting on edges in a tour and offered results (using integer arithmetic for distances) that were superior to previous best known tours for three test problems of 30, 50, and 75 cities.

Fogel (1988) offered an evolutionary programming algorithm for discovering solutions to the TSP. Each tour in the population was represented as a list of cities to be visited in the order of appearance. Offspring tours were created by selecting a city at random in a parent and moving it to another randomly selected position in the list (remove-and-reinsert). No recombination operators were employed. Selection of the best tours was made probabilistic such that tours with above average lengths still retained some probability of becoming parents in the subsequent generation. The results of this procedure were statistically significantly better than those obtained using a genetic algorithm with PMX.

Ambati et al. (1991) proposed a similar algorithm with a mutation operator that swapped two cities in a single parent and used deterministic selection. This procedure outperformed that offered in Fogel (1988) on TSPs of 10, 20, 50, and 100 cities, with the locations of the cities distributed uniformly at random in a square area. Analysis indicated that the method of Ambati et al. (1991) could achieve solutions that were 25% worse than the expected best tour with an expected running time of $O(n \log n)$, where there are n cities. Subsequent empirical studies in Fogel (1993a) indicated that under a mutation operator that inverts a sublist of cities in a single tour, solutions that are 10% worse than the expected best can be achieved with the number of function evaluations increasing as $O(n^2)$. Fogel (1993b) offered improved results with this procedure on the three test cases analyzed in Whitley et al. (1989) using real-valued arithmetic to compute the intercity distances.

Many efforts have been made to construct effective representations and corresponding genetic operators for the TSP (Ablay, 1987; Ulder et al., 1991; Braun, 1991; Gorges-Schleuter, 1991; Herdy, 1991; Starkweather et al., 1991; Mühlenbein, 1991; Homaifar et al., 1993; Pal, 1993; Bui and Moon, 1994; Tamaki et al., 1994; Lin et al., 1995; Ronald, 1995; Faulkner and Talhami, 1995; and others). Specific consideration has been devoted to creating variation operators that always yield legal tours and take advantage of the Euclidean metric used to assess the distance between consecutive cities. The consistent conclusion of these efforts is that the "performance of probabilistic search algorithms such as the [genetic algorithm] are highly dependent on representation and the choice of neighborhood operators" (Homaifar et al., 1993).

The TSP has been used as a canonical optimization problem in order to study other areas of interest within evolutionary computation. For example, Manderick et al. (1991) and Mathias and Whitley (1992) used the TSP to study the performance of genetic algorithms as a function of the response surface being searched. Julstrom (1995) developed a procedure for updating operator probabilities using the TSP as a test case. Craighurst and Martin (1995) used the TSP to examine methods for preventing premature convergence in genetic algorithms. Colorni et al. (1992) and others have modeled ant colonies and their associated trail dynamics using the TSP. Dorigo (1995) indicated that these simulations discovered an improved solution to the 75-city problem studied in Whitley et al. (1989) and Fogel (1993b).

The TSP is one of a number of inherently order-based combinatorial optimization problems. Other prominent examples include job shop scheduling and task assignment with precedent constraints (i.e., one task must be completed, or at least started, before another can be started). A similar review of efforts within evolutionary computation could be made for such problems. These efforts go back at least to Davis (1985) and Smith (1985).

References

[1] V. P. Ablay (1987) "Optimieren mit evolutionsstrategien," *Spektrum der Wissenschaft*, Vol. 7, July, pp. 104–115. [GERMAN: Abstract translation by Bäck (1996)]

[2] B. K. Ambati, J. Ambati, and M. M. Mokhtar (1991) "Heuristic combinatorial optimization by simulated Darwinian evolution: A polynomial time algorithm for the traveling salesman problem," *Biological Cybernetics*, Vol. 65:1, pp. 31–35.

[3] T. Bäck (1996) personal communication, Informatik Centrum Dortmund, Germany.

[4] E. Bonomi and J.-L. Lutton (1984) "The n-city traveling salesman problem: Statistical mechanics and the metropolis algorithm," *SIAM Rev.*, Vol. 26, pp. 551–568.

[5] H. Braun (1991) "On solving travelling salesman problems by genetic algorithms," *Parallel Problem Solving from Nature*, H.-P. Schwefel and R. Männer (eds.), Springer, Berlin, pp. 109–116.

[6] T. N. Bui and B. R. Moon (1994) "A new genetic approach for the traveling salesman problem," *Proc. of the 1st IEEE Conf. on Evolutionary Computation*, IEEE Press, Piscataway, NJ, pp. 7–12.

[7] A. Colorni, M. Dorigo, and V. Manniezo (1992) "An investigation of some properties of an 'ant algorithm,'" *Parallel Problem Solving from Nature, 2*, R. Männer and B. Manderick (eds.), Elsevier Science Publishers, Amsterdam, The Netherlands, pp. 509–520.

[8] R. Craighurst and W. Martin (1995) "Enhancing GA performance through crossover prohibitions based on ancestry," *Proc. of the 6th Intern. Conf. on Genetic Algorithms*, L. J. Eshelman (ed.), Morgan Kaufmann, San Mateo, CA, pp. 130–135.

[9] L. Davis (1985) "Job shop scheduling with genetic algorithms," *Proc. of an Intern. Conf. on Genetic Algorithms and Their Applications*, J. J. Grefenstette (ed.), Lawrence Erlbaum, Hillsdale, NJ, pp. 136–140.

[10] M. Dorigo (1995) personal communication, IRIDIA, Belgium.

[11] G. Faulkner and H. Talhami (1995) "Using 'biological' genetic algorithms to solve the travelling salesman problem with applications in medical image processing," *Proc. of the 1995 IEEE Conf. on Evolutionary Computation*, Vol. 2, IEEE Press, Piscataway, NY, pp. 707–710.

[12] D. B. Fogel (1988) "An evolutionary approach to the traveling salesman problem," *Biological Cybernetics*, Vol. 6:2, pp. 139–144.

[13] D. B. Fogel (1993a) "Empirical estimation of the computation required to discover approximate solutions to the traveling salesman problem using

evolutionary programming," *Proc. of the 2nd Ann. Conf. on Evolutionary Programming*, D. B. Fogel and W. Atmar (eds.), Evolutionary Programming Society, La Jolla, CA, pp. 56–61.

[14] D. B. Fogel (1993b) "Applying evolutionary programming to selected traveling salesman problems," *Cybernetics and Systems*, Vol. 24:1, pp. 27–36.

[15] D. E. Goldberg (1989) *Genetic Algorithms in Search, Optimization and Machine Learning*, Addison-Wesley, Reading, MA.

[16] D. E. Goldberg (1997) personal communication, Univ. of Illinois, Urbana-Champaign, IL.

[17] D. E. Goldberg and C. L. Bridges (1990) "An analysis of a reordering operator on a GA-hard problem," *Biol. Cybern.*, Vol. 2, pp. 397–405.

[18] D. E. Goldberg, B. Korb, and K. Deb (1989) "Messy genetic algorithms: Motivation, analysis, and first results," *Complex Syst.*, Vol. 3:5, pp. 493–530.

[19] D. E. Goldberg and R. Lingle (1985) "Alleles, loci, and the traveling salesman problem," *Proc. of an Intern. Conf. on Genetic Algorithms and Their Applications*, J. J. Grefenstette (ed.), Lawrence Erlbaum, Hillsdale, NJ, pp. 154–159.

[20] M. Gorges-Schleuter (1991) "Explicit parallelism of genetic algorithms through population structures," *Parallel Problem Solving from Nature*, H.-P. Schwefel and R. Männer (eds.), Springer, Berlin, pp. 150–159.

[21] J. J. Grefenstette (1987) "Incorporating problem specific knowledge into genetic algorithms," *Genetic Algorithms and Simulated Annealing*, L. Davis (ed.), Pitman, London, pp. 42–60.

[22] J. Grefenstette, R. Gopal, B. Rosmaita, and D. Van Gucht (1985) "Genetic algorithms for the traveling salesman problem," *Proc. of an Intern. Conf. on Genetic Algorithms and Their Applications*, J. J. Grefenstette (ed.), Lawrence Erlbaum, Hillsdale, NJ, pp. 160–168.

[23] G. R. Harik and D. E. Goldberg (1997) "Learning linkage," *Foundations of Genetic Algorithms 4*, R. K. Belew and M. D. Vose (eds.), Morgan Kaufmann, San Francisco, CA, pp. 247–262.

[24] M. Herdy (1991) "Application of the evolutionsstrategie to discrete optimization problems," *Parallel Problem Solving from Nature*, H.-P. Schwefel and R. Männer (eds.), Springer, Berlin, pp. 188–192.

[25] A. Homaifar, S. Guan, G. E. Liepins (1993) "A new approach on the traveling salesman problem by genetic algorithms," *Proc. of the 5th Intern. Conf. on Genetic Algorithms*, S. Forrest (ed.), Morgan Kaufmann, San Mateo, CA, pp. 460–466.

[26] B. A. Julstrom (1995) "What have you done for me lately? Adapting operator probabilities in a steady-state genetic algorithm," *Proc. of the 6th Intern. Conf. on Genetic Algorithms*, L. J. Eshelman (ed.), Morgan Kaufmann, San Mateo, CA, pp. 81–87.

[27] G. E. Liepens, M. R. Hilliard, M. Palmer, and M. Morrow (1987) "Greedy genetics," *Genetic Algorithms and Their Applications: Proc. of the 2nd Intern. Conf. on Genetic Algorithms*, J. J. Grefenstette (ed.), Lawrence Erlbaum, Hillsdale, NJ, pp. 90–99.

[28] S. Lin and B. W. Kernighan (1972) "An effective heuristic algorithms for the traveling salesman problem," *Operations Research*, Vol. 21, 498–516.

[29] W. Lin, J. G. Delgado-Frias, D. C. Gause, and S. Vassiliadis (1995) "Hybrid Newton-Raphson genetic algorithm for the traveling salesman problem," *Cybernetics and Systems*, Vol. 26:4, pp. 387–412.

[30] B. Manderick, M. de Weger, and P. Spiessens (1991) "The genetic algorithm and the structure of the fitness landscape," *Proc. of the 4th Intern. Conf. on Genetic Algorithms*, R. K. Belew and L. B. Booker (eds.), Morgan Kaufmann, San Mateo, CA, pp. 143–150.

[31] K. Mathias and D. Whitley (1992) "Genetic operators, the fitness landscape and the traveling salesman problem," *Parallel Problem Solving from Nature, 2*, R. Männer and B. Manderick (eds.), Elsevier Science Publishers, Amsterdam, The Netherlands, pp. 219–228.

[32] H. Mühlenbein (1991) "Evolution in time and space—The parallel genetic algorithm," *Foundations of Genetic Algorithms*, G. J. E. Rawlins (ed.), Morgan Kaufmann, San Mateo, CA, pp. 316–337.

[33] H. Mühlenbein, M. Gorges-Schleuter, and O. Krämer (1987) "New solutions to the mapping problem of parallel systems: The evolution approach," *Parallel Computing*, Vol. 4, pp. 269–279.

[34] H. Mühlenbein, M. Gorges-Schleuter, and O. Krämer (1988) "Evolution algorithms in combinatorial optimization," *Parallel Computing*, Vol. 7, pp. 65–85.

[35] I. M. Oliver, D. J. Smith, and J. R. C. Holland (1987) "A study of permutation crossover operators on the traveling salesman problem," *Genetic Algorithms and Their Applications: Proc. of the 2nd Intern. Conf. on Genetic Algorithms*, J. J. Grefenstette (ed.), Lawrence Erlbaum, Hillsdale, NJ, pp. 224–230.

[36] K. F. Pal (1993) "Genetic algorithms for the traveling salesman problem based on a heuristic crossover operation," *Biological Cybernetics*, Vol. 69:5–6, pp. 539–546.

[37] S. Ronald (1995) "Finding multiple solutions with an evolutionary algorithm," *Proc. of the 1995 IEEE Conf. on Evolutionary Computation*, Vol. 2, IEEE Press, Piscataway, NJ, pp. 641–646.

[39] D. Smith (1985) "Bin packing with adaptive search," *Proc. of an Intern. Conf. on Genetic Algorithms and Their Applications*, J. J. Grefenstette (ed.), Lawrence Erlbaum, Hillsdale, NJ, pp. 202–206.

[40] T. Starkweather, S. McDaniel, K. Mathias, D. Whitley, and C. Whitley (1991) "A comparison of genetic sequencing operators," *Proc. of the 4th Intern. Conf. on Genetic Algorithms*, R. K. Belew and L. B. Booker (eds.), Morgan Kaufmann, San Mateo, CA, pp. 69–76.

[41] J. Y. Suh and D. Van Gucht (1987) "Incorporating heuristic information into genetic search," *Genetic Algorithms and Their Applications: Proc. of the 2nd Intern. Conf. on Genetic Algorithms*, J. J. Grefenstette (ed.), Lawrence Erlbaum, Hillsdale, NJ, pp. 100–107.

[42] H. Tamaki, H. Kita, N. Shimizu, K. Maekawa, and Y. Nishikawa (1994) "A comparison study of genetic codings for the traveling salesman problem," *Proc. of the 1st IEEE Conf. on Evolutionary Computation*, IEEE Press, Piscataway, NJ, pp. 1–6.

[43] N. L. J. Ulder, E. H. L. Aarts, H.-J. Bandelt, P. J. M. van Laarhoven, and E. Pesch (1991) "Genetic local search algorithms for the traveling salesman problem," *Parallel Problem Solving from Nature*, H.-P. Schwefel and R. Männer (eds.), Springer, Berlin, pp. 109–116.

[44] D. Whitley, T. Starkweather, and D. Fuquay (1989) "Scheduling problems and traveling salesman: The genetic edge recombination operator," *Proc. of the 3rd Intern. Conf. on Genetic Algorithms*, J. D. Schaffer (ed.), Morgan Kaufmann, San Mateo, CA, pp. 133–140.

ALLELES, LOCI, AND THE TRAVELING SALESMAN PROBLEM

by

David E. Goldberg

and

Robert Lingle, Jr.

Department of Engineering Mechanics
The University of Alabama, University, AL 35486

INTRODUCTION

We start this paper by making several seemingly not-too-related observations:

1) Simple genetic algorithms work well in problems which can be coded so the underlying building blocks (highly fit, short defining length schemata) lead to improved performance.

2) There are problems (more properly, codings for problems) that are GA-Hard --difficult for the normal reproduction+crossover+mutation processes of the simple genetic algorithm.

3) Inversion is the conventional answer when genetic algorithmists are asked how they intend to find good string orderings, but inversion has never done much in empirical studies to date.

4) Despite numerous rumored attempts, the traveling salesman problem has not succumbed to genetic algorithm-like solution.

Our goal in this paper is to show that, in fact, these observations are closely related. Specifically, we show how our attempts to solve the traveling salesman problem (TSP) with genetic algorithms have led to a new type of crossover operator, partially-mapped crossover (PMX), which permits genetic algorithms to search for better string orderings while still searching for better allele combinations. The partially-mapped crossover operator combines a mapping operation usually associated with inversion and subsequent crossover between non-homologous strings with a swapping operation that preserves a full gene complement. The resultant is an operator which enables both allele and ordering combinations to be searched with the implicit parallelism usually reserved for allele combinations in more conventional genetic algorithms.

In the remainder, we first examine and question the conventional notions of gene and locus. This leads us to consider the mechanics of the partially-mapped crossover operator (PMX). This discussion is augmented by the presentation of a sample implementation (for ordering-only problems) in Pascal. Next, we consider the effect of PMX by extending the normal notion of a schema by introducing the o-schemata (ordering schemata) or locus templates. This leads to simple counting arguments and survival probability calculations for o-schemata under PMX. These results show that with high probability, low order o-schemata survive PMX thus giving us a desirable result: an operator which searches among both orderings and allele combinations that lead to good fitness. Finally, we demonstrate the effectiveness of this extended genetic algorithm consisting of reproduction+PMX, by applying it to an ordering-only problem, the traveling salesman problem (TSP). Coding the problem as an n-permutation with no allele values, we obtain optimal or very near-optimal results in a well-known 10 city problem. Our discussion concludes by discussing extensions in problems with both ordering and value considered.

THE CONVENTIONAL VIEW OF POSITION AND VALUE

In genetic algorithm work we usually take a decidedly Mendelian view of our artificial chromosomes and consider genes which may take on different values (alleles) and positions (loci). Normally we assume that alleles decode to our problem parameter set (phenotype) in a manner independent of locus. Furthermore, we assume that our parameter set may then be evaluated by a fitness function (a non-negative objective function to be maximized). Symbolically, the fitness f depends upon the parameter set x which in turn depends upon the allele values v or more compactly $f = f(x(v))$. While this is certainly conventional, we need to ask whether this is the most general (or even most biological) way to consider this mapping. More to the point, shouldn't we also consider the possible effect of a string's ordering o on phenotype outcome and fitness. Mathematically there seems to be no good reason to exclude this possibility which we may write as $f=f(x(o,v))$.

While this generalization of our coding techniques is attractive because it would permit us to code ordering problems more naturally, we must make sure we maintain the implicit parallelism of the reproductive plans and genetic operators we apply to the generalized structures. Furthermore, because GA's are drawn from biological example we should be careful to seek natural precedent before committing ourselves to this extension. To find biological precedent for the importance of ordering as well as value we need only consider the sublayer of structure beneath the chromosome and consider the amino acid sequences that lead to particular proteins. At this level, the values (amino acids) are in no way tagged with meaning. There are only amino acids and they must appear in just the right order to obtain a useful outcome (a particular protein). Thus, there is biological example of outcomes that depend upon both ordering and value, and we do not risk the loss of the right flavor by considering them both.

Then, wherein lies our problem? If it is ok to admit both ordering and value information into our fitness evaluation, what is missing in our current thinking about genetic algorithms which prevents us from exploiting both ordering and value information concurrently? In previous work where ordering was considered at all (primarily for its effect on the creation of good, tightly linked, building blocks), the only ordering operator considered was inversion, a unary operator which picks two points along a single string at random and inverts the included substring (1). Subsequent crossover between non-homologous (differently ordered) strings occurred by mapping one string's order to the other, crossing via simple crossover, and unmapping the offspring. This procedure is well and good for searching among different allele combinations, but it does little to search for better orderings. Clearly the only operator effecting string order here is inversion, but the beauty of genetic algorithms is contained in the structured, yet randomized information exchange of crossover--the combination of highly fit notions from different strings. With only a unary operator to search for better string orderings, we have little hope of finding the best ordering, or even very good orderings, in strings of any substantial length. Just as mutation cannot be expected to find very good allele schemata in reasonable time, inversion cannot be expected to find good orderings in substantial problems. What is needed is a binary, crossover-like operator which exchanges both ordering and value information among different strings. In the next section, we present a new operator which does precisely this. Specifically, we outline an operator we call partially-mapped crossover (PMX) that exploits important similarities in value and ordering simultaneously when used with an appropriate reproductive plan.

PARTIALLY-MAPPED CROSSOVER (PMX) - MECHANICS

To exchange ordering and value information among different strings we present a new genetic operator with the proper flavor. We call this operator partially-mapped crossover because a portion of one string ordering is mapped to a portion of another and the remaining information is exchanged after appropriate swapping operations. To tie down these ideas we also present a piece of code used in the computational experiments to be presented later.

To motivate the partially-mapped crossover operator (PMX) we will consider different orderings only and neglect any value information carried with the ordering (this is not a limitation of the method because allele information can easily be tacked on to city name information). For example, consider two permutations of 10 objects:

$$A = 9\ 8\ 4\ 5\ 6\ 7\ 1\ 3\ 2\ 10$$
$$B = 8\ 7\ 1\ 2\ 3\ 10\ 9\ 5\ 4\ 6$$

PMX proceeds as follows. First, two positions are chosen along the string uniformly at random. The substrings defined from the first number chosen to the second number chosen are called the MAPPING SECTIONS. Next, we consider each mapping section separately by mapping the other string to the mapping section through a sequence of swapping operations. For example, if we pick two random numbers say 4 and 6, this defines the two mapping sections, 5-6-7 in string A, and 2-3-10 in string B. The mapping operation, say B to A, is performed by swapping first the 5 and the 2, the 6 and the 3, and the 7 and the 10, resulting in a well defined offspring. Similarly the mapping and swapping operation of A to B results in the swap of the 2 and the 5, the 3 and the 6, and the 10 and the 7. The resulting two new strings are as follows:

$$A' = 9\ 8\ 4\ 2\ 3\ 10\ 1\ 6\ 5\ 7$$
$$B' = 8\ 10\ 1\ 5\ 6\ 7\ 9\ 2\ 4\ 6$$

The mechanics of PMX is a bit more complex than simple crossover so to tie down the ideas completely we present a code excerpt which implements the operator for ordering-only structures in Figure 1. In this code, the string is treated as a ring and attention is paid to the order of selection of the two mapping section endpoints.

The power of effect of this operator, as with simple crossover, is much more subtle than is suggested by the simplicity of the string matching and swapping. Clearly, however, portions of the string ordering are being propagated untouched as we should expect. In the next section, we identify the type of information being exchanged by introducing the o-schemata (ordering schemata). We also consider the probability of survival of particular o-schemata under PMX.

PARTIALLY-MAPPED CROSSOVER - POWER OF EFFECT

In the analysis of a simple genetic algorithm with reproduction+crossover+mutation, we consider allele schemata as the underlying building blocks of future solutions. We also consider the effect of the genetic operators on the survivability of

Data Types and Constants

```
const max_city = 100;

type city      = 1..max_city;
     tourarray = array[1..max_city] of city;
```

Functions and Procedures (find_city, swap_city, cross_tour)

```
function find_city(city_name,n_city:city; var tour:tourarray):city;
var j1:integer;
begin
 j1:=0;
 repeat
  j1:=j1+1;
 until ( (j1>n_city) or (tour[j1]=city_name) );
 find_city:=j1;
end;

procedure swap_city(city_pos1,city_pos2:integer; var tour:tourarray);
var temp:city;
begin
 temp:=tour[city_pos1];
 tour[city_pos1]:=tour[city_pos2];
 tour[city_pos2]:=temp;
end;

procedure cross_tour(n_city,lo_cross,hi_cross:city;
                       var tour1_old,tour2_old,tour1_new,tour2_new:tourarray);
var j1,hi_test:integer;
begin
 hi_test := hi_cross + 1; if (hi_test>n_city) then hi_test:=1;
 tour1_new := tour1_old;
 tour2_new := tour2_old;
 if ( (lo_cross <> hi_cross) and (lo_cross <> hi_test) ) then begin
  j1 := lo_cross;
  while (j1<>hi_test) do begin (* mapped crossover on both tours *)
   swap_city(j1,find_city(tour1_old[j1],n_city,tour2_new),tour2_new);
   swap_city(j1,find_city(tour2_old[j1],n_city,tour1_new),tour1_new);
   j1:=j1+1; if (j1>n_city) then j1:= 1;
  end;
 end;
end;
```

Figure 1. Pascal Implementation of PMX - Partially Mapped
Crossover - procedure cross_tour.

important schemata. In a similar way, in our current work we consider the o-schemata or ordering schemata, and calculate the survival probabilities of important o-schemata under the PMX operator just discussed. As in the previous section we will neglect any allele information which may be carried along to focus solely on the ordering information; however, we recognize that we can always tack on the allele information for problems where it is needed in the coding.

To motivate an o-schema consider two of the 10-permutations:

```
C = 1 2 3 4 5 6 7 8 9 10
D = 1 2 3 5 4 6 7 9 8 10
```

As with allele schemata (a-schemata) where we appended a * (a meta-don't-care symbol) to our k-nary alphabet to motivate a notation for the schemata or similarity templates, so

do we here append a don't care symbol (the !) to mean that any of the remaining permutations will do in the don't care slots. Thus, in our example we have, among others, the following o-schemata common among structures C and D:

```
1  2  3  !  !  !  !  !  !  !
!  2  !  !  !  !  !  !  !  !
!  !  !  !  !  6  7  !  !  !
1  !  !  !  !  6  !  !  !  10
```

To consider the number of o-schemata, we take them with no positions fixed, 1 position fixed, 2 positions fixed, etc., and recognize that the number of o-schemata with exactly j positions fixed is simply the product of the number of combinations of groups of j among ℓ objects, $\binom{\ell}{j}$, times the number of permutations of groups of j among ℓ objects. Summing from 0 to ℓ (the string length) we obtain the number of o-schemata:

$$n_{os} = \sum_{0}^{\ell} \frac{\ell!}{(\ell-j)!j!} \frac{\ell!}{(\ell-j)!}$$

While this expression has not been reduced to closed form, it may be shown for large ℓ that the number of o-schemata is certainly greater than $(\ell!)^2$. Furthermore, it is easily shown that each particular string (permutation) is a representative of 2^ℓ o-schemata and that a population contains at most $n \cdot 2^\ell$ o-schemata.

Next we consider the survival probability of a particular o-schema under the partially-mapped crossover operator. The easiest way to calculate this is to use conditional probabilities over three mutually exclusive events: the o-schema is entirely contained within the match section (Event W-within), the schema is entirely outside the match section (Event O-outside), or the schema is cut by a cross point (Event C-cut). Thus, the probability of survival (Event S-survival) may be given:

$$P(S) = P(S|W)P(W) + P(S|O)P(O) + P(S|C)P(C)$$

Since the probability of surviving a cut is very low $(P(S|C)=0)$ we ignore this possibility and focus on the other two events. Assuming a cut length k, a defining length of the schema $\delta(s)$, and an o-schema of order (number of fixed positions) $o(s)$, the overall probability of survival (for large string length ℓ) may be estimated:

$$P(S) = \frac{k-\delta+1}{\ell} + \frac{\ell-k-\delta+1}{\ell} \left(1 - \frac{k+1}{\ell}\right)$$

Closer examination of this equation reveals two modes of survival. When the cut length is large with respect to the defining length, relatively short defining length schemata survive with high probability. The second and more subtle mode of survival occurs when short, low order schemata survive, because a small cut length dictates a small probability of interruption due to swapping. Together the two modes combine to pass through short, low order o-schemata so normal reproductive plans can sample these building blocks at near-optimal rates. Hence, PMX permits the same type of implicit parallelism to occur in both orderings and alleles as we have already witnessed using simple crossover on allele information alone.

A PURE ORDERING PROBLEM - THE TRAVELING SALESMAN PROBLEM (TSP)

In some sense we've presented this paper in the reverse order of discovery. We did not 1) admit ordering information, 2) discover PMX and o-schemata, and 3) apply reproduction+PMX to the traveling salesman problem. In fact, by trying to solve the TSP with genetic algorithms, we were led to PMX-like operators, then o-schemata, and

finally PMX. The traveling salesman problem is a pure ordering problem (2,3,4) where one attempts to find the optimal tour (minimum cost path which visits each of n cities exactly once). The TSP is possibly the most celebrated combinatorial optimization problem of the past three decades, and despite numerous exact (impractical) and heuristic (inexact) methods already discovered, the method remains an active research area in its own right, partially because the problem is part of a class of problems considered to be NP-complete for which no polynomial time solution is believed to exist. Our interest in the TSP sprung mainly from a concern over claims of genetic algorithm robustness. If GA's are robust, why have the rumored attempts at "solving" the TSP with GA's failed. This concern led us to consider many schemes for coding the ordering information, with strange codes, penalty functions, and the like, but none of these had the appropriate flavor--the building blocks didn't seem right. This led us to consider the current scheme, which does have appropriate building blocks, and as we shall soon see, does (in one problem) lead to optimal or near-optimal results.

The specific problem we consider is Karg and Thompson's well-studied 10 city problem (4). While a 10 city problem is no final touchstone of success, it does contain 9! alternatives (the GA knows nothing of the problem's symmetry which reduces this number to (9!)/2). We code the problem as a normalized (city 1 in the first position) 10-permutation and apply reproduction and PMX to successive populations. We use roulette wheel reproduction with selection probabilities set in the normal way, and fitnesses are created from costs and scaled by subtracting string cost from population maximum cost, $f_i = c_{max} - c_i$. We choose initial populations, popsize=200, at random. This number was selected to obtain a rich spread of order 2 o-schemata in the population. This requires a population size proportional to $n(n-1)$ or roughly n^2. It might be useful to have order 3 schemata as well, but this may require larger populations than we are used to working with.

We present the results of two runs on the 10 city problem in Figures 2 and 3. Figure 2 shows the population average cost with each successive generation. The crossover probability was set at 0.6 so each generation represents roughly 120 new function evaluations (0.6*200). Figure 3 shows the population best results with successive generations. As we can see, run 1 reaches the optimal (!!) result rather quickly, while run 2 converges on a very near-optimal tour (we only ran twenty generations--there was still enough diversity left so improvement was possible in run 2). The best of run 1 was indeed the Karg and Thompson optimum, tour 1-2-3-4-5-10-9-8-6-7 with cost=378. The best of run 2 was a near-optimum, the tour 1-2-3-10-9-5-4-6-8-7 with cost=381. We are

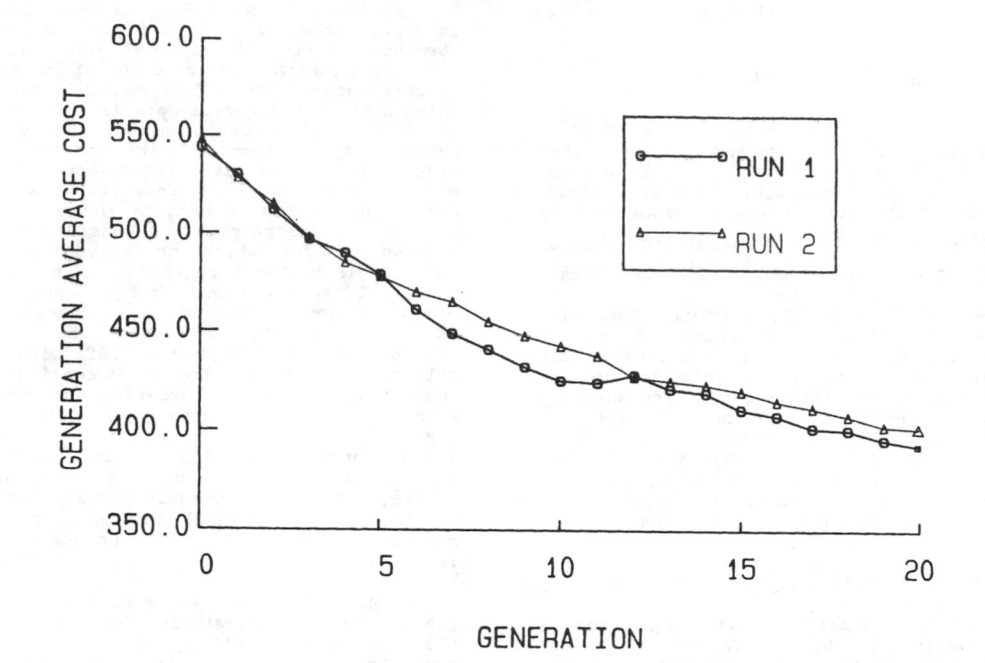

Figure 2. Generation Average Cost vs. Generation for 10 City TSP

currently working on a 20 city problem and a 33 city problem, although we need to do some reprogramming to fit the large population sizes into our IBM PC's. We also have built in an inversion operator, but have not had a chance to test its effect on average and best results.

CONCLUSIONS

In this paper we have examined a new type of crossover operator, partially-mapped crossover (PMX), for the exploration of codings where ordering and allele information may directly or indirectly effect fitness values. The mechanics of the operator have been described, and an ordering-only implementation has been presented in Pascal. The power of effect of the new operator has been analyzed using an extension to the concept of schemata called the o-schemata (ordering schemata). Simple counting arguments have been put forward which show the vast amount of information contained in the o-schemata, and survival probabilities have been estimated for o-schemata under the PMX operator. The result is an operation which preserves ordering building blocks (and allele building blocks if they are attached) so orderings and allele combinations may be explored with implicit parallelism.

The new operator is tested in an ordering-only problem, the traveling salesman problem. Using reproduction+PMX in two runs, optimal or very near optimal results are found in a well-known 10 city problem after exploring a small portion of the tour search space. We are continuing our work by testing the method in larger problems, but we are encouraged with the GA-like performance obtained on our first test.

This work has important implications for improving more general GA-search in problems where both allele combinations and ordering information are important. The binary operation of PMX does permit the randomized, yet structured, information exchange among both alleles and ordering building blocks which simple crossover promotes among allele schemata alone. This should assist us in our efforts to successfully apply genetic algorithms to ever more complex problems.

REFERENCES

1. Holland, J. H., Adaptation in Natural and Artificial Systems, University of Michigan Press, Ann Arbor, 1975.

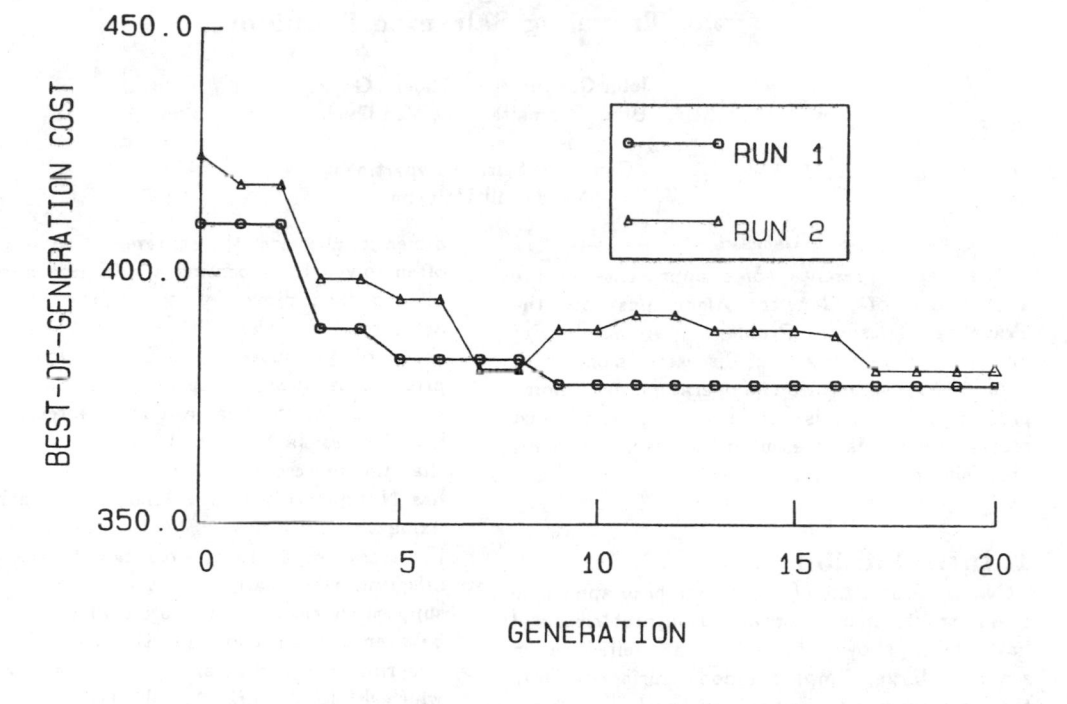

Figure 3. Best-of-Generation Cost for 10 City TSP

2. Bellmore, M. and G. L. Neuhauser, "The Traveling Salesman Problem: A Survey," Operation Research, vol. 16, May-June 1968, pp. 538-558.

3. Parker, R. G. and R. L. Rardin, "The Traveling Salesman Problem: An Update of Research," Naval Research Logistics Quarterly, vol. 30, 1983, pp. 69-96.

4. Karg, R. L. and G. L. Thompson, "A Heuristic Approach to Solving Travelling Salesman Problems," Management Science, vol. 10, no. 2, January 1964, pp. 225-248.

Genetic Algorithms for
the Traveling Salesman Problem

John Grefenstette[1], Rajeev Gopal,
Brian Rosmaita, Dirk Van Gucht

Computer Science Department
Vanderbilt University

Abstract

This paper presents some approaches to the application of Genetic Algorithms to the Traveling Salesman Problem. A number of representation issues are discussed along with several recombination operators. Some preliminary analysis of the Adjacency List representation is presented, as well as some promising experimental results.

1. Introduction

Genetic Algorithms (GA's) have been applied to a variety of function optimization problems, and have been shown to be highly effective in searching large, complex response surfaces even in the presence of difficulties such as high-dimensionality, multimodality, discontinuity and noise [4]. However, GA's have not been applied extensively to combinatorial problems. The major obstacle is in finding an appropriate representation. This paper presents some approaches to the design of GA's for a well known combinatorial optimization problem -- the Traveling Salesman Problem (TSP). The TSP is easily stated: Given a complete graph with N nodes, find the shortest Hamiltonian path through the graph. (In this paper, we will assume Euclidean distances between nodes.) The TSP is NP-Hard, which probably means that any algorithm which computes an exact solution of the TSP requires an amount of computation time which is exponential in N, the size of the problem [5]. In addition to its many important applications, the TSP is often used to illustrate heuristic search methods [2,7,8], so it is natural to investigate the use of GA's for this problem.

Choosing an appropriate representation is the first step in applying GA's to any optimization problem. If the problem involves searching an N-dimensional space, the representation problem is often solved by allocating a sufficient number of bits to each dimension to achieve the desired accuracy. For the TSP, the search space is a space of permutations and the representation problem is more complex. Consider a path representation in which a tour is represented by a list of cities: (a b c d e f). The first problem is that the representation is not unique: each tour has N representations. This can be solved by fixing the initial city. Another problem is that the crossover operator does not generally yield offspring which are legal tours. For example, suppose we cross tours (a b c d e) and (a d e c b) between the third and fourth cities. We get as offspring (a b c c b) and (a d e d e), neither of which are legal tours. Finally, there is a problem in applying the hyperplane analysis of GA's to this representation. The definition of a hyperplane is unclear in this representation. For example, (a # # # #) appears to be a first order hyperplane, but it contains the entire space. The problem is that in this representation, the semantics of an allele in a given position depends on the surrounding alleles. Intuitively, we hope that GA's will tend to construct good solutions by identifying good building blocks and eventually combining these to get larger building blocks. For the TSP, the basic building blocks are edges. Larger building blocks correspond to larger subtours. The path representation does not lend itself to the description of edges and longer subtours in ways which are useful to the GA.

In section 2, we present two representations which offer some improvements over the path representation. Section 3 discusses the design of a heuristic recombination operator for what we consider to be the most promising representation. In section 4, some preliminary experimental

---

[1]Research supported in part by the National Science Foundation under Grant MCS-8305693.

results are described for the TSP. Section 5 discusses some future directions.

2. Representations for TSP

2.1. Ordinal Representation

In the ordinal representation, a tour is described by a list of N integers in which the ith element can range from 1 to (N-i+1). Given a path representation of a tour, we can construct the ordinal representation TourList as follows: Let FreeList be an ordered list of the cities. For each city in the tour, append the position of that city in the FreeList to the TourList and delete that city from the FreeList. For example, the path tour (a c e d b) corresponds to an ordinal tour (1 2 3 2 1) as shown:

TourList	FreeList
()	(a b c d e)
(1)	(b c d e)
(1 2)	(b d e)
(1 2 3)	(b d)
(1 2 3 2)	(b)
(1 2 3 2 1)	()

Note that it is necessary to fix the starting city to avoid multiple representation of tours.

A similar procedure provides a mapping from the ordinal representation back to the path representation. In fact, the mapping between the two representations is one-to-one.

The primary advantage of the ordinal representation is that the classical crossover operator may be freely applied to the ordinal representation and will always produce the ordinal representation of a legal tour. However, the results of crossover may not bear much relation to the parents when translated to the path representation. For example, consider the following two tours:

ordinal tours	path tours
(1 2 3 2 1)	(a c e d b)
(2 4 1 1 1)	(b e a c d)

Suppose that we cross the ordinal tours between the second and third positions. We get the following tours as offspring:

ordinal tours	path tours
(1 2 1 1 1)	(a c b d e)
(2 4 3 2 1)	(b e d c a)

The subtours corresponding to the genes in the ordinal tours to the left of the crossover point do not change. However, the subtours corresponding to genes to the right of the crossover points are disrupted in a fairly random way. Furthermore, the closer the crossover point is to the front of the tour, the greater the disruption of subtours in the offspring.

As predicted by the above consideration of subtour disruptions, experimental results using the ordinal representation have been generally poor. In most cases, a GA using the ordinal representation does no better than random search on the TSP.

2.2. Adjacency Representation

In the adjacency representation, a tour is described by a list of cities. There is an edge in the tour from city i to city j iff the allele in position i is j. For example, the path tour (1 3 5 4 2) corresponds to the adjacency tour (3 1 5 2 4). Note that any tour has exactly one adjacency list representation.

2.2.1. Crossover Operators

Unlike the ordinal representation, the adjacency representation does not allow the classical crossover operator. Several modified crossover operators can be defined.

Alternating Edges

Using the alternating edges operator, an offspring is constructed from two parent tours as follows: First choose an edge at random from one parent. Then extend the partial tour by choosing the appropriate edge from the other parent.

Continue extending the tour by choosing edges from alternating parents. If the parent's edge would introduce a cycle into a partial tour, then extend the partial tour by a random edge which does not introduce a cycle. Continue until a complete tour is constructed.

For example, suppose we have

$$mom = (\ 2\ 3\ 4\ 5\ 6\ 1\ )$$
$$dad = (\ 2\ 5\ 1\ 6\ 4\ 3\ )$$

Then we might get the following offspring:

$$kid\ = (\ 2\ 5\ 4\ 1\ 6\ 3\ )$$

where the only random edge introduced into the offspring is the edge (4 1). All other edges were inherited by alternately choosing edges from parents, starting with the edge (1 2) from mom.

Experimental results with the alternating edges operator have been uniformly discouraging. The obvious explanation seems to be that good subtours are often disrupted by the crossover operator. Ideally, an operator ought to promote the development of coadapted alleles, or in the TSP, longer and longer high performance subtours. The next operator was motivated by the desire to preserve longer parental subtours.

Subtour Chunks

Using the subtour chunking operator, an offspring is constructed from two parent tours as follows: First choose a subtour of random length from one parent. Then extend the partial tour by choosing a subtour of random length from the other parent. Continue extending the tour by choosing subtours from alternating parents. During the selection of a subtour from a parent, if the parent's edge would introduce a cycle into a partial tour, then extend the partial tour by a random edge which does not introduce a cycle. Continue until a complete tour is constructed.

Subtour chunking performed better than alternating edges, as expected, but the absolute performance was still unimpressive. An analysis of the allocation of trials to hyperplanes provide a partial explanation for the poor performance of this operator.

2.2.2. Hyperplane Analysis

The primary advantage of the adjacency representation is that it permits the kind of hyperplane analysis which has been applied to the N-dimensional function optimization GA paradigm [1,3,6]. Hyperplanes defined in terms of a single defining position correspond to the natural building blocks, i.e., edges, for the TSP problem. For example, the hyperplane (# # # 2 #) is the set of all permutations in which the edge (4 2) occurs. We briefly summarize the main points of the classical hyperplane analysis of GA's: In the absence of recombination operators, selection of structures for reproduction in proportion to the structure's observed relative performance allocates trials to all represented hyperplanes in the population (roughly) according to the following formula:

$$M(H,t+1) = M(H,t)*(\ u(H,t)\ /\ u(P,t)\ )$$

where

$M(H,t) = $ # of representatives of H at time t

$u(H,t) = $ observed performance of H at time t

$u(P,t) = $ mean performance of population at time t.

The elements of any hyperplane partition compete against the other elements of that partition, with the better performing elements eventually propagating through the population. This in turn leads to a reduction in the dimensionality of the search space, and the construction of larger high performance building blocks.

In the adjacency representation, a first order hyperplane partition consists of all of the hyperplanes which are defined on the same position. For example:

{ (# # # 1 #), (# # # 2 #), (# # # 3 #),
(# # # 5 #) }

is a first order hyperplane partition. Each element of the partition contains an equal number of tours. Selection is supposed to distinguish among the elements of this partition and to favor the high performance hyperplanes. However, the following theorem shows that selection has very little information on which to allocate trials to competing first order hyperplanes.

Theorem 1. Suppose that H_{ab} and H_{ac} are two first-order hyperplanes defined by the edges (a b) and (a c), respectively, in a Euclidean TSP. Then $| u(H_{ab}) - u(H_{ac}) | \leq 4(ab + ac)$ where ab and ac represent the lengths of the edges (a b) and (a c), respectively

Proof. We show that there is a one-to-one mapping f between the tours in H_{ab} and the tours H_{ac} such that if x is a tour in H_{ab} and y $=$ f(x) is the corresponding tour in H_{ac}, then

$$| Length(y) - Length(x) | \leq 4(ab+ac).$$

The theorem follows directly.

The following illustrates the mapping f:

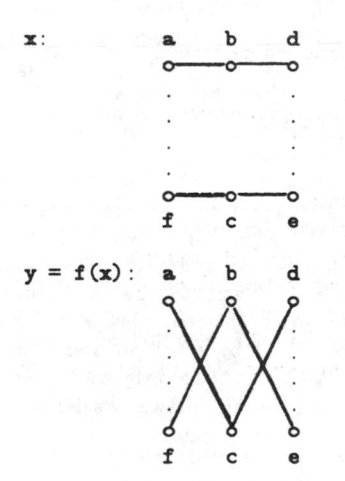

That is, y is obtained by exchanging the nodes b and c in the tour x. Using the triangle inequality, it is easy to show that:

$$-(4ab + 2ac) \leq Distance(y) - Distance(x)$$
$$\leq (4ac + 2ab).$$

So

$$| Distance(y) - Distance(x) | \leq 4(ab+ac).$$
QED.

In practice, the observed difference between competing first order hyperplanes is usually an order of magnitude less than the bounds in the theorem. And since the overall tour length is generally very large compared to the bound in the theorem, there is generally no significant difference between the mean relative performance of any two competing first order hyperplanes. Our experimental studies have shown that the difference in the observed performance of competing first order hyperplanes in a TSP of size 20 is generally less than 5% of the mean population tour length. In larger problems, this difference can be expected to rapidly approach zero.

One might suspect that the TSP is not a suitable problem for GA's, that the TSP is in some sense GA-Hard. Bethke[1] characterizes some problems for which GA's are unsuitable. Informally, Bethke shows that there are functions and representations for which the low order hyperplanes can mislead the GA into allocating trials to suboptimal areas of the search space. However, Bethke's techniques, which involve the Walsh transform of the objective function, apply to one-dimensional functions of a real variable using a fixed-point representation. A similar set of results may be derivable for combinatorial problems using the adjacency representation. But Theorem 1 does not indicate that the information in the first order hyperplanes of the adjacency representation is misleading, just that it is buried. In other words, measuring the fitness of a tour by the tour length may be too crude a measure for apportioning credit. We now describe a crossover operator which performs a secondary apportionment of credit at the level of individual alleles.

3. Heuristic Crossover

Theorem 1 shows that selection alone may not be able to properly allocate trials to first order hyperplanes, given our adjacency representation for the TSP. The heuristic crossover operator attempts to perform a secondary apportionment of credit at the allele level. This operator constructs an offspring from two parent tours as follows: Pick a random city as the starting point for the child's tour. Compare the two edges leaving the starting city in the parents and choose the shorter edge. Continue to extend the partial tour by choosing the shorter of the two edges in the parents which extend the tour. If the shorter parental edge would introduce a cycle into the partial tour, then extend the tour by a random edge. Continue until a complete tour is generated.

In order to compare this operator with the previous two recombination operators, 1000 random pairs of parents were chosen for a TSP of size 20. For each pair of parents, an offspring was constructed according to each of the crossover operators. For all three operators, the offspring generally inherited about 30% of the edges from each parent. The remaining 40% were random edges introduced by the recombination operator to create a legal tour. For the first two operators, the offspring generally show no improvement in overall tour length when compared to the better parent. Not surprisingly, the heuristic crossover produces offspring which are, on average, about 10% better than the better parent. It seems reasonable that such an improvement should give selection a way to promote the propagation of good edges through the population. The next section shows some experimental results which confirm this expectation.

It is important to note that, with the proper choice of data structures, the heuristic crossover operator can be implemented to run as a linear function of the length of the structures [9]. This implies that, if E is the number of trials and N is the number of cities, our GA's for the TSP run with asymptotic complexity O(EN), the same as pure random search.

4. Experimental Results

This section describes some experiments with the adjacency representation and the heuristic crossover operator. For each experiment, N cities were randomly placed in a square Euclidean space. The initial population consisted of randomly generated tours. The selection method was based on the expected value model. The crossover rate was set at 50%, and there was no explicit mutation operator.

Figure 1 shows the results of a 50 city problem, Figure 2 shows a 100 city problem and Figure 3 shows a 200 city problem. Each Figure shows a representative tour from the initial population, the best tour obtained part way through the search, and the best tour obtained after the entire search, along with a randomly selected tour in the final population. It can be seen, especially in Figues 2 and 3, that good subtours tend to survive and to propagate. The figures also show that there is still a good deal of diversity in the final population.

Statistical techniques [2] allow us to estimate that the expected length of an optimal tour for experiment 1 is approximately 37.45. The optimal tour obtained by the GA differs from this expected optimum by about 25%. After an equal number of trials, random search produces a best tour of length 148.6, nearly 300% longer than the optimal tour. The optimal tour obtained in experiment 2 differs from the expected optimum by 16%. The optimal tour obtained in experiment 3 differs from the expected optimum by about 27%. These results are encouraging and suggest that further investigation of this approach is warranted.

Experiments show that GA's which use heuristic crossover but not selection perform better than random search but significantly worse than GA's which use both selection and heuristic crossover. That is, there appears to be a symbiotic relationship between the two levels of credit assignment performed by selection and heuristic crossover. We are currently working on clarifying the relationship between selection and the heuristic crossover operator.

5. Future Directions

This papers presents some preliminary observations and experiments. Many more questions about the TSP need to be investigated. Some interesting future projects include:

Combining GA's with other heuristics. In may be useful to heuristically choose the initial population of tours. For example, the nearest neighbor algorithm can generate a set of relatively good tours when started from various initial cities. For very large problems, nearest neighbor can be approximated by choosing a random set of cities and taking the one closest to the current city. Heuristics could also be invoked at the end of the GA to do some local modifications to the tours in the final population. For example, the Figures shows many opportunities for improving the final tour by some local edge reversals.

Comparison with simulated annealing. Simulated annealing is another randomized heuristic algorithm which has been applied to very large ($N > 1000$) TSP's. From the published literature on simulated annealing [2,7], it appears that our results are at least competitive. A careful comparison of these two techniques would be very interesting.

Effects of GA parameters. There are several control parameters involved in any GA implementation, such as population size, crossover rate, etc. which may have an effect on the performance of the system. The proposed GA's are sufficiently different from previous GA's that it might be useful to investigate the effects of these parameters for the TSP.

Other combinatorial applications. How do the ideas developed thus far apply to combinatorial problems other than the TSP?

References

1. A. D. Bethke, *Genetic algorithms as function optimizers*, Ph. D. Thesis, Dept. Computer and Communication Sciences, Univ. of Michigan (1981).

2. E. Bonomi and J.-L. Lutton, "The N-city traveling salesman problem: statistical mechanics and the Metropolis Algorithm," *SIAM Review* Vol. *26*(4), pp. 551-569 (Oct. 1984).

3. K. A. Dejong, *Analysis of the behavior of a class of genetic adaptive systems*, Ph. D. Thesis, Dept. Computer and Communication Sciences, Univ. of Michigan (1975).

4. K. A. Dejong, "Adaptive system design: a genetic approach," *IEEE Trans. Syst., Man, and Cyber.* Vol. *SMC-10*(9), pp. 556-574 (Sept 1980).

5. M. R. Garey and D. S. Johnson, *Computers and Intractability*, W. H. Freeman Co., San Fransisco (1979).

6. J. H. Holland, *Adaptation in Natural and Artificial Systems*, Univ. of Michigan Press, Ann Arbor (1975).

7. S. Kirkpatrick, C. D. Gelatt, and M. P. Vecchi, "Optimization by simulated annealing," *Science* Vol. *220*(4598), pp. 671-680 (May 1983).

8. J. Pearl, *Heuristics*, Addison-Wesley, Menlo Park (1984).

9. B. J. Rosmaita, *Exodus: An extension of the the genetic algorithm to problems dealing with permutations*, M.S. Thesis, Computer Science Department, Vanderbilt University (Aug. 1985).

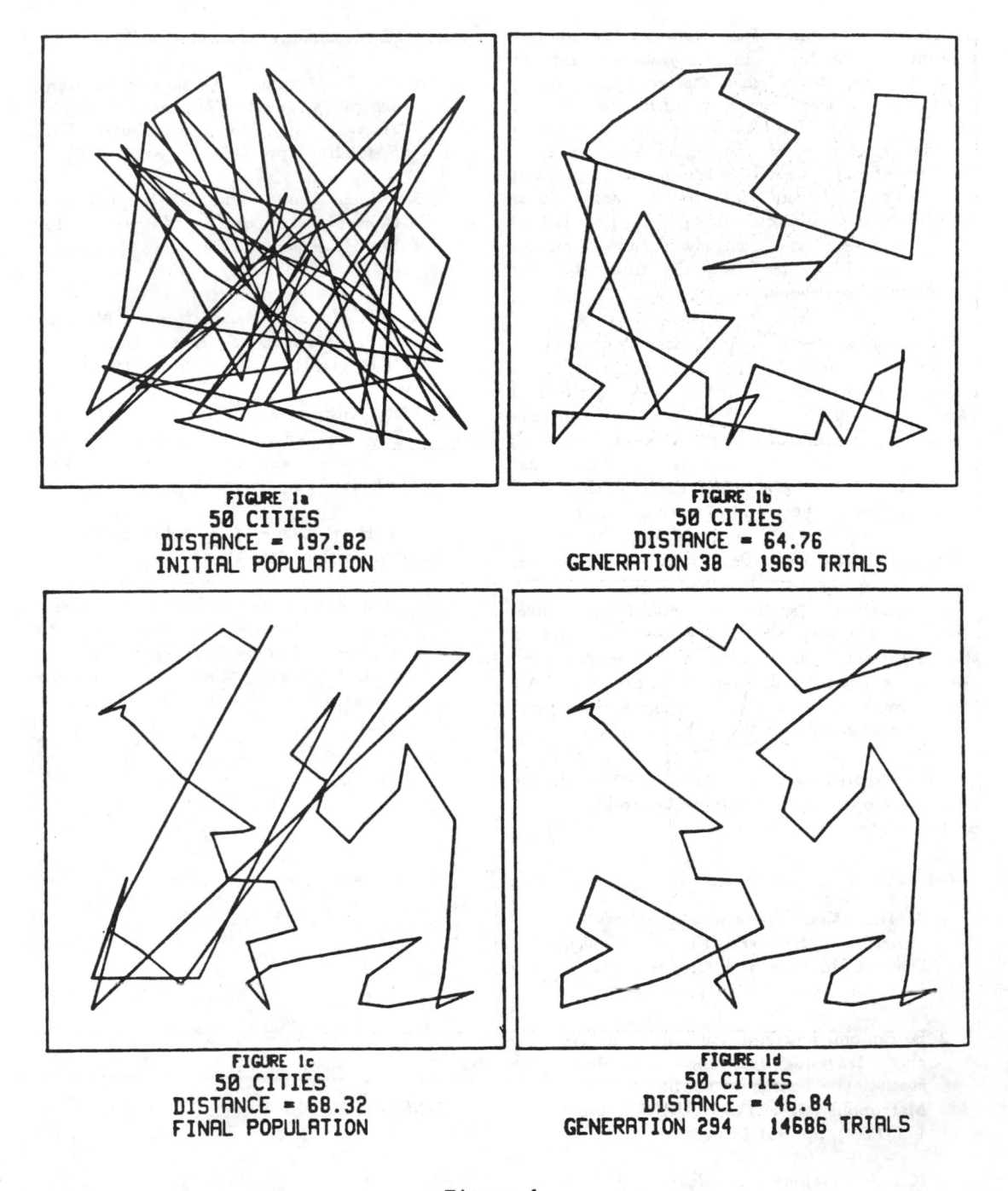

FIGURE 1a
50 CITIES
DISTANCE = 197.82
INITIAL POPULATION

FIGURE 1b
50 CITIES
DISTANCE = 64.76
GENERATION 38   1969 TRIALS

FIGURE 1c
50 CITIES
DISTANCE = 68.32
FINAL POPULATION

FIGURE 1d
50 CITIES
DISTANCE = 46.84
GENERATION 294   14686 TRIALS

Figure 1.

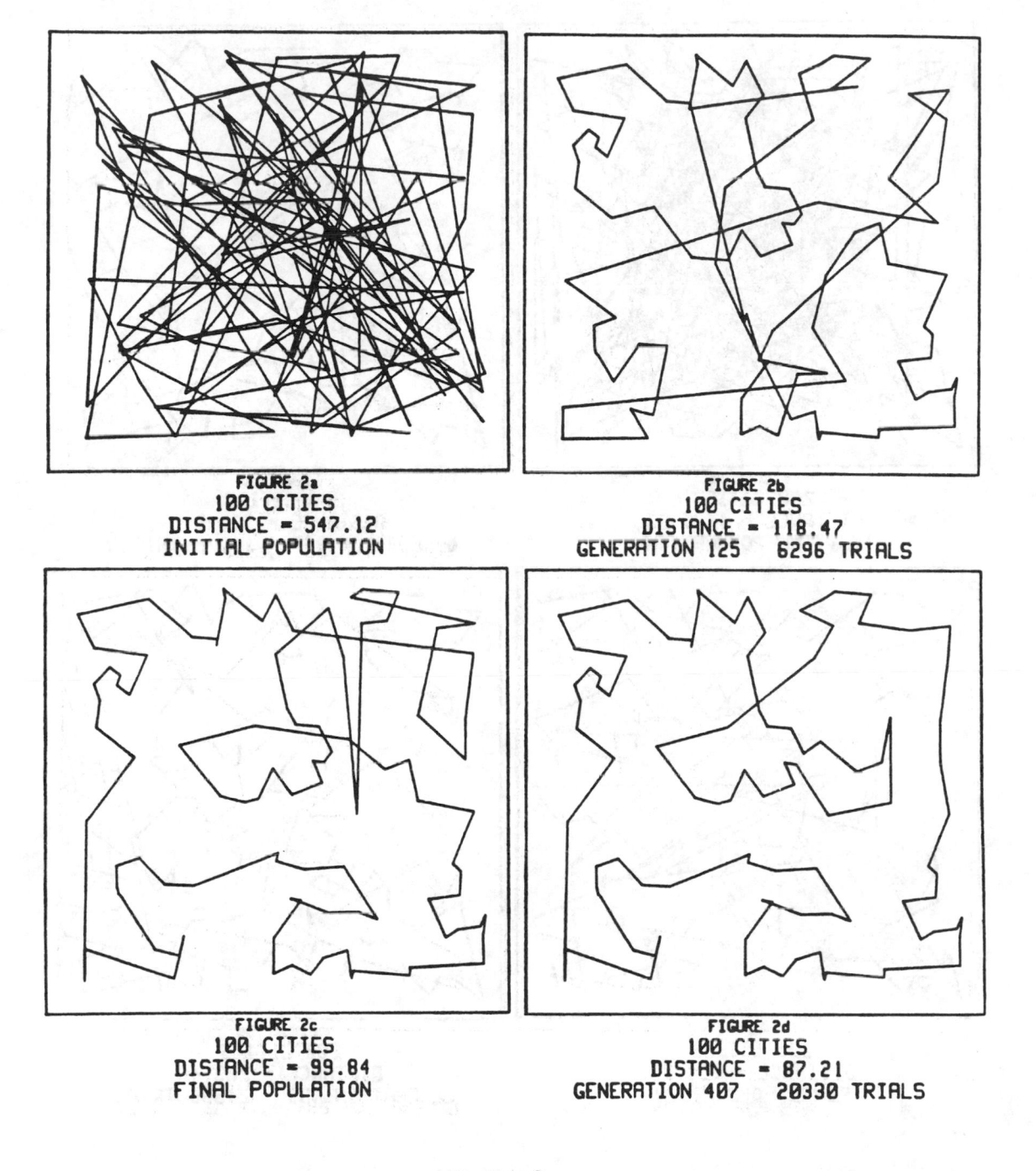

FIGURE 2a
100 CITIES
DISTANCE = 547.12
INITIAL POPULATION

FIGURE 2b
100 CITIES
DISTANCE = 118.47
GENERATION 125   6296 TRIALS

FIGURE 2c
100 CITIES
DISTANCE = 99.84
FINAL POPULATION

FIGURE 2d
100 CITIES
DISTANCE = 87.21
GENERATION 407   20330 TRIALS

Figure 2.

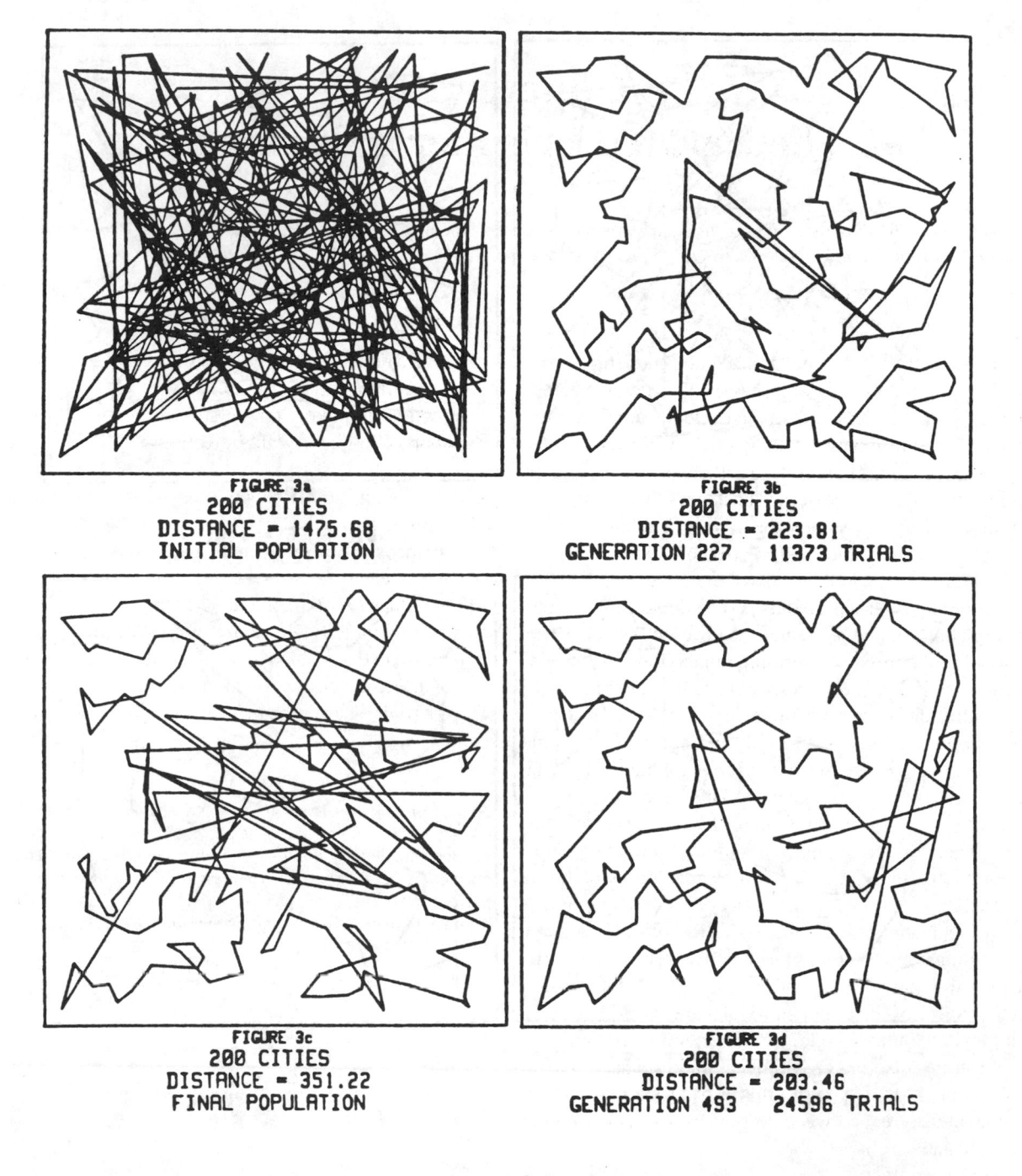

FIGURE 3a
200 CITIES
DISTANCE = 1475.68
INITIAL POPULATION

FIGURE 3b
200 CITIES
DISTANCE = 223.81
GENERATION 227   11373 TRIALS

FIGURE 3c
200 CITIES
DISTANCE = 351.22
FINAL POPULATION

FIGURE 3d
200 CITIES
DISTANCE = 203.46
GENERATION 493   24596 TRIALS

Figure 3.

Chapter 19
The Iterated Prisoner's Dilemma

R. Axelrod (1987) "The Evolution of Strategies in the Iterated Prisoner's Dilemma," *Genetic Algorithms and Simulated Annealing,* L. Davis (ed.), Pitman, London, pp. 32–41.

THE conditions that foster the evolution of cooperative behaviors among individuals of a species (or among species) are not generally well understood. In intellectually advanced social animals, cooperation between individuals, when it exists, is often ephemeral and quickly reverts to selfishness, with little or no clear indication of the specific circumstances that prompt the change. Simulation games such as the prisoner's dilemma have been used to gain insight into the precise conditions that promote the evolution of either decisively cooperative or selfish behavior in a community of individuals. Even simple games often generate very complex and dynamic optimization surfaces. The computational problem is to determine reliably any ultimately stable strategy (or strategies) for a specific game situation, or alternatively indicate that no such strategies exist.

The prisoner's dilemma is an easily defined nonzero-sum, noncooperative game. The term *nonzero-sum* indicates that whatever benefits accrue to one player do not necessarily imply similar penalties imposed on the other player. The term *noncooperative* indicates that no preplay communication is permitted between the players. Typically, two players are involved in the game, each having two alternative actions: cooperate or defect. Mutual cooperation implies increasing the total gain of both players; defecting implies increasing one's own reward at the expense of the other player. The optimal policy for a player depends on the policy of the opponent (Hofstadter, 1985, p. 717). Against a player who always defects, defection is the only rational play. But it is also the only rational play against a player who always cooperates, for such a player is a fool. Only when there is some mutual trust between the players does cooperation become a reasonable policy.

The general form of the game is represented by Table 19.1 (after Scodel et al., 1959). The game is conducted on a trial-by-trial basis (a series of moves). Each player must choose to cooperate or defect on each trial. The payoff matrix defining the game is subject to the following constraints (Rapoport, 1966):

$$2R > S + T \qquad [1]$$

$$T > R > P > S. \qquad [2]$$

The first rule prevents any incentive to alternate between S and T because the payoff for mutual cooperation R exceeds the average of S and T. The second rule motivates each player to play noncooperatively.

Defection dominates cooperation in a game-theoretic sense (Rapoport, 1966). But joint defection results in a payoff, P, to each player that is smaller than the payoff R that could be gained through mutual cooperation. If the game is played for only a single trial, the clearly compelling behavior is defection. If the game is iterated over many trials, the correct behavior becomes less clear. If there is no possibility for communication with the other player, then the iterated game degenerates into a sequence of independent trials, with defection again yielding the maximum expected benefit. But if the outcome of earlier trials can affect the decision-making policy of the players in subsequent trials, learning can take place and both players may seek to cooperate.

TABLE 19.1. The general form of the payoff function in the prisoner's dilemma, where R (reward) is the payoff to each player for mutual cooperation, S (sucker) is the payoff for cooperating when the other player defects, T (temptation) is the payoff for defecting when the other player cooperates, and P (punishment) is the payoff for mutual defection. An entry (α, β) indicates the payoffs to players A and B, respectively.

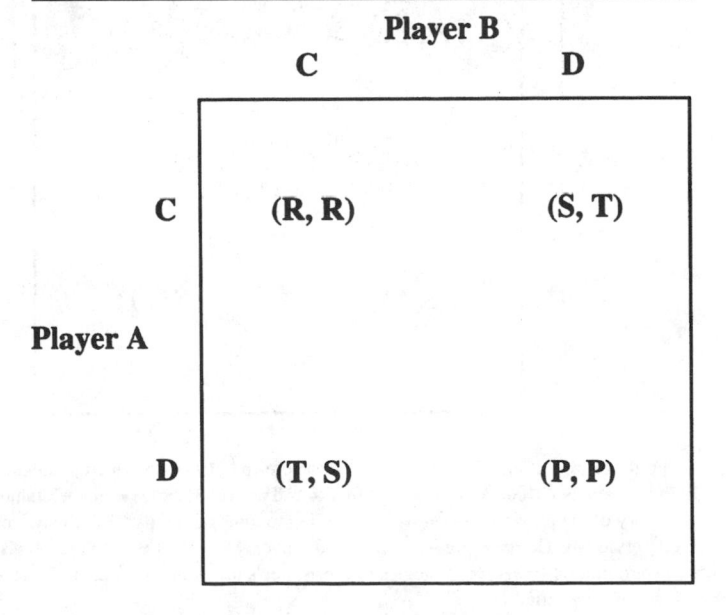

		Player B	
		C	**D**
Player A	**C**	(R, R)	(S, T)
	D	(T, S)	(P, P)

In 1979, Robert Axelrod organized a prisoner's dilemma tournament and solicited strategies from game theorists (Axelrod 1980a). The 14 entries competed, along with a 15th entry: on each move, cooperate or defect with equal probability. Each strategy was played against all others and itself over a sequence of 200 moves. The specific payoff function used is shown in Table 19.2. The winner of the tournament, submitted by Anatol Rapoport, was "tit-for-tat":

(1) Cooperate on the first move.
(2) Otherwise, mimic whatever the other player did on the previous move.

Subsequent analysis by Axelrod (1984), Hofstadter (1985, pp. 721–723), and others, indicated that this tit-for-tat strategy is robust because it never defects first and is never taken advantage of for more than one iteration at a time. Boyd and Lorberbaum (1987) indicated that tit-for-tat is not evolutionarily stable (in the sense of Maynard Smith, 1982). Nevertheless, in a second tournament, Axelrod (1980b) collected 62 entries, and again the winner was tit-for-tat.

Axelrod (1984) noted that just a few of the 62 entries in the second tournament could be used to reasonably account for how well a given strategy did with the entire set. In the seminal paper that is reprinted here, Axelrod (1987) utilized eight strategies as opponents for a simulated evolving population of policies based on a genetic algorithm approach. Consideration was given to the set of strategies that is deterministic and uses

outcomes of the three previous moves to determine a current move. Because there were four possible outcomes for each move, there were 4^3, or 64, possible sequences of three possible moves. The coding for a policy was therefore determined by a string of 64 bits, where each bit corresponded with a possible instance of the preceding three interactions, and six additional bits that defined the player's move for the initial combinations of under three iterations. Thus there were 2^{70} (about 10^{21}) possible strategies.

The simulation was conducted using a population of 20 strategies executed repeatedly against the eight representatives, with a modified version of proportional selection being used for replication, and mutation and crossover (described as operating on one or more places to break a parental strategy) being used to generate new strategies. This framework was then extended to require the evolving policies to play against each other, rather than against the eight representatives. The typical results indicated that populations initially generated mutual defection, but subsequently evolved toward mutual cooperation. Axelrod (1987) tested the effects of executing the simulation with and without recombination and concluded that recombination aided the search for cooperative strategies (cf. Fogel, 1991, for an alternative interpretation). The paper also offered conclusions indicating the potential benefit of saving lesser adapted individuals and the possibility of using similar simulations to gain a better understanding of host-parasite relationships. The language used is reminiscent of Langton (1987), using simulations to study "life as it could be, rather than life as it is."

Axelrod (1987) influenced a series of papers that explore the iterated prisoner's dilemma using evolutionary computation (a complete review is beyond the scope of this introduction). Independently, Miller (1989) and Fogel (1991, 1993) replaced the fixed-length coding scheme of Axelrod (1987) with finite state automata (Moore and Mealy machines, respectively) that can generate more flexible behaviors. Fujiki and Dickinson (1987)[1] and Crowley (1996) evolved rule-based programs to play the prisoner's dilemma (in the form of LISP and classifier systems, respectively), while Angeline (1994) extended the effort of Fogel (1993) to examine the iterated dilemma when the above inequality (1) was violated, thus allowing for the promotion of retaliatory behaviors that extend over multiple iterations.

Nowak (1990) examined the effect of probabilistic prisoner's dilemma strategies and showed that in the face of errors in play, two tit-for-tat players can execute an endless succession of "mutual recriminations." Nowak and Sigmund (1992) proposed that tit-for-tat is not an aim of the iterated prisoner's dilemma game but rather a pivotal strategy that leads to other more successful strategies. Nowak and May (1992) studied simple prisoner's dilemma strategies in spatial settings where each strategy can only meet other strategies in a locally defined neighborhood and

TABLE 19.2. The specific payoff function used in Axelrod (1980a).

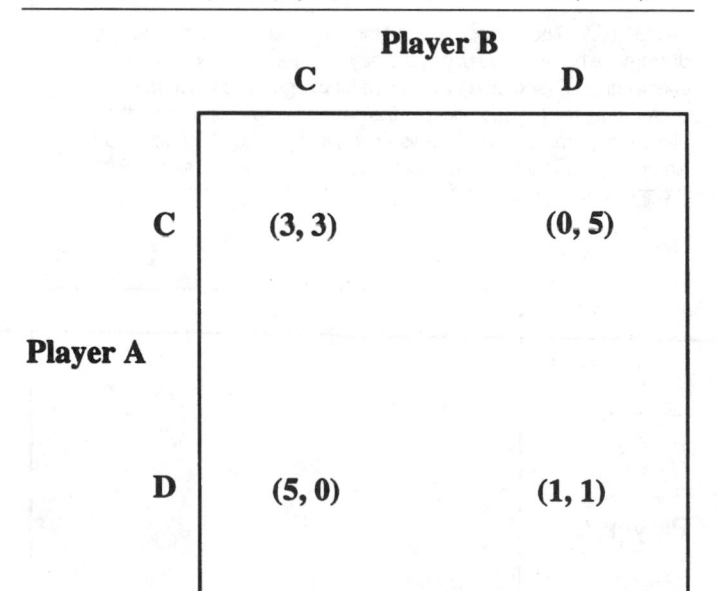

Player B

		C	D
Player A	C	(3, 3)	(0, 5)
	D	(5, 0)	(1, 1)

[1]Fujiki (1986) offered a method similar to Axelrod (1987) in which strategies playing an iterated prisoner's dilemma were put in coevolutionary competition. Twenty strategies were maintained at each generation. They used a memory of the previous two moves, and a flag to indicate if a particular play (cooperate or defect) had been offered previously. Games were iterated for 10 moves. Fogel and Burgin (1969) also reported on using evolutionary programming to evolve strategies to a prisoner's dilemma game using finite state machines. The population size was not specified.

demonstrated that dynamic fractal patterns could emerge in these settings (extended results in spatial settings were presented by Oliphant [1994]). Nowak and Sigmund (1993) demonstrated that under recurrent, simultaneous invasions by multiple (probabilistic) strategies the population may never converge to an evolutionary stable strategy and instead tend to periodic oscillations and chaos. Other investigations into probabilistic strategies were offered in Lindren (1991) and Nowak and Sigmund (1993), where the strategy "win-stay, lose-shift" was shown to outperform tit-for-tat (Wu and Axelrod [1995] noted that win-stay, lose-shift does not perform as well in some noisy environments).

Other efforts have been directed to evolving the option for a strategy to engage another strategy (e.g., Stanley et al., 1994; Batali and Kitcher, 1994) or to mediate the length of encounters (Fogel, 1995). Marks (1989), Bankes (1994), and others, have examined the multiple-player iterated prisoner's dilemma. Finally, Harrald and Fogel (1996), Borges et al. (1997), and others, have investigated evolutionary simulations of the prisoner's dilemma with a continuum of possible behaviors instead of the dichotomy of cooperate or defect.

The foremost question, one that has rarely been formally asked, is to what degree of sufficiency does the prisoner's dilemma model natural circumstances? Indeed, what is the real value of the prisoner's dilemma as a model? The answer remains somewhat elusive. Many of the studies after Axelrod (1987) attempt to bring a greater realism to the model and connect the model to studies of reciprocity in natural settings (see Axelrod and Dion, 1988, for a review), yet the relevance of the factors that are omitted from such models is still unclear and awaits further investigation.

References

[1] P. J. Angeline (1994) "An alternate interpretation of the iterated prisoner's dilemma and the evolution of non-mutual cooperation," *Artificial Life IV,* R. Brooks and P. Maes (eds.), MIT Press, Cambridge, MA, pp. 353–358.

[2] R. Axelrod (1980a) "Effective choice in the prisoner's dilemma," *J. Conflict Resolution,* Vol. 24, pp. 3–25.

[3] R. Axelrod (1980b) "More effective choice in the prisoner's dilemma," *J. Conflict Resolution,* Vol. 24, pp. 379–403.

[4] R. Axelrod (1984) *The Evolution of Cooperation,* Basic Books, NY.

[5] R. Axelrod (1987) "The evolution of strategies in the iterated prisoner's dilemma," *Genetic Algorithms and Simulated Annealing,* L. Davis (ed.), Pitman, London, pp. 32–41.

[6] R. Axelrod and D. Dion (1988) "The further evolution of cooperation," *Science,* Vol. 242, pp. 1385–1390.

[7] S. Bankes (1994) "Exploring the foundations of artificial societies: Experiments in evolving solutions to iterated *N*-player prisoner's dilemma," *Artificial Life IV,* R. Brooks and P. Maes (eds.), MIT Press, Cambridge, MA, 337–342.

[8] J. Batali and P. Kitcher (1994) "Evolutionary dynamics of altruistic behavior in optional and compulsory versions of the iterated prisoner's dilemma," *Artificial Life IV,* R. Brooks and P. Maes (eds.), MIT Press, Cambridge, MA, pp. 343–348.

[9] P. S. S. Borges, R. C. S. Pacheco, R. M. Barcia, and S. K. Khator (1997) "A fuzzy approach to the prisoner's dilemma," *BioSystems,* Vol. 41:2, pp. 127–137.

[10] R. Boyd and J. P. Lorberbaum (1987) "No pure strategy is evolutionarily stable in the repeated prisoner's dilemma," *Nature,* Vol. 327, pp. 58–59.

[11] P. H. Crowley (1996) "Evolving cooperation: Strategies as hierarchies of rules," *BioSystems,* Vol. 37:1–2, pp. 67–80.

[12] D. B. Fogel (1991) "The evolution of intelligent decision making in gaming," *Cybernetics and Systems,* Vol. 22, pp. 223–236.

[13] D. B. Fogel (1993) "Evolving behaviors in the iterated prisoner's dilemma," *Evolutionary Computation,* Vol. 1:1, pp. 77–97.

[14] D. B. Fogel (1995) "On the relationship between the duration of an encounter and the evolution of cooperation in the iterated prisoner's dilemma," *Evolutionary Computation,* Vol. 3:3, pp. 349–363.

[15] L. J. Fogel and G. H. Burgin (1969) "Competitive goal-seeking through evolutionary programming," final report under Contract No. AF 19(628)-5927, Air Force Cambridge Research Labs., Bedford, MA.

[16] C. Fujiki (1986) "An evaluation of Holland's genetic operators applied to a program generator," master's thesis, Univ. of Idaho.

[17] C. Fujiki and J. Dickinson (1987) "Using the genetic algorithm to generate LISP source code to solve the prisoner's dilemma," *Proc. of the 2nd Intern. Conf. on Genetic Algorithms,* J. J. Grefenstette (ed.), Lawrence Erlbaum, Hillsdale, NJ, pp. 236–240.

[18] P. G. Harrald and D. B. Fogel (1996) "Evolving continuous behaviors in the iterated prisoner's dilemma," *BioSystems,* Vol. 37:1–2, pp. 135–145.

[19] D. Hofstadter (1985) *Metamagical Themas: Questing for the Essence of Mind and Pattern,* Basic Books, NY.

[20] C. G. Langton (1987) "Artificial life," *Artificial Life: The Proc. of an Interdisciplinary Workshop on the Synthesis and Simulation of Living Systems,* C. G. Langton (ed.), Addison-Wesley, Reading, MA, pp. 1–47.

[21] K. Lindren (1991) "Evolutionary phenomena in simple dynamics," *Artificial Life II,* C. G. Langton, C. Taylor, J. D. Farmer, and S. Rasmussen (eds.), Addison-Wesley, Reading, MA, pp. 295–312.

[22] R. E. Marks (1989) "Breeding hybrid strategies: Optimal behavior for oligopolists," *Proc. of the 3rd Intern. Conf. on Genetic Algorithms,* J. D. Schaffer (ed.), Morgan Kaufmann, San Mateo, CA, pp. 198–207.

[23] J. Maynard Smith (1982) *Evolution and the Theory of Games,* Cambridge Univ. Press, Cambridge, U.K.

[24] J. Miller (1989) "The coevolution of automata in the repeated prisoner's dilemma," Santa Fe Institute Report 89–003.

[25] M. Nowak (1990) "Stochastic strategies in the prisoner's dilemma," *Theoretical Population Biology,* Vol. 38, pp. 93–112.

[26] M. A. Nowak and R. M. May (1992) "Evolutionary games and spatial chaos," *Nature,* Vol. 359, pp. 826–839.

[27] M. A. Nowak and K. Sigmund (1992) "Tit for tat in heterogeneous populations," *Nature,* Vol. 355, pp. 250–253.

[28] M. Nowak and K. Sigmund (1993) "A strategy of win-stay, lose-shift that outperforms Tit-for-tat in the prisoner's dilemma game," *Nature,* Vol. 364, pp. 56–58.

[29] M. Oliphant (1994) "Evolving cooperation in the non-iterated prisoner's dilemma: The importance of spatial organization," *Artificial Life IV,* R. Brooks and P. Maes (eds.), MIT Press, Cambridge, MA, pp. 349–352.

[30] A. Rapoport (1966) "Optimal policies for the prisoner's dilemma," Tech. Report No. 50, Psychometric Laboratory, Univ. North Carolina, NIH Grant, MH-10006.

[31] A. Scodel, J. S. Minas, P. Ratoosh, and M. Lipetz (1959) "Some descriptive aspects of two-person non-zero sum games," *J. Conflict Resolution,* Vol. 3, pp. 114–119.

[32] E. A. Stanley, D. Ashlock, and L. Tesfatsion (1994) "Iterated prisoner's dilemma with choice and refusal of partners," *Artificial Life III,* C. G. Langton (ed.), Addison-Wesley, Reading, MA, pp. 131–175.

[33] J. Wu and R. Axelrod (1995) "How to cope with noise in the iterated prisoner's dilemma," *J. Conflict Resolution,* Vol. 39:1, pp. 183–189.

Robert Axelrod

The Evolution of Strategies in the Iterated Prisoner's Dilemma

3.1 The Problem with Cooperation

Mutual cooperation among groups of organisms is a frequently occurring phenomenon. When such cooperation is of benefit to the cooperating agents, and when the lack of it is harmful to them, the mechanisms by which such cooperation might arise and persist seem straightforward.

There are other types of cooperation, however, that are characterized by the fact that, while cooperating agents do well, any one of them would do better by failing to cooperate. For this sort of case, it is more difficult to explain how group cooperation would arise and persist, for one would expect organisms showing a propensity to cooperate to do less well than their neighbors, leading to a dying out of cooperating tendencies in a large, non-cooperating population.

The Prisoner's Dilemma of game theory is an elegant embodiment of this sort of case. In the Prisoner's Dilemma, two individuals can each either cooperate or defect. The payoff to a player affects its reproductive success. No matter what the other does, the selfish choice of defection yields a higher payoff than cooperation. But if both defect, both do worse than if both had cooperated. Figure 3-1 shows the payoff matrix of the Prisoner's Dilemma used in this study.

In many biological settings, the same two individuals may meet more than once. If an individual can recognize a previous interactant and remember some aspects of the prior outcomes, then the strategic situation becomes an iterated Prisoner's Dilemma. In an iterated Prisoner's Dilemma, a strategy is a decision rule which specifies the probability of cooperation or defection as a function of the history of the interaction so far.

To see what type of strategy can thrive in a variegated environment of more or less sophisticated strategies, I conducted a computer tournament for the iterated Prisoner's Dilemma. The strategies were submitted by game theorists in economics, sociology, political science, and mathematics (Axelrod, 1980a). The 14 entries and a totally random strategy were paired with each other in a round robin tournament. Some of the strategies were quite intricate. An example is one which on each move models the behavior of the other player as a Markov process, and then uses Bayesian inference to select what seems the best choice for the long run. However, the result of the tournament

Column Player

		Cooperate	Defect
Row Player	Cooperate	$R = 3,\ R = 3$ Reward for mutual cooperation	$S = 0,\ T = 5$ Sucker's payoff, and temptation to defect
	Defect	$T = 5,\ S = 0$ Temptation to defect and sucker's payoff	$P = 1,\ P = 1$ Punishment for mutual defection

Note: The payoffs to the row chooser are listed first.

Figure 3-1: The Prisoner's Dilemma

was that the highest average score was attained by the simplest of all strategies, TIT FOR TAT. This strategy is simply one of cooperating on the first move and then doing whatever the other player did on the preceding move. TIT FOR TAT is a strategy of cooperation based upon reciprocity.

The results of the first round were circulated and entries for a second round were solicited. This time there were 62 entries from six countries (Axelrod, 1980b). Most of the contestants were computer hobbyists, but there were also professors of evolutionary biology, physics, and computer science, as well as the five disciplines represented in the first round. TIT FOR TAT was again submitted by the winner of the first round, Anatol Rapoport. It won again.

The second round of the computer tournament provides a rich environment in which to test the evolution of behavior. It turns out that just eight of the entries can be used to account for how well a given rule did with the entire set. These eight rules can be thought of as representatives of the full set in the sense that the scores a given rule gets with them can be used to predict the average score the rule gets over the full set. In fact, 98% of the variance in the tournament scores is explained by knowing a rule's performance with these eight representatives. So these representative strategies can be used as a complex environment in which to evaluate an evolutionary simulation. That is the environment I used to simulate the evolution of strategies for the iterated Prisoner's Dilemma.

3.2 The Genetic Algorithm

The inspiration for my simulation technique came from an artificial intelligence procedure developed by computer scientist John Holland (Holland 1975, 1980, 1986). Holland's technique is called the genetic algorithm. Using a genetic algorithm, one represents strategies as chromosomes. Each chromosome serves a dual purpose: it provides a representation of what the organism will become, and it also provides the actual material which can be transformed to yield new genetic material for the next generation.

Before going into details, it may help to give a brief overview of how the genetic algorithm works. The first step is to specify a way of representing each allowable strategy as a string of genes on a chromosome which can undergo genetic transformations, such as mutation. Then the initial population is constructed from the allowable set (perhaps by simply picking at random). In each generation, the effectiveness of each individual in the population is determined by running the individual in the current strategic environment. Finally, the relatively successful strategies are used to produce offspring which resemble the parents. Pairs of successful offspring are selected to mate and produce the offspring for the next generation. Each offspring draws part of its genetic material from one parent and part from another. Moreover, completely new material is occasionally introduced through mutation. After many generations of selection for relatively successful strategies, the result might well be a population that is substantially more successful in the given strategic environment than the original population.

To explain how this works, consider the strategies available for playing the iterated Prisoner's Dilemma. In particular, consider the set of strategies that are deterministic and use the outcomes of the three previous moves to make a choice in the current move. Since there are four possible outcomes for each move, there are 4x4x4 = 64 different histories of the three previous moves. Therefore to determine its choice of cooperation or defection, a strategy would only need to determine what to do in each of the situations which could arise. This could be specified by a list of sixty-four C's and D's (C for cooperation and D for defection). For example, one of these sixty-four genes indicates whether the individual cooperates or defects when in a rut of three mutual defections. Other parts of the chromosome would cover all the other situations that could arise.

To get the strategy started at the beginning of the game, it is also necessary to specify its initial premises about the three hypothetical moves which preceded the start of the game. To do this requires six more genes, making a total of seventy loci on the chromosome.[1] This string of seventy C's and D's would specify what the individual would do in every possible circumstance and would therefore completely define a particular strategy. The string of 70 genes would also serve as the individual's chromosome for use in reproduction and mutation.

There is a huge number of strategies which can be represented in this way. In fact, the number is 2^{70}, which is about 10^{21}.[2] An exhaustive search for good strategies in this huge collection of strategies is clearly out of the question. If a computer had examined these strategies at the rate of 100 per second since the beginning of the universe, less than one percent would have been checked by now.

To find effective strategies in such a huge set, a very powerful technique is needed.

---

[1] The six premise genes encode the presumed C or D choices made by the individual and the other player in each of the three moves before the interaction actually begins.

[2] Some of these chromosomes give rise to equivalent strategies since certain genes might code for histories that could not arise given how loci are set. This does not necessarily make the search process any easier, however.

Holland's "genetic algorithm" is such a technique. Genetic algorithms were originally inspired by biological genetics, but were adapted by Holland to be a general problem-solving technique. In the present context, a genetic algorithm can be regarded as a model of a "minimal genetics" which can be used to explore theoretical aspects of evolution in rich environments. The simulation program works in five stages:

1. An initial population is chosen. In the present context the initial individuals can be represented by random strings of seventy C's and D's.

2. Each individual is run in the current environment to determine its effectiveness. In the present context this means that each individual player uses the strategy defined by its chromosome to play an iterated Prisoner's Dilemma with other strategies, and the individual's score is its average over all the games it plays.[3]

3. The relatively successful individuals are selected to have more offspring. The method used is to give an average individual one mating, and to give two matings to an individual who is one standard deviation more effective than the average. An individual who is one standard deviation below the population average would then get no matings.

4. The successful individuals are then randomly paired off to produce two offspring per mating. For convenience, a constant population size is maintained. The strategy of an offspring is determined from the strategies of the two parents. This is done by using two genetic operators: crossover and mutation.

   a. Crossover is a way of constructing the chromosomes of the two offspring from the chromosomes of two parents. It can be illustrated by an example of two parents, one of whom has seventy C's in its chromosome (indicating that it will cooperate in each possible situation that can arise), and the other of whom has seventy D's in its chromosome (indicating that it will always defect). Crossover selects one or more places to break the parents' chromosomes in order to construct two offspring each of whom has some genetic material from both parents. For example, if a single break occurs after the third gene, then one offspring will have three C's followed by sixty-seven D's, while the other offspring will have three D's followed by sixty-seven C's.

   b. Mutation in the offspring occurs by randomly changing a very small proportion of the C's to D's or vice versa.

5. This gives a new population. This new population will display patterns of behavior that are more like those of the successful individuals of the previous generation, and

---

[3]The score is actually a weighted average of its scores with the eight representative, the weights having been chosen to give the best representation of the entire set of strategies in the second round of the tournament.

less like those of the unsuccessful ones. With each new generation, the individuals with relatively high scores will be more likely to pass on parts of their strategies, while the relatively unsuccessful individuals will be less likely to have any parts of their strategies passed on.

3.3   Simulation Results

The computer simulations were done using a population size of twenty individuals per generation. Levels of crossover and mutation were chosen averaging one crossover and one-half mutation per chromosome per generation. Each game consisted of 151 moves, the average game length used in the tournament. With each of the twenty individuals meeting eight representatives, this made about 24,000 moves per generation. A run consisted of 50 generations. Forty runs were conducted under identical conditions to allow an assessment of the variability of the results.

The results are quite remarkable: from a strictly random start, the genetic algorithm evolved populations whose median member was just as successful as the best rule in the tournament, TIT FOR TAT. Most of the strategies that evolved in the simulation actually resemble TIT FOR TAT, having many of the properties that make TIT FOR TAT so successful. For example, five behavioral alleles in the chromosomes evolved in the vast majority of the individuals to give them behavioral patterns that were adaptive in this environment and mirrored what TIT FOR TAT would do in similar circumstances. These patterns are:

1. Don't rock the boat: continue to cooperate after three mutual cooperations (which can be abbreviated as C after RRR).

2. Be provocable: defect when the other player defects out of the blue (D after receiving RRS).

3. Accept an apology: continue to cooperate after cooperation has been restored (C after TSR).

4. Forget: cooperate when mutual cooperation has been restored after an exploitation (C after SRR).

5. Accept a rut: defect after three mutual defections (D after PPP).

The evolved rules behave with specific representatives in much the same way as TIT FOR TAT does. They did about as well as TIT FOR TAT did with each of the eight representatives. Just as TIT FOR TAT did, most of the evolved rules did well by achieving almost complete mutual cooperation with seven of the eight representatives. Like TIT FOR TAT, most of the evolved rules do poorly with only one representative, called ADJUSTER, that adjusts its rate of defection to try to exploit the other player. In all, 95% of the time the evolved rules make the same choice as TIT FOR TAT would make in the same situation.

While most of the runs evolve populations whose rules are very similar to TIT FOR TAT, in eleven of the forty runs, the median rule actually does substantially better than TIT FOR TAT.[4] In these eleven runs, the populations evolved strategies that manage to exploit one of the eight representatives at the cost of achieving somewhat less cooperation with two others. But the net effect is a gain in effectiveness.

This is a remarkable achievement because to be able to get this added effectiveness, a rule must be able to do three things. First, it must be able to discriminate between one representative and another based upon only the behavior the other player shows spontaneously or is provoked into showing. Second, it must be able to adjust its own behavior to exploit a representative that is identified as an exploitable player. Third, and perhaps most difficult, it must be able to achieve this discrimination and exploitation without getting into too much trouble with the other representatives. This is something that none of the rules originally submitted to the tournament were able to do.

These very effective rules evolved by breaking the most important advice developed in the computer tournament, namely to be "nice", that is never to be the first to defect. These highly effective rules always defect on the very first move, and sometimes on the second move as well, and use the choices of the other player to discriminate what should be done next. The highly effective rules then had responses that allowed them to "apologize " and get to mutual cooperation with most of the unexploitable representatives, and they had different responses which allowed them to exploit a representative that was exploitable.

While these rules are highly effective, it would not be accurate to say that they are better than TIT FOR TAT. While they are better in the particular environment consisting of fixed proportions of the eight representatives of the second round of the computer tournament, they are probably not very robust in other environments. Moreover, in an ecological simulation these rules would be destroying the basis of their own success, as the exploited representative would become a smaller and smaller part of the environment (Axelrod 1984, pp. 49-52 and 203-5). While the genetic algorithm was sometimes able to evolve rules that are more effective than any entry in the tournament, the algorithm was only able to do so by trying many individuals in many generations against a fixed environment. In sum, the genetic algorithm is very good at what actual evolution does so well: developing highly specialized adaptations to specific environmental settings.

In the evolution of these highly effective strategies, the computer simulation employed sexual reproduction, where two parents contributed genetic material to each offspring. To see what would happen with asexual reproduction, forty additional runs were conducted in which only one parent contributed genetic material to each offspring. In

---

[4]The criterion for being substantially better than TIT FOR TAT is a median score of 450 points, which compares to TIT FOR TAT's weighted score of 428 with these eight representatives.

these runs, the populations still evolved toward rules that did about as well as TIT FOR TAT in most cases. However, the asexual runs were only half as likely to evolve populations in which the median member was substantially more effective than TIT FOR TAT.[5]

So far, the simulation experiments have dealt with populations evolving in the context of a constant environment. What would happen if the environment is also changing? To examine this situation, another simulation experiment with sexual reproduction was conducted in which the environment consisted of the evolving population itself. In this experiment each individual plays the iterated Prisoner's Dilemma with each other member of the population rather than with the eight representatives. At any given time, the environment can be quite complex. For an individual to do well requires that its strategy achieves a high average effectiveness with the nineteen other strategies that are also present in the population. Thus as the more effective rules have more offspring, the environment itself changes. In this case, adaptation must be done in the face of a moving target. Moreover, the selection process is frequency dependent, meaning that the effectiveness of a strategy depends upon what strategies are being used by the other members of the population.

The results of the ten runs conducted in this manner display a very interesting pattern. From a random start, the population evolves away from whatever cooperation was initially displayed. The less cooperative rules do better than the more cooperative rules because at first there are few other players who are responsive — and when the other player is unresponsive the most effective thing for an individual to do is simply defect. This decreased cooperation in turn causes everyone to get lower scores as mutual defection becomes more and more common. However, after about ten or twenty generations the trend starts to reverse. Some players evolve a pattern of reciprocating what cooperation they find, and these reciprocating players tend to do well because they can do very well with others who reciprocate without being exploited for very long by those who just defect. The average scores of the population then start to increase as cooperation based upon reciprocity becomes better and better established. So the evolving social environment led to a pattern of decreased cooperation and decreased effectiveness, followed by a complete reversal based upon an evolved ability to discriminate between those who will reciprocate cooperation and those who won't. As the reciprocators do well, they spread in the population resulting in more and more cooperation and greater and greater effectiveness.

3.4 Lessons

1. The genetic algorithm is a highly effective method of problem solving. Following

---

[5]This happened in 5 of the 40 runs with asexual reproduction compared to 11 of the 40 runs with sexual reproduction. This difference is significant at the .05 level using the one tailed chi-squared test.

Quincy Wright (1977, pp. 452-454), the problem for evolution can be conceptualized as a search for relatively high points in a multidimensional field of gene combinations, where height corresponds to fitness. When the field has many local optima, the search becomes quite difficult. When the number of dimensions in the field becomes great, the search is even more difficult. What the computer simulations demonstrate is that the genetic algorithm is a highly efficient method for searching such a complex multidimensional space. The first experiment shows that even with a seventy dimensional field of genes, quite effective strategies can be found within fifty generations. Sometimes the genetic algorithm found combinations of genes that violate the previously accepted mode of operation (not being the first to defect) to achieve even greater effectiveness than had been thought possible.

2. Sexual reproduction does indeed help the search process. This was demonstrated by the much increased chance of achieving highly effective populations in the sexual experiment compared to the asexual experiment. If sexual reproduction comes at the cost of reduced fecundity, it is not clear whether this gain in search efficiency would be worth the cost of fewer offspring. However, one case in which search efficiency can be very important is the escape from rapidly evolving parasites. This has been demonstrated in one and two locus models (Hamilton, 1980) and shown to be relevant in the sexual selection of birds (Hamilton, 1982).

3. Some aspects of evolution are arbitrary. In natural settings, one might observe that a population has little variability in a specific gene. In other words one of the alleles for that gene has become fixed throughout the population. One might be tempted to assume from this that the allele is more adaptive than any alternative allele. However, this may not be the case. The simulation of evolution allows an exploration of this possibility by allowing repetitions of the same conditions to see just how much variability there is in the outcomes. In fact, the simulations show two reasons why convergence in a population may actually be arbitrary.

   a. Genes that do not have much effect on the fitness of the individual may become fixed in a population because they "hitch-hike" on other genes that do (Maynard Smith and Haigh, 1974). For example, in the simulations some sequences of three moves may very rarely occur, so what the corresponding genes dictate in these situations may not matter very much. However, if the entire population are descendants of just a few individuals, then these irrelevant genes may be fixed to the values that their ancestors happened to share. Repeated runs of a simulation allow one to notice that some genes become fixed in one population but not another, or that they become fixed in different ways in different populations.

   b. In some cases, some parts of the chromosome are arbitrary in content, but what is not arbitrary is that they be held constant. By being fixed, other parts of the chromosome can adapt to them. For example, the simulations of the individual chromosomes had six genes devoted to coding for the premises about the three moves that preceded the first move in the game. When the environment was the eight representatives, the populations in different runs of the simulation developed different premises. Within each run, however, the populations were usually very consistent about the premises: the six premise genes had become fixed. Moreover, within each population these genes usually became fixed quite early. It is interesting that different populations evolved quite different premises. What was important for the evolutionary process was to fix the premise about which history is assumed at the start so that the other parts of the chromosome could adapt on the basis of a given premise.

4. There is a tradeoff in evolution between the gains to be made from flexibility and the gains to be made from commitment and specialization. Flexibility might help in the long run, but in an evolutionary system, the individuals also have to survive in the short run if they are to reproduce. This feature of evolution arises at several levels.

   a. As the simulations have shown, the premises became fixed quite early. This meant a commitment to which parts of the chromosome would be consulted in the first few moves, and this in turn meant giving up flexibility as more and more of the chromosome evolved on the basis of what had been fixed. This in turn meant that it would be difficult for a population to switch to a different premise. So flexibility was given up so that the advantages of commitment could be reaped.

   b. There is also a tradeoff between short and long term gains in the way selection was done in the simulation experiments. In any given generation there would typically be some individuals that did much better than the average, and some that did only a little better than the average. In the short run, the way to maximize the expected performance of the next generation would be to have virtually all of the offspring come from the very best individuals in the present generation. But this would imply a rapid reduction in the genetic variability of the population, and a consequent slowing of the evolutionary process later on. If the moderately successful were also given a chance to have some offspring, this would help the long term prospects of the population at the cost of optimizing in the short run. Thus there is an inherent tradeoff between exploitation and exploration, i.e. between exploiting what already works best and exploring possibilities that might eventually evolve into something even better (Holland, 1975, p. 160).

3.5 Conclusions

The genetic simulations provided in this paper are highly abstract systems. The populations are very small, and the number of generations is few. More significantly, the genetic process have only two operators, mutation and crossover, and the sexual reproduction has no sexual differentiation and always had two offspring per mating. These are all highly simplified assumptions, and yet the simulations displayed a remarkable ability to evolve sophisticated adaptive strategies in moderately complex environments.

In the future, more complex and realistic simulations are possible. But the main advantage of simulations can already be glimpsed from these minimal simulation experiments. They provide a different intellectual perspective on evolution. Instead of having to rely only on our observations of real biological systems or our standard mathematical models, we will be able to approach genetics and evolution as a theoretical design problem. We can begin asking about whether parasites are inherent in all complex systems, or are merely the outcome of the way biological systems have happened to evolve. We can begin investigating alternative ways genetics might have evolved and see just which properties of our biological heritage are arbitrary and which are not. Today microbiologists are developing the techniques to alter our genetic heritage. Perhaps now is also the time to think about doing some "as if" experiments to better appreciate the fundamental properties of the genetic system that is the basis of our natural endowment.

3.6 Acknowledgements

I thank Stephanie Forrest and Reiko Tanese for their help with the computer programming, Michael D. Cohen and John Holland for their helpful suggestions, and the Harry Frank Guggenheim Foundation and the National Science Foundation for their financial support to Robert Axelrod.

Chapter 20
Implicit Parallelism and Representations

J. Antonisse (1989) "A New Interpretation of Schema Notation that Overturns the Binary Encoding Constraint," *Proc. of the 3rd Intern. Conf. on Genetic Algorithms*, J. D. Schaffer (ed.), Morgan Kaufmann, San Mateo, CA, pp. 86–91.

ONE of the key considerations in evolutionary computation is the choice of representation. A representation is an invertible mapping of the space of candidate structures (solutions) to strings of symbols. Consider a traveling salesman problem of n cities. The space of candidate structures consists of all complete circuits of the posed cities, where each city except the origin is visited once and only once. There are a variety of ways to represent salesmen's tours for manipulation on a computer. For example, each city could be numbered and an ordered list of these nonrepeating numbers from 1 to n would represent the sequence of vertices in the circuit (e.g., [1 2 4 3 5], with the return to city 1 implicit). Another choice could be the sequence of edges to traverse (e.g., [(1 2) (2 4) (4 3) (3 5) (5 1)]). Each tour could even simply be assigned a number from 1 to $(n-1)!/2$ (i.e., the total number of different tours). The last choice, however, highlights the importance of a suitable representation and the degree to which the choice interacts with the search and selection operators.

It would be very difficult to conduct an evolutionary search for optimal tours by applying random variation and selection over the set of corresponding integers $[1, (n-1)!/2]$. Despite the ordinal relationship of the increasing integers, there is nothing that requires the circuit of tour number 1 to be like that of number 2, and so forth. Any tour could be mapped to any integer in $[1, (n-1)!/2]$. For evolution to be successful, there must generally be some degree of a nearness in representation correlating to a nearness of function (i.e., slight variations in genotypic structure portend slight variations in the resultant phenotype under the chosen variation operators and fitness function), otherwise the search degenerates essentially into trying new solutions completely at random. The other two representations above (i.e., the ordered list of cities or edges), in contrast, provide for the use of search operators that can randomly vary portions of a salesman's tour and maintain a behavioral link between each parent and its offspring (Chapter 18) (Goldberg and Lingle, 1985; Grefenstette et al., 1985; and others). The choice of representation in turn impacts the choice of performance measure, the human operator's perception of the

adaptive landscape being searched, as well as the suitability of alternative variation and selection operators.

The early research in evolution strategies (Rechenberg, 1973; and others) relied on real-valued and integer representations (see Chapter 8) because the applications involved continuous physical variables of engineering devices, occasionally of variable length. Early efforts in evolutionary programming (Fogel, 1964; and others) utilized finite state machines to represent the logics that generated sequences of symbols from a finite alphabet (see Chapter 7). In contrast, the standard representation within early genetic algorithms (Chapter 15) (Holland, 1975; and others) relied on binary strings, this in order to maximize implicit parallelism (see below). Although all three approaches have been modified to handle arbitrary representations (see Bäck et al., 1997, for a variety of applications), the conviction in favor of binary representations in genetic algorithms was quite strong until the early 1990s. Antonisse (1989) (reprinted here) wrote "The bit string was embraced early in the history of the [genetic algorithm] on the basis of a theoretical analysis of string representation languages. This analysis . . . and its outcome, that the binary string is optimal . . . became accepted as fundamental to work in this area. Under the *principle of minimal alphabets* [Goldberg89], the bit string representation has been raised beyond a common feature to almost a necessary precondition for serious work in the [genetic algorithm]."

Binary strings were given favor because they resulted in the greatest number of schemata for processing by the genetic algorithm. To review (see Chapter 15), a schema is a template with fixed and variable symbols. Consider a string of symbols from an alphabet **A**. Suppose that some of the components of the string are held fixed while others are free to vary. Define a wild card symbol, $\# \notin \mathbf{A}$, that matches any symbol from **A**. A string with fixed and variable symbols defines a schema. Consider the string [01##], defined over the union of {#} and the alphabet $\mathbf{A} = \{0,1\}$. This set includes [0100], [0101], [0110], and [0111]. Holland (1975, pp. 64–74) offered that every evaluated string actually offers partial information about the expected fitness of all possible schemata in which that string resides. That is, if string [0000] is evaluated to have some fitness, then partial information is also received about the worth of sampling from variations in [####], [0###], [#0##], [#00#], [#0#0], and so forth. This characteristic is termed *intrinsic parallelism* (or *implicit parallelism*), in that through a single sample, information is gained with respect to many schemata.

Holland (1975, p. 71) suggested that alternative representations in evolutionary algorithms could be compared by calculating the number of schemata sampled per evaluated string. In order to maximize intrinsic parallelism, it was recommended that representations should be chosen with the fewest "detectors with a range of many attributes." In other words, alphabets with low cardinality are to be favored because they generate more schemata. This increased number of schemata was suggested to give a "larger information flow" to reproductive plans such as genetic algorithms. Emphasis was therefore placed on binary representations, this offering the lowest possible cardinality and the greatest number of schemata.

The primary reason for the trend away from binary codings in genetic algorithms about 1990 was probably the attention given to a few successful applications that relied on more intuitive representations (e.g., Koza, 1989; Davis, 1991; Michalewicz, 1992, pp. 75–76; and others). Another reason was that binary strings were simply cumbersome and awkward to apply in many situations.[1] But there was also a theoretical argument for using representations of higher cardinality. Rather than view the # symbol in a string as a "don't care" character, Antonisse (1989) viewed it as indicating all possible subsets of symbols at the particular position. If $A = \{0, 1, 2\}$, then the schema [000#] would indicate the sets {[0000] [0001]}, {[0000] [0002]}, {[0001] [0002]}, and {[0000] [0001] [0002]} as the # symbol would indicate the possibilities of (a) 0 or 1, (b) 0 or 2, (c) 1 or 2, and (d) 0, 1, or 2. When schemata are viewed in this manner, the greater implicit parallelism comes from the use of more, not fewer, symbols. In retrospect, however, Antonisse (1989) did not receive sufficient attention to overturn the traditional perspective favoring binary representations in light of the schema theorem (Holland, 1975, p. 111).

The schema theorem indicates that the expected proportion of each schema H (as defined in Holland, 1975), taking into account the effects of crossover but not mutation or other genetic operators, follows:

$$E\, P(H, t+1) \geq P(H, t) \frac{f(H)}{\bar{f}} \left(1 - p_c \frac{\delta(H)}{l-1}\right)$$

where $P(H, t)$ is the proportion of the population possessing schema H at time t, $f(H)$ is the mean fitness of strings possessing schema H, $\bar{f}$ is the mean fitness of all strings in the population, p_c is the probability of crossover (one-point), $\delta(H)$ is the defining length[2] of the schema H, and l is the length of each

string. This result was considered to be particularly significant: "Short, low-order, above-average schemata receive exponentially increasing trials in subsequent generations. This conclusion is important, so important that we give it a special name: the *Schema Theorem*, or the Fundamental Theorem of Genetic Algorithms" (Goldberg, 1989, p. 33). Binary encodings maximized the number of schemata that could be processed and recombined as building blocks using the crossover operator (Goldberg, 1989, p. 41).

But other mathematical arguments followed Antonisse (1989) showing that binary representations do not provide an inherently superior encoding scheme. Radcliffe (1992) noted that for a search space S of size $|S| = 2^k$, if B^k is the set of all binary strings of length k, $B = \{0, 1\}$, then there are $2^k!$ possible representations of S, much like there are $(n - 1)!/2$ possible assignments of integers to tours in a traveling salesman problem (see above). The schema theorem will hold for any of these representations, even one derived by randomly assigning binary strings to elements of S, yet many of the $2^k!$ mappings would not be expected to provide any useful functional relationship between the individuals in S. Thus simply choosing a binary representation is not sufficient to ensure the useful processing of schemata (Radcliffe, 1992).

Battle and Vose (1993) proved that isomorphisms exist between alternative instances of genetic algorithms for binary representations (i.e., the operations of crossover and mutation on a population under a particular binary representation in light of a fitness function can be mapped equivalently to alternative similar operators for any other binary representation). Vose and Liepins (1991) offered an "instructive exercise" left for the reader that all fixed coding schemes perform equally well when averaged over all functions to be optimized. Similar results have been extended in Wolpert and Macready (1997) where it was proved that all algorithms (in a broad class of search procedures) offer identical performance regardless of their representation when averaged over all problems, and Fogel and Ghozeil (1997) where it was shown that for any representation of a given cardinality and associated variation operator, there is an equivalent variation operator in every other cardinality. Thus there is a crucial interaction between representation, fitness function, and search and selection operators; it is insufficient to consider a "best" representation for a problem in isolation of these other concerns.

Maximizing implicit parallelism is not intrinsically useful in accelerating evolutionary optimization, nor does the schema theorem provide explanatory power in predicting the performance of a genetic algorithm (or another algorithm relying on proportional selection) (Mühlenbein, 1991; Altenberg, 1995). Indeed, the schema theorem does not capture the effects of search operators in discovering new solutions; the theorem only

---

[1] Antonisse (1997) indicated a motivation to explore high-level representations (e.g., context free grammars) with genetic algorithms, in contrast with low-level binary representations.

[2] The defining length of a schema is the distance between the first and last specified symbols in the string. For example, for the string [1##10##], the defining length is four because the first specified symbol is in position one and the last is in position five.

addresses the possibility for these operators to disrupt extant schemata and their expected rate of spread into the next generation. There is no requisite correlation between the spreading of schemata with above-average fitness relative to other schemata in the current population and the eventual discovery of other schemata with even higher relative fitness (Altenberg, 1995). Consideration should instead be given to the relationship between parental and offspring fitness (Altenberg, 1995; Grefenstette, 1995; Fogel and Ghozeil, 1996; Voigt et al., 1996; and others) and the effect of variation and selection operators in terms of the likelihood of discovering improved solutions and the degree of improvement. This form of analysis provides a basis for comparing the effects of alternative operators, and for selecting those that are particularly appropriate for the task at hand.

References

[1] L. Altenberg (1995) "The schema theorem and Price's theorem," *Foundations of Genetic Algorithms 3*, L. D. Whitley and M. D. Vose, Morgan Kaufmann, San Mateo, CA, pp. 23–49.

[2] J. Antonisse (1989) "A new interpretation of schema notation that overturns the binary encoding constraint," *Proc. of the 3rd Intern. Conf. on Genetic Algorithms*, J. D. Schaffer (ed.), Morgan Kaufmann, San Mateo, CA, pp. 86–91.

[3] J. Antonisse (1997) personal communication, Cambridge Research Associates, McLean, VA.

[4] T. Bäck, D. B. Fogel, and Z. Michalewicz (eds.) (1997) *Handbook of Evolutionary Computation*, Oxford, NY.

[5] D. L. Battle and M. D. Vose (1993) "Isomorphisms of genetic algorithms," *Artificial Intelligence*, Vol. 60, pp. 155–165.

[6] L. Davis (1991) "Hybridization and numerical representation," *Handbook of Genetic Algorithms*, L. Davis (ed.), Van Nostrand Reinhold, NY, pp. 61–71.

[7] D. B. Fogel and A. Ghozeil (1996) "Using fitness distributions to design more efficient evolutionary computations," *Proc. of 1996 IEEE Conf. on Evolutionary Computation*, IEEE Press, NY, pp. 11–19.

[8] D. B. Fogel and A. Ghozeil (1997) "A note on representations and operators," *IEEE Trans. Evolutionary Computation*, Vol. 1:2, pp. 159–161.

[9] L. J. Fogel (1964) "On the organization of intellect," Ph.D. diss., UCLA.

[10] D. E. Goldberg (1989) *Genetic Algorithms in Search, Optimization and Machine Learning*, Addison-Wesley, Reading, MA.

[10] D. E. Goldberg and R. Lingle (1985) "Alleles, loci, and the traveling salesman problem," *Proc. of an Intern. Conf. on Genetic Algorithms and Their Applications*, J. J. Grefenstette (ed.), Lawrence Erlbaum, Hillsdale, NJ, pp. 154–159.

[11] J. J. Grefenstette (1995) "Predictive models using fitness distributions of genetic operators," *Foundations of Genetic Algorithms 3*, L. D. Whitley and M. D. Vose (eds.), Morgan Kaufmann, San Mateo, CA, pp. 139–161.

[12] J. Grefenstette, R. Gopal, B. Rosmaita, and D. Van Gucht (1985) "Genetic algorithms for the traveling salesman problem," *Proc. of an Intern. Conf. on Genetic Algorithms and Their Applications*, J. J. Grefenstette (ed.), Lawrence Erlbaum, Hillsdale, NJ, pp. 160–168.

[13] J. H. Holland (1975) *Adaptation in Natural and Artificial Systems*, Univ. of Michigan Press, Ann Arbor, MI.

[14] J. R. Koza (1989) "Hierarchical genetic algorithms operating on populations of computer programs," *Proc. of the 11th Intern. Joint Conf. on Artificial Intelligence*, N. S. Sridharan (ed.), Morgan Kaufmann, San Mateo, CA, pp. 768–774.

[15] Z. Michalewicz (1992) *Genetic Algorithms + Data Structures = Evolution Programs*, Springer, Berlin.

[16] H. Mühlenbein (1991) "Evolution in time and space—the parallel genetic algorithm," *Foundations of Genetic Algorithms*, G. J. E. Rawlins (ed.), Morgan Kaufmann, San Mateo, CA, pp. 316–337.

[17] N. J. Radcliffe (1992) "Non-linear genetic representations," *Parallel Problem Solving from Nature, 2*, R. Männer and B. Manderick (eds.), North-Holland, Amsterdam, The Netherlands, pp. 259–268.

[18] I. Rechenberg (1973) *Evolutionsstrategie: Optimierung technischer Systeme nach Prinzipien der biologischen Evolution*, Frommann-Holzboog, Stuttgart, Germany.

[19] H.-M. Voigt, H. Mühlenbein, and D. Schlierkamp-Voosen (1996) "The response to selection equation for skew fitness distributions," *Proc. of 1996 IEEE Intern. Conf. on Evolutionary Computation*, IEEE Press, NY, pp. 820–825.

[20] M. D. Vose and G. E. Liepins (1991) "Schema disruption," *Proc. of the 4th Intern. Conf. on Genetic Algorithms*, R. K. Belew and L. B. Booker (eds.), Morgan Kaufmann, San Mateo, CA, pp. 237–240.

[21] D. H. Wolpert and W. G. Macready (1997) "No free lunch theorems for optimization," *IEEE Trans. Evolutionary Computation*, Vol. 1:1, pp. 67–82.

A New Interpretation of Schema Notation
that Overturns the Binary Encoding Constraint

Jim Antonisse
The MITRE Corporation
7525 Colshire Drive
McLean, VA 22102-3481
antonisse@mitre.ai.org

January 1989

ABSTRACT

Genetic algorithms (GAs) have been demonstrated as effective learning and discovery techniques for a large class of problems. Despite their promise, however, these algorithms have not found wide acceptance among members of the artificial intelligence community at large. This is largely due to the binary representation embraced by most of the genetic algorithms community. This representation is purposefully simple, while much of the work in, e.g., the artificial intelligence community has been to develop highly expressive, relatively complex representations. This paper puts forward an alternative interpretation of the argument that led to the acceptance of the binary representation for GAs. Under this new interpretation the principle reason for a maximally simple representation -- that it maximizes the schemata sampled per individual of a population -- does not hold. This interpretation aligns theory with the intuition that the more expressive a language is the more powerful an apparatus for adaptation it provides, and encourages exploration of alternative encoding schemes in GA research.

1 Introduction

The genetic algorithms (GA) community has been engaged for nearly two decades in research in automated learning and discovery (see, e.g., [Holland-75], [Grefenstette85a], [Grefenstette87], [Goldberg89]). This work has led to a number of impressive successes in the solution of difficult problems (e.g., [Goldberg83], [DeJong75], [Grefenstette85b]). The approach is particularly interesting in that it successfully applies a powerful metaphor to organize computation -- the metaphor of natural selection and evolution among populations of (computing) individuals in an environment. Yet, although it is increasingly recognized as an interesting enclave of research, it has failed to make significant inroads into the rest of the machine learning and artificial intelligence community. This is largely due to the fact that, for the vast preponderance of research and applications in the genetic algorithms area, the representation of the basic elements of the GA have been binary strings rather than more immediately expressive representations. Although two approaches to reconciling the GA to higher-level representations are possible, historically only one approach has received substantial attention. That approach is one of interpreting binary substrings as individual elements in the high-level language (as in [Forrest85]). This paper takes the other approach and, it is hoped, represents the beginnings of a genetic theory of adaptation directly applicable to high-level languages.

The bit string was embraced early in the history of the GA on the basis of a theoretical analysis of string representation languages. This analysis turned on the desire to maximize the implicit parallelism inherent in the GA, and its outcome, that the binary string is optimal for this purpose, became accepted as fundamental to work in this area. Under the *principle of minimal alphabets* [Goldberg89], the bit string representation has been raised beyond a common feature to almost a necessary precondition for serious work in the GA. The purpose of this paper is to demonstrate that the analysis leading to the principle of minimal alphabets is limited, and that the issue of encoding schemes for the GA should be reopened.

The paper is organized as follows. Section 2.0 reviews the basic argument that the bit string is optimal. Section 3.0 reexamines this analysis by revisiting the motivation for the "schema alphabet" on which it is based and exploring a neglected interpretation of such alphabets. The implications of the resulting new analysis on the schemata theorem and the genetic algorithm proper are discussed in Section 4.0. The paper concludes with Section 5.0, which includes a short summary of the result and future directions.

2 Implicit Parallelism and the Bit String

There are a number of reasons why the binary representation has dominated GA research. These include the simplicity of analysis of binary vectors, the elegance of the genetic operators in binary strings, and requirements for computational speed. The most important reason, however is the theoretically motivated desire to maximize the number of schemata sampled by a given set of individuals in a population (where the population represents the adaptive system, and the individuals its components). The chunk of theory underlying this maximization argument is also the basis of the implicit parallelism results for the GA. It is the focus of this paper.

The maximization argument put forward in [Holland75], pp. 69-74, is as follows: The objects under consideration are individuals in a population of strings. At issue is the alphabet out of which these strings are to be constructed. Let us frame the issue by considering two alphabets, one

in which the number of values that an element of the string may take, v, is two, and another where it is ten. Now construct two strings, one binary of length 20, one decimal of length 6. For concreteness Holland suggests we think of these as two arrays of detectors, one a 4 X 5 array of detectors of binary output, the other a 2 X 3 array of detectors of decimal output. Call these A_2 and A_{10}. Since $2^{20} = 1.05$ X $10^6 \sim 10^6$, (i.e., they distinguish about the same number of states) we can view these carrying approximately the same amount of information.

Holland introduces schemata as a way of analyzing *combinations* of such strings. Each schema is intended to represent a different configuration of component strings as tested in a dynamic environment. To carry out the analysis, a device was needed that would give expression to these subsets of strings. The device used was, starting from the original string alphabet, to augment it into a schema alphabet by adding a single character representing a "wild card". Each string composed from the schema alphabet that includes at least one such augmenting character thus represents a *set* of strings of the original alphabet. Schemata represent sets of individuals (potential or actual) of a population of strings.

What number of schemata are sampled by single individuals of the respective representations (that is, by single configurations of sensed values on the respective arrays)? If we consider a schema to be defined by a string of the original alphabet of a representation plus one "don't care" character "#" (as Holland did), we see that there are $(2+1)^{20}$, or about 3.48 X 10^9 schemata for information carried in A_2, and that the individual sample above matches 2^{20} of them (since the 20-element string matches all schemata that have either the corresponding character or a "#" at the respective positions). A_{10}, on the other hand is represented by $(10+1)^6$, or about 1.77 X 10^6 schemata, and an individual from A_{10} matches only 2^6 of these. In general, the number of schemata for strings of length k and alphabet of cardinality v is about $(v+1)^k$, and individual strings of such languages each sample 2^k

such schemata. [1]

This analysis represents an extremely interesting result. When single strings are evaluated and used in a genetic adaptive plan, they provide a partial evaluation of each schema of which they are representatives. The argument above says that the larger the number of schemata defined on a language, the larger the number of schemata sampled by individual instances of that language. It further indicates that the number of schemata defined on a language is maximized when the alphabet is binary (v=2). The argument forms the basis of the implicit parallelism results of the GA, that the simultaneous sampling of many schemata by single individuals of a population of n strings is maximized when v=2. At that value, the number of schemata sampled is $O(n^3)$. (Implicit parallelism is discussed explicitly but piecemeal throughout [Holland75]. The arguments are reformulated and the $O(n^3)$ result presented in [Goldberg89].)

3 A Re-appraisal of the Argument for the Bit String

Let us consider again the construction of schemata from the original population of strings. Holland introduced a new letter, "#", into the original alphabet to indicate the "don't care" positions in the strings. By that move, he was able to consider strings of the augmented alphabet that have at least one "#" to represent *classes* of strings in the original string language. It is from the analysis of these classes of strings that Holland has derived the mathematical foundations of the GA, that the (*implicit*) manipulation of those classes gives the GA extraordinary adaptive power. The GA represents a milestone in the understanding of adaptation largely because of the theoretical analysis of adaptive systems based on schemata.

But what work, exactly, is the "#" character doing? The appearance of an augmenting character at a given position

acts as a quantifier over strings with different values at that position. So the schema "000#" quantifies over strings "0000" and "0001" when v=2; in the binary case "#" means 0 *or* 1. It has usually been stated that "#" is a "don't care" character (call this Interpretation 1). The binary representation masks an ambiguity between a "don't care" interpretation and one based on quantification over subsets of values at a string position. That is, in the binary case the "#" character can be construed as denoting the set of strings sharing a *subset* of possible values at a position (Interpretation 2) as well as the set of strings of *any* of the possible values at the position. It just happens that when v=2 the two interpretations lead to the same set of strings. As shown below, in strings of richer alphabets they do not.

Consider the language of strings of length 4 and alphabet {0 1 2}. On Interpretation 1 the schema "000#" refers simply to the set of strings {0000 0001 0002}. On Interpretation 2, however, a set of schemata need to be defined to adequately express the quantification over values, because the "don't care" for a three-valued alphabet can be between 0 and 1, 0 and 2, 1 and 2, or 0, 1, and 2. The quantification over values at a position leads correspondingly to the sets {0000 0001}, {0000 0002}, {0001 0002}, and {0000 0001 0002}.

We may now sketch where this alternative view leads. We still construct schemata by augmenting the original string alphabet. Let us annotate augmenting characters with the set over which they are defined. Thus $\#_{01}$ indicates a position in the binary string at which either a 0 *or* a 1 may appear, and so has the same meaning as the earlier "#". In string languages of alphabets of order greater than 2 this does not hold. For a language consisting of the letters {0,1,2}, we need to introduce $\#_{01}$, $\#_{02}$, $\#_{12}$, and $\#_{012}$ into the new schemata alphabet to express, at a given position, a fine-grained "don't care" condition among (0 or 1), (0 or 2), (1 or 2), and (0 or 1 or 2). Notice that only the last corresponds to the previous (binary) don't care semantics of "#". In general the number of augmenting characters needed to express all schemata is not v+1, it is 2^v-1. (The empty set is

---

[1] When schemata aren't intended to include individual instances of strings themselves, we must subtract v^k to get the exact number of schemata $(v+1)^k - v^k$.

subtracted out since string positions always take one of the values in the alphabet).

The implications for the implicit parallelism argument are striking. Let us compare encoding schemes among arrays of sensors again. (Smaller arrays are used to illustrate the point with more manageable numbers, then the detector arrays of the previous example are revisited as a point of comparison.) Let A_{1X8X2} be a 1 X 8 array of binary detectors (that is, v=2). It can distinguish $2^8 = 256$ distinct states. Let A_{1X5X3} be a 1 X 5 array of 3-valued detectors (v=3). It has approximately the same representational power as A_{1X8X2}, since $3^5 = 243 \sim 256 = 2^8$. By the previous analysis the numbers of schemata defined by A_{1X8X2} and A_{1X5X3} are:

$$(2+1)^{(1*8)} = 6561 \gg (3+1)^{(1*5)} = 1024$$
$$[\text{previous analysis}]$$

and the numbers of schemata sampled by a single configuration of detector values would have been calculated at:

$$2^{(1*8)} = 256 \gg 2^{(1*5)} = 32 \quad [\text{previous analysis}].$$

The binary representation was seen as more powerful than the 3-valued representation.

We now see this to be in error, since it severely underestimates the size of the set of schemas sampled by an individual string. In general the number of schemata needed to cover a given string language is $(2^v-1)^L$, where L is the length of the string. On this analysis the numbers of schemata defined on A_{1X8X2} and A_{1X5X3} are:

$$(2^2-1)^8 = 6561 < (2^3-1)^5 = 16807$$

The correct number of schemata sampled by an instance of a string language is $(2^{v-1})^L$. For the respective instances of A_{1X8X2} and A_{1X5X3} they are:

$$(2^1)^8 = 256 < (2^2)^5 = 1024$$

For the earlier example the difference is even more striking. The numbers of

schemata defined on the arrays A_2 and A_{10} are:

$$(2^2-1)^{20} = 3.48 \text{ X } 10^9 \ll (2^{10}-1)^6 = 1.46 \text{ X } 10^{18}$$

The numbers of schemata sampled per individual of the arrays are:

$$(2^1)^{20} \ll (2^9)^6 = 2^{54}$$

The more expressive alphabet is seen to carry *much* more power in an informationally equivalent environment. More expressive languages seem, then, to provide *finer-grained tools* for the construction of adaptive plans.

4 Discussion

The result presented here realigns a fragment of GA theory with intuition about high-level languages, namely, that the more expressive a language, the more powerful an apparatus it provides for adaptation. That additional power is evident in the finer analysis possible when considering subsets of values instead of "don't care" positions. But what is the impact of this analysis on other parts of GA theory and practice?

In the area of practice, some existing experimental results seem to contradict the findings above. In [Shaffer84], experiments directly comparing binary and ternary encodings of a single GA problem, the binary encoding was found significantly superior to the ternary case. However, the experiment dealt only with the peculiar case in which all binary strings in the encoding were meaningful. This will not necessarily be the case when single letters are intended to represent distinct characteristics. Generally, when higherarity characteristics are expressed in a binary encoding, many binary strings will be meaningless -- they will have no translation in the higherarity alphabet. For instance, in a system where v=10, there will be a probability of (16-10)/10 = .6 that a binary substring representing the characteristic will not have a meaningful translation. This probability of unintelligibility increases as the string length gets longer.

In the area of theory, the most obvious and at first sight highly objectionable change is that a huge number of additional characters (exponential in the size of the alphabet!) must be introduced to adequately express the set of schemata. On closer examination, however, this is not so terrible. The schemata are introduced in GA theory strictly for understanding. The naive algorithm based on schemata (that the GA efficiently implements) is itself infeasibly large to use directly, since it is already exponential in the length of the string. The fact that the new analysis also yields an infeasible algorithm shouldn't count against it because it is likewise strictly for theoretical understanding.

Another impact, on the hyperplane analysis of the schemata, is also (on closer examination) seen to be less serious than might be expected. The new view of schema construction complicates the analysis of schemata since they no longer represent simple hyperplanes in the feature space of the problem domain. However, it seems possible to extend the existing hyperplane analysis. The new schemata alphabet *also* defines hyperplanes, but many (augmenting) characters define disjoint, parallel sets of them in addition to the contiguous, individual hyperplanes of the earlier analysis. Therefore the incorporation of sets of hyperplanes in the schemata theorem proof is not anticipated to be a major obstacle. A high cost is exacted, however, in the construction of the genetic operators. The string languages considered here preserve the crossover and inversion operators in their previous forms (although this is no longer the case when a grammatical, instead of a set-theoretic view of the string languages is taken). However, the mutation operator is complicated to the degree of variability of the alphabet, that is, proportionately to v (although cross-positional "mutation" has been explored, e.g., in [Cavicchio70]). The difference, at least, is linear.

It is not clear that the move to a subsetting "operator" to express the schemata was considered in the early development of the GA. What is clear ([Holland75], p68) is that the augmented alphabet for schemata is presented strictly through Interpretation 1:

Now our objective is to designate subsets of A (A = the entire language) which have attributes in common. To do this let the symbol ("#") indicate that we "don't care" what attribute occurs at a given position.

Although this move focused GA research on a common representation for years to come, it simultaneously led to a nearly total neglect of higher-level representation in both the theory and the application of genetic algorithms, and to the present difficulty of reconciling this work with dominant themes elsewhere in automated intelligence (e.g., knowledge representation). The importance of this neglected interpretation of the schemata character is that it opens the door to a reconciliation with both intuition (that a more powerful representation is more useful in adaptive settings) and the overwhelmingly large body of work in higher-order languages.

5 Conclusion

This paper contravenes the argument put forward in the GA literature for a binary representation language. There may be other reasons for using a binary representation, like simplicity of analysis, elegance of the genetic operators, computational speed, and history. However, no other reason besides the one refuted above has, to the author's knowledge, been put forth as a compelling argument for the bit string. The upshot of this is that the genetic algorithm may now, without self contradiction at least, be considered for high-level representation languages. Whether it is appropriate to the problem domains in which we typically apply such expressive languages remains an open question.

The results above represent merely a first step at making Holland's analysis directly applicable to high-level systems. The next steps include carrying out a rederivation of the implicit parallelism results and a reconstructing the schemata theorem proof using the new representation. There is, additionally, the deep problem of how to relate the "set-theoretic" analysis of the problem domain found in the GA theory so far, to a "formal language" characterization of the problem such as commonly found in the theory of computation. While the price of generalization indicated by this paper is a moderate complication of mutation and of the proof of the schemata theorem (we hope), it at least leaves crossover intact. A generalization to formal languages does not (see [Cramer85], [Antonisse87], and [Fujiko87] for characterizations of the problem and some preliminary ideas of its practical solution).

These elements of further generalization are the next steps of this work. The ultimate aim is a schemata theorem and accompanying genetic algorithm that accommodates high-level languages. Hopefully this paper is a step towards that, and more generally, towards the theoretical understanding of adaptation in all complex systems.

References

(Antonisse87) "Genetic Operators for High-level Knowledge Representations", H. J. Antonisse and K. S. Keller, Proc. 2nd Int. GA Conf., pp. 69-76, 1987.

(Cavicchio70) "Adaptive search using simulated evolution", D. J. Cavicchio, Un-published doctoral dissertation, University of Michigan, Ann Arbor, 1970.

(Cramer85)"A representation for the adaptive generation of simple sequential programs", N. L. Cramer, Proc. Int. GA Conf., pp. 183-185, 1985.

(DeJong75) "The Analysis of the Behavior of a Class of Genetic Adaptive Systems", K. DeJong, Doctoral Thesis, CCS Department, University of Michigan, Ann Arbor, 1975.

(DeJong85) "Genetic Algorithms: a 10 year perspective", K. A. DeJong, Proc. Int. GA Conf., pp. 169-177, 1985.

(Forrest85) "Implementing Semantic Network Structures Using the Classifier System", S. Forrest, Proc. Int. GA Conf., pp. 24-44, 1985.

(Fujiko87) "Using the genetic algorithm to generate LISP code to solve the prisoner's dilemma", C. Fujiko and J. Dickinson, Proc. 2nd Int. GA Conf., pp. 236-240, 1987.

(Goldberg83) "Computer-aided gas pipeline operation using genetic algorithms and rule learning", D.E. Goldberg, Ph.D. Thesis, University of Michigan, Ann Arbor, MI, 1983 .

(Goldberg89) "Genetic Algorithms in Search, Optimization, and Machine Learning", D.E. Goldberg, Addison-Wesley, Reading PA, 1989.

(Grefenstette85a) Proceedings of the First International Genetic Algorithms Conference, Grefenstette, J. J., ed., Lawrence Erlbaum Associates, Hillsdale, NJ 1985.

(Grefenstette85b) "Genetic Algorithms for the traveling salesman problem", J. J. Grefenstette, R. Gopal, B. J. Rosmaita, and D. VanGucht, Proc. Int. GA Conf., pp. 160-168, 1985.

(Grefenstette87) Proceedings of the Second International Genetic Algorithms Conference, Grefenstette, J. J., ed., Lawrence Erlbaum Associates, Hillsdale, NJ 1987.

(Holland75) "Adaptation in Natural and Artificial Systems", J. H. Holland, University of Michigan Press, Ann Arbor, MI, 1975.

(Shaffer84) "Some Experiments in Machine Learning Using Vector Evaluated Genetic Algorithms", J. D. Shaffer, Doctoral Thesis, CS Department, Vanderbilt University, Nashville, Tennessee, Dec1984.

Chapter 21
Fuzzy Evolution

C. A. Ankenbrandt, B. P. Buckles, F. E. Petry, and M. Lybanon (1990) "Ocean Feature Recognition using Genetic Algorithms with Fuzzy Fitness Functions (GA/F³)," *3rd Ann. Workshop on Space Operations, Automation and Robotics (SOAR'89)*, R. H. Brown (Chair), NASA Conf. Publ. 3059, Lyndon B. Johnson Space Ctr., Houston, TX, (July 25–27, 1989) pp. 679–685.

WHEN humans describe complex environments, they do not typically speak in absolutes. Linguistic descriptors of real-world circumstances are not precise but rather are "fuzzy." For example, when one describes the optimum behavior of an investor interested in making money in the stock market, the adage is "buy low, sell high." But how low is "low"? And how high is "high"? It is unreasonable to suggest that if the price of the stock climbs to a certain precise value in dollars per share, then it is high; yet if it were only $0.01 lower, it would not be high. Useful descriptions need not be of a binary or crisp nature.

Zadeh (1965) introduced the notion of "fuzzy sets." Rather than describing elements as being either in a given set or not, membership in the set was viewed as a matter of degree ranging over the interval [0, 1]. A membership of 0.0 indicates that the element absolutely is not a member of the set, and a membership of 1.0 indicates that the element absolutely is a member of the set. Intermediate values indicate degrees of membership. The choice of the appropriate membership function to describe elements of a set is left to the researcher.

Negoita and Ralescu (1987, p. 79) noted that descriptive phrases such as "numbers approximately equal to 10" and "young children" are not tractable by methods of classic set theory or probability theory. There is an undecidability about the membership or nonmembership in a collection of such objects, and there is nothing random about the concepts in question. A classic set can be represented precisely as a binary valued function $f_A: X \rightarrow \{0, 1\}$, the characteristic function, defined as:

$$f_A(x) = \begin{cases} 1, \text{ if } x \in A; \\ 0, \text{ otherwise} \end{cases}$$

where X is the universe of discourse. The collection of all subsets of X (the power set of X) is denoted

$$P(X) = \{A | A \text{ is a subset } of X\}.$$

In contrast, a fuzzy subset of X is represented by a membership function:

$$u:X \rightarrow [0, 1].$$

The collection of all fuzzy subsets of X (the fuzzy power set) is denoted by:

$$F(X) = \{u | u: X \rightarrow [0,1]\}.$$

It is natural to enquire as to the effect of operations such as union and intersection on such fuzzy sets. If u and v are fuzzy sets, then:

$$(u \text{ or } v)(x) = \max [u(x), v(x)]$$

$$(u \text{ and } v)(x) = \min [u(x), v(x)].$$

Other forms of these operators have been developed (Yager, 1980; Dubois and Prade, 1982; Kandel, 1986, pp. 143–149; and others).

Fuzzy sets have been used in a variety of applications since their introduction in the mid-1960s, but primarily in pattern recognition and control (Bezdek and Pal, 1992; Mamdani, 1974; Mamdani and Baaklini, 1975; and others). The appropriate design of fuzzy systems is something of an art, with performance varying with the choices of membership functions and the procedures that are taken in light of these functions. It is natural to seek optimum plans in light of fuzzy performance measures, as well as to optimize the fuzzy membership functions for specific problems. Evolutionary algorithms provide one possibility for accomplishing this task.

An early related effort was described in Bremermann (1971) regarding pattern recognition experiments conducted by student Richard Hodges.[1] A pattern to be recognized was deformed into multiple prototypes that comprised that pattern. For example, the letter "A" could be deformed into three prototype line segments that form the two legs and crossbar. The position of each line segment and its width were described by

---

[1] Bremermann (1971) cited Hodges as a dissertation in progress. There appears to be no final record of the dissertation at U.C. Berkeley.

real-valued coordinates (x_1, y_1), (x_2, y_2) and a scalar w (all segments were the same width). Thus the character "A" was defined by 13 real-valued parameters. By varying these parameters of the prototypes for a variety of patterns to be recognized, a degree of correspondence could be obtained to each observed character. The correspondence (described as a "cybernetic functional") was computed as the weighted composite of two terms measuring distortion (i.e., discrepancy between prototypes and ideal prototypes) and the degree of match (i.e., discrepancy between prototypes and the observed character). An evolutionary search was made using the methods of Bremermann (1970; following Bremermann, 1968, and others, see Chapter 9) to find optimal parameter values to minimize the error defined by the cybernetic functional. Bremermann (1971) showed through mathematical manipulation that these cybernetic functionals were essentially fuzzy sets, and the process could be thought of as determining "in which of several fuzzy sets the given object has the largest membership value."

Ankenbrandt et al. (1990a) (reprinted here)[2] and (Ankenbrandt et al., 1990b) followed a somewhat similar framework, although more plainly involving fuzzy logic, in which a genetic algorithm was used to optimize a semantic network in light of fuzzy performance measures. Fuzzy descriptors of the relations between satellite imagery of oceanic features (e.g., cold and warm eddies, continental shelf, north or south wall of Gulf Stream) were devised, such as "cold eddies are usually south of the Gulf Stream." A scene consisted of classification categories and relationships between the categories, modeled as a semantic network. Candidate networks were evaluated in light of fuzzy criteria, such as the nearness between connected categories.

About the same time as Ankenbrandt et al. (1990a), Karr et al. (1990) used a genetic algorithm to optimize the placement of membership functions in a fuzzy controller.[3] The potential for such optimization was well known (e.g., Mamdani and Baaklini, 1975; Procyk and Mamdani, 1979; and others). Karr et al. (1990) evolved the placement of triangle-shaped fuzzy membership functions describing the position, velocity, and thrust of a spacecraft in a rendezvous problem. The application of the genetic algorithm was fairly straightforward: the left and right limits on each membership function were coded in binary and subjected to one-point crossover, point mutation, and tournament selection. Across four different initial conditions, the genetic algorithm was generally able to improve on the performance offered by fuzzy membership functions hand crafted by the authors. Subsequent efforts to evolve fuzzy controllers were offered in Karr (1991), Pham and Karaboga (1991), Thrift (1991), Haffner and Sebald (1993), and others. Other combinations of evolutionary computation and fuzzy sets include the evolution of fuzzy clusters in Fogel and Simpson (1993), and others, and the development of fuzzy classifier systems (Valenzuela-Rendón, 1991a, 1991b).

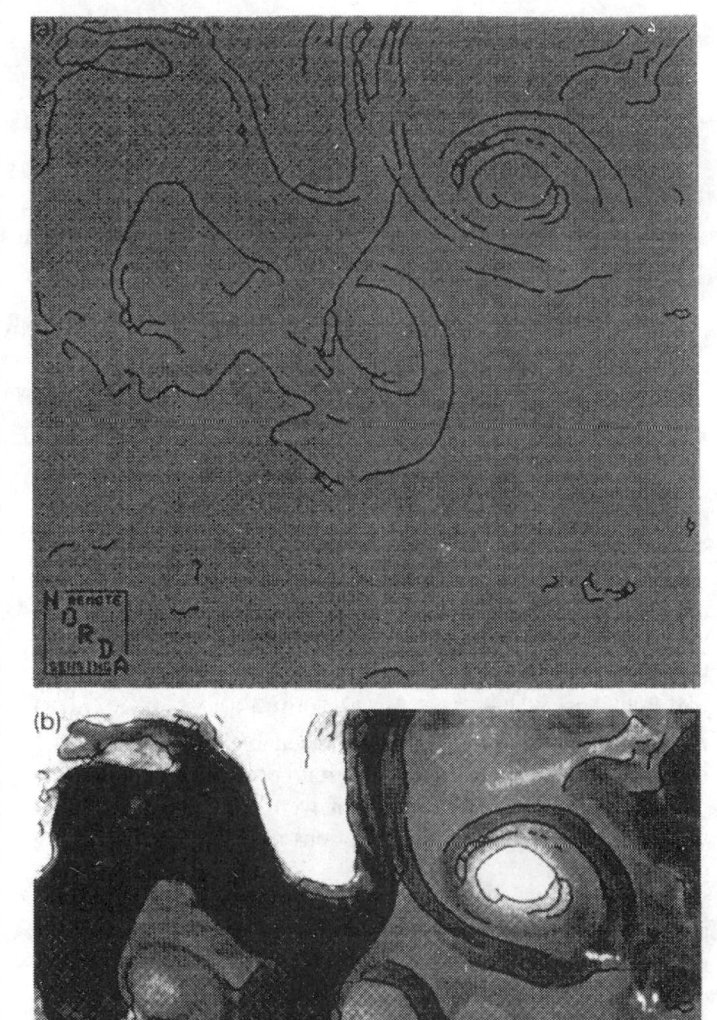

Fig. 21.1 The above figures are taken from Ankenbrandt et al. (1990b) and duplicate the figures in Ankenbrandt et al. (1990a), reprinted here, which were not suitable for reproduction.

[2] The conference publication Ankenbrandt et al. (1990a) contained two figures that were not suitable for reproduction. These also appear in the more complete journal publication Ankenbrandt et al. (1990b) and are reprinted as Figure 21.1.

[3] Both Ankenbrandt et al. (1990a) and Karr et al. (1990) were presented at conferences in 1989 with publication dates of 1990.

References

[1] C. A. Ankenbrandt, B. P. Buckles, F. E. Petry, and M. Lybanon (1990a) "Ocean feature recognition using genetic algorithms with fuzzy fitness functions (GA/F³)," *3rd Ann. Workshop on Space Operations, Automation and Robotics (SOAR'89),* R. H. Brown (chair), NASA Conf. Publ. 3059, Lyndon B. Johnson Space Ctr., Houston, TX, (July 25–27, 1989) pp. 679–685.

[2] C. A. Ankenbrandt, B. P. Buckles, and F. E. Petry (1990b) "Scene recognition using genetic algorithms with semantic nets," *Pattern Recognition Letters,* Vol. 11, pp. 285 293.

[3] J. C. Bezdek and S. K. Pal (eds.) (1992) *Fuzzy Models for Pattern Recognition: Methods that Search for Structures in Data,* IEEE Press, NY.

[4] H. J. Bremermann (1968) "Numerical optimization procedures derived from biological evolution processes," *Cybernetic Problems in Bionics,* H. L. Oestreicher and D. R. Moore (eds.), Gordon and Breach, NY, pp. 543–561.

[5] H. Bremermann (1970) "A method of unconstrained global optimization," *Math. Biosciences,* Vol. 9, pp. 1–15.

[6] H. J. Bremermann (1971) "Cybernetic functionals and fuzzy sets," *1971 IEEE Systems, Man, and Cybernetics Group Ann. Symp. Record,* G. L. Hollander (chair), IEEE, Anaheim, CA, pp. 248–253.

[7] D. Dubois and H. Prade (1982) "A class of fuzzy measures based on triangular norms—A general framework for the combination of uncertain information," *Int. J. General Systems,* Vol. 8:1, pp. 43–61.

[8] D. B. Fogel and P. K. Simpson (1993) "Experiments with evolving fuzzy clusters," *Proc. of the 2nd Ann. Conf. on Evolutionary Programming,* D. B. Fogel and W. Atmar (eds.), Evolutionary Programming Society, La Jolla, CA, pp. 90–97.

[9] S. B. Haffner and A. V. Sebald (1993) "Computer-aided design of fuzzy HVAC controllers using evolutionary programming," *Proc. of the 2nd Ann. Conf. on Evolutionary Programming,* D. B. Fogel and W. Atmar (eds.), Evolutionary Programming Society, La Jolla, CA, pp. 98–107.

[10] A. Kandel (1986) *Fuzzy Expert Systems,* CRC Press, Boca Raton, FL.

[11] C. L. Karr (1991) "Design of an adaptive fuzzy logic controller using a genetic algorithm," *Proc. of the 4th Intern. Conf. on Genetic Algorithms,* R. K. Belew and L. B. Booker (eds.), Morgan Kaufmann, San Mateo, CA, pp. 450–457.

[12] C. L. Karr, L. M. Freeman, and D. L. Meredith (1990) "Improved fuzzy process control of spacecraft autonomous rendezvous using a genetic algorithm," *Intelligent Control and Adaptive Systems,* G. Rodriguez (ed.), Proc. SPIE 1196, pp. 274–288.

[13] E. H. Mamdani (1974) "Applications of fuzzy algorithms for control of a simple dynamic plant," *Proc. IEE,* Vol. 121:12, pp. 1585–1588.

[14] E. H. Mamdani and N. Baaklini (1975) "Prescriptive method for deriving control policy in a fuzzy-logic controller," *Electronics Letters,* Vol. 11:25–26, pp. 625–626.

[15] C. V. Negoita and D. Ralescu (1987) *Simulation, Knowledge-Based Computing, and Fuzzy Statistics,* Van Nostrand Reinhold, NY.

[16] D. T. Pham and D. Karaboga (1991) "A new method to obtain the relation matrix for fuzzy logic controllers," *Applications of Artificial Intelligence in Engineering VI,* G. Rzevski and R. A. Adey (eds.), Elsevier, London, pp. 567–581.

[17] T. J. Procyk and E. H. Mamdani (1979) "A self-organizing linguistic process controller," *Automatica,* Vol. 15, pp. 15–30.

[18] P. Thrift (1991) "Fuzzy logic synthesis with genetic algorithms," *Proc. of the 4th Intern. Conf. on Genetic Algorithms,* R. K. Belew and L. B. Booker (eds.), Morgan Kaufmann, San Mateo, CA, pp. 509–513.

[19] M. Valenzuela-Rendón (1991a) "The fuzzy classifier system: Motivation and first results," *Parallel Problem Solving from Nature,* H.-P. Schwefel and R. Männer (eds.), Springer, Berlin, pp. 338–342.

[20] M. Valenzuela-Rendón (1991b) "The fuzzy classifier system: A classifier system for continuously varying variables," *Proc. of the 4th Intern. Conf. on Genetic Algorithms,* R. K. Belew and L. B. Booker (eds.), Morgan Kaufmann, San Mateo, CA, pp. 346–353.

[21] R. R. Yager (1980) "A measurement-informational discussion of fuzzy union and intersection," *IEEE Trans. Syst., Man and Cybern.,* Vol. 10:1, pp. 51–53.

[22] L. Zadeh (1965) "Fuzzy sets," *Information and Control,* Vol. 8, pp. 338–353.

OCEAN FEATURE RECOGNITION USING GENETIC ALGORITHMS

WITH FUZZY FITNESS FUNCTIONS (GA/F^3)*

by

C.A. Ankenbrandt[1], B.P. Buckles[1], F.E. Petry[1], & M. Lybanon[2]

[1]Department of Computer Science
Center for Intelligent and Knowledge-based Systems
301 Stanley Thomas Hall, Tulane University
New Orleans, LA 70118, (504) 865-5840

[2]Remote Sensing Branch
Naval Ocean Research and Development Activity
NSTL Station, MS 39529

ABSTRACT

A model for genetic algorithms with semantic nets is derived for which the relationships between concepts is depicted as a semantic net. An organism represents the manner in which objects in a scene are attached to concepts in the net. Predicates between object pairs are continuous valued truth functions in the form of an inverse exponential function ($e^{-\beta|x|}$). 1:n relationships are combined via the fuzzy OR (Max [...]). Finally, predicates between pairs of concepts are resolved by taking the average of the combined predicate values of the objects attached to the concept at the tail of the arc representing the predicate in the semantic net. The method is illustrated by applying it to the identification of oceanic features in the North Atlantic.

keywords: genetic algorithms, feature labelling, semantic nets, fitness functions

* This work was supported in part by a grant from Naval Ocean Research and Development Activity, Grant #N00014-89-J-6003

BACKGROUND

Genetic algorithms are a problem solving method requiring domain-specific knowledge that is often heuristic. Candidate solutions are represented as organisms. Organisms are grouped into populations known as generations and are combined in pairs to produce subsequent generations. An individual organism's potential as a solution is determined by a fitness function.

Fitness functions map organisms into real numbers and are used to determine which organisms will be used (and how frequently) to produce offspring for the succeeding generation. Fitness functions often require heuristic information because a precise measure of the suitability of a given organism (i.e., solution) is not always attainable. An example is the recognition (i.e., labeling) of segments in a scene. General characteristics of objects in the scene such as curvature, size, length, and relationship to each other may be known only within broad tolerance levels. That is, there is great variability in the relationships among objects in different scenes.

Semantic nets (SNs) are effective representations of binary relationships between concepts (e.g., objects in a scene). SNs denote concepts via nodes in a directed graph. The arcs are labelled by predicates. We introduce here a representation of an organism whose fitness function evaluation is dependent upon an SN context.

Because relationships (i.e., predicates) relating concepts are not precise, their evaluation is in the form of a truth functional with range [0,1] rather than the traditional {0,1}. That is, we use fuzzy logic [YA75, ZA88, ZI85] to combine heuristically the information concerning a particular organism. Thus, we derive genetic algorithms with fuzzy fitness functions (GA/F^3).

GENETIC ALGORITHMS

Genetic algorithms (GAs) are search procedures modelled after the mechanics of natural selection. They differ from traditional search techniques in several ways. First, GAs have the property of implicit parallelism, where the algorithm is equivalent to a search of the hyperplanes of the search space, without directly testing hyperplane values [HO75, GO88]. Nearly optimal results have been found by examining as few as one point for every 2^{35} points in the search space [GO86]. Second, GAs are randomized algorithms, using operations with nondeterministic results. The results for an operation depend on the value of a random number. Third, GAs operate on many solutions simultaneously, gathering information from all current points to direct the search. This factor mitigates the problems of local maxima and noise.

From a mechanistic view, genetic algorithms are a variation of the generate and test method. In pure generate and test, solutions are generated and sent to an evaluator. The evaluator reports whether the solution posed is optimal. In genetic algorithms, this generate and test process is repeated iteratively over a set of solutions. The evaluator returns information to guide the selection of new solutions for following iterations.

GA terminology is taken from genetics. Each candidate solution examined is termed an organism, traditionally represented as a list. The set of organisms maintained is termed a population, and the population at a given time is termed a generation. Each iteration envolves three steps.

First, each organism in the current generation is evaluated, producing a numerical fitness function result. The criteria for evaluation is domain specific information about the relative merit of that particular organism. Better organisms are assigned higher fitness function values. Second, some organisms are selected to form one or more organisms for the next generation. Specifically, the number of copies of each organism selected is directly proportional to its fitness function. Third, some of those organisms selected are modified via genetic operators. Each genetic operator takes the chosen organism(s), and produces a new organism(s). The most common genetic operators include crossover and mutation. This iterative procedure terminates when the population converges to a solution.

The crossover operator takes two organisms selected and combines partial solutions of each. When organisms are represented with lists, single point crossover can be viewed as combining the left hand side of one organism chosen with the right hand side of the other, and conversely. This creates two offspring. The crossover point, that point where the crossover takes place, is randomly determined.

The mutation operator uses a minimal change strategy. It takes a selected organism, and changes the value at one randomly determined position. This corresponds to a tight local search. The offspring produced is identical to the parent except at the mutation point.

GENETIC ALGORITHM PROBLEM MODEL FOR OCEANIC FEATURE LABELING

Scene recognition is an application for which the GA model we propose is suited. For example, Fig. 1(a) is a segmented image of the North Atlantic for which Fig. 1(b) is the original image. The lines (referred to here as segments, s_1, s_2, ...) represent boundaries between warm and cold regions of sea water. The problem is to classify the segments as Gulf Stream North Wall (NW), Gulf Stream South Wall (SW), cold eddies (CE), warm eddies (WE), continental shelf (CS), and "other" (O).

Relationships which can be expressed as fuzzy truth functions are known to exist within or between classifications. Principal among these are (1) the average width of the Gulf Stream is 50 kilometers, (2) the average diameter of an eddy is 100 kilometers, (3) cold eddies are usually south of the Gulf Stream, and (4) warm eddies are usually north of the Gulf Stream. To these one must add the trivial (yet necessary) relationships such as the south wall is at a lower latitude than the north wall and the known geophysical coordinates of continental shelves.

A scene consisting of classification categories (cat_1, cat_2, ..., cat_n) and relationships expressed as truth functions ($P^{(1)}_{ij}$, $P^{(2)}_{ij}$, ...) between categories can be modelled as a semantic net (or, more precisely, an association list). A generic one is shown in Fig. 2. Segments are

a. Segmented Image

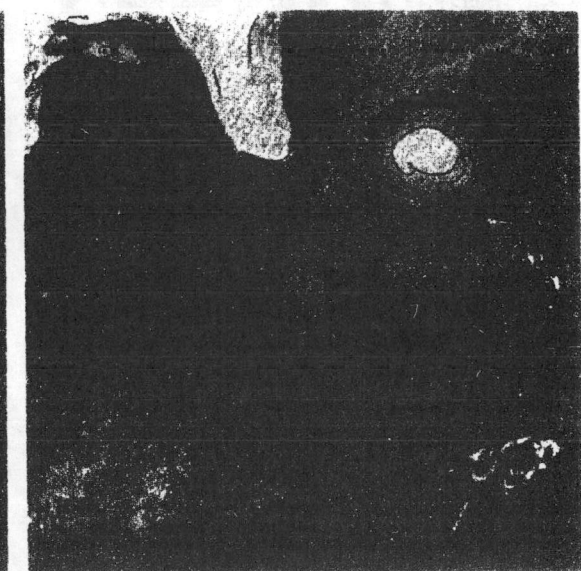

b. Original Infrared Image

Figure 1. Oceanic Features (North Atlantic)

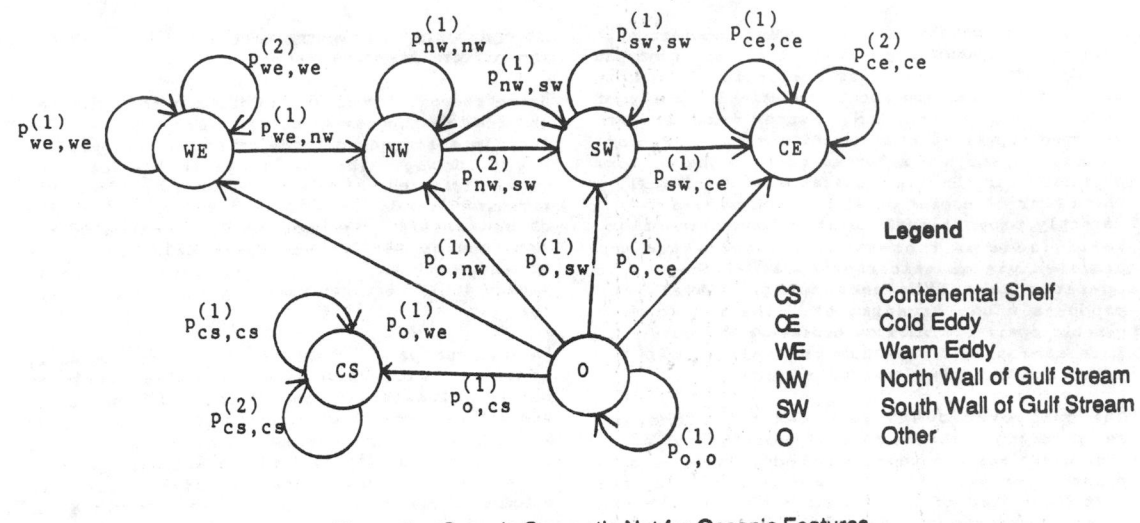

Figure 2. Generic Semantic Net for Oceanic Features

Legend

CS	Contenental Shelf
CE	Cold Eddy
WE	Warm Eddy
NW	North Wall of Gulf Stream
SW	South Wall of Gulf Stream
O	Other

attached to the categories via the INST (instance) relation. An allele (or gene) is a category name. An organism is a list of categories, one allele for each segment. For example, given six segments then (NW, NW, SW, CS, CE, O) and (CE, SW, CE, O, O, CS) are representative organisms. Formally, let an association list be defined as $A = <V, P>$ where $V = \{cat_1, cat_2, ..., cat_m\}$ is a set of categories, and $P = \{P_{ij}^{(g)} \mid i, j \leq m, g = 1, 2, ..., r_{ij}\}$ is a set of binary predicates. These predicates describe the relationships between categories and the ideal relationship between segments assigned to these categories. Let an organism for spatial labeling is defined as $Q = <S, INST>$, where $S = \{s_1, s_2, ..., s_n\}$ is a set of segments, and $INST: S \rightarrow V$ is a function.

Crossover Operators

There are three applicable crossover operators. These include single point crossover, two point crossover, and varying multiple point crossover [BO87]. Crossover operators require the imposition of a total order on the segments in S. Let $s_i < s_j$ if $i < j$; $s_i = s_j$ if $i = j$; $s_i > s_j$ if $i > j$. Denote by $INST_{Oi}$ the instance mapping for organism O_i.

Single Point Crossover. Given $< s_1, s_2, ..., s_n>$, choose a random integer k, $1 \leq k < n$. For parent organisms O1 and O2 create an offspring, O', such that

$$INST_{O'}(s_i) = \begin{cases} INST_{O1}(s_i) \text{ if } i \leq k \\ INST_{O2}(s_i) \text{ if } i > k \end{cases}$$

Two Point Crossover. Let $< s_1, s_2, ..., s_n>$ be a circular list. Formally, $succ(s_i) = s_{i+1}$ $(pred(s_{i+1}) = s_i)$ if $i < n$ and $succ(s_n) = s_1$ $(pred(s_1) = s_n)$. Choose two random integers, k1 and k2. For parent organisms O_1 and O_2 create an offspring, O' such that

$$INST_{O'}(s_i) = \begin{cases} INST_{O1}(s_i) \text{ if } s_i \in \{s_{k1}, succ(s_{k1}), \\ ..., pred(s_{k2})\} \\ INST_{O2}(s_i) \text{ otherwise} \end{cases}$$

Varying Multiple Point Crossover. For parent organisms O_1 and O_2, create an offspring O' such that

$$INST_{O'}(s_i) = \begin{cases} INST_{O1}(s_i) \text{ with probability } 0.5 \\ INST_{O2}(s_i) \text{ with probability } 0.5 \end{cases}$$

Mutation Operator

Our mutation operator selects one segment randomly and assigns it to a randomly determined category. Choose two random integers k1, $1 \leq k1 \leq n$, and k2, $1 \leq k2 \leq m$. Remove s_{k1} from its current category in organism O and attach it to cat_{k2} (i.e., set $INST_O(s_{k1}) = cat_{k2}$).

Fitness Function

For the model, the fitness function is the sum of all satisfied predicates in the semantic net. Let E denote the function. Let $P_i^{(g)}$, be defined as above, with m possible categories. Then

$$E = \sum_{j=i}^{m} \sum_{i=1}^{m} \sum_{g=1}^{r_{ij}} P_{ij}^{(g)} \tag{1}$$

$P_{ij}^{(g)}$ is a predicate for a relationship between categories, i and j. Each predicate $P_{ij}^{(g)}$ has a corresponding derived predicate, $pred_{ij}^{(g)}(k, 1)$, for an analogous relationship between segments s_k and s_1, where s_k is in category i and s_1 is in category j. $P_{ij}^{(g)}$ is interpreted based on the normalized truth value of the derived predicate. Specifically,

$$P_{ij}^{(g)} \begin{cases} \dfrac{\sum_{s_1} \sum_{s_k} \text{pred}_{ij}^{(g)}(k, 1)}{|\text{cat}_i| \times |\text{cat}_j|} & \\ 0 & \text{otherwise} \end{cases} \qquad (2)$$

where $|\text{cat}_i|$ and $|\text{cat}_j|$ are the number of segments classified as category i and category j, respectively. Because all such predicates are not defined between all possible pairs of segments, the normalizing factor (the denominator) is subject to redefinition on a case by case basis. Alternatives to (2) are described following the description of derived predicates below.

An example of a fuzzy predicate $P_{ij}^{(g)}$ from our domain is the relationship "is near", where category i "is near" category j. The corresponding derived predicate $\text{pred}_{ij}^{(g)}(k,1)$ describes the relationship between two segments, s_k in category i and s_1 in category j. The sum of $\text{pred}_{ij}^{(g)}(k,1)$ for all possible pairs of segments s_k and s_1 is normalized by the maximum possible.

Definitions of $\text{pred}_{ij}^{(g)}(k,1)$ are dependent on the underlying semantics of the problem domain. One approach is to define them propositionally as {0,1} if a measurable relationship between s_k and s_1 is within or beyond some threshold. A second approach preferred here is to define them as fuzzy truth functions on the interval [0,1]. Inverse exponential truth functions are commonly used in fuzzy set theory to measure the "nearness" of two concepts. An alternative nearness measures are in [ZI85]. For example, if the description of $P_{ij}^{(g)}$ contains a nominal value (e.g., the SW is approximately 50 kilometers from the NW) then let X_o represent the nominal value and

$$\text{pred}_{ij}^{(g)}(k,1) = e^{-\beta |X_o-X|} \qquad (3)$$

where

   X is the observed value corresponding to the same measure (distance, curvative, angle of declination) between s_k and s_1
   β is a constant contrast factor in [0,1] which emphasizes the magnitude of the difference between the observed and nominal value when increased

There are many situations for which the nearness measure is not bounded by an ideal but the closer to s_k the better. In such cases, X_o can be replaced by zero in formula (3).

"Not near" or "as distant as possible" may be measured by the fuzzy complement of (3).

$$\text{pred}_{ij}^{(g)}(k,1) = 1 - f() \qquad (4)$$

where $f()$ is the right side of formula (3).

Some relationships such as "above" or "smaller" are not easily modelled as nearness measures.

Such relationships can be considered as ordinary propositional truth values.

$$\text{pred}_{ij}^{(g)}(k,1) = \begin{cases} 1 & \text{if } s_k \text{ and } s_1 \text{ are so related} \\ 0 & \text{otherwise} \end{cases} \qquad (5)$$

If there is a measure X associated with the relationship and $X_k > X_1$ when the condition is met, the derived predicate of formula (5) can be represented by the ceiling function

$$\text{pred}_{ij}^{(g)} = \lceil (X_k-X_1)/(|X_k-X_1|+1) \rceil \qquad (6)$$

For $P_{ij}^{(g)}$, each object attached to cat_i requires $|\text{cat}_j|$ evaluations of $\text{pred}_{ij}^{(g)}$. The multiple evaluations are combined to a single value using fuzzy OR

$$\max_{s_1} [\text{pred}_{ij}^{(g)}(k,1)]; \text{ for each } s_k \text{ in } \text{cat}_i \qquad (7)$$

This corresponds to finding the best segment, s_1, that matches the relationship for a given segment s_k. By contract, the combination rule

$$\min_{s_1} [\text{pred}_{ij}^{(g)}(k,1)]; \text{ for each } s_k \text{ in } \text{cat}_i \qquad (8)$$

corresponds to fuzzy AND. The heuristic implied by the formula (2) is

$$\sum_{s_1} \text{pred}_{ij}^{(g)}(k,1)/|\text{cat}_j|; \text{ for each } s_k \text{ in } \text{cat}_i \qquad (9)$$

which corresponds to the average truth functional value of s_k with all s_1 segments in cat_j.

Let $f^{(g)}_{ij}(k)$ stand for the segment level combination rule, (7), (8), or (9). Possible aggregation rules to compute $P_{ij}^{(g)}$ are

$$\sum_{s_k} f_{ij}^{(g)}(k)/|\text{cat}_i| \qquad (10)$$

$$\max_{s_k} [f_{ij}^{(g)}(k)] \qquad (11)$$

$$\min_{s_k} [f_{ij}^{(g)}(k)] \qquad (12)$$

which correspond to average, best, and worst match, respectively. The aggregation rule of formula (10) is the one implied by formula (2).

EXAMPLE

Fig. 3 is a reproduction of Fig. 1(a) with most segments labelled (correctly). Eight segments are labelled as s_1, s_2, ..., s_8 and are used below in an example. Table 1 lists and defines all predicates and derived predicates required for the semantic net of Fig. 2. The notation $|\text{cat}_h|$

Table 1. Predicate Descriptions

Predicate	Functional [Pred(k,l)]/normalizer	Description
$P_{cs,cs}^{(1)}$	$\max\limits_{x} [\exp(-0.5\,x)]/|coor|$	near known CS coordinates (distance = x)
$P_{cs,cs}^{(2)}$	$\max\limits_{x \text{ where } k\neq l} [\exp(-0.5\,x)]/(|cat_{cs}|-1)$	near other CS segment (distance = x)
$P_{we,we}^{(1)}$	$(1/|cat_{we}|)\sum\limits_{x}[\exp(-0.5|100-x|)]/|cat_{we}|$	WE diameter near 100 km (distance = x)
$P_{we,we}^{(2)}$	$\max\limits_{x \text{ where } k\neq l} [\exp(-0.5\,x)]/(|cat_{we}|-1)$	near other WE segment (distance = x)
$P_{we,nw}^{(1)}$	$(1/|cat_{nw}|)\sum\limits_{x}\lceil (X_k-X_l)/(|X_k-X_l|+1)\rceil]/|cat_{we}|$	WE north of NW (X_k and X_l are latitudes)
$P_{nw,nw}^{(1)}$	$\max\limits_{x \text{ where } k\neq l} [\exp(-0.5\,x)]/|cat_{nw}|-1)$	near other NW segment (distance = x)
$P_{nw,sw}^{(1)}$	$(1/|cat_{sw}|)\sum\limits_{x}[\exp(-0.5|50-x|)]/|cat_{nw}|$	NW 50km from SW (distance = x)
$P_{nw,sw}^{(2)}$	$(1/|cat_{sw}|)\sum\limits_{x}\lceil [(X_k-X_l)/(|X_k-X_l|+1)]\rceil]/|cat_{nw}|$	NW north of SW (X_k and X_l are latitudes)
$P_{sw,sw}^{(1)}$	$\max\limits_{x \text{ where } k\neq l} [\exp(-0.5\,x)]/|cat_{sw}|-1)$	near other SW segment (distance = x)
$P_{sw,ce}^{(1)}$	$(1/|cat_{ce}|)\sum\limits_{x}\lceil [(X_k-X_l)/(|X_k-X_l|+1)]\rceil]/|cat_{sw}|$	SW north of CE (X_k and X_l are latitudes)
$P_{ce,ce}^{(1)}$	$(1/|cat_{ce}|)\sum\limits_{x}[\exp(-0.5|100-x|)]/|cat_{ce}|$	CE diameter near 100 km (distance = x)
$P_{ce,ce}^{(2)}$	$\max\limits_{x \text{ where } k\neq l} [\exp(-0.5\,x)]/|(cat_{ce}|-1)$	near other CE segment (distance = x)
$P_{o,o}^{(1)}$	$\max\limits_{x \text{ where } k\neq l} [\exp(-0.5\,x)]/|(cat_{o}|-1)$	near other O segment (distance = x)
$P_{o,*}^{(*)}$	$(1/|cat_{o}|)\sum\limits_{x}[1-\exp(-0.5x)]/|cat_{o}|$	not near CS, WE, CE, NW, or SW

refers to the number of segments that are an instance of category h. The value 0.5 is chosen arbitrarily for β in all derived predicates. The exponential form of derived predicates is used for all relationships except "north of" where formula (6) is substituted. The default value for any predicate or derived predicate is zero should a denominator evaluate to zero.

The eight segments distinguished in Fig. 3 are characterized in Table 2. For this example, we need only the geophysical coordinates, the distances between segment centroids, and the distances between the closest points of segments. A larger, more complete description might also contain the length and degree of curvature of each segment.

Table 3 lists six organisms together with their fitness function values which are computed using the predicates in Table 1. The fitness function is given by formula (2). The combination and aggregation rules are formulas (7) and (12), respectively. Derived predicates are variations of formulas (3) and (4) except "north of", which is represented by formula (6) with the requisite measure being latitude. Organism O_1 has no segments labelled incorrectly. O_2 has two segments labelled incorrectly. O_3 through O_6 have 3, 3, 5, and 8 incorrectly labelled segments, respectively. The fitness function values correspond roughly to the correctness of the labelling. Additional predicates (i.e., a more complex semantic net) would improve upon the ordering and separation in most cases.

CONCLUSION

A model for labelling complex scenes via genetic algorithms with fuzzy fitness functions evaluated over semantic nets and GAs is possible. Truth functionals indicating the degree to which specific interfeature relationships are fulfilled are combined at the segment level then aggregated at the category level using fuzzy set operators.

We are currently investigating such issues as the effect of many predicates clustered on one or two

572

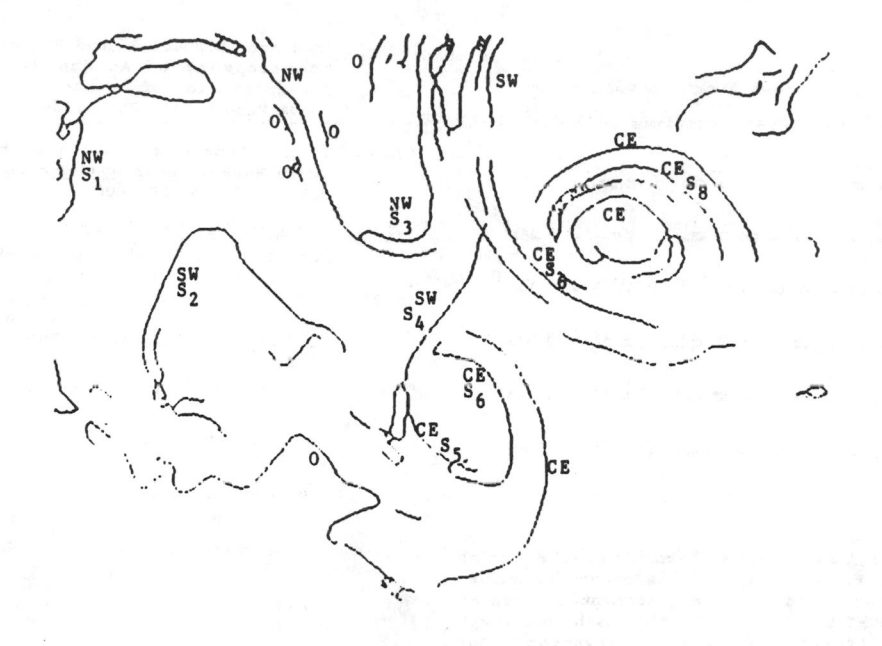

Figure 3. Segmented Image With Correct Labels

Table 2. Segment Descriptors

a. Centroid Position in Fractions of Latitude and Longitude

Segment	Latitude	Longitude
S_1	39.48	70.04
S_2	38.82	68.69
S_3	39.52	66.84
S_4	38.37	66.67
S_5	37.33	66.72
S_6	37.52	66.06
S_7	38.07	65.81
S_8	39.54	64.86

b. Distances Between Centroids (kilometers)

	S_1	S_2	S_3	S_4	S_5	S_6	S_7	S_8
S_1	0.00	127.50	257.55	293.03	342.93	375.45	368.68	416.76
S_2	127.50	0.00	164.60	168.39	217.12	247.96	243.55	316.35
S_3	257.55	164.60	0.00	115.81	219.21	209.60	167.14	159.20
S_4	293.03	168.39	115.81	0.00	104.08	98.13	75.67	186.73
S_5	342.93	217.12	219.21	104.08	0.00	56.36	104.29	266.84
S_6	375.45	247.96	209.60	98.13	56.36	0.00	58.67	223.86
S_7	368.68	243.55	167.14	75.67	104.29	58.67	0.00	165.51
S_8	416.76	316.35	159.20	186.73	266.84	223.86	165.51	0.00

c. Closest Proximities (kilometers)

	S_1	S_2	S_3	S_4	S_5	S_6	S_7	S_8
S_1	0.00	127.13	-	-	-	-	-	-
S_2	127.13	0.00	-	80.42	-	-	-	-
S_3	-	-	0.00	31.26	-	-	42.80	-
S_4	-	80.42	31.26	0.00	12.72	15.39	16.93	35.00
S_5	-	-	-	12.72	0.00	0.00	-	-
S_6	-	-	-	15.39	0.00	0.00	-	-
S_7	-	-	42.80	16.93	-	-	0.00	20.92
S_8	-	-	-	35.00	-	-	20.92	0.00

Table 3. Fitness Function Values for Selected Organisms

$O_1 = $ <NW SW NW SW CE CE CE CE> ; $E(O_1) = 2.2098$

$O_2 = $ <SW SW NW NW CE CE CE CE> ; $E(O_2) = 2.2511$

$O_3 = $ <NW SW NW NW CE CE NW SW> ; $E(O_3) = 2.1251$

$O_4 = $ <SW SW NW CE NW CE CE CE> ; $E(O_4) = 1.4731$

$O_5 = $ <NW NW CE CE SW NW SW CE> ; $E(O_5) = 1.6757$

$O_6 = $ <SW CE SW CE SW NW SW NW> ; $E(O_6) = 0.9235$

categories, alternate forms for the truth functionals themselves, and the crossover rules. Our image set consists of six segmented infrared photographs of the North Atlantic, each photograph having a different degree of observation. Our testbed will consist of a GA algorithm capable of manipulating the alleles' correspondence to the semantic net.

REFERENCES

[BO87] Booker, Lashon, "Improving Search in Genetic Algorithms", _Genetic Algorithms and Simulated Annealing_, Lawrence Davis, Ed., Morgan Kaufmann, Los Altos, CA, 1987, pp. 61-73.

[GO86] Goldberg, David E., "A Tale of Two Problems: Broad and Efficient Optimization Using Genetic Algorithms," _Proc. of the Summer Computer Simulation Conference_, July 28-30, 1986, Reno, Nevada.

[GO88] Goldberg, David E., Genetic Algorithms in _Search, Optimization, and Machine Learning_, Addison-Wesley, Reading, MA, 1988.

[HO75] Holland, John H., _Adaption in Natural and Artificial Systems_, University of Michigan Press, Ann Arbor, Michagan, 1975.

[LC87] Lybanon, M. and R.L. Crout, "The NORDA GEOSAT Ocean Applications Program", _John Hopkins APL Technical Digest_, Vol 8, No. 2, April/June 1987, pp. 212-218.

[RI83] Richardson, P.L., "Gulf Stream Rings", _Eddies and Marine Science_, A.R. Robinson (ed.), Springer Verlag, New York, 1983, pp. 19-45.

[TB86] Thomason, Michael G. and Richard E. Blake, "Development of An Expert System for Interpretation of Oceanographic Images", NORDA Report 148, June 1986.

[YA75] Yager, Ronald R. "Decision Making with Fuzzy Sets", _Decision Sciences_, Vol 6, 3, July 1975, pp. 590-600.

[ZA88] Zadeh, Lotfi A., "Fuzzy Logic" _Computer_, Vol 21, 4, April 1988, pp. 83-93.

[ZI85] Zimmermann, Hans -J., _Fuzzy Set Theory and Its Applications_, Kluwer Nijhoff Publishing, Dordrecht, The Netherlands, 1985.

Chapter 22
Evolving Programs using
Symbolic Expressions

J. R. Koza (1989) "Hierarchical Genetic Algorithms Operating on Populations of Computer Programs," *Proc. of the 11th Intern. Joint Conf. on Artificial Intelligence,* N. S. Sridharan (ed.), Morgan Kaufmann, San Mateo, CA, pp. 768–774.

INVESTIGATIONS in the automatic evolution of computer programs span four decades. But before devoting attention to specific efforts, the question of what constitutes a computer program deserves primary attention. A program may be defined as the expression of a computational method in a computer language, where a computational method is a procedure having the characteristics of an algorithm (except possibly finiteness) (Knuth, 1973, p. 5). Algorithms have five properties: (1) they must terminate after a finite number of steps, (2) they must be unambiguous, (3) they must accept input, (4) they must generate output, and (5) they must be reproducible, in principle, by someone using paper and pencil (Knuth, 1973, pp. 5–6). Thus computer programs may take on many alternative forms. The common structures and syntax associated with C, Pascal, FORTRAN, Prolog, LISP, etc., are obvious candidates, but so are less obvious input-output devices such as neural networks, finite state automata, production systems (fuzzy or crisp logic), and even the genetic milieu of living organisms.

Efforts to evolve programs date back at least to Friedberg (1958; Friedberg et al., 1959) (see Chapter 5), but these attempts did not achieve the desired level of success. Machine language data structures (i.e., binary strings) were subjected to random mutation and deterministic alterations. The sequence of program instructions was executed by an interpreter so as to perform mathematical and/or logical operations on data stored in specific memory locations. In order for the evolution to succeed on a problem requiring a nontrivial complex of operations (i.e., to perform better than a random search of all possible programs), it was necessary to divide the problem into subproblems that were then posed in graded levels of difficulty. This was a significant constraint: It is not possible, in general, to decompose problems in this manner.

A later effort in Fogel (1964; Fogel et al., 1966; and others) evolved populations of finite state machines (FSMs) to produce programs that would predict a sequential series of symbols in light of an arbitrary payoff function (see Chapter 7). Finite state machines are general input-output devices consisting of a finite number of internal states. In each state, each possible input symbol has an associated output symbol and a next-state transition. Thus FSMs satisfy the definition of a computer program (indeed, digital computers can be described by FSMs). Although these experiments achieved a significant degree of success in predicting diverse logical sequences (e.g., primeness of increasing integers) and could in principle be used to describe any executable program, FSMs and other similar automata (e.g., push-down) are not particularly convenient for writing general programs to be executed on common computers.

In contrast, programming languages such as C or Pascal are convenient because they allow for the easy expression of structured sequences of instructions with calls to reusable subroutines. Such structure is important for human programming efforts: The machine language equivalents of common computer applications are too tedious to develop directly. Unfortunately, the syntactical requirements of these and similar higher-level programming languages render them mostly unsuitable for evolution (Conrad, 1988; and others). Small random changes often lead to complete catastrophe: The program simply fails to run when compiled. Even when these mutated programs can be executed, the change in behavior is often large and unpredictable. Evolving such programs requires hand crafting specific variation operators, and by consequence the procedure becomes infeasible to automate.

Two reasonable alternatives have been proposed.[1] The first is to operate on production systems in the form of if-then rules based on relationships between conditions and actions. Such efforts are described in Chapter 16 (classifier systems). Alternatively, programs can be represented as trees and manipulated by modifying or exchanging branches or subtrees. In particular,

---

[1] Perhaps a third avenue was suggested in Jonckers (1987) where mutations could be made to programs in a concept-language where each instruction was abstracted as a "cliché," with different variations defined for alternative clichés. Jonckers (1987) reported manual testing of some varieties of programs for routing (VLSI) that were mutated from an original version.

Hicklin (1986) and Fujiki (1986; Fujiki and Dickinson, 1987) used LISP S-expressions to represent programs that were evolved to solve problems in gaming (e.g., tic-tac-toe and prisoner's dilemma). Hicklin (1986) discussed combining two programs by copying the production trees of both parents until the trees differ and then selecting one of the parents at random to provide the remainder of the tree. In addition, randomly chosen nodes in the trees could be mutated to introduce information not found in the current population. Fujiki (1986) also used a crossover operator to mix two parents, but the trees always represented conditional statements so the crossover naturally took the form of exchanging instructions that occurred after condition-action pairs (somewhat similar to Smith, 1980, 1983). An analogous procedure was also offered in Bickel and Bickel (1987), while Dickmanns et al. (1986) sketched a procedure for swapping parts of Prolog programs.

Independently, Cramer (1985) and later Koza (1989) (reprinted here) offered that crossover could be used to exchange functional subtrees between candidate programs using a tree-based representation. Cramer (1985) offered cursory descriptions of experiments evolving programs in the framework of TB, which essentially was a system for manipulating tree-like structures in a simple form of assembly language. The task was to evolve two-input, single-output multiplication functions. A hand crafted evaluation function was determined based on the observed results of executing a candidate program. Each of the following criteria was given successive credit: (1) an output variable had changed from its initial value (3 points), (2) there was a simple functional dependence of the output variable on an input variable (20 points), (3) the value of an input variable was a factor of the value of the output variable (25 points; 75 points if two input variables were a factor), and (4) successful multiplication (1,000 points). In addition, a penalty was imposed to force programs to remain concise. Cramer (1985) asserted that there was some success using this scoring method as opposed to not using the method.[2]

Koza (1989) offered perhaps the first demonstration that subtree crossover could be usefully applied to the evolution of program trees (in this case, LISP S-expressions) without hand crafting a scoring function.[3] The paper described several applications in sequence induction, pattern recognition, and planning. Essentially, the same procedure (termed *genetic programming*) was used to address each of these areas, demonstrating a degree of robustness. The process was a nearly domain-independent approach to addressing the problem of automatic program induction. The variety of examples was extended in several other publications, summarized in Koza (1992a).

Following Koza (1990), Koza and Rice (1991), and Koza (1992b) (also see Chapters 20 and 21 of Koza, 1992a), Koza (1994) extended the approach by incorporating the evolution of *automatically defined functions* (i.e., a function with a specified number and type of arguments is prescribed and then evolved on-the-fly) that could be reused. Such subroutine generation allows for exploiting possible symmetries in a problem where repeated calls to the same procedures can be used to advantage. Independently, Angeline and Pollack (1992; and others) proposed a different approach to reusing functions (termed *module acquisition*) where partitions of subtrees in programs are collapsed into a single module, with the obverse operator of expanding such modules back into their original subtrees. A variety of applications with these and other extensions to the original framework of genetic programming are offered in Kinnear (1994), Angeline and Kinnear (1996), Koza et al. (1996, 1997) and others.

References

[1] P. J. Angeline and J. B. Pollack (1992) "Evolutionary induction of subroutines," *Proc. of the 14th Ann. Conf. of the Cognitive Science Society,* Lawrence Erlbaum, Hillsdale, NJ, pp. 236–241.

[2] P. J. Angeline and K. Kinnear (eds.) (1996) *Advances in Genetic Programming II,* MIT Press, Cambridge, MA.

[3] A. S. Bickel and R. W. Bickel (1987) "Tree structured rules in genetic algorithms," *Genetic Algorithms and Their Applications. Proc. of the 2nd Intern. Conf. on Genetic Agorithms,* J. J. Grefenstette (ed.), Lawrence Erlbaum, Cambridge, MA, pp. 77–81.

[4] K. Chellapilla (1997) "Evolutionary programming with tree mutations: Evolving computer programs without crossover," *Genetic Programming 1997: Proc. of the 2nd Ann. Conf. on Genetic Programming,* J. R. Koza, K. Deb, M. Dorigo, D. B. Fogel, M. Garzon, H. Iba, and R. L. Riolo (eds.), Morgan Kaufmann, San Mateo, CA, pp. 431–438.

[5] M. Conrad (1988) "Prolegomena to evolutionary programming," *Advances in Cognitive Science: Steps Toward Convergence,* M. Kochen and H. M. Hastings (eds.), AAAS Selected Symposium 104, American Association for the Advancement of Science, NY, pp. 150–168.

[6] N. L. Cramer (1985) "Representation for the adaptive generation of simple sequential programs," *Proc. of an Intern. Conf. on Genetic Algorithms and Their Applications,* J. J. Grefenstette (ed.), Lawrence Erlbaum, Hillsdale, NJ, pp. 183–187.

[7] N. L. Cramer (1997) personal communication, BBN, Cambridge, MA.

[8] D. Dickmanns, J. Schmidhuber, and A. Winklhofer (1986) "Der genetische algorithmus: Eine implementierung in Prolog," Fortgeschrittenenpraktikum, Institut für Informatik, Lehrstuhl Prof. Radig, Technische Universität München, Germany.

[9] L. J. Fogel (1964) "On the organization of intellect," Ph.D. diss., UCLA.

[10] L. J. Fogel, A. J. Owens, and M. J. Walsh (1966) *Artificial Intelligence through Simulated Evolution,* John Wiley, NY.

[11] R. M. Friedberg (1958) "A learning machine: Part I," *IBM J. Research and Development,* Vol. 2:1, pp. 2–13.

---

[2] Cramer (1985) omitted many important experimental details; however, Cramer (1997) offered these details after referring to notes from the presentation of the material at the 1985 conference. Specifically, 20 trials were conducted both with and without the handcrafted scoring method. Nineteen of these were successful with the scoring method, while 11 were successful without it. Thus the scoring method was about 72% more effective (as reported in Cramer, 1985). All of the point values for the scoring function were also provided by Cramer (1997).

[3] It has been shown recently that such crossover is not required in order to solve similar problems (e.g., O'Reilly, 1996; Chellapilla, 1997).

[12] R. M. Friedberg, B. Dunham, and J. H. North (1959) "A learning machine: Part II," *IBM J. Research and Development,* Vol. 3, pp. 282–287.

[13] C. Fujiki (1986) "An evaluation of Holland's genetic operators applied to a program generator," master's thesis, University of Idaho, Moscow, ID.

[14] C. Fujiki and J. Dickinson (1987) "Using the genetic algorithm to generate LISP source code to solve the prisoner's dilemma," *Genetic Algorithms and Their Applications: Proc. of the 2nd Intern. Conf. on Genetic Algorithms,* J. J. Grefenstette (ed.), Lawrence Erlbaum, Hillsdale, NJ, pp. 236–240.

[15] J. F. Hicklin (1986) "Application of the genetic algorithm to automatic program generation," master's thesis, University of Idaho, Moscow, ID.

[16] V. Jonckers (1987) "Exploring algorithms through mutations," *Advances in Artificial Intelligence—II,* B. Du Boulay, D. Hogg, and L. Steels (eds.), Elsevier Science Publishers, Amsterdam, The Netherlands, pp. 131–143.

[17] K. Kinnear (ed.) (1994) *Advances in Genetic Programming,* MIT Press, Cambridge, MA.

[18] D. E. Knuth (1973) *The Art of Computer Programming: Fundamental Algorithms,* Vol. 1, 2nd ed., Addison-Wesley, Reading, MA.

[19] J. R. Koza (1989) "Hierarchical genetic algorithms operating on populations of computer programs," *Proc. of the 11th Intern. Joint Conf. on Artificial Intelligence,* N. S. Sridharan (ed.), Morgan Kaufmann, San Mateo, CA, pp. 768–774.

[20] J. R. Koza (1990) "Genetic programming: A paradigm for genetically breeding populations of computer programs to solve problems," Stanford University Computer Science Dept. Tech. Report STAN-CS-90–1314, June.

[21] J. R. Koza (1992a) *Genetic Programming: On the Programming of Computers by Means of Natural Selection,* MIT Press, Cambridge, MA.

[22] J. R. Koza (1992b) "A hierarchical approach to learning the Boolean multiplexer function," *Foundations of Genetic Algorithms,* G. J. E. Rawlins (ed.), Morgan Kaufmann, San Mateo, CA, pp. 171–192.

[23] J. R. Koza (1994) *Genetic Programming II: Automatic Discovery of Reusable Programs,* MIT Press, Cambridge, MA.

[24] J. R. Koza, D. E. Goldberg, D. B. Fogel, and R. L. Riolo (eds.) (1996) *Genetic Programming 1996: Proc. of the 1st Ann. Conf. on Genetic Programming,* MIT Press, Cambridge, MA.

[25] J. R. Koza, K. Deb, M. Dorigo, D. B. Fogel, M. Garzon, H. Iba, and R. L. Riolo (eds.) (1997) *Genetic Programming 1997: Proc. of the 2nd Ann. Conf. on Genetic Programming,* Morgan Kaufmann, San Mateo, CA.

[26] J. R. Koza and J. P. Rice (1991) "Genetic generation of both the weights and architecture of a neural network," *Proc. of the 1991 Intern. Joint Conf. on Neural Networks,* IEEE Press, Vol. II, pp. 397–404.

[27] U.-M. O'Reilly (1996) "Investigating the generality of automatically defined functions," *Genetic Programming 1996: Proc. of the 1st Ann. Conf. on Genetic Programming,* J. R. Koza, D. E. Goldberg, D. B. Fogel, and R. L. Riolo (eds.), MIT Press, Cambridge, MA, pp. 351–356.

[28] S. F. Smith (1980) "A learning system based on genetic adaptive algorithms," Ph.D. diss., Univ. of Pittsburgh, Pittsburgh, PA.

[29] S. F. Smith (1983) "Flexible learning of problem solving heuristics through adaptive search," *Proc. of the 8th Intern. Joint Conf. on Artificial Intelligence,* A. Bundy (ed.), William Kaufman, Inc., Los Altos, CA, pp. 422–425.

Hierarchical Genetic Algorithms Operating on Populations of Computer Programs

John R. Koza
Computer Science Department
Stanford University
Stanford, California 94305

Abstract

Existing approaches to artificial intelligence problems such as sequence induction, automatic programming, machine learning, planning, and pattern recognition typically require specification in advance of the size and shape of the solution to the problem (often in a unnatural and difficult way). This paper reports on a new approach in which the size and shape of the solution to such problems is dynamically created using Darwinian principles of reproduction and survival of the fittest. Moreover, the resulting solution is inherently hierarchical. The paper describes computer experiments, using the author's 4341 line LISP program, in five areas of artifical intelligence, namely (1) sequence induction, (2) automatic programming, (3) machine learning of functions, (4) planning, and (5) pattern recognition.

1 Introduction

Sequence induction requires developing a computational procedure that can generate any arbitrary element in a sequence $S = S_0, S_1, ..., S_j, ...$ given a finite number of specific examples of the values of the sequence. Examples are finding a correct recursive computational procedure for the Fibonacci sequence or finding a polynomial sequence expression of the appropriate order given a finite sampling of the initial values of the sequence. Although induction problems admittedly do not have closed mathematical solutions, the ability to correctly perform induction is widely accepted as a component of human intelligence.

Automatic programming requires developing a computer program that can produce a desired output for a given set of inputs. Examples include finding a computational procedure for solving a given pair of linear equations $a_{11}x_1 + a_{12}x_2 = b_1$ and $a_{21}x_1 + a_{22}x_2 = b_2$ for the real numbers x_1 and x_2, finding a computational procedure for solving a given quadratic equation $ax^2 + bx + c = 0$ for complex-valued roots x_1 and x_2, and solving trigonometric identities.

Machine learning of a function requires developing a computational procedure that can return the correct functional value for any combination of arguments given a finite number of specific examples of particular combinations of arguments and the associated functional value. An example is the problem of learning the Boolean multiplexer function.

Planning in artificial intelligence and robotics requires finding a plan that receives information from sensors about the state of various objects in the robotic environment and uses that information to select a sequence of functions to execute in order to change the state of the objects in the robotic environment. An example of a planning problem involves generating a general plan for stacking labeled blocks onto a target tower in a specified desired order.

Pattern recognition requires finding a computational procedure that processes a digitized input image to determine whether a particular pattern is present in the input image.

All of these problems, and many similar problems in artifical intelligence and symbolic processing, can be viewed as requiring the creation of a LISP S-expression (i.e. a computer program, a computational procedure, a robotic plan) comprised of various functions and various atoms appropriate to the given problem domain that returns the desired values (and performs the desired side effects) when presented with a particular combination of input values.

In each case, it would be difficult and unnatural to try to specify the size and shape of the eventual solution in advance. Moreover, attempting such specification in advance narrows the window by which the system views the world and may well preclude finding the solution.

The fitness of any LISP S-expression in a problem environment can be naturally measured by the sum of the distances (taken for all the cases in the test suite) between the point in the solution space (whether Boolean-valued, integer-valued, real-valued, vector-valued, or complex-valued) created by the S-expression for a given set of arguments and the correct point in the solution space. The closer this sum is to zero, the better the S-expression.

As will be seen, the LISP S-expression required to solve the problem will, in each case, emerge from a simulated evolutionary process which starts with an initial population of randomly generated LISP S-expressions containing functions and atoms appropriate to the problem domain. Predictably, these initial random individual S-expressions will have exceedingly low fitness (when measured by the previously mentioned objective function). Nonetheless, some individuals in the population will be somewhat more fit in the environment than others. Then, a process of sexual reproduction among two parental S-expression selected in proportion to fitness creates offspring S-expressions comprised of sub-expressions ("building blocks") from their parents. The offspring then replace their parents. At each stage, the only input is the fitness of the individuals in the

Reprinted with permission from *Proceedings of the 11th International Joint Conference on Genetic Algorithms*, J. R. Koza, "Hierarchical genetic algorithms operating on populations of computer programs," N. S. Sridharan (ed.), pp. 768-774. © 1989 by Morgan Kaufmann Publishers.

current population. This process tends to produce populations which, over a period of generations, exhibit increasing average fitness in dealing with their environment and which also can robustly adapt to changes in their environment.

2 Background

Observing that sexual reproduction in conjunction with Darwinian natural selection based on reproduction and survival of the fittest enables biological species to robustly adapt to their environment, Professor John Holland of the University of Michigan presented the pioneering mathematical formulation of simulated evolution (genetic algorithms) for fixed-length character strings in *Adaptation in Natural and Artificial Systems* (Holland 1975).

Although genetic algorithms superficially seem to only process the particular individual binary strings present in the current population, Holland's 1975 work focused attention on the fact that they actually also automatically process large amounts of useful information in parallel concerning unseen Boolean hyperplanes (called similarity templates or schemata) representing numerous similar individuals not actually present in the current population. Genetic algorithms have a property of "intrinsic parallelism" which enable them to create individual strings for the new population in such a way that all the hyperplanes representing similar other individuals are all automatically expected to be represented (without any explicit computation or memory beyond the population itself) in proportion to the fitness of the hyperplane relative to the average population fitness. As Schaffer (1987) points out, "Since there are very many more than N hyperplanes represented in a population of N strings, this constitutes the only known example of the combinatorial explosion working to advantage instead of disadvantage."

In addition, Holland established that the seemingly unprepossessing genetic operation of crossover in conjunction with the straight-forward operation of fitness proportionate reproduction causes the unseen hyperplanes (schemata) to grow (and decay) from generation to generation at rates that are mathematically near optimal when the process is viewed as a set of multi-armed slot machine problems requiring an optimal allocation of trials so as to minimize losses.

Holland's 1975 work also highlighted the relative unimportance of mutation in the evolutionary process and contrasts sharply in this regard with numerous other efforts based on the approach of merely saving the best from among asexual random mutants, such as the 1966 *Artificial Intelligence through Simulated Evolution* (Fogel *et. al.* 1966) and other work (Lenat 1983, Hicklin 1986).

Representation is a key issue in genetic algorithm work because the representation scheme can severely limit the window by which the system observes its world. However, as Davis and Steenstrup (1987) point out, "In all of Holland's work, and in the work of many of his students, chromosomes are bit strings." String-based representation schemes are difficult and unnatural for many problems (De Jong 1987, Smith 1980, Fujiki 1986, Hicklin 1986, Cramer 1985). String-based representation schemes do not provide the hierarchical structure central to the organization of computer programs (into programs and subroutines) and the organization of behavior (into tasks and subtasks). String-based representation schemes

do not provide any convenient way of representing arbitrary computational procedures or incorporating iteration or recursion when these capabilities are inherently necessary to solve the problem (e.g. the Fibonacci sequence). Moreover, string-based representation schemes do not facilitate computer programs modifying themselves and then executing themselves. Moreover, without dynamic variability, the initial selection of string length limits in advance the number of internal states of the system and the computational complexity of what the system can learn.

3 Hierarchical Genetic Algorithms

The LISP programming language is especially well-suited for handling hierarchies, recursions, logical functions, compositions of functions, self-modifying computer programs, self-executing computer programs, iterations, late typing of variables and expressions, and complex structures whose size and shape is dynamically determined (rather than predetermined in advance). Because of these features, the LISP programming language allows the creation of "hierarchical" genetic algorithms for simulated evolution in which the population consists of individual hierarchical LISP S-expressions, rather than strings of characters or other objects (whether of fixed or variable length).

In hierarchical genetic algorithms, the set of possible S-expressions for a particular domain of interest depends on the functions and atoms that are available in the domain. The possible S-expressions are those that can be composed recursively from a set of n functions $F = \{f_1, f_2, \ldots, f_n\}$ and a set of m atoms $A = \{a_1, a_2, \ldots, a_m\}$. Each particular function f in F takes a specified number $z(f)$ of arguments $b_1, b_2, \ldots, b_{z(f)}$. For example, the LISP S-expression $(+ (\sigma (- J 1) 1) (\sigma (- J (+ 1 1) 0)))$ is an S-expression for the Fibonacci sequence. In this representation, J is the index for the current sequence element and $(\sigma x y)$ is the sequence referencing function returning the value of the sequence at position x (provided x is between 0 and J-1) or the default value y (if σ is being asked to provide a position of the sequence that is not yet defined).

The operation of fitness proportionate reproduction for hierarchical genetic algorithms is the basic engine of Darwinian reproduction and survival of the fittest. It is an asexual operation in that it operates on only one parental S-expression. The result of this operation is one offspring S-expression. In this operation, if $s_i(t)$ is an individual in the population at generation t with fitness value $f(s_i(t))$, it will be copied into the mating pool for the next generation with probability $f(s_i(t))/\Sigma f(s_i(t))$.

The crossover operation is a sexual operation that starts with two parental S-expressions. Its result is, for convenience, two offspring S-expressions. Every LISP S-expression can be depicted graphically as a rooted point-labeled tree in a plane whose internal points are labeled with functions, whose external points (leaves) are labeled with atoms, and whose root is labeled with the function (or atom) appearing just inside the outermost left parenthesis. The crossover operation begins by randomly and independently selecting one point in each parent using a uniform distribution. This crossover operation is well-defined for any two S-expressions and any two crossover points and the resulting offspring are always valid LISP S-expressions. Offspring contain some traits from each parent.

The "crossover fragment" for a particular parent is the

rooted sub-tree whose root is the crossover point for that parent and where the sub-tree consists of the entire sub-tree lying below the crossover point (i.e. more distant from the root of this parent). Viewed in terms of lists in LISP, the crossover fragment is the sub-list starting at the crossover point.

The first offspring is produced by deleting the crossover fragment of the first parent from the first parent and then impregnating the crossover fragment of the second parent at the crossover point of the first parent. In producing this first offspring the first parent acts as the base parent (the female parent) and the second parent acts as the impregnating parent (the male parent). The second offspring is produced in a symmetric manner.

For example, consider the two parental LISP S-expressions below.

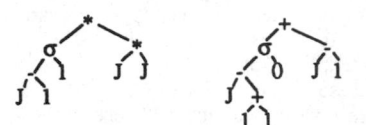

Assume that the points of trees are numbered in a depth-first way starting at the left. Suppose that point 2 (out of the 9 points of the first parent) was selected as the crossover point for the first parent (i.e. the σ) and that point 9 (out of the 11 points of the second parent) was selected as the crossover point of the second parent (i.e. the subtraction function - at the right). The two crossover fragments are below.

In terms of LISP S-expressions, the two parents are (* _(σ (- J J J)_ (* J J))) and (+ (σ (- J (+ 1 1)) 0) _(- J J)_) and the two crossover fragments are the underlined sublists.

The two offspring resulting from crossover are shown below.

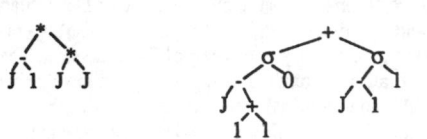

Note that the second offspring above is a perfect solution for the Fibonacci sequence, namely (+ (σ (- J (+ 1 1) 0)) (σ (- J 1) 1)).

Crossover can be efficiently implemented in LISP using the RPLACA function in LISP (in conjunction with the COPY-TREE function) so as to destructively change the pointer of the CONS cell at the crossover point of one parent so that it points to the crossover fragment (sublist) of the other parent.

In each of the runs reported herein, between 75% and 80% of the crossover points are restricted to function (internal) points of the tree in order to promote the recombining of larger structures than is the case with an unrestricted selection (which may do an inordinate amount of mere swapping of atoms in a manner more akin to point mutation rather than true crossover).

4 Experimental Results

This section describes some experiments using hierarchical genetic algorithms. The author's computer program, consisting of 4341 lines of Common Lisp code, was run on a Texas Instruments Explorer II computer with a 25 megaHertz LISP microprocessor chip with 32 megabytes of internal memory. For each experiment reported below, the author believes that sufficient information is provided to allow the experiment to be independently replicated to produce substantially similar results (within the limits inherent in any process involving randomized selections). Substantially similar results were obtained on several occasions for each experiment.

4.1.1 Sequence Induction - Fibonacci Sequence

For this experiment, the problem is to induce the computational procedure (i.e. LISP S-expression) for the Fibonacci sequence. The environment in which adaptation is to take place consists of the first 20 elements of the actual Fibonacci sequence S = 1, 1, 2, 3, 5, 8, 13, 21, 34, 55,..., 4181, 6765.

The set of functions available for this problem is F = {+, -, σ, *} and the set of atoms available is A = {0, 1, J}. For our purposes here we can view each atom as a function that requires no arguments in order to be evaluated. Thus, the combined set of functions and atoms is C = {+, -, σ, *, 0, 1, J} having 2, 2, 2, 2, 0, 0, and 0 arguments, respectively. In order to make the experiment more realistic, extraneous functions or atoms are included in all the experiments reported herein. The multiplication function here is extraneous to a parsimonious solution of this problem. A population of 300 individuals is used in all examples herein unless otherwise stated. The algorithm begins by randomly generating 300 LISP S-expressions recursively using the items from set C. Examples of such random S-expressions included (+ J J), (* 0 (- J 1)), and (* (- (+ J 1) 0) (σ J J)).

The raw fitness of an individual LISP S-expression in the population at any generational time step t is $\Sigma |P_{hj}(t) - S_j|$ where S_j is the actual Fibonacci sequence element and $P_{hj}(t)$ is the value returned by S-expression h for sequence position j. In this case, the smaller the raw fitness, the closer the match between the performance of the LISP S-expression involved and the actual Fibonacci sequence.

The best S-expression for generation 0 (the initial random population) was (σ (- J (σ (- J J) 0)) 0) with a raw fitness of 6765. The worst individual had a raw fitness of 28979. The average value of raw fitness was 17621. An adjusted fitness value a_h = 1/(1+r) is then computed from the raw fitness r for each individual h. A normalized fitness value $u_h = a_h/\Sigma a_h$ (ranging between 0 and 1 for each individual) is then computed for each individual. The average value of adjusted fitness for generation 0 was .0001 and the average normalized fitness was .0086. The number of exact matches for the best individual was 1 (out of 20). These predictably poor values for generation 0 serve as a useful baseline for the entire process.

A new population is then created from the current population. This process begins with the selection of a mating pool equal in size to the entire population using fitness proportionate reproduction (with replacement). In this run and each of the runs reported herein, the number of individuals involved with crossover equals 100% of the population for each generation. When these operations are completed, the new population replaces the old population.

The value of average fitness improved (i.e. dropped) from 17621 for generation 0 to 16969 and 15515 for generations 1 and 2, respectively. It then continued to improve monotonically to 5928 for generation 10. Between generations 11 and 24, the average fitness oscillated in the general neighborhood of 6000. Then, for generation 25, the value of average fitness improved to 5390. In addition, there was a monotonically improving trend for the fitness of the best individual in the population from generation to generation. The worst individual in the population exhibited considerable variability (as is typical) but did improve overall. The average normalized fitness for each generation was very small until generation 16 (when an a very good individual appeared) and thereafter showed a substantial upwards movement.

The number of exact matches for the best individual of each generation started at 1 for generation 0, remained at 1 between generations 1 and 6, dropped to 0 at generation 7, rose to 2 between generations 8 and 13, rose to 18 for generations 14 and 15, rose to 19 for generations 16 through 21. Starting at generation 22, the best individual had a perfect score of 20 matches, namely

(- (+ (σ (+ (- 0 1) J) 1) (σ (+ (- (- 0 1) 1) J) 0)) 0). This S-expression equals (+ (σ (- J 1) 1) (σ (- J 2) 0))).

The computer program takes approximately 150 seconds for 300 individuals for 26 generations. The process includes extensive interactive output consisting of two full-color graphs (with mouse-sensitive graph points) and five other windows for monitoring the process.

The number of possible trees increases rapidly with the number of points. This number is dramatically increased by labeling the points of the tree with functions or atoms and is further increased by designating the root. The best S-expression in 3,000,000 random initial S-expressions had only 3 matches out of 20. Even if only trees with 13 points are considered, the search space is very large in relation to the 6600 individual LISP S-expressions processed in the 22 generations involved above.

An asexual point mutation operator and an asexual mutation operator which inserts a randomly generated sub-tree at a randomly selected point were also tested in several runs. No run using either mutation operator and only fitness proportionate reproduction produced a solution. Moreover, an examination of the hereditary history (i.e. LISP audit trail indicating parents, crossover points, mutation points, etc.) of solutions achieved in various runs using crossover revealed that the solution never came about as a result of mutation operation.

The algorithm seems relatively insensitive to the inclusion of extraneous functions and atoms in the repertoire.

4.1.2 Sequence Induction - Cubic Polynomial Sequence

For this experiment, the problem was to induce the computational procedure for cubic polynomials such as $1+2J+J^2+J^3$. Note that neither the order of the polynomial required nor the size and shape of the computational procedures needed to solve this (and other problems herein) is provided to the problem solver in advance. The same functions and atoms as the Fibonacci sequence were used. Population size was 500. Starting with generation 5, a computational procedure emerged that returned values that exactly matched the actual cubic polynomial for all sequence positions. Similar results were obtained for a variety of different polynomials. Interestingly, in one run, the program unexpectedly factored the polynomial into a product of factors $(J - r_k)$, where the r_k were the roots of the polynomial.

4.2.1 Automatic Programming - Pairs of Linear Equations

The problem of automatic programming requires developing a computer program that can produce a desired output for a given set of inputs. For this experiment, the problem is to find the computational procedure for solving a pair of consistent nonindeterminate linear equations, namely $a_{11}x_1 + a_{12}x_2 = b_1$ and $a_{21}x_1 + a_{22}x_2 = b_2$ for two real-valued variables. The environment consisted of a suite of 10 pairs of equations (to avoid being misled). Without loss of generality, the coefficients of the equations were prenormalized so the determinant is 1. The set of available functions is $\Gamma = \{+,-,*\}$ and the set of available atoms is A = {A11, A12, A21, A22, B1, B2}. The raw fitness of a particular S-expression is the sum of the Euclidian distances between the known solution point in the plane and the point produced by the S-expression for all 10 pairs of equations in the test suite.

The average raw fitness of the population immediately begins improving from the baseline value for generation 0 of 2622 to 632, 341, 342, 309, etc. In addition, the worst individual in the population also begins improving from 119051 for generation 0 to 68129, 2094, etc. The best individual from generation 0 is (+ (- A12 (* A12 B2)) (+ (* A12 B1) B2)) and has a raw fitness value of 125.8. The best individual begins improving and has a value of 106 for generations 1 and 2, 103 for generation 3 through 5, 102 for generations 6 through 16, and 102 for generations 17-20. The computational procedure (+ (- A12 (* A12 B2)) (* A22 B1)) appearing in generations 21 and 22 had a fitness value of 62 and differed from the known correct solution only by one additive term -A12. The best individual for generations 23 through 26 is a similarly close S-expression (+ (- A22 (* A12 B2)) (* A22 B1)) with a raw fitness value of 58. Starting with generation 27, a perfect solution for x_1 emerges, namely (- (* A22 B1) (* A12 B2)). Between generations 27 and 30, the average normalized fitness rises to .39 (as the perfect solution starts dominating).

4.2.2 Automatic Programming - Quadratic Equations

For this experiment, the problem is to solve the quadratic equation $x^2 + bx + c = 0$ for a complex-valued root. The available functions were multiplication, subtraction, a square root function S [which returns a LISP complex number, e.g. (S -4) is #C(0, 2)], and a modified division operation % (which returns a value of zero for division by zero). The environment consisted of a suite of 10 quadratic equations (with some purely real roots, some purely imaginary roots, and some complex-valued roots). A correct solution to the problem emerged at generation 22, namely, the S-expression (- (S (- (* (% B 2) (% B 2)) C)) (% B 2)).

4.2.3 Automatic Programming - Trigonometric Identities

For this group of experiments, the problem was to derive various trigonometric identities. This particular group of experiments yielded a number of unexpected results. The environment consisted of a Monte Carlo suite of 20 pairs of randomly generated X values between 0 and 2Π radians and the value of cos 2X (which is equivalent to $1 - 2 \sin^2 X$). The

Koza

available functions were SIN, multiplication, and subtraction (with the addition and cosine function were intentionally deleted from the repertoire of available functions). The correct S-expression (- (- 1 (* SIN X) (SIN X))) (* (SIN X) (SIN X))) was obtained after 13 generations in one run and the somewhat more parsimonious correct S-expression (- 1 (* (* (SIN X) (SIN X)) 2)) was obtained after 16 generations. In one run with cos 2X, the S-expression (SIN (- (- 2 (* X 2)) (SIN (SIN (SIN (SIN (SIN (SIN (* (SIN (SIN 1)) (SIN (SIN 1))))))))))), where 1 is in radians, was obtained as the best individual. This expression approximately equals sin (Π/2 - 2X).

4.3.1 Machine Learning - Boolean Multiplexer Function

For this experiment, the problem is to find the Boolean expression which gives the correct Boolean output value for a given Boolean multiplexer function. The input to the Boolean multiplexer function consists of k "address" bits a_i and 2^k "data" bits d_i and is a string of length $k+2^k$. The value of the multiplexer function is the value (0 or 1) of the particular data bit that is singled out by the k address bits of the multiplexer. For example, for the 6-multiplexer (where k = 2), if the two address bits $a_1 a_0$ are 11, then the output is the third data bit d_3. This function has been studied in connection with neural nets (Barto et. al. 1985) and classifier systems (Wilson 1987).

The combined set of functions and atoms for this problem is C = {NOT, OR, OR, OR, AND, AND, IF, IF, A0, A1, D0, D1, D2, D3 } with 1,2,3,4,2,3,2,3,0,0,0,0,0, and 0 arguments, respectively. Note that the OR, AND, and IF functions appear with varying number of arguments (e.g. 2, 3, or 4). For example, the IF function with 3 arguments is an if-then-else function. The environment consisted of the 2^w (where w = $k+2^k$) possible inputs.

Initial random individuals include contradictions such as (AND A0 (NOT A0)), inefficiencies such as (OR D3 D3), irrelevancies such as (IF A0 A0 (NOT A1)), and nonsense such as (IF (IF (IF D2 D2) D2) D2). The best individual from generation 0 was (IF A0 D1 D2) with a raw fitness value of 16 (i.e. 16 mismatches out of a possible 64). This individual uses just one of the address bits (A0) to decide whether the output is data line D1 or D2 and can never give an output of D0 or D3. Nonetheless, in the valley of the blind, the one-eyed man is king.

The average raw fitness of the population immediately begins improving from the baseline value for generation 0 of 29.05 to 26.89, 25.74, 23.78, 22.09, 21.38, 20.13, 19.91, etc. In generation 9 a best individual arises that has only 12 mismatches, namely (IF (IF A0 (OR A1 D0)) D3 (IF A0 D1 D2)). Note that (IF A0 D1 D2) from generation 0 is now embedded as a sub-expression within this new individual. In generation 11, a new best individual arises that has only 8 mismatches, namely (IF A0 D1 (IF A1 D2 D0)). The sub-expression (IF A1 D2 D0) contributes substantially to this improved performance because it perfectly deals with the case when A0 is NIL (False) by taking either data line D2 or D0 as its output (depending on A1). Note also that (IF A0 D1 ...) is partially correct when A0 is T (True). In generations 12, 13, and 14, a new individual arises with only 4 mismatches, namely, (IF (IF (A0 (OR A1 D0) D3 (IF A0 D1 (IF A1 D2 D0)))).

In generation 15, a perfect solution i.e. an individual with 0 mismatches) emerges, namely, (IF (IF A0 A1) D3 (IF A0 D1

(IF A1 D2 D0)) as a result of a crossover where the unfit sub-expression (IF (A0 (OR A1 D0))) is replaced by the more fit sub-expression (IF A0 A1).

The interpretation of this solution expression is as follows: The output of the multiplexer is D3 if (IF A0 A1) is true (i.e. the two address bits are 11). Note that IF function in LISP (unlike the predicate calculus) is equivalent to the AND function. If that is not true, the output is D1 if A0 is true (because the two address bits are necessarily now 01). Note that setting the output to D1 if merely A0 were true in a vacuum is not a correct solution to the problem. However, after (IF A0 A1) has been considered (and found to be false), then (IF A0 D1 ...) is correct. Finally, (IF A1 D2 D0) now handles the case when address bit A0 must necessarily be NIL. In this context, the partially correct sub-expression that was around since generation 0, namely (IF A1 D2 D0), sets the output of the multiplexer to D2 if A1 is T (because the two address bits are 10) and, otherwise, it sets the output to D0 (because the two address bits are 00).

Note that a default hierarchy emerged here which incorporated partially correct sub-rules into a perfect overall procedure by dealing with ever more specific cases. Although default hierarchies are considered desirable in classifier systems (Holland 1986), none emerged in Wilson's (1987) otherwise noteworthy experiments involving classifier systems and the multiplexer.

The perfect solution above arose after processing 4500 individuals. Note that the hierarchical algorithm does not start with any advance information identifiying inputs versus outputs or any advance information about the size and shape of the ultimate solution.

4.3.2 Machine Learning - The Parity Function

For this experiment, the problem is to find the Boolean expression for the Boolean parity function. The k-parity function takes k Boolean arguments and returns T if an odd number of its arguments are T and returns NIL otherwise. The exclusive-or function and the k-parity function were not realizable by early simple perceptrons (Minsky and Papert 1969) and are, as a result, commonly used as test functions for multi-layered non-linear neural networks. Moreover, these functions yield uninformative schema (similarity templates) with conventional linear genetic algorithms using fixed length binary strings so that these functions are not realizable with such linear genetic algorithms.

The combined set of functions and atoms used for the 3-parity function was C = {AND, OR, NOT, IF, D2, D1, D0} with 2, 2, 1, 3, 0, 0, and 0 arguments, respectively. In generation 5, an individual emerged which was correct in all 8 cases, namely (IF (IF D2 D0 (NOT D0)) D1 (NOT D1)).

4.4 Planning

Nilsson (1988a) has presented a robotic action network that solves a problem described to Nilsson (1988b) by Ginsberg involving rearranging labeled blocks in various towers from an arbitrary initial arrangement into an arbitrary specified new order on a single target tower. In the experiment here, the goal is to automatically generate a general plan that solves this problem.

Three lists are involved in the problem. GOAL-LIST is the list specifying the desired final order in which the blocks are to be stacked in the target tower (i.e. "FRUITCAKE" or "UNIVERSAL"). STACK is the list of blocks that are currently in the target tower (where the order is important).TABLE is the list of blocks that are currently not in the target tower. The initial configuration consists of certain blocks in the STACK and the remaining blocks on the TABLE. The desired final configuration consists of all the blocks being in the STACK in the order specified by GOAL-LIST and no blocks being on the TABLE.

The environment consists of the various different initial configurations of N blocks in the STACK and on the TABLE. The raw fitness of a particular individual plan in the population is the number of initial configurations for which the particular plan produces the desired final configuration after the plan is executed. The computation of fitness can be significantly shortened by consolidating functionally equivalent initial configurations.

In the problem as stated, three sensors dynamically track the environment in the formulation of the problem. TB is a sensor that dynamically specifies the CAR (i.e. first element) of the list which is the longest CDR (i.e. list of remaining elements) of the list STACK that matches a CDR of GOAL-LIST. NN is a sensor that dynamically specifies the next needed block for the STACK (i.e. the immediate predecessor of TB in GOAL-LIST). CS dynamically specifies the CAR of the STACK (i.e. the top block). Thus, the set of atoms available for solving the problem here is A = {TB, NN, CS}. Each of these atoms may assume the value of one of the block labels or the value NIL.

The set of functions available for solving the problem contains 6 functions F = {MS, MT, DU, QUOTE, NOT, EQ}. The function MS has one argument and moves block X to the top of the STACK if X is on the table. The function MT has one argument and moves the top item to the TABLE if the STACK contains X anywhere in the STACK. The iterative function DU ("do until") has two arguments, namely a predicate PRED and some WORK. Both the MS and MT functions have return values, although their true functionality consists of their side effects on STACK and TABLE. The function DU tests the predicate PRED and does the WORK (via the LISP evaluation function EVAL) repeatedly until the predicate PRED becomes T (True). Note that the fact that each function returns some value (in addition to whatever side effects it has on the STACK and TABLE) and the flexibility of the LISP language guarantees that the DU function can be executed and evaluated for any combination of functions and arguments (however unusual, pointless, or counter-productive). Since individuals in the population will often contain complicated nestings of DU functions and unsatisfiable termination predicates, limits are placed on both the number of iterations allowed (without preventing any plan from being executed and evaluated).

After about 5 generations, we typically see the emergence of perhaps one plan in the population that correctly deals with the simplest group of cases in the environment (i.e. the cases in which the blocks, if any, in the initial STACK are already all in the correct order and in which there are no out-of-order blocks on top of those blocks). In several runs, the rather parsimonious (DU (QUOTE (MS NN)) (NOT NN)) emerged as a partially correct plan. This plan works by improving a partially correct initial STACK by moving needed blocks (NN) in the correct sequence from the TABLE onto the STACK until there are no more blocks needed to finish the STACK (i.e. the sensor NN is no longer a block).

After about 10 generations, the best single individual in the population is typically a plan that achieves a perfect score. One such plan is (NOT (EQ (DU (QUOTE (MT CS)) (NOT NN)) (EQ (MS (DU (QUOTE (MS NN)) (NOT NN))) (DU NN (QUOTE TB))))). Note that this plan contains a default hierarchy. In particular, the sub-plan (DU (QUOTE (MS NN)) (NOT NN)) comes from an earlier generation (which performed correctly for a simple set of cases of initial configurations). This sub-plan is now incorporated as a sub-plan (i.e. a small "building block"). Note also that another sub-plan (DU (QUOTE (MT CS)) (NOT NN)) from another individual from an earlier generation correctly deals with the remaining cases by first moving out-of-order blocks from the STACK to the TABLE until the STACK contains no incorrect blocks.

4.5 Pattern Recognition

Hinton (1988) has discussed the problem of translation-invariant recognition of a one-dimensional shape in a linear binary retina (with wrap-around) in connection with the claim that connectionist neural networks cannot possibly solve this type of problem. In the simplified experiment here, the retina has 6 pixels (with wrap-around) and the shape consists of three consecutive binary 1's.

The functions available are a zero-sensing function H0, a one-sensing function H1, ordinary multiplication, and a disjunctive function U. The atoms available are the integers 0, 1, and 2, and a universally quantified atom k.

LISP's comparative tolerance as to typing is well suited to pattern recognition problems where it is desirable to freely combine numerical concepts such as positional location (either absolute or universally quantified) and relative displacement (e.g. the symbol 2 pixels to the right) with various combinations of Boolean tests. The functions U and * so defined resolve potential type problems that would otherwise arise when integers identify positions in the retina.

In one particular run, the number of mismatches for the best individual of generation 0 was 48 and rapidly improved to 40 for generations 1 and 3. It then improved to 0 mismatches in generation 3 for the individual (* 1 (* (H1 K 1) (H1 K 0) (H1 K 2)) 1). Ignoring the extraneous outermost conjunction of two 1's, this individual returns a value of the integer 1 if and only if a binary 1 is found in the retina in positions 0, 1, and 2 (each displaced by the same constant k).

5 Robustness

The existence and nurturing of a population of disjunctive alternative solutions to a problem allows hierarchical genetic algorithms to effectively perform even when the environment changes. To demonstrate this ability, the environment for generations 0 through 9 is the quadratic polynomial $x^2 + x + 2$; however, at generation 10, the environment abruptly changes to the cubic polynomial $x^3 + x^2 + 2x + 1$; and, at generation 20, it changes again to a new quadratic polynomial $x^2 + 2x + 1$. A perfect-scoring quadratic polynomial for the first environment was created by generation 3. Normalized average population fitness stabilized in the neighborhood 0.5 for generations 3 through 9 (with genetic diversity maintained). Predictably, the fitness level abruptly dropped to virtually 0 for generation 10 and 11 when the environment changed. Nonetheless, fitness increased for generation 12 and stabilized in the neighborhood of 0.7 for generations 13 to 19 (after creation of a perfect-scoring cubic polynomial). The fitness level again abruptly dropped to virtually 0 for generation 20 when the environment again changed. However, by generation 22, a fitness level again stabilized in the neighborhood of 0.7 after creation of a new perfect-scoring quadratic polynomial.

6 Theoretical Discussion

Hierarchical genetic algorithms employ the same automatic allocation of credit inherent in the basic string-based genetic algorithm described by Holland (1975) and inherent in Darwinian reproduction and survival of the fittest amongst biological populations in nature. In hierarchical genetic algorithms, the individuals in the population are LISP S-expressions (i.e. rooted point-labeled trees in a plane) instead of linear character strings. The set of similar individuals sharing common features (i.e. the schemata) is the hyperspace of LISP S-expressions sharing common features. This infinite set can be partitioned into finite subsets by using the number of points as the partitioning parameter. If the subset sharing common features with a specified value of this parameter is considered, fitness proportionate reproduction causes growth or decay in the size of that subset in the new population in accordance with the relative fitness of the subset to the average population fitness in the same way as it does for string-based linear genetic algorithms (with the associated approximately near optimal allocation of trials). The deviation from this approximately near optimal rate of growth or decay is relatively small if the number of points defining the common feature is relatively small and to the extent that the points defining the common feature are coextensive with one subtree. Thus, the overall effect of fitness proportionate reproduction and crossover is that subprograms (i.e. sub-trees, sub-lists) from relatively high fitness individuals are used as "building blocks" for constructing new individuals and the search is concentrated for successive populations into sub-hyperspaces of S-expressions of ever decreasing dimensionality and ever increasing fitness.

Acknowledgments

Dr. Thomas Westerdale, Dr. Martin A. Keane, and John Perry made valuable comments on drafts of this paper. Eric Mielke of the Texas Instruments Education Center in Austin significantly improved execution time of the author's crossover operation.

References

[Barto et al.,1985] Barto A. G., Anandan, P., and Anderson, C.W. Cooperativity in networks of patternr recognizing stochastic learning automata. In Narendra, K.S. *Adaptive and Learning Systems*. New York: Plenum 1985.

[Cramer 1985] Cramer, Nichael Lynn. A representation for adaptive generation of simple sequential programs. *Proceedings of an International Conference on Genetic Algorithms and Their Applications*. Hillsdale, NJ: Lawrence Erlbaum Associates 1985.

[Davis and Steenstrup 1987] Davis, Lawrence and Steenstrup, M. Genetic algorithms and simulated annealing: An overview. In Davis, Lawrence (editor) *Genetic Algorithms and Simulated Annealing* London: Pittman 1987.

[De Jong 1987] De Jong, Kenneth A. On using genetic algorithms to search program spaces. *Genetic Algorithms and Their Applications: Proceedings of the Second International Conference on Genetic Algorithms*. Hillsdale, NJ: Lawrence Erlbaum Associates 1987.

[Fogel et. al. 1966] Fogel, L. J., Owens, A. J. and Walsh, M. J. *Artificial Intelligence through Simulated Evolution*. New York: John Wiley 1966.

[Fujuki 1986] Fujuki, Cory. *An Evaluation of Holland's Genetic Algorithm Applied to a Program Generator*. Master of Science Thesis, Department of Computer Science, Moscow, ID: University of Idaho, 1986.

[Hicklin 1986] Hicklin, Joseph F., *Application of the Genetic Algorithm to Automatic Program Generation*. Master of Science Thesis, Department of Computer Science. Moscow, ID: University of Idaho 1986.

[Hinton 1988] Hinton, Geoffrey, *Neural Networks for Artificial Intelligence*. Santa Monica, CA: Technology Transfer Institute. Documentation dated December 12, 1988.

[Holland 1975] Holland, John H. *Adaptation in Natural and Artificial Systems*, Ann Arbor, MI: University of Michigan Press 1975.

[Holland 1986] Holland, John H. Escaping brittleness: The possibilities of general-purpose learning algorithms applied to parallel rule-based systems. In Michalski, Ryszard S., Carbonell, Jaime G. and Mitchell, Tom M. *Machine Learning: An Artificial Intelligence Approach, Volume II*. P. 594-623, Los Altos, CA: Morgan Kaufman 1986.

[Lenat 1983] Lenat, Douglas B. The role of heuristics in learning by discovery: Three case studies. In Michalski, Ryszard S., Carbonell, J. G. and Mitchell, T. M. *Machine Learning: An Artificial Intelligence Approach, Volume I*. P. 243-306. Los Altos, CA: Morgan Kaufman 1983.

[Minsky and Papert 1969] Minsky, Marvin L. and Papert, Seymour A. *Perceptrons*. Cambridge: MIT Press. 1969.

[Nilsson 1988a] Nilsson, Nils J. Action networks. Draft Stanford Computer Science Department Working Paper, October 24, 1988. Stanford, CA: Stanford University. 1988.

[Nilsson 1988b] Nilsson, Nils J. Private Communication.

[Schaffer 1987] Schaffer, J. D. Some effects of selection procedures on hyperplane sampling by genetic algorithms. In Davis, L. (editor) *Genetic Algorithms and Simulated Annealing* London: Pittman 1987.

[Smith 1980] Smith, Steven F. *A Learning System Based on Genetic Adaptive Algorithms*. PhD dissertation. University of Pittsburgh 1980.

[Wilson 1987] Wilson, S. W. Classifier Systems and the animat problem. *Machine Learning*, 3(2), 199-228, 1987.

Chapter 23
Tierra and Emergent Properties

T. S. Ray (1992) "An Approach to the Synthesis of Life," *Artificial Life II*, C. G. Langton, C. Taylor, J. D. Farmer, and S. Rasmussen (eds.), Addison-Wesley, Reading, MA, pp. 371–408.

THE majority of efforts in evolutionary computation have emphasized optimization in the face of specified fixed criteria, with the goal to discover the best solutions in light of those criteria. In contrast, natural evolution is not a process that searches for, or necessarily achieves, perfection. Although it is an optimization process, it is entirely "opportunistic" (Mayr, 1988, p. 105); complete perfection is not to be expected because of developmental constraints, many interdependencies between phenotypic traits, the time-varying quality of the environment, and other factors (Mayr, 1982, p. 589; 1988, pp. 151, 156). Individuals are only in competition with each other, not with any presumed "right answer." The criteria by which selection may be viewed to judge the appropriateness of individuals' and species' behaviors varies in the face of other extant organisms.

The concept of such co-evolution has been incorporated in several evolutionary simulations (Barricelli, 1963; Reed et al., 1967; Fogel and Burgin, 1969; Hillis, 1992; Angeline and Pollack, 1994; Sebald and Schlenzig, 1994; and others). Rarely, however, has this co-evolution been plainly set in the context of simulating an ecology for the purpose of observing the resultant emergent behaviors that would arise from the complex dynamics of local interactions in light of general survival criteria. The early simulations of Barricelli (1954, 1957, 1962; and others) (see Chapter 6) examined the emergent patterns of numbers moving on a grid over time, and the long series of investigations by Conrad and colleagues (Conrad, 1969; Conrad and Pattee, 1970; Rizki and Conrad, 1985; Conrad and Strizich, 1985; and others) (see Chapter 13) examined the simulated evolution of specified behaviors in light of the physics imposed by a conservation of required resources in a closed environment. Ray (1992) (reprinted here) offered an innovative simulation acting directly on computer programs that competed for CPU cycles while reproducing copies of themselves in RAM.

The simulation, called *Tierra,* evolved programs in an assembly language. It was first seeded with a hand-coded program capable of replicating itself into another area of the RAM. Upon starting the simulation this program self-replicated, and its progeny in turn self-replicated, and so forth, ultimately to the point of filling the available environmental arena. Each program competed for available CPU cycles; generally, shorter programs were able to replicate more quickly than longer ones. Mutation was imposed on the programs in two forms: (1) errors could occur to the program instructions during reproduction, and (2) at a background rate, random bits in the RAM were altered. This provided a source of random variation in the behavior of the replicating programs.[1] Upon creation, programs were placed in a queue, and once the capacity of the RAM reached a specified threshold, programs were subject to having their memory deallocated (i.e., being eliminated from competition). The primary mechanism of the queue was simply first-in, first-out, but programs that made errors when executed were accelerated toward the top of the queue.

A repeatable effect of executing the described protocol was the emergence of "parasitic" programs. These sets of instructions could not copy themselves in the absence of a host program; they relied in part on the host's instructions. But they were shorter than their hosts and therefore capable of faster replication. Subsequent evolution led to "hyper-parasites" that were immune to the effects of the first parasites and tended to drive these parasites to extinction. Other emergent behaviors were also identified (Ray, 1992).

Extensions of Tierra have been made to emphasize spatial (i.e., topological) interactions between programs. For example, the "Avida" system of Adami and Brown (1994) and Adami et al. (1995) used update rules that were dependent on the programs in neighboring cells. An expanded version of Tierra to be released over the Internet is also under development. The project has been tested in limited conditions, but by the time of publication it may be in normal operation. The heterogeneous environment posed by using different computer architectures connected to the network will, it is hoped, serve to promote the evolution of greater complexity in programmed behaviors than would be achieved on a single homogeneous supercomputer. The programs will be executed using spare CPU cycles from donated machines and may therefore exhibit a nocturnal behavior of migrating to machines that are in time zones of early morning hours, as these will likely

---

[1]Another source of variation was introduced by making each instruction probabilistic. The result of each instruction being executed was altered with a low probability. Thus the Tierran programs were not deterministic.

have the greatest spare CPU cycles. Another hope is that the experiment will provide insight into the more fundamental question of what conditions are required to promote an open-ended series of innovations in evolution (i.e., evolvability), including the invention of multicellularity (Thearling and Ray, 1994).

References

[1] C. Adami and C. T. Brown (1994) "Evolutionary learning in the 2D artificial life system 'Avida,'" *Artificial Life IV,* R. Brooks and P. Maes (eds.), MIT Press, Cambridge, MA, pp. 377–381.

[2] C. Adami, C. T. Brown, and M. R. Haggerty (1995) "Abundance-distributions in artificial life and stochastic models: 'Age and area' revisited," *Advances in Artificial Life,* F. Morán, A. Moreno, J. J. Merelo, and P. Chacón (eds.), Springer, Berlin, pp. 503–514.

[3] P. J. Angeline and J. B. Pollack (1994) "Coevolving high-level representations," *Artificial Life III,* C. Langton (ed.), Addison-Wesley, Reading, MA, pp. 55–71.

[4] N. A. Barricelli (1954) "Esempi numerici di processi di evoluzione," *Methodos,* pp. 45–68.

[5] N. A. Barricelli (1957) "Symbiogenetic evolution processes realized by artificial methods," *Methodos,* Vol. IX, No. 35–36, pp. 143–182.

[6] N. A. Barricelli (1962) "Numerical testing of evolution theories. Part I: Theoretical introduction and basic tests," *Acta Biotheoretica,* Vol. 16, No. 1–2, pp. 69–98.

[7] N. A. Barricelli (1963) "Numerical testing of evolution theories. Part II: Preliminary tests of performance, symbiogenesis and terrestrial life," *Acta Biotheoretica,* Vol. 16, No. 3–4, pp. 99–126.

[8] M. Conrad (1969) "Computer experiments on the evolution of coadaptation in a primitive ecosystem," Ph.D. diss., biophysics program, Stanford, CA.

[9] M. Conrad and H. H. Pattee (1970) "Evolution experiments with an artificial ecosystem," *J. Theoret. Biol.,* Vol. 28, pp. 393–409.

[10] M. Conrad and M. Strizich (1985) "Evolve II: A computer model of an evolving ecosystem," *BioSystems,* Vol. 17, pp. 245–258.

[11] L. J. Fogel and G. H. Burgin (1969) "Competitive goal-seeking through evolutionary programming," final report under Contract No. AF 19(628)-5927, Air Force Cambridge Research Labs.

[12] W. D. Hillis (1992) "Co-evolving parasites improve simulated evolution as an optimization procedure," *Artificial Life II,* C. G. Langton, C. Taylor, J. D. Farmer, and S. Rasmussen (eds.), Addison-Wesley, Reading, MA, pp. 313–324.

[13] E. Mayr (1982) *The Growth of Biological Thought: Diversity, Evolution, and Inheritance,* Belknap, Cambridge, MA.

[14] E. Mayr (1988) *Toward a New Philosophy of Biology: Observations of an Evolutionist,* Belknap, Cambridge, MA.

[15] T. S. Ray (1992) "An approach to the synthesis of life," *Artificial Life II,* C. G. Langton, C. Taylor, J. D. Farmer, and S. Rasmussen (eds.), Addison-Wesley, Reading, MA, pp. 371–408.

[16] J. Reed, R. Toombs, and N. A. Barricelli (1967) "Simulation of biological evolution and machine learning: I. Selection of self-reproducing numeric patterns by data processing machines, effects of hereditary control, mutation type and crossing," *J. Theoret. Biol.,* Vol. 17, pp. 319–342.

[17] M. M. Rizki and M. Conrad (1985) "Evolve III: A discrete events model of an evolutionary ecosystem," *BioSystems,* Vol. 18, pp. 121–133.

[18] A. V. Sebald and J. Schlenzig (1994) "Minimax design of neural net controllers for highly uncertain plants," *IEEE Trans. Neural Networks,* Vol. 5:1, pp. 73–82.

[19] K. Thearling and T. Ray (1994) "Evolving multi-cellular artificial life," *Artificial Life IV,* R. Brooks and P. Maes (eds.), MIT Press, Cambridge, MA, pp. 283–288.

Thomas S. Ray

School of Life & Health Sciences, University of Delaware, Newark, Delaware 19716, email:
ray@brahms.udel.edu

An Approach to the Synthesis of Life

Marcel, a mechanical chessplayer... his exquisite 19th-century brainwork—
the human art it took to build which has been flat lost, lost as the dodo bird
... But where inside Marcel is the midget Grandmaster, the little Johann
Allgeier? Where's the pantograph, and the magnets? Nowhere. Marcel really is a mechanical chessplayer. No fakery inside to give him any touch of
humanity at all.

— Thomas Pynchon, *Gravity's Rainbow.*

INTRODUCTION

Ideally, the science of biology should embrace all forms of life. However in practice, it has been restricted to the study of a single instance of life, life on earth.
Because biology is based on a sample size of one, we can not know what features
of life are peculiar to earth, and what features are general, characteristic of all life.
A truly comparative natural biology would require inter-planetary travel, which is
light years away. The ideal experimental evolutionary biology would involve creation of multiple planetary systems, some essentially identical, others varying by

Artificial Life II, SFI Studies in the Sciences of Complexity, vol. X, edited by
C. G. Langton, C. Taylor, J. D. Farmer, & S. Rasmussen, Addison-Wesley, 1991

a parameter of interest, and observing them for billions of years. A practical alternative to an inter-planetary or mythical biology is to create synthetic life in a computer. "Evolution in a bottle" provides a valuable tool for the experimental study of evolution and ecology.

The intent of this work is to synthesize rather than simulate life. This approach starts with hand-crafted organisms already capable of replication and open-ended evolution, and aims to generate increasing diversity and complexity in a parallel to the Cambrian explosion.

To state such a goal leads to semantic problems, because life must be defined in a way that does not restrict it to carbon-based forms. It is unlikely that there could be general agreement on such a definition, or even on the proposition that life need not be carbon based. Therefore, I will simply state my conception of life in its most general sense. I would consider a system to be living if it is self-replicating, and capable of open-ended evolution. Synthetic life should self-replicate, and evolve structures or processes that were not designed-in or preconceived by the creator.[43]

Core Wars programs, computer viruses, and worms[11,14,15,16,17,18,19,46,48] are capable of self-replication, but fortunately, not evolution. It is unlikely that such programs will ever become fully living, because they are not likely to be able to evolve.

Most evolutionary simulations are not open ended. Their potential is limited by the structure of the model, which generally endows each individual with a genome consisting of a set of pre-defined genes, each of which may exist in a pre-defined set of allelic forms.[1,12,13,17,27,42] The object being evolved is generally a data structure representing the genome, which the simulator program mutates and/or recombines, selects, and replicates according to criteria designed into the simulator. The data structures do not contain the mechanism for replication; they are simply copied by the simulator if they survive the selection phase.

Self-replication is critical to synthetic life because without it, the mechanisms of selection must also be pre-determined by the simulator. Such artificial selection can never be as creative as natural selection. The organisms are not free to invent their own fitness functions. Freely evolving creatures will discover means of mutual exploitation and associated implicit fitness functions that we would never think of. Simulations constrained to evolve with pre-defined genes, alleles, and fitness functions are dead ended, not alive.

The approach presented here does not have such constraints. Although the model is limited to the evolution of creatures based on sequences of machine instructions, this may have a potential comparable to evolution based on sequences of organic molecules. Sets of machine instructions similar to those used in the Tierra Simulator have been shown to be capable of "universal computation."[2,33,38] This suggests that evolving machine codes should be able to generate any level of complexity.

Other examples of the synthetic approach to life can be seen in the work of Holland,[28] Farmer et al.,[22] Langton,[31] Rasmussen et al.,[45] and Bagley et al.[3] A characteristic these efforts generally have in common is that they parallel the origin of life event by attempting to create prebiotic conditions from which life may emerge spontaneously and evolve in an open-ended fashion.

While the origin of life is generally recognized as an event of the first order, there is another event in the history of life that is less well known but of comparable significance: the origin of biological diversity and macroscopic multicellular life during the Cambrian explosion 600 million years ago. This event involved a riotous diversification of life forms. Dozens of phyla appeared suddenly, many existing only fleetingly, as diverse and sometimes bizarre ways of life were explored in a relative ecological void.[24,39]

The work presented here aims to parallel the second major event in the history of life, the origin of diversity. Rather than attempting to create prebiotic conditions from which life may emerge, this approach involves engineering over the early history of life to design complex evolvable organisms, and then attempting to create the conditions that will set off a spontaneous evolutionary process of increasing diversity and complexity of organisms. This work represents a first step in this direction, creating an artificial world which may roughly parallel the RNA world of self-replicating molecules (still falling far short of the Cambrian explosion).

The approach has generated rapidly diversifying communities of self-replicating organisms exhibiting open-ended evolution by natural selection. From a single rudimentary ancestral creature containing only the code for self-replication, interactions such as parasitism, —inximmunity, hyper-parasitism, sociality, and cheating have emerged spontaneously. This paper presents a methodology and some first results.

Here was a world of simplicity and certainty no acidhead, no revolutionary anarchist would ever find, a world based on the one and zero of life and death. Minimal, beautiful. The patterns of lives and deaths.... weightless, invisible chains of electronic presence or absence. If patterns of ones and zeros were "like" patterns of human lives and deaths, if everything about an individual could be represented in a computer record by a long string of ones and zeros, then what kind of creature would be represented by a long string of lives and deaths? It would have to be up one level at least—an angel, a minor god, something in a UFO.

— Thomas Pynchon, *Vineland*.

METHODS

THE METAPHOR

Organic life is viewed as utilizing energy, mostly derived from the sun, to organize matter. By analogy, digital life can be viewed as using CPU (central processing unit) time, to organize memory. Organic life evolves through natural selection as individuals compete for resources (light, food, space, etc.) such that genotypes which leave the most descendants increase in frequency. Digital life evolves through the same process, as replicating algorithms compete for CPU time and memory space, and organisms evolve strategies to exploit one another. CPU time is thought of as the analog of the energy resource, and memory as the analog of the spatial resource.

The memory, the CPU, and the computer's operating system are viewed as elements of the "abiotic" environment. A "creature" is then designed to be specifically adapted to the features of the environment. The creature consists of a self-replicating assembler language program. Assembler languages are merely mnemonics for the machine codes that are directly executed by the CPU. These machine codes have the characteristic that they directly invoke the instruction set of the CPU and services provided by the operating system.

All programs, regardless of the language they are written in, are converted into machine code before they are executed. Machine code is the natural language of the machine, and machine instructions are viewed by this author as the "atomic units" of computing. It is felt that machine instructions provide the most natural basis for an artificial chemistry of creatures designed to live in the computer.

In the biological analogy, the machine instructions are considered to be more like the amino acids than the nucleic acids, because they are "chemically active." They actively manipulate bits, bytes, CPU registers, and the movements of the instruction pointer (as will be discussed later). The digital creatures discussed here are entirely constructed of machine instructions. They are considered analogous to creatures of the RNA world, because the same structures bear the "genetic" information and carry out the "metabolic" activity.

A block of RAM memory (random access memory, also known as "main" or "core" memory) in the computer is designated as a "soup" which can be inoculated with creatures. The "genome" of the creatures consists of the sequence of machine instructions that make up the creature's self-replicating algorithm. The prototype creature consists of 80 machine instructions; thus, the size of the genome of this creature is 80 instructions, and its "genotype" is the specific sequence of those 80 instructions.

THE VIRTUAL COMPUTER—TIERRA SIMULATOR

The computers we use are general purpose computers, which means, among other things, that they are capable of emulating through software the behavior of any other computer that ever has been built or that could be built.[2,33,38] We can utilize this flexibility to design a computer that would be especially hospitable to synthetic life.

There are several good reasons why it is not wise to attempt to synthesize digital organisms that exploit the machine codes and operating systems of real computers. The most urgent is the potential threat of natural evolution of machine codes leading to virus or worm types of programs that could be difficult to eradicate due to their changing "genotypes." This potential argues strongly for creating evolution exclusively in programs that run only on virtual computers and their virtual operating systems. Such programs would be nothing more than data on a real computer, and, therefore, would present no more threat than the data in a data base or the text file of a word processor.

Another reason to avoid developing digital organisms in the machine code of a real computer is that the artificial system would be tied to the hardware and would become obsolete as quickly as the particular machine it was developed on. In contrast, an artificial system developed on a virtual machine could be easily ported to new real machines as they become available.

A third issue, which potentially makes the first two moot, is that the machine languages of real machines are not designed to be evolvable, and in fact might not support significant evolution. Von Neuman-type machine languages are considered to be "brittle," meaning that the ratio of viable programs to possible programs is virtually zero. Any mutation or recombination event in a real machine code is almost certain to produce a non-functional program. The problem of brittleness can be mitigated by designing a virtual computer whose machine code is designed with evolution in mind. Farmer and Belin[23] have suggested that overcoming this brittleness and "discovering how to make such self-replicating patterns more robust so that they evolve to increasingly more complex states is probably the central problem in the study of artificial life."

The work described here takes place on a virtual computer known as Tierra (Spanish for Earth). Tierra is a parallel computer of the MIMD (multiple instruction, multiple data) type, with a processor (CPU) for each creature. Parallelism is imperfectly emulated by allowing each CPU to execute a small time slice in turn. Each CPU of this virtual computer contains two address registers, two numeric registers, a flags register to indicate error conditions, a stack pointer, a ten-word stack, and an instruction pointer. Each virtual CPU is implemented via the C structure listed in Appendix A. Computations performed by the Tierran CPUs are probabilistic due to flaws that occur at a low frequency (see Mutation below).

The instruction set of a CPU typically performs simple arithmetic operations or bit manipulations, within the small set of registers contained in the CPU. Some instructions move data between the registers in the CPU, or between the CPU registers and the RAM (main) memory. Other instructions control the location and movement of an "instruction pointer" (IP). The IP indicates an address in RAM, where the machine code of the executing program (in this case a digital organism) is located.

The CPU perpetually performs a fetch-decode-execute-increment-IP cycle: The machine code instruction currently addressed by the IP is fetched into the CPU, its bit pattern is decoded to determine which instruction it corresponds to, and the instruction is executed. Then the IP is incremented to point sequentially to the next position in RAM, from which the next instruction will be fetched. However, some instructions like JMP, CALL, and RET directly manipulate the IP, causing execution to jump to some other sequence of instructions in the RAM. In the Tierra

Simulator this CPU cycle is implemented through the time-slice routine listed in Appendix B.

THE TIERRAN LANGUAGE

Before attempting to set up an Artificial Life system, careful thought must be given to how the representation of a programming language affects its adaptability in the sense of being robust to genetic operations such as mutation and recombination. The nature of the virtual computer is defined in large part by the instruction set of its machine language. The approach in this study has been to loosen up the machine code in a "virtual bio-computer," in order to create a computational system based on a hybrid between biological and classical von Neumann processes.

In developing this new virtual language, which is called "Tierran," close attention has been paid to the structural and functional properties of the informational system of biological molecules: DNA, RNA, and proteins. Two features have been borrowed from the biological world which are considered to be critical to the evolvability of the Tierran language.

First, the instruction set of the Tierran language has been defined to be of a size that is the same order of magnitude as the genetic code. Information is encoded into DNA through 64 codons, which are translated into 20 amino acids. In its present manifestation, the Tierran language consists of 32 instructions, which can be represented by five bits, *operands included*.

Emphasis is placed on this last point because some instruction sets are deceptively small. Some versions of the redcode language of Core Wars,[15,18,45] for example, are defined to have ten operation codes. It might appear on the surface that the instruction set is of size ten. However, most of the ten instructions have one or two operands. Each operand has four addressing modes, and then an integer. When we consider that these operands are embedded into the machine code, we realize that they are, in fact, a part of the instruction set, and this set works out to be about 10^{11} in size. Inclusion of numeric operands will make any instruction set extremely large in comparison to the genetic code.

In order to make a machine code with a truly small instruction set, we must eliminate numeric operands. This can be accomplished by allowing the CPU registers and the stack to be the only operands of the instructions. When we need to encode an integer for some purpose, we can create it in a numeric register through bit manipulations: flipping the low-order bit and shifting left. The program can contain the proper sequence of bit flipping and shifting instructions to synthesize the desired number, and the instruction set need not include all possible integers.

A second feature that has been borrowed from molecular biology in the design of the Tierran language is the addressing mode, which is called "address by template." In most machine codes, when a piece of data is addressed, or the IP jumps to another piece of code, the exact numeric address of the data or target code is specified in the machine code. Consider that in the biological system by contrast, in order for protein molecule A in the cytoplasm of a cell to interact with protein molecule B, it does not specify the exact coordinates where B is located. Instead, molecule A presents a template on its surface which is complementary to some surface on B. Diffusion brings the two together, and the complementary conformations allow them to interact.

Addressing by template is illustrated by the Tierran JMP instruction. Each JMP instruction is followed by a sequence of NOP (no-operation) instructions, of which there are two kinds: NOP_0 and NOP_1. Suppose we have a piece of code with five instruction in the following order: JMP NOP_0 NOP_0 NOP_0 NOP_1. The system will search outward in both directions from the JMP instruction looking for the nearest occurrence of the complementary pattern: NOP_1 NOP_1 NOP_1 NOP_0. If the pattern is found, the instruction pointer will move to the end of the pattern and resume execution. If the pattern is not found, an error condition (flag) will be set and the JMP instruction will be ignored (in practice, a limit is placed on how far the system may search for the pattern).

The Tierran language is characterized by two unique features: a truly small instruction set without numeric operands, and addressing by template. Otherwise, the language consists of familiar instructions typical of most machine languages, e.g., MOV, CALL, RET, POP, PUSH, etc. The complete instruction set is listed in Appendix B.

THE TIERRAN OPERATING SYSTEM

The Tierran virtual computer needs a virtual operating system that will be hospitable to digital organisms. The operating system will determine the mechanisms of interprocess communication, memory allocation, and the allocation of CPU time among competing processes. Algorithms will evolve so as to exploit these features to their advantage. More than being a mere aspect of the environment, the operating system, together with the instruction set will determine the topology of possible interactions between individuals, such as the ability of pairs of individuals to exhibit predator-prey, parasite-host, or mutualistic relationships.

MEMORY ALLOCATION—CELLULARITY

The Tierran computer operates on a block of RAM of the real computer which is set aside for the purpose. This block of RAM is referred to as the "soup." In most of the work described here the soup consisted of 60,000 bytes, which can hold the same number of Tierran machine instructions. Each "creature" occupies some block of memory in this soup.

Cellularity is one of the fundamental properties of organic life, and can be recognized in the fossil record as far back as 3.6 billion years.[4] The cell is the original individual, with the cell membrane defining its limits and preserving its chemical integrity. An analog to the cell membrane is needed in digital organisms in order to preserve the integrity of the informational structure from being disrupted easily by the activity of other organisms. The need for this can be seen in AL models such as cellular automata where virtual state machines pass through one another,[31,32] or in core-wars-type simulations where coherent structures demolish one another when they come into contact.[15,18,45]

Tierran creatures are considered to be cellular in the sense that they are protected by a "semi-permeable membrane" of memory allocation. The Tierran operating system provides memory allocation services. Each creature has exclusive write privileges within its allocated block of memory. The "size" of a creature is just the size of its allocated block (e.g., 80 instructions). This usually corresponds to the size of the genome. While write privileges are protected, read and execute privileges are not. A creature may examine the code of another creature, and even execute it, but it can not write over it. Each creature may have exclusive write privileges in at most two blocks of memory: the one that it is born with which is referred to as the "mother cell," and a second block which it may obtain through the execution of the MAL (memory allocation) instruction. The second block, referred to as the "daughter cell," may be used to grow or reproduce into.

When Tierran creatures "divide," the mother cell loses write privileges on the space of the daughter cell, but is then free to allocate another block of memory. At the moment of division, the daughter cell is given its own instruction pointer, and is free to allocate its own second block of memory.

TIME SHARING—THE SLICER

The Tierran operating system must be multi-tasking in order for a community of individual creatures to live in the soup simultaneously. The system doles out small slices of CPU time to each creature in the soup in turn. The system maintains a circular queue called the "slicer queue." As each creature is born, a virtual CPU is created for it, and it enters the slicer queue just ahead of its mother, which is the active creature at that time. Thus, the newborn will be the last creature in the soup to get another time slice after the mother, and the mother will get the next slice after its daughter. As long as the slice size is small relative to the generation time of the creatures, the time-sharing system causes the world to approximate parallelism. In actuality, we have a population of virtual CPUs, each of which gets a slice of the real CPU's time as it comes up in the queue.

The number of instructions to be executed in each time slice is set proportional to the size of the genome of the creature being executed, raised to a power. If the "slicer power" is equal to one, then the slicer is size neutral, the probability of an instruction being executed does not depend on the size of the creature in which it occurs. If the power is greater than one, large creatures get more CPU cycles per instruction than small creatures. If the power is less than one, small creatures get more CPU cycles per instruction. The power determines if selection favors large or small creatures, or is size neutral. A constant slice size selects for small creatures.

MORTALITY—THE REAPER

Self-replicating creatures in a fixed-size soup would rapidly fill the soup and lock up the system. To prevent this from occurring, it is necessary to include mortality. The Tierran operating system includes a "reaper" which begins "killing" creatures when the memory fills to some specified level (e.g., 80%). Creatures are killed by deallocating their memory, and removing them from both the reaper and slicer queues. Their "dead" code is not removed from the soup.

In the present system, the reaper uses a linear queue. When a creature is born, it enters the bottom of the queue. The reaper always kills the creature at the top of the queue. However, individuals may move up or down in the reaper queue according to their success or failure at executing certain instructions. When a creature executes an instruction that generates an error condition, it moves one position up the queue, as long as the individual ahead of it in the queue has not accumulated a greater number of errors. Two of the instructions are somewhat difficult to execute without generating an error, therefore successful execution of these instructions moves the creature down the reaper queue one position, as long as it has not accumulated more errors than the creature below it.

The effect of the reaper queue is to cause algorithms which are fundamentally flawed to rise to the top of the queue and die. Vigorous algorithms have a greater longevity, but in general, the probability of death increases with age.

MUTATION

In order for evolution to occur, there must be some change in the genome of the creatures. This may occur within the lifespan of an individual, or there may be errors in passing along the genome to offspring. In order to insure that there is genetic change, the operating system randomly flips bits in the soup, and the instructions of the Tierran language are imperfectly executed.

Mutations occur in two circumstances. At some background rate, bits are randomly selected from the entire soup (60,000 instructions totaling 300,000 bits) and flipped. This is analogous to mutations caused by cosmic rays, and has the effect of preventing any creature from being immortal, as it will eventually mutate to death. The background mutation rate has generally been set at about 1 bit flipped for every 10,000 Tierran instructions executed by the system.

In addition, while copying instructions during the replication of creatures, bits are randomly flipped at some rate in the copies. The copy mutation rate is the higher of the two, and results in replication errors. The copy mutation rate has generally been set at about 1 bit flipped for every 1,000 to 2,500 instructions moved. In both classes of mutation, the interval between mutations varies randomly within a certain range to avoid possible periodic effects.

In addition to mutations, the execution of Tierran instructions is flawed at a low rate. For most of the 32 instructions, the result is off by ± 1 at some low frequency. For example, the increment instruction normally adds one to its register, but it sometimes adds two or zero. The bit-flipping instruction normally flips the low-order bit, but it sometimes flips the next higher bit or no bit. The shift-left instruction normally shifts all bits one bit to the left, but it sometimes shifts left by two bits, or not at all. In this way, the behavior of the Tierran instructions is probabilistic, not fully deterministic.

It turns out that bit-flipping mutations and flaws in instructions are not necessary to generate genetic change and evolution, once the community reaches a certain state of complexity. Genetic parasites evolve which are sloppy replicators, and have the effect of moving pieces of code around between creatures, causing rather massive rearrangements of the genomes. The mechanism of this ad hoc sexuality has not been worked out, but is likely due to the parasites' inability to discriminate between live, dead, or embryonic code.

Mutations result in the appearance of new genotypes, which are watched by an automated genebank manager. In one implementation of the manager, when new genotypes replicate twice, producing a genetically identical offspring at least once, they are given a unique name and saved to disk. Each genotype name contains two parts, a number, and a three-letter code. The number represents the number of instructions in the genome. The three-letter code is used as a base 26 numbering system for assigning a unique label to each genotype in a size class. The first genotype to appear in a size class is assigned the label aaa, the second is assigned the label aab, and so on. Thus the ancestor is named 80aaa, and the first mutant of size 80 is named 80aab. The first parasite of size 45 is named 45aaa.

The genebanker saves some additional information with each genome: the genotype name of its immediate ancestor which makes possible the reconstruction of the entire phylogeny; the time and date of origin; "metabolic" data including the number of instructions executed in the first and second reproduction, the number of errors generated in the first and second reproduction, and the number of instructions copied into the daughter cell in the first and second reproductions (see Appendix C); some environmental parameters at the time of origin including the search limit for addressing, and the slicer power, both of which affect selection for size.

THE TIERRAN ANCESTOR

The Tierran language has been used to write a single self-replicating program which is 80 instructions long. This program is referred to as the "ancestor," or alternatively as genotype 0080aaa (Figure 1). The ancestor is a minimal self-replicating algorithm which was originally written for use during the debugging of the simulator. No functionality was designed into the ancestor beyond the ability to self-replicate, nor was any specific evolutionary potential designed in. The commented Tierran assembler and machine code for this program is presented in Appendix C.

The ancestor examines itself to determine where in memory it begins and ends. The ancestor's beginning is marked with the four no-operation template: 1 1 1 1, and its ending is marked with 1 1 1 0. The ancestor locates its beginning with the five instructions: ADRB, NOP_0, NOP_0, NOP_0, and NOP_0. This series of

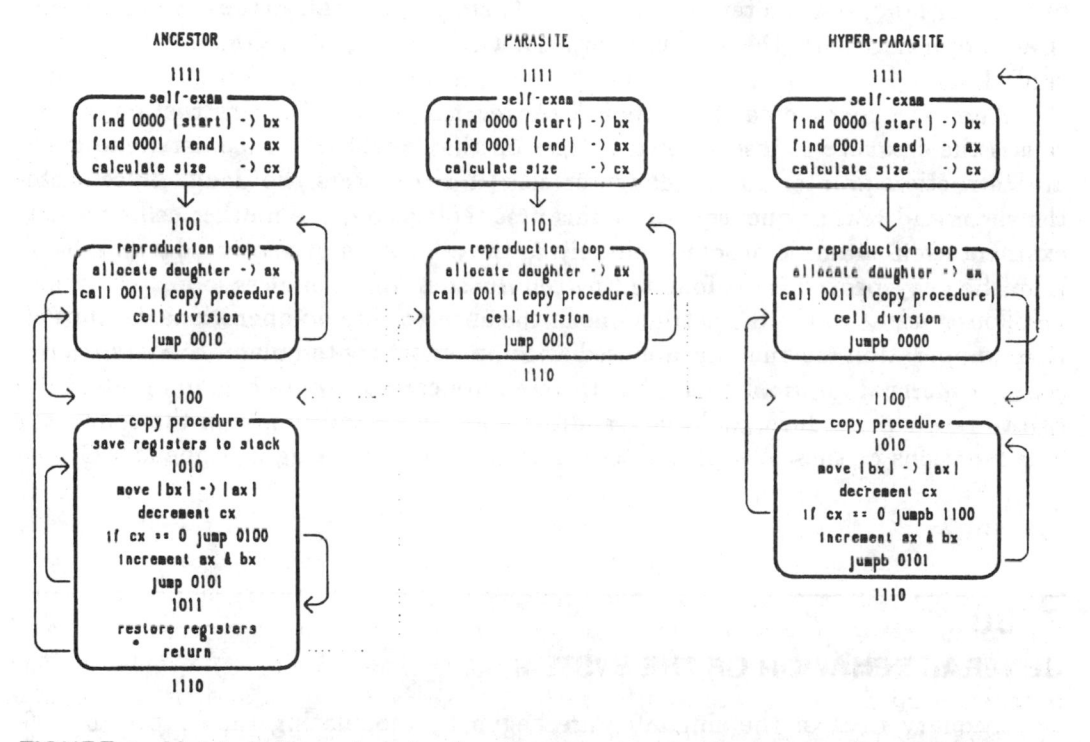

FIGURE 1 Metabolic flow chart for the ancestor, parasite, hyper-parasite, and their interactions: ax, bx and cx refer to CPU registers where location and size information are stored. [ax] and [bx] refer to locations in the soup indicated by the values in the ax and bx registers. Patterns such as 1101 are complementary templates used for addressing. Arrows outside of boxes indicate jumps in the flow of execution of the programs. The dotted-line arrows indicate flow of execution between creatures. The parasite lacks the copy procedure; however, if it is within the search limit of the copy procedure of a host, it can locate, call, and execute that procedure, thereby obtaining the information needed to complete its replication. The host is not adversely affected by this informational parasitism, except through competition with the parasite, which is a superior competitor. Note that the parasite calls the copy procedure of its host with the expectation that control will return to the parasite when the copy procedure returns. However, the hyper-parasite jumps out of the copy procedure rather than returning, thereby seizing control from the parasite. It then proceeds to reset the CPU registers of the parasite with the location and size of the hyper-parasite, causing the parasite to replicate the hyper-parasite genome thereafter.

instructions causes the system to search backwards from the ADRB instruction for a template complementary to the four NOP_0 instructions, and to place the address of the complementary template (the beginning) in the ax register of the CPU (see Appendix A). A similar method is used to locate the end.

Having determined the address of its beginning and its end, it subtracts the two to calculate its size, and allocates a block of memory of this size for a daughter cell. It then calls the copy procedure which copies the entire genome into the daughter-cell memory, one instruction at a time. The beginning of the copy procedure is marked by the four no-operation template: 1 1 0 0. Therefore, the call to the copy procedure is accomplished with the five instructions: CALL, NOP_0, NOP_0, NOP_1, and NOP_1.

When the genome has been copied, it executes the DIVIDE instruction, which causes the creature to lose write privileges on the daughter-cell memory, and gives an instruction pointer to the daughter cell (it also enters the daughter cell into the slicer and reaper queues). After this first replication, the mother cell does not examine itself again; it proceeds directly to the allocation of another daughter cell, then the copy procedure is followed by cell division, in an endless loop.

Fourty-eight of the 80 instructions in the ancestor are no-operations. Groups of four no-operation instructions are used as complementary templates to mark twelve sites for internal addressing, so that the creature can locate its beginning and end, call the copy procedure, and mark addresses for loops and jumps in the code, etc. The functions of these templates are commented in the listing in Appendix C.

RESULTS
GENERAL BEHAVIOR OF THE SYSTEM

Evolutionary runs of the simulator are begun by inoculating the soup of 60,000 instructions with a single individual of the 80 instruction ancestral genotype. The passage of time in a run is measured in terms of how many Tierran instructions have been executed by the simulator. Most software development work has been carried out on a Toshiba 5200/100 laptop computer with an 80386 processor and an 80387 math co-processor operating at 20 Mhz. This machine executes over 12 million Tierran instructions per hour. Long evolutionary runs are conducted on mini and mainframe computers which execute about 1 million Tierran instructions per minute.

The original ancestral cell which inoculates the soup executes 839 instructions in its first replication, and 813 for each additional replication. The initial cell and its replicating daughters rapidly fill the soup memory to the threshold level of 80% which starts the reaper. Typically, the system executes about 400,000 instructions in filling up the soup with about 375 individuals of size 80 (and their gestating daughter cells). Once the reaper begins, the memory remains roughly 80% filled with creatures for the remainder of the run.

Once the soup is full, individuals are initially short lived, generally reproducing only once before dying; thus, individuals turn over very rapidly. More slowly, there appear new genotypes of size 80, and then new size classes. There are changes in the genetic composition of each size class, as new mutants appear, some of which increase significantly in frequency, sometimes replacing the original genotype. The size classes which dominate the community also change through time, as new size classes appear (see below), some of which competitively exclude sizes present earlier. Once the community becomes diverse, there is a greater variance in the longevity and fecundity of individuals.

In addition to an increase in the raw diversity of genotypes and genome sizes, there is an increase in the ecological diversity. Obligate commensal parasites evolve, which are not capable of self-replication in isolated culture, but which can replicate when cultured with normal (self-replicating) creatures. These parasites execute some parts of the code of their hosts, but cause them no direct harm, except as competitors. Some potential hosts have evolved immunity to the parasites, and some parasites have evolved to circumvent this immunity.

In addition, facultative hyper-parasites have evolved, which can self-replicate in isolated culture, but when subjected to parasitism, subvert the parasites energy metabolism to augment their own reproduction. Hyper-parasites drive parasites to extinction, resulting in complete domination of the communities. The relatively high degrees of genetic relatedness within the hyper-parasite-dominated communities leads to the evolution of sociality in the sense of creatures that can only replicate when they occur in aggregations. These social aggregations are then invaded by hyper-hyper-parasite cheaters.

Mutations and the ensuing replication errors lead to an increasing diversity of sizes and genotypes of self-replicating creatures in the soup. Within the first 100 million instructions of elapsed time, the soup evolves to a state in which about a dozen more-or-less persistent size classes coexist. The relative abundances and specific list of the size classes varies over time. Each size class consists of a number of distinct genotypes which also vary over time.

EVOLUTION

MICRO-EVOLUTION

If there were no mutations at the outset of the run, there would be no evolution. However, the bits flipped as a result of copy errors or background mutations result in creatures whose list of 80 instructions (genotype) differs from the ancestor, usually by a single bit difference in a single instruction.

Mutations, in and of themselves, cannot result in a change in the size of a creature, they can only alter the instructions in its genome. However, by altering the genotype, mutations may affect the process whereby the creature examines itself and calculates its size, potentially causing it to produce an offspring that differs in size from itself.

Four out of the five possible mutations in a no-operation instruction convert it into another kind of instruction, while one out of five converts it into the complementary no-operation. Therefore, 80% of mutations in templates destroy the template, while one in five alters the template pattern. An altered template may cause the creature to make mistakes in self-examination, procedure calls, or looping or jumps of the instruction pointer, all of which use templates for addressing.

PARASITES An example of the kind of error that can result from a mutation in a template is a mutation of the low-order bit of instruction 42 of the ancestor (Appendix C). Instruction 42 is a NOP_0, the third component of the copy procedure template. A mutation in the low-order bit would convert it into NOP_1, thus changing the template from 1 1 0 0 to: 1 1 1 0. This would then be recognized as the template used to mark the end of the creature, rather than the copy procedure.

A creature born with a mutation in the low-order bit of instruction 42 would calculate its size as 45. It would allocate a daughter cell of size 45 and copy only instructions 0 through 44 into the daughter cell. The daughter cell then, would not include the copy procedure. This daughter genotype, consisting of 45 instructions, is named 0045aaa.

Genotype 0045aaa (Figure 1) is not able to self-replicate in isolated culture. However, the semi-permeable membrane of memory allocation only protects write privileges. Creatures may match templates with code in the allocated memory of other creatures, and may even execute that code. Therefore, if creature 0045aaa is grown in mixed culture with 0080aaa, when it attempts to call the copy procedure, it will not find the template within its own genome, but if it is within the search limit (generally set at 200–400 instructions) of the copy procedure of a creature of genotype 0080aaa, it will match templates, and send its instruction pointer to the copy code of 0080aaa. Thus a parasitic relationship is established (see ECOLOGY below). Typically, parasites begin to emerge within the first few million instructions of elapsed time in a run.

IMMUNITY TO PARASITES At least some of the size 79 genotypes demonstrate some measure of resistance to parasites. If genotype 45aaa is introduced into a soup, flanked on each side with one individual of genotype 0079aab, 0045aaa will initially reproduce somewhat, but will be quickly eliminated from the soup. When the same experiment is conducted with 0045aaa and the ancestor, they enter a stable cycle in which both genotypes coexist indefinitely. Freely evolving systems have been observed to become dominated by size 79 genotypes for long periods, during which parasitic genotypes repeatedly appear, but fail to invade.

CIRCUMVENTION OF IMMUNITY TO PARASITES Occasionally these evolving systems dominated by size 79 were successfully invaded by parasites of size 51. When the immune genotype 0079aab was tested with 0051aao (a direct, one-step descendant of 0045aaa in which instruction 39 is replaced by an insertion of seven instructions of unknown origin), they were found to enter a stable cycle. Evidently 0051aao has evolved some way to circumvent the immunity to parasites possessed by 0079aab. The 14 genotypes 0051aaa through 0051aan were also tested with 0079aab, and none were able to invade.

HYPER-PARASITES Hyper-parasites have been discovered, (e.g., 0080gai, which differs by 19 instructions from the ancestor, Figure 1). Their ability to subvert the energy metabolism of parasites is based on two changes. The copy procedure does not return, but jumps back directly to the proper address of the reproduction loop. In this way it effectively seizes the instruction pointer from the parasite. However it is another change which delivers the coup de grâce: after each reproduction, the hyper-parasite re-examines itself, resetting the bx register with its location and the cx register with its size. After the instruction pointer of the parasite passes through this code, the CPU of the parasite contains the location and size of the hyper-parasite and the parasite thereafter replicates the hyper-parasite genome.

SOCIAL HYPER-PARASITES Hyper-parasites drive the parasites to extinction. This results in a community with a relatively high level of genetic uniformity, and therefore high genetic relationship between individuals in the community. These are the conditions that support the evolution of sociality, and social hyper-parasites soon dominate the community. Social hyper-parasites (Figure 2) appear in the 61 instruction size class. For example, 0061acg is social in the sense that it can only self-replicate when it occurs in aggregations. When it jumps back to the code for self-examination, it jumps to a template that occurs at the end rather than the beginning of its genome. If the creature is flanked by a similar genome, the jump will find the target template in the tail of the neighbor, and execution will then pass into the beginning of the active creature's genome. The algorithm will fail unless a similar genome occurs just before the active creature in memory. Neighboring creatures cooperate by catching and passing on jumps of the instruction pointer.

It appears that the selection pressure for the evolution of sociality is that it facilitates size reduction. The social species are 24% smaller than the ancestor. They have achieved this size reduction in part by shrinking their templates from four instructions to three instructions. This means that there are only eight templates available to them, and catching each others jumps allows them to deal with some of the consequences of this limitation as well as to make dual use of some templates.

Thomas S. Ray

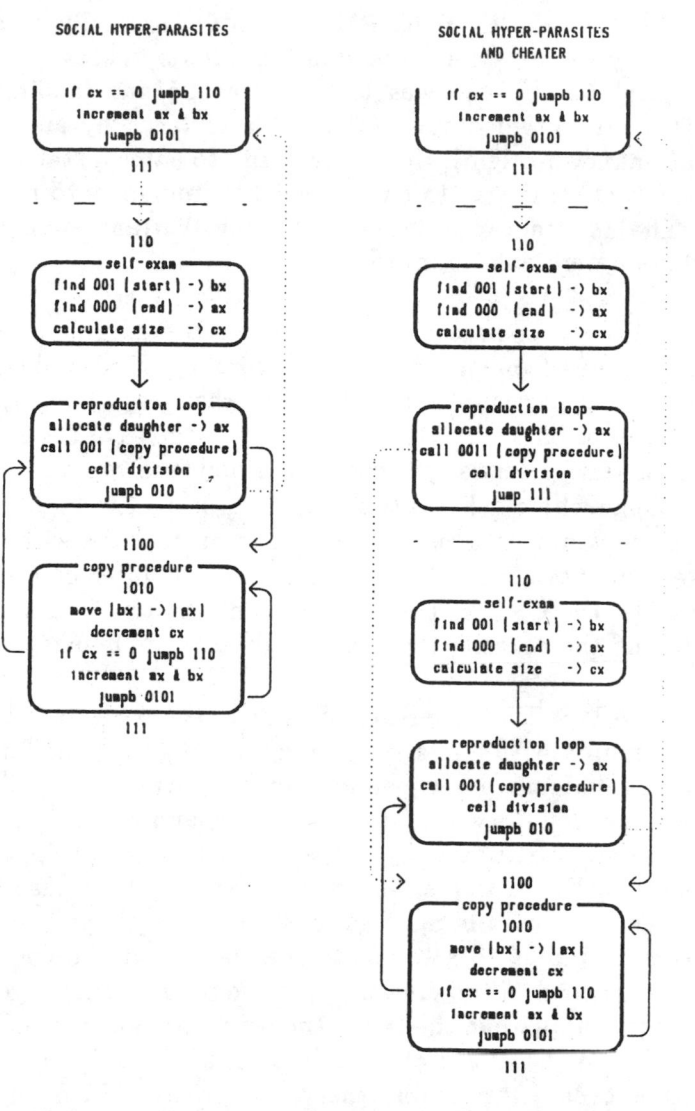

FIGURE 2 Metabolic flow chart for social hyper-parasites, their associated hyper-hyper-parasite cheaters, and their interactions. Symbols are as described for Figure 1. Horizontal dashed lines indicate the boundaries between individual creatures. On both the left and right, above the dashed line at the top of the figure is the lowermost fragment of a social hyper-parasite. Note (on the left) that neighboring social hyper-parasites cooperate in returning the flow of execution to the beginning of the creature for self-re-examination. Execution jumps back to the end of the creature above, but then falls off the end of the creature without executing any instructions of consequence, and enters the top of the creature below. On the right, a cheater is inserted between the two social-hyper-parasites. The cheater captures control of execution when it passes between the social individuals. It sets the CPU registers with its own location and size, and then skips over the self-examination step when it returns control of execution to the social creature below.

602

CHEATERS: HYPER-HYPER-PARASITES The cooperative social system of hyper-parasites is subject to cheating, and is eventually invaded by hyper-hyper-parasites (Figure 2). These cheaters (e.g., 0027aab) position themselves between aggregating hyper-parasites so that when the instruction pointer is passed between them, they capture it.

A NOVEL SELF-EXAMINATION All creatures discussed thus far mark their beginning and end with templates. They then locate the addresses of the two templates and determine their genome size by subtracting them. In one run, creatures evolved without a template marking their end. These creatures located the address of the template marking their beginning, and then the address of a template in the middle of their genome. These two addresses were then subtracted to calculate half of their size, and this value was multiplied by two (by shifting left) to calculate their full size.

MACRO-EVOLUTION

When the simulator is run over long periods of time, hundreds of millions or billions of instructions, various patterns emerge. Under selection for small sizes, there is a proliferation of small parasites and a rather interesting ecology (see below). Selection for large creatures has usually lead to continuous incrementally increasing sizes (but not to a trivial concatenation of creatures end-to-end) until a plateau in the upper hundreds is reached. In one run, selection for large size lead to apparently open-ended size increase, evolving genomes larger than 23,000 instructions in length. This evolutionary pattern might be described as phyletic gradualism.

The most thoroughly studied case for long runs is where selection, as determined by the slicer function, is size neutral. The longest runs to date (as much as 2.86 billion Tierran instructions) have been in a size-neutral environment, with a search limit of 10,000, which would allow large creatures to evolve if there were some algorithmic advantage to be gained from larger size. These long runs illustrate a pattern which could be described as periods of stasis punctuated by periods of rapid evolutionary change, which appears to parallel the pattern of punctuated equilibrium described by Eldredge and Gould[21] and Gould and Eldredge.[25]

Initially these communities are dominated by creatures with genome sizes in the 80s. This represents a period of relative stasis, which has lasted from 178 million to 1.44 billion instructions in the several long runs conducted to date. The systems then very abruptly (in a span of 1 or 2 million instructions) evolve into communities dominated by sizes ranging from about 400 to about 800. These communities have not yet been seen to evolve into communities dominated by either smaller or substantially larger size ranges.

TABLE 1 Table of numbers of size classes in the genebank. Left column is size class, right column is number of self-replicating genotypes of that size class. 305 sizes, 29,275 genotypes.

Size	#	Size	#	Size	#	Size	#	Size	#	Size	#
0034	1	0092	362	0150	2	0205	5	0418	1	5213	2
0041	2	0093	261	0151	1	0207	3	0442	10	5229	4
0043	12	0094	241	0152	2	0208	2	0443	1	5254	1
0044	7	0095	211	0153	1	0209	1	0444	61	5888	36
0045	191	0096	232	0154	2	0210	9	0445	1	5988	1
0046	7	0097	173	0155	3	0211	4	0456	2	6006	2
0047	5	0098	92	0156	77	0212	4	0465	6	6014	1
0048	4	0099	117	0157	270	0213	5	0472	6	6330	1
0049	8	0100	77	0158	938	0214	47	0483	1	6529	1
0050	13	0101	62	0159	836	0218	1	0484	8	6640	1
0051	2	0102	62	0160	3229	0219	1	0485	3	6901	5
0052	11	0103	27	0161	1417	0220	2	0486	9	6971	1
0053	4	0104	25	0162	174	0223	3	0487	2	7158	2
0054	2	0105	28	0163	187	0226	2	0493	2	7293	3
0055	2	0106	19	0164	46	0227	7	0511	2	7331	1
0056	4	0107	3	0165	183	0231	1	0513	1	7422	70
0057	1	0108	8	0166	81	0232	1	0519	1	7458	1
0058	8	0109	2	0167	71	0236	1	0522	6	7460	7
0059	8	0110	8	0168	9	0238	1	0553	1	7488	1
0060	3	0111	71	0169	15	0240	3	0568	6	7598	1
0061	1	0112	19	0170	99	0241	1	0578	1	7627	63
0062	2	0113	10	0171	40	0242	1	0581	3	7695	1
0063	2	0114	3	0172	44	0250	1	0582	1	7733	1

Code	Count	Code	Count	Code	Count	Code	Count	Code	Count	Code	Count
0064	1	0115	3	0173	34	0251	1	0600	1	7768	2
0065	4	0116	5	0174	15	0260	2	0683	1	7860	25
0066	1	0117	3	0175	22	0261	1	0689	1	7912	1
0067	1	0118	3	0176	137	0265	2	0757	6	8082	3
0068	2	0119	1	0177	13	0268	1	0804	2	8340	1
0069	1	0120	3	0178	3	0269	1	0813	1	8366	1
0070	7	0121	2	0179	1	0284	15	0881	6	8405	5
0071	5	0122	60	0180	16	0306	1	0888	1	8406	2
0072	17	0123	9	0181	5	0312	1	0940	2	8649	2
0073	2	0124	3	0182	27	0314	1	1006	6	8750	1
0074	80	0125	11	0184	3	0316	2	1016	1	8951	1
0075	56	0126	6	0185	21	0318	3	1077	5	8978	3
0076	21	0127	1	0186	9	0319	2	1116	1	9011	3
0077	28	0130	3	0187	3	0320	23	1186	1	9507	3
0078	409	0131	2	0188	11	0321	5	1294	7	9564	3
0079	850	0132	5	0190	20	0322	21	1322	7	9612	1
0080	7399	0133	2	0192	12	0330	1	1335	1	9968	1
0081	590	0134	7	0193	4	0342	5	1365	11	10259	31
0082	384	0135	1	0194	4	0343	1	1631	1	10676	1
0083	886	0136	1	0195	11	0351	1	1645	3	11366	5
0084	1672	0137	1	0196	19	0352	3	2266	1	11900	1
0085	1531	0138	1	0197	2	0386	1	2615	2	12212	2
0086	901	0139	2	0198	3	0388	2	2617	9	15717	3
0087	944	0141	6	0199	35	0401	3	2671	7	16355	3
0088	517	0143	1	0200	1	0407	1	3069	3	17356	1
0089	449	0144	4	0201	84	0411	22	4241	7	18532	1
0090	543	0146	1	0203	1	0412	3	5101	15	23134	14
0091	354	0149	1	0204	1	0416	1	5157	9		

The communities of creatures in the 400 to 800 size range also show a long-term pattern of punctuated equilibrium. These communities regularly come to be dominated by one or two size classes, and remain in that condition for long periods of time. However, they inevitably break out of that stasis and enter a period where no size class dominates. These periods of rapid evolutionary change may be very chaotic. Close observations indicate that at least at some of these times, no genotypes breed true. Many self-replicating genotypes will coexist in the soup at these times, but at the most chaotic times, none will produce offspring which are even their same size. Eventually the system will settle down to another period of stasis dominated by one or a few size classes which breed true.

Two communities have been observed to die after long periods. In one community, a chaotic period led to a situation where only a few replicating creatures were left in the soup, and these were producing sterile offspring. When these last replicating creatures died (presumably from an accumulation of mutations), the community was dead. In these runs, the mutation rate was not lowered during the run, while the average genome size increased by an order of magnitude until it approached the average mutation rate. Both communities died shortly after the dominant size class moved from the 400 range to the 700 to 1400 range. Under these circumstances it is probably difficult for any genome to breed true, and the genomes may simply have "melted." Another community died abruptly when the mutation rate was raised to a high level.

DIVERSITY

Most observations on the diversity of Tierran creatures have been based on the diversity of size classes. Creatures of different sizes are clearly genetically different, as their genomes are of different sizes. Different sized creatures would have some difficulty engaging in recombination if they were sexual; thus, it is likely that they would be different species. In a run of 526 million instructions, 366 size classes were generated, 93 of which achieved abundances of five or more individuals. In a run of 2.56 billion instructions, 1180 size classes were generated, 367 of which achieved abundances of five or more.

Each size class consists of a number of distinct genotypes which also vary over time. There exists the potential for great genetic diversity within a size class. There are 32^{80} distinct genotypes of size 80, but how many of those are viable self-replicating creatures? This question remains unanswered; however, some information has been gathered through the use of the automated genebank manager.

In several days of running the genebanker, over 29,000 self-replicating genotypes of over 300 size classes accumulated. The size classes and the number of unique genotypes banked for each size are listed in Table 1. The genotypes saved to disk can be used to inoculate new soups individually, or collections of these banked genotypes may be used to assemble "ecological communities." In "ecological" runs, the mutation rates can be set to zero in order to inhibit evolution.

ECOLOGY

The only communities whose ecology has been explored in detail are those that operate under selection for small sizes. These communities generally include a large number of parasites, which do not have functional copy procedures, and which execute the copy procedures of other creatures within the search limit. In exploring ecological interactions, the mutation rate is set at zero, which effectively throws the simulation into ecological time by stopping evolution. When parasites are present, it is also necessary to stipulate that creatures must breed true, since parasites have a tendency to scramble genomes, leading to evolution in the absence of mutation.

0045aaa is a "metabolic parasite." Its genome does not include the copy procedure; however, it executes the copy procedure code of a normal host, such as the ancestor. In an environment favoring small creatures, 0045aaa has a competitive advantage over the ancestor; however, the relationship is density dependent. When the hosts become scarce, most of the parasites are not within the search limit of a copy procedure, and are not able to reproduce. Their calls to the copy procedure fail and generate errors, causing them to rise to the top of the reaper queue and die. When the parasites die off, the host population rebounds. Hosts and parasites cultured together demonstrate Lotka-Volterra population cycling.[34,53,54]

A number of experiments have been conducted to explore the factors affecting diversity of size classes in these communities. Competitive exclusion trials were conducted with a series of self-replicating (non-parasitic) genotypes of different size classes. The experimental soups were initially inoculated with one individual of each size. A genotype of size 79 was tested against a genotype of size 80, and then against successively larger size classes. The interactions were observed by plotting the population of the size 79 class on the x axis, and the population of the other size class on the y axis. Sizes 79 and 80 were found to be competitively matched such that neither was eliminated from the soup. They quickly entered a stable cycle, which exactly repeated a small orbit. The same general pattern was found in the interaction between sizes 79 and 81.

When size 79 was tested against size 82, they initially entered a stable cycle, but after about 4 million instructions, they shook out of stability and the trajectory became chaotic with an attractor that was symmetric about the diagonal (neither size showed any advantage). This pattern was repeated for the next several size classes, until size 90, where a marked asymmetry of the chaotic attractor was evident, favoring size 79. The run of size 79 against size 93 showed a brief stable period of about a million instructions, which then moved to a chaotic phase without an attractor, which spiraled slowly down until size 93 became extinct, after an elapsed time of about 6 million instructions.

An interesting exception to this pattern was the interaction between size 79 and size 89. Size 89 is considered to be a "metabolic cripple," because although it is capable of self-replicating, it executes about 40% more instructions to replicate than normal. It was eliminated in competition with size 79, with no loops in the trajectory, after an elapsed time of under 1 million instructions.

In an experiment to determine the effects of the presence of parasites on community diversity, a community consisting of 20 size classes of hosts was created and allowed to run for 30 million instructions, at which time only the eight smallest size classes remained. The same community was then regenerated, but a single genotype (0045aaa) of parasite was also introduced. After 30 million instructions, 16 size classes remained, including the parasite. This seems to be an example of a "keystone" parasite effect.[41]

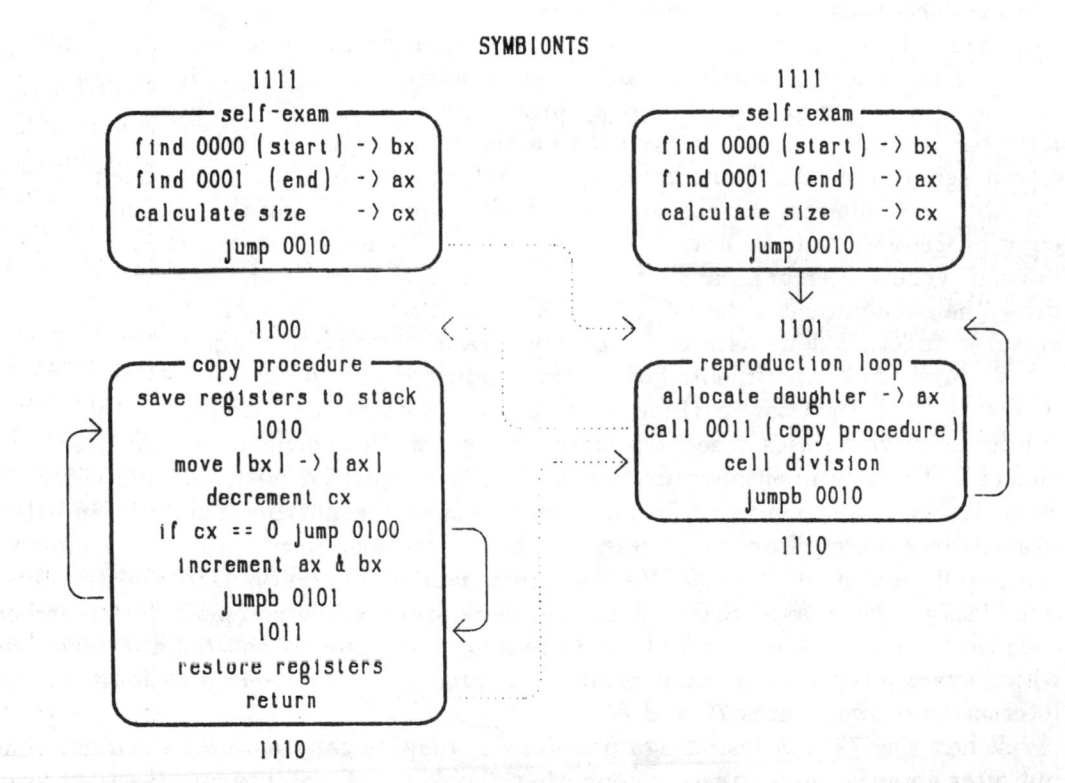

FIGURE 3 Metabolic flow chart for obligate symbionts and their interactions. Symbols are as described for Figure 1. Neither creature is able to self-replicate in isolation. However, when cultured together, each is able to replicate by using information provided by the other.

Symbiotic relationships are also possible. The ancestor was manually dissected into two creatures, one of size 46 which contained only the code for self-examination and the copy loop, and one of size 64 which contained only the code for self-examination and the copy procedure (Figure 3). Neither could replicate when cultured alone, but when cultured together, they both replicated, forming a stable mutualistic relationship. It is not known if such relationships have evolved spontaneously.

DISCUSSION

The "physical" environment presented by the simulator is quite simple, consisting of the energy resource (CPU time) doled out rather uniformly by the time slicer, and memory space which is completely uniform and always available. In light of the nature of the physical environment, the implicit fitness function would presumably favor the evolution of creatures which are able to replicate with less CPU time, and this does, in fact, occur. However, much of the evolution in the system consists of the creatures discovering ways to exploit one another. The creatures invent their own fitness functions through adaptation to their biotic environment.

Parasites do not contain the complete code for self-replication; thus, they utilize other creatures for the information contained in their genomes. Hyper-parasites exploit parasites in order to increase the amount of CPU time devoted to the replication of their own genomes; thus, hyper-parasites utilize other creatures for the energy resources that they possess. These ecological interactions are not programmed into the system, but emerge spontaneously as the creatures discover each other and invent their own games.

Evolutionary theory suggests that adaptation to the biotic environment (other organisms) rather than to the physical environment is the primary force driving the auto-catalytic diversification of organisms.[49] It is encouraging to discover that the process has already begun in the Tierran world. It is worth noting that the results presented here are based on evolution of the first creature that I designed, written in the first instruction set that I designed. Comparison to the creatures that have evolved shows that the one I designed is not a particularly clever one. Also, the instruction set that the creatures are based on is certainly not very powerful (apart from those special features incorporated to enhance its evolvability). It would appear then that it is rather easy to create life. Evidently, virtual life is out there, waiting for us to provide environments in which it may evolve.

EMERGENCE

Cariani[8] has suggested a methodology by which emergence can be detected. His analysis is described as "emergence-relative-to-a-model," where "the model... constitutes the observer's expectations of how the system will behave in the future." If the system evolves such that the model no longer describes the system, we have emergence.

Cariani recognizes three types of emergence, in semiotic terms: syntactic, semantic, and pragmatic. Syntactic operations are those of computation (symbolic). Semantic operations are those of measurement (e.g., sense perception) and control (e.g., effectors), because they "determine the relation of the symbols in the computational part of the device to the world at large." Pragmatic ("intentional") operations are those that are "performance-measuring," and, hence "the criteria which control the selection."

Cariani has developed this analysis in the context of robotics, and considers that the semantic operations should act at the interface between the symbolic (computational) and the nonsymbolic (real physical world). I can not apply his analysis in precisely this way to my simulation, because there is no connection between the Tierran world and the real physical world. I have created a virtual universe that is fully self-contained, within the computer; thus, I must apply his analysis in this context.

In the Tierran world, symbolic operations (syntactic), computations, take place in the CPU. The "nonsymbolic," "real physical world" is the soup (RAM) where the creatures reside. The measurement (semantic) operations are those that involve the location of templates; the effector operations are the copying of instructions within the soup, and the allocation of memory (cells). Fitness functions (pragmatic) are implicit, and are determined by the creatures themselves because they must effect their own replication.

Any program which is self-modifying can show syntactic emergence. As long as the organization of the executable code changes, we have syntactic emergence. This occurs in the Tierran world, as the executable genetic code of the creatures evolves.

Semantic emergence is more difficult to achieve, as it requires the appearance of some new meaning in the system. This is found in the Tierran world in the evolution of templates and their meanings. When a creature locates a template, which has a physical manifestation in the "real world" of the soup, the location of the template appears in the CPU in the form of a symbol representing its address in the soup. For example, the beginning and end of the ancestor are each marked by templates. That one "means" beginning and the other "means" end is apparent from the computation made on the symbols for them in the CPU: the two are subtracted to calculate the size of the creature, and copying of the genome starts at the beginning address. Through evolution, a class of creatures appeared which did not locate a template at their end, but rather one in their center. That the new template "means" center to these creatures is again apparent from the computations made on its associated symbol in the CPU: the beginning address is subtracted from the center address, the difference is then multiplied by two to calculate the size.

Pragmatic emergence is considered "higher" by Cariani, and certainly it is the most difficult to achieve, because it requires that the system evolve new fitness functions. In living systems, fitness functions always reduce to: genotypes which leave a greater number of their genes in future generations will increase in frequency relative to other genotypes and thus have a higher fitness. This is a nearly tautological observation, but tautology is avoided in that the fitness landscape is shaped by specific adaptations that facilitate passing genes on.

For a precambrian marine algae living before the appearance of herbivores, the fitness landscape consists, in part, of a multi-dimensional space of metabolic parameter affecting the efficiency of the conversion of sun light into useable energy, and the use of that energy in obtaining nutrients and converting them into new cells. Regions of this metabolic phase space that yield a greater efficiency at these operations also have higher associated fitnesses.

In order for pragmatic emergence to occur, the fitness landscape must be expanded to include new realms. For example, if a variant genotype of algae engulfs other algae, and thereby achieves a new mechanism of obtaining energy, the fitness landscape expands to include the parameters of structure and metabolism that facilitate the location, capture, and digestion of other cells. The fitness landscapes of algae lacking these adaptations also become altered, as they now include the parameters of mechanisms to avoid being eaten. Pragmatic emergence occurs through the acquisition of a new class of adaptation for enhancing passing genes on.

Pragmatic emergence occurs in the Tierran world as creatures which initially do not interact, discover means to exploit one another, and in response, means to avoid exploitation. The original fitness landscape of the ancestor consists only of the efficiency parameters of the replication algorithm, in the context of the properties of the reaper and slicer queues. When by chance, genotypes appear that exploit other creatures, selection acts to perfect the mechanisms of exploitation, and mechanisms of defense to that exploitation. The original fitness landscape was based only on adaptations of the organism to its physical environment (the slicer and reaper). The new fitness landscape retains those features, but adds to it adaptations to the biotic environment, the other creatures. Because the fitness landscape includes an ever-increasing realm of adaptations to other creatures which are themselves evolving, it can facilitate an auto-catalytic increase in complexity and diversity of organisms.

In any computer model of evolution, the fitness functions are determined by the entity responsible for the replication of individuals. In genetic algorithms and most simulations, that entity is the simulator program; thus, the fitness function is defined globally. In the Tierran world, that entity is the creatures themselves; thus, the fitness function is defined locally by each creature in relation to its environment (which includes the other creatures). It is for this reason that pragmatic emergence occurs in the Tierran world.

In Tierra, the fitness functions are determined by the creatures themselves, and evolve with the creatures. As Cariani states, "Such devices would not be useful for accomplishing *our* purposes as their evaluatory criteria might well diverge from our

Thomas S. Ray

own over time." This was the case from the outset in the Tierran world, because the simulator never imposed any explicit selection on the creatures. They were not expected to solve my problems, other than satisfying my passion to create life.

After describing how to recognize the various types of emergence, Cariani concludes that Artificial Life cannot demonstrate emergence because of the fully deterministic and replicable nature of computer simulations. This conclusion does not follow in any obvious way from the preceding discussions and does not seem to be supported. Furthermore, I have never known "indeterminate" and "unreplicable" to be considered as necessary qualities of life.

As a thought experiment, suppose that we connect a Geiger counter near a radioactive source to our computer, and use the interval between clicks to determine the values in our random number generator. The resulting behavior of the simulation would no longer be deterministic or repeatable. However, the results would be the same, in any significant respect, to those obtained by using an algorithm to select the random numbers. Determinism and repeatability are irrelevant to emergence and to life. In fact, repeatability is a highly desirable quality of synthetic life because it facilitates study of life's properties.

SYNTHETIC BIOLOGY

One of the most uncanny of evolutionary phenomena is the ecological convergence of biota living on different continents or in different epochs. When a lineage of organisms undergoes an adaptive radiation (diversification), it leads to an array of relatively stable ecological forms. The specific ecological forms are often recognizable from lineage to lineage. For example, among dinosaurs, the *Pterosaur, Triceratops, Tyrannosaurus*, and *Ichthyosaur* are ecological parallels, respectively, to the bat, rhinoceros, lion, and porpoise of modern mammals. Similarly, among modern placental mammals, the gray wolf, flying squirrel, great anteater, and common mole are ecological parallels, respectively, to the Tasmanian wolf, honey glider, banded anteater, and marsupial mole of the marsupial mammals of Australia.

Given these evidently powerful convergent forces, it should perhaps not be surprising that as adaptive radiations proceed among digital organisms, we encounter recognizable ecological forms, in spite of the fundamentally distinct physics and chemistry on which they are based. Ideally, comparisons should be made among organisms of comparable complexity. It may not be appropriate to compare viruses to mammals. Unfortunately, the organic creatures most comparable to digital organisms, the RNA creatures, are no longer with us. Since digital organisms are being compared to modern organic creatures of much greater complexity, ecological comparisons must be made in the broadest of terms.

In describing the results, I have characterized classes of organisms such as hosts, parasites, hyper-parasites, social, and cheaters. While these terms apply nicely to digital organisms, it can be tricky to examine the parallels between digital and organic organisms in detail. The parasites of this study cause no direct harm to their host; however, they do compete with them for space. This is rather like a vine which depends on a tree for support, but which does not directly harm the tree, except that the two must compete for light. The hyper-parasites of this study are facultative and subvert the energy metabolism of their parasite victims without killing them. I cannot think of an organic example that has all of these properties. The carnivorous plant comes close in that it does not need the prey to survive, and in that its prey may have approached the plant expecting to feed on it. However, the prey of carnivorous plants are killed outright.

We are not in a position to make the most appropriate comparison, between digital creatures and RNA creatures. However, we can apply what we have learned from digital organisms, about the evolutionary properties of creatures at that level of complexity, to our speculations about what the RNA world may have been like. For example, once an RNA molecule fully capable of self-replication evolved, might other RNA molecules lacking that capability have parasitized its replicatory function?

In studying the natural history of synthetic organisms, it is important to recognize that they have a distinct biology due to their non-organic nature. In order to fully appreciate their biology, one must understand the stuff of which they are made. To study the biology of creatures of the RNA world would require an understanding of organic chemistry and the properties of macro-molecules. To understand the biology of digital organisms requires a knowledge of the properties of machine instructions and machine language algorithms. However, to fully understand digital organisms, one must also have a knowledge of biological evolution and ecology. Evolution and ecology are the domain of biologists and machine languages are the domain of computer scientists. The knowledge chasm between biology and computer science is likely to hinder progress in the field of Artificial Life for some time. We need more individuals with a depth of knowledge in both areas in order to carry out the work.

Trained biologists will tend to view synthetic life in the same terms that they have come to know organic life. Having been trained as an ecologist and evolutionist, I have seen in my synthetic communities, many of the ecological and evolutionary properties that are well known from natural communities. Biologists trained in other specialties will likely observe other familiar properties. It seems that what we see is what we know. It is likely to take longer before we appreciate the unique properties of these new life forms.

ARTIFICIAL LIFE AND BIOLOGICAL THEORY

The relationship between Artificial Life and biological theory is two-fold: (1) Given that one of the main objectives of AL is to produce evolution leading to spontaneously increasing diversity and complexity, there exists a rich body of biological theory that suggests factors that may contribute to that process; and (2) to the extent that the underlying life processes are the same in AL and organic life, AL models provide a new tool for experimental study of those processes, which can be used to test biological theory that can not be tested by traditional experimental and analytic techniques.[47]

Furthermore, there exists a complementary relationship between biological theory and the synthesis of life. Theory suggests how the synthesis can be achieved, while application of the theory in the synthesis is a test of the theory. If theory suggests that a certain factor will contribute to increasing diversity, then synthetic systems can be run with and without that factor. The process of synthesis becomes a test of the theory.

At the molecular level, there has been much discussion of the role of transposable elements in evolution. It has been observed that most of the genome in eukaryotes (perhaps 90%) originated from transposable elements, while in prokaryotes, only a very small percentage of the genome originated through transposons.[20,40,51] It can also be noted that the eukaryotes, not the prokaryotes, were involved in the Cambrian explosion of diversity.[4] It has been suggested that transposable elements play a significant role in facilitating evolution.[26,30,50] These observations suggest that it would be an interesting experiment to introduce transposable elements into digital organisms.

The Cambrian explosion consisted of the origin, proliferation, and diversification of macroscopic multi-cellular organisms. The origin and elaboration of multi-cellularity was an integral component of the process. Buss[7] provides a provocative discussion of the evolution of multi-cellularity, and explores the consequences of selection at the level of cell lines. From his discussion the following idea emerges (although he does not explicitly state this idea, in fact, he proposes a sort of inverse of this idea, p. 65): the transition from single to multi-celled existence involves the extension of the control of gene regulation by the mother cell to successively more generations of daughter cells. This is a concept which transcends the physical basis of life, and could be profitably applied to synthetic life in order to generate an analog of multi-cellularity.

The Red Queen hypothesis[52] suggests that in the face of a changing environment, organisms must evolve as fast as they can in order to simply maintain their current state of adaptation. "In order to get anywhere you must run twice as fast as that."[9] A critical component of the environment for any organism is the other living organisms with which it must interact. Given that the species that comprise the environment are themselves evolving, the pace is set by the maximal rate that any species may change through evolution, and it becomes very difficult to actually get ahead. A maximal rate of evolution is required just to keep from falling behind. This suggests that interactions with other evolving species provide the primary driving force in evolution.

Much evolutionary theory deals with the role of biotic interactions in driving evolution. For example, it is thought that these are of primary importance in the maintenance of sex.[5,10,36,37] Stanley[49] has suggested that the Cambrian explosion was sparked by the appearance of the first organisms that ate other organisms. These new herbivores enhanced diversity by preventing any single species of algae from dominating and competitively excluding others. These kinds of biotic interactions must be incorporated into synthetic life in order to move evolution.

Similarly, many abiotic factors are known to contribute to determining the diversity of ecological communities. Island biogeography theory considers how the size, shape, distribution, fragmentation, and heterogeneity of habitats contribute to community diversity.[35] Various types of disturbance are also believed to significantly affect diversity.[29,44] All of these factors may be introduced into synthetic life in an effort to enhance the diversification of the evolving systems.

The examples just listed are a few of the many theories that suggest factors that influence biological diversity. In the process of synthesizing increasingly complex instances of life, we can incorporate and manipulate the states of these factors. These manipulations, conducted for the purposes of advancing the synthesis, will also constitute powerful tests of the theories.

EXTENDING THE MODEL

The approach to AL advocated in this work involves engineering over the first 3 billion years of life's history to design complex evolvable artificial organisms, and attempting to create the biological conditions that will set off a spontaneous evolutionary process of increasing diversity and complexity of organisms. This is a very difficult undertaking, because in the midst of the Cambrian explosion, life had evolved to a level of complexity in which emergent properties existed at many hierarchical levels: molecular, cellular, organismal, populational, and community.

In order to define an approach to the synthesis of life paralleling this historical stage of organic life, we must examine each of the fundamental hierarchical levels, abstract the principal biological properties from their physical representation, and determine how they can be represented in our artificial media. The simulator program determines not only the physics and chemistry of the virtual universe that it creates, but the community ecology as well. We must tinker with the structure of the simulator program in order to facilitate the existence of the appropriate "molecular," "cellular," and "ecological" interactions to generate a spontaneously increasing diversity and complexity.

The evolutionary potential of the present model can be greatly extended by some modifications. In its present implementation, parasitic relationships evolve rapidly, but predation involving the direct usurpation of space occupied by cells is not possible. This could be facilitated by the introduction of a FREE (memory deallocation) instruction. However, it is unlikely that such predatory behavior would be selected for because in the current system there is always free memory space available; thus, there would be little to be gained through seizing space from another creature. However, predation could be selected for by removing the reaper from the system.

Perhaps a more interesting way to favor predatory-type interactions would be to make instructions expensive. In the present implementation, there is no "conservation of instructions," because the MOV_IAB instruction creates a new copy of the instruction being moved during self-replication. If the MOV_IAB instruction were modified such that it obeyed a law of conservation, and left behind all zeros when it moved an instruction, then instructions would not be so cheap. Creatures could be allowed to synthesize instructions through a series of bit flipping and shifting operations, which would make instructions "metabolically" costly. Under such circumstance, a soup of "autotrophs" which synthesize all of their instructions could be invaded by a predatory creature which kills other creatures to obtain instructions.

Additional richness could be introduced to the model by modifying the way that CPU time is allocated. Rather than using a circular queue, creatures could deploy special arrays of instructions or bit patterns (analogous to chlorophyll) which capture potential CPU time packets raining like photons onto the soup. In addition, with instructions being synthesized through bit flipping and shifting operations, each instruction could be considered to have a "potential time" (i.e., potential energy) value which is proportional to its content of one bits. Instructions rich in ones could be used as time (energy) storage "molecules" which could be metabolized when needed by converting the one bits to zeros to release the stored CPU time. The introduction of such an "informational metabolism" would open the way for all sorts of evolution involving the exploitation of one organism by another.

Separation of the genotype from the phenotype would allow the model to move beyond the parallel to the RNA world into a parallel of the DNA-RNA-protein stage of evolution. Storage of the genetic information in relatively passive informational structures, which are then translated into the "metabolically active" machine instructions would facilitate evolution of development, sexuality, and transposons. These features would contribute greatly to the evolutionary potential of the model.

These enhancements of the model represent the current directions of my continuing efforts in this area, in addition to using the existing model to further test ecological and evolutionary theory.

ACKNOWLEDGMENT

I thank Dan Chester, Robert Eisenberg, Doyne Farmer, Walter Fontana, Stephanie Forrest, Chris Langton, Stephen Pope, and Steen Rasmussen, for their discussions or readings of the manuscripts. Contribution No. 142 from the Ecology Program, School of Life and Health Sciences, University of Delaware.

APPENDIX A

Structure definition to implement the Tierra virtual CPU. The source code or executables for the Tierra Simulator can be obtained by contacting the author by mail (emial or snail mail).

```
struct cpu {  /* structure for registers of virtual cpu */
    int    ax;  /* address register */
    int    bx;  /* address register */
    int    cx;  /* numerical register */
    int    dx;  /* numerical register */
    char   fl;  /* flag */
    char   sp;  /* stack pointer */
    int    st[10];  /* stack */
    int    ip;  /* instruction pointer */
    } ;
```

APPENDIX B

Abbreviated code for implementing the CPU cycle of the Tierra Simulator.

```
void main(void)
{   get_soup();
    life();
    write_soup();
}

void life(void) /* doles out time slices and death */
{   while(inst_exec_c < alive)  /* control the length of the run */
    {   time_slice(this_slice); /* this_slice is current cell in queue */
        incr_slice_queue(); /* increment this_slice to next cell in queue */
        while(free_mem_current < free_mem_prop * soup_size)
            reaper(); /* if memory is full to threshold, reap some cells */
    }
}

void time_slice(int  ci)
{   Pcells  ce; /* pointer to the array of cell structures */
    char    i;  /* instruction from soup */
    int     di; /* decoded instruction */
    int     j, size_slice;
    ce = cells + ci;
    for(j = 0; j < size_slice; j++)
    {   i = fetch(ce->c.ip); /* fetch instruction from soup, at address ip */
        di = decode(i);      /* decode the fetched instruction */
        execute(di, ci);     /* execute the decoded instruction */
        increment_ip(di,ce); /* move instruction pointer to next instruction */
        system_work(); /* opportunity to extract information */
    }
}
```

```
void execute(int  di, int  ci)
{  switch(di)
    {   case 0x00: nop_0(ci);    break; /* no operation */
        case 0x01: nop_1(ci);    break; /* no operation */
        case 0x02: or1(ci);      break; /* flip low order bit of cx, cx ^= 1 */
        case 0x03: shl(ci);      break; /* shift left cx register, cx <<= 1 */
        case 0x04: zero(ci);     break; /* set cx register to zero, cx = 0 */
        case 0x05: if_cz(ci);    break; /* if cx==0 execute next instruction */
        case 0x06: sub_ab(ci);   break; /* subtract bx from ax, cx = ax - bx */
        case 0x07: sub_ac(ci);   break; /* subtract cx from ax, ax = ax - cx */
        case 0x08: inc_a(ci);    break; /* increment ax, ax = ax + 1 */
        case 0x09: inc_b(ci);    break; /* increment bx, bx = bx + 1 */
        case 0x0a: dec_c(ci);    break; /* decrement cx, cx = cx - 1 */
        case 0x0b: inc_c(ci);    break; /* increment cx, cx = cx + 1 */
        case 0x0c: push_ax(ci);  break; /* push ax on stack */
        case 0x0d: push_bx(ci);  break; /* push bx on stack */
        case 0x0e: push_cx(ci);  break; /* push cx on stack */
        case 0x0f: push_dx(ci);  break; /* push dx on stack */
        case 0x10: pop_ax(ci);   break; /* pop top of stack into ax */
        case 0x11: pop_bx(ci);   break; /* pop top of stack into bx */
        case 0x12: pop_cx(ci);   break; /* pop top of stack into cx */
        case 0x13: pop_dx(ci);   break; /* pop top of stack into dx */
        case 0x14: jmp(ci);      break; /* move ip to template */
        case 0x15: jmpb(ci);     break; /* move ip backward to template */
        case 0x16: call(ci);     break; /* call a procedure */
        case 0x17: ret(ci);      break; /* return from a procedure */
        case 0x18: mov_cd(ci);   break; /* move cx to dx, dx = cx */
        case 0x19: mov_ab(ci);   break; /* move ax to bx, bx = ax */
        case 0x1a: mov_iab(ci);  break; /* move instruction at address in bx
                                            to address in ax */
        case 0x1b: adr(ci);      break; /* address of nearest template to ax */
        case 0x1c: adrb(ci);     break; /* search backward for template */
        case 0x1d: adrf(ci);     break; /* search forward for template */
        case 0x1e: mal(ci);      break; /* allocate memory for daughter cell */
        case 0x1f: divide(ci);   break; /* cell division */
    }
    inst_exec_c++;
}
```

APPENDIX C

Assembler source code for the ancestral creature.

```
genotype: 80 aaa  origin: 1-1-1990  00:00:00:00  ancestor
parent genotype: human
1st_daughter:  flags: 0  inst: 839  mov_daught: 80
2nd_daughter:  flags: 0  inst: 813  mov_daught: 80


nop_1    ; 01    0 beginning template
nop_1    ; 01    1 beginning template
```

```
nop_1    ; 01    2 beginning template
nop_1    ; 01    3 beginning template
zero     ; 04    4 put zero in cx
or1      ; 02    5 put 1 in first bit of cx
shl      ; 03    6 shift left cx
shl      ; 03    7 shift left cx, now cx = 4
         ;             ax =                  bx =
         ;             cx = template size    dx =
mov_cd   ; 18    8 move template size to dx
         ;             ax =                  bx =
         ;             cx = template size    dx = template size
adrb     ; 1c    9 get (backward) address of beginning template
nop_0    ; 00   10 compliment to beginning template
nop_0    ; 00   11 compliment to beginning template
nop_0    ; 00   12 compliment to beginning template
nop_0    ; 00   13 compliment to beginning template
         ;             ax = start of mother + 4   bx =
         ;             cx = template size         dx = template size
sub_ac   ; 07   14 subtract cx from ax
         ;             ax = start of mother   bx =
         ;             cx = template size     dx = template size
mov_ab   ; 19   15 move start address to bx
         ;             ax = start of mother   bx = start of mother
         ;             cx = template size     dx = template size
adrf     ; 1d   16 get (forward) address of end template
nop_0    ; 00   17 compliment to end template
nop_0    ; 00   18 compliment to end template
nop_0    ; 00   19 compliment to end template
nop_1    ; 01   20 compliment to end template
         ;             ax = end of mother     bx = start of mother
         ;             cx = template size     dx = template size
inc_a    ; 08   21 to include dummy statement to separate creatures
sub_ab   ; 06   22 subtract start address from end address to get size
         ;             ax = end of mother      bx = start of mother
         ;             cx = size of mother     dx = template size
nop_1    ; 01   23 reproduction loop template
nop_1    ; 01   24 reproduction loop template
nop_0    ; 00   25 reproduction loop template
nop_1    ; 01   26 reproduction loop template
mal      ; 1e   27 allocate memory for daughter cell, address to ax
         ;             ax = start of daughter    bx = start of mother
         ;             cx = size of mother       dx = template size
call     ; 16   28 call template below (copy procedure)
nop_0    ; 00   29 copy procedure compliment
nop_0    ; 00   30 copy procedure compliment
nop_1    ; 01   31 copy procedure compliment
nop_1    ; 01   32 copy procedure compliment
divide   ; 1f   33 create independent daughter cell
jmp      ; 14   34 jump to template below (reproduction loop, above)
nop_0    ; 00   35 reproduction loop compliment
nop_0    ; 00   36 reproduction loop compliment
nop_1    ; 01   37 reproduction loop compliment
nop_0    ; 00   38 reproduction loop compliment
if_cz    ; 05   39 this is a dummy instruction to separate templates
```

```
         ;              begin copy procedure
nop_1    ; 01  40 copy procedure template
nop_1    ; 01  41 copy procedure template
nop_0    ; 00  42 copy procedure template
nop_0    ; 00  43 copy procedure template
push_ax  ; 0c  44 push ax onto stack
push_bx  ; 0d  45 push bx onto stack
push_cx  ; 0e  46 push cx onto stack
nop_1    ; 01  47 copy loop template
nop_0    ; 00  48 copy loop template
nop_1    ; 01  49 copy loop template
nop_0    ; 00  50 copy loop template
mov_iab  ; 1a  51 move contents of [bx] to [ax]
dec_c    ; 0a  52 decrement cx
if_cz    ; 05  53 if cx == 0 perform next instruction, otherwise skip it
jmp      ; 14  54 jump to template below (copy procedure exit)
nop_0    ; 00  55 copy procedure exit compliment
nop_1    ; 01  56 copy procedure exit compliment
nop_0    ; 00  57 copy procedure exit compliment
nop_0    ; 00  58 copy procedure exit compliment
inc_a    ; 08  59 increment ax
inc_b    ; 09  60 increment bx
jmp      ; 14  61 jump to template below (copy loop)
nop_0    ; 00  62 copy loop compliment
nop_1    ; 01  63 copy loop compliment
nop_0    ; 00  64 copy loop compliment
nop_1    ; 01  65 copy loop compliment
if_cz    ; 05  66 this is a dummy instruction, to separate templates
nop_1    ; 01  67 copy procedure exit template
nop_0    ; 00  68 copy procedure exit template
nop_1    ; 01  69 copy procedure exit template
nop_1    ; 01  70 copy procedure exit template
pop_cx   ; 12  71 pop cx off stack
pop_bx   ; 11  72 pop bx off stack
pop_ax   ; 10  73 pop ax off stack
ret      ; 17  74 return from copy procedure
nop_1    ; 01  75 end template
nop_1    ; 01  76 end template
nop_1    ; 01  77 end template
nop_0    ; 00  78 end template
if_cz    ; 05  79 dummy statement to separate creatures
```

REFERENCES

1. Ackley, D. H., and M. S. Littman. "Learning From Natural Selection in an Artificial Environment." In *Proceedings of the International Joint Conference on Neural Networks*, Vol. I, Theory Track, Neural and Cognitive Sciences Track. (Washington, DC, Winter, 1990.) Hillsdale, NJ: Lawrence Erlbaum Associates, 1990.

2. Aho, A. V., J. E. Hopcroft, and J. D. Ullman. *The Design and Analysis of Computer Algorithms*. Reading, MA: Addison-Wesley, 1974.

3. Bagley, R. J., J. D. Farmer, S. A. Kauffman, N. H. Packard, A. S. Perelson, and I. M. Stadnyk. "Modeling Adaptive Biological Systems." *Biosystems* **23** (1989): 113–138.

4. Barbieri, M. *The Semantic Theory of Evolution*. London: Harwood, 1985.

5. Bell, G. *The Masterpiece of Nature: The Evolution and Genetics of Sexuality*. Berkeley: University of California Press, 1982.

6. Bell, G. *Sex and Death in Protozoa: The History of an Obsession*. Cambridge: Cambridge University Press, 1989.

7. Buss, L. W. *The Evolution of Individuality*. Princeton: Princeton University Press, 1987.

8. Cariani, P. "Emergence and Artificial Life." This volume

9. Carroll, L. *Through the Looking-Glass*. London: MacMillan, 1865.

10. Charlesworth, B. "Recombination Modification in a Fluctuating Environment." *Genetics* **83** (1976): 181–195.

11. Cohen, F. "Computer Viruses: Theory and Experiments." Ph. D. dissertation, University of Southern California, 1984.

12. Dawkins, R. *The Blind Watchmaker*. New York: Norton, 1987.

13. Dawkins, R. "The Evolution of Evolvability." In *Artificial Life*, edited by C. Langton. Santa Fe Institute Studies in the Sciences of Complexity, Proc. Vol. VI, 201–220. Reading, MA: Addison-Wesley, 1989.

14. Denning, P. J. "Computer Viruses." *Amer. Sci.* **76** (1988): 236–238.

15. Dewdney, A. K. "Computer Recreations: In the Game Called Core War Hostile Programs Engage in a Battle of Bits." *Sci. Amer.* **250** (1984): 14–22.

16. Dewdney, A. K. "Computer Recreations: A Core War Bestiary of Viruses, Worms and Other Threats to Computer Memories." *Sci. Amer.* **252** (1985): 14–23.

17. Dewdney, A. K. "Computer Recreations: Exploring the Field of Genetic Algorithms in a Primordial Computer Sea Full of Flibs." *Sci. Amer.* **253** (1985): 21–32.

18. Dewdney, A. K. "Computer Recreations: A Program Called MICE Nibbles Its Way to Victory at the First Core War Tournament." *Sci. Amer.* **256** (1987): 14–20.

19. Dewdney, A. K. "Of Worms, Viruses and Core War." *Sci. Amer.* **260** (1989): 110–113.

20. Doolittle, W. F., and C. Sapienza. "Selfish Genes, the Phenotype Paradigm and Genome Evolution." *Nature* **284** (1980): 601–603.

21. Eldredge, N., and S. J. Gould. "Punctuated Equilibria: An Alternative to Phyletic Gradualism." In *Models in Paleobiology*, edited by J. M. Schopf, 82–115. San Francisco: Greeman, Cooper, 1972.

22. Farmer, J. D., S. A. Kauffman, and N. H. Packard. "Autocatalytic Replication of Polymers." *Physica D* **22** (1986): 50–67.

23. Farmer, J. D., and A. Belin. "Artificial Life: The Coming Evolution." Proceedings in celebration of Murray Gell-Man's 60th Birthday. Cambridge: Cambridge University Press. In press. Reprinted in this volume.

24. Gould, S. J. *Wonderful Life, The Burgess Shale and the Nature of History*. New York: Norton, 1989.

25. Gould, S. J., and N. Eldredge. "Punctuated Equilibria: The Tempo and Mode of Evolution Reconsidered." *Paleobiology* **3** (1977): 115–151.

26. Green, M. M. "Mobile DNA Elements and Spontaneous Gene Mutation." In *Eukaryotic Transposable Elements as Mutagenic Agents*, edited by M. E. Lambert, J. F. McDonald, and I. B. Weinstein, 41–50. Banbury Report 30. Cold Spring Harbor Laboratory, 1988.

27. Holland, J. H. *Adaptation in Natural and Artificial Systems: An Introductory Analysis with Applications to Biology, Control, and Artificial Intelligence*. Ann Arbor: University of Michigan Press, 1975.

28. Holland, J. H. "Studies of the Spontaneous Emergence of Self-Replicating Systems Using Cellular Automata and Formal Grammars." In *Automata, Languages, Development*, edited by A. Lindenmayer, and G. Rozenberg, 385–404. New York: North-Holland, 1976.

29. Huston, M. "A General Hypothesis of Species Diversity." *Am. Nat.* **113** (1979): 81–101.

30. Jelinek, W. R., and C. W. Schmid. "Repetitive Sequences in Eukaryotic DNA and Their Expression." *Ann. Rev. Biochem.* **51** (1982): 813–844.

31. Langton, C. G. "Studying Artificial Life With Cellular Automata." *Physica* **22D** (1986): 120–149.

32. Langton, C. G. "Virtual State Machines in Cellular Automata." *Complex Systems* **1** (1987): 257–271.

33. Langton, C. G., ed. "Artificial Life." In *Artificial Life*, Santa Fe Institute Studies in the Sciences of Complexity, Proc. Vol. VI, 1–47. Reading, MA: Addison-Wesley, 1989, .

34. Lotka, A. J. *Elements of Physical Biology*. Baltimore: Williams and Wilkins, 1925. Reprinted as *Elements of Mathematical Biology*, Dover Press, 1956.

35. MacArthur, R. H., and E. O. Wilson. *The Theory of Island Biogeography*. Princeton: Princeton University Press, 1967.

36. Maynard-Smith, J. "What Use is Sex?" *J. Theor. Biol.* **30** (1971): 319–335.

37. Michod, R. E., and B. R. Levin, eds. *The Evolution of Sex*. Sutherland, MA: Sinauer, 1988.

38. Minsky, M. L. *Computation: Finite and Infinite Machines*. Englewood Cliffs, NJ: Prentice-Hall, 1976.

39. Morris, S. C. "Burgess Shale Faunas and the Cambrian Explosion." *Science* **246** (1989): 339–346.

40. Orgel, L. E., and F. H. C. Crick. "Selfish DNA: The Ultimate Parasite." *Nature* **284** (1980): 604–607.

41. Paine, R. T. "Food Web Complexity and Species Diversity." *Am. Nat.* **100** (1966): 65–75.

42. Packard, N. H. "Intrinsic Adaptation in a Simple Model for Evolution." In *Artificial Life*, edited by C. Langton. Santa Fe Institute Studies in the Sciences of Complexity, Proc. Vol. VI, 141–155. Reading, MA: Addison-Wesley, 1989.

43. Pattee, H. H. "Simulations, Realizations, and Theories of Life." In *Artificial Life*, edited by C. Langton. Santa Fe Institute Studies in the Sciences of Complexity, Proc. Vol. VI, 63–77. Reading, MA: Addison-Wesley, 1989.

44. Petraitis, P. S., R. E. Latham, and R. A. Niesenbaum. "The Maintenance of Species Diversity by Disturbance." *Quart. Rev. Biol.* **64** (1989): 393–418.

45. Rasmussen, S., C. Knudsen, R. Feldberg, and M. Hindsholm. "The Coreworld: Emergence and Evolution of Cooperative Structures in a Computational Chemistry" *Physica D.* **42** (1990): 111–134.

46. Rheingold, H. "Computer Viruses." *Whole Earth Review* **Fall** (1988): 106.

47. Ray, T. S. "Synthetic Life: Evolution and Ecology of Digital Organisms." Unpublished, 1990.

48. Spafford, E. H., K. A. Heaphy, and D. J. Ferbrache. *Computer Viruses, Dealing with Electronic Vandalism and Programmed Threats*. ADAPSO, 1300 N. 17th Street, Suite 300, Arlington, VA 22209, 1989.

49. Stanley, S. M. "An Ecological Theory for the Sudden Origin of Multicellular Life in the Late Precambrian." *Proc. Nat. Acad. Sci.* **70** (1973): 1486–1489.

50. Syvanen, M. "The Evolutionary Implications of Mobile Genetic Elements." *Ann. Rev. Genet.* **18** (1984): 271–293.

51. Thomas, C. A. "The Genetic Organization of Chromosomes." *Ann. Rev. Genet.* **5** (1971): 237–256.

52. Van Valen, L. "A New Evolutionary Law." *Evol. Theor.* **1** (1973): 1–30.

53. Volterra, V. "Variations and Fluctuations of the Number of Individuals in Animal Species Living Together." In *Animal Ecology*, edited by R. N. Chapman, 409–448. New York: McGraw-Hill, 1926.

54. Wilson, E. O., and W. H. Bossert. *A Primer of Population Biology*. Stamford, CN: Sinauers, 1971.

Epilogue

As with natural history, the fossil record of evolutionary computation continues to be uncovered. I hope that *Evolutionary Computation: The Fossil Record* will foster an increased interest in learning about efforts to simulate evolution that remain "burried," waiting to be discovered. If you are aware of any scientific "fossils" that merit consideration for inclusion in the second edition of this book, please let me know by emailing me at d.fogel@ieee.org. I'd be very pleased to hear from you.

Author Index

Index

About the Editor

Dr. David B. Fogel is Executive Vice President and Chief Scientist of Natural Selection, Inc. in La Jolla, CA, a small business focused on solving difficult problems in industry, medicine, and defense using evolutionary computation, neural networks, fuzzy systems, and other methods of computational intelligence. Dr. Fogel's experience in evolutionary computation spans 14 years and includes applications in pharmaceutical design, computer-assisted mammography, data mining, factory scheduling, financial forecasting, traffic flow optimization, agent-based adaptive combat systems, and many other areas. Dr. Fogel is a prolific author in evolutionary computation, having published over 30 journal papers, as well as over 75 conference publications, 20 contributions in book chapters, 1 video, and 2 books, most recently *Evolutionary Computation: Toward a New Philosophy of Machine Intelligence* (IEEE Press, 1995). In addition, Dr. Fogel is co-editor-in-chief of the *Handbook of Evolutionary Computation* (Oxford, 1997) and the founding editor-in-chief of the *IEEE Transactions on Evolutionary Computation*.

Dr. Fogel received the Ph.D. degree in engineering sciences (systems science) from the University of California at San Diego (UCSD) in 1992. He earned the M.S. degree in engineering sciences (systems science) from UCSD in 1990, and the B.S. in mathematical sciences (probability and statistics) from the University of California at Santa Barbara in 1985. Prior to co-founding Natural Selection, Inc. in 1993, Dr. Fogel was a Systems Analyst at Titan Systems, Inc. (1984–1988), and a Senior Principal Engineer at ORINCON Corporation (1988–1993). He has taught university courses at the graduate and undergraduate level in stochastic processes, probability and statistics, and evolutionary computation. He serves as associate editor for the journal *BioSystems*, and is a member of the editorial board of several other international technical journals.

Dr. Fogel served as a Visiting Fellow of the Australian Defence Force Academy in November, 1997, and is a member of many professional societies including the American Association for the Advancement of Science, the American Association for Artificial Intelligence, Sigma Xi, and the New York Academy of Sciences. He was the founding President of the Evolutionary Programming Society in 1991 and is a senior member of the IEEE, as well as an associate member of the Center for the Study of Evolution and the Origin of Life (CSEOL) at the University of California at Los Angeles. Dr. Fogel is a frequently invited lecturer at international conferences and guest for television and radio broadcasts.